MICROSCALE COMBUSTION AND POWER GENERATION

MICROSCALE COMBUSTION AND POWER GENERATION

**YIGUANG JU, CHRISTOPHER CADOU
AND KAORU MARUTA**

 MOMENTUM PRESS

MOMENTUM PRESS, LLC, NEW YORK

Microscale Combustion and Power Generation
Copyright © Momentum Press®, LLC, 2015.

First published by Momentum Press®, LLC
222 East 46th Street, New York, NY 10017
www.momentumpress.net

ISBN-13: 978-1-60650-306-5 (print)
ISBN-13: 978-1-60650-308-1 (e-book)

Momentum Press Mechanical Engineering Collection

DOI: 10.5643/9781606503081

Cover and interior design by Exeter Premedia Services Private Ltd., Chennai, India
Cover page photo for Swiss roll combustor was taken by Prof. NAM IL KIM at KAIST, Korea.

10 9 8 7 6 5 4 3 2 1

Printed in the United States of America

ABSTRACT

Recent advances in microfabrication technologies have enabled the development of entirely new classes of small-scale devices with applications in fields ranging from biomedicine (portable defibrillators, drug delivery systems, etc.), to wireless communication and computing (cell phones, laptop computers, etc.), to reconnaissance (unmanned air vehicles, microsatellites etc.), and to augmentation of human function (exoskeletons etc.). In many cases, however, what these devices can actually accomplish is limited by the low energy density of their energy storage and conversion systems.

This breakthrough book brings together in one place the information necessary to develop the high energy density combustion-based power sources that will enable many of these devices to realize their full potentials. Engineers and scientists working in energy-related fields will find here:

- An overview of the fundamental physics and phenomena of microscale combustion;
- Presentations of the latest modeling and simulation techniques for gas-phase and catalytic micro-reactors;
- The latest results from experiments in small-scale liquid film, micro-tube, and porous combustors, micro-thrusters, and micro heat engines;
- An assessment of the additional research necessary to develop compact and high energy density energy conversion systems that are truly practical

KEYWORDS

microscale combustion, flameless combustion, combustion limits, combustion instability, excess enthalpy combustion, small-scale liquid film combustors, micro-tubes and porous combustors, Swiss-roll combustors, catalytic reactors, micro-heat engines, micro-reactors, micro-power generators, micro-thrusters, model aircraft engines, 2-stroke engines, piston engines, heterogeneous combustion, catalytic combustion, conjugate heat transfer, scale-effects on combustion, thermoelectric power generation, micro gas turbine engine, micro-rotary engine, micro-rockets, microfabrication, MEMS

CONTENTS

LIST OF FIGURES

LIST OF TABLES

LIST OF CONTRIBUTORS

Yiguang Ju
Department of Mechanical and Aerospace Engineering, Princeton University, Princeton, NJ 08544, USA, Email: yju@princeton.edu

Kaoru Maruta
Institute of Fluid Science, Tohoku University, 2-1-1 Katahira, Aoba, Sendai 980-8577, Japan, Email: maruta@ifs.tohoku.ac.jp

International Combustion and Energy (ICE) Laboratory, School of Engineering, Far Eastern Federal University, 8, Sukhanova Str., Vladivostok 690950, Russia

Christopher Cadou
University of Maryland, Department of Aerospace Engineering, College Park, MD 20816, Email: cadou@umd.edu

Paul D. Ronney
Department of Aerospace and Mechanical Engineering, University of Southern California, Los Angeles, CA, USA 90089-1453, Email: ronney@usc.edu

John Mantzaras
Paul Scherrer Institute, Combustion Research, Switzerland, Email: ioannis.mantzaras@psi.ch

Yei-Chin Chao
Institute of Aeronautics & Astronautics, National Cheng Kung University, Taiwan, Email: ycchao@mail.ncku.edu.tw

Chih-Peng Chen
Missile & Rocket Systems Research Division, National Chung-Shan Institute of Science & Technology, Longtan, Taoyuan, Taiwan 325, Taiwan, Email: 0238cpc@gmail.com

David P. Tse
Los Alamos National Laboratory, Los Alamos, NM 87547, USA, Email: idavidtse@gmail.com

Dimitrios C. Kyritsis
Department of Mechanical Engineering, Khalifa University, United Arab Emirates,
Email: dimitrios.kyritsis@kustar.ac.ae

Marco Schultze
Combustion Research Laboratory, Paul Scherrer Institute CH-5232 Villigen PSI, Switzerland,
Email: marco.schultze@psi.ch

Guan-Bang Chen
Research Center for Energy Technology and Strategy, National Cheng Kung University,
Tainan, Taiwan 701, Taiwan, Email: gbchen26@gmail.com

Ming-hsun Wu
Department of Mechanical Engineering, National Cheng Kung University, Tainan City,
Taiwan, Email: minghwu@mail.ncku.edu.tw

Richard A. Yetter
Department of Mechanical Engineering, The Pennsylvania State University, University Park,
PA, USA, Email: ray8@psu.edu

Vigor Yang
School of Aerospace Engineering, Georgia Institute of Technology, Atlanta, GA, USA,
Email: vigor.yang@aerospace.gatech.edu

Carlos Fernandez-Pello
Department of Mechanical Engineering, University of California, Berkeley, CA 94720-1740,
Email: ferpello@me.berkeley.edu

Shyam Menon
Energy Systems Laboratory, Oregon State University-Cascades, Ponderosa Hall, 2600 NW
College Way, Bend OR 97701, USA, Email: smenon@caltech.edu

Elaine S. Oran
Department of Aerospace Engineering, College Park, University of Maryland, USA,
Email: eoran@umd.edu

Vadim N. Gamezo
Naval Research Laboratory, Lab for Computational Physics & Fluid Dynamics, Washington,
DC 203755321, USA, Email: gamezo@lcp.nrl.navy.mil

Ryan Houim
Department of Aerospace Engineering, College Park, University of Maryland, USA,
Email: rhouim@umd.edu

Sang Hee Won
Department of Mechanical and Aerospace Engineering, Princeton University, Princeton,
New Jersey 08544, USA, Email: sangwon@princeton.edu

Derek Dunn-Rankin
Department of Mechanical and Aerospace Engineering. University of California, Irvine, CA
92697-3975, USA, Email: ddunnran@uci.edu

William A. Sirignano
Mechanical and Aerospace Engineering, University of California Irvine, CA 92697-3975,
USA, Email: sirignan@uci.edu

Yueh-Heng Li
Department of Aeronautics and Astronautics, National Cheng Kung University,
Tainan, 70101, Taiwan, Email: Gyueheng@mail.ncku.edu.tw

Christopher M. Spadaccini
Lawrence Livermore National Laboratory, 7000 East Avenue, Livermore, CA 94550,
Email: spadaccini2@llnl.gov

Ian A. Waitz
Department of Aeronautics and Astronautics, Massachusetts Institute of Technology,
77 Massachusetts Avenue, Cambridge, Massachusetts 02139, USA, Email: iaw@mit.edu

PREFACE

In 1998, Chang-Lin Tien came to Caltech to give a talk on microscale boiling heat transfer [1]. He had just stepped down after a distinguished term as Chancellor of UC Berkeley and it was clear that he was really looking forward to returning to research. He began by recounting some advice that Llewellen M. K. Boelter, another luminary in heat transfer, had given him many years prior at the beginning his academic career. The gist of Boelter's advice was this: "Go to the extremes. That is where the interesting things will be and don't worry about the applications yet. They will come." Professor Tien went on to explain how his career in high flux heat transfer (which soon became relevant because of the development of nuclear power plants) had begun and that now, for his return to research, he was moving to the other extreme: heat transfer at very small scales. Unknown to everyone in the room then, but true to Boelter's hypothesis, was that rapid advances in microfabrication technology would soon make microscale heat transfer an extremely important topic. Zhigang Suo, a Harvard Professor and a world renowned expert on micromechanics, gave similar advice in 2000: "go microcombustion!"

In some sense then, this book is a result of Boelter's and Zhigang's visionary advice given more than a decade ago. The first two sections focus on what happens to various types of flames when one attempts to stabilize them in small passages. A more careful discussion of "microcombustion" and what "small" means in the context of flames is presented in the introduction. The last section focuses on applications of miniature combustion systems. While there are some applications in chemical processing (in situ generation of hazardous reagents for example), the overwhelming majority is in energy conversion: that is the construction of miniature heat engines and propulsion systems.

That we can think about microcombustion-powered systems at all is due in large part to advances in manufacturing technologies like chemical etching, deep reactive ion etching, chemical vapor deposition, wafer bonding, and so on that allow us to make things with micron-scale dimensions. These manufacturing tools have enabled the miniaturization of all sorts of devices (cameras, telephones, automotive sensors, etc.) and have spawned the general field of "MEMS" which stands for microelectro mechanical systems and "microsystems." Advances in this field enable the functional integration of microelectronic and micromechanical systems on one chip. One successful example is the microchip-based drug delivery system [2]. Advances in integration are also occurring at somewhat larger "mesoscales" with one especially important, power-limited, example being exoskeletons for augmenting human function [3, 4]. Given these successes, it is logical to think about miniaturizing the power/energy source too. But can it be done?

Richard Feynman is often credited for launching the field of MEMS through his now famous talk "There's plenty of room at the bottom." While combustion is mentioned, it is in only one sentence and it is not encouraging: "This rapid heat loss [associated with miniaturization] would prevent the gasoline from exploding, so an internal combustion engine is impossible" [5]. These are concerning words indeed for advocates of microcombustion—especially because they are true! The key question, however, is at what scale they are true since we all know that engines work at the scales of model airplanes and larger. As a result, one important objective of this book is to help developers of combustion-powered microsystems choose sensible sizes for their devices. Another is to determine under what conditions it makes sense to develop completely self-contained combustion-powered devices on a single chip and under what conditions somewhat larger less integrated devices make more sense.

This volume focuses on combustion at small scales—that is in passages with dimensions of a few millimeters or smaller. The main applications are in miniature power sources for devices like those described above. Examples include heat engines [6–9], fuel processors for fuel cells [10–13], and thermoelectric generators [14, 15]. There are also other applications in chemical processing like in situ generation of hazardous reagents [16, 17] but these will not be the focus here.

The distinguishing aspect of this volume compared to others in microsystems [18, 19] is its exclusive focus on fundamentals of microcombustion and thermo-chemical energy conversion. While other books have chapters that address microscale combustion, it is always in the context of much broader applications like power generation or rotating machines and the level of detail is relatively limited. That this book on microcombustion follows those on its broader applications is consistent with how the field developed. The classical concept of the "quenching diameter" as a limiting passage size prevented many combustion researchers from taking the idea of microscale combustion seriously until microfabricators actually built devices at smaller scales and showed that it was in fact possible to sustain reaction. However, because the early efforts were largely fabrication driven, sufficient resources were not available to address the combustion problems (incomplete combustion, flame instability, and excessive heat loss) that were encountered. As a result, another objective of this volume is to give developers of combustion-powered microsystems enough physical insight into the microscale combustion processes to develop suitably efficient and stable microcombustors.

Finally, a volume like this is not produced in a vacuum and there are many people to thank. First and foremost, we would like to thank the authors of each chapter who made time in their busy schedules to assemble their chapters. This is much easier said than done and we truly appreciate their willingness to support this project and the quality of their contributions. Thanks go as well to the many un-named graduate students and research assistants who performed a lot of the foundational work upon which these chapters are based. Second, we would like to thank our external reviewers Drs. Derek Dunn-Rankin and Paul Ronney for making sure that the material measures up technically and for offering their valuable insights. We would also like to thank Dr. Dunn-Rankin and William Sirignano for championing the idea of a book on microcombustion and for making the connection to Momentum Press. Third, we would like to thank Kurt Annen and Aerodyne Research Inc., Josh Collins and Sun Power Inc., Werner Dahm, Alan Epstein, Carlos Fernandez-Pello, Paul Ronney, and Rich Yetter for either supplying or permitting us to use images of their technology in the introduction. Fourth, many thanks go to Exeter team and especially to Joel Stein, Millicent Treloar, Cindy Durand, Destiny Hadley, and the rest of the staff at Momentum Press for their advice, support, and understanding as this project stretched far longer than intended. We truly appreciate your patience and hope that the wait was worth it! Finally, we would like to thank our families and home institutions for their support during this effort. Without you, we would not have been able to complete this project.

Chris Cadou, Yiguang Ju, and Kaoru Maruta Oct. 20, 2013

REFERENCES

[1] C.-L. Tien, Recent Developments in Microscale Thermophysical Engineering, Pasadena, California: Sixth James R. and Shirley A. Kleigel Lecture in Engineering and Applied Science, 1998.

[2] J. T. Santini, A. C. Richards, R. Scheidt, M. H. Cima and R. Langer, "Microchips as controlled drug delivery devices," Angew. Chem. Int., vol. 39, pp. 2396–2407, 2000.

[3] EKSO Bionics, [Online]. Available: http://eksobionics.com/. [Accessed 10 July 2013].

[4] U. S. O. Command, USSOCOM RFI ST Tactical Light Operator Suit (TALOS), 2013.

[5] R. Feynman, "There's Plenty of Room at the Bottom," Caltech Engineering and Science, vol. 23:5, p. 26, February 1960.

[6] A. a. S. S. D. Epstein, Science, vol. 276, p. 1211, 23 May 1997.

[7] J. Piers, D. Deynaerts and Verplaetsen, "A microturbine for electric power generation," Sensors and Actuators A, vol. 113, pp. 86–93, 2004.

[8] A. C. Fernandez-Pello, "Micropower generation using combustion: Issues and approaches," Proceedings of the Combustion Institute, vol. 29, pp. 883–889, 2002.

[9] K. Annen, D. Stickler and J. Woodroffe, "Miniature Internal Combustion Engine-Generator for High Energy Density Portable Power," in Proceedings of the Army Science Conference (26th), Orlando, FL, December 1–4, 2008.

[10] J. D. Holladay, E. O. Jones, M. Phelps and J. L. Hu, "Microfuel processor for use in a miniature power supply," Journal of Power Sources, vol. 108, pp. 21–27, 2002.

[11] J. C. Ganley, E. G. Seebauer and R. I. Masel, "Porous anodic alumina microreactors for production of hydrogen from ammonia," AICHE Journal, vol. 50, pp. 829–834, 2004.

[12] A. V. Pattekar and M. V. Kothare, "A microreactor for hydrogen production in micro fuel cell applications," Journal of Microelectromechanical Systems, vol. 13, pp. 7–18, 2004.

[13] S. R. Deshmukh and D. G. Vlachos, "Effect of flow configuration on the operation of coupled combustor/reformer microdevices for hydrogen production," Chemical Engineering Science, vol. 60, pp. 5718–5728, 2005.

[14] W. M. Yang, S. K. Chou, C. Shu, Z. W. Li and H. Xue, "A prototype microthermophotovoltaic power generator," Applied Physics Letters, vol. 84, pp. 3864–3866, 2004.

[15] A. L. Cohen, P. Ronney, U. Frodis, L. Sitzki, E. Meiburg and S. Wussow, "Microcombustor and combustion-based thermoelectric microgenerator." US Patent 6613972, 2 September 2003.

[16] S. J. Haswell and P. Watts, "Green chemistry: synthesis in micro reactors," Green Chem, vol. 5, pp. 240–249, 2003.

[17] K. F. Jensen, "Microreaction engineering—is small better?," Chemical Engineering Science, vol. 56, pp. 293–303, 2001.

[18] J. H. Lang, Ed., Multi-Wafer Rotating MEMS Machines, New York: Springer, 2009.

[19] A. Mitsos and P. I. Barton, Eds., Microfabricated Power Generation Devices, Weinheim: Wiley-VCH, 2009.

[20] "Data, data everywhere," The Economist, 25 Feb 2010.

[21] Honeywell Inc., 5 July 2012. [Online]. Available: http://aerospace.honeywell.com/markets/defense/unmanned-systems/2012/07-July/t-hawk. [Accessed 10 July 2013].

[22] Air Force Research Laboratory, December 2005. [Online]. Available: http://www.kirtland.af.mil/shared/media/document/AFD-070404-108.pdf. [Accessed 10 July 2013].

INTRODUCTION

MICROSCALE COMBUSTION AND POWER GENERATION

Hardly a year passes without the introduction of another small power-consuming device that has the potential to fundamentally transform people's lives. Examples of the past include mobile computers, telephones, and "smart" phones. That the impact of these devices needs no explaining is perhaps the best testament to their significance. Examples of the present include miniature satellites and unmanned air vehicles (Figure 1) which have moved beyond the military realm to serve as mobile sensor platforms for monitoring everything from traffic flow [1, 2] to agricultural performance [3] to environmental conditions in forests [4, 5]. The data that are collected—but one component of what is sometimes termed "big data"—will help us to understand and improve everything from weather prediction to security [6]. More transformative technologies are on the horizon. For example, in another 10 years it could be common for people to wear electronic devices like Google Glass (Figure 2a) that augment human senses and link them in a more biologically natural way to cyber-data sources. Similarly, people may wear external mechanisms like exoskeletons (Figure 2b) that enhance physical capabilities or permit disabled people to function more normally. In 20 years, large fleets of miniature spacecraft could be launched that increase our reach into the vastness of space in a far more pervasive way than larger manned vehicles ever could. This could increase the probability of success in the somewhat stochastic process of discovering new sources of critical materials—and possibly new habitats.

Whether or not these technologies live up to their potential depends to a large degree on the availability of appropriate power/energy sources. They must be compact (i.e., of the scale of the device or smaller) and lightweight enough to be worn or carried, rapidly rechargeable, long-lived, provide power output ranging from several milliwatts to hundreds of watts [7–9] (depending on the application), mass-producible/inexpensive, safe reliable, and power/energy dense. Note that the latter also means thermodynamically efficient.

At present, batteries are the only power sources available at scales suitable for many of these devices. While batteries are efficient (if discharged properly), their relatively low energy densities and long recharge times place significant limits on what the systems they power can ultimately accomplish. For example, the exoskeleton pictured in Figure 2b can operate for only a few hours before needing to recharge. While this is not bad—especially for a person who would otherwise be unable to walk—imagine the difference that being able to walk for a full day as opposed to 2 hours would make to a disabled person.

Interest in miniature combustion-based power systems derives from two factors: the enormous mass-specific energy advantage of liquid hydrocarbon fuels (e.g., gasoline, diesel, and JP-8) over electrochemical materials and advances in MEMS fabrication techniques that are

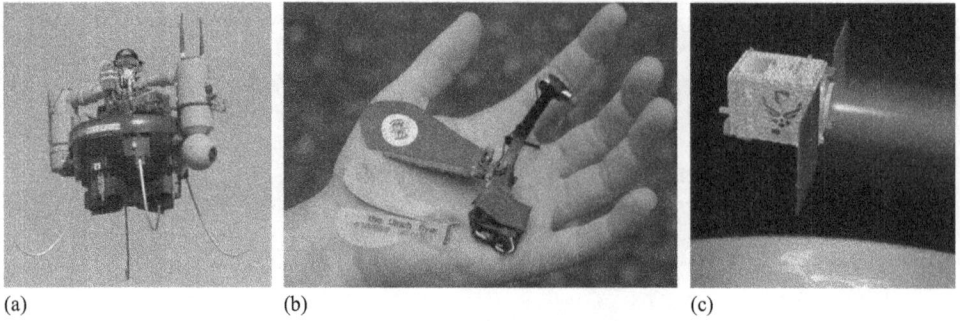

(a)   (b)   (c)

Figure 1. Examples of small unmanned air vehicles (UAVs) and satellites. (a) Honeywell T-Hawk UAV intended for battlefield surveillance (from Honeywell Inc. [77]) (b) University of Maryland's powered/controllable Samara intended for surveillance and general environmental monitoring (c) US Air Force XSS-11 microsatellite developed to re-supply and service larger satellites (from Air Force Research Laboratory [78]).

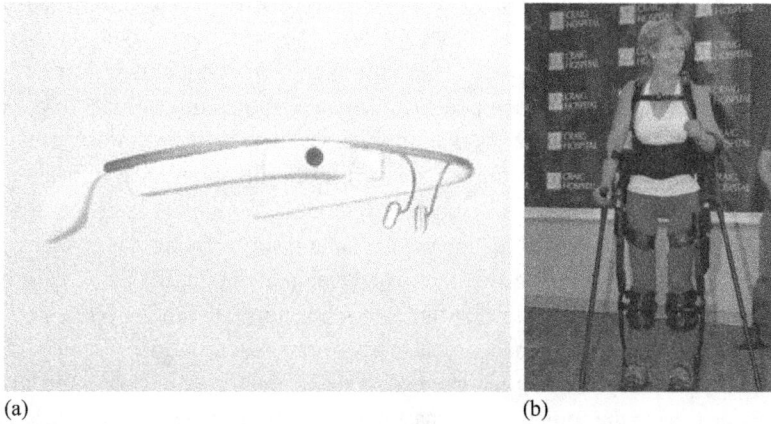

(a)   (b)

Figure 2. Examples of wearable devices for enhancing human perception and mobility. (a) Google Glass headset that lets the wearer view various types of information on a miniature display, can take pictures and video from the wearer's perspective, and can connect to the internet. (b) A lower extremity exoskeleton developed by Ekso Bionics to assist people with limited lower body function (from EKSO Bionics [79]).

capable of making "engine-like" components at micron scales.[1] Figure 3 shows that the energy density of a typical liquid hydrocarbon fuel (~44 MJ/kg [10]) is more than 60 times that of the most advanced rechargeable lithium-ion batteries (0.72 MJ/kg [11]). While the very best (and most exotic/expensive) primary batteries can achieve energy densities as high as 1.9 MJ/kg [11], this is still more than a factor of 20 lower than what is offered by a typical liquid hydrocarbon. Therefore, even a relatively low-performing (10 percent efficient) liquid hydrocarbon-based system will outperform the most exotic primary batteries by more than a factor of 2 and

---

[1]Note that while no group has succeeded in building a working MEMS engine to date, manufacturing processes for building very small machines out of an increasingly wide range of materials are available and continue to evolve.

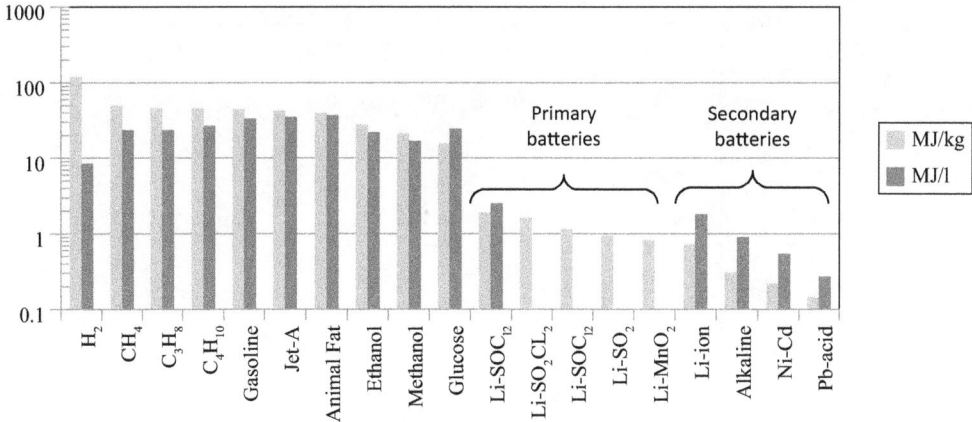

Figure 3. Gravimetric and volumetric energy densities of various energy storage materials [9–11, 80–82].

all secondary batteries by factors of 6 or more. However, to fully realize this benefit, the energy conversion system must be small enough so that it weighs little more than the non-electrochemical components (electrodes etc.) of the battery that it replaces. A similar tradeoff applies to chemical rockets except that the ratio of propellant to engine mass is often much larger than in other vehicles or power systems. For example, a single stage rocket using H_2–O_2 propellants needs a fuel mass fraction of at least 10 to achieve low earth orbit and would need larger mass fractions for propellants with lower specific energies. Reducing this mass fraction is also one reason that rockets are staged.

The tradeoff between fuel and conversion system mass/volume is more easily appreciated with the help of a Ragone diagram (Figure 4) which shows how the power and energy density of a generic power system (consisting of fuel storage and an energy conversion device) depends on the performance of the conversion device (engine), the energy content of the fuel, and the size of the fuel tank relative to the engine. The horizontal limit is the power/mass ratio of the conversion device alone while the vertical limit is the product of the overall conversion efficiency and the fuel energy density. Practical systems lie somewhere between these limits where increasing endurance (diagonal contours) comes at the price of increased fuel mass fraction and hence weight. When system weight is fixed, there are only two ways to increase endurance. One is to reduce the mass of the energy conversion system so that more energy (fuel) can be stored. The other is to increase the product of the energy density of the fuel itself with the thermodynamic conversion efficiency. This usually involves looking for higher energy density (Q_F) fuels (since the weight of batteries' electrodes relative to the electrochemical material is already very small) but consideration must also be given to the efficiency with which they are converted or else the energy density improvement will not be realized.

An analogous tradeoff between fuel and conversion system mass applies to chemical rockets although it is usually expressed in terms of thrust-to-weight ratio (T/W) and specific impulse (I_{sp}) as opposed to power and energy density. These can be converted to power and energy density, respectively. Doing so, one can show for a chemical rocket (single stage) that the horizontal asymptote (propulsion power per unit mass of system) on the Ragone diagram is $TI_{sp} g/M_e$ (where M_e is the empty mass of the propulsion system) and the vertical asymptote is $(I_{sp}g)^2/2$. The total Δv of the system is given by $I_{sp}g\ln(1+\zeta)$ where ζ is the ratio of fuel to

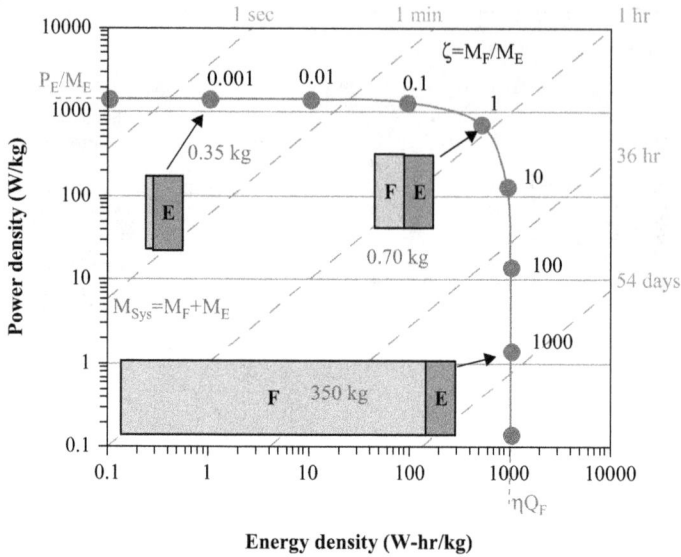

Figure 4. Power and energy density of a generic power system consisting of an engine or energy conversion system (E) of fixed size (500 W, 350 g) and a fuel tank (F) of varying size. The energy density of the fuel is assumed to be 25 MJ/kg and the overall thermodynamic efficiency of the conversion system is assumed to be 15 percent. These numbers are representative of glow-fuel powered model aircraft engines.

conversion system (engine) mass. Taken together, Figure 4 shows that the essence of the micro-power challenge is building miniaturized heat engines that consume energy-dense fuels and achieve thermodynamic efficiencies that are comparable to the much larger-scale heat engines found in automobiles, aircraft, and chemical rockets.

Figure 5 compares the levels of performance achieved by various types of power and energy systems available today (heat engines, fuel cells, and rockets) to targets associated with chemical rocket propulsion and power/energy density objectives for various military applications [9, 12, 13]. The figure shows that while conventional-scale technologies are able to access most target areas, decreasing scale generally shifts performance down and to the left with the smallest-scale fall devices falling short of the targets. This illustrates a fundamental challenge for fuel or chemical propellant-based power and energy systems: miniaturization generally comes at the price of reduced power and energy density regardless of the type of system. We will see why in the next section. Mitigating this effect is the key to realizing the advantage of fuel or propellant-based systems over electrochemical ones.

SCALE AND EFFICIENCY

Early advocates of engine miniaturization argued that building suitably small power systems was possible because of the "cube-square law" [14]. It says that if the power per unit air mass flow is constant (because it is set by the mass burning rate which depends on chemical rates that are scale-invariant), then the power per unit mass should increase as the scale of the device is

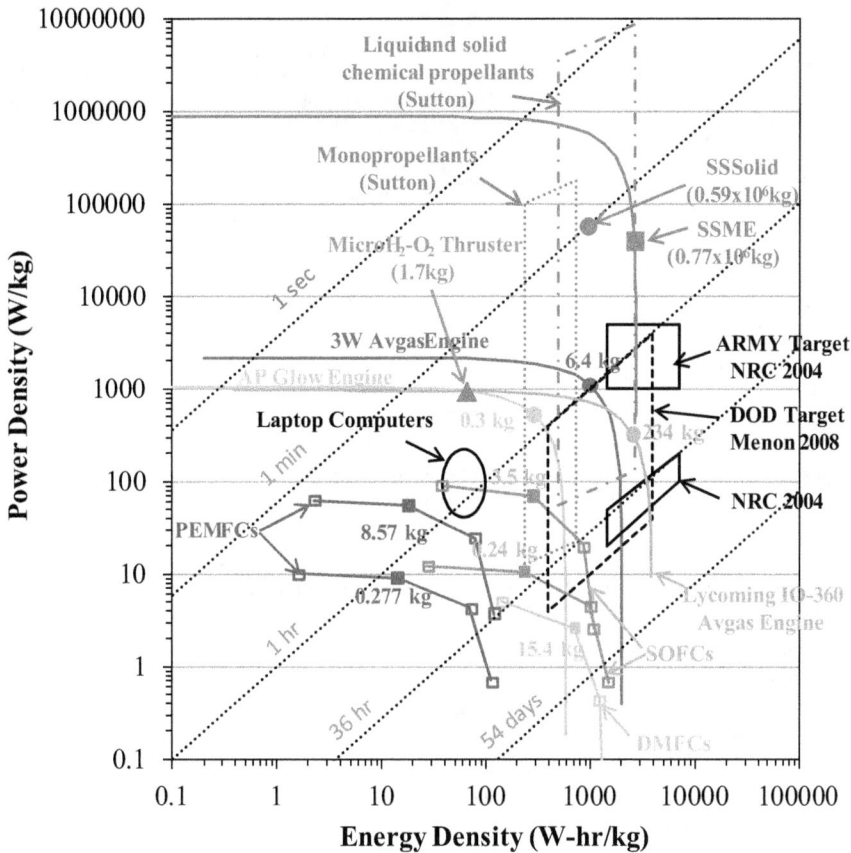

Figure 5. Ragone plot showing the relationship of various types of chemical energy conversion systems (fuel cells, laptop computers, piston engines, and chemical rockets) to DOD performance targets. SOFC stands for solid oxide fuel cell. PEMFC stands for proton exchange membrane fuel cell. DMFC stands for direct methanol fuel cell. The targets correspond to objectives identified by the National Research Council (NRC) (from Lewis [9]) and the demands of existing DOD systems (from Menon [12]).

reduced because the power per unit mass of air flow scales with the cross-sectional area (length squared) whereas the mass of the device scales with the volume (length cubed). As a practical matter, however, maintaining constant power per unit air mass flow is difficult as engine size is reduced. The reason is that the relative importance of losses associated with fluid friction, heat transfer, and radical quenching at surfaces scales with the surface to volume ratio of the device which typically increases as the device is miniaturized.[2]

As an example, consider flow through the simple tubular combustor pictured in Figure 6. The solid line shows power density when convective losses are included while the dotted line

_____

[2]An additional important constraint that is not considered in this simple analysis is the fact that the length scale of the device is also constrained by the product of flow velocity and the characteristic reaction time of the fuel. If the characteristic reaction time is fixed, smaller engines will need to operate at proportionally smaller air flow rates in order to ensure that combustion is complete.

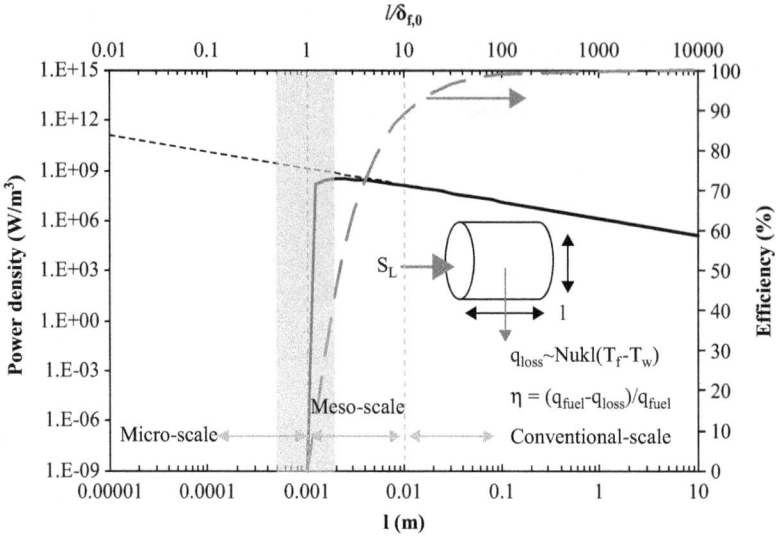

Figure 6. Power density (left axis) and efficiency (right axis) of a notional combustor as a function of its characteristic length l (lower axis) and the ratio of the characteristic length to the freely-propagating flame thickness $\delta_{f,0}$ (upper axis). $\delta_{f,0}$ is assumed to be ~1 mm. Also illustrated are definitions of microscale, mesoscale, and conventional-scale combustion regimes based on physical scale and scale relative to flame thickness. The shading highlights the somewhat ambiguous transition between "Mesoscale" and "Microscale" regimes. Stoichiometric, atmospheric pressure combustion of a "typical" liquid hydrocarbon fuel with a specific energy of 44 MJ/kg, a laminar flame speed of 42 cm/s, and a laminar flame thickness of 1 mm are assumed.

shows the power/volume predicted by the "cube-square" law. The dashed line shows the associated thermal efficiency. The "cube-square" law for power density holds until the rate of heat loss approaches the rate of heat generation. Near this point but to the right of it, reducing the size of the device no longer increases power density and comes at the price of significantly reduced efficiency. Not far to the left of this point, thermal losses outstrip production and the device is no longer thermodynamically viable. While this example's extreme simplicity does not allow one to make reliable estimates of maximum achievable power density, minimum combustor volume, or optimum combustor configuration—all of which are important to designers of combustion-based microengines—it clearly demonstrates the essential challenges of miniaturizing a heat engine: thermal management of the combustor and developing fundamental understanding of the combustion process under conditions where losses of heat and radicals to the walls are significant.

WHAT IS "MICROSCALE" COMBUSTION?

Over the past 10 or so years, various devices have been developed which have been termed "microthrusters," "microengines," or "microreactors" by their creators [7, 8, 15–27]. However, there is no consistent standard for applying this terminology. Part of this is probably due to the fact that the choice of reference length scale is somewhat arbitrary and part is probably due to the particular interests/perspectives of individual researchers. Regardless, the net result is that it is often difficult to know what people are really talking about when they refer to "microscale," "mesoscale," or "conventional-scale" combustion. This section describes the main approaches to classification that have been taken to date and proposes a standard for future classification.

Three approaches have generally been taken for defining "microscale" combustion. The first is based on the physical dimensions of the combustor: If the characteristic physical length scale is less than 1 or 2 mm, then the combustion occurs at the "microscale" and is called "microcombustion." If the physical length scale is of the order of a few millimeters, the combustion is called "mesoscale" combustion. This method is often used in the context of micro-engines [8, 15] where combustion is expected to occur in a region of fixed size. The second approach is based on some reference length scale of the flame itself. This can be the quenching diameter [28] or the freely-propagating flame thickness [29, 30]. Combustion is termed "microscale" if the combustor size is smaller than these length scales and is termed "mesoscale" if the combustor size is of the order of a few of these length scales. This method is sensible when thinking phyically (i.e., in terms of flame regimes) and thus is favored in fundamental studies. However, it too can be ambiguous as the quenching diameter can be a few times the flame thickness or more and is a strong function of mixture composition, temperature, and wall properties (thermal conductivity, thermal conductance, and surface reactivity). The third approach is based on device size relative to a "conventional" size device with a similar function. For example, a thruster on a microsatellite (which is defined as a satellite weighing between 10 and 100 kg [31]) may be termed a "microthruster" even though its (and its combustor's) dimensions would not qualify as "micro" under the previous two criteria [16, 32]. This definition is sometimes used by developers of devices for specific applications.

Figure 6 proposes a new classification scheme that has a physical basis while remaining consistent with the original conception of a "microcombustor" as a device with dimensions ~ a few millimeters or less. The lower horizontal axis shows the physical dimension of the combustor while the upper horizontal axis shows the ratio of the physical dimension to a reference flame thickness $\delta_{f,0}$. The latter is assumed to be 1 mm here which is reasonable for atmospheric pressure hydrocarbon-air flames. Two extremes are apparent when thinking in terms of flame regimes (upper axis). On the far left are flames that are much thicker than the passage dimension and on the far right are flames that are much thinner than the passage dimension. These are termed "microscale" and "conventional scale," respectively. The limits are analogous to the flame sheet and distributed reaction zone approximations used in the modeling of turbulent flames. Between these two extremes lies a transition region where the flame is neither thick nor thin with respect to the characteristic passage dimension. This is the "mesoscale" region. The key question, however, is deciding where the boundaries of the "mesoscale region" should be.

A strategy that makes physical sense is to choose $1/N < 1/\delta_{f,0} < N$ where N is sufficiently large that the distributed reaction or flame sheet assumptions hold with a reasonable level of accuracy at the boundaries. However, there are two problems with this approach even if we choose N to be relatively small—say 10: People could be confused because most systems that used to be called "microscale" would have to be re-classified as "mesoscale" and truly "microscale" systems would be so small that their chances of achieving thermodynamic viability would be poor (see the efficiency curve in Figure 6). Therefore, we propose defining a "microcombustor" to be a device with a physical length scale (passage height) < 1 mm and a "mesoscale combustor" to have dimensions in the range 1 mm < l < 1 cm. This makes the

Table 1. Proposed classification methods for reacting flows in small passages

Basis	Regime	Criteria	Examples
Physical length (L)	Mesoscale	$L \sim 1\text{–}10$ mm	MEMS power [19, 21, 33], Microreactor [27]
	Microscale	$L < 1$ mm	
Reference quenching dia. ($\delta_{q,0}$)	Mesoscale	$L/\delta_{q,0} \sim (1\text{–}10)$	Swiss roll combustor [18, 34], Fuel cells
	Microscale	$L < \delta_{q,0}$	
Reference flame thickness ($\delta_{f,0}$)	Mesoscale	$L/\delta_{f,0} \sim (1\text{–}10)\delta_f$	Thrusters/energy conversion [29, 30]
	Microscale	$L < \delta_{f,0}$	
Reference device (L_R)	Mesoscale	$L \sim L_R/(10\text{–}100)$	Microsatellites, Micro UAVs [16, 32]
	Microscale	$\sim L_R/1000$	

left boundary consistent with the concept of microcombustors as they were envisioned by the first practitioners in the field [14, 21] and the right boundary consistent with the physics of the problem where a freely propagating flame is thin ($\sim 1/10\times$) compared to the passage dimension. These criteria along with some examples are summarized in Table 1. Note that the limits for the "Reference device" in the last row of the table are inspired by the third classification approach described above but remains mostly arbitrary. We feel that this is not a good basis by which to classify combustion systems.

Finally, some researchers may prefer to define microcombustion using the Peclet number (Pe) instead of reference quenching diameters or flame thicknesses. While Pe has the advantage of being nondimensional, it only represents the ratio of convective to diffusive timescales and completely neglects the chemistry that defines the flame propagation limit. Although the minimum Pe may contain some of this information implicitly, other losses (e.g., radiation and radical quenching), fuel heating value, and boundary conditions like the wall temperature also play critical roles near the kinetic limit for flame propagation. Therefore, Pe is not a more "universal" parameter than the reference quenching diameter or flame thickness.

SCALE AND COMBUSTION

A central objective of this book is to explore the influence of scale on the combustion process. While the previous section established boundaries for micro, meso, and conventional-scale combustion with the help of some simple physical reasoning, combustion is an extremely complex process involving gas phase reaction, surface reaction, molecular transport (convective and diffusive), and thermal transport (convective, diffusive, and radiative). Therefore, it is useful to consider how other length and time scales vary with the physical scale of the device so that one can understand how physical scale influences device performance and modeling choices. Some length and time scales are presented in Tables 2 and 3 respectively.

Nondimensional parameters are also useful for understanding the effect of scale and several potentially important ones along with their definitions are presented in Table 4. They are plotted as functions of length scale in Figure 7 assuming flow in a closed channel with square

Table 2. Length scales associated with combustion in small passages

Description	Symbol	Notes
Combustor diameter	D	
Combustor length	L	
Combustor wall thickness	t_w	
Characteristic loss length	L_c	L_c=Volume/Surface area
Freely propagating flame thickness	$\delta_{f,o}$	
Flame thickness	δ_f	$\delta_f = \delta_{f,o}\beta^{1/2}$ *
Mean free path	λ	6.84×10^{-8} m (air)
Quenching distance	δ_o	

*Beta (β) is a parameter that accounts for the influence of wall proximity and composition on the effective thermal diffusivity. See [29] for more details.

Table 3. Time scales associated with combustion in small passages

Description	Symbol	Notes
Combustion time	t_c	S_L/δ_f
Residence (convective) time	t_r	S_L/L
Gas phase thermal diffusion	$t_{\alpha,g}$	$D^2/4\alpha_e$
Solid phase thermal diffusion	$t_{\alpha,s}$	$D^2/4\alpha_s$ ‡
Mass diffusion	t_d	$D^2/4D_{AB}$ §
Acoustic wave	t_a	D/a **
Heat loss to environment	$t_{h,e}$	$\rho_s C_s L_c^2/Nu_e k_e$ ††
Heat loss to wall	$t_{h,w}$	$L_c^2/Nu_i \alpha_e$

*S_L is the laminar flame speed. It is given by $S_L = S_{L,o}\beta^{1/2}$ where $S_{L,o}$ is the freely propagating laminar flame speed.
†α_e is the "effective" thermal diffusivity. It is given by $\beta\alpha_g$ where α_g is the thermal diffusivity of the gas.
‡α_s is the thermal diffusivity of the structure material.
§D_{AB} is the binary diffusivity of the fuel into air in m²/s.
**a is the local speed of sound a=$(\gamma RT)^{1/2}$.
††ρ_s is the density of the structure material, C_s is the heat capacity of the structure material, L_c is a characteristic length for heat transfer from the structure to the environment (typically the ratio of volume/surface area ratio), Nu_e is the Nusselt number for heat transfer from the structure to the environment (~3.7 for free convection), k_e is the thermal conductivity of the gaseous environment.

Table 4. Non-dimensional parameters associated with combustion in small passages

Description	Symbol	Notes
Biot number	Bi	$N_{u,e}k_eL_c/(Dk_s)$ *
Damköhler number (convection)	Da_{res}	t_r/t_c
Damköhler number (diffusion)	Da_{diff}	t_d/t_c
Lewis number	Le	α_e/D_{AB}
Reynolds number	Re	S_LD/ν †
Mach number	M	S_L/a
Fourier number	Fo	$t_c/t_{\alpha,s}$
Knudsen number	Kn	λ/D
Peclet number for wall heat transfer	$Pe_{h,w}$	$t_{h,w}/t_r$
Peclet number for mixing	Pe_{mix}	t_d/t_r
Non-dimensional flame thickness	D/δ_f	
Acoustic to chemical time ratio	t_a/t_c	
External heat loss to chemical time ratio	$t_{h,e}/t_c$	

*k_s is the thermal conductivity of the structure material.
†ν is the kinematic viscosity of the gas flow through the combustor.

cross-section, inner dimension D, length L (=10D), and wall thickness t (=D/10). The velocity in the channel is taken to be the laminar burning velocity. Two qualifications are in order. First, this is not an "all-inclusive" list of non-dimensional parameters and there are undoubtedly others that are important too. Second, the combustor model upon which the following analysis is based is extremely simple and takes no account of flame shape or turbulence that could be present at larger scales. Its predictions are only qualitative and therefore the important things are the *trends* which can help us understand how and why combustion stabilized in small passages differs from conventional-scale combustion.

As the combustor size is reduced, the surface to volume ratio increases dramatically leading to strong thermal and chemical interactions between the reacting flow and the combustor structure. This is the distinguishing aspect of meso and microscale combustion and has a number of important consequences. First, the increase in surface/volume ratio makes the flow more susceptible to heat losses as illustrated by the behavior of $t_{h,e}/t_c$ in Figure 7 and ultimately the efficiency curve in Figure 6.[3] However, some of the first work in microscale combustion showed that both heat loss from the outer structure *and* near wall flame quenching associated with flame curvature (i.e., shearing at the wall) play important roles in determining extinction limits in small channels [35–37]. Second, the *effective* thermal diffusivity (α_e) of the reactant

---

[3]The miniaturization limit identified in Figure 6 corresponds to the ratio of the thermal loss time scale to the combustion time scale ~ unity.

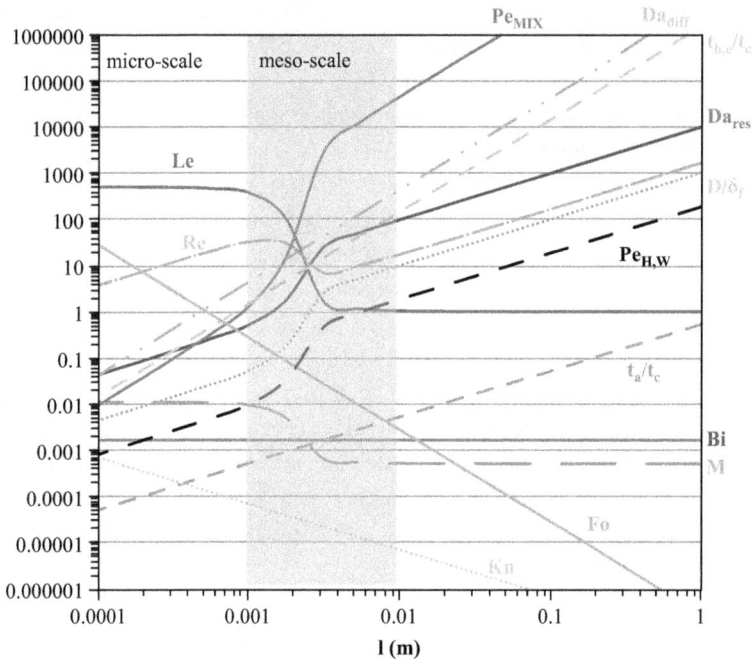

Figure 7. Scale dependence of various nondimensional parameters associated with reacting flow in a channel.

mixture increases significantly as scale is reduced. This is because reducing the passage dimension facilitates heat transfer to the wall and most structural materials have thermal conductivities that are at least two orders of magnitude greater than those of most gases. The result is a transition from gas-dominated to wall-dominated thermal diffusivity which causes the "kinks" in some of the curves in Figure 7 and a number of other interesting effects.

One important consequence of the transition from gas-dominated to wall-dominated thermal diffusivity is an increase in the laminar flame speed over the value it would have as an unconfined (i.e., freely propagating) laminar flame. This changes the chemical time scale and begins at scales of the order of the freely propagating flame thickness ($\delta_{f,o}$) [29, 30, 38]. A related consequence is flame broadening in which the reaction zone becomes thicker than in a freely propagating flame (i.e., $\delta_f > \delta_{f,o}$) [29, 30]. While this increases the required combustor volume, the flame speed increases more so the power density that can be achieved actually increases with reduced scale until a thermal quenching limit is reached [29]. Other consequences of enhanced thermal feedback through the structure are multiple combustion regimes in which both normal and weak flames exist [38, 39] and broadened flammability limits. The latter was shown analytically for heat-recirculating combustors by Ronney [40] and Ju [41] and explains the experimental findings of Weinberg [42]. Similar studies in rectangular channels by Ju [43, 44] and Maruta [45, 46] also show a broadening of flammability limits as scale is reduced that is associated with a new "weak" flame regime.

The increase in effective Lewis and Fourier numbers as scale is reduced suggests that combustion instability could also be a challenge. And, indeed, a wide variety have been observed including extinction and re-ignition instability [45, 47, 48], spinning instability [49], cellular instability [50], and flame "streets" [51]. Some computational studies have also been able to reproduce re-ignition and pulsating flame instabilities [52].

The only parameter that is scale-invariant (to the extent that the Nusselt number for external heat loss, Nu_e, is scale-invariant) is the Biot number and its relatively low value ($\sim 10^{-3}$) suggests that temperature gradients within the combustor's walls are small. Thus, it may be appropriate to assume that walls are spatially isothermal. However, the fact that the Fourier number increases as scale is reduced suggests that there will be scales where the walls are no longer temporally isothermal and it will be necessary to account for the unsteady temperature distribution in the solid phase. The strong gas-wall coupling also leads to wall temperatures that are higher than in conventional systems. While this can lower the ignition delay time, it places a proportionally larger burden on the materials used to construct micro- and mesoscale combustors. At a minimum, they must retain their strength at higher temperatures but depending on the structure's function they may also have to maintain other properties like dielectric constant or permanent magnetism too.

Figure 7 shows that the Mach number changes depending on whether one is in a wall-dominated or gas-dominated regime but overall it remains very low indicating that compressibility effects are negligible. In addition, the decrease in the acoustic time scale relative to the chemical one suggests that acoustic effects could become even less important at small scales. The Reynolds number decreases with scale as one expects, but its relatively low value across the range of scales serves as a reminder that trends rather than magnitudes are the most useful things to extract from this analysis. The Reynolds numbers reported here should be thought of as lower bounds because practical combustors have flameholders that hold the reaction zone at an angle relative to the incoming flow enabling operation at velocities far exceeding the laminar flame speed. Nevertheless, the decrease in Reynolds number with decreasing scale has several important consequences. First, it suggests that turbulence will be less available (or even non-existent) as a means for wrinkling the reaction zone, increasing the flame area, and thus the burning rate. While this is offset somewhat by the increase in flame speed associated with thermal feedback through the structure, it remains an important limitation by driving the Damköhler number down. This, in turn, can lead to problems with incomplete combustion. Second, mixing of fuel and oxidizer will be more challenging as turbulence may not be available to maintain steep concentration gradients [53]. Thus, the distance that must be allocated to mix fuel and oxidizer may need to be proportionally larger than in conventional-scale systems. This further complicates efforts to achieve high power density.

Finally, Figure 7 suggests that the most significant property changes occur near the interface between the mesoscale and microscale regions and thus this is where one should see the strongest effects of scale and the richest range of behaviors. Coincidentally, this is also where most devices built to date operate. Figure 7 also indicates that most practical devices (i.e., those with $t_{h,e}/t_c \gg 1$) should lie well within the continuum region (Kn < 0.01). However, recent work suggests that non-equilibrium effects involving thermal and concentration slip and radical quenching at the walls can be important [54–56] in certain situations. Under these conditions, simplified representations of heat, mass, and momentum transfer at walls break down and different methods appropriate for non-continuum flows are required.

A BRIEF SURVEY OF MICROPOWER GENERATION DEVICES

The idea of micropower generation began with the pioneering work of Epstein et al. [22] at MIT who sought to construct a miniature gas turbine with integrated electrical generator. This work catalyzed many other innovative projects to develop microcombustors for propulsion, power generation, chemical sensing, and heating. The MIT engine (Figure 8) was a true MEMS device (although a very large one at $20 \times 20 \times 8$ mm) fabricated out of silicon using various etching, deposition, and wafer bonding techniques. The final performance target was 50 W of electrical power on 7 grams of jet fuel per hour. While it never operated as a stand-alone device, the contributions that the project made to our understanding of the design of miniature power systems and MEMS fabrication techniques are wide-ranging and significant. Similarly influential efforts at UC Berkeley by Fernandez-Pello and co-workers [33] focused on developing liquid-fueled rotary engines at both the meso- and microscales. The mesoscale version (Figure 9) was manufactured using conventional EDM technology while the microscale version was a true MEMS device. Target performance levels were 30 watts and a few milliwatts, respectively. Honeywell [57], Georgia Tech [58], University of Michigan [24] (Figure 10), Aerodyne Research [26, 59] (Figure 11), and a Yale/Sunpower team [60, 61] (Figure 12) focused on free

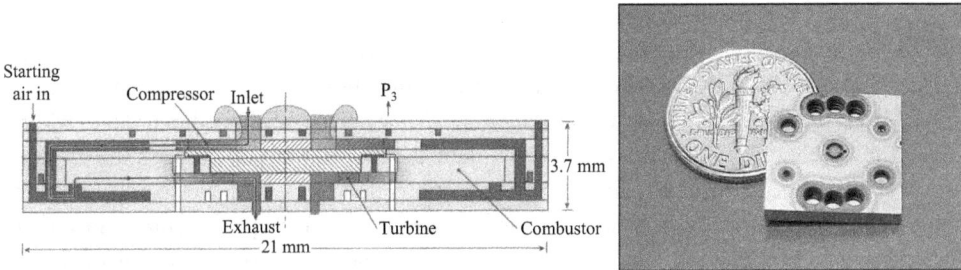

Figure 8. MIT microengine. Cross-sectional view (left, from Spadaccini [83]) and prototype "bearing rig" (right, from Knapp [84]).

Figure 9. UC Berkeley micro Wankel engine (from Zandonella [85]).

Figure 10. University of Michican Micro Internal Combustion Swing Engine (MICSE). Cutaway view showing oscillating rotor and generator (left) and a prototype device next to a propane fuel canister (from Dahm [86]).

piston designs with some using homogeneous charge compression ignition (HCCI). Three of these efforts led to true stand-alone devices [24, 59, 61] and work continues on one but at larger scales suitable for UAVs (K. Annen, pers. comm.) However, most of these designs suffered from problems related to thermal loss and the use of moving parts—friction and sealing/leakage—with the net result being low energy conversion efficiency. Other challenges include material degradation and the delivery, atomization, and ignition of liquid fuel.

Similar efforts have been made to miniaturize chemical thrusters. "Digital" monopropellant [18, 20] and bi-propellant microthrusters [19] have been constructed using MEMS fabrication technologies. Recently, mesoscale and microscale thrusters using high melting temperature metals [16] (Figure 13), ceramics [28], and quartz [62] have been developed in order to address engine life problems associated with silicon's poor thermal stability at combustion temperatures (1600–1800 K). These have improved combustion efficiency and flame stability considerably but the wall heat loss is still large and so the overall specific impulse remains low. Other challenges include liquid fuel injection, mixing, ignition, and cold start. Fabrication of ceramic microcombustors also remains difficult.

Swiss Roll microcombustors [63–65] (Figure 14) have been developed as a way to address problems with thermal efficiency and combustion instability by transferring thermal energy in the burned gas back to the unburned mixture. This heat recirculation can enable the reaction zone to reach temperatures that are at least two times higher than the adiabatic flame temperature (although the equilibrium adiabatic flame temperature itself is a purely thermodynamic quantity that does not change). This makes the flame more resistant to thermal quenching and enables combustion well beyond the standard flammability limits [40, 41]. The concept, based on previous work by Weinberg and co-workers [42, 66, 67] and Takeno et al. [68], addresses two very important challenges of microscale combustion: thermal management and flame stability. An important challenge that remains, however, is how to integrate such a burner into a thermal cycle. One approach is to use the burner as a heat source for a thermoelectric generator. Another may be to use it to sustain high temperatures for use with a single-chamber solid-oxide fuel cell [69].

Finally, catalysts can also be used to enhance flame stability and to broaden flammability limits so various meso- and microscale catalytic reactors have been developed for power generation [62, 64, 70, 71]. Thermal management can be more of a challenge in these devices, however, because chemical reaction occurs on the surfaces of flow passages making it easier for

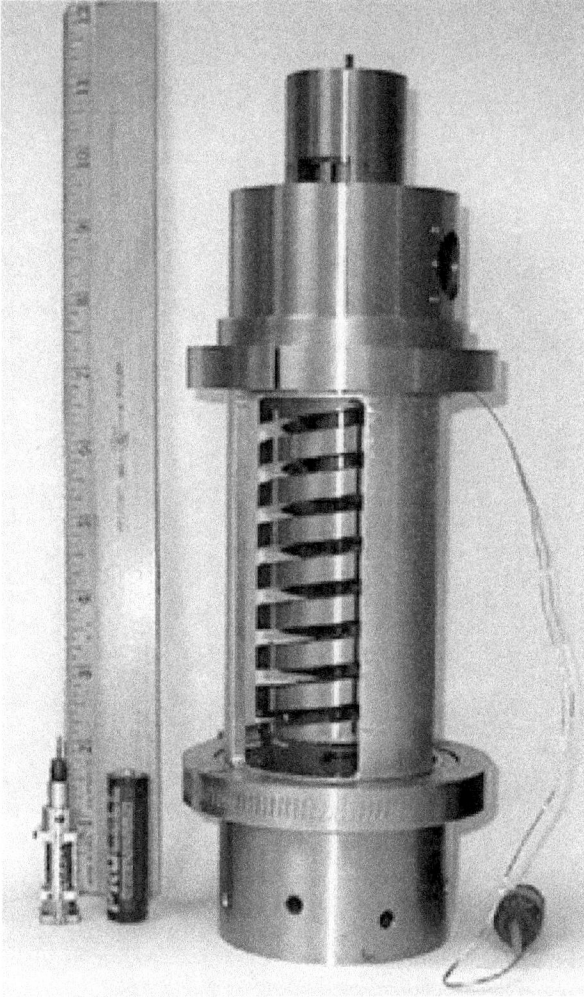

Figure 11. Two of Aerodyne's Miniature Internal Combustion engine-generators (MICE): 10 W (left, funded by the Defense Advanced Projects Administration) and 500 W (right, funded by the U.S. Army) (from Haswell [87]).

heat to be transferred to the structure and escape to the environment. Table 5 summarizes the attributes of a variety of microthrusters and engines that have been developed.

WHERE DO WE GO FROM HERE? ENABLING RESEARCH FOR MICROSCALE POWER SYSTEMS

In light of the challenges identified in above, a number of research topics are important for the realization of practical meso- and microscale combustion-based power systems. At the most basic level, much more needs to be understood about the fundamental processes of flame stabilization and chemical reaction at the meso- and microscales—especially the roles

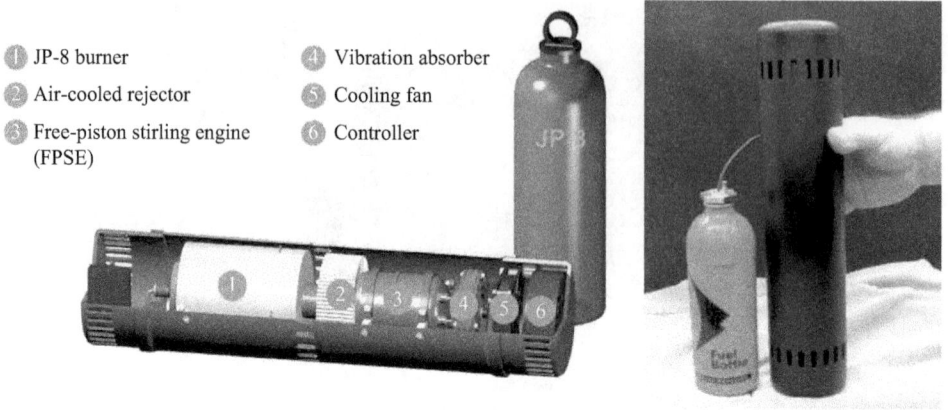

① JP-8 burner
② Air-cooled rejector
③ Free-piston stirling engine (FPSE)
④ Vibration absorber
⑤ Cooling fan
⑥ Controller

Figure 12. Yale-Sun Power free piston Stirling engine: cutaway view (left) and weight optimized prototype (right) (from Huth and Collins [61]).

Figure 13. PSU Inconel microthruster (right) installed on thrust stand (left).

Figure 14. Example of a USC swiss roll combustor (from Ronney [88]).

of surface chemistry, flame-wall thermal coupling, chemical–electrical interactions involving plasma discharge, and the suppression (or possible exploitation) of the various thermo-diffusive instabilities that have been observed. Understanding the influence of these additional

Table 5. Examples of various micro and mesoscale chemical energy conversion systems

Technology	Material	Scale	Design thrust/power	Propellant	Pressure (atm)	Challenges
Digital rocket chip (Caltech)	Silicon	0.5 mm³	0.1 N, 100W	$C_6H_3N_3O_8Pb$	~1	Large heat loss $d/\delta_f \sim O(10)$
Bi-propellant (MIT)	Silicon	705 mm³	1N, 750 W	O_2/CH_4	12.3	Low flame temperature and small $Da=t_{res}/t_c$
Staged thruster (PU)	Quartz	2000 mm³	~500 W	Methanol, Butene/air	2	Wall-flame coupling $t_c/t_s= O(1)$
Vortex flow combustor (PSU)	Inconel	10–50 mm³	~500 W	methane/oxygen	1	Small $Da=t_{res}/t_c$
Electrolytic thruster (PSU)	ceramic	322 mm³	197 mN	HAN	1	Cracking, heat loss, $d/\delta_f \sim O(10)$
Non-premixed combustor (UIUC)	Quartz, alumina	0.1–2 mm	10–50 W	H_2, Methane	1	Thermal, radical quenching $Da=t_{res}/t_c \sim 1$
Micro gas turbine (MIT, Tohoku)	Silicon	60–200 mm³	50W	Jet fuel	1	Friction, sealing
Rotary engine (UCB)	Silicon, Stainless	1–1000 mm³	0.01–30 W	Hydrocarbon fuel	1	Fuel delivery, friction, sealing, $Da=t_{res}/t_c \sim 1$
Free-piston engine (Honeywell, UM, Georgia Tech)	Stainless	1 cm³	~10W	heptane, butane, Jet fuel	1–5	Mass loss, sealing, ignition, $t_{ig}/t_{res}=O(1)$

interactions on the combustion processes is challenging because they greatly increase the complexity of the problem. For example, properly representing surface reactions requires accounting for the composition and morphology of the active surface (which may change over the course of a device's operation) in addition to the chemical reaction and heat transfer within the structure itself. However, meeting these challenges will yield real benefits by, for example, helping us to understand how to use wall–thermal coupling to create low pressure-drop combustors with extraordinarily high turndown ratios or how to use plasmas to reduce activation energy, broaden flammability limits, and control misfire in small UAV engines [72, 73]—a significant problem.

Another huge barrier to understanding the operation of meso- and microscale power systems is making the measurements that are necessary to validate model predictions. The combustion environment is already extremely hostile to physical probes and protection strategies like water cooling used at the conventional-scale are virtually impossible to implement in a way that does not alter the phenomena of interest. Therefore, non intrusive methods are essential. However, modifying the device to permit optical access is often as disruptive as installing some kind of physical probe. One promising approach is to exploit Silicon's transmissivity in the infrared which makes it possible to perform spectroscopic measurements directly through a Silicon MEMS structure [74, 75]. However, this may not be possible in structures made of heavily doped Silicon or integrated with other devices so better techniques are required. The challenge remains significant even at the mesoscale where there are no pressure transducers that are small enough to install in the cylinder of the smallest commercial engines. Some pressure and flame luminescence measurements have been made in engines at somewhat larger scales [76] but better techniques are needed to probe the smallest engines where the really interesting things may be happening.

Going hand-in-hand with miniaturization is an increased level of integration across the device. Thus, accounting for 'communication' between components is very important. We have already seen that accounting for the thermal properties of the wall is critical in understanding the performance of meso- and microscale combustors. Therefore, we should expect that understanding the performance of meso- and microscale *devices* will require even more complex, multi-disciplinary models that account for many different types of interactions between components (i.e., mechanical, thermal, electrical).

Miniature combustion-based energy conversion devices cannot exist without similarly compact systems for delivering fuel and mixing it with oxidizer. Such systems are not presently available. The first step in developing them is learning more about the mixing process at the meso- and microscale in order to establish the length of the flow path that needs to be devoted to mixing. At present there are little to no data in the open literature for liquid–gas mixing at the Reynolds numbers expected in with meso and microscale power systems. The second step is to devise some kind of compact fuel injection (i.e., metering and mixing) system based on what was learned in step one. It should also be easily scalable in order to accommodate a range of engine sizes. The smallest fuel injection systems available today are intended as retrofits for motorcycles and are much too large for meso- and microscale power systems. Related to the need for miniature fuel injection systems is the need for miniature valves that are capable of switching pressures ~ tens of atmospheres with very low power consumption.

Another very important challenge is the development of new high temperature materials. These must resist oxidation, retain their properties at combustion temperatures (1600–1800 K), and should have low thermal conductivity. However, we have seen above that some of meso- and microscale combustion's advantages (higher flame speeds and wider stability limits) stem from this thermal feedback. So, the ideal material (for non-counter-current heat recirculating combustors) could be one that is thermally anisotropic: conductive in the streamwise direction while insulating in the transverse direction. While it may not be possible to achieve this with a single material, it may be possible to do so using hybrid structures formed by bonding highly conductive materials to insulating ones. Another critically important need is for low thermal conductivity, heat-resistant materials that are easy to machine.

While polymers like Vespel have very low thermal conductivities, they are only viable for temperatures up to 500°C. Ceramics can tolerate higher temperatures but they have much higher thermal conductivities, are much harder to machine, and have many more problems with thermal stress. New manufacturing technologies are also required. For example, better techniques are needed for micromachining existing refractory materials like Silicon Carbide and Alumina. To date, most 'true' microcombustors have been machined from Silicon simply because so much more is known about Silicon microfabrication than other materials. Future efforts to build meso and microscale combustion power systems could be a lot more successful if more were known about how to micromachine refractory materials like silicon carbine and alumina. An interim strategy is to develop more precise ways of cutting, aligning, and bonding green ceramic tapes. Finally, traditional MEMS fabrication is limited to planar structures. The ability to create three-dimensional MEMS structures (without excessive surface roughness) would be truly revolutionary.

Finally, a number of practical problems need to be solved. First among them is devising strategies to mitigate or eliminate mismatches in coefficients of thermal expansion (CTE) that produce high internal stresses and lead to cracking and mechanical failure. This may be accomplished either through design (i.e., by designing components capable of deforming at interfaces) or devising ways to adjust a particular material's CTE to match that of adjacent materials. Fluidic connections for supplying fuel are another "annoying" but critically important challenge as existing glass frit technologies remain finicky and fragile. Better methods also need to be developed for sealing moving components like pistons and rotors without incurring large frictional losses. While compression is needed to improve the thermal efficiency of most heat engines, it could be possible to bypass the aforementioned sealing and friction problems by developing compression techniques that do not rely on traditional pistons and turbomachinery and thus are more appropriate for the meso- and microscale.

OVERVIEW OF THIS VOLUME

The objective of this book is to provide an overview of the technological developments and advances in the fundamental understanding of micro and mesoscale combustion that has occurred over the last two decades. It consists of 16 chapters which are grouped into three parts. All chapters are contributed by scientists who have long and distinguished records in this area.

Part 1 addresses fundamental physical aspects of microscale combustion. Chapters 1 and 2 discuss the flame propagation limits in mesoscale channels and the dynamics of flame acceleration in a boundary layer. Chapter 3 summarizes phenomena and mechanisms of instabilities of premixed flames due to flame-structure interaction. Chapter 4 presents numerical modeling results associated with the work described in Chapters 1–3. Chapter 5 discusses research efforts in non-premixed microscale combustion, and Chapter 6 investigates diffusion flame instability and cell formation of non-premixed microscale combustion.

Part 2 presents combined theoretical and experimental investigations of a number of different types of miniature reactors. Chapter 7 discusses the effects of wall geometry on repetitive extinction/re-ignition and sound emission. Chapter 8 addresses fundamental aspects of the development of catalytic reactors. Chapter 9 focuses on liquid film combustion as a possible means to mitigate thermal losses. Chapter 10 presents the "Swiss roll" combustor as another strategy for mitigating thermal losses. Chapter 11 presents a detailed analysis of catalytic reactors and their construction for the purposes of power generation. Chapter 12 investigates the effect of varying the external wall temperature distribution on the combustion process occurring in a small quartz tube.

Part 3 focuses on the development of microengines and thrusters. It includes four chapters which discuss the design and performance of microthrusters (Chapter 13), microrotary engines (Chapter 14), mesoscale reciprocating engines (Chapter 15), and micro gas-turbine engines (Chapter 16).

REFERENCES

[1] K. Ro, J.-S. Oh and L. Dong, "Lessons Learned: Application of Small UAV for Urban Highway Traffic Monitoring," in 45th AIAA Aerospace Sciences Meeting and Exhibit, Reno, 2007.

[2] B. Coifman, M. McCord, R. G. Mishalani, M. Iswlat and Y. Ji, "Roadway Traffic Monitoring from an Unmanned Aeroal Vehicle," IEEE Proceedings of Intelligent Transportation System, vol. 153, no. 1, March 2006.

[3] L. F. Johnson, S. Herwitz, S. Dunagan, B. Lobitz, D. Sullivan and R. Slye, "Collection of Ultra High Spatial and Spectral Resolution Image Data Over California Vineyards with a Small UAV," in Proceedings of the 30th International Symposium on Remote Sensing of Environment, Honolulu, HA, November 10–14, 2003.

[4] D. W. Casbeer, R. W. Beard, T. W. McLain, S. Li and R. K. Mehra, "Forest Fire Monitoring with Multiple Small UAVs," in American Control Conference, Portland, OR, June 8–10, 2005.

[5] L. Wallace, A. Lucieer, C. Watson and D. Turner, "Development of a UAV-LiDAR System with Application to Forest Inventory," Remote Sensing, vol. 4, no. 6, pp. 1519–1543, 2001.

[6] "Data, Data Everywhere," The Economist, 25 February 2010.

[7] D. Dunn-Rankin, E. M. Leal and D. C. Walther, "Personal Power Systems," Progress in Energy and Combustion Science, vol. 31, no. 5–6, pp. 422–465, 2005.

[8] Y. Ju and K. Maruta, "Microscale Combustion: Technology Development and Fundamental Research," Progress in Energy and Combustion Science, vol. 37, no. 6, pp. 669–715, 2011.

[9] N. Lewis, "Portable Energy for the Dismounted Soldier," MITRE Corporation, 2002.

[10] A. F. Mills, "Basic Heat and Mass Transfer," in Table A.25, Chicago, IL: Richard D. Irwin Inc., 1995, p. 879.

[11] M. Broussely and G. Archdale, "Li-ion Batteries and Portable Power Source Prospects for the Next 5-10 years," Journal of Power Sources, vol. 136, no. 2, pp. 386–394, 2004.

[12] S. Menon, Ph.D Thesis: The Scaling of Performance and Losses in Miniature Internal Combustion Engines, University of Maryland, 2011, pp. 188–194.

[13] P. F. Flynn, "Meeting the Energy Needs of Future Warriors," Washington, DC: National Academies Press, 2004.

[14] A. H. Epstein and S. D. Senturia, "Macro power from Micro Machinery," Science, vol. 276, no. 5316, p. 1211, 1997.

[15] A. C. Fernandez-Pello, "MicropowerGeneration Using Combustion: Issues and Approaches," Proceedings of the Combustion Institute, vol. 29, pp. 883–889, 2002.

[16] R. A. Yetter, V. Yang, M. H. Wu, Y. Want, D. Milius, I. A. Aksay and F. L. Dryer, "Combustion Issues and Approaches for Chemical Microthrusters," International Journal of Energetic Materials and Chemical Propulsion, vol. 6, no. 4, pp. 393–424, 2007.

[17] H. H. Wu, R. A. Yetter and V. Yang, "Development and Characterization of Cermaic Micro Chemical Propulsion and Combustion Systems," in 46th AIAA Aerospace Sciences Meeting and Exhibit, 2008.

[18] K. Zhang, S. Chou and S. Ang, "MEMS-Based Solid Propellant Microthruster Design, Simulation, Fabrication, and Testing," Journal of Microelectromechanical Systems, vol. 13, no. 2, pp. 165–175, 2004.

[19] A. London, A. Ayon, A. Epstein, S. Spearing, T. Harrison, Y. Peles and J. L. Kerrebrock, "Microfabrication of a High Pressure Bipropellant Rocket Engine," Sensors and Actuators A—Physical, vol. 92, no. 1–3, pp. 351–357, 2001.

[20] D. Lewis, S. Janson, R. Cohen and E. Antonsson, "Digital Micropropulsion," Sensors and Actuators A-Physical, vol. 80, no. 2, pp. 143–154, 2000.

[21] A. Mehra, X. Zhang, A. A. Ayon, I. A. Waitz, M. A. Schmidt and C. M. Spadaccini, "A Six-Wafer Combustion System for a Silicon Micro Gas Turbine Engine," Journal of Microelectromechanical Systems, vol. 9, no. 4, pp. 517–527, 2000.

[22] A. H. Epstein, S. D. Senturia, I. Al-Midani, G. Anathasuresh, A. Ayon, K. Breuer, K. S. Chen, F. F. Ehrich, E. Esteve, L. Frechette et al., "Micro-Heat Engines, Gas Turbines, and Rocket Engines—The MIT Microengine Project," in 28th AIAA Fluid Dynamics Conference, Snowmass Village, CO, 1997.

[23] C. M. Spadaccini, X. Zhang, C. P. Cadou, N. Miki and I. A. Waitz, "Preliminary Development of a Hydrocarbon-Fueled Catalytic Micro Combustor," Sensors and Acutators A, vol. 103, no. 1–2, pp. 219–224, 2003.

[24] W. A. Dahm, J. Ji, R. Mayor, G. Qiao, S. Dyer, A. Benjamin, Y. Gu, Y. Lei and M. Papke, "Micro Internal Combustion Swing Engine (MISCE) for Portable Power Generation Systems," in 40th AIAA Aerospace Sciences Meeting and Exhibit, Reno, NV, 2002.

[25] H. T. Aichlmayr, D. B. Kitteson and M. R. Zachariah, "Free-Piston Homogeneous Charge Compression Ignition Compressor Concept—Part I: Performance Estimation and Design Considerations Unique to Small Dimensions," Chemical Engineering Science, vol. 57, no. 19, pp. 4161–4171, 2002.

[26] K. D. Annen, D. B. Stickler and P. I. Kebabian, "Miniature Motor Generator." US Patent 6349683, February 2002.

[27] C. M. Miesse, R. I. Masesl, C. D. Jensen, M. A. Shannon and M. Short, "Submillimeter-scale combustion," AIChE Journal, vol. 50, no. 12, pp. 3206–3214, 2004.

[28] K. Zhang, S. Chou and S. Ang, "Development of a Low Temperature Co-Fired Ceramic Solid Propellant Microthruster," Journal of Micromechanics and Microengineering, vol. 15, no. 5, pp. 944–952, 2005.

[29] T. T. Leach and C. P. Cadou, "The Role of Structural Heat Exchange and Heat Loss in the Design of Efficient Silicon Micro-Combustors," Proceedings of the Combustion Institute, vol. 30, pp. 2437–2444, 2005.

[30] T. T. Leach, C. P. Cadou and G. S. Jackson, "Effect of Structural Conduction and Heat Loss on Combustion in Micro-channels," Combustion Theory and Modeling, vol. 10, no. 1, pp. 85–103, 2006.

[31] J. Mueller, "Thruster Options for Microspacecraft: A Review and Evaluation of State-of-the-Art and Emerging Technologies," in Micropropulsion for Small Spacecraft, M. Micci and A. D. Ketsdever, Eds., Reston, VA: American Institute of Aeronautics and Astronautics, 2000, pp. 45–137.

[32] N. Chigier and T. Gemci, "A Review of Micro Propulsion Technology," in 41st AIAA Aerospace Sciences Meeting and Exhibit, Reno, 2003.

[33] K. Fu, A. Knobloch, F. Martinez, D. Walther, C. Fernandez-Pello, D. Liepmann, K. Miyaska and K. Maruta, "Design and experimental results of small-scale rotary engines," in Proceedings of 2001 ASME International Mechanical Engineering Congress and Exposition, New York, NY, November, 11–16, 2001.

[34] A. L. Cohen, P. Ronney, U. Frodis, L. Sitzki, E. Meiburg and S. Wussow, "Microcombustor and Combustion-Based Thermoelectric Microgenerator." US Patent 6613972, 2 September 2003.

[35]   J. Daou and M. Matalon, "Flame Propagation in Poiseuille Flow Under Adiabatic Conditions," Combustion and Flame, vol. 124, no. 3, pp. 337–349, 2001.

[36]   J. Daou and M. Matalon, "Influence of Conductive Heat Losses on the Propagation of Premixed Flames in Channels," Combustion and Flame, vol. 128, no. 4, pp. 321–339, 2002.

[37]   J. Daou, J. Dold and M. Matalon, "The Thick Flame Asymptotic Limit and Damkohler's Hypothesis," Combustion Theory and Modeling, vol. 6, no. 1, pp. 141–153, 2002.

[38]   A. Veeraragavan and C. Cadou, "The Influence of Heat Recirculation on Flame Speed and Stabilization in Micro/Mesoscale Combustion," Combustion and Flame, vol. 158, no. 11, pp. 2178–2187, 2011.

[39]   Y. Ju and B. Xu, "Effects of Channel Width and Lewis Number on the Multiple Flame Regimes and Propagation Limits in Mesoscale," Combustion Science and Technology, vol. 178, no. 10–11, pp. 1723–1753, 2006.

[40]   P. D. Ronney, "Analysis of Non-adiabatic Heat Recirculating Combustors," Combustion and Flame, vol. 135, no. 4, pp. 421–439, 2003.

[41]   Y. Ju and C. W. Choi, "An Analysis of Sub-limit Flame Dynamics Using Opposite Propagating Flames in Mesoscale Channels," Combustion and Flame, vol. 133, no. 4, pp. 483–493, 2003.

[42]   F. J. Weinberg, "He first half-million years of combustion research and today's burning problems," Symposium (International) on Combustion, vol. 15, no. 1, pp. 1–17, 1975.

[43]   Y. Ju and S. Minaev, "Dynamics and Flammability Limit of Stretched Premixed Flames Stabilized by a Hot Wall," Proceedings of the Combustion Institute, vol. 29, pp. 949–956, 2002.

[44]   Y. Ju and B. Xu, "Theoretical and Experimental Studies on Mesoscale Flame Propagation and Extinction," Proceedings of the Combustion Institute, vol. 30, no. 2, pp. 2445–2453, 2005.

[45]   K. Maruta, T. Kataoka, N. I. Kim, S. Minaev and R. Fursenko, "Characteristics of Combustion in a Narrow Channel with a Temperature Gradient," Proceedings of the Combustion Institute, vol. 30, no. 2, pp. 2429–2436, 2005.

[46]   H. Nakamura, A. Fan, H. Minamizono, K. Maruta, H. Kobayashi and T. Niioka, "Bifurcations of Stretched Premixed Flames Stabilized by a Hot Wall," Proceedings of the Combustion Institute, vol. 32, no. 1, pp. 1367–1374, 2009.

[47]   K. Maruta, J. K. Park, K. C. Oh, T. Fujimpori, S. S. Minaev and R. V. Furshenko, "Characteristics of Microscale Combustion in a Narrow Heated Channel," Combustion, Explosion and Shock Waves, vol. 40, no. 5, pp. 516–523, 2004.

[48]   F. Richecoeur and D. C. Kyritsis, "Experimental Study of Flame Stabilization in Low Reynolds and Dean Number Flows in Curved Mesoscale Ducts," Proceedings of the Combustion Institute, vol. 30, no. 2, pp. 2419–2427, 2005.

[49]   B. Xu and Y. Ju, "Experimental Study of Spinning Combustion in a Mesoscale Divergent Channel," Proceedings of the Combustion Institute, vol. 31, no. 2, pp. 3285–3292, 2007.

[50]   C. Miesse, R. Masel, M. Short and M. A. Shannon, "Diffusion Flame Instabilities in a 0.75 mm Non-premixed Microburner," Proceedings of the Combustion Institute, vol. 30, no. 2, pp. 2499–2507, 2005.

[51]   B. Xu and Y. Ju, "Studies on Non-premixed Flame Streets in a Mesoscale Channel," Proceedings of the Combustion Institute, vol. 32, no. 1, pp. 1375–1382, 2009.

[52]   T. Jackson, J. Buckmaster, Z. Lu, D. C. Kyritsis and I. Massa, "Flames in Narrow Circular Tubes," Proceedings of the Combustion Institute, vol. 31, no. 1, pp. 955–962, 2007.

[53]   K. J. Dellimore, A. Veeraragavan and C. P. Cadou, "Modeling and Simulation of Fuel-Oxidizer Mixing in Micro-Power Systems," AIAA Journal, vol. 50, no. 5, pp. 1090–1102, 2012.

[54]   B. Xu and Y. Ju, "Concentration Slip and its Impact on Heterogeneous Combustion in a Micro Scale Chemical Reactor," Chemical Engineering Science, vol. 60, no. 13, pp. 955–962, 2005.

[55]   B. Xu and Y. Ju, "Theoretical and Numerical Studies of Non-equilibrium Slip Effects on a Catalytic Surface," Combustion Theory and Modeling, vol. 10, no. 6, pp. 961–979, 2006.

[56] P. Aghalayam, P. A. Bui and D. G. Vlachos, "The Role of Radical Wall Quenching in Flame Stability and Wall Heat Flux: Hydrogen-Air Mixtures," Combustion Theory and Modeling, vol. 2, no. 4, pp. 515–530, 1998.

[57] W. Yang, U. Bonne and B. R. Johnson, "MicroCombustion Engine/Generator." US Patent 6276313, 2001.

[58] J. Jagoda, "The development and investigation of a small high aspect ratio two-stroke engine," in 41st Aerospace Sciences Meeting and Exhibit, Reno, NV, January 6–9, 2003.

[59] K. Annen, D. Stickler and J. Woodroffe, "Miniature Internal Combustion Engine-Generator For High Energy Density Portable Power," in Proceedings of the Army Science Conference (26th), Orlando, FL, December 1–4, 2008.

[60] A. Gomez, J. J. Berry, S. Roychoudhury, B. Coriton and J. Huth, "From Jet Fuel to Electric Power Using a Mesoscale, Efficient StirlingCycle," Proceedings of the Combustion Institute, vol. 31, no. 2, pp. 3251–3259, 2007.

[61] J. Huth and J. Collins, "Diesel Fuel-to-Electric Conversion Using Compact, Portable, Stirling Engine-Based Systems," in 13th International Stirling Engine Conference, Tokyo, 2007.

[62] P. Gaile, B. Xu and Y. Ju, "Kinetic Enhancement of Mesoscale Combustion by Using an Novel Nested Doll Combustor," in 45th AIAA Aerospace Sciences Meeting and Exhibit, 2007.

[63] J. Vican, B. F. Gajdeczko, F. L. Dryer, D. I. Milius, I. A. Aksay and R. A. Yetter, "Development of a Microreactor as a Thermal Source for Microelectromechanical System Power Generation," Proceedings of the Combustion Institute, vol. 29, no. 1, pp. 909–916, 2002.

[64] L. Sitzki, K. Borer, S. Wussow, E. Schuster, K. Maruta and P. D. e. a. Ronney, "Combustion in Microscale Heat-recirculating Burners," in 39th Aerospace Sciences Meeting and Exhibit, Reno, 2001.

[65] N. I. Kim, S. Kato, T. Kataoka, T. Yokomori, S. Maruyama, T. Fujimori and K. Maruta, "Flame Stabilization and Emission of Small Swiss-roll Combustors as Heaters," Combustion and Flame, vol. 141, pp. 229–240, 2005.

[66] S. A. Lloyd and F. J. Weinberg, "A Birner for Mixtures of Very Low Heat Content," Nature, vol. 251, no. 5470, pp. 47–49, 1974.

[67] A. R. Jones, S. A. Lloyd and F. J. Weinberg, "Combustion in Heat Exchangers," Proceedings of the Royal Society of London A, Mathematical and Physical Sciences, vol. 360, no. 1700, pp. 97–115, 1978.

[68] T. Takeno and K. Sato, "An Excess Enthalpy Flame Theory," Combustion Science and Technology, vol. 20, no. 1–2, pp. 73–84, 1979.

[69] J. Ahn, P. Ronney, Z. Shao and S. Haile, "A Thermally Self-Sustaining Miniature Solid Oxide Fuel Cell," Journal of Fuel Cell Science and Technology, vol. 6, no. 4, 2009.

[70] K. Maruta, K. Takeda, L. Sitzki, K. Borer, P. Ronney, S. Wussow and O. Deutschmann, "Catalytic Combustion in Microchannel for MEMS Power Generation," in Third Asia-Pacific Conference on Combustion, 2001.

[71] D. C. Kyritsis, I. Guerrero-Arias, S. Roychoudhury and A. Gomez, "MesoscalePower Generation by a Catalytic Combustor Using Electrosprayed Liquid Hydrocarbons," Proceedings of the Combustion Institute, vol. 29, no. 1, pp. 965–971, 2002.

[72] W. Sun, M. Uddi, T. Ombrello, S. H. Won, C. Carter and Y. Ju, "Effects of non-equilibrium plasma discharge on counerflow diffusion flame extinction," Proceedings of the Combustion Institute, vol. 33, no. 2, pp. 3211–3218, 2009.

[73] J. Lefkowitz, Y. Ju, R. Tsuruoka and Y. Ikeda, "Studies of Plasma-Assisted Ignition in a Small Internal Combustion Engine," in 50th AIAA Aerospace Sciences Conference and Exhibit, Nashville, TN, 2012.

[74] S. Heatwole, C. P. Cadou and S. G. Buckley, "In-situ Infrated Diagnostics in a Silicon-walled Microscale Combustion Reactor: Initial Measurements," Combustion Science and Technology, vol. 16, no. 6, pp. 481–492, 2005.

[75] S. Heatwole, A. Veeraragavan, S. Buckley and C. Cadou, "In-situ Species and temperature Measurements in a Micro-combustor," Journal of Nanoscale and Microscale Thermophysical Engineering, vol. 13, no.1, pp. 1–23, 2009.

[76] S. Menon and C. P. Cadou, "Investigation of Combustion Processes in Miniature Internal Combustion Engines," Combustion Science and Technology, vol. 185, no. 11, pp. 1667–1695, 2013.

[77] Honeywell Inc., 5 July 2012. [Online]. Available: http://aerospace.honeywell.com/markets/defense/unmanned-systems/2012/07-July/t-hawk [Accessed 10 July 2013].

[78] Air Force Research Laboratory, December 2005. [Online]. Available: http://www.kirtland.af.mil/shared/media/document/AFD-070404-108.pdf [Accessed 10 July 2013].

[79] EKSO Bionics, [Online]. Available: http://eksobionics.com/ [Accessed 10 July 2013].

[80] K. Blaxter, Energy Metabolism in Animals and Man, Cambridge, 1989, pp. Table A1.1, 297.

[81] T. Edwards and L. Maurice, "Surrogate Mixtures to Represent Complex Aviation and Rocket Fuels," AIAA Journal of Propulsion and Power, vol. 17, no. 2, pp. 461–466, 2001.

[82] International Energy Agency, "Oil Information Documentation for Beyond 2020 Files," 2010. [Online]. Available: http://wds.iea.org/wds/pdf/documentation_oil_2010.pdf [Accessed 29 July 2013].

[83] C. M. Spadaccini, A. Mehra, J. Lee, X. Zhang, S. Lukachko and I. A. Waitz, "High Power Density Silicon Combustion Systems for Micro Gas Turbine Engines," Journal of Engineering for Gas Turbines and Power, vol. 125, pp. 709-719, 2003.

[84] L. Knapp, "The Little Engine That Could Be," Wired.com, 26 November 2001. [Online]. Available: http://archive.wired.com/science/discoveries/news/2001/11/48400?currentPage=all. [Accessed 8 July 2014].

[85] C. Zandonella, "UC Berkeley researchers create world's smallest rotary internal combustion engine," University of California Berkeley, 2 April 2001. [Online]. Available: http://www.berkeley.edu/news/media/releases/2001/04/02_engin.html. [Accessed 8 July 2014].

[86] W. A. Dahm, Courtesy of W. J. A. Dahm, 2014.

[87] K. D. Annen, Photograph courtesy of Aerodyne Research Inc. and K. D. Annen., 2014.

[88] P. D. Ronney, Photo courtesy of P. D. Ronney, 2014.

CHAPTER 1

MESO- AND MICROSCALE COMBUSTION AND FLAMMABILITY LIMITS

*Yiguang Ju, Kaoru Maruta, and
Christopher Cadou*

In meso- and microscale combustion, the increase in surface area-to-volume ratio and the reduction of the characteristic thermal inertia time of the combustor will significantly enrich the combustion phenomena via flame–structure coupling. This chapter provides an overview of the quenching diameter, flammability limit, sublimit combustion, and the effects of heat recirculation, heat loss, radical quenching as well as catalytic reactions on meso- and microscale combustion. At first, it will introduce the concept of the propagation limit of a premixed flame. This is followed by the description of analytical and computational results of the quenching diameter. Then, the flame–structure thermal coupling in a mesoscale reactor is analyzed and the phenomena of sublimit combustion and flame regimes will be discussed. Finally, the catalytic effect and the non-equilibrium transport effect on meso- and microscale combustion will be presented.

1.1 PREMIXED FLAMES IN MESO- AND MICROSCALE COMBUSTION

A premixed flame is an exothermic self-propagating wavefront in a premixed mixture. The propagation speed of an adiabatic, one-dimensional, unstretched premixed flame has a unique value and is a function of fuel chemistry and transport properties of the reactants. However, for a non-adiabatic flame, either a heat loss or a radical loss will slowdown the flame speed or even quench the flame, leading to a flammability limit or extinction limit. On the other hand, if external energy and radicals are added to the flame, the flame becomes super-adiabatic or excess enthalpy, and the flame propagation speed will increase. For meso- and microscale combustion, as seen below, the effects of flame-wall thermal and kinetic coupling on the propagation of a premixed flame can be either positive or negative. The positive feedback via heat recirculation and catalytic reactions will strongly change the burning regime and propagation limits of flames in meso- and microscale combustion.

1

1.2 FLAMMABILITY LIMIT AND QUENCHING DIAMETER

The flammability limits of a mixture are defined as the lean and rich flame propagation limits of a one-dimensional, planar, and unstretched premixed flame. The theoretical and experimental determinations of flammability limits have been conducted for many decades [1–14]. Recently, it has been concluded that radiation heat loss from the emitting species in combustion products such as H_2O and CO_2 is the cause of the flammability limit via temperature sensitive flame chemistry. Thus, the flame speed of a non-adiabatic flame (U) normalized by the adiabatic flame speed depends on the radiative heat loss, H, normalized by the total chemical heat release of the flame and can be given as

$$U^2 \ln U^2 = -\beta H,\qquad(1.1)$$

where β is the reduced Zeldovich number of the global chemical reaction [3, 14].

As shown in Figure 1.1, with the increase of normalized heat loss (H) (i.e., the increase of heat loss or the decrease of fuel concentration), the normalized flame speed decreases. As the normalized heat loss increases to a critical value,

$$H = e^{(-1)}/\beta\qquad(1.2)$$

the normalized flame speed will reach a critical value ($U = e^{-1/2}$), beyond which a leaner or richer mixture with lower flame heat release does not have a finite flame speed. This critical fuel concentration is the so-called flammability limit.

For microscale combustion, due to the increased flame surface-to-volume ratio, the heat loss does not just occur from the radiation of the flame, the heat loss due to the convection to the wall structures becomes much greater than that of thermal radiation. Using a simple

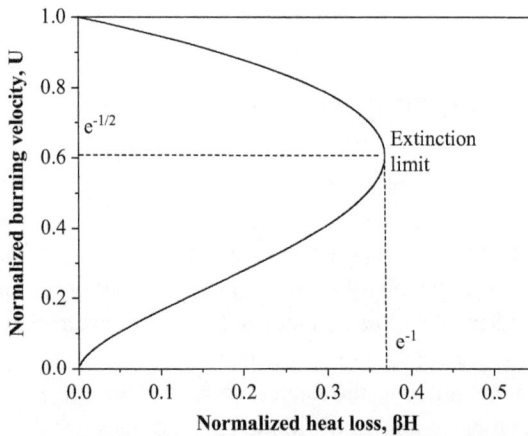

Figure 1.1. The dependence of the normalized burning velocity ($U = S_L/S_{L,ad}$) on the Zeldovich number weighted heat loss (H) normalized by the chemical heat release of fuel of a premixed flame (S_L is the laminar flame speed, $S_{L,ad}$ is the adiabatic flame speed).

normalization [14], the convective heat loss normalized by the chemical heat release of a flame in a cylindrical channel with an inner diameter of d can be given as

$$H = \frac{\text{Heat loss to the wall}}{\text{Total chemical heat release}} = \frac{4\delta_f^2}{d^2} Nu \qquad (1.3)$$

where δ_f is the flame thickness and Nu is the Nusselt number. For a laminar flow in a channel with a constant wall temperature, the Nusselt number is approximately constant [15]. Therefore, the normalized heat loss from the flame to the wall is proportional to the inverse of the channel diameter square (d^2). From Equation (1.1), the normalized flame speed can be given as,

$$U^2 \ln U^2 = -\beta \frac{4\delta_f^2}{d^2} Nu . \qquad (1.4)$$

Figure 1.1 shows that the flame propagation limit occurs at a critical channel diameter,

$$d_0 = 2\sqrt{e\beta Nu}\, \delta_f . \qquad (1.5)$$

This critical tube diameter, d_0, is the so-called quenching diameter [1, 2], below which a flame fails to propagate into a channel. If the Nusselt number for convective heat transfer in a laminar cylindrical flow takes a value between 3.66 and 4.36 (at a constant wall surface temperature or heat flux) [15] and the Zeldovich number is about 10, Equation (1.5) shows that the quenching diameter is about 16 times of the flame thickness. Since flame thickness depends on the thermal diffusivity and flame speed (i.e., fuel concentration and the mean mixture molecular weight), the quenching diameter is a function of fuel concentration and the thermochemical properties of the reactants.

The earliest measurement of the quenching diameter was made by Davy [1] and Lewis and Von Elbe [2]. Since the observed flames in the experimental tubes were not strictly one-dimensional, the measured quenching diameter was a function of flow rate, ranging between $d/\delta_f = 10$ and 30. For example, for stoichiometric hydrogen, propane, and methane/air mixtures, the measured dimensional quenching diameters were between 2 mm and 5 mm, depending on the experimental conditions. Recently, flame propagation in a narrow channel has been computed numerically [16, 17]. Figure 1.2 shows the dependence of the normalized flame propagation speed in a quiescent channel on the channel width (d) for various intensities of the convective heat loss (k), which is proportional to the Nusselt number (Nu) in Equation (1.4). Figure 1.2 shows that for an adiabatic wall ($Nu = k = 0$), there is no convective heat loss from the flame to the wall and the flame speed is independent of the channel width. With the increase of heat loss, the flame speed decreases and there is a critical channel width (the smallest value of d), below which the flame fails to propagate. This critical channel width is equivalent to the quenching diameter at a given heat loss. As the convective heat transfer rate becomes infinite ($k = \infty$) and the wall temperature is kept at the room temperature, the results show that the flame extinguishes at $U = 0.6$ and the resulting quenching channel width is approximately, $d_0/\delta_f \approx 15$, which is consistent to the theoretical prediction by Equation (1.5) and the experimental observation. The above studies imply that when there is a wall heat loss, near-wall flame quenching occurs in a region within $7.5\delta_f$ from the wall. Furthermore, it was also shown that the flow velocity and flow direction also affected the quenching diameter. If the flow is directed

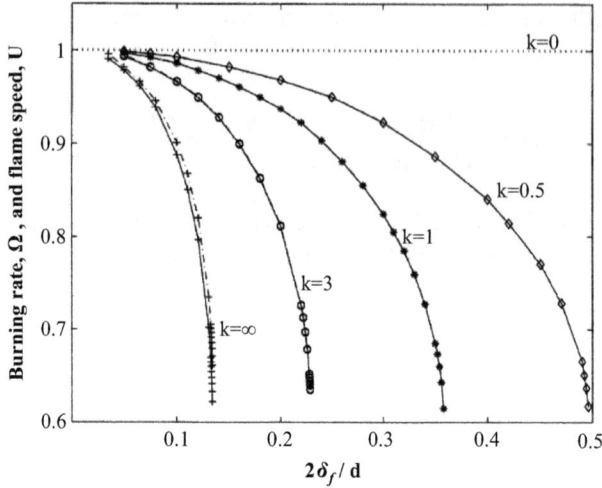

Figure 1.2. Burning rate (solid line) and normalized flame propagation speed U (dash-dotted line) plotted against the ratio of flame thickness to channel width (d) for selected values of reduced heat transfer coefficient (k) in a quiescent, two-dimensional channel flow (reproduced by permission from Daou and Matalon [16]).

from the unburned side to the burned side, the flame is more vulnerable to heat loss. In addition, the quenching diameter increases as the mean flow rate increases.

The effect of the transport properties (e.g., the Lewis number) on the flame geometry and flame speed in a circular channel was computed by Kurdyumov et al. [18] using a three-dimensional model. The results showed that when *Le* is less than unity, the flame speed was greater than that of the adiabatic flame.

As seen from the above results, for a quite long time, researchers had believed that combustion in a scale below the "standard" quenching diameter was not possible. Fortunately, the idea of heat recirculation or excess enthalpy via flame-structure coupling made the standard quenching diameter or flammability limit less relevant and the sublimit combustion possible.

1.3 HEAT RECIRCULATION

Since the pioneering work of Weinberg and co-workers on the Swiss-roll burners [19, 20], extensive experiments and numerical simulations [21–23] have been conducted to study the effect of heat recirculation on flame speed and flammability limit. As shown in Figure 1.3a and 1.3b, the combustor chamber of the Swiss-roll burner is located in the center of the Swill-roll. The cold unburned mixture flows into the combustion chamber tangentially. In parallel to the unburned mixture, the burned gas flows out tangentially from the center of the combustion chamber. As a result, the burned gas will transfer its heat (enthalpy) to the unburned mixture (Figure 1.3c and 1.3d) via convective and conductive heat transfer and leads to excess enthalpy combustion in the combustion chamber.

In order to understand the dynamics of flames with heat recirculation, Ronney [24] and Ju and Choi [25] conducted theoretical analyses independently by using a simplified U-shaped and

(a)

(b)

(c) U shaped flame

(d) Opposite propagating flames

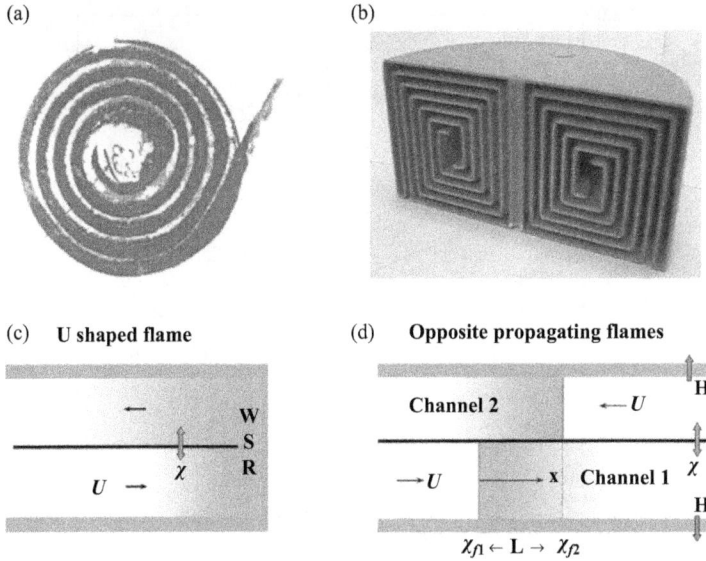

Figure 1.3. Swiss-roll burners and simplified heat recirculation models: (a) two-dimensional swiss-roll (reproduced by permission from Lloyd and Weinberg [20]); (b) three dimensional Swiss-roll (reproduced by permission from Ahn et al. [21]); (c) the U-shaped flame model [25]; and (d) the opposite propagation flames (reproduced by permission from Ju and Choi [25]).

the opposite propagating Swiss-roll flame models (Figure 1.3c and 1.3d), respectively. In the opposite propagating Swiss-roll flame model, the unburned mixtures in two parallel channels flow in the opposite directions, so that the burned gas in one channel will preheat the other. Since both flames are exactly the same, the mutual heat recirculation can be considered as self-heat recirculation in a Swiss-roll. Therefore, the latter model is more general and allows the change of flame location and consideration of heat transfer in both the stream-wise and vertical directions. Therefore, the flame speed and flame separation distance between the two flames ($L = x_{f1} - x_{f2}$) will be an eigenvalue of the mixture concentration, heat transfer rates across the wall (χ), and heat losses (H) to the environment. As such, the opposite propagating Swiss-roll flame model can be used to examine the flame dynamics of sublimit combustion in a Swiss-roll with heat recirculation.

Figure 1.4 shows the dependence of flame speed on the convective heat loss (H) to the outside wall at different heat recirculation factors across the inside wall (χ). When there is no heat recirculation ($\chi = 0$), extinction occurs at a critical heat loss, H_c. For a heat loss less than H_c, as already shown in Figure 1.1, there are two flame speeds, a fast flame mode and a slow flame mode. For a heat loss larger than H_c, Figure 1.4 shows that no flame can be held. However, by allowing heat recirculation across the inner wall ($\chi > 0$), it is seen that the normalized critical heat loss to quench the flame increases significantly with the increase of heat recirculation. In addition, the flame speed increases dramatically, particularly in near adiabatic conditions (small H_c). This figure clearly demonstrates that an increase of heat recirculation (excess enthalpy) will reduce the fuel concentration (increase of normalized heat loss) at the extinction limit and thus enable sub-limit combustion. As a result, a mixture can be burned below the standard flammability limits or the classical quenching diameter if heat recirculation is encouraged.

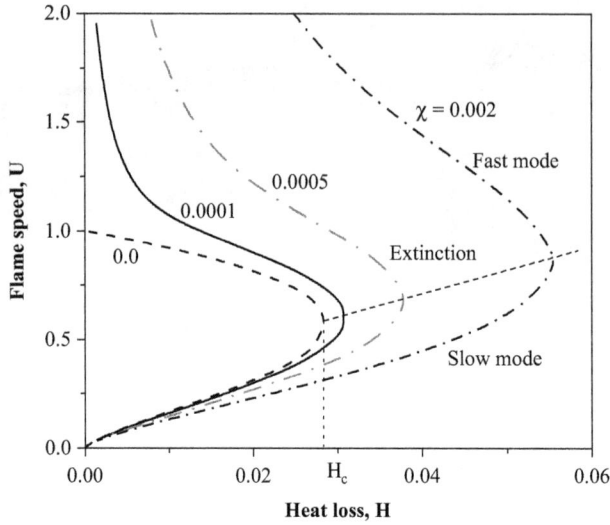

Figure 1.4. The dependence of flame speed on the external heat loss at different intensity of heat recirculation for normalized fuel mass fraction of 0.0365 and flame separation distance of $L = 0$ (reproduced by permission from Ju and Choi [25]).

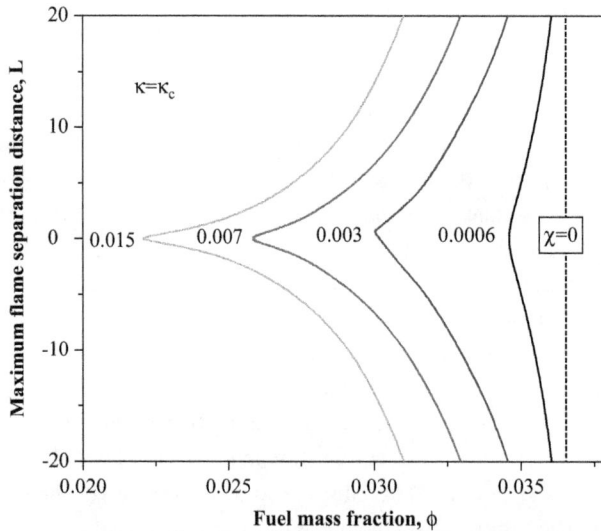

Figure 1.5. The diagram of flammable region: dependence of maximum flame separation distance on fuel mass fraction (reproduced by permission from Ju and Choi [25]).

To further demonstrate the effect of heat recirculation on the extension of the lean flammability limit, Figure 1.5 shows the dependence of the flame separation distance (L) on the fuel mass fraction with a heat loss at $H = H_c$. Therefore, the fuel mass fraction at $\chi = 0$ defines the standard lean flammability limit without heat recirculation ($\varphi = 0.0365$). At the flammability limit, both flames can propagate independently so that the flame separation distance (L) can take any large value. However, by allowing heat recirculation ($\chi > 0$), the flame can be held in much leaner conditions, indicating the existence of sub-limit flames. Moreover, due to the need of heat recirculation for a sub-limit flame, the flame separation distance (L) is bounded to a given region. As the heat recirculation intensity increases, the flammable region of the sub-limit flames is dramatically extended and the flame separation distance becomes smaller, which reduces to the case of the well-stirred reactor (WSR) model in Figure 1.3c. By using these analytical formulations, the flame speed and the extended flammability limit for sublimit flames via heat recirculation can be appropriately estimated.

1.4 FLAME AND WALL STRUCTURE COUPLING IN MICROSCALE COMBUSTION

In Figures 1.3–1.5, the analysis of heat recirculation emphasized only on the direct enthalpy transfer from the burned gas to the unburned cold mixture. There is another indirect enthalpy transfer from the burned gas to the unburned gas through the heat loss from the burned gas to the structure and then from the structure to the unburned gas (Figure 1.6). Because the rate of heat recirculation depends on the rate of heat conduction in the burner structure, the indirect heat recirculation via a combustor wall structure will cause a strong thermal coupling between the flame and the combustor wall. Figure 1.6 schematically shows how the flame–wall structure thermal coupling affects the flame propagation. When a flame propagates at a velocity of U_f relative to a channel with a channel width of d and a mean flow velocity of u, the burned gas loses heat to the wall structure at a rate coefficient of H. A fraction of the heat loss H will go to the environment at a rate of H_0, and the rest of it (Q) will be transferred to upstream and given back to the unburned gas. Therefore, the heat recirculation strongly depends on the heat transfer rates at the inner and outer wall surfaces, and the thermal diffusivity and heat capacity of the wall structure. In addition, since the direction of heat conduction in the solid phase is parallel to that of the thermal diffusion of the gas phase from the burned zone to the unburned mixture, the indirect heat recirculation will change the effective Lewis number of the flame and consequently the flame dynamics.

To understand the flame dynamics with flame–wall structure coupling, a theoretical analysis was carried by Ju et al. [26] by using the simplified model shown in Figure 1.6. By assuming a constant wall temperature at far upstream (T_w), a uniform flow velocity, constant thermal properties, a thin flame model, and the one-dimensional quasi-steady state flame propagation,

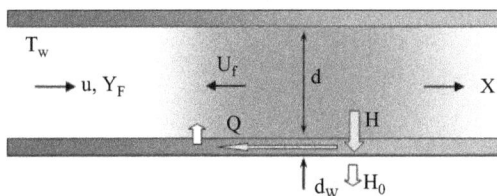

Figure 1.6. Schematic of flame propagation in a mesoscale channel with indirect heat recirculation through flame–wall structure thermal coupling.

the normalized governing equations for energy of gas phase, energy in the solid phase, and the fuel mass fraction in the coordinate attached to the flame front can be rewritten as

$$\left(u-U_f\right)\frac{d\theta}{dx}=\frac{d^2\theta}{dx^2}-H\left(\theta-\theta_w\right)+\exp\left(\beta\left(\theta_f-1\right)/2\right)\delta\left(x-x_f\right)$$

$$\left(u-U_f\right)\frac{dY_F}{dx}=\frac{1}{Le}\frac{d^2Y_F}{dx^2}-\exp\left(\beta\left(\theta_f-1\right)/2\right)\delta\left(x-x_f\right) \quad\quad (1.6)$$

$$-U_f\frac{d\theta_w}{dx}=\frac{d^2\theta_w}{dx^2}\frac{\alpha_w}{\alpha_g}+HC(\theta-\theta_w)-H_0C\theta_w/\beta$$

The boundary conditions are

$$x\rightarrow-\infty,\quad\theta=\theta_w=0,\quad Y_F=1;\quad x\rightarrow+\infty,\quad\frac{d\theta}{dx}=\frac{d\theta_w}{dx}=0,\quad Y_F=0,\quad(1.7)$$

where u, θ, and Y are, respectively, the flow velocity, temperature, and mass fraction; x is the coordinate in the flow direction, U_f is the flame propagation speed relative to the channel, H and H_0 are the normalized convective heat transfer coefficients on the inner and outer walls of the channel, C is the ratio of heat capacity of gas phase to the solid structure, Le is the Lewis number, α_g and α_w are the thermal diffusivities of gas and solid structures, respectively, δ is the Dirac delta function, and the subscripts f and w represent variables at flame front and on the channel wall, respectively.

By carrying the asymptotic analysis in the limit of large activation energy (β), the normalized flame speed ($m = u - U_f$) can be given as [26],

$$m=\exp\left(\frac{\beta(\theta_f-1)}{2(1+\sigma(\theta_f-1))}\right),\quad\quad\theta_f=\theta_f^0+\theta_f^1/\beta,\quad\quad(1.8)$$

where θ_f^0 and θ_f^1 are, respectively, the equilibrium and perturbed flame temperatures, which can be obtained from the following algebraic equations:

$$-m=F(\lambda_3-\lambda_1)+A(\lambda_1-\lambda_2)-B\lambda_3 \quad\quad (1.9)$$

$$F(\theta_f^0-A)+DA=E(\theta_f^0-B)+B \quad\quad (1.10)$$

$$B=m/(m-U_f/C), \quad\quad (1.11)$$

$$\theta_f^1=-\frac{H_0}{m-U_f/C}\left[\frac{\beta(1+\alpha_w/\alpha_gC)}{m-U_f/C}+\frac{F(\theta_f^0-A)}{\lambda_1}+\frac{AD}{\lambda_2}\right], \quad\quad (1.12)$$

in which A and B are constants, and λ_1, λ_2, and λ_3 are, respectively, the eigenvalues of the following third-order ordinary differential equations:

$$\frac{\alpha_w}{\alpha_gC}D^3-\left(\frac{\alpha_w}{\alpha_gC}m+\frac{n}{C}\right)D^2+\left(-\frac{\alpha_w}{\alpha_gC}H-H+\frac{nm}{C}\right)D-\frac{U_f}{C}H+mH=0, \quad\quad (1.13)$$

and D, E, and F are respectively,

$$D = 1 + \frac{m\lambda_2 - \lambda_2^2}{H}, \quad E = 1 + \frac{m\lambda_3 - \lambda_3^2}{H}, \quad F = 1 + \frac{m\lambda_1 - \lambda_1^2}{H}. \tag{1.14}$$

In the limit case of zero thermal conductivity of the wall ($\alpha_w = 0$), that is, the wall becomes a heat sink with no thermal conduction in the flow direction to allow heat recirculation, the problem reduces to the flame propagation in a particle-laden mixture, in which particles are the heat sink of the gaseous flame. Equations (1.8) reduces to

$$\ln m^2 = -\frac{\beta H(2m^2 + CH)}{(m^2 + CH)^2}, \tag{1.15}$$

which is exactly the same as the solution of a homogeneous particle-laden flame obtained by Joulin et al. [27]. Equation (1.15) shows that there is a maximum particle loading (or structure mass) heat capacity (C) beyond which the flame extinguishes. In the limit of zero wall heat capacity ($C = 0$), Equation (1.15) further reduces to

$$m^2 \ln m^2 = -2\beta H, \quad m = u - U_f, \tag{1.16}$$

which is exactly in the same form as Equation (1.1). A factor of two appears in (1.16) is because the heat losses to both walls of the channel in Figure 1.6 was considered.

Figure 1.7 shows the effect of flow velocity on the flame propagation speed for the normalized wall heat capacity of $C = 0.01$ and thermal diffusivity ratio of $\alpha_w/\alpha_g = 0.1$. It is seen that at $u = 0$, the extinction curve is similar to Figure 1.1. The extinction limit defines the minimum channel width at a given heat loss rate and fuel concentration (see the definition of H in Equation (1.3)).

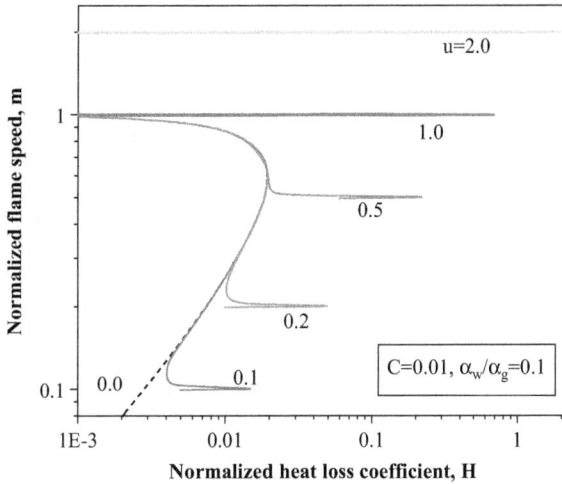

Figure 1.7. The effect of flow velocity on flame speeds, flame regimes, flame regime transition, and extinction limits (quenching channel width) (reproduced by permission from Ju and Xu [26]).

However, when the flow speed is increased to $u = 0.1$, different from the results of zero flow rate, it is seen that there exist two steady-state flame solutions and two extinction limits. The first extinction limit at a higher flame speed is the conventional quenching limit. However, the second extinction limit at a lower flame speed is a limit of a new flame branch, which has a flame speed almost equal to the flow speed ($m \rightarrow 0$)). This new flame regime is called the "weak flame," which is induced by the flame–wall structure thermal coupling. It is seen that at $u = 0.1$, the new flame regime has a quenching diameter larger than the standard quenching diameter of the normal flame. As the flow velocity further increases to $u = 0.2$, it is interesting to note that the weak flame regime extinguishes at a larger normalized heat loss coefficient (H) than that of the normal flame, indicating that the weak flame regime can be stabilized at a combustor scale smaller than the standard quenching diameter. In addition, it is noted that with the increase of flow velocity, the normal flame extinction limit does not change. The flame transition from the normal flame to the weak flame occurs via extinction transition. As the normalized flow speed increases to $u = 0.5$, it is seen that the extinction quenching diameter of the weak flame further decreases (larger H) and the flame transition from the normal flame to the weak flame is a smooth transition without an extinction limit. The smooth transition means that it can be observed in experiment. It is also interesting to note that when the flow velocity is higher than the adiabatic flame speed ($u > 1$), the flame speed with flame–wall structure thermal coupling can be higher than the adiabatic flame speed. This above theoretical result strongly suggests that microscale combustion is only possible when strong flame wall combustion is enabled.

Figure 1.8 shows the dependence of the flame speed on the normalized heat loss coefficient for various wall thermal diffusivities. It is seen that the wall thermal diffusivity affects the flame speed dependence on heat loss significantly. At a small wall thermal diffusivity (weak flame–wall thermal coupling), similar to Figure 1.7, there are two flame regimes and two extinction limits. However, at a high wall thermal diffusivity, the two flame regimes combine into a new

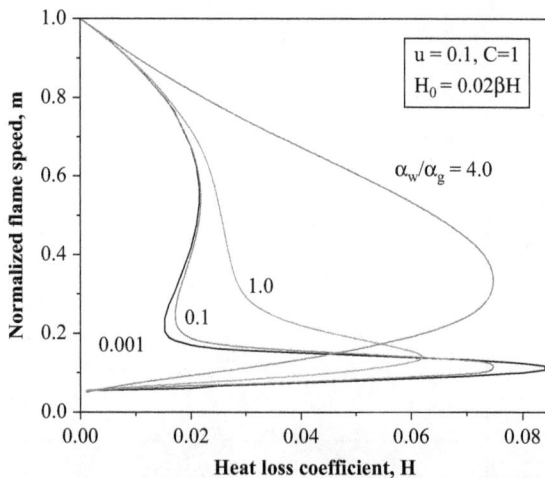

Figure 1.8. The effect of flow velocity on flame speeds, flame regimes, flame regime transition, and extinction limits (quenching channel width) with different thermal diffusivities of the wall (reproduced by permission from Ju and Xu [26]).

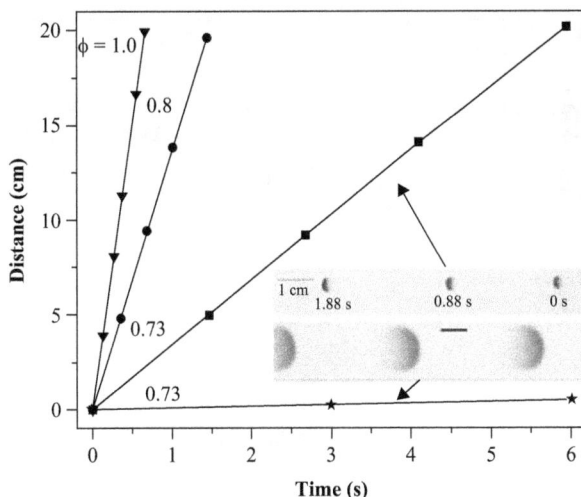

Figure 1.9. Time history of flame location for various equivalence ratios and flame regimes for C_3H_8–air flame with the tube diameter d = 5 mm; upper image: fast flame recorded by high speed camera; lower image: slow flame recorded by CCD camera (reproduced by permission from Ju and Xu [26]).

flame branch, leading to a dramatic reduction of the quenching diameter (increase of *H*). Therefore, when flame–wall thermal coupling is encouraged, materials with a higher thermal diffusivity lead to more extended flammability limit in microscale combustion.

Experimental observation of the two different flame regimes and the increase of flame speed with flow velocity in a mesoscale (d = 5 mm) was also conducted [26]. Figure 1.9 shows the measured time history of flame fronts propagating in the channel at various equivalence ratios for C_3H_8–air mixtures. It is seen that at all equivalence ratios, flame location changes linearly with time. However, at an equivalence ratio of 0.73, the imbedded flame images show that there exist two different flame trajectories (flame regimes). One flame (upper) propagates faster (at 3.5 cm/s relative to the channel wall) and the other weak flame (lower) propagates only at a speed of less than one millimeter per second. These two flames correspond, respectively, to the fast and slow flame regimes predicted in Figures 1.7 and 1.8.

The measured flame speed of methane–air mixtures as a function of equivalence ratio is shown in Figure 1.10 for flow rates at 5.3, 5.69, and 6.09 cm^3/s, respectively. The results show interesting transitions between the two flame regimes, which were predicted by theory (Figure 1.7). The results show that there are two flame regimes and their transition depends on the flow rate. For a low flow rate of 5.3 cm^3/s, the transition from the normal flame to weak flame is extinction transition. However, for a high flow rate the transition is a smooth transition. In addition, it is seen that the flame speed of weak flames increases with the flow rate. The existence of different flame regimes was also observed by Maruta et al. [28]. Earlier experiments by Zamashchikov [29] also showed the existence of weak flames in a quartz tube at very low propagating velocities for nitrogen-diluted methane flames although the fundamental mechanism was not explored.

A one-dimensional numerical model for reacting flow in a micro-channel with flame–structure thermal coupling has been developed by Leach et al. [30] to identify optimum

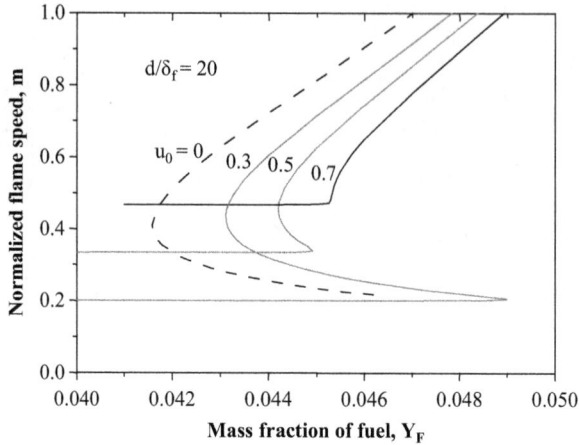

Figure 1.10. Dependence of flame speed on equivalence ratio for CH_4–air flame with d = 5 mm.

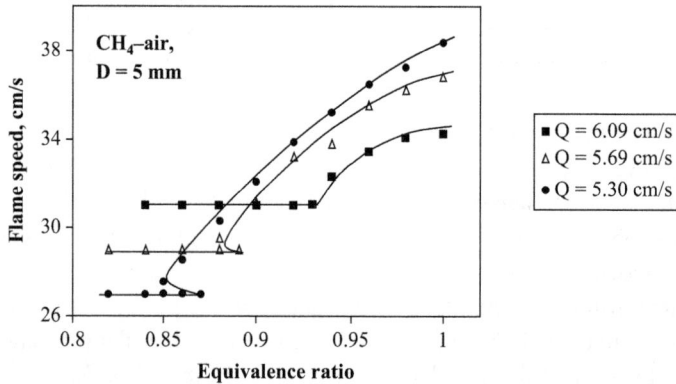

Figure 1.11. Dependence of simulated flame speed on fuel concentration in a two-dimensional channel with a parabolic flow distribution; the channel is 20 times of flame thickness (reproduced by permission from Ju and Xu [26]).

configurations for microcombustors. The model shows that as the size of a combustor is reduced, axial heat transfer through the structure from the postflame to the preflame plays an increasingly important role in determining the combustor performance by broadening the reaction zone and increasing the burning rate. The model also shows that heat transfer from the structure to the environment places a lower limit on the combustor volume that tends to decrease with increasing pressure. The combustor design that maximizes power density generally does not maximize efficiency. This model is in a good agreement with the theoretical analysis presented above.

Two-dimensional simulations of the flame–wall thermal coupling were also conducted to examine the effects of near-wall flame quenching and flame structure on the flame regimes and propagation speed [17, 31, 32]. Figure 1.11 shows the dependence of flame speed on the fuel mass fraction of flame propagation in a two-dimensional channel with a parabolic flow distribution for methane–air mixtures. Constant thermal properties and a one-step reaction model were employed. The channel width is 20 times of flame thickness. It is seen that similar to the one-dimensional theory and experimental observation, two flame regimes and two different kinds of flame transitions exist. In addition, the weak flame velocity increases with the flame rate. However, different from the one-dimensional theory, the normal flame speed decreases with the increase of flow velocity. Moreover, the flame speed of the weak flame is significantly lower than the mean flow velocity because of the flame curvature and near-wall extinction.

The comparison of flame structures (contours of temperature, reaction rate, and fuel mass fraction) without (left) and with (right) flame–wall thermal coupling for flame propagation in a channel with ($d/2 = 10\delta_f$) is shown in Figure 1.12. It is seen that without flame–wall coupling, the wall temperature ($y > 10$) is held constant and flame quenching occurs near the wall surface.

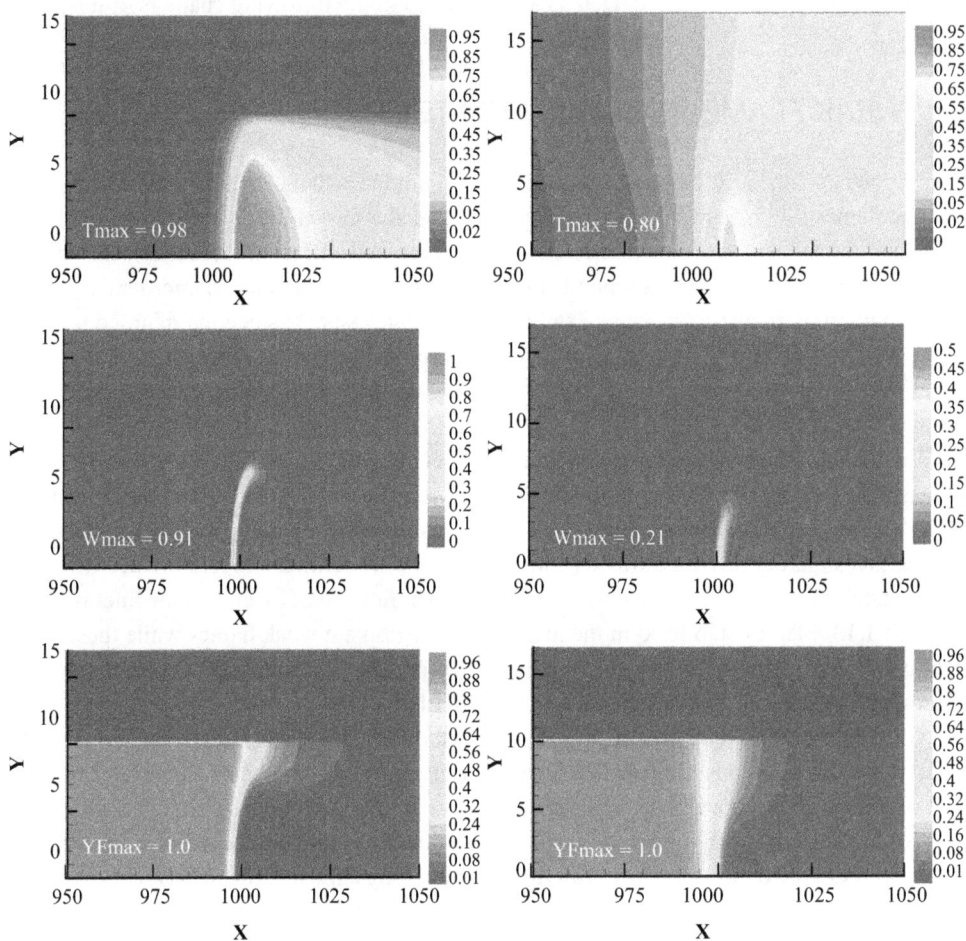

Figure 1.12. Comparison of contours of flame structures without (left) and with (right) flame–structure thermal coupling, temperature (top), reaction rate (middle), and fuel concentration (bottom) at $Y_F = 0.044$ for u = 0.3. The half channel width is ten times of the flame thickness, y = 0 is the centerline and y = 10 is the wall surface of the channel, respectively.

In addition, the flame surface is curved and has a smaller surface area than the cross-section area of the channel. This is why the normal flame speed is lower than the mean flow velocity (Figure 1.11). For the flame–wall coupling case (Figure 1.12, right), the temperature contours show that the heat conduction in the wall structure is faster than that of thermal diffusion in the gas flow, leading to an indirect heat recirculation. This is the reason why the flame–wall thermal coupling can stabilize a weak flame at a fuel concentration or a channel width far below that of the normal flame.

Flame propagation in microchannels (around 100 μm) was also modeled by Vlachos and co-workers [33–35] using two-dimensional parabolic equations with detailed multi-component transport and gas-phase chemistry. The results showed that the near-entrance heat and radical losses at the wall are key issues in controlling flame propagation. More recently, multi-dimensional direct numerical simulations with detailed chemistry and transport for fuel lean hydrogen/air flames in planar microchannels with prescribed wall temperature was conducted [32]. Comprehensive flame responses including oscillating, chaotic, and ignition/extinction flame dynamics are obtained. Details are discussed in the following chapters of this book.

1.5   WEAK FLAME REGIMES WITH TEMPERATURE GRADIENTS

As described above, flame–structure thermal coupling induces flame bifurcations, which exhibits weak flames. Such flames are also observed under the gas–solid thermal coupling in particle-laden flames and strained flames under the combined effects of radiation and unequal diffusive-thermal transport (Lewis number effect) in counterflow flames in microgravity. Weak flames in micro combustion systems exhibit various novel aspects since the microcombustion device is under the strong thermal coupling in nature. In this section, characteristics of weak flames in temperature gradient resulting from thermal coupling of flames with thermally thick wall will be discussed.

Thermal coupling of flames with a micro-channel having a controlled temperature profile is investigated using a fine, straight quartz tube by Maruta et al. [28, 36]. Stationary monotonic positive temperature gradient by an external heater is formed along the inner surface of the tube wall for investigating micro-flame behavior under the well-defined temperature profiles at the flow channel. Measured flame velocity to varying mean flow velocity at the tube inlet is shown in Figure 1.13. Flames stabilized in the low velocity region are weak flames while those in the high and middle velocity branch respectively correspond to experimental normal flames and Flames with Repetitive Extinction and Ignition (FREI).

Since the tube wall temperature is kept by an external heater, weak flame extinction induced by heat loss due to its insufficient heat generation is not likely to occur. However, the lower limit of weak flame is identified experimentally at very low mixture flow velocity of methane/air mixture [37]. Although a weak flame at a mean flow velocity of 0.2 cm/s was observed, no flame was observed below that flow velocity even with long exposure time. This result indicates that the existence of a lower limit of weak flame, even with heat compensation by external heating. Thus, gas-phase and wall temperature measurement in such conditions are carefully conducted. The temperature difference between the flame and the tube wall at the flame position, that is, temperature increase in the weak flame, is shown in Figure 1.14. Figure 1.14 shows the gradual decrease of temperature difference with the decrease of the mean flow velocity and it is almost zero at a mean flow velocity of 0.2 cm/s, where the wall temperature is around 1225 K regardless of mixture composition. Based on the two characteristics of weak flame, that is,

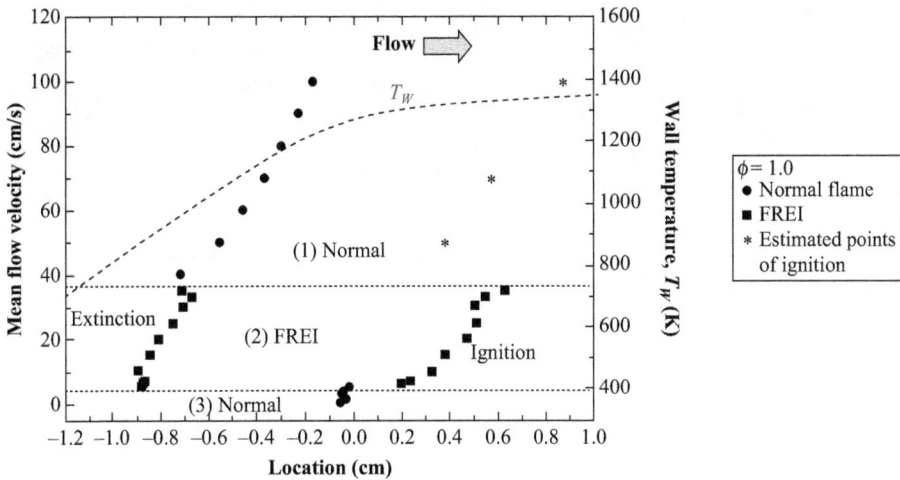

Figure 1.13. Measured flame positions and extinction and ignition points of FREI for variable mean flow velocity at equivalence ratio of 1.0. Estimated ignition locations in upper normal flame regime are also indicated.

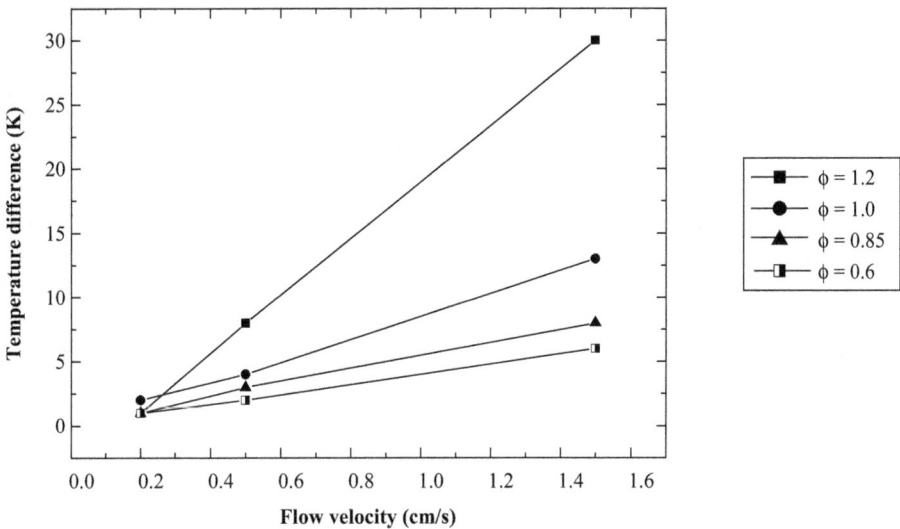

Figure 1.14. Measured temperature difference between the flame and the inner surface of the tube wall at the flame position.

(a) small temperature increase and (b) flame location close to the ignition point, flame temperature at the lower limit of weak flame is considered to correspond to the ignition temperature of the given mixture. Although ignition and flame propagation have been considered to be separated phenomena in general, the stationary propagating flame near the ignition point in the weak flame is successfully established in the experiment. Hence, ignition and flame propagation would merge with each other at this extreme condition. This fact also indicates that the general ignition temperature of given mixture, which is independent of the experimental apparatus and methodology, can be expected to be identified with the position of the weakest flame by this methodology.

One-dimensional computations with detailed chemistry and transport properties are conducted for examining the mechanism of the lower weak flame limit [37]. The energy equation, which has a convective heat transfer between gas and wall, is as follows:

$$\dot{M}\frac{dT}{dx} - \frac{1}{c_p}\frac{d}{dx}\left(\lambda A\frac{dT}{dx}\right) + \frac{A}{c_p}\sum_{k=1}^{K}\rho Y_k V_k c_{pk}\frac{dT}{dx} + \frac{A}{c_p}\sum_{k=1}^{K}\dot{\omega}_k h_k W_k - \frac{A}{c_p}\frac{4\lambda Nu}{d^2}(T_W - T) = 0. \quad (1.17)$$

Because of the fully developed flow in a circular tube for the case between constant wall heat flux and constant wall temperature, the Nusselt number of the inner wall surface is assumed to be constant ($Nu = 4$). Figure 1.14 shows the computed flame response to the mean flow velocity at equivalence ratio $\phi = 1$. The location of the CH peak is considered to be the flame position. The dashed line with open circles in the figure is the wall temperature profile. The computational flame response showed an e-shaped curve, which has an additional lower velocity branch with the S-shaped curve. Based on the stability analysis for the S-shaped curve, the lowest velocity branch here is considered to be unstable. If this assumption is true, the existence of two stable and two unstable solutions in the four regimes can be inferred. Therefore, the existence of the lower limit of the experimental weak flame can be interpreted as the lower limit of the region (3) in Figure 1.15.

To further consider the experimental lower limit of the weak flames, the variation of the computed temperature difference between the flame and the fixed inner wall surface temperature at the flame position is shown in Figure 1.16. Flame temperature is defined as the temperature of gas at the CH peak. The computed temperature difference becomes smaller with the decrease of mean flow velocity, which agrees well with findings of the experiment. The temperature difference becomes almost zero ($T_g - T_w < 1$ K) at a mean flow velocity of 0.1 cm/s when wall temperature is 1230 K, which quantitatively agrees with the experimental results (1225 K). The lower limit of stable weak flame is considered to be at the midpoint of the lowest velocity branch (regime (4)) in Figure 1.15. Since the boundary between the stable and unstable branches is unresolved, stability analysis should be conducted. It should be noted that the extremely small temperature increase does not directly correspond to flame quenching in the low velocity

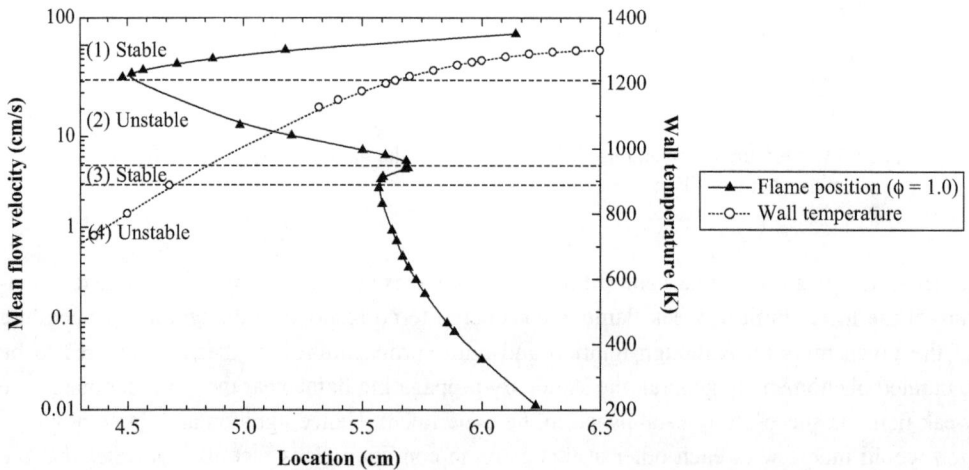

Figure 1.15. Computed flame position with mean flow velocity ($\phi = 1.0$).

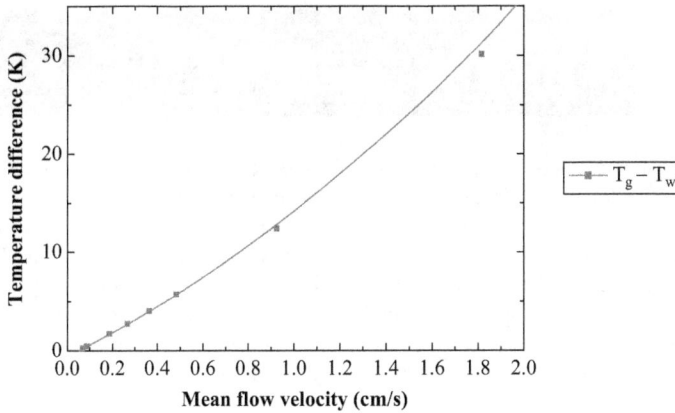

Figure 1.16. Computed temperature difference between the flame and the inner surface of the tube wall at the flame position ($\phi = 1.0$).

regime; however, the conventional reaction with intense heat release no longer occurs in regime (4), even though the heat loss from the flame zone is compensated by the external heating.

If thermal quenching does not occur at the lower limit of the stable weak flame, what is the possible mechanism for the extinction of weak flame? The authors consider the cause of such limit to be related to the dominance of diffusive mass dissipation over the convective mass transfer in the extremely low velocity regime. Existence of the weakest flame and its mechanism is also investigated by flames under the reduced pressure [37, 38].

So far, weak flames in temperature gradient are found to be an effective way for identifying the lower limit of weak flames. More practically, it is equivalent to the lowest ignition temperature of given mixture at reactor pressure and given residence time. In fact, some studies are conducted for elucidating comprehensive ignition characteristics of other hydrocarbon fuels using weak flame regimes in the microflow reactor with controlled temperature profile. Oshibe et al. [39] examined ignition and combustion characteristics of stoichiometric dimethyl ether (DME)/air mixture using the microreactor and demonstrated that the existence of the steady double luminous flames in the weak flame region as well as stable flames in the high velocity region and unstable flames in the middle velocity region. Based on gas sampling analysis and computation with detailed chemistry and transport, it is concluded that these double weak flames are separated multi-stage oxidations of the DME/air mixture (Figure 1.17). It is expected that these steady-state multiple flame phenomena can be utilized for examining a general transient ignition process in modern compression ignition engines since multi-stage oxidations in compression ignition can be visualized as stabilized multiple flames in the microflow reactor.

1.6 COUPLING OF THERMAL AND KINETIC QUENCHING IN MICROSCALE COMBUSTION

The diffusion time scale in a microscale combustor is proportional to the square of the combustor length scale,

$$t_D \propto d^2/D \tag{1.18}$$

Figure 1.17. CH-filtered, long-exposure images of double weak flames in a channel with an inner diameter of 2 mm for stoichiometric DME/air mixture at channel inlet velocity = 2 cm/s. Flow direction is from the left to the right. Channel wall temperature is increasing from the room temperature up to 1300 K in a flow direction. Dotted lines are inner surface of the channel wall. Dual weak flames are stabilized at the wall temperatures 1054 K and 1198 K. Gas sampling analysis confirms cool flame reaction starts further upstream of the first weak flame where the wall temperature is around 600 K–700 K. Bright luminescence in the downstream channel wall is radiation from the high-temperature wall due to an external heat source. (reproduced by permission from Oshibe et al. [39].)

where D is the thermal or mass diffusivity. As the length scale of the combustor decreases, the diffusion time of chemical radicals from the flame zone to the wall will become comparable to the characteristic formation time of radicals, leading to a strong radical quenching on the wall surface. The radical quenching will be coupled with thermal quenching and render simulations of microscale combustion inapplicable without considering radical (e.g., H and OH) recombination in the wall, particularly at high wall temperatures [40, 41].

In order to understand qualitatively the impact of wall temperature, radical quenching, and radical transport on the flame propagation speed in a microchannel with a parabolic flow, numerical computations were made by using a skeletal kinetic model,

Initiation reactions

$$F \rightarrow R \tag{I}$$

Branching reaction

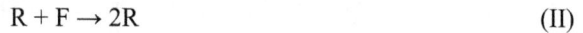

$$R + F \rightarrow 2R \tag{II}$$

Termination reaction

$$R + R\,(+M) \rightarrow 2P\,(+M) + Q \tag{III}$$

Radical quenching on the wall

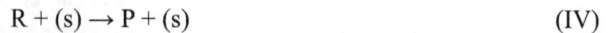

$$R + (s) \rightarrow P + (s) \tag{IV}$$

where F, R, and P represent reactants, radicals, and products, respectively, and s denotes the surface sites. The chain initiation, branching, and termination reaction rates are given as

$$\omega_I = \rho Y_F \exp\left[\frac{E_{aI}}{R_0}\left(\frac{1}{T_I} - \frac{1}{T}\right)\right]$$

$$\omega_{II} = \rho^2 Y_F Y_R \exp\left[\frac{E_{aB}}{R_0}\left(\frac{1}{T_B} - \frac{1}{T}\right)\right] \tag{1.19}$$

$$\omega_{III} = \rho^2 Y_R Y_R$$

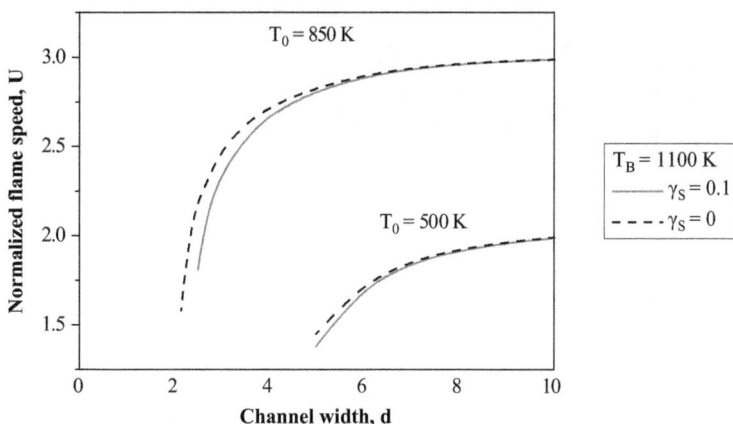

Figure 1.18. Effect of wall temperature on flame speed in a micro-channel and the competition between thermal quenching and radical quenching.

where the activation energies of the initiation and branching reactions are, respectively, $Ea_I =$ 90 Kcal/mol and $E_{aB} = 30$ Kcal/mol. The cross-over temperatures for initiation and branching reactions are, respectively, $T_I = 3000$ K and $T_B = 1100$ K. The radical quenching reaction rate of reaction (IV) can be simplified by using a radical quenching coefficient (γ_s) for the molecular collision flux on the wall,

$$\omega_s = \gamma_s n_R c_R / 4 \tag{1.20}$$

where n_R is the radical number density and c_R is the radical mean molecular velocity. All calculations were made by assuming constant properties.

Figure 1.18 shows the effect of wall temperature on the flame propagation speed for different channel widths. At a low temperature (500 K), which is far below the chain-branching crossover temperature (1100 K), the thermal quenching is dominant. However, as the wall temperature increases to 850 K, the radical quenching plays a more important role to accelerate the flame quenching. This effect becomes more significant when the wall temperature approaches the chain-branching cross-over temperature.

Figure 1.19 shows the effect of radical quenching coefficient on the normalized flame speed for different channel widths and radical Lewis numbers. For a large channel width ($d/2 = 3.25$), flame speed decreases with the increase of radical quenching coefficient indicating that the decrease of flame speed is limited by the rate of radical recombination. However, when the channel width is small ($d/2 = 2.7$), the results show that the flame speed decreases with an initial increase of radical quenching coefficient; however, a further increase of radical quenching rate does not affect the flame speed. This implies that the radical quenching is limited by diffusion transport.

Therefore, in microscale combustion, thermal quenching is dominant when the wall temperature is low. However, radical quenching plays an increasingly important role in affecting flame propagation as the wall temperature increases close to the chain-branching crossover temperature. In addition, radical quenching via surface recombination can be dominated by the diffusion process as the channel width decreases. Accurate modeling of microscale combustion at high temperatures needs to include rigorous models of radical quenching mechanism and radical diffusion transport.

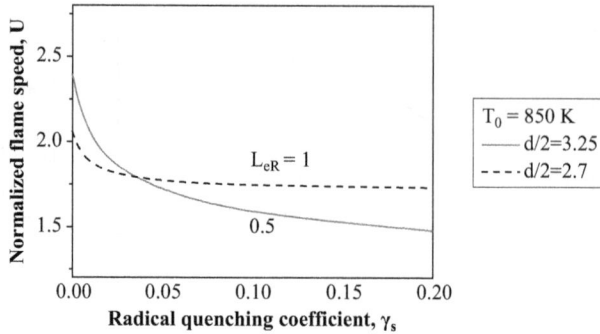

Figure 1.19. Effect of radial quenching coefficient on flame speed in a micro-channel with different channel widths and radical Lewis numbers.

The feasibility study of hydrocarbon combustion in a small burner was experimentally conducted by paying attention to thermal and radical quenching [40–42]. Several wall materials that are not expected to trap radicals are selected. Results show that quenching distance did not depend on wall materials when the wall temperature is below 500 K, whereas those at wall temperature near 1273 K are strongly dependent on the wall materials [42]. Therefore, it was concluded that thermal quenching is dominant at low temperature wall while radical quenching is dominant at high temperature wall. The experimental results are consistent with the analytical results (Figures 1.18 and 1.19).

1.7 NON-EQUILIBRIUM COMBUSTION

As the reactor and combustor scale further decreases and becomes comparable to a magnitude (e.g., the pore size in solid fuel cells is in the order of 1–10 micrometers) of two orders of mean free path (~0.1 μm for air at one atmospheric pressure), the diffusion transport becomes non-equilibrium, that is, the near-wall molecular collisions are not sufficient enough (less than 10^3 collisions) to be modeled by using a continuous model such as the Navier–Stokes equations. In this case, the rarefied gas effect becomes significant [43]. The flow near the solid wall cannot reach thermodynamic equilibrium with the wall, yielding large slips of thermodynamic properties between the wall and the gas in the near-wall boundary. In the slip flow regime ($10^{-3} < Kn < 0.1$), the flow outside of the Knudsen layer can still be governed by the Navier–Stokes equations. However, corrections of non-continuum boundary conditions for discontinuities in macroscopic variables such as velocity, temperature, and concentrations are needed to consider the non-equilibrium effects (Figure 1.20). Flows in the microchemical reactors fall mostly into this regime. Therefore, the slip effects in modeling the heterogeneous reaction or combustion systems like catalytic reactors [44], solid oxide fuel cells [45], fuel reforming, and combustion synthesis of thin films become important especially at low pressures or small scales.

As shown in Figure 1.21, when the length scale of the flow approaches 10^3 the gaseous mean free path, the wall temperature, $T(s)$, and the wall temperature outside of the Knudsen layer $T(0^+)$ are not equal ($T(0^+) \neq T(s)$). The Knudsen layer is called the slip regime ($Kn = 0.001 - 0.1$). For example, a catalytic microreactor working at 1000 K and 1 atm, the Knudsen number of 10^{-3} approximately corresponds to 300 μm. In this regime, velocity, temperature,

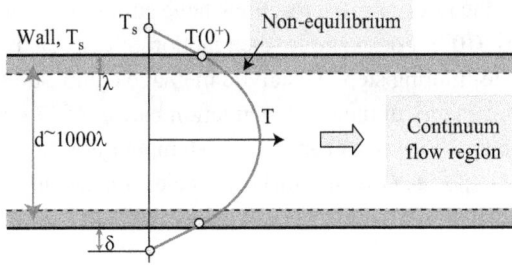

Figure 1.20. Non-equilibrium model in slip regime.

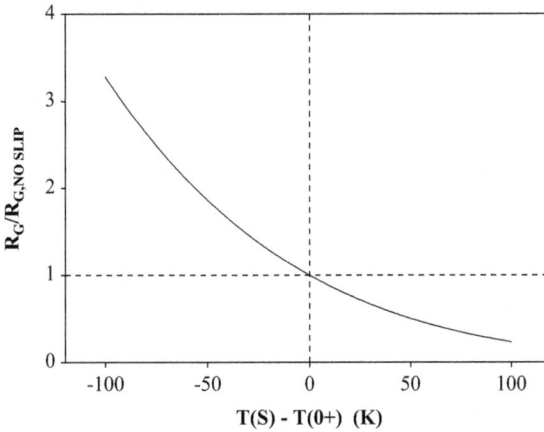

Figure 1.21. The actual reaction rate constant R_G, normalized by the rate for the case without temperature slip, for a gas/surface reaction with $E_A = 27$ kcal/mol on a 1000 K surface, as a function of temperature slip (reproduced by permission from Glumac and Ju, 2005).

and species slips across the Knudsen layer have to be corrected in applications of existing continuum models. The existence of velocity and temperature slips in heat transfer systems has been extensively studied for the development of micro-fluidic systems in microelectromechanical systems (MEMS) [46]. However, very limited work emphasizing non-equilibrium transport has been done in microscale reactive systems, especially for concentration slips (N. Glumac, pers. comm.) [47, 48]. In fact, the temperature slip in microreactive systems will not only affect the temperature in reactants transported onto the surface, but also change the rates of transport and the local species concentration.

A surface kinetic mechanism typically consists of three temperature-dependent process types: reactions of surface species with other surface species, surface species diffusion, and reactions of gas species with surface species. Sensitivity analysis has shown that the dissociative adsorption and abstraction reactions (N. Glumac, pers. comm.) play a dominant role in catalytic fuel oxidation. However, for these reactions, a substantial fraction of the temperature dependence will come from the gas temperature at $T(0^+)$, which provides most of the energy required for a gas-species to overcome a large activation energy barrier to reaction with the surface species. As such, the rates of these reactions have strong dependence on kinetic energy of the incident species at $T(0^+)$. For example, the reaction of dissociative adsorption of methane on platinum is the rate-limiting step in catalytic methane combustion [49] and has a strong dependence on the kinetic energy of methane (activation barrier of 27 kcal/mol) [50].

As a first-order estimate of the rate change due to temperature slip, we assume that the rate of reaction of a gas-phase species G with a surface species S takes the form (N. Glumac, pers. comm.):

$$R_G = \begin{pmatrix} \text{surface} \\ \text{flux of } G \end{pmatrix} \cdot \begin{pmatrix} \text{reactive} \\ \text{site fraction} \end{pmatrix} \cdot \begin{pmatrix} \text{sticking} \\ \text{coefficient} \end{pmatrix} \cdot \begin{pmatrix} \text{fraction of } G \\ \text{with } E > E_A \end{pmatrix}$$

$$= \frac{n_G \langle c \rangle_G}{4} \theta_S S_0 e^{-E_A/RT} = S_0 X_G \frac{n \langle c \rangle_G}{4} \theta_S e^{-E_A/RT} = S_0 X_G P \theta_S \left(\frac{1}{2kT\pi m_G} \right)^{1/2} e^{-E_A/RT} \qquad (1.21)$$

where T is the gas temperature, $T = T(0^+)$, θ_S is the surface coverage of species S, n and n_G are the total number density and the number density of species G, respectively, $\langle c \rangle_G$ is the mean thermal speed of species G, X_G is the mole fraction of species G, and S_0 is the sticking coefficient for this reaction.

By assuming a constant pressure process, and neglecting for now the temperature dependence of X_G and θ_S, we can estimate the effect of the gas temperature T on R_G as $R_G \sim T^{-1/2} e^{-EA/RT}$. Figure 1.21 shows the normalized reaction rate, $R_G/R_{G,\text{No Slip}}$, at 1000 K for the case of a reaction with $E_A = 27$ kcal/mol for different temperature jumps $T(0^+) - T(s)$. It is seen that a temperature slip of only 6% leads to nearly a factor of two difference between the actual reaction rate and that calculated with the assumption that $T(s) = T(0^+)$. For reactions with larger activation energies or larger temperature slips [51] measured a slip of 20 percent in atmospheric pressure CVD flame reactors), an order of magnitude error could be introduced by neglecting the temperature slip. Therefore, including non-equilibrium transport in microscale combustion is necessary to predict the reaction rates correctly.

The effect of temperature slip in low-pressure catalytic combustion systems were investigated experimentally [52] by using low-pressure combustion systems. The results showed that the temperature jump was 34 K at a pressure of 2.5 Torr and had a very significant effect on the methane oxidation. Numerical modeling of temperature discontinuity in a 100 μm microcombustor at one atmospheric pressure was conducted by Raimondeau et al. [41]. The results showed that the temperature slip at the wall was negligible.

In addition to the temperature slip, the mole fraction gradients of gas species near the wall will also yield a concentration slip. The phenomenon of composition slip has been detected in simulations of gas mixtures by Bird [43], Papadopoulos and Rosner [53], Kramers and Kistemaker [54]. More detailed reviews of the concentration slip were made by Scott et al. [55] and Gupta et al. [56]. Recently, Rosner and Papadopoulos [57] derived a simple expression for concentration slip based on the similarity between the transport of mass and energy. More recently, Xu and Ju [47, 48] derived a more general expression to consider the concentration slip in a non-equilibrium flow based on the gas kinetic theory,

$$\left(\frac{n_{i,s}}{n_{i,0+}} \right) \left(\frac{T_s}{T_{0+}} \right)^{1/2} = \left\{ 1 + \frac{4}{3\pi \bar{v}} \lambda \left(\frac{\partial u}{\partial y} \right)_{0+} + \frac{\alpha_{i,m} - 2}{\alpha_{i,m}} \lambda \left(\frac{\partial \ln(X_i)}{\partial y} \right)_{0+} \right\} \qquad (1.22)$$

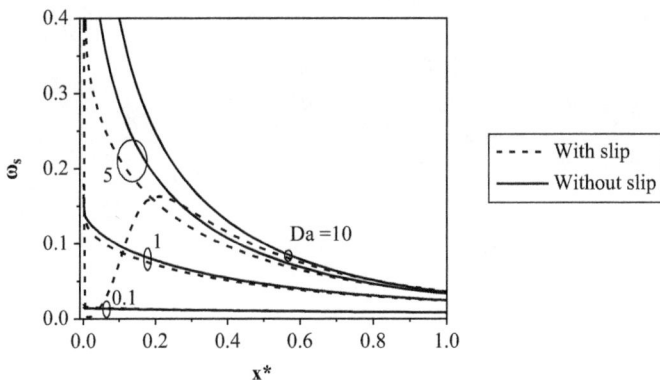

Figure 1.22. Surface reaction rate at different Damköhler numbers along the channel (T_{in} = 600 K, Kn = 0.1) (reproduced by permission from Xu and Ju [47]).

where n is the concentration, u is the velocity, T is the temperature, α is the surface accommodation coefficient, and X_i is the mole fraction.

Equation (1.22) shows that the concentration slip is coupled with the temperature slip. Unfortunately, in most of the previous work, the coupling of concentration slips with temperature and velocity slips was not examined. Figure 1.22 shows the slip effect on the surface reaction rate at different reduced Damköhler numbers defined by the ratio of diffusion time to the catalytic reaction time [47]. The results show that the slip effect becomes increasingly significant when the Damköhler number is larger than unity. When the Damköhler number is large, the reaction process is diffusion controlled and a faster surface reaction will cause a larger concentration gradient and thus a larger slip effect. On the other hand, if the Damköhler number is small, the problem becomes reaction controlled. As a result, the relatively faster diffusion process yields a more uniform distribution of temperature and species near the reaction surface. Therefore, the slip effect becomes smaller at small Damköhler numbers. In practical applications, depending on the reactors' working condition, the system Damköhler number can be either greater or less than unity. As the Damköhler number is increased from 0.1 to 5, Figure 1.22 shows that both the slip model and the non-slip model show a monotonic increase of the reaction rates. However, if the Damköhler number is increased to 10, the reaction rate will be dominated by diffusion transport. It is seen that a non-monotonic distribution of the reaction rate appears near the entrance of the reaction channel. This phenomenon is caused by the competition between the chemical reaction and species diffusion represented by the Damköhler number. It is also seen that because the coupling between temperature and concentration slips, it is not accurate to predict the surface reaction rate by only considering a temperature slip.

Numerical simulations including both temperature and concentration slips were also conducted by using detailed chemical mechanisms [48]. The results show that the concentration slip is proportional to the product of the Knudsen number and the species Damköhler number. It is also shown that the non-equilibrium process has a dramatic effect on the radical reaction rates and mass fractions at a large Damköhler number and Knudsen number. It is demonstrated that at the leading edge of the microchannel, the temperature slip has a greater effect on the fuel oxidation rate. However, as the reaction proceeds, the effect of concentration slip on radicals dominates. In addition, as shown in Figure 1.23, the concentration slip has a larger effect on

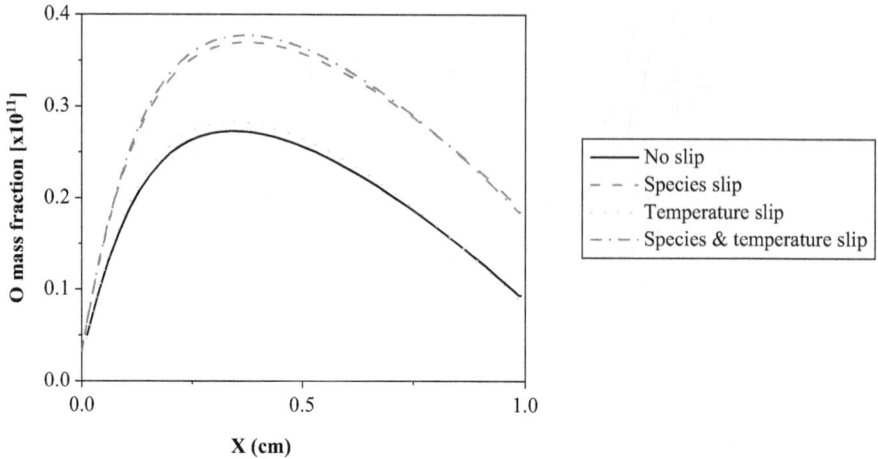

Figure 1.23. The distribution of O on the surface along the streamwise direction (T_{in} = 300 K, T_w = 1000 K).

the reaction rates and mass fractions of OH, H, and O radicals than on those of the reactants. Therefore, the non-equilibrium concentration slip and temperature slip are important issues in microscale combustion and need to be addressed.

Therefore, in mesoscale and microscale combustion, many new phenomena such as the sublimit combustion, kinetic extinction, weak flames, flame cells, flame instability, spiral flames, oscillating and spinning combustion, and temperature and concentration slips, which will be discussed in details in the following chapters, arise due to the reduced timescales of combustor, increased surface-to-volume ratio, flame–flame and flame–structure thermal coupling, flame–wall kinetic coupling, and the non-equilibrium transport. This chapter provides an overview of the fundamental mechanisms of meso- and microscale combustion for the rest of the chapters in this book.

REFERENCES

[1] Davy, H (1817). "Some researches on flame." Philosophical Transactions of the Royal Society of London 107(1): 45–76. doi: http://dx.doi.org/10.1098/rstl.1817.0008.

[2] Lewis, B and Von Elbe, G (1987). Combustion, Flames and Explosions of Gases. Orlando, FL: Academic Press.

[3] Zeldovich, Y B, Barenblatt, G I, Librovich, V B and Makhviladze, G M (1985). The Mathematical Theory of Combustion and Explosions. New York, NY: Consultants Bureau.

[4] Spalding, D B (1957). "A theory of inflammability limits and flame-quenching." Proceedings of the Royal Society of London Series A 240(1220): 83–100. doi: http://dx.doi.org/10.1098/rspa.1957.0068.

[5] Joulin, G and Clavin, P (1979). "Linear stability analysis of nonadiabatic flames: Diffusional-thermal model." Combustion and Flame 35: 139–153. doi: http://dx.doi.org/10.1016/0010-2180(79)90018-X.

[6] Strehlow, R A, Noe, K A and Wherley, B L (1988). "The effect of gravity on premixed flame propagation and extinction in a vertical standard flammability tube." Symposium (International) on Combustion 21(1): 1899–1908. doi: http://dx.doi.org/10.1016/S0082-0784(88)80426-0.

[7] Ronney, P D (1988). "On the mechanisms of flame propagation limits and extinguishment-processes at microgravity." Symposium (International) on Combustion 22(1): 1615–1623. doi: http://dx.doi.org/10.1016/S0082-0784(89)80173-0.

[8] Sibulkin, M and Frendi, A (1990). "Prediction of flammability limit of an unconfined premixed gas in the absence of gravity." Combustion and Flame 82(3–4): 334–345. doi: http://dx.doi.org/10.1016/0010-2180(90)90006-D.

[9] Lakshmisha, K N, Paul, P J and Mukunda, H S (1990). "On the flammability limit and heat loss in flames with detailed chemistry." Symposium (International) on Combustion 23(1): 433–440. doi: http://dx.doi.org/10.1016/S0082-0784(06)80288-2.

[10] Platt, J and T'ien, J (1990). "Flammability of a weakly stretched premixed flame: The effect of radiation loss." Fall Technical Meeting, Eastern Section, Combustion Institute.

[11] Guo, H, Ju, Y, Maruta, K, Niioka, T and Liu, F (1997). "Radiation extinction limit of counter flow premixed lean methane-air flames." Combustion and Flame 109(4): 639–646. doi: http://dx.doi.org/10.1016/S0010-2180(97)00050-3.

[12] Ju, Y, Guo, H, Maruta, K and Liu, F (1997). "On the extinction limit and flammability limit of non-adiabatic stretched methane-air premixed flames." Journal of Fluid Mechanics 342(1): 315–334. doi: http://dx.doi.org/10.1017/S0022112097005636.

[13] Ju, Y, Maruta, K and Niioka, T (2001). "Combustion limits." Applied Mechanics Reviews 54(3): 257–277. doi: http://dx.doi.org/10.1115/1.3097297.

[14] Ju, Y and Maruta, K (2011). "Microscale combustion." Progress in Energy and Combustion Science 37(6): 669–715, 2011. doi: http://dx.doi.org/10.1016/j.pecs.2011.03.001.

[15] Holman, J P (2001). Heat Transfer. New York: McGraw Hill Higher Education.

[16] Daou, J and Matalon, M (2002). "Influence of conductive heat-losses on the propagation of premixed flames in channels." Combustion and Flame, 128(4): 321–339. doi: http://dx.doi.org/10.1016/S0010-2180(01)00362-5.

[17] Ju, Y and Xu, B (2006). "Effects of channel width and Lewis number on the multiple flame regimes and propagation limits in mesoscale." Combustion Science and Technology 178(10–11): 1723–1753. doi: http://dx.doi.org/10.1080/00102200600788643.

[18] Kurdyumov, V N and Fernandez-Tarrazo, E (2002). "Lewis number effect on the propagation of premixed laminar flames in narrow open ducts." Combustion and Flame 128(4): 382–394. doi: http://dx.doi.org/10.1016/S0010-2180(01)00358-3.

[19] Lloyd, S A and Weinberg, F J (1974). "A burner for mixtures of very low heat content." Nature 251(5470): 47–49. doi: http://dx.doi.org/10.1038/251047a0.

[20] Weinberg, F J (1975). "The first half-million years of combustion research and today's burning problems." Symposium (International) on Combustion 15(1): 1–17. doi: http://dx.doi.org/10.1016/S0082-0784(75)80280-3.

[21] Ahn, J, Eastwood, C, Sitzki, L, Ronney, P D, Agrawal, A K, Cadou, C, et al. (2005). "Gas-phase and catalytic combustion in heat-recirculating burners." Proceedings of the Combustion Institute 30(2): 2463–2472. doi: http://dx.doi.org/10.1016/j.proci.2004.08.265.

[22] Kuo, C H and Ronney, P D (2007). "Numerical modeling of non-adiabatic heat-recirculating combustors." Proceedings of the Combustion Institute 31: 3277–3284. doi: http://dx.doi.org/10.1016/j.proci.2006.08.082.

[23] Yetter, R A, Yangt, V, Wang, Z, Wang, Y, Milius, D, Peluse, M, Aksay, I A, Angioletti, M and Dryer, F L (2003, January 6–9). "Development of meso and micro scale liquid propellant thrusters." 41st Aerospace Sciences Meeting and Exhibit, AIAA-2003-0676, Reno, Nevada.

[24] Ronney, P D (2003). "Analysis of non-adiabatic heat-recirculating combustors." Combustion and Flame 135(4): 421–439. doi: http://dx.doi.org/10.1016/j.combustflame.2003.07.003.

[25] Ju, Y and Choi, C W (2003). "An analysis of sub-limit flame dynamics using opposite propagating flames in mesoscale channels." Combustion and Flame 133(4): 483–493. doi: http://dx.doi.org/10.1016/S0010-2180(03)00058-0.

[26] Ju, Y and Xu, B (2005). "Theoretical and experimental studies on mesoscale flame propagation and extinction." Proceedings of the Combustion Institute 30(2): 2445–2453. doi: http://dx.doi.org/10.1016/j.proci.2004.08.234.

[27] Joulin, G and Deshaies, B (1986). "On radiation-affected flame propagation in gaseous-mixtures seeded with inert particles." Combustion Science and Technology 47(5–6): 299–315. doi: http://dx.doi.org/10.1080/00102208608923879.

[28] Maruta, K, Kataoka, T, Kim, N I, Minaev, S and Fursenko, R (2005). "Characteristics of combustion in a narrow channel with a temperature gradient." Proceedings of the Combustion Institute 30(2): 2429–2436. doi: http://dx.doi.org/10.1016/j.proci.2004.08.245.

[29] Zamashchikov, V (1997). "Experimental investigation of gas combustion regimes in narrow tubes." Combustion and Flame 108(3): 357–359. doi: http://dx.doi.org/10.1016/S0010-2180(96)00169-1.

[30] Leach, T T and Cadou, C P (2005). "The role of structural heat exchange and heat loss in the design of efficient silicon micro-combustors." Proceedings of the Combustion Institute 30(2): 2437–2444. doi: http://dx.doi.org/10.1016/j.proci.2004.08.229.

[31] Jackson, T L, Buckmaster, J, Lu, Z, Kyritsis, D C and Massa, L (2007). "Flames in narrow circular tubes." Proceedings of the Combustion Institute 31(1): 955–962. doi: http://dx.doi.org/10.1016/j.proci.2006.07.032.

[32] Pizza, G, Frouzakis, C E, Mantzaras, J, Tomboulides, A G and Boulouchos, K (2008). "Dynamics of premixed hydrogen/air flames in mesoscale channels." Combustion and Flame 155(1–2): 2–20. doi: http://dx.doi.org/10.1016/j.combustflame.2008.08.006.

[33] Norton, D G and Vlachos, D G (2003). "Combustion characteristics and flame stability at the microscale: A CFD study of premixed methane/air mixtures." Chemical Engineering Science 58(21): 4871–4882. doi: http://dx.doi.org/10.1016/j.ces.2002.12.005.

[34] Norton, D G and Vlachos, D G (2004). "A CFD study of propane/air microflame stability." Combustion and Flame 138(1–2): 97–107. doi: http://dx.doi.org/10.1016/j.combustflame.2004.04.004.

[35] Kaisare, N S and Vlachos, D G (2007). "Optimal reactor dimensions for homogeneous combustion in small channels." Catalysis Today 120(1): 96–106. doi: http://dx.doi.org/10.1016/j.cattod.2006.07.036.

[36] Maruta, K, Park, J K, Oh, K C, Fujimori, T, Minaev, S S and Fursenko, R V (2004). "Characteristics of microscale combustion in a narrow heated channel." Combustion, Explosion and Shock Waves 40(5): 516–523. doi: http://dx.doi.org/10.1023/B:CESW.0000041403.16095.a8.

[37] Tsuboi, Y, Yokomori, T and Maruta, K (2008). "Extinction characteristics of premixed flame in heated microchannel at reduced pressures." Combustion Science and Technology 180(10–11): 2029–2045. doi: http://dx.doi.org/10.1080/00102200802269723.

[38] Tsuboi, Y, Yokomori, T and Maruta, K (2009). "Lower limit of weak flame in a heated channel." Proceedings of the Combustion Institute 32: 3075–3081. doi: http://dx.doi.org/10.1016/j.proci.2008.06.151.

[39] Oshibe, H, Nakamura, H, Tezuka, T, Hasegawa, S and Maruta, K (2009). "Two stage reactions of dimethyl ether-air mixture in micro flowreactor with controlled temperature profile." Proceedings of Seventh Asia-Pacific Conference on Combustion, ASPACC 09, Taipei, Taiwan.

[40] Aghalayam, P, Bui, P A and Vlachos, D G (1998). "The role of radical wall quenching in flame stability and wall heat flux: Hydrogen-air mixtures." Combustion Theory and Modelling 2(4): 515–530. doi: http://dx.doi.org/10.1088/1364-7830/2/4/010.

[41] Raimondeau, S, Norton, D, Vlachos, D G and Masel, R I (2002). "Modeling of high-temperature microburners." Proceedings of the Combustion Institute 29(1): 901–907. doi: http://dx.doi.org/10.1016/S1540-7489(02)80114-6.

[42] Miesse, C, Masel, R I, Short, M and Shannon, M A (2005). "Diffusion flame instabilities in a 0.75 mm non-premixed microburner." Proceedings of the Combustion Institute 30(2): 2499–2507. doi: http://dx.doi.org/10.1016/j.proci.2004.08.140

[43] Bird, G A (1994). Molecular Gas Dynamics and the Direct Simulation of Gas Flows. New York, NY: Oxford University Press.

[44] Norton, D G, Vlachos, D G and Chen, J H (2005). "Hydrogen assisted self-ignition of propane/air mixtures in catalytic microburners." Proceedings of the Combustion Institute 30(2): 2473–2480. doi: http://dx.doi.org/10.1016/j.proci.2004.08.188.

[45] Kee, R J, Zhu, H and Goodwin, D G. (2005). "Solid-oxide fuel cells with hydrocarbon fuels." Proceedings of the Combustion Institute 30(2): 2379–2404. doi: http://dx.doi.org/10.1016/j.proci.2004.08.277.

[46] McNenly, M J, Gallis, M A, Boyd, I (2003). "Slip model performance for micro-scale gas flows." AIAA paper-2003-4050. doi: http://arc.aiaa.org/doi/abs/10.2514/6.2003-4050.

[47] Xu, B and Ju, Y (2005). "Concentration slip and its impact on heterogeneous combustion in a micro scale chemical reactor." Chemical Engineering Science 60(13): 3561–3572. doi: http://dx.doi.org/10.1016/j.ces.2005.01.022.

[48] Xu, B and Ju, Y (2006). "Theoretical and numerical studies of non-equilibrium slip effects on a catalytic surface." Combustion Theory and Modelling 10(6): 961–979. doi: http://dx.doi.org/10.1080/13647830600792313.

[49] Chou, C P, Chen, J Y, Evans, G H and Winters, W S (2000). "Numerical studies of methane catalytic combustion inside a monolith honeycomb reactor using multi-step surface reactions." Combustion Science and Technology 150(1): 27–57. doi: http://dx.doi.org/10.1080/00102200008952116.

[50] Lee, M B, Yang, Q Y and Ceyer, S T (1987). "Dynamics of the activated dissociative chemisorption of CH4 and implication for the pressure gap in catalysis: A molecular beam-high resolution electron energy loss study." Journal of Chemical Physics 87(5): 2724. doi: http://dx.doi.org/10.1063/1.453060.

[51] Bertagnolli, K E and Lucht, R P (1996). "Temperature profile measurements in stagnation-flow, diamond-forming flames using hydrogen cars spectroscopy." Symposium (International) on Combustion 26(2): 1825–1833. doi: http://dx.doi.org/10.1016/S0082-0784(96)80003-8.

[52] Shankar, N, Glumac, N (2003). "Experimental investigations into the effect of temperature slip on catalytic combustion." Eastern States Section Meeting of the Combustion Institute, Pennsylvania State University.

[53] Papadopoulos, D and Rosner, D (1996). "Direct simulation of concentration creep in a binary gas-filled enclosure." Physics of Fluids 8(11): 3179–3193. doi: http://dx.doi.org/10.1063/1.869094.

[54] Kramers, H A and Kistemaker, J (1943). "On the slip of a diffusing gas mixture along a wall." Physica 10(8): 699–713. doi: http://dx.doi.org/10.1016/S0031-8914(43)80018-5.

[55] Scott, C D (1992). Wall Catalytic Recombination and Boundary Conditions in Nonequilibrium Hypersonic Flows—With Applications. In Advances in Hypersonics (pp. 176–250). Boston: Birkhäuser.

[56] Gupta, R N, Scott, C D and Moss, J N (1985). Slip-boundary equations for multicomponent non-equilibrium airflow. NASA STI/Recon Technical Report N, 86, 14530, Houston, TX.

[57] Rosner, D E and Papadopoulos, D H (1996). "Jump, slip, and creep boundary conditions at nonequilibrium gas/solid interfaces." Industrial and Engineering Chemistry Research 35(9): 3210–3222. doi: http://dx.doi.org/10.1021/ie9600351.

CHAPTER 2

BOUNDARY-ACCELERATED FLAMES IN MICROCHANNELS

A NEW APPROACH TO CHEMICAL MICROPROPULSION

Elaine S. Oran, Vadim N. Gamezo, and Ryan Houim

A boundary layer is a static or slow-moving layer of fluid that forms next to a material interface. The fluid near the interface is slowed at the boundary by viscous forces (friction) that effectively hold back the flow nearest the surfaces. Properties of boundary layers are determined by the gradient in fluid velocity near the boundary and material properties of the adjacent surfaces. The fluid nearest the surfaces is the most affected, and fluid far enough away from the surface can be relatively unaffected. In general, and from a macroscopic fluid-dynamics point of view, predicting what happens in and around a boundary layer requires understanding how viscous forces distort flow patterns and how these, in turn, can distort the flow away from the boundary.

In many considerations of combustion phenomena, we can neglect boundary layers between solid and fluid surfaces, and so ignore their effects on the flow and the combustion process itself. This is valid when flow dynamics are controlled by events far from walls. For many applications, we design systems in which boundary layer effects on the main body of the flow are small or essentially nonexistent. In opposite situations, such as in microflow devices, we direct the flow to the walls to take advantage of long residence times, wall reactions, or catalysis.

The issue of boundary layers and flames is revisited here. Due to current trends toward smaller fluid and combustion systems, boundary effects can now become significant or even dominant. This may seem natural and expected for the case of small, low-velocity flow systems, such as those characterized by laminar flame speeds, low Reynolds numbers, and the absence of shocks or detonations. The importance of boundary layers is more of a surprise, however, for small systems that evolve to high-speed flows, in which shocks, turbulent deflagrations, and detonations *can* develop.

The work described in this chapter began with the Ph.D. dissertation of James Ott in 1999 [1–3] and subsequent papers by Gamezo et al. [4, 5] that applied and greatly extended the original study to show its use for propulsion. The body of work explores some of the basic concepts of the boundary-layer and flame interaction, a laminar propagates from the closed to the open end of a narrow channel. The flame drives a flow downstream in the cold gas. The interaction of this flow with the channel walls creates a boundary layer on the downstream walls. The boundary layer grows in time, and the flame itself must interact with this growing boundary layer. This process was studied through numerical simulations of an idealized exothermic reactive flow. The most important results for micropropulsion are achieved by suppressing or at least retarding the development of turbulence and heating the walls. Then, if the channel size is adjusted properly with respect to properties of the flame, significant thrust can be obtained as the flow exits the channel. If the channel is long enough, the interactions lead to detonations and sometimes to turbulent flames and then detonations.

The flame acceleration phenomenon described in this chapter is different from the well-known case of flame propagation in a channel with cold walls, where the flame quenches near the walls and the flame velocity oscillates. Such oscillating flames were originally reported in the classical work of Mallard and Le Chatelier [6], and recently studied in experiments [7, 8]. The flame propagation in adiabatic and nonadaibatic channels was also studied in recent theoretical and numerical works [9–11].

In addition, the work presented here is for narrow channels, with an emphasis on how this might occur for laminar flow, without turbulence present. The effects of turbulence, which could significantly alter the process, and which should be delayed as long as possible, could have a significant effect on the results. That turbulence can be suppressed for enough time after spark ignition was shown in early experiments [12] and is discussed in later sections of this chapter. Here, we are not concerned with one of the possible outcomes of flame acceleration, which is a deflagration-to-detonation transition, although this is briefly discussed in the text below.

As pointed out in Bychkov et al. [13], the effects we are seeing in the simulations were discussed in early work by Shelkin [14] who related flame acceleration both to the presence of the boundary layer increasing the surface area of the flame and to expansion of burning material. After the work of Ott, further theoretical work replicated and extended concepts, as presented here [13, 15].

The discussion begins with a description of the early results, which are a series of two-dimensional solutions of the reactive Navier–Stokes equations. This lead to a series of next questions: What happens in three dimensions? What happens for nonadiabatic boundary conditions? What happens when the flame is allowed to accelerate to sonic or supersonic speeds? Finally, applications to micropropulsion applications and navigation control in low-gravity environments are considered.

2.1   PHYSICAL AND NUMERICAL MODELS

The problem is addressed by solving the fully compressible, explicit, second-order, reactive Navier–Stokes equations in the form:

$$\frac{\partial \rho}{\partial t} + \nabla \cdot (\rho \mathrm{U}) = 0, \qquad (2.1)$$

$$\frac{\partial(\rho \mathbf{U})}{\partial t} + \nabla \cdot (\rho \mathbf{U}\mathbf{U}) + \nabla P + \nabla \cdot \tau = 0, \tag{2.2}$$

$$\frac{\partial E}{\partial t} + \nabla \cdot ((E + P)\mathbf{U}) + \nabla \cdot (\mathbf{U} \cdot \tau) + \nabla \cdot (K\nabla T) + \rho q \dot{\omega} = 0, \tag{2.3}$$

$$\frac{\partial(\rho Y)}{\partial t} + \nabla \cdot (\rho Y \mathbf{U}) + \nabla \cdot (\rho D \nabla Y) = \rho \dot{\omega}. \tag{2.4}$$

Here ρ is the mass density, U is the fluid velocity vector, E is the energy density, P is the pressure, Y is the unburned mass fraction, K is the thermal conduction coefficient, D is the mass diffusion coefficient, q is the chemical energy release, and $\dot{\omega}$ is the reaction source term. The viscous stress tensor is defined as

$$\tau = \rho v \left(\frac{2}{3} (\nabla \cdot \mathbf{U}) \mathbf{I} - (\nabla \mathbf{U}) - \nabla \mathbf{U})^\dagger \right), \tag{2.5}$$

where v is the kinematic shear viscosity, I is a unit matrix, and superscript $\dagger$ indicates the matrix transposition operation.

The equation of state is that of an ideal gas

$$P = \frac{\rho R_u T}{M_u}, \quad E = \frac{P}{(\gamma - 1)} + \frac{\rho U^2}{2}, \tag{2.6}$$

where γ is the adiabatic index (ratio of specific heats), R_u is the universal gas constant, and M_u is the molecular weight. The chemical source term describes first-order Arrhenius kinetics,

$$\frac{dY}{dt} \equiv \dot{\omega} = -A \rho Y \exp\left(-\frac{Q}{R_u T} \right), \tag{2.7}$$

where A is the pre-exponential factor and Q is the activation energy. Taking the reaction rate $\dot{\omega}$ proportional to ρ accounts for the binary nature of chemical reactions taking place in real combustion systems.

Molecular transport coefficients are functions of temperature,

$$v = v_0 \frac{T^n}{\rho}, \quad D = D_0 \frac{T^n}{\rho}, \quad \frac{K}{\rho C_p} = \kappa_0 \frac{T^n}{\rho}, \tag{2.8}$$

where v_0, D_0, and κ_0 are constants, $C_p = \gamma R/M(\gamma - 1)$ is the specific heat at constant pressure, and $n = 0.7$ models the temperature dependence typical of these coefficients in reactive hydrocarbon systems.

There are two main ways to include a boundary layer in a numerical simulation, either to "capture" or to model it. It can be captured by solving the Navier–Stokes equations with no-slip boundary conditions and enough computational cells near the walls to resolve velocity and other gradients. It can be modeled, and a model used at the walls by any one of a number of boundary-layer parameterization methods, such as subtracting momentum from the region near the wall, or using an approximate analytic solution and enforcing it on the computational domain. The approach used here is to resolve the boundary layer so that it appears self-consistently as part of the solution of the equations. With this approach, it is possible to compute, and hopefully understand feedback effects between the boundary layer and other parts of the flow. Even though it is more expensive to capture a boundary layer, doing so in this case is critical for understanding the evolution of the reactive flow in the presence of boundary layers.

The original computations [1–3] solved the coupled two-dimensional reactive continuity equations described above by a high-order flux-corrected transport method on a structured, uniformly spaced grid. The results were of high accuracy everywhere on the computational domain, even in regions that probably did not require it. This was done because one focus of that work was on acoustic waves and how they were important in the early stages of flow development. These results are summarized in Section 2.4 below.

The calculations were subsequently repeated using a lower-order Godunov method and a dynamically adapting, Cartesian, structured mesh. The adaptive mesh refinement (AMR) increases numerical resolution in the vicinity of important flow features, such as flame fronts, shocks, and boundary layers, and catches and resolves incipient reactive regions. It also reduces the computing time enough for two-dimensional simulations to become routine, and even to perform a limited number of three-dimensional calculations in a reasonable amount of time [16, 17]. The AMR code has been extensively tested on a variety of combustion problems [17–21], including flame–shock interactions, boundary-layer effects in turbulent combustion, and the deflagration-to-detonation transition. The most recent code used for these calculations replaced the lower-order Godunov method with another high-order monotone method, and used a block-structured AMR algorithm. These are referenced as the work is described.

The model reactive gas used in the simulations presented in this chapter was homogeneously premixed stoichiometric acetylene and air at low pressure [17]. The input parameters for the model are given in Table 2.1, which also shows the values of laminar flame velocity and

Table 2.1. Material and chemistry parameters for stoichiometric acetylene–air mixture

Input		
T_0	293 K	Initial temperature
P_0	1.33×10^5 ergs/cm^3	Initial pressure
ρ_0	1.58×10^{-4} g/cm^3	Initial density
γ	1.25	Adiabatic index
M	29 g/mol	Molecular weight
A	10^{12} cm^3/g-s	Pre-exponential factor
Q	$29.3\,RT_0$	Activation energy
q	$35.0\,RT_0/M$	Chemical energy release
$v_0 = \mu_0 = D_0$	1.3×10^{-6} g/s-cm-K$^{0.7}$	Transport constants in Equation (2.5)
Output		
x_l	0.012 cm	Laminar flame thickness
s_l	153 cm/s	Laminar flame velocity

thickness obtained by solving the model with the inputs. Similar parametric chemical-diffusive models now exist for low-pressure stoichiometric ethylene in air [19], or stoichiometric hydrogen in air [20], and lean, stoichiometric or rich methane in air [21].

2.2 THE BOUNDARY-LAYER ACCELERATED FLAME

2.2.1 EARLY SIMULATIONS

The early studies of boundary-layer accelerated flames [1–3] were carried out with a fully compressible reactive fluid-dynamics code that used a high-order flux-corrected transport algorithm on a fine, uniform, Cartesian mesh. The code was implemented on a massively parallel computer with a special architecture (the CM 5) that has since become obsolete. It can be argued that there was a large waste of computing resources in those calculations because so much of the computational grid was assigned to regions of space where nothing of particular note was happening. At the time, however, part of the focus of the research was on acoustic waves and flame-acoustic interactions. In order to explore these phenomena, it was necessary to resolve as much of the acoustic field as possible and its effects on the structure of the laminar flame.

The computational cell size was then chosen so that there would always be 5–15 cells in a laminar one-dimensional flame width. As was demonstrated then and has been demonstrated by many recent numerical tests, this small number of computational cells is adequate when the reaction mechanism is as simple as the one-step Arrhenius kinetics shown in Equation (2.7). When a detailed reaction mechanism is used, however, this resolution is inadequate because the important rise and fall of the production and consumption of individual intermediate chemical species cannot be resolved. When these features are diffused because there is not adequate numerical resolution, the flame structure and energy release rate depend on numerical diffusion instead of the physical diffusion effects included in the model. Prior studies of laminar flames indicated that, if all of the constituent species are included and their profiles resolved, at least 30–50 computational cells in a hydrogen-flame thickness were required to model the laminar flame. Some researchers have claimed that they need as many as 150–200 cells.

The schematic in Figure 2.1 shows a laminar flame propagating toward the open end of a narrow channel filled with a gaseous combustible mixture. The propagating flame creates an outflow in the unburned fluid ahead of it, and this outflow produces a boundary layer along the walls. Early experimental studies showed that this flame could accelerate, oscillate, or die out, depending on the material, the wall temperature, and the channel width [12]. There are two main questions to ask now. *What exactly is the structure of the flame at the boundary layer? What are the dynamics of a flame propagating into a boundary layer?*

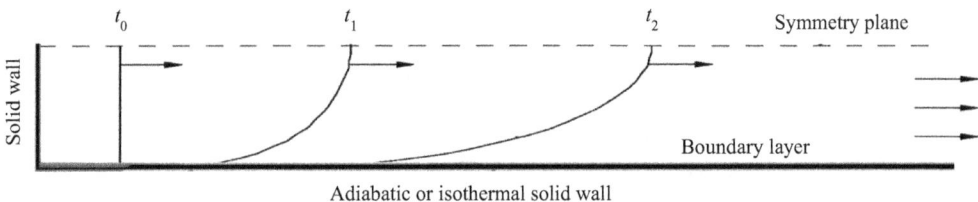

Figure 2.1. Schematic of general problem and the boundary conditions. The diagram shows that initially planar wave at t_0 becomes more and more curved (t_1, t_2) due to the presence of the boundary layer [1].

The Ott et al. computations for a low-pressure acetylene–air mixture (Table 2.1) considered both adiabatic (no heat losses) and isothermal walls. The results were strikingly different for these two different boundary conditions. When the walls were adiabatic, the flame continuously accelerated. When the walls were held at constant temperature, the flame speed oscillated around the laminar burning velocity as the flame moved down the channel. Figure 2.2 shows the flame speed and position in laboratory coordinates for both cases. Although this figure does summarize the results of the calculations, it does not present any explanations of what was happening or why. Physical explanations were found by examining instantaneous profiles of some of the key physical variables, such as those shown in Figures 2.3 to 2.5.

First consider the adiabatic-wall case shown in Figure 2.3, which is a composite of instantaneous flame profiles. These profiles show that as the flow develops and the flame propagates downstream, the boundary layer grows, and the channel effectively becomes narrower. Figure 2.4 also shows that there is additional momentum added to the flow around the centerline as unburned boundary-layer material reacts, jets *toward the center*, and then is forced to turn toward the flame. Jetting material turns toward the flame because the previously burned confined material behind it behaves like a wall or a plug. The jet can only turn downstream. Throughout all of this, the laminar burning velocity of the flame does not change much, but the flame velocity in the laboratory coordinate changes greatly. Thus the flame along the centerline is accelerated by two effects: the effective narrowing of the channel due to the presence of the boundary layer, and the extra "push" created by additional jetting material.

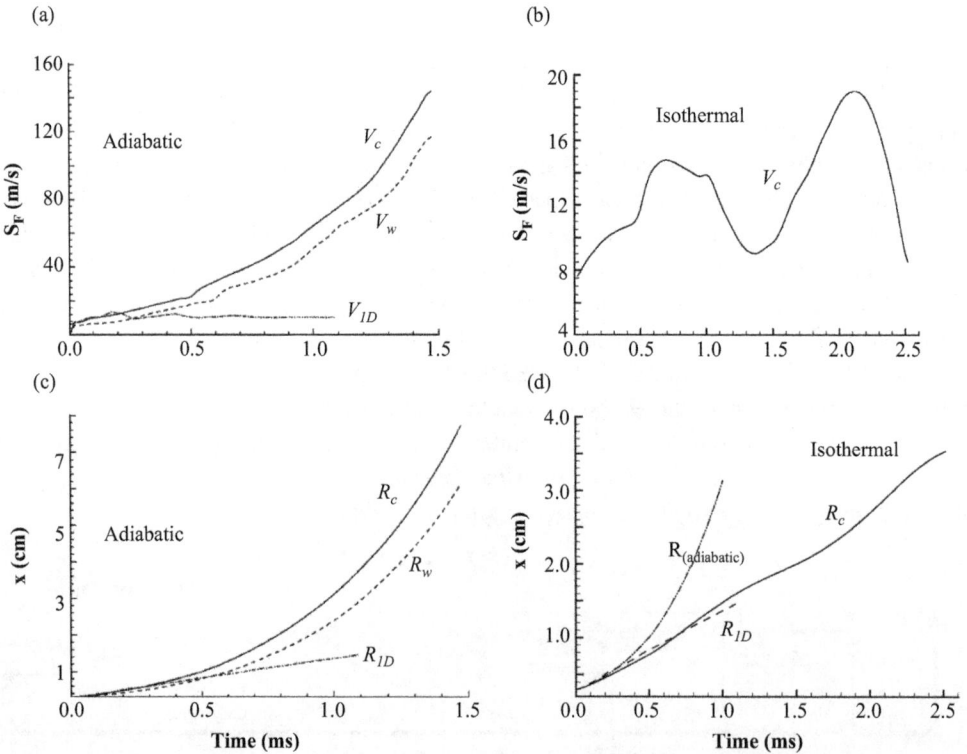

Figure 2.2. Flame velocities S_F (a) and flame positions R (c) in the laboratory coordinate system for adiabatic walls and isothermal walls (b) and (d). Subscripts c and w indicate that the variable is shown for the centerline and walls, respectively, and ID indicates the one-dimensional solution. Subscript adiabatic in figure (d) indicates the adiabatic solution from figure (c), shown for comparison [1].

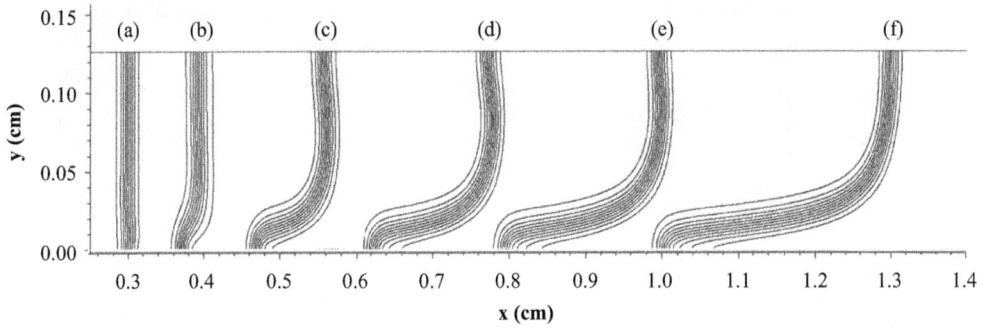

Figure 2.3. Composite taken from a two-dimensional simulation [1–3] of temperatures for the case of an adiabatic wall, at the flame front at selected early times (a–f): 0.02138, 0.10798, 0.26006, 0.39012, 0.49843, and 0.60671 ms. Contours are evenly spaced values of temperature, ranging from 400 K on the right to 2400 K on the left [1].

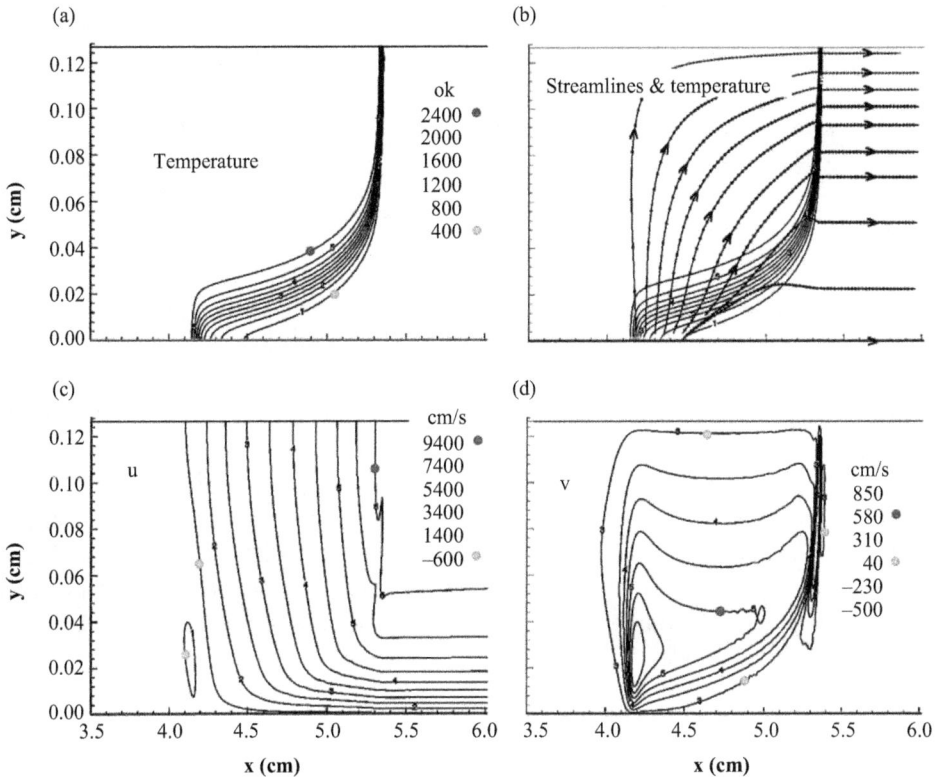

Figure 2.4. Selected variables for the case with adiabatic walls (Figure 2.2 a, c) at 1.28 ms. (a) Temperature. (b) Temperature with superimposed instantaneous streamlines. (c) Velocity u. (d) Velocity v [1].

The explanation for the flame oscillation that occurs when there are isothermal walls is more complex. Figure 2.5 shows that the oscillation of the flame speed is due to changing vortical flow structures that develop in the boundary layer itself. The direction of the flow can reverse locally, and this affects the flame velocity. Eventually, enough energy is lost to the walls and the flame can be quenched. In early extensions of this work, the channel height was varied.

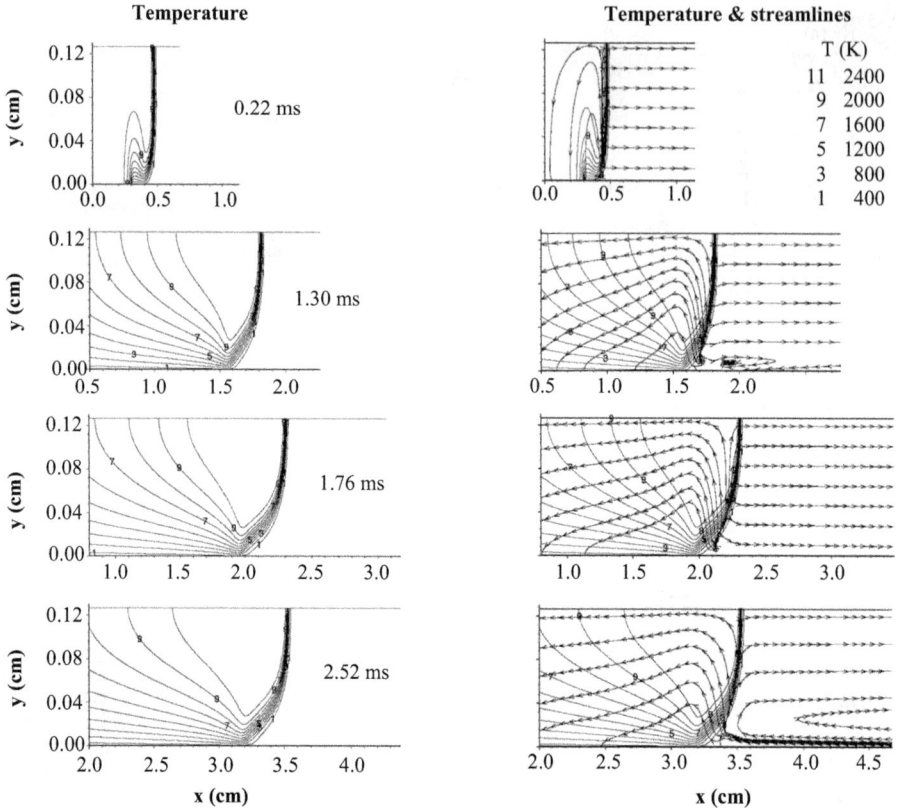

Figure 2.5. For selected times, contours of temperature (left column) and temperature imposed on instantaneous streamlines (right column) for the isothermal case [1].

This showed that acceleration and quenching effects are considerably reduced as the channel height increases [3, 4].

Now consider Figure 2.2. This figure shows histories of the position of the two-dimensional flames at the centerline of the channel, R_c, and at the wall, R_w, and compares these with the position of a one-dimensional flame with no boundary effects, R_{1D}. The curve R_c closely follows R_{1D} until about 0.3 ms. Initially, R_w is less than R_{1D}, but this reverses at about 0.55 ms. These quantities may be converted into flame speeds in the laboratory system, V_c (Figure 2.2 a, b). The one-dimensional velocity V_{1D} levels off to a constant value, but both V_c and V_w are still increasing by the time the flame reaches at the end of the channel. An analysis of the boundary-layer development is given in by Ott et al. [1–3]. Their conclusion is that when the channel is relatively narrow (that is, thick flames in small channels), and the walls are insulated or heated, there can be substantial flow acceleration.

When the laminar flame accelerates as it moves down the channel, the unburned fluid ahead of the flame must also accelerate. Then from a propulsion standpoint, the next question is whether this accelerated outflow creates any substantial thrust.

2.2.2. CONFIRMATION FROM NEW TESTS

Ott's original work was not published until 2003. After that, significantly improved computational resources became available, and the concept was revisited with the intention of examining whether it could be used to create a micropropulsion device [4, 5]. More specifically, could the flame–boundary layer interaction be used to create significant thrust? The simulations described in the remainder of this chapter were done with the second code, which used the same chemical-diffusive model, a lower-order fluid algorithm to solve the Navier–Stokes equations, but added the computational advantage of AMR to resolve local flow features.

As before, the simulations started by modeling a half of a channel with a no-slip reflecting boundary condition at $x = 0$ and a symmetry condition at $d/2$, where d is the channel width. Two channel lengths L were considered: 1.024 cm and 8.192 cm. The channel width d was varied from 0.0128 cm to 1.024 cm, or from 1 to 80 in units of the laminar flame thickness (defined as the distance between the $Y = 0.2$ and $Y = 0.8$ planes).

A planar laminar flame was ignited near the closed end of the channel by placing a 0.01 cm layer of hot burned material at adiabatic flame temperature near the left boundary. The initial discontinuity between burned and unburned materials was spread by molecular diffusion, heat conduction, and chemical reaction. During the first several thousand time steps ($\sim 0.2 \times 10^{-4}$ s), the reactive structure evolved into a propagating flame.

Figure 2.6 shows details of the flame evolution for the shorter channel ($L = 1.024$ cm) and height $d/x_l = 10$. As before, a boundary layer develops in the flow of unburned material downstream of the burning, expanding gas. As the boundary layer forms, the flow velocity changes across the channel, so that it is zero at the wall and reaches a maximum in the center of the channel. This non-uniform flow stretches the flame, so that its shape reflects the velocity profile across the channel. This means that the surface area of the flame increases, and this increases the rate of energy release. This does not have a major effect on the laminar flame velocity, S_L,

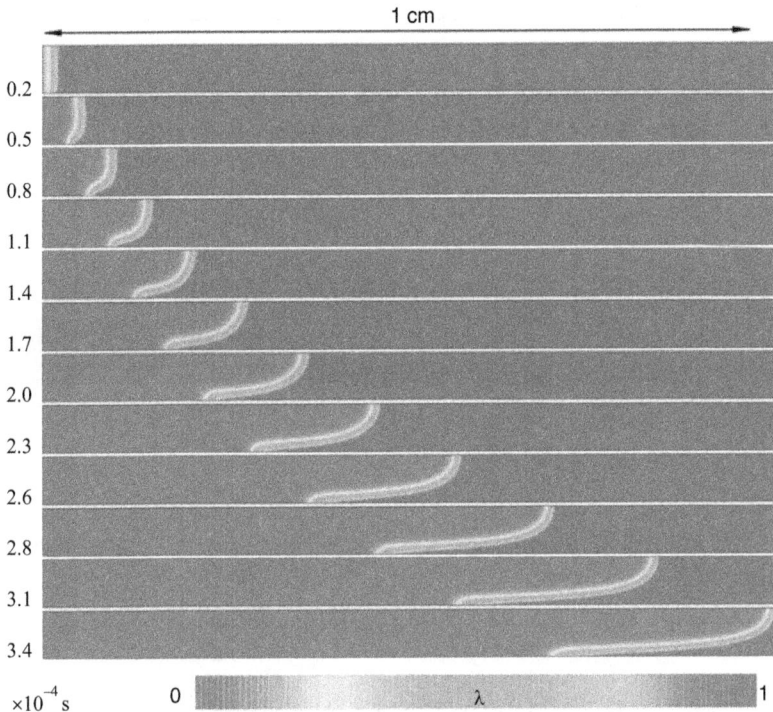

Figure 2.6. Time sequence of unburned mass fraction plots showing the evolution of a laminar flame in 2D half-channel for $L = 1.024$ cm and $d/2 = 0.064$ cm. The minimum computational cell size is 0.001 cm [4]. Color figure in E-book.

which remains on the order of a few tens of centimeters per second. As the flow accelerates, the flame speed in the laboratory frame of reference increases as well, and eventually becomes much higher than S_L. Figure 2.7 shows that the outflow velocity reaches ~55 m/s by the time the flame arrives at the end of the channel. The flame speed in the laboratory frame of reference at that time is practically the same. Further, the result of resolution tests shown in that figure indicates that numerical resolution has only a minor effect on the computed outflow velocity. A typical flow field in the vicinity of the accelerating flame is shown in Figure 2.8.

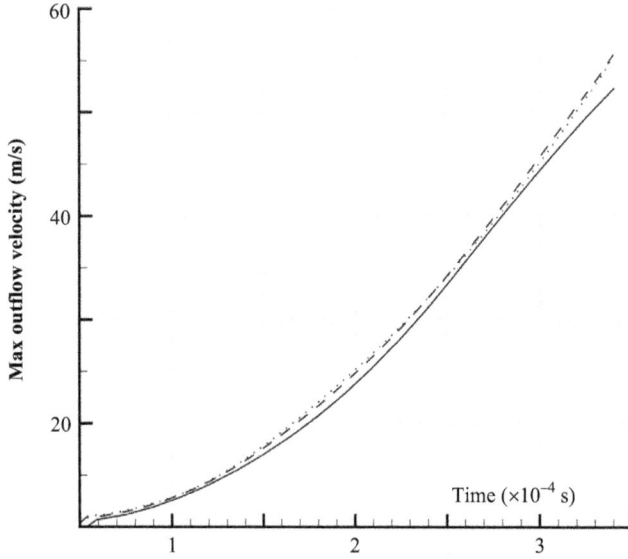

Figure 2.7. Maximum outflow velocity at the right boundary calculated for 2D half-channel 1.024×0.064 cm for different numerical resolutions. Solid, dashed, and dotted lines correspond to minimum computational cell sizes 0.002, 0.001, and 0.0005 cm, respectively [4].

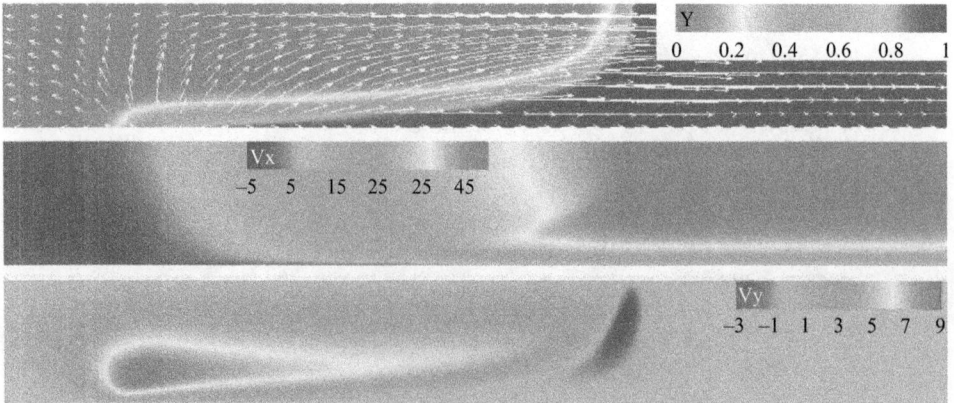

Figure 2.8. Unburned mass fraction Y with superimposed velocity vectors, longitudinal velocity Vx, and transverse velocity Vy in the vicinity of an accelerating flame in 2D half-channel 1.024×0.064 cm. The part of the computational domain shown is 0.5×0.064 cm adjacent to the right (outflow) boundary, at 3.1×10^{-4} seconds after ignition [4]. Color figure in E-book.

The most important point of results shown above are that these newer computations, using a new generation of computers and very different algorithms for solving the governing equations, show the same mechanism of flame acceleration induced by a boundary layer as was reported earlier. That is, they show: the curved flame surface, the boundary layer near the bottom wall, the fast-moving unburned material ahead of the flame, the relatively slow-moving burned material behind the flame, and the velocity gradient inside the tip of the flame that is responsible for the growth of the flame surface area.

2.3 EFFECTS OF VARYING CHANNEL DIMENSIONS

2.3.1 CHANGING THE CHANNEL WIDTH

A series of simulations were performed to study the effect of varying the channel width, while keeping the channel length constant ($L = 1.024$ cm). Physically, we are looking at the effects of increasing the ratio of channel height to flame thickness.

Figure 2.9 shows the maximum outflow velocity as a function of scaled channel width. For channel widths comparable to the flame thickness, $d \sim x_l$, the flame remains practically one-dimensional for the entire time it travels down the channel. As d becomes larger, however, so that $d > 2x_l$, boundary layers begin to affect the flow, and the flame and flow velocities in the channel increase. The maximum flame acceleration occurs for $d/x_l = 5$, where, for this case, the outflow velocity reaches 100 m/s. As the channels become wider, the flame acceleration decreases, and thus the outflow velocity increases because the part of the flame surface initially affected by the developing boundary layer is smaller.

In longer channels, the flame and the boundary layer have more time to develop. For such cases, the flow can eventually accelerate to the sound speed. For example, simulations for

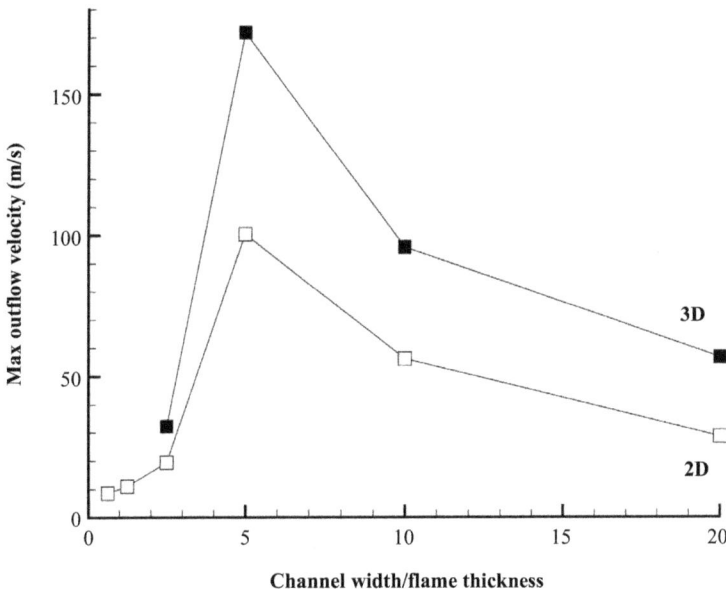

Figure 2.9. Maximum outflow velocity calculated for the fixed channel length 1.024 cm as a function of channel width scaled by the laminar flame thickness 0.0128 cm [4].

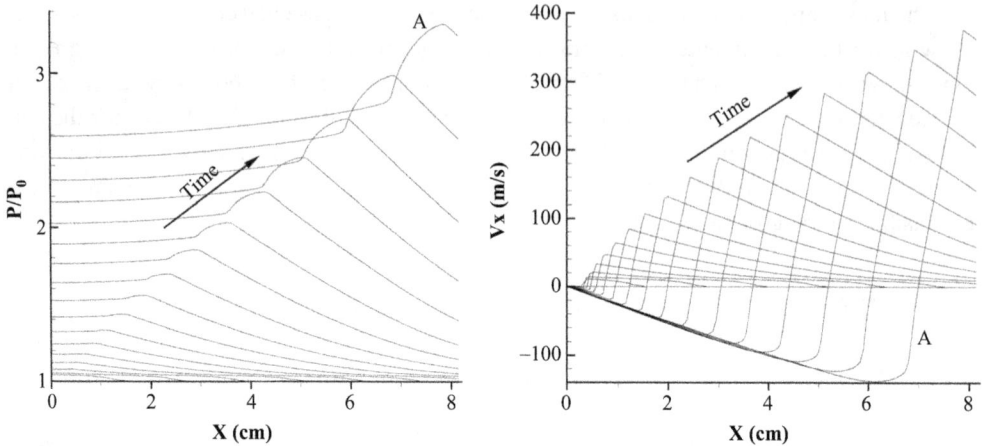

Figure 2.10. Time sequence of pressure and longitudinal velocity profiles at the symmetry line of a 2D half-channel 8.192×0.032 cm. Time changes from 0 to 5.66×10^{-4} s with increment $(2.5 – 3.5) \times 10^{-6}$ s between profiles. Profiles A correspond to 5.66×10^{-4} s [4].

$L = 8.192$ cm and $d/x_l = 5$ predict a maximum outflow velocity of 370 m/s, which is close to the sound speed in the unburned material. The sequence of profiles in Figure 2.10 show that, as the system evolves, the pressure and velocity increase between the open end of the channel ($x = 8.192$ cm) and the flame, and then drop in the reaction zone. The velocity becomes negative, and then increases to 0 at the closed end ($x = 0$), as the boundary condition requires. In the reaction zone and behind the flame, the pressure slightly decreases. It is still high enough however to reverse the flow in the positive direction at later times, after the flame reaches the open end.

Ahead of the flame, the pressure, velocity, and their gradients increase in time. This is caused by compression waves generated by the energy release. In a longer channel or in a more energetic system, these compression waves can converge to form a strong shock. This does not occur for the reactive mixture and channel sizes and for gas considered here.

2.3.2 THREE-DIMENSIONAL SIMULATIONS

For the three-dimensional (3D) simulations, the computational domain covered a quarter of a square channel. The boundary conditions were no-slip adiabatic walls at $y = 0$ and $z = 0$, and symmetry at $y = d/2$ and $z = d/2$. Four channel lengths were considered: 1.024, 2.048, 4.096, and 8.192 cm. The scaled channel width, d/x_l, thus varied from 2.5 to 20. Simulations were terminated when the flame reached the open end of the channel. The same reactive acetylene–air mixture was used. The initial condition is a planar flame.

Figure 2.11 shows the evolution of the full 3D flame surface for $L = 1.024$ cm and $d/x_l = 5$, 10, and 20. When the channel is narrow, the flame surface first becomes convex and then, as the flow accelerates and interacts with boundaries, the flame shape generally follows the shape of the longitudinal velocity profile. For larger channels, the flame surface becomes slightly concave in the middle of the channel and then becomes convex. For the largest channel ($d/x_l = 20$), the flame surface develops into a concave cavity near the symmetry planes, and this persists until the flame reaches the end of the channel. Similar concave flames were observed in 2D simulations and analyzed in [1]. They appear at initial stages of flame development in

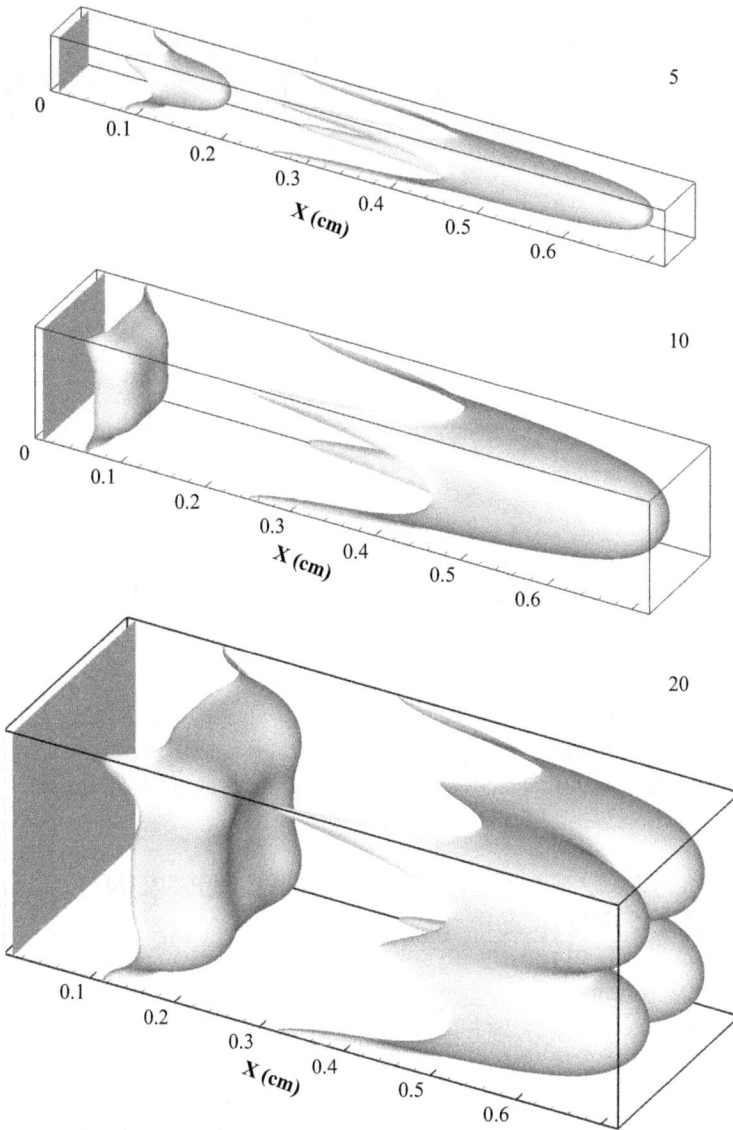

Figure 2.11. Evolution of the flame surface in full 3D channels of different width and $L = 1.024$ cm. Numbers 5, 10, and 20 are scaled channel widths d/x_l. Flame surfaces are shown at 0, 0.98×10^{-4}, and 1.59×10^{-4} s for $d/x_l = 5$; at 0, 0.74×10^{-4}, and 2.19×10^{-4} s for $d/x_l = 10$; and at 0, 1.50×10^{-4}, and 2.97×10^{-4} s for $d/x_l = 20$ [4].

larger channels, when the boundary-layer thickness is small compared to the channel width, and disappear with time if the channel is long enough.

The flame-induced flow acceleration in 3D channels is significantly greater than in 2D channels, which is due to the extra boundary-layer effect in the third dimension. Because the distorted surface area of a 3D flame grows faster than a 2D flame, the energy-release rate and the flow velocity increase faster. Figures 2.9 and 2.12 show this effect rather clearly. The maximum outflow velocity reached in 3D channels was approximately a factor of two higher than

(a)

(b)

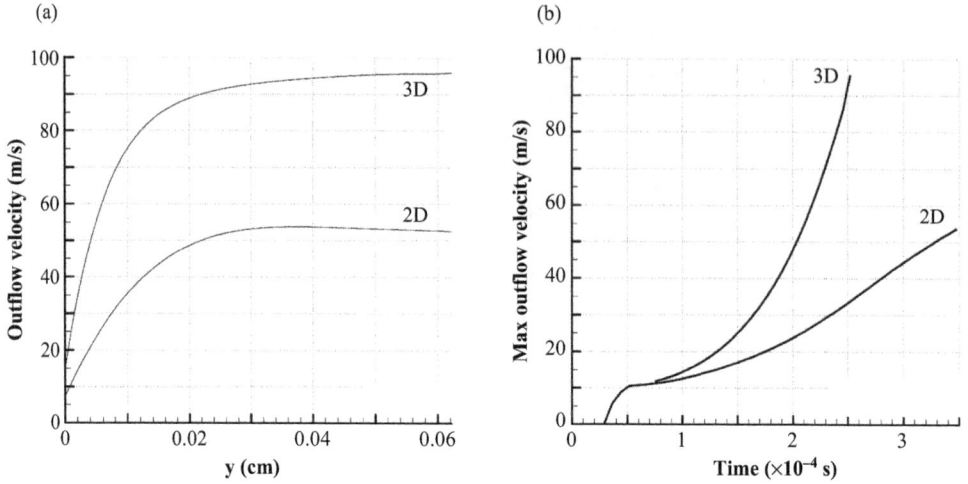

Figure 2.12. Outflow velocity profiles at the time when the flame reaches the open boundary (left) and maximum outflow velocity as a function of time (right) for 2D and 3D channels. The channel length is 1.024 cm, and the channel half-width is 0.064 cm ($d/x_l = 10$). The 3D velocity profile is shown for the symmetry plane. Maximum velocities in profiles are the same [4].

that in 2D channels of the same dimension. Figure 2.12 also shows another important effect. For both 2D and 3D, the outflow speed is high through a substantial cross-section of the channel (~0.7 for 2D and ~0.5 for 3D). As shown below, this produces significant thrust.

2.4 OUTFLOW ACCELERATION AND PROPULSION CHARACTERISTICS

Figures 2.13 and 2.14 summarize the propulsion characteristics for 3D channels. Figure 2.13 shows the propulsion characters for a 3D channel of fixed length, 1.024 cm, and for a range of cross-sections. Figure 2.14 shows the propulsion characteristics of the cases of constant channel half-width $d/2 = 0.064$ cm ($d/x_l = 10$) and four channel lengths: 1.024, 2.048, 4.096, and 8.192 cm. The quantities shown are the maximum outflow velocity, mass outflow, thrust, ratio of the ejected mass to the burned mass, and the specific impulse, I_{sp}. In these figures, I_{sp} defined in two ways [4]: total impulse scaled by an ejected mass and total impulse scaled by a burned mass.

Figure 2.13 shows that the maximum outflow velocity, the mass outflow, and the thrust are approximately proportional to each other. The maximum thrust was for $d/2 = 0.032$ cm ($d/x_l = 5$), where it reached 158 g/cm^2 (or 0.65 g for the full channel), with an outflow velocity of 172 m/s and the flame near the open end of the channel. The ratio of the maximum thrust to the initial mass of the material in the channel is about 10^6.

Figure 2.14 shows that at a fixed physical time, the outflow velocity in a longer channel is lower than that in a shorter channel. There are several explanations for this. First, the mass of the unburned material accelerated ahead of the flame is larger for longer channels. Second, additional viscous losses in longer channels slow the flow acceleration. Finally, it takes longer for compression waves to reach the outflow boundary in a longer channel. In longer channels,

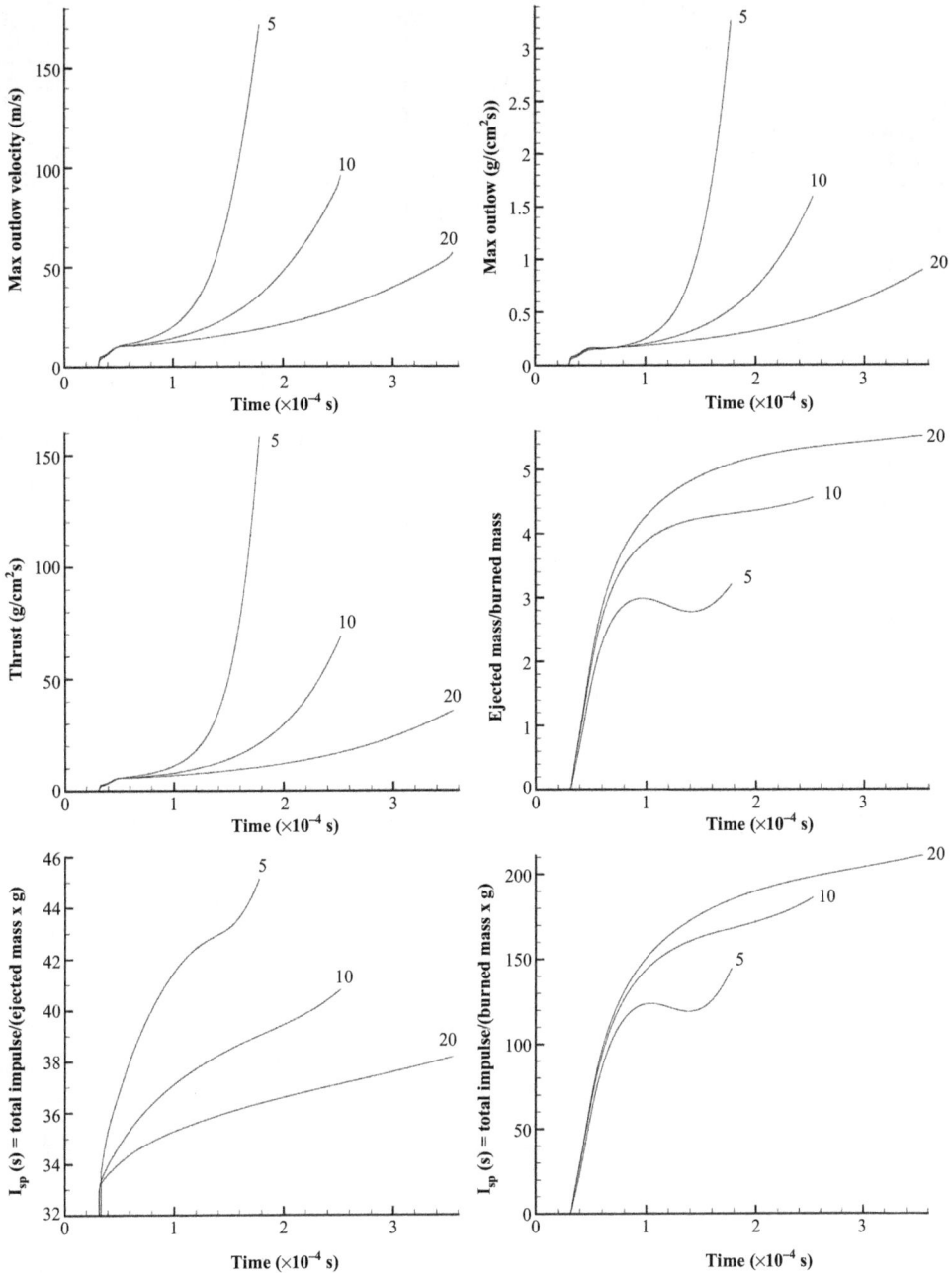

Figure 2.13. Integral propulsion characteristics calculated for 1.024 cm long 3D channels. Numbers 5, 10, and 20 are scaled channel widths [4].

however, the flame interacting with the boundary layer has more time to develop and eventually burns more material at higher rate. Outflow velocities compared at the times when the flame reaches the open end of each channel increase dramatically with the channel length. When the channel length increases from 1.024 to 8.192 cm, the maximum outflow velocity increases from 95 to 360 m/s, and the maximum thrust increases from 70 to 700 g/cm^2.

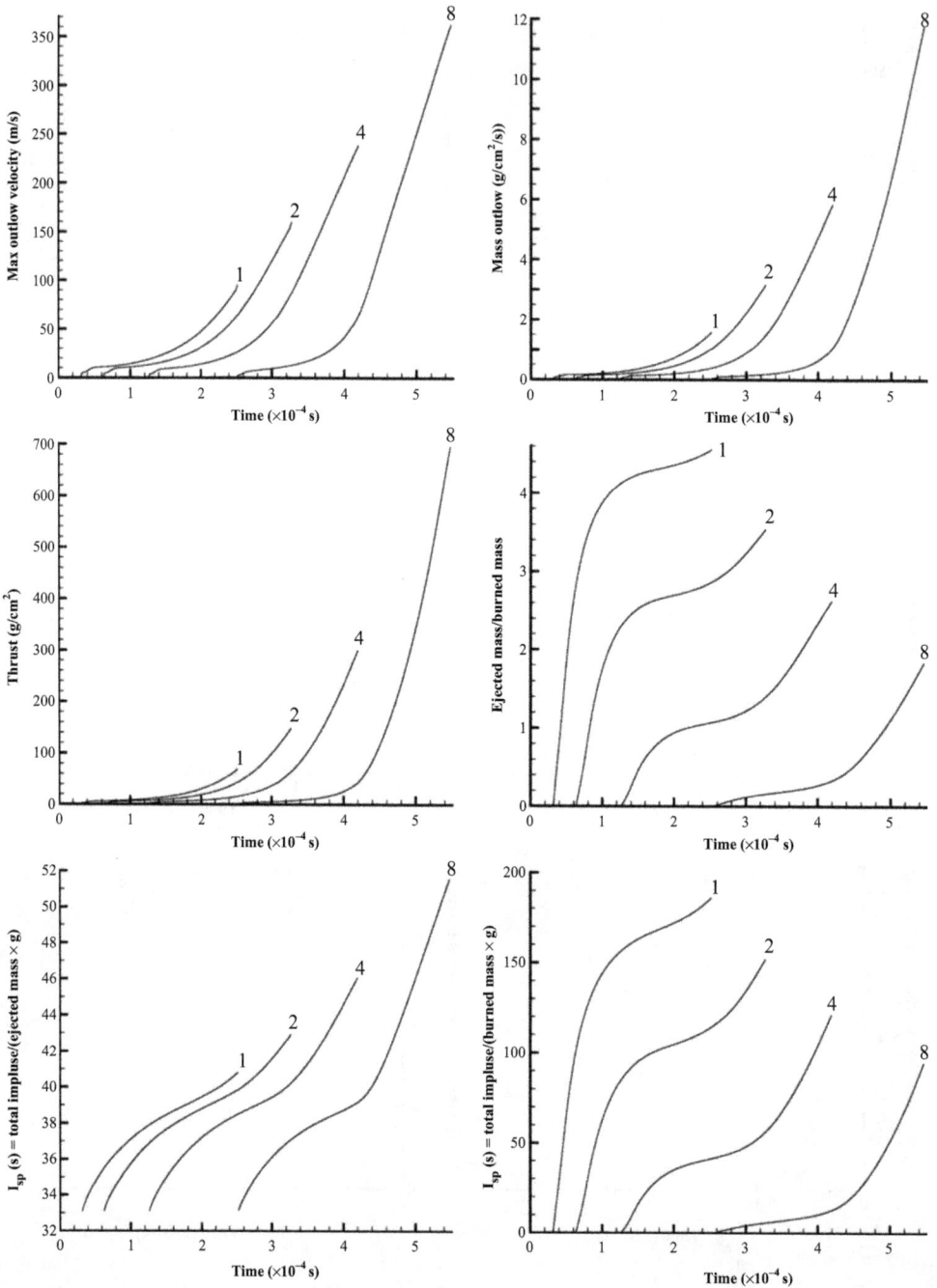

Figure 2.14. Integral propulsion characteristics calculated for channel half-width 0.064 cm ($d/x_l = 10$). Numbers 1, 2, 4, and 8 correspond to channel lengths 1.024, 2.048, 4.096, and 8.192 cm [4].

The evolution of the flame surface area, S_F, and the total energy generation rate, W_q, were computed for the 3D simulations, as shown in Figure 2.15. The curves are scaled by S_F^0, the initial flame surface area, and W_0, the corresponding initial energy-release rate. These curves show that the flame evolution is independent of the channel length, that is, given a

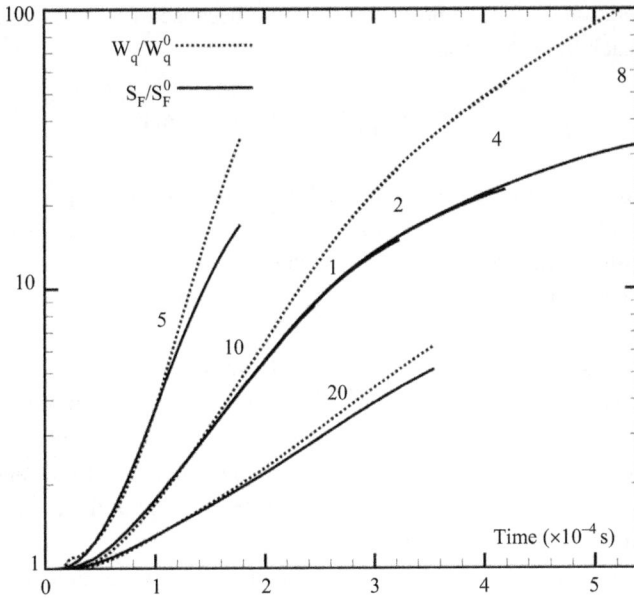

Figure 2.15. Scaled flame surface area S_F/S_F^0 and total energy generation rate W_q/W_q^0 as functions of time for all 3D cases shown in Figures 2.9 and 2.10. Numbers 5, 10, and 20 are scaled channel widths d/x_l for 1.024 cm long channels. Numbers 1, 2, 4, and 8 correspond to channel lengths 1.024, 2.048, 4.096, and 8.192 cm for 0.064 cm wide channels ($d/x_l = 10$). Numbers 1, 2, 4, and 8 are located at the ends of the corresponding curves, which practically coincide for their effective time intervals. The scales S_F^0 and W_q^0 correspond to the planar laminar flame [4].

fixed width and variable length, scaled values of S_F and W_q, almost coincide for their effective time intervals.

2.5   EFFECTS OF WALL TEMPERATURE

It is expected that the wall boundaries would be important for any practical flow. Therefore, three types of wall boundaries were tested: adiabatic, isothermal, and with a limited heat loss coefficient that was varied to model different levels of material insulation.

For adiabatic walls, the outflow velocity at the open end of the tube can reach hundreds of meters per second and provide a significant thrust. This can be achieved without creating strong shocks or detonations inside the channel. Because the material ahead of the flame is ejected without burning, the channel can be partially filled with an inert gas or combustion products from a previous pulse without affecting the energy release. This case gives upper limits for outflow velocities and thrusts.

In contrast, channels with isothermal walls give a lower limit in which energy losses to the walls can cause the flame to propagate more slowly, oscillate, or even extinguish. For practical systems, energy losses are not as fast as for isothermal-wall limit, and a laminar flame can still accelerate to high velocities.

Energy losses to the walls were parameterized by introducing a heat flux at the boundary cells that is proportional to the difference between the current temperature in a cell T and the ambient temperature T_w The temperature decrease because of heat losses is computed for boundary cells at each time step n as

$$T^n = T_w + \left(T^{n-1} - T_w\right)\exp(-\alpha\Delta t)$$

where T^{n-1} is the temperature in a boundary cell at a previous time step, Δt is the time increment for the current time step, and α is the heat loss coefficient. Adiabatic walls correspond to $\alpha = 0$; isothermal walls are described by $\alpha \to \infty$. For the minimum computational cell size 0.001 cm used here, which corresponds to approximately 13 computational cells per laminar flame thickness, a typical value of Δt is about 10^{-8} s.

The sequence of temperature maps in Figure 2.16 shows the flame evolution for a narrow channel, $d/2 = 0.064$ cm, and $\alpha = 6 \times 10^4$ s^{-1}. The flame propagation in this channel is influenced by the velocity gradient in the downstream boundary layer in the way similar to what is described for adiabatic channels. In this case, however, the speed of the flow ahead of the flame has increased to only 30 m/s by the time the flame reached the end of the channel, in contrast to the 55 m/s reached when the walls were adiabatic.

Figure 2.17 shows the maximum outflow velocity as a function of time for a range of channel widths and heat-loss rates. As expected, flame propagation is more affected by heat losses in smaller channels. For $d/2 = 0.032$ cm and α above 5×10^4 s^{-1}, the flow decelerates after some initial acceleration, and the flame eventually is quenched. Isothermal walls have the most severe effect on the flame propagation here, limiting the outflow velocity below 13 m/s for all the cases considered.

Cold walls affect the flow and flame acceleration mostly through energy losses from the hot burned material behind the flame to the wall, as shown by temperature fields in Figure 2.18.

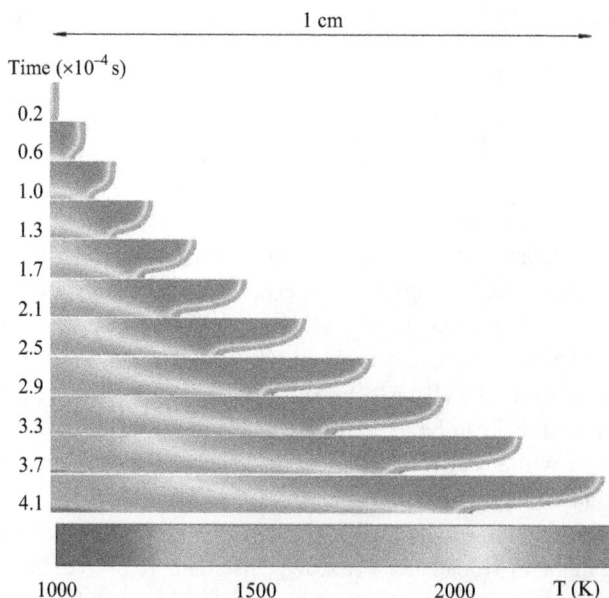

Figure 2.16. Time sequence of temperature fields showing the evolution of a laminar flame in 2D half-channel with heat losses. $L = 1.024$ cm, $d/2 = 0.064$ cm, and $\alpha = 6 \times 10^4$ s^{-1}. The minimum computational cell size is 0.001 cm [5]. Color figure in E-book.

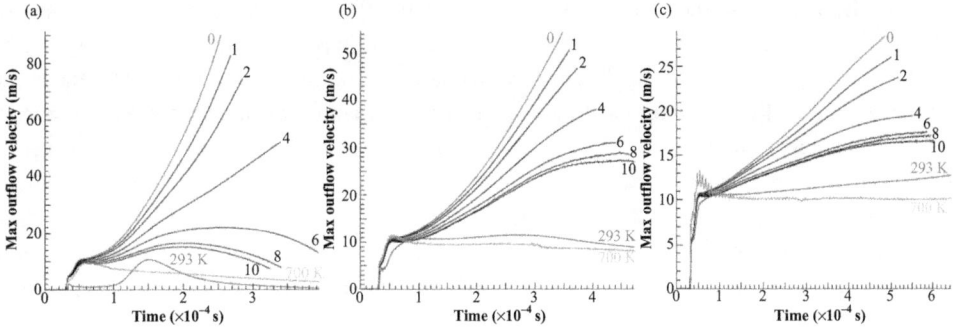

Figure 2.17. Maximum outflow velocity (near the symmetry plane) as a function of time calculated for (a) $d/2 = 0.032$ cm, (b) $d/2 = 0.064$ cm, and (c) $d/2 = 0.128$ cm. Red curves correspond to adiabatic walls and blue and green curves correspond to isothermal walls with $T_w = 293$ K and 700 K, respectively. Numbers near black curves show the heat loss coefficient $\alpha \times 10^{-4}$ s^{-1}. The channel length 1.024 cm is fixed [5].

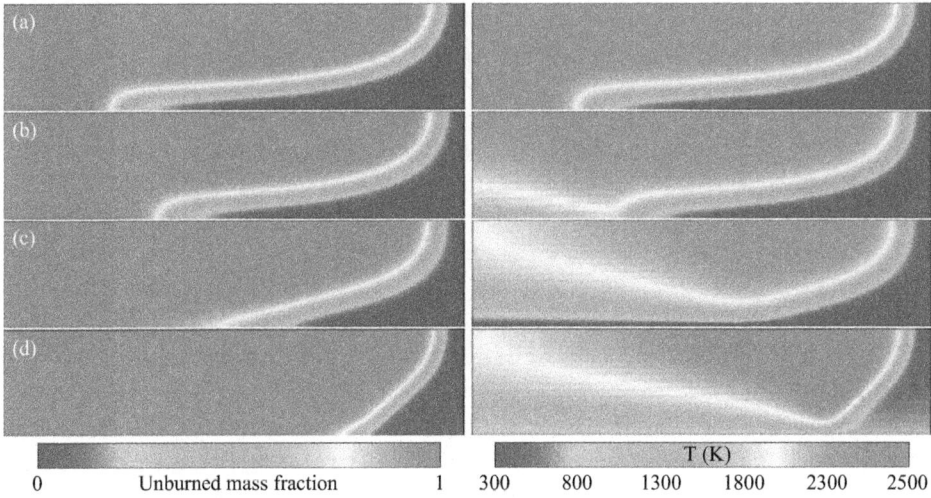

Figure 2.18. Unburned mass fraction and temperature plots showing typical flame shapes for (a) adiabatic walls, (b) $\alpha = 1 \times 10^5$ s^{-1}, (c) isothermal walls with $T_w = 293$ K, and (d) isothermal walls with $T_w = 700$ K. $d/2 = 0.064$ cm [5]. Color figure in E-book.

This reduces the expansion rate of the material, and therefore the flow acceleration. These energy losses are proportional to $T_b - T_w$, where $T_b \sim 2350$ K is the temperature of burned material. Increasing the wall temperature T_w does not improve the flame acceleration in channels with isothermal walls because T_w should be below the ignition temperature, which is close to 700 K for our system and is still "cold" compared to T_b.

Outflow velocities computed for channels with isothermal walls for two different wall temperatures, $T_w = 293$ K and $T_w = 700$ K, are shown in Figure 2.17. For $d/2 = 0.128$ and 0.064 cm, the flame propagates slightly faster in channels with colder walls because the wall temperature affects the flame structure near the wall, and therefore the shape of the flame and flame surface area. The effect of wall temperature is different for the smaller channel ($d/2 = 0.032$ cm), where the flame quenches for $T_w = 293$ K, but propagates if the wall is heated to $T_w = 700$ K.

Thus, heat losses to the walls do have a significant effect on flame propagation in narrow channels. Even though for a limited heat loss coefficient the flame near the wall may not be quenched, energy losses from the hot burned material behind the flame to the wall reduce the flow acceleration. For a fixed heat loss coefficient, the maximum flame acceleration occurs in larger channels compared to the case with adiabatic walls. The maximum outflow velocity reached in a channel of a given width is lower than for a similar channel with adiabatic walls. These effects increase with heat losses and prevent any significant acceleration of laminar flames in channels with isothermal walls. From a practical point of view, this means that thermo-insulating properties of wall material will affect the efficiency of a propulsion device that uses the laminar flame acceleration.

2.6 LAMINAR-FLAME TRANSITION TO DETONATION IN LONG CHANNELS

If the channel is long enough, compression waves generated by accelerating flames converge to form a shock. The channel length needed for this depends on the energetics of the reactive system considered.

Consider a two-dimensional channel 8.192 cm long and 0.064 cm wide containing a reactive gas that is 2.6 times more energetic than the stoichiometric acetylene–air mixture. The pressure profiles in Figure 2.19 show that a shock forms early and then gradually increases in strength as it propagates in the channel. The temperature of the unreacted material between the shock and the flame also increases, and eventually reaches the point of self-ignition in the boundary layer near the wall. The result is a spontaneous reaction wave that propagates through the temperature gradient along the wall. This spontaneous wave can evolve into a detonation, as shown by the temperature profiles in Figure 2.19c.

The evolution of the spontaneous wave and transition to a detonation are also shown by the time sequence of the density fields in Figure 2.20. The spontaneous wave develops in the boundary layer, and this generates a compression wave that propagates along the boundary layer and triggers the energy release in the hot material. The compression wave evolves into a shock, and the shock-reaction complex eventually becomes a self-sustained detonation wave

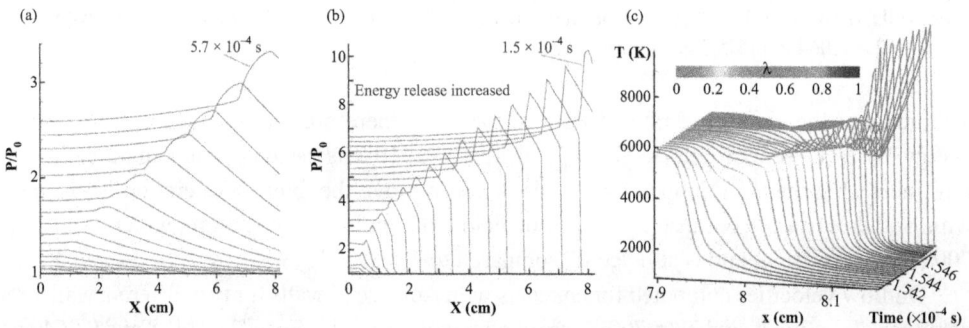

Figure 2.19. Evolution to a detonation. (a) Pressure waves generated for the original system. (b) Pressure waves generated for the more energetic system, showing the development of a shock wave. (c) Temperature profiles near the channel wall for the more energetic system, with the unburned mass fraction, λ, superimposed in color.

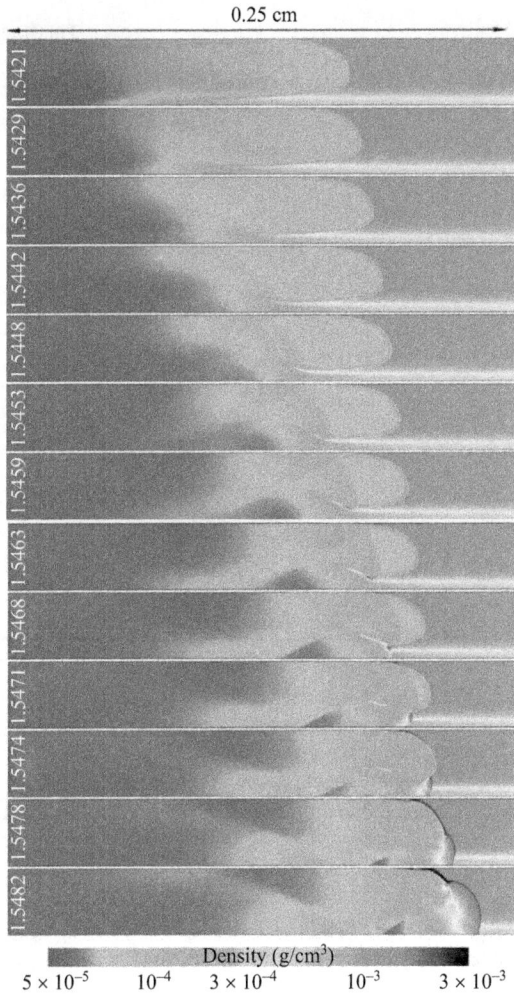

Figure 2.20. Density profiles showing the evolution of the laminar flame interacting with a boundary layer, and a transition to a detonation. Note that the spontaneous wave emerges at the bottom wall, in the boundary layer. It evolves into a shock-reaction complex that propagates through the boundary layer and emerges finally as an overdriven detonation. Color figure in E-book.

that spreads from the boundary layer to the bulk flow in the middle of the channel. The problem considered here is similar in form to that studied by Kagan and Sivashinsky [15], but differs in the way in which the transition to detonation occurs.

Flames propagating in narrow tubes can accelerate to the point where products are choked with respect to velocity of the flame tip, the Chapman–Jouget deflagration condition. Flames propagating at this condition are inherently unstable and eventually cause a transition to detonation. The details of this transition are dependent on whether the flame structure as a whole it choked or if the tip of the flame itself is choked. Pressure waves form if the overall flame structure is choked and interact with the unburned reactants in the heated boundary layer. One of these pressure waves causes a transition to detonation similar to the mechanism described above. Pressure builds locally at the flame tip if it approaches the choking condition, as shown

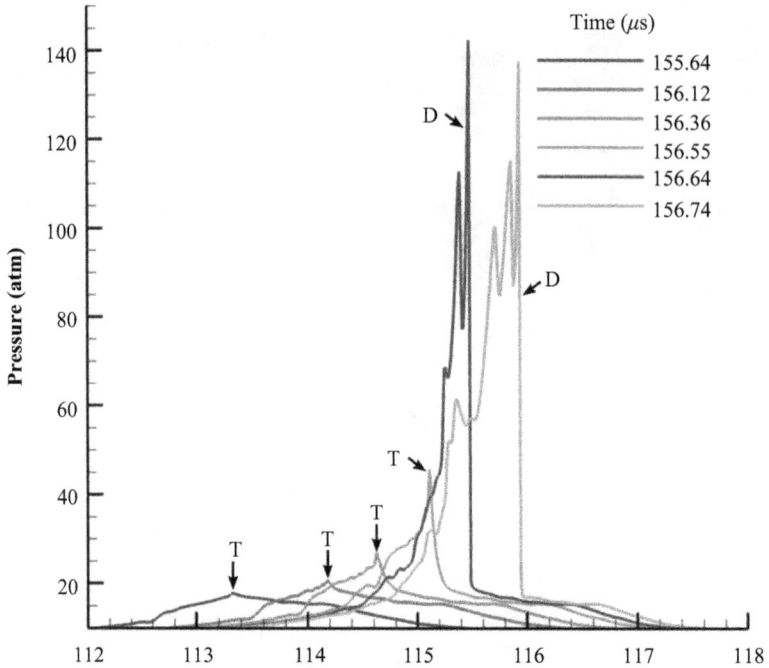

Figure 2.21. Pressure profiles at various times near the tip of a hydrogen–oxygen flame propagating in 0.5-mm-diameter circular tube during transition to detonation. The flame tip is marked *T* and the overdriven detonation is marked by the D.

in Figure 2.21 for a stoichiometric hydrogen and oxygen flame propagating in a 0.5-mm-diameter tube. This produces a runaway process where increasing pressure increases the local rate of energy release. Compression waves produced from this process coalesce to form a shock. The shock eventually becomes of sufficient strength to ignite a detonation near the flame tip. Similar mechanisms have been observed in the simulations by other investigators [22].

2.7 DISCUSSION

The concepts explored in this chapter have focused on determining how to get as much acceleration as possible from a flame in very narrow channel, with an application to micropropulsion in mind. The flow we deal with is laminar, and we are interested in how it behaves in the absence of any signification shock development or a transition to turbulence. If such a flame propagation could be achieved in a very small channel, then by controlling the length of the region filled with fuel and tuning the size of the channel to the properties of the laminar flame in the specific fuel mixture, flame acceleration and thrust could be adjusted. This micropropulsion concept is of particular interest for low-gravity environments, where relatively small, controlled thrusts could be used for navigational corrections. In such cases, there might not be a need for the high repetition rates required for pulse combustion and pulse detonation engines.

It is well known that the presence of turbulence leads to flame acceleration, and that more intense turbulence increases the turbulent flame velocity. In a narrow tube, however, we have seen that when the system becomes turbulent, the turbulece actually slows down the flame. Even though turbulent wrinkling increases the surface area and allows the overall flame structure to move more quickly, the laminar mechanism of flame acceleration in narrow channels is more efficient and results in higher flame and flow velocities. Thus we would want to avoid turbulence to obtain the largest, most controlled thrust. This might not be easy experimentally, and it is a topic that requires further research.

Much of the recent work on flame acceleration in small channels has focused on possible transitions to detonation and the mechanisms for these transitions. As we showed above, the laminar flame could, in principle, go directly to detonation, without the presence of turbulence, if the acceleration is high enough, and certainly if strong enough shocks form. We have shown one mechanism for this, where a spontaneous wave forms in the boundary layer. Other mechanisms have been suggested. Again, it is not clear that this is realistic without accounting for significant turbulence in the system, although it has been shown in experiments by Thomas (private communication) that expansion effects could delay its onset, perhaps long enough for the effects described below to hold quantitatively as well as qualitatively. Propane–air flames observed in experiments [12] in $109 \times 2.4 \times 2.4$ cm^3 channels look laminar for the first 30–40 cm, develop the shape similar to that shown in Figure 2.1 or 2.11, and accelerate to about 50 m/s for a stoichiometric mixture before the turbulence develops.

There is always an issue of heat losses to walls. It is known, and was shown above, that cold walls extract enough energy to cause the flame to oscillate and eventually die. Nevertheless, recent experiments [23, 24] have shown that detonations can be initiated from accelerating ethylene–oxygen flames in tubes with diameters as small as 0.5 mm and gaps as small as 0.26 mm. Details and timing of the detonation initiation and propagation mode were found to be heavily dependent on various factors including the flame initiation mechanism and the tube diameter. Detonations that form can either survive, fail, or propagate in a galloping mode depending on the tube diameter. Using detonations to develop micropropulsion devices is an area of active research [25].

In summary, for laminar flames in narrow channels:

- A laminar flame propagating toward the open end of a narrow channel filled with a gaseous combustible mixture can accelerate or oscillate, depending on the wall temperature and the channel width.
- The maximum flame acceleration and outflow velocity occur for adiabatic walls. As heat losses to the walls increase, the flame acceleration and outflow velocity are reduced.
- For isothermal walls, there is no significant flame acceleration and the flame may be quenched.
- The maximum flame acceleration in channels with adiabatic walls occurs when the channel was about five times larger than the reaction zone of a laminar flame.
- The accelerating flame generates weak compression waves that propagate with a local sound speed and can accelerate the unreacted material ahead of the flame to the velocities close to the sound speed without creating strong shocks.
- High-seed flows generated by accelerating flames have the potential to provide significant thrust.

- In principle, if the channel is long enough, the system can transition from a flame to a detonation. It appears that there are a number of mechanisms by which this can happen, some involving the boundary layer and others resulting from flame acceleration due to the presence of the boundary layer.

ACKNOWLEDGMENTS

The original ideas and work presented here are based on theoretical and computational work done in collaborations with James D. Ott, John D. Anderson, Jr., and Alexei M. Khokhlov. Subsequent work related to micropropulsion was supported by experimental work based on collaborations with Geraint O. Thomas (University of Wales, Aberystwyth) and Uday Hegde (NASA-Glenn) and theoretical work by Gregory I. Sivashinsky. This work was partially funded by NASA through the Combustion and Reacting Systems and Fluid Physics Program in the Office of Biological and Physical Research, partly by the Naval Research Laboratory through the Office of Naval Research, and currently by the University of Maryland through Minta Martin Endowment Funds in the Department of Aerospace Engineering, and the Glenn L. Martin Institute Chaired Professorship at the A. James Clark School of Engineering.

REFERENCES

[1] Ott, J.D., 1999, "*The Interaction of a Flame and Its Self-Induced Boundary Layer*," Ph.D. Dissertation, Department of Aerospace Sciences, University of Maryland.

[2] NASA report; Ott, J.D., Oran, E.S., and Anderson, J.D., Jr., 1999, "*The Interaction of a Flame and Its Self-Induced Boundary Layer*," NASA CR-1999-209401, Naval Research Lab., Washington, DC.

[3] Ott, J.D, Oran, E.S., and Anderson, J.D., Jr., 2003, "A mechanism for flame acceleration in narrow tubes," *AIAA Journal*, vol. 41, 1391–1396. doi: http://dx.doi.org/10.2514/2.2088.

[4] Gamezo, V.N., and Oran, E.S., 2006, "Flame acceleration in narrow tubes: Applications for micropropulsion in low-gravity environments," *AIAA Journal*, vol. 44, 329–336. doi: http://dx.doi.org/10.2514/1.16446.

[5] Gamezo, V.N., and Oran, E.S., 2006, "Flame Acceleration in Narrow Tubes: Effect of Wall Temperature on Propulsion Characteristics," *AIAA Paper AIAA-2006-1134*, American Institute of Aeronautics and Astronautics, Reston, VA.

[6] Mallard, E., and Le Chatelier, H.L., 1883, "Recherches expérimentales et théoriques sur la Combustion des mélanges gazeux explosifs," *Les Annales des Mines, 8th Series*, vol. 4, 274–376.

[7] Kerampran, S., Desbordes, D., and Veyssiére, B., 2000, "Study of the mechanisms of flame acceleration in a tube of constant cross-section," *Combustion Science and Technology*, vol. 158, 71–83. doi: http://dx.doi.org/10.1080/00102200008947328.

[8] Kerampran, S., Desbordes, D., and Veyssiére, B., and Bauwens, L., "Flame propagation in a tube from closed to open end," *AIAA Paper AIAA-2006-1082*, 2001, American Institute of Aeronautics and Astronautics, Reston, VA.

[9] Daou, J., and Matalon, M., 2001, "Flame propagation in Poiseuille flow under adiabatic conditions," *Combustion and Flame*, vol. 124, 337–349. doi: http://dx.doi.org/10.1016/S0010-2180(00)00209-1.

[10] Daou, J., and Matalon, M., 2002, "Influence of conductive heat-losses on the propagation of premixed flames in channels," *Combustion and Flame*, vol. 128, 321–339. doi: http://dx.doi.org/10.1016/S0010-2180(01)00362-5.

[11] Cui, C., Matalon, M., and Jackson, T.L., 2005, "Pulsating mode of flame propagation in two-dimensional channels," *AIAA Journal*, vol. 43, 1284–1292. doi: http://dx.doi.org/10.2514/1.9691.

[12] Schmidt, E.H.W., Steinicke, H., and Neubert, U., 1952, "Flame and schlieren photographs of combustion waves in tubes," *Proceedings of the Combustion Institute*, vol. 4, 658–667. doi: http://dx.doi.org/10.1016/S0082-0784(53)80086-X.

[13] Bychkov, V., Petchenko, A., Akkerman, V., and Eriksson, L.E., 2005, "Theory and modeling of accelerating flames in tubes," *Physical Review E*, vol. 72, 046307. doi: http://dx.doi.org/10.1103/PhysRevE.72.046307.

[14] Shelkin, K.I., 1940, "Influence of tube walls on detonation ignition," *Zhurnal Eksperimental'noi i Teoreticheskoi Fiziki (Journal of Experimental and Theoretical Physics)*, vol. 10, 823–837.

[15] Kagan, L., and Sivashinsky, G., 2003, "The transition from deflagration to detonation in thin channels," *Combustion and Flame*, vol. 134, 389–397. doi: http://dx.doi.org/10.1016/S0010-2180(03)00138-X.

[16] Khokhlov, A.M., 1998, "Fully threaded tree algorithms for adaptive refinement fluid dynamics simulations," *Journal of Computational Physics*, vol. 143, 519–543. doi: http://dx.doi.org/10.1006/jcph.1998.9998.

[17] Khokhlov, A.M., Oran, E.S., and Thomas, G.O., 1999, "Numerical simulation of deflagration-to-detonation transition: the role of shock-flame interactions in turbulent flames," *Combustion and Flame*, vol. 117, 323–339. doi: http://dx.doi.org/10.1016/S0010-2180(98)00076-5.

[18] Gamezo, V.N., Khokhlov, A.M., Oran, E.S., Chtchelkanova, A.Y., Rosenberg, R.O., 2003, "Thermonuclear Supernova: Simulations of the Deflagration State and Their Implications," *Science*, vol. 299, 77–81.

[19] Gamezo, V.N., Khokhlov, A.M., and Oran, E.S., 2001, "The influence of shock bifurcations on shock-flame interactions and DDT," *Combustion and Flame*, vol. 126, 181–1826. doi: http://dx.doi.org/10.1016/S0010-2180(01)00291-7.

[20] Gamezo, V.N., Ogawa, T., and Oran, E.S., 2007, "Numerical simulations of flame propagation and DDT in obstructed channels filled with hydrogen–air mixture," *Proceedings of the Combustion Institute*, vol. 31, 2463–2471. doi: http://dx.doi.org/10.1016/j.proci.2006.07.220.

[21] Kessler, D.A., Gamezo, V.N., and Oran, E.S., 2010, "Simulations of flame acceleration and deflagration to detonation transitions in methane-air systems," *Combustion and Flame*, vol. 157, 2063–2077. doi: http://dx.doi.org/10.1016/j.combustflame.2010.04.011.

[22] Ivanov, M.F., Kiverin, A.D., and Liberman, M.A., "Flame acceleration and DDT of hydrogen-oxygen gaseous mixtures in channels with no-slip walls," *International Journal of Hydrogen Energy*, vol. 36, 7714–7727.

[23] Wu, M., Burke, M.P., Son, S.F., and Yetter, R.A., 2007, "Flame acceleration and the transition to detonation of stoichiometric ethylene/oxygen in microscale tubes," *Proceedings of the Combustion Institute*, vol. 31, 2429–2436. doi: http://dx.doi.org/10.1016/j.proci.2006.08.098.

[24] Wu, M. and Kuo, W., 2013, "Accelerative expansion and DDT of stoichiometric ethylene/oxygen flame rings in micro-gaps," *Proceedings of the Combustion Institute*, vol. 34, 2017–2024. doi: http://dx.doi.org/10.1016/j.proci.2012.07.008.

[25] Wu, M. and Lu, T., 2012, "Development of a chemical microthruster based on pulsed detonation," *Journal of Micromechanics and Microengineering*, vol. 22, 105040. doi: http://dx.doi.org/10.1088/0960-1317/22/10/105040.

CHAPTER 3

FLAME INSTABILITY IN MICROSCALE COMBUSTION

Kaoru Maruta, Yiguang Ju, and Dimitrios Kyritsis

The large heat loss and strong flame–wall thermal interaction in microscale and mesoscale combustion result in rich phenomena of various flame instability such as the repetitive ignition and extinction [1–5], split flame [6, 7], spinning combustion [8], and spiral and other flame patterns. These phenomena are inherent for microscale combustion and have not been observed in the conventional combustion system at a large scale.

3.1 REPETITIVE EXTINCTION AND RE-IGNITION INSTABILITY

When the local wall temperature of a mesoscale channel is increased above the auto-ignition temperature either by flame–wall thermal coupling or by an external heat source, ignition will occur and the resulting flame may propagate upstream (Figure 3.1). If the wall temperature in upstream is reduced or the upstream channel width is below the quenching diameter of the normal propagating flame, extinction will occur at upstream. Since the downstream wall remains hot, when the mixture after flame extinction reaches a location where the wall temperature is higher than the ignition temperature, auto-ignition will occur again. This cycle repeats and forms a repetitive ignition and extinction instability.

The experiments to observe ignition and extinction instability were conducted by Maruta and co-workers [1, 2], Kyritsis and co-workers [3, 5], and Fan et al. [9, 10]. As shown in Figure 3.2, a cylindrical quartz glass tube with an inner diameter of 2 mm was used as a model channel [1, 2]. The results are replicated from Maruta et al. [1]. The inner diameter of the channel was slightly smaller than the standard quenching diameter of the employed fuel/air mixtures. Methane/air and propane/air mixtures were used for those studies. The channel was set either between two parallel plate-heaters or above flat flame burner. A temperature gradient was formed in the middle of the tube. The measured wall temperature is shown in the figure. After being ignited, the flame was stabilized at a certain location within the temperature gradient. The average mixture velocities were varied up to 100 cm/s.

Figure 3.3 shows direct photographs of the two different flames, a stable flame at a high flow rate (A) and a repetitive extinction and ignition flame at a low flow rate (B). The existence

Low wall temperature ◄── High wall temperature

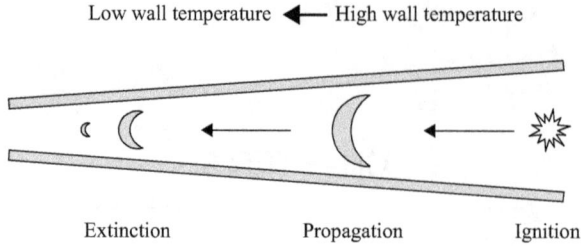

Extinction   Propagation   Ignition

Figure 3.1. Schematic of repetitive ignition and extinction of mesoscale combustion.

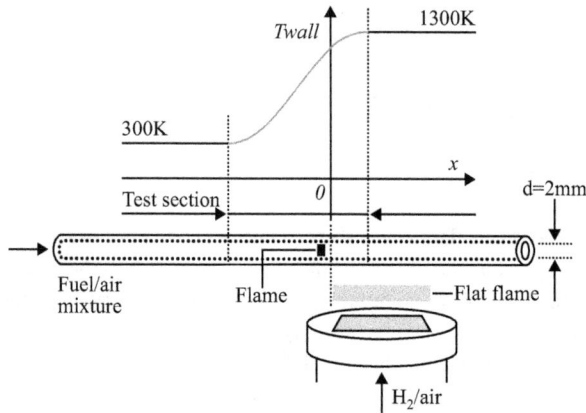

Figure 3.2. A cylindrical quartz tube with a prescribed wall temperature profile. A tube with an inner diameter of 2 mm, which is smaller than the ordinary quenching diameter, was chosen. Though hydrogen/air flat-flame burner was shown as a heat source here, twin flat panel heaters were also used in the earlier studies [1].

of a stable flame at a high flow rate is because the flame speed with a high wall temperature is balanced by the high flow rate. The existence of the repetitive extinction and ignition flame is due to the mechanism shown in Figure 3.1. Similar observations of the two different flame regimes were also made by Kyritsis and co-workers [3, 5] (Figure 3.4) in a curved channel. The results are replicated from Richecoeur and Kyritsis [3].

The histories of the locations of the stable and unstable flames are shown in Figure 3.5. The results are replicated from Maruta et al. [1]. It is seen that the flame location of the stable flame (Figure 3.5a) does not change. Figure 3.5b shows the temporal motion of the reaction zone obtained from the flame shown in Figure 3.3b. The oscillating flames were termed as *flames with repetitive extinction and ignition* (FREI) by their nature. The frequency of the oscillating flame is about 50 Hz. In addition to the two kinds of combustion modes, other flame modes that exhibit different dynamics were also observed (Figure 3.5c and 3.5d). The so-called 1D pulsating flame (Figure 3.5c) exhibited regular periodical motions with almost constant luminescence. Furthermore, flames that possess the characteristics of both pulsating flame and FREI were also observed (Figure 3.5d). These flames exhibit combined characteristics of the periodical motions of pulsating flame with small amplitude and repetitive ignition and extinction with

(a)

(b)

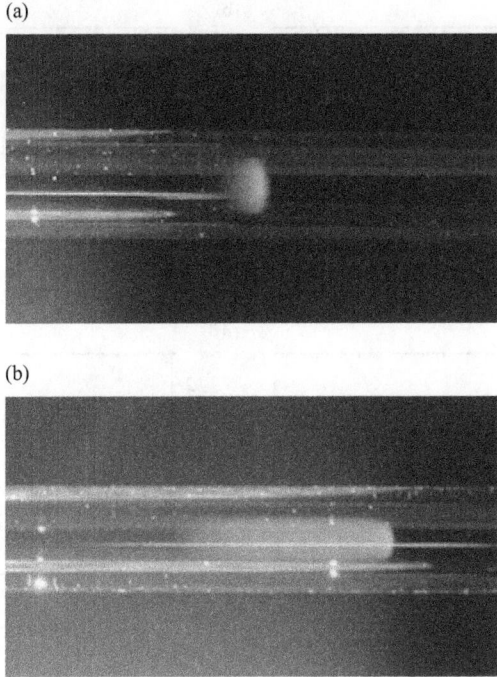

Figure 3.3. Direct images of (a) a stable flame and (b) a repetitive extinction and ignition flame. The mixture flow direction is from right to left [1].

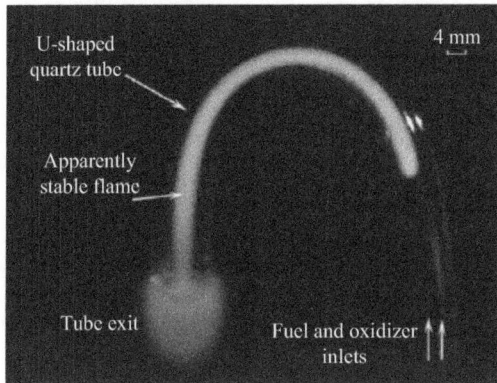

Figure 3.4. An oscillating flame in a curved tube [3].

large amplitude. The intrinsic mechanism for the occurrence of different flame modes is due to the existence of multiple flame regimes and extinction limits.

These non-stationary flame behaviors are described analytically by proposing one-dimensional (1D) nonlinear evolutionary equation of flame front, which considers flame front acceleration and rate of flame temperature variation [4]. Figure 3.6 indicates analytically derived amplitude of flame oscillations, that is, length between upstream and downstream turning points during FREI together with S-shaped steady-state solutions, which show flame location depending on the mixture flow velocity. The results are replicated from Minaev et al. [4].

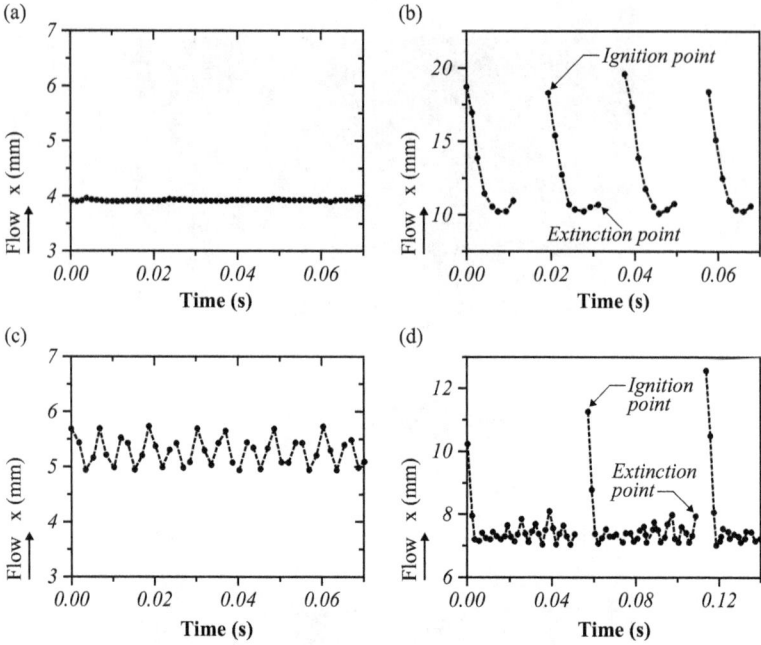

Figure 3.5. Temporal motions of various flames of propane/air mixtures at an average mixture velocity of 30 cm/s and an equivalence ratio of 0.5: (a) stable flame in the case of maximum wall temperature, T_c = 1320 K, (b) flame with repetitive extinction and ignition, T_c = 1130 K, (c) pulsating flame, T_c = 1270 K, and (d) flame with a combination of pulsating flame and FREI, T_c = 1200 K [1].

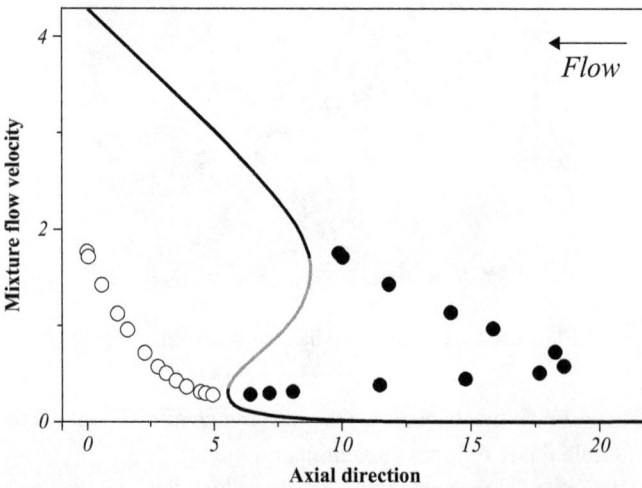

Figure 3.6. S-shaped steady-state solutions depending on mixture flow velocity and amplitude of flame oscillation indicated by the upstream and downstream turning points of flames during FREI. Closed and open circles represent upstream and downstream turning points, respectively [4].

The length of the flame brush during FREI shortens when mixture flow velocity decreased toward the transition point from the FREI regime (middle branch of steady-state solution) to the weak flame regime (lower branch). This implies that weak flames are stabilized on or very close to the ignition point of the given mixture.

Numerical simulations of oscillating flames in small-scale combustors were conducted by Norton et al. [11], Jackson et al. [12], and Kessler and Short [13]. Figure 3.7 shows the limit cycle of self-sustained oscillations in cases of a high heat loss of microscale combustor and the temperature evolution as a function of time. The results are replicated from Norton and Vlachos [11].

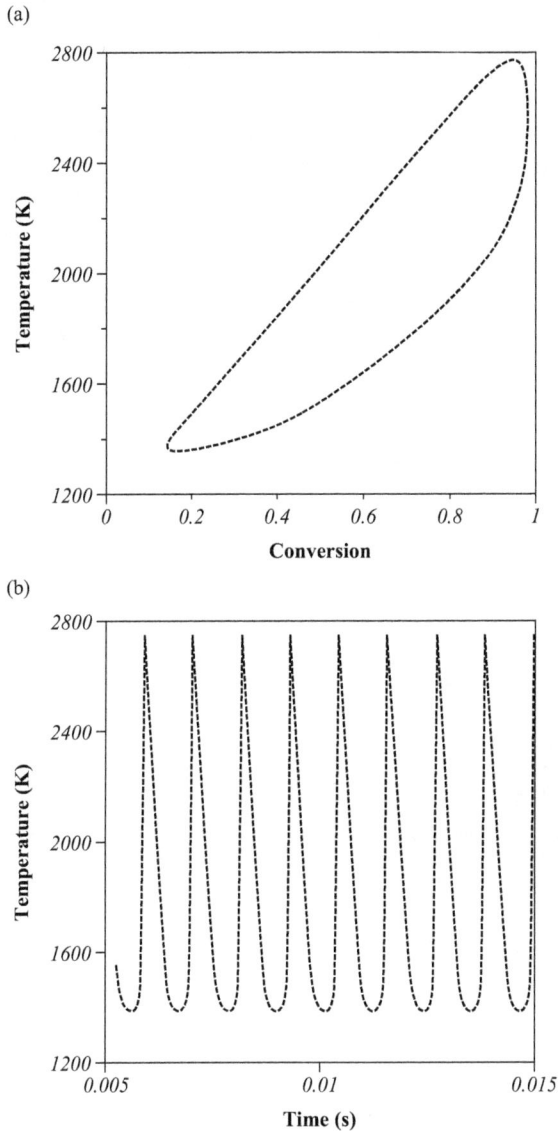

(a)

(b)

Figure 3.7. (a) Limit cycle for a point on the centerline of channel, indicating self-sustained oscillations at high heat loss near extinction. (b) The time history of temperature of this point. The oscillations occur with a period of 1 ms. The channel width and wall thickness are 600 and 200 μm [11].

The results showed that the stationary flames became unstable in cases of large heat loss. The limit cycle frequency is about 1000 Hz. Regular flame oscillations and repetitive quenching and re-ignition instabilities were observed.

Two-dimensional (2D) direct numerical simulation with detailed chemistry and transport for fuel lean ($\phi = 0.5$) hydrogen/air flames in planar microchannels with prescribed wall temperature [14, 15] captured the details of the repetitive ignition/extinction phenomena and was predicting rich phenomena of various flame instabilities (See Chapter 4 for the details). One of the notable flame features among these predicted phenomena is the splitting flame [15]. Indeed, the splitting flame was experimentally observed for methane/air mixture independently [6, 7] and further splitting phenomena due to different time scales in chemical reactions with and without significant heat generations were shown numerically as shown in Figure 3.8. The results are replicated from Nakamura et al. [16]. FREI described in the above starts at $t = 20.4$ ms and reaction front P1 propagated to upstream while the split flame front P2 moved to downstream. After finishing significant heat release, P1 still survived and further splitting phenomena occur twice.

Further detailed experiments using the rectangular quartz microchannel heated by the infrared lamp were conducted [9, 10]. Extinction and reignition phenomena were also observed and planar laser-induced fluorescence (PLIF) measurements for CH* and OH* were conducted. Movements of these excited species during extinction and reignition were captured. It is noted that OH* kernel during the FREI exhibited nearly one order higher propagation speed than the laminar flame speed of the same mixture [10], which typically represents ignition kernel speed in homogeneous mixture with a temperature gradient. The same order of the high-speed reaction front propagation was also shown numerically [16].

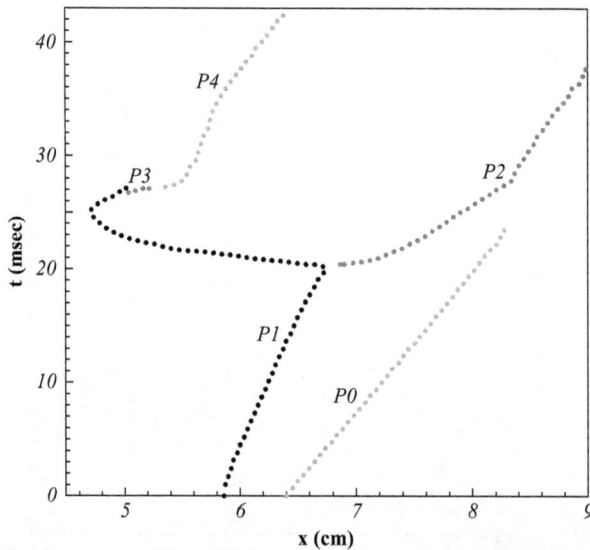

Figure 3.8. Reaction front bifurcation and their movements as a function of time. FREI starts at $t = 20.4$ ms and rapid reaction front propagations toward both upstream and downstream can be seen. Main reaction front P1 exhibited further bifurcation in its weak reaction regime [16].

3.2 SPINNING INSTABILITY

When there is a thermal coupling between the flame and the wall, the thermal transport in the flame structure augments the enthalpy transport in the flame via the indirect heat recirculation and leads to a multi-dimensional spinning instability in mesoscale combustion.

The spinning instability was observed by Xu and Ju [8] in a convergent and divergent channel (Figure 3.9, the results are replicated from Xu and Ju [8]). The channel was vertically mounted and had two equal constant-area sections A and D, a converging section B, and a diverging section C. The inner diameter of the constant-area sections was 10 mm and that of at the throat was 4 mm. The wall thickness was 1 mm. The lengths of the diverging and converging parts were 50 mm and that of the A section was 600 mm ($L/d > 60$) to ensure a fully developed flow entering section B. The convergent and divergent angles were all 6.9 degrees. A mesh grid was placed near the inlet of the tube to eliminate any flow disturbance. The divergent cross-section was introduced to ensure that the final stabilization flame was always in the weak flame regime with a strong flame–wall coupling. The flame propagation history was recorded by a high-speed camera at 500 frames per second.

In the experiments, four different flame modes, the propagating flame, a self-extinguished flame, the stabilized planar flame, and the spinning flame, were observed. For a given equivalence ratio, when the mixture was ignited at the tube exit, a propagating flame is formed in the D section and propagates into the C section. Figure 3.10 shows the flame trajectory histories of transitions from a propagating flame to either a stabilized planar flame at equivalence ratio $\varphi = 1.5$ and flow rate $Q = 3$ cm^3/s (left) and a spinning flame at $\varphi = 1.5$ and $Q = 5$ cm^3/s (right).

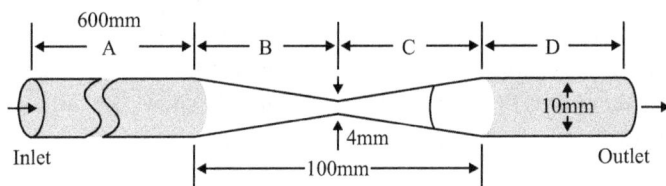

Figure 3.9. Schematic of convergent and divergent mesoscale channel [8].

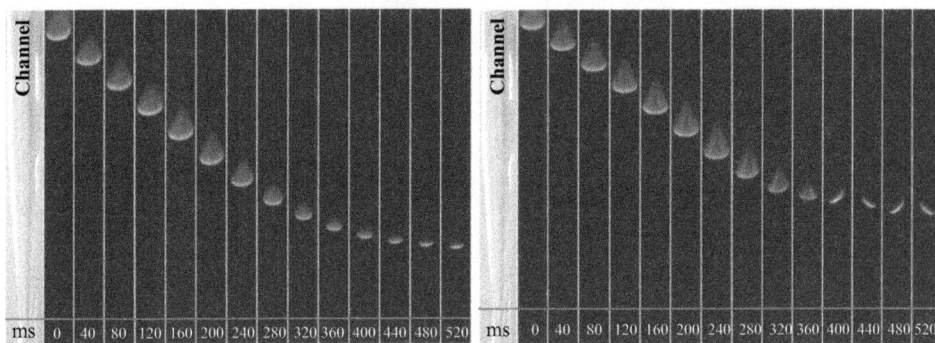

Figure 3.10. Flame evolution histories for C$_3$H$_8$–air flames. Left: $\varphi = 1.5$, $Q = 3$ cm^3/s; right: $\varphi = 1.5$, $Q = 5$ cm^3/s [8].

The results are replicated from Xu and Ju [8]. As shown in Figure 3.10, the flame propagation has three stages. First, as the flame propagates in the section of the constant cross-sectional area, the flame speed is almost constant. Second, as the flame enters the divergent section, the decrease of tube diameter resulted in an increase of the local flow velocity and thus a decrease of flame propagation speed relative to the wall. Third, as the flame further moves upstream, it is stabilized in the divergent section around $t = 500$ ms (Figure 3.10, left). However, with an increase of flow rate (Figure 3.10, right), the propagating flame slows down first in the divergent section and then slowly transforms to a spinning flame around $t = 400$ ms at the same location with a constant spinning frequency. This slow flame transition process indicates that the onset of flame spin requires a certain time to initiate flame–wall coupling. Interestingly, it was also found that the transitions between spinning flames and stabilized planar flames can be controlled by varying the flame–wall coupling externally via external cooling or heating of the wall. For example, if the wall is cooled, the spinning flame will stop and become a stable flame. This observation further confirmed that the flame instability is controlled by flame–wall thermal coupling. In addition, spinning flames were observed for both methane and propane flames at lean and rich conditions with both Le > 1 and Le < 1. This spinning irrelevance with the Lewis number contradicts with the classical theory of spinning combustion [17]. Moreover, the flame spin direction was randomly selected with equal probability in clockwise and counterclockwise directions. The random selection of spinning direction implies that the instability is not governed by flow motion but by the thermal diffusion process. Furthermore, the spinning flame was also confirmed in tubes with only one divergent section and observed when the tube was set horizontally or upside down. Therefore, it was concluded that this spinning flame is not produced by buoyancy and is governed by the flame–wall thermal coupling.

It was also found that when the flow rate was decreased, the propagating flame transformed to a stabilized planar flame at a smaller diameter where the flow velocity balanced with the flame speed. If the flow rate was further reduced, the propagating flame would either extinguish at a smaller tube diameter or pass through the throat. Figure 3.11 shows the different flame regimes for self-extinguished flames, stabilized planar flames, and spinning flames. It is seen that for

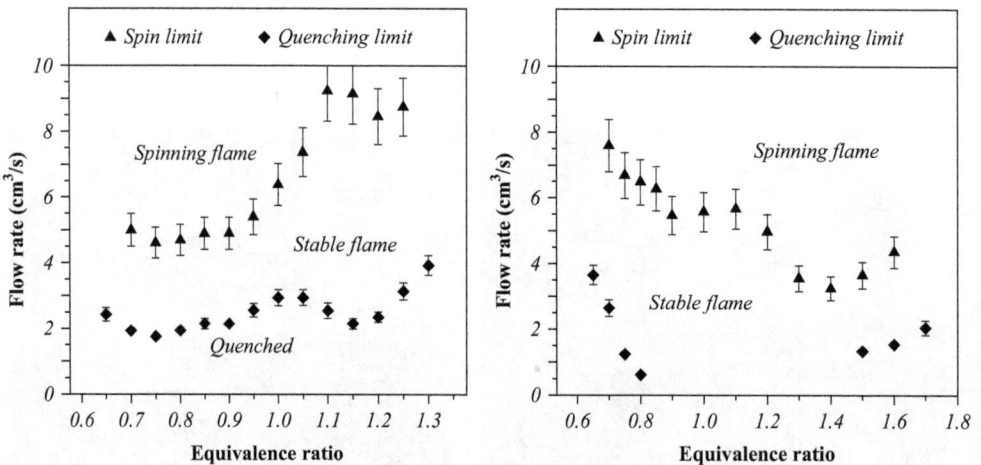

Figure 3.11. Flame regime diagram for different flow rates equivalence ratios (left: methane–air mixture; right: propane–air mixtures) [8].

both methane–air (left) and propane–air (right) flames, there exist critical flow rates, above which spinning flames exist and below which flames are stable. For methane flames, there is a quenching limit at low flow rates. For propane flames, the quenching limit exists for lean and rich mixtures at low flow rates. At near stoichiometric conditions and low flow rates, propane flames will pass the throat area without extinction. The results are replicated from Xu and Ju [8].

The spinning frequencies of methane–air flames at different equivalence ratios and flow rates are shown in Figure 3.12. The results are replicated from Xu and Ju [8]. Experiments were made by varying flow velocity at a given mixture concentration. It is seen that the spin frequency is roughly proportional to the average flame speed and only slightly affected by the flow rate.

In addition, the experiments showed that although the onset of flame spin was not governed by the mixture Lewis number, the mixture Lewis number had a strong effect on the shape of the spinning flame. Depending on Le, two different spinning flame shapes, an "L"-shaped flame (flame tail is open) and an "S"-shaped flame (flame tail closed), were observed for both methane and propane flames. In the conventional theory of thermal-diffusion instability, the onset of spinning combustion only happens for a mixture Lewis number larger than a critical value, $\beta(Le - 1) > 11$ [18]. However, in small-scale combustion, the enthalpy is also transported in the solid phase although the mass diffusion only happens in the gas phase. In order to understand the onset of flame instability, the estimation of an effective Lewis number for flame propagation in small-scale combustion with flame–wall thermal coupling is necessary. Since the laminar flame speed is a function of Lewis number for the same boundary condition, the effective Lewis number can be simply extracted by comparing the flame speed calculated with heat recirculation with that obtained without heat recirculation. The calculated effective Lewis number is shown in Figure 3.13. The results are replicated from Xu and Ju [8]. It is seen that the

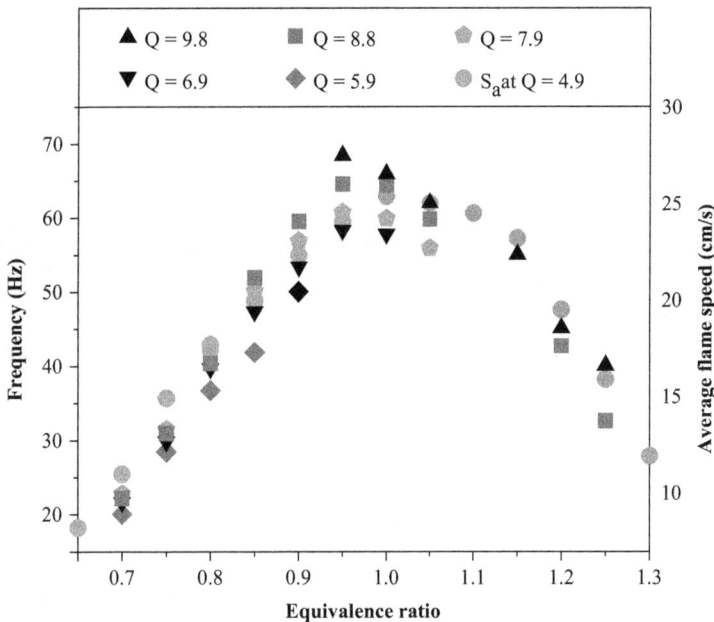

Figure 3.12. Frequencies of methane and propane spin flames at different equivalence ratios for methane–air flames [8].

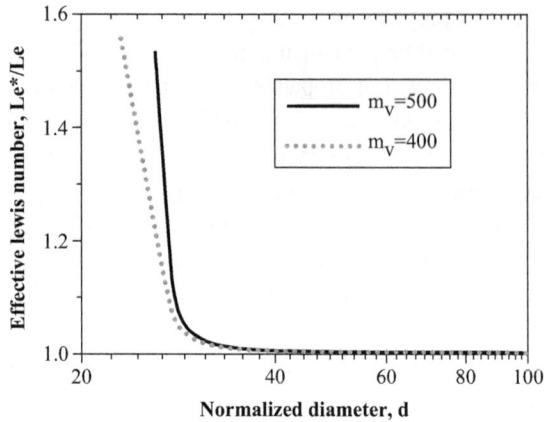

Figure 3.13. The effective Lewis number calculated from theoretical analysis by matching the flame speeds with and without heat recirculation for methane–air mixture [8].

effective Lewis number increases as the diameter decreases. In particular, when the flame starts to couple with the wall, the effective Lewis number increases dramatically. In addition, the increase of flow rate results in an increase of the effective Lewis number at the same tube diameter. For methane–air flames that have Lewis number around unity, the effective Lewis number with flame–wall coupling can be as high as 1.5. If the Zeldovich number is around $\beta = 10$, the reduced Lewis number $\beta(Le - 1)$ will be larger than the critical value for the onset of spinning flames. This explains why the flow rate is important to cause the onset of the spinning flame.

Spinning flames in a heated channel was identified also by three-dimensional (3D) DNS for hydrogen/air flames [19]. (See Chapter 4 for details.)

3.3 SPIRAL FLAMES AND PATTERN FORMATIONS

Spiral and other patterns, commonly observed in excitable media, have been reported in combustion systems [20–22]. In addition to them, pattern formations were observed in micro- and mesoscale flame systems. Repetitive extinction and ignition instability in a straight tube with an external heat source exhibit various instability and oscillatory phenomena such as regular cyclic oscillation, pulsation, and other chaotic oscillations that described in Section 3.1. While these phenomena occurred in a spatially 1D geometry, studies on the 2D radial-flow system have demonstrated other chaotic instabilities and pattern formations. Here, such phenomena observed in radial microchannel [23–25] and rectangular channel between two parallel plates [14, 15] both with external heat sources are described.

A straight tube with an external heat source, which is modeled as 1D plug flow geometry, was extended to a spatially 2D radial-flow system, that is, the radial microchannel between two circular plates with a temperature gradient as shown in Figure 3.14. An external heater for establishing a positive temperature gradient in the radial direction from room temperature to ~850 K was employed. Repetitive extinction and ignition instability in a system with additional degree of freedom in angular direction are likely to occur in the radial-flow system. In fact,

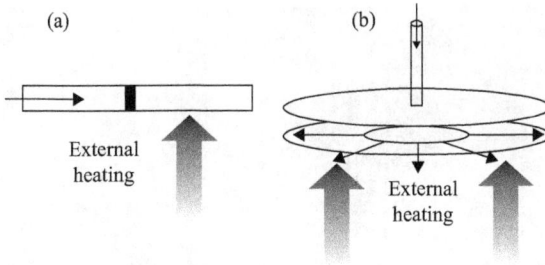

Figure 3.14. Combustion systems in a straight tube (a) and in a radial channel (b): the bold curve indicates the flame position and the arrows show the gas-flow direction.

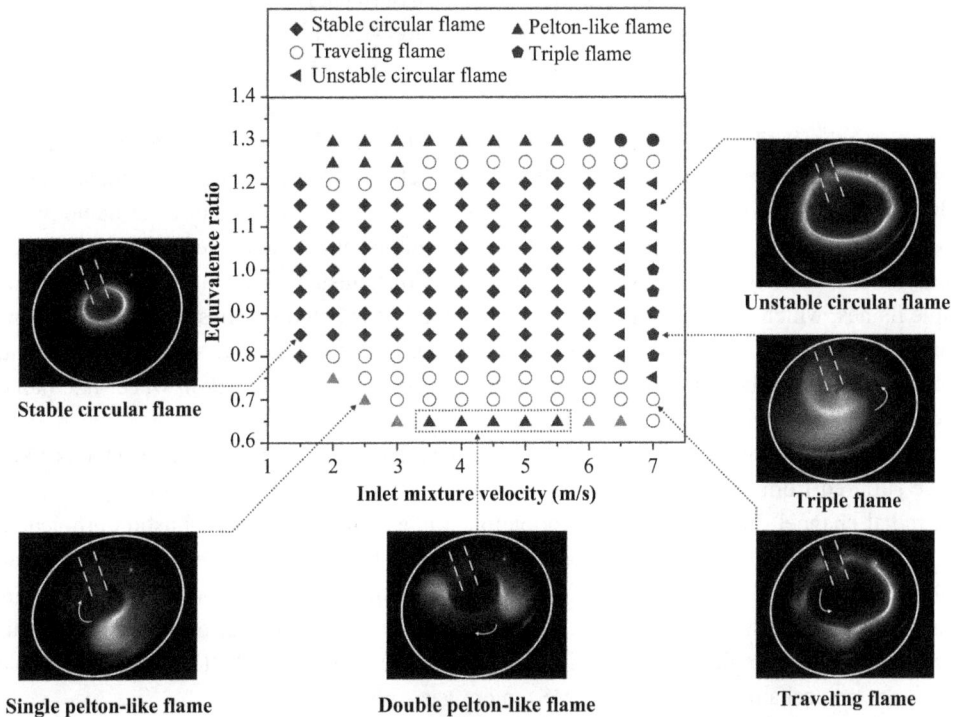

Figure 3.15. Flame patterns observed in the radial microchannel and their regime diagram for methane–air mixtures at the gap between two plates is 2 mm [23].

various flame patterns are observed and each pattern can be transformed one to another with the variations of parameters such as the mixture flow rate, equivalence ratio, and gap between two plates as shown in Figure 3.15. The results are replicated from Fan et al. [23]. One can expect the stable circular flame in the radial microchannel from the steady normal flame in a straight tube. As expected, the most probable flame pattern in the regime diagram is the stable circular flame when the gap between two plates is 2 mm as shown in the figure. However, single or double separated flames rotating around the center of the channel, which are termed

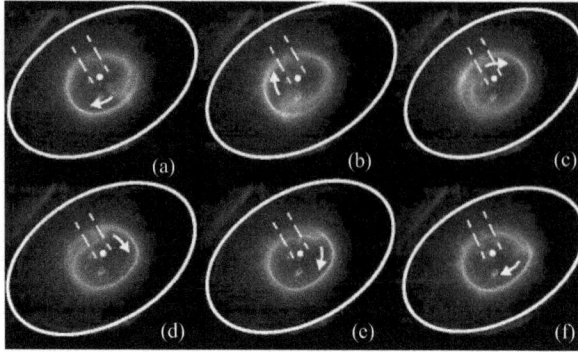

Figure 3.16. Evolution of spiral flames shown as sequential images (a)–(f) in a 2 ms time interval. The mean flow velocity in 4 mm i.d. delivery tube (white dashed line) was 5.5 m/s, $\phi = 1.25$, and plate distance was 2 mm [25].

as pelton (wheel)-like flames [26], are also observed at fuel lean and rich conditions. At the region between those for circular stable flames and pelton (wheel)-like flames, circular flames, which are termed traveling flame, are observed. The traveling flame is a circular flame, which is divided into some fractions where each fraction exhibits tangential movement [27]. At higher inlet mixture velocity and near stoichiometric conditions, unstable circular flames and rotating triple flames, which have three branches are observed. It is noted that observed flame patterns are sensitive, in particular, to the gap between two plates. For instance, no stable circular flames are observed when the gap is equal to 1.5 mm. Instead, traveling and pelton wheel-like flames occupy 80 and 20 percent of regime diagram, respectively.

This implies that a small-scale combustion chamber with radial flow geometry is likely to be suffered from these combustion instabilities. Accordingly, combustion completeness in the radial channel during the pattern formation was examined [24]. Combustion efficiencies estimated from the burned gas composition were 0.82–0.88 for double pelton flames whereas 0.48–0.82 for a single pelton flame. Significant amount of unburned fuel and CO were flowing out without completing combustion in low combustion efficiency conditions. One can imagine the difficulties in completing combustion at small-scale combustor with 2D geometries, such as microelectromechanical systems (MEMS) micro gas-turbines.

Besides the above flame patterns, spiral flames were observed at a high velocity region at the gap between two plates larger than 2 mm [23, 25]. Figure 3.16 shows spiral flame evolution recorded by a high-speed camera with 2 ms time interval at a mean flow velocity of 5.5 m/s in 4 mm i.d. delivery tube, $\phi = 1.25$, and plate distance of 2 mm. The results are replicated from Kumar et al. [25]. Target patterns were also observed in the vicinity of pattern transition from the stable flame to the bifurcated structure.

Mechanisms of these pattern formations have been investigated, and some aspects are clarified by multi-dimensional computations. Inner and outer edges of the pelton and traveling flames are correlated with the locations of ignition and extinction of FREI phenomena based on experimental results [28]. Single/double pelton flames were reproduced with 2D transient computations [28], and the spiral flame structure was clarified by the 3D numerical simulations [29].

REFERENCES

[1] K. Maruta, T. Kataoka, N. Il Kim, S. Minaev, and R. Fursenko, *Characteristics of combustion in a narrow channel with a temperature gradient.* Proceedings of the Combustion Institute, 2005. **30**(2): p. 2429–2436. doi: http://dx.doi.org/10.1016/j.proci.2004.08.245.

[2] K. Maruta, J.K. Parc, K.C. Oh, T. Fujimori, S.S. Minaev, and R.V. Fursenko, *Characteristics of microscale combustion in a narrow heated channel.* Combustion. Explosion and Shock Waves, 2004. **40**(5): p. 516–523. doi: http://dx.doi.org/10.1023/B:CESW.0000041403.16095.a8.

[3] F. Richecoeur and D.C. Kyritsis, *Experimental study of flame stabilization in low Reynolds and Dean number flows in curved mesoscale ducts.* Proceedings of the Combustion Institute, 2005. **30**(2): p. 2419–2427. doi: http://dx.doi.org/10.1016/j.proci.2004.08.015.

[4] S. Minaev, K. Maruta, and R. Fursenko, *Nonlinear dynamics of flame in a narrow channel with a temperature gradient.* Combustion Theory and Modelling, 2007. **11**(2): p. 187–203. doi: http://dx.doi.org/10.1080/13647830600649364.

[5] C.J. Evans and D.C. Kyritsis, *Operational regimes of rich methane and propane/oxygen flames in mesoscale non-adiabatic ducts.* Proceedings of the Combustion Institute, 2009. **32**(2): p. 3107–3114. doi: http://dx.doi.org/10.1016/j.proci.2008.06.089.

[6] S. Minaev, E.V. Sereshchenko, R.V. Fursenko, A. Fan, and K. Maruta, *Splitting flames in a narrow channel with a temperature gradient in the walls.* Combustion, Explosion, and Shock Waves, 2009. **45**(2): p. 119–125. doi: http://dx.doi.org/10.1007/s10573-009-0016-6.

[7] A. Fan, S.S. Minaev, E.V. Sereshchenko, Y. Tsuboi, H. Oshibe, H. Nakamura, and K. Maruta, *Dynamic behavior of splitting flames in a heated channel.* Combustion, Explosion, and Shock Waves, 2009. **45**(3): p. 245–250. doi: http://dx.doi.org/10.1007/s10573-009-0032-6.

[8] B. Xu and Y. Ju, *Experimental study of spinning combustion in a mesoscale divergent channel.* Proceedings of the Combustion Institute, 2007. **31**(2): p. 3285–3292. doi: http://dx.doi.org/10.1016/j.proci.2006.07.241.

[9] Y. Fan, Y. Suzuki, and N. Kasagi, *Experimental study of micro-scale premixed flame in quartz channels.* Proceedings of the Combustion Institute, 2009. **32**(2): p. 3083–3090. doi: http://dx.doi.org/10.1016/j.proci.2008.06.219.

[10] Y. Fan, Y. Suzuki, and N. Kasagi, *Quenching mechanism study of oscillating flame in micro channels using phase-locked OH-PLIF.* Proceedings of the Combustion Institute, 2011. **33**(2): p. 3267–3273. doi: http://dx.doi.org/10.1016/j.proci.2010.05.041.

[11] D.G. Norton and D.G. Vlachos, *Combustion characteristics and flame stability at the microscale: A CFD study of premixed methane/air mixtures.* Chemical Engineering Science, 2003. **58**(21): p. 4871–4882. doi: http://dx.doi.org/10.1016/j.ces.2002.12.005.

[12] T.L. Jackson, J. Buckmaster, Z. Lu, D.C. Kyritsis, and L. Massa, *Flames in narrow circular tubes.* Proceedings of the Combustion Institute, 2007. **31**(1): p. 955–962. doi: http://dx.doi.org/10.1016/j.proci.2006.07.032

[13] D.A. Kessler and M. Short, *Ignition and transient dynamics of sub-limit premixed flames in microchannels.* Combustion Theory and Modelling, 2008. **12**(5): p. 809–829. doi: http://dx.doi.org/10.1080/13647830801956295.

[14] G. Pizza, C.E. Frouzakis, J. Mantzaras, A.G. Tomboulides, and K. Boulouchos, *Dynamics of premixed hydrogen/air flames in microchannels.* Combustion and Flame, 2008. **152**(3): p. 433–450. doi: http://dx.doi.org/10.1016/j.combustflame.2007.07.013.

[15] G. Pizza, C.E. Frouzakis, J. Mantzaras, A.G. Tomboulides, and K. Boulouchos, *Dynamics of premixed hydrogen/air flames in mesoscale channels.* Combustion and Flame, 2008. **155**(1–2): p. 2–20. doi: http://dx.doi.org/10.1016/j.combustflame.2008.08.006.

[16] H. Nakamura, A. Fan, S. Minaev, E. Sereshchenko, R. Fursenko, Y. Tsuboi, and K. Maruta, *Bifurcations and negative propagation speeds of methane/air premixed flames with repetitive extinction*

and ignition in a heated microchannel. Combustion and Flame, 2012. **159**(4): p. 1631–1643. doi: http://dx.doi.org/10.1016/j.combustflame.2011.11.004.

[17] G. Joulin and P. Clavin, *Linear stability analysis of nonadiabatic flames: Diffusional-thermal model.* Combustion and Flame, 1979. **35**: p. 139–153. doi: http://dx.doi.org/10.1016/0010-2180(79)90018-x.

[18] F.A. Williams, *Combustion Theory.* Second edition 1985: Benjamin/Cummings.

[19] G. Pizza, C.E. Frouzakis, J. Mantzaras, A.G. Tomboulides, and K. Boulouchos, *Three-dimensional simulations of premixed hydrogen/air flames in micro tubes.* Journal of Fluid Mechanics, 2010. **658**: p. 463–491. doi: http://dx.doi.org/10.1017/S0022112010001837.

[20] V. Nayagam and F.A. Williams, *Rotating spiral edge flames in von Karman swirling flows.* Physical Review Letters, 2000. **84**(3): p. 479. doi: http://dx.doi.org/10.1103/PhysRevLett.84.479.

[21] H.G. Pearlman and P.D. Ronney, *Self-organized spiral and circular waves in premixed gas flames.* Journal of Chemical Physics, 1994. **101**(3): p. 2632. doi: http://dx.doi.org/10.1063/1.467636.

[22] K. Robbins, M. Gorman, J. Bowers, and R. Brockman, *Spiral dynamics of pulsating methane–oxygen flames on a circular burner.* Chaos, 2004. **14**: p. 467. doi: http://dx.doi.org/10.1063/1.1688532.

[23] A. Fan, S. Minaev, S. Kumar, W. Liu, and K. Maruta, *Regime diagrams and characteristics of flame patterns in radial microchannels with temperature gradients.* Combustion and Flame, 2008. **153**(3): p. 479–489. doi: http://dx.doi.org/10.1016/j.combustflame.2007.10.015.

[24] A. Fan, S. Minaev, S. Kumar, W. Liu, and K. Maruta, *Experimental study on flame pattern formation and combustion completeness in a radial microchannel.* Journal of Micromechanics and Microengineering, 2007. **17**(12): p. 2398–2406. doi: http://dx.doi.org/10.1088/0960-1317/17/12/002.

[25] S. Kumar, K. Maruta, S. Minaev, and R. Fursenko, *Appearance of target pattern and spiral flames in radial microchannels with CH4-air mixtures.* Physics of Fluids, 2008. **20**: p. 024101. doi: http://dx.doi.org/10.1063/1.2836670.

[26] S. Kumar, K. Maruta, and S. Minaev, *On the formation of multiple rotating pelton-like flame structures in radial microchannels with lean methane–air mixtures.* Proceedings of the Combustion Institute, 2007. **31**(2): p. 3261–3268. doi: http://dx.doi.org/10.1016/j.proci.2006.07.174.

[27] S. Kumar, K. Maruta, and S. Minaev, *Pattern formation of flames in radial microchannels with lean methane–air mixtures.* Physical Review E, 2007. **75**: p. 016208. doi: http://dx.doi.org/10.1103/PhysRevE.75.016208.

[28] A. Fan, S. Minaev, E. Sereshchenko, R. Fursenko, S. Kumar, W. Liu, and K. Maruta, *Experimental and numerical investigations of flame pattern formations in a radial microchannel.* Proceedings of the Combustion Institute, 2009. **32**(2): p. 3059–3066. doi: http://dx.doi.org/10.1016/j.proci.2008.06.092.

[29] S. Minaev, R. Fursenko, E. Sereshchenko, A. Fan, and S. Kumar, *Oscillating and rotating flame patterns in radial microchannels.* Proceedings of the Combustion Institute, 2013. **34**(2): p. 3427–3434 doi: http://dx.doi.org/10.1016/j.proci.2012.06.029.

CHAPTER 4

MICROSCALE COMBUSTION MODELING

John Mantzaras

Microreactor modeling encompasses intricate interactions between fluid flow, chemical kinetics, species transport, heat conduction in the solid, and external heat loss mechanisms. Numerical models are reviewed for stationary and transient microreactor simulation, focusing on both catalytic and non-catalytic systems. One-dimensional models and subsequently multidimensional simulations with detailed chemistry and transport are introduced and applied to investigate the combustion stability and start-up of microreactors, flame dynamics in catalytic and non-catalytic channels, and turbulent hetero-/homogeneous channel-flow combustion. Key comparisons of simulations with experiments are provided, demonstrating the capacity of the employed models. One-dimensional and two-dimensional simplified models based on perturbation theory (activation energy asymptotics) are also presented, providing closed-form solutions in micro-channel combustion (ignition criteria, flame propagation speeds). Limitations of the quasisteady-state assumption for the gas-phase in transient microreactor modeling and finite Knudsen number models based on the Lattice Boltzmann Method (LBM) are finally outlined.

4.1 INTRODUCTION

Microreactors used for portable power generation and fuel processing encompass a wide range of geometrical configurations and fuel conversion methodologies. The latter involve homogeneous (gas phase), heterogeneous (catalytic), and combined hetero-/homogeneous combustion processes. Microreactors for fuel processing exclusively employ catalytic approaches to achieve demanding product yields and selectivities. On the other side, all three aforementioned fuel conversion methods can be utilized for power generation applications, whereby total oxidation is required. The fundamental physicochemical processes in each microreactor type are outlined next so as to pave the way for the presentation of appropriate mathematical models.

Hetero-/homogeneous microreactors entail the largest complexity and are discussed first. Although various reactor configurations have been established, heterogeneous conversion in many small-scale power and chemical engineering processes is achieved in honeycomb structures comprising straight channels (Figure 4.1a), with each channel having a hydraulic diameter ranging from 0.5 to 2.0 mm. Such reactors are also used for large-scale power generation [1, 2], the main difference being the higher mass throughput per channel and the larger number of channels per honeycomb structure as compared to microreactor applications [3]. Depending on the power requirements, single-channel geometries with large cross-flow aspect ratios have also been constructed (Figure 4.1b). Moreover, excess enthalpy (recirculation type) catalytic microreactors with complex three-dimensional geometries have been tested (configurations similar to the non-catalytic Swiss-roll microreactor shown in Figure 4.2a). The underlying physicochemical processes are shown in Figure 4.1c within the context of straight catalytic channel geometries.

Fuel and oxidizer are mixed and subsequently enter the reactor at inlet velocities that yield, for most power-generation and chemical processing applications, laminar channel flows [6, 7]. Both reagents diffuse transversely to the channel walls and react on the catalytic surfaces. Heterogeneous ignition (light-off) is achieved at a certain distance from the channel entry, which depends not only on the operating conditions (inlet temperature, velocity and pressure, fuel type, stoichiometry) and the particular catalyst formulation, but also on key in-channel heat

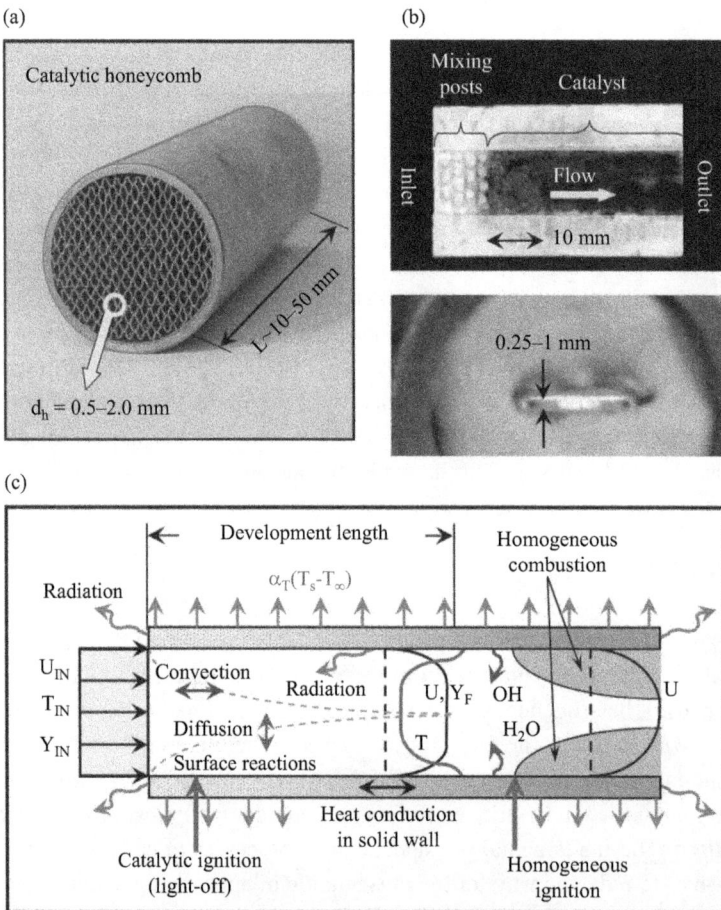

Figure 4.1. (a) Typical honeycomb catalytic structure made of FeCr alloy and coated with Pt, similar to the one used in Karagiannidis et al. [3] for mesoscale power generation, (b) rectangular-shaped, single-channel, Pt-coated catalytic microreactor (reproduced by permission from Norton et al. [4]), and (c) physicochemical processes within a catalytic channel.

(a)

(b)

$D = 2$ mm

$D_0 = 2$ mm

Burned

D_c

$D = 2$ mm

Flame

Unburned

$D_i = 1$ mm

(c)

Heat gain/loss $\alpha_T(T_W - T_\infty)$

U_{IN}
T_{IN}
Y_{IN}

Radiation
Heat conduction
in solid

U
T

Flame

Radiation

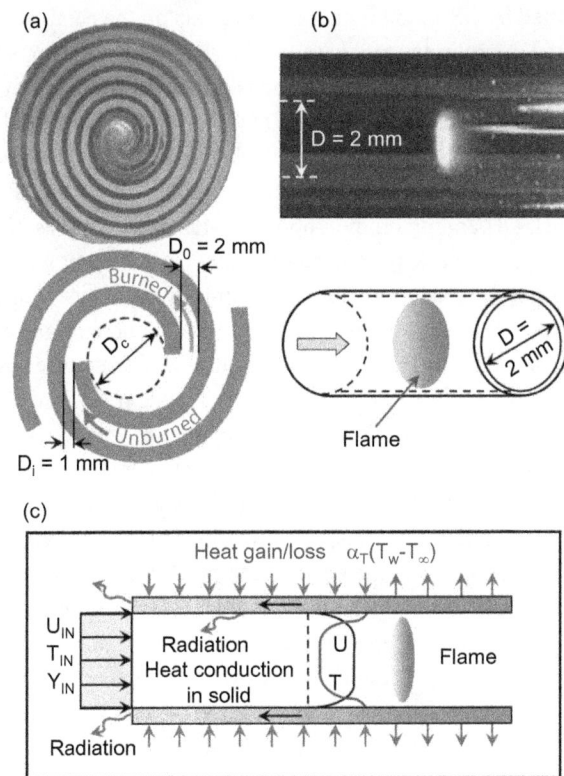

Figure 4.2. (a) Swiss-roll recirculation-type homogeneous microreactor [5], (b) single-channel tubular homogeneous microreactor [26], and (c) physicochemical processes within a homogeneous combustion microreactor channel.

transfer mechanisms, such as upstream heat transfer via conduction inside the solid channel walls and surface radiation heat transfer [8–12]. The surface temperature at the light-off position is high enough to ensure a shift of the catalytic conversion from the kinetically controlled to the transport-controlled regime [8, 13, 14]. This is manifested by the practically zero concentration of the limiting reactant on the catalyst surface, as shown from the profile of the fuel mass fraction (Y_F) in Figure 4.1c.

Downstream of the light-off location, fuel and oxidizer react vigorously at the catalyst surface creating a *degenerate* diffusion reaction sheet [15]. This terminology reflects the facts that the reaction zone is fixed in space (channel walls) and that both reagents diffuse from the same side of the reaction zone in contrast to gaseous diffusion flames. In catalytic combustion, heat and combustion products diffuse back to the flow and, if conditions are appropriate, a homogeneous (gas-phase) combustion zone may also be established (Figure 4.1c). Owing to the lower effective activation energies of catalytic reactions compared to those of homogeneous reactions [8, 16], gas-phase combustion (when present) is typically located farther downstream of the light-off position. Four major coupling routes between the two reaction pathways affect the likelihood of homogeneous ignition. The near-wall depletion of the limiting reactant induced by the catalyst (see Figure 4.1c) inhibits homogeneous ignition, whereas the heat transfer from the hot catalytic walls to the reacting gas promotes ignition [14]. Gas-phase ignition is, in most cases, modestly inhibited by the recombination of radicals on the catalyst [17–19] and, finally, it can either be inhibited or promoted by heterogeneously produced major species (such as H_2O of CO_2), depending on the fuel type and operating conditions [17, 18, 20].

Homogeneous ignition within the catalytic module may not always be avoided, even at the large microreactor geometrical confinements (sub-millimeter channel hydraulic diameters), especially at modest pressures relevant to microturbine-based mesoscale power-generation concepts [3, 10, 21, 22]. Depending on the imposed external heat losses, the onset of homogeneous combustion may be of concern for reactor thermal management when using hydrocarbon fuels. However, this is not the case for diffusionally imbalanced fuels with Lewis numbers less than unity, such as hydrogen or hydrogen-rich syngas fuels; therein, gas-phase combustion is actually advantageous since it moderates the reactor surface temperatures as will also be shown in Section 4.2 [8, 17, 20, 23]. Whenever thermal management (in terms of both reactor and catalyst thermal stability) and product selectivity (for fuel-processing microreactors) are not overriding issues, gaseous combustion in catalytic microreactors has the added advantage of completing the fuel conversion at shorter reactor lengths since the homogeneous reaction pathway does not exhibit the strong mass transport limitations of the catalytic pathway. Moreover, as will be elaborated in Section 4.8.4, the catalyst can be very effective in suppressing intrinsic gas-phase combustion instabilities, which are driven by flame–wall interactions [24, 25].

Homogeneous combustion microreactors utilize heat recirculation (e.g., excess flue gas enthalpy concepts) as the Swiss roll reactor in Figure 4.2a. Such concepts are used more often in homogeneous microreactors rather than in their catalytic counterparts due to the considerably lower gaseous reactivity of hydrocarbons and hydrogen when compared to their corresponding heterogeneous reactivity. Straight channel reactors, however, can also be used (Figure 4.2b). The underlying processes in homogeneous microreactors are depicted in Figure 4.2c, again within the context of straight channel geometry. Energy recirculation of the flue gases can be accounted for via imposed external heat loss or gain at different reactor parts (see Figure 4.2c), while other heat transfer mechanisms in the channel (heat conduction, surface radiation heat transfer) are similar to the corresponding ones in the catalytic channel (Figure 4.1c). Although the channel in Figure 4.2c is non-catalytic, radical recombination reactions on the surfaces should be considered, depending on the particular fuel, wall material properties, and operating temperature. It is finally emphasized that flames established inside a homogeneous microchannel reactor are only stable for limited operating parameters as will be discussed in Section 4.8.4.

4.2 MICROREACTOR THERMAL MANAGEMENT

In both catalytic and non-catalytic microreactors, the attained surface temperature is an important design parameter that numerical models should be capable of accurately reproducing. The wall temperature directly impacts the reactor integrity and, moreover, for heterogeneous systems it affects the catalyst stability and the product selectivity when considering microreformers. Starting from catalytic microreactors, even in the absence of external heat losses, the coupling of transport and chemistry can give rise to surface temperatures noticeably different from the corresponding adiabatic equilibrium temperatures, thus exemplifying the importance of numerical models adequately describing this coupling. There are three mechanisms responsible for such temperature deviations: diffusional imbalance of the limiting reactant (Lewis number, $Le \neq 1$), upstream recirculation of heat via heat conduction in the solid (see Figure 4.1c), and non-equilibration of the combustion products due to short reactor residence times [7]. The former two are controlling in fuel-lean combustion, whereas the latter two are important in fuel-rich (reforming) processes.

The constraint of the catalytic reaction zone on the channel wall, when coupled to non-equal heat and mass diffusivities, gives rise to super- or under-adiabatic surface temperatures. The surface temperature, T_w, in channel-flow catalytic combustion under infinitely fast (mass transport limited) conversion of the deficient fuel reactant is [17]

$$T_w = T_g + Le_f^{\beta-1} \Delta T, \tag{4.1}$$

where $\Delta T = T_{ad} - T_g = Y_{f,g} q/c_p$, T_{ad} is the adiabatic equilibrium temperature, T_g is the bulk gas temperature, $Y_{f,g}$ is the fuel mass fraction in the bulk of the gas, q is the heat release per unit mass of fuel, c_p is the mixture heat capacity, Le_f is the Lewis number of the fuel, and $\beta = 0$ or $1/3$ for fully developed and developing channel-flows, respectively. Fully developed flows are rarely realized in microreactors due to the short channel aspect ratio and the combustion-induced flow acceleration. Fuels with $Le < 1$ ($Le > 1$) lead to super-adiabatic (under-adiabatic) surface temperatures under conditions of infinitely fast catalytic chemistry.

Predicted wall temperatures are presented in Figure 4.3a for the combustion of a preheated ($T_{IN} = 600$ K), fuel-lean ($\varphi = 0.24$) H_2/air mixture at atmospheric pressure in a tubular, Pt-coated catalytic microreactor, with internal diameter $D = 1.2$ mm, length $L = 50$ mm, solid wall thickness of 100 μm, surface emissivity $\varepsilon = 0.6$ and solid thermal conductivity $k_s = 16$ W/(mK); the latter value pertains to an FeCr-alloy metal typically used for constructing honeycomb structures similar to that in Figure 4.1a. A 2D model for the gas and the solid accounting for

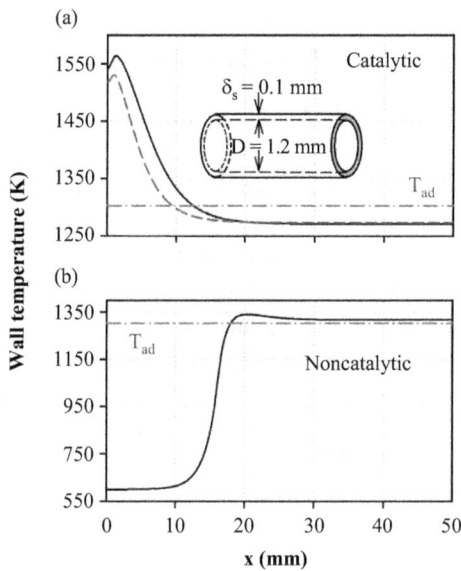

Figure 4.3. (a) Predicted surface temperature profile in a catalytic tubular microreactor with a length of 50 mm and a diameter of 1.2 mm. Combustion of H_2/air ($\varphi = 0.24$, $T_{IN} = 600$ K, p = 1 bar, $U_{IN} = 10$ m/s). Solid lines: simulations with only catalytic reactions; dashed lines: simulations with combined heterogeneous and homogeneous reactions. (b) Predicted surface temperatures in a non-catalytic tubular microreactor having the same geometry and operating conditions as in (a).

all processes in Figure 4.1c is employed (model details will be presented in Section 4.8.1.1), with detailed heterogeneous [27] and homogeneous [28] chemical reaction schemes. The maximum surface temperature predicted using only catalytic reactions is 1562 K (solid line in Figure 4.3a), exceeding by 260 K the adiabatic equilibrium temperature; this is due to the diffusional imbalance of the limiting reactant ($Le_{H_2} \sim 0.3$) according to Equation (4.1). The addition of gaseous reactions reduces the catalytically induced super-adiabaticity to 225 K (dashed lines in Figure 4.3a), the reason being that homogeneous combustion shields the catalyst from the hydrogen-rich channel core [17, 29], thus reducing catalytic conversion. Wall temperatures as high as the ones depicted in Figure 4.3a may cause reactor meltdown and/or catalyst deactivation. Therefore, even though the presence of a catalyst can significantly enhance the stability limits of a microreactor [10, 22], careful design with appropriate external cooling is needed when dealing with hydrogen-containing fuels in order to avoid the creation of hot spots.

Finally, corresponding simulations in a non-catalytic channel with the same geometry and externally adiabatic surfaces ($\alpha_T = 0$, see schematic in Figure 4.1c) are provided in Figure 4.3b. In this case, the smaller degree of super-adiabaticity (38 K above T_{ad}) is due to heat recirculation inside the 100 μm thick solid wall and also due to upstream surface radiation heat transfer from the hot channel rear to the colder front.

4.3   HETEROGENEOUS AND HOMOGENEOUS CHEMISTRY MODELING

Indispensable inputs in any microreactor model are validated heterogeneous and homogeneous chemical reaction schemes. This section summarizes the current understanding of kinetics at microreactor-relevant operating conditions.

4.3.1   HETEROGENEOUS KINETICS

The reaction rate is specific to the catalyst formulation and every catalyst exhibits a unique rate expression depending not only on the catalyst active material but also on the support material, the type and structure of the washcoat, and finally the preparation method. Generally, different catalyst surface structures vary in their reaction pathways and kinetic expressions. Moreover, the temporal history of the reaction, recrystallization due to changes in species concentration and temperature, and diffusion of adsorbed species into the catalyst bulk may also modify the local reaction rates.

Global catalytic chemical steps have been the preferred modeling choice for many years [13, 16, 30, 31], with the reaction rate expressed on catalyst mass, catalyst volume, reactor volume, or catalyst external surface area. A frequently used approach for establishing rate expressions is the Langmuir–Hinshelwood–Hougen–Watson (LHHW) methodology [32], which is based on Langmuir adsorption, surface reaction between adsorbed intermediates and finally desorption, under the key assumption that one of the involved steps is rate limiting. Such an approach cannot account for the complex variety of the aforementioned surface phenomena and the rate parameters must be evaluated experimentally for each new catalyst and different operating conditions.

The current modeling direction is towards development of detailed reaction mechanisms from elementary steps on the catalyst surface. The mean-field approach is the most popular methodology, treating heterogeneous reactions in a way similar to that of gas-phase reactions. This entails the assumption that all adsorbates are randomly distributed on the surface, which is in turn considered uniform. The state of the catalytic surface is described by the temperature and a set of surface species coverages, θ, both depending on the macroscopic position in the reactor, which is an average over microscopic local fluctuations. Reaction rates depend not only on the gas phase but also on surface and bulk species. Surface species are those adsorbed on the top atomic layer of the solid, whereas bulk species are those diffusing in the inner solid material. Steric effects of adsorbed species and various resulting configurations, that is the type of the chemical bonds between adsorbates and solid, can be modeled as follows. Each surface structure (there can be several surface structures representing different materials, crystal structures, or reconstructions of the same crystal structure) is associated with a surface site density Γ (units of mol/cm^2), defining the maximum number of species that can adsorb on a unit area of catalyst material. The surface site density can be computed from the molecular structure of the catalyst material and should not be confused to the active catalytic surface area describing the particular catalyst loading. Surface site densities are of the order of 10^{-9} mol/cm^2, corresponding roughly to 10^{15} adsorption sites per cm^2. A coordination number σ_m is associated with each surface species m, accounting for the number of surface sites it occupies. Under the aforementioned assumptions, multi-step (quasi-elementary) reaction mechanisms can be set up [33]. The local net production rate $\dot{s}_k$ ($= $ mol/cm^{2}s) of the k-th gaseous species due to adsorption or desorption on the catalyst surface is then described in a manner analogous to gas-phase chemical reactions:

$$\dot{s}_k = \sum_{i=1}^{N_R} \nu_{ki} \left\{ k_{f_i} \prod_{j=1}^{K} \left[X_j \right]^{v'_{ji}} - k_{r_i} \prod_{j=1}^{K} \left[X_j \right]^{v''_{ji}} \right\}, \tag{4.2}$$

with N_R the total number of surface reactions including adsorption and desorption, K the number of all species (gas, surface, and bulk), $v_{ji} \left(= v''_{ji} - v'_{ji} \right)$ denotes the stoichiometric coefficients, k_{f_i} and k_{r_i} the forward and reverse rate coefficients for the i-th reaction, and $[X_j]$ the concentration of species j (for surface species in mol/cm^2 and for gaseous species in mol/cm^3). Due to local variations in temperature and gaseous species concentrations, the surface coverage varies spatially. Nonetheless, lateral interactions between surface species at different locations on the catalytic surface are ignored in the mean field approach. This assumption is justified, since the computational cells used in a reactor simulation are usually much larger than the range of lateral surface interactions. Thus, for any reactor configuration, the variations of surface coverage do not include spatial gradients and can therefore be modeled as

$$\frac{d\theta_m}{dt} = \sigma_m \frac{\dot{s}_m}{\Gamma} - \frac{\theta_m}{\Gamma} \frac{d\Gamma}{dt}, \quad m = 1, \ldots, M_s, \tag{4.3}$$

with M_s the number of surface species. The second term on the right side of Equation (4.3) signifies a non-conserved surface site density, although in most catalytic applications Γ is usually conserved. Because the adsorption binding energies on the surface vary with surface coverage, the expression for the rate coefficients in Equation (4.3) can be considerably more complex compared to gaseous rate coefficients [33]:

$$k_{f_k} = A_k T^{\beta_k} \exp\left[\frac{-E_{a_k}}{R^\circ T} \right] \prod_{i=1}^{M_s} \theta_i^{\mu_{ik}} \exp\left[\frac{\varepsilon_{i_k} \theta_i}{R^\circ T} \right]. \tag{4.4}$$

The parameters μ_{i_k} and ε_{i_k} describe the dependence of the rate coefficients on the surface coverage of the i-th species. For adsorption reactions, sticking coefficients are commonly used and are converted to conventional rate coefficients as

$$k_{f_i}^{ads} = \left(\frac{\gamma_i}{1 - \gamma_i / 2} \right) \frac{1}{\Gamma^m} \sqrt{\frac{R^\circ T}{2\pi W_i}} \tag{4.5}$$

with γ_i the sticking coefficient of the i-th gaseous species and m the number of sites occupied by the adsorbing species. The denoninator $(1-\gamma_i/2)$ in Equation (4.5) is the Motz correction factor [34] accounting for non-Maxwellian velocity distribution due to species adsorption. Improvement of this factor to $(1 - \theta_{free}\gamma_i/2)$, with θ_{free} the free sites, has been proposed in Dogwiler et al. [9].

The assumption of a uniform surface in the mean-field approach is in most cases questionable. Site heterogeneities are present because a technical catalyst is characterized by terraces of different crystal structures, steps, edges, impurities, and defects. In the mean field approach, the site heterogeneity is averaged out by mean rate coefficients. If the distribution of the different types of adsorption sites and the reaction kinetics on those sites are known, the mean-field concept can be used to construct a reaction mechanism, which consists of several sub-mechanisms for the different surface structures. This approach has been applied in the framework of a "two-adsorption site model" for the simulation of CO oxidation on polycrystalline Pt [35]. The site heterogeneity can be described by the probability that an arbitrary site is characterized by the associated reaction kinetics. This probability function is usually a sum over a finite number of surface structures.

Effects resulting from lateral adsorbate interactions are inherently more difficult to treat. In the mean-field approximation they are either neglected or incorporated by mean rate coefficients. However, a variety of adsorbate–adsorbate interactions have been experimentally observed, including two-dimensional phase transitions depending on temperature and coverage, island formation of adsorbed species, dependence of sticking probability and surface reaction rate on the local environment (number and nature of species occupying the adjacent sites), and so forth. When the specific surface interaction mechanisms are known, Monte Carlo (MC) simulation of the surface chemistry can be carried out at the nanometer scale [36]. The simulation starts with a particular configuration of adsorbed species on a preferably large array of surface sites, on which various reactive processes occur, such as adsorption, diffusion, surface reactions, desorption, phase transitions, and so forth. The rates of these processes are expressed in terms of the local environment. After a sufficient number of recorded events, an overall reaction rate can be calculated. Although MC calculations are quite promising for understanding complex surface phenomena, their application to catalytic combustion of hydrocarbons is not yet possible, due to both computer power limitations and incomplete knowledge of the specific surface interactions. Owing to their inherent transient nature, MC simulations have been used to investigate dynamic oscillatory phenomena driven by surface kinetic processes, including surface–restructuring, substrate–substrate, and substrate–adsorbate interactions [37–39]. Such dynamics may also be relevant in catalytic microreactors and will be discussed in Section 4.8.4.3.

Modern experimental methodologies used to develop mean-field-based kinetic schemes are outlined next. Elementary heterogeneous chemical reaction schemes have relied primarily on ultra high vacuum (UHV) surface science experiments. Notwithstanding recent advances in *in situ* surface science diagnostics [40–42], surface science measurements at practical operating conditions are still not possible. Therefore, extension of the reaction schemes to realistic pressures and technical catalysts has necessitated appropriate kinetic rate modifications in order to bridge the well-known *pressure and materials gap*. These modifications have been aided by kinetic measurements in a variety of laboratory-scale reactors [43]. Two basic reactor configurations have been singled out in the last years for kinetic measurements: the nearly isothermal, low-temperature ($T \leq 500°C$), gradientless tubular or annular flow reactor [44, 45] (fed with highly diluted fuel/oxidizer mixtures so as to maintain a small temperature rise), and the stagnation flow reactor [46–49]. Although modeling of the former reactor is straightforward, its low-to-moderate temperature operation poses inherent limitations. Stagnation flow configurations, on the other hand, have provided a wealth of data on catalytic ignition/extinction, steady fuel conversion and product selectivity under realistic temperatures and mixture compositions. The measurements, in conjunction with numerical predictions from well-established one-dimensional stagnation-flow reacting codes [50], have aided considerably the refinement of surface reaction mechanisms. Nonetheless, despite their amenable geometry, most stagnation-flow experiments provided measurements of global rather than local quantities.

More recently, a new methodology involving *in situ* spatially resolved Raman measurements of major gas-phase species concentrations across the boundary layer of a channel-flow catalytic reactor has been introduced [17], providing a direct way to assess the catalytic reactivity as well as the gaseous reactivity when combined with laser-induced fluorescence (LIF) of radical species. This methodology has been used to study lean catalytic combustion of methane, propane, and hydrogen at pressures up to 16 bar (encompassing microreactor and gas-turbine operating pressures) and the rich catalytic combustion of methane at pressures up to 10 bar over noble metals [17, 18, 51–57]. A similar approach, based on gas-phase Raman spectroscopy, was used to investigate the catalytic partial oxidation of methane over Pt in a stagnation flow reactor and to assess the performance of various heterogeneous kinetic schemes [58]. An example from the validation of catalytic schemes for propane total oxidation over Pt is illustrated in Figure 4.4 for a rectangular channel with 7 mm height, at moderate pressures up

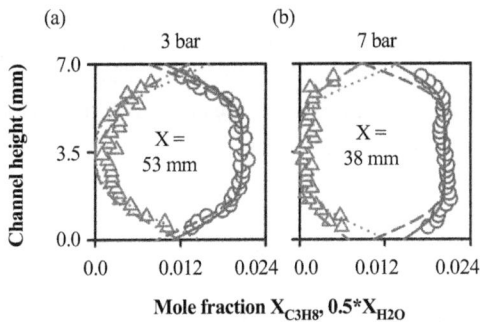

Figure 4.4. Predicted (lines) and Raman-measured (symbols) transverse profiles of propane and water mole fractions in a Pt-coated catalytic rectangular channel with a height of 7 mm: (a) $p = 3$ bar, $\varphi = 0.27$, and (b) $p = 7$ bar, $\varphi = 0.26$. Measurements: C_3H_8 (circles), H_2O (triangles). Predictions: dashed lines (C_3H_8) and dotted lines (H_2O) with original reaction scheme of Garetto et al. [59]; solid lines (C_3H_8) and H_2O (dashed-dotted lines) are predicted with a pressure-corrected reaction mechanism (reproduced by permission from Karagiannidis et al. [52]).

to 7 bar relevant to microturbine-based catalytic microreactors [3]. It is evident in Figure 4.4 that the atmospheric-pressure catalytic step by Garetto et al. [59] must be corrected at elevated pressures to account for the O_2-induced inhibition of propane adsorption due to the rise in O_2 partial pressure. Such corrections are of the utmost importance when examining the combustion stability of catalytic microreactors at modest pressures (see Section 4.8.2).

Detailed kinetic schemes for the oxidation of H_2 over noble metals have been developed in previous studies [27, 46, 48, 60–65]. CO oxidation schemes are reported in [66–69] and corresponding mechanisms for oxidation of fuel-lean and fuel-rich methane/air mixtures over noble metals have been proposed in [27, 46, 70–75].

4.3.2 HOMOGENEOUS KINETICS

Gas-phase chemistry is required in both catalytic and non-catalytic microreactors; for the former, gas-phase chemistry cannot always be neglected, particularly at elevated pressures [55]. Even though homogeneous combustion chemistry of low hydrocarbons and hydrogen has been extensively studied over the last decades, uncertainties remain about low-temperature and ultra-low stoichiometry gas-phase chemistry associated with either catalytic or non-catalytic microreactors. Although many C_1 mechanisms are available in the literature for lean CH_4 combustion, only a few of them encompass the regimes of low equivalence ratios and temperatures ($\varphi \leq 0.50$, $T \leq 1100°C$) applicable to catalytic microreactor applications relevant to power generation. The validity of the gaseous mechanisms can be assessed with homogeneous ignition studies and measurements of radical concentrations (notably OH) over the catalyst boundary layer. Radical concentrations above the catalyst have been routinely measured with LIF, point or planar [17, 54, 62, 76–82]. Using planar LIF of the OH radical in a channel-flow catalytic reactor along with detailed numerical simulations, validated gas-phase mechanisms have been provided for fuel-lean CH_4/air combustion at pressures up to 16 bar [18, 54, 56, 76, 80] and for fuel-lean H_2/air mixtures at pressures up to 10 bar [17, 29, 53].

Figure 4.5 provides LIF-measured and numerically predicted (using a 2D numerical model discussed in Section 4.8.1.1) distributions of the OH radical during gas-phase combustion of fuel-lean CH_4/air mixtures over Pt, in a 300 mm long and 7 mm high rectangular reactor [55, 56]. Simulations have been carried out using three different gas-phase mechanisms [28, 83, 84]. The onset of homogeneous ignition in Figure 4.5 has been defined as the location where OH rises to 5 percent of its maximum value in the channel and is marked with vertical arrows. For pressures 6 bar $\leq p \leq 16$ bar, the scheme of Warnatz et al. [28] provided good agreement between measured and predicted homogeneous ignition distances (see e.g., the comparisons in Figure 4.5(3a, 3b) pertaining to 6 bar). For pressures less than 6 bar, which are relevant to microreactors, no gas-phase scheme was able to reproduce the onset of homogeneous ignition (see Figure 4.5 (1a–1d, 2a–2d)). Moreover, the differences between experiments and predictions in terms of ignition delays (determined by using the ignition distance and the flow velocity) were up to 100 ms, that is considerably longer than the residence times of typical microreactors in Figure 4.1. These differences, which are particular to the low equivalence ratios and temperatures of catalytic microreactors, exemplify the importance of suitable gas-phase kinetic mechanisms in numerical models. A corrected gas-phase scheme is proposed in Reinke et al. [56] by modifying the scheme of Warnatz in the single-chain branching reaction CHO + M = CO + H + M, following updated literature kinetic rate measurements. The resulting modified Warnatz scheme

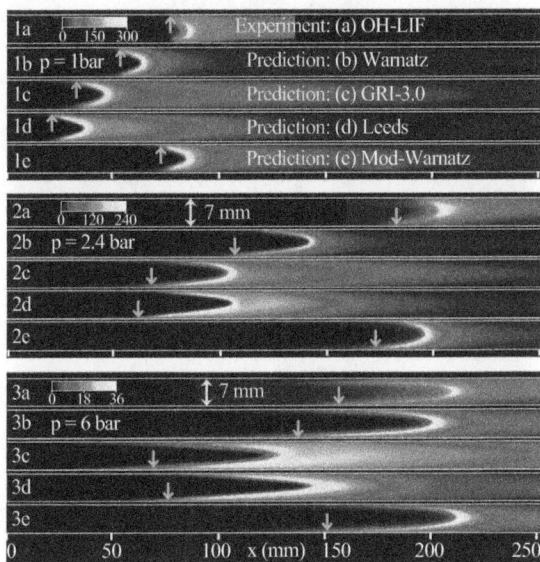

Figure 4.5. (a) OH LIF measurements and (b, c, d, e) predicted OH distributions, in methane/air combustion over Pt: (1) $p = 1$ bar, $\varphi = 0.31$, $T_{IN} = 754$ K, (2) $p = 2.4$ bar, $\varphi = 0.40$, $T_{IN} = 556$ K and (3) $p = 6$ bar, $\varphi = 0.36$, $T_{IN} = 569$ K. Predictions with the catalytic scheme of Deutschmann et al. [27] and the gas-phase schemes of: (b) Warnatz et al. [28], (c) GRI-3.0 [83], (d) Leeds [84] and (e) modified-Warnatz [56]. The arrows denote the onset of homogeneous ignition and the color bars give the OH in ppmv [56].

yielded good agreement between experiments and predictions at pressures up to 16 bar, thus encompassing operational ranges of microreactors and large-scale gas turbine reactors. Comparisons in Figure 4.5 manifest the aptness of the new mechanism at microreactor relevant pressures.

In a similar fashion, C_2 gas-phase mechanisms pertinent to fuel-rich catalytic combustion of methane (catalytic micro reformers) have been investigated [51, 57]. Finally, validated heterogeneous and homogeneous reaction schemes can be reduced jointly, accounting for pathway coupling: reduction methodologies for combined hetero-/homogeneous reaction schemes have been provided in [55, 85, 86].

In non-catalytic reactors, apart from a suitable gas-phase mechanism, a set of radical recombination reactions on the reactor walls may also be required. Such reactions have been proposed in [87] (see Table 4.1), and their inclusion in numerical modeling of methane-fueled tubular microreactors with a diameter of 0.2 mm [88] showed that wall radical quenching could suppress flame propagation. Table 1 draws on ideas in Raimondeau et al. [87]. Non-catalytic channel reactors operating at moderate prescribed wall temperatures in the so-called *weak combustion mode* have further been used to validate the low temperature chemistry of higher hydrocarbons (e.g., *n*-heptane studies in [89] and discussion in Section 4.6.3).

Table 4.1. Radical wall quenching mechanism[a]

1	$CH_3 + (s) \rightarrow CH_3(s)$	0 to 1
2	$H + (s) \rightarrow H(s)$	0 to 1
3	$OH + (s) \rightarrow OH(s)$	0 to 1
4	$O + (s) \rightarrow O(s)$	0 to 1
5	$2CH_3(s) \rightarrow C_2H_6(s) + 2(s)$	10^{13}
6	$2H(s) \rightarrow H_2 + 2(s)$	10^{13}
7	$2OH(s) \rightarrow H_2O + O(s) + (s)$	10^{13}
8	$2O(s) \rightarrow O_2 + 2(s)$	10^{13}
9	$CH_3(s) + H(s) \rightarrow CH_4 + 2(s)$	10^{13}
10	$OH(s) + H(s) \rightarrow H_2O + 2(s)$	10^{13}

[a]Kinetic parameters are sticking coefficients in reactions (1–4) and pre-exponentials (s^{-1}) in reactions (5–10); (s) indicates a surface site.

4.4 MODELS USED IN MICROREACTOR RESEARCH

Models capable of treating catalytic surface reactions and gaseous volumetric reactions have been constructed during the last years, leading to a better understanding of the underlying hetero-/homogeneous coupled processes and to the optimization of catalytic reactors. Modeling and numerical simulation of microreactor devices requires the description of the flow field, the heat transfer mechanisms, and the chemical reactions (see Figures 4.1c and 4.2c). However, the development of advanced models poses severe challenges due to different physicochemical processes with disparate spatio-/temporal scales: fluid mechanical transport (laminar in most microreactor applications) surface and gas-phase chemistry (the latter being of particular importance in catalytic microreactors at elevated pressures [18, 56]), intraphase diffusion within the catalyst washcoat, heat conduction in the solid channel walls, and finally radiation heat transfer. For example, in a catalytic microreactor, there are various time scales leading to complex interactions between the solid heat conduction (time scale of seconds), the flow inside the channels (residence times down to $\sim 10^{-3}$ s), interphase diffusion from the gas to the catalytic channel wall and intraphase diffusion inside the washcoat (time scales of $\sim 10^{-4}$ s), chemical reactions on the catalyst ($\sim 10^{-9}$ s) and possibly in the gas phase, and finally surface and gas radiation heat transfer exchange (ca. 10^{-10} s for typical microreactor sizes).

The most accurate way for modeling microreactor systems is the solution of the 3D continuum Navier–Stokes (NS) equations in conjunction with detailed homogeneous and heterogeneous reaction schemes and composition/temperature-dependent transport coefficients. The models should consider all related physicochemical phenomena, namely fluid mechanical transport, realistic gas-phase and surface chemistry (the latter accounting also for catalyst-support interactions), intraphase diffusion in the washcoat, heat conduction in the solid substrate, and radiation exchange between the hot channel walls. Because such simulations are computationally very intensive, appropriate approximations and simplifications depending on the flow conditions will also be introduced in Section 4.7.

Microreactor models, steady or transient, are classified based on their spatial dimensionality in Sections 4.5, 4.6, and 4.7, starting with simple zero-dimensional models in Section 4.5. One-dimensional models with lumped heat and mass transport coefficients (Nusselt and Sherwood numbers) were the first used in microreactor research [90] and are reviewed in Section 4.6. The models are appealing for their simplicity; however, unique lumped heat and mass transport coefficients for the entry channel-flow problem (see Figures 4.1c and 4.2c) are not always warranted, particularly in the presence of hetero-/homogeneous chemical reactions [91]. Even though problem-specific transport correlations can lead to acceptable 1D predictions for pure catalytic reaction applications, a spatial dimensionality of at least two is required to correctly describe homogeneous combustion in catalytic microreactors. This is because gas-phase combustion is strongly dependent on the reactant and temperature boundary layer profiles [14, 92]. Two- and three-dimensional models, presented in Section 4.7 (with applications in Section 4.8), reproduce realistically the in-channel fluid mechanical transport and, when coupled with detailed hetero-/homogeneous chemistry, they can provide a powerful tool for both reactor design and fundamental research. Although 3D effects play a role even in straight-channel catalytic microreactors [93, 94, 95], the large 3D computational effort is not always justified.

Elliptic (NS) 2D models with simplified chemistry [96–98] and nowadays with detailed hetero-/homogeneous chemistry [1, 9, 17, 27, 54, 55, 57, 75, 76, 99, 100] have become common research tools. On the other hand, parabolic (boundary-layer) 2D models [101, 102], again with detailed hetero-/homogeneous chemistry, are computationally very efficient by utilizing a marching solution algorithm. Elliptic and parabolic models are presented in Sections 4.7.1 and 4.7.2, respectively. Briefly, the applicability of parabolic models has been demonstrated in pure catalytic combustion at Reynolds numbers as low as 20 [99]. Comparisons between elliptic and parabolic solvers have also delineated the regimes of applicability of the parabolic approach in combined hetero-/homogeneous combustion [100]. It was shown that for low laminar propagation speeds (in comparison to the inflow velocities), the parabolic model can adequately describe homogeneous combustion.

Interphase transport is typically laminar in catalytic microreactors; however, in power-generation applications the increased pressure requirements of new generation turbines (~30 bar) have resulted in incoming Reynolds numbers, based on the individual catalytic channel hydraulic diameter, of up to 30,000 at full machine load. Each channel in such reactors falls in the microscale range, rendering turbulence modeling of general interest for microreactors. Moreover, turbulence can also be created via the strong mixing of high-velocity fuel and air streams upon injection in the volume of practical microreactors. Turbulence modeling issues will be addressed in Section 4.9. Despite recent advances in laminar modeling, turbulent reacting channel-flow combustion has not yet received proper attention. Recent turbulent model developments for catalytic reacting channel flows have been presented in [6, 103, 104]. Comparisons with in-channel measurements of velocity and thermo-scalars in catalytic mesoscale reactors have shown that crucial for the model performance is the capture of the strong flow laminarization induced by the heat transfer from the hot catalytic plates [6].

Transient laminar flow models are needed for two classes of microreactor combustion problems. The first deals with conventional dynamic operation, arising during reactor start-up (ignition), shutdown (including extinction), or change in reactor operating conditions. The other class of problems deals with intrinsic dynamic oscillatory phenomena appearing as bifurcations when a system parameter (e.g., velocity) is altered. Therein, two subclasses are identified: dynamic oscillatory phenomena driven solely by kinetics [105–107] or by the interaction of gas-phase kinetics and solid walls [24, 108, 109]. Modeling of these two distinct transient processes will be addressed in Sections 4.8.1 and 4.8.4. Transient 1D models with a global catalytic chemistry step were initially used to investigate light-off of catalytic microreactors [110]. The light-off time scale (order of a few seconds to tens of seconds) renders fully transient 2D or 3D channel-flow models with detailed hetero-/homogeneous chemistry very demanding. Two-dimensional transient models for a single channel usually employ the quasi-steady approximation for the flow and gas-phase chemistry [111, 112]. Although the convective and chemical reaction time scales are usually shorter than the solid heat conduction time scale, the chemical time scales may become comparable to the solid time scale during light-off, thereby invalidating at specific times and spatial locations the quasi-steady gas-phase assumption [111]. The requirements for the validity of the quasi-steady assumption will be outlined in Section 4.8.1.2. Fundamental studies can also be carried out using fully transient stagnation-flow 1D codes [113, 114], which accommodate detailed chemistry at an affordable cost; however, stagnation flow geometries are remote to those of practical microreactors.

In addition to single-channel models, the *continuum* approach is commonly used for meso-scale catalytic honeycomb reactors, when the incoming flow is non-uniform and when external heat losses are present in the periphery of the structure. It is clarified that the terminology *continuum* should not be confused with the non-continuum models discussed in Section 4.10, referring to finite Knudsen number flows wherein the NS equations break down. The continuum approach for the honeycomb structure allows for 2D or 3D modeling, steady or transient, with detailed chemistry. The approach has been established in Aris and co-workers [115, 116]. Two time-dependent, coupled partial differential equations describe the heat transfer in the solid-phase structure and gas-phase, respectively. The quasi-steady assumption for the gas-phase is further applied in a manner analogous to that of single-channel modeling [110, 111]. The heat balance for the solid is then coupled to steady-state single-channel simulations via given axial wall temperature profiles (representing the energy boundary condition for the single-channel calculations) and via heat source terms derived from the gaseous convective heat transfer and heat release in the single channel [117–119]. Continuum models for the honeycomb structure will be laid down in Section 4.7.3 and an application will be provided in Section 4.8.5.

Analytical studies based either on activation energy asymptotics or on simplified description of the flow (thermodiffusive constant density models) will be dealt in Sections 4.6.3 and 4.8.3. Such models can reveal the controlling processes in microreactor combustion and provide closed-form criteria for ignition or extinction. Although these results may not be directly used for microreactor design, they can aid elaborate computational fluid dynamics (CFD) simulations by reducing the parameter space under investigation. Finally, finite Knudsen number effects (i.e., non-continuum flows, manifested by the presence of velocity, concentration and temperature slips at the gas–wall interface) are discussed in Section 4.10. Although for atmospheric pressure applications most microreactors fall in the continuum regime, certain microreactor components, such as electrodes in micro-solid oxide fuel cells (SOFCs), may operate in the slip regime. Lattice Boltzmann (LB) models are particularly suited for the slip regime; the basic LB theory with microreactor applications is also presented in Section 4.10.

4.5 ZERO-DIMENSIONAL MODELS

The simplest approach is the spatially zero-dimensional model for a perfectly stirred reactor (PSR) and for a constant pressure batch reactor. Although the idealization of perfect mixing is not relevant to practical microreactors, the computational efficiency of such models renders them amenable to fundamental kinetic studies with complex chemical reaction schemes (e.g., sensitivity analyses, investigation of hetero-/homogeneous chemistry coupling, impact of operating pressure and temperature on kinetics, effect of reactor flow residence time and surface-to-volume ratio on catalytic and gaseous species conversion, etc. [55, 120–124]). The time-dependent governing equations for the K_g gaseous species and the energy in a PSR with heterogeneous and homogeneous chemical reactions are [125]

$$\frac{dY_k}{dt} = -\frac{1}{\tau}(Y_k - Y_{k,IN}) + \frac{1}{\rho}\dot{\omega}_k W_k + \frac{A}{\rho V}\dot{s}_k W_k, \ k = 1, \dots K_g \ \text{and} \tag{4.6}$$

$$c_p \frac{dT}{dt} = \frac{1}{\tau}\sum_{k=1}^{K_g} Y_{k,IN}(h_{k,IN} - h_k) - \frac{Q}{\rho V} - \frac{1}{\rho}\sum_{k=1}^{K_g} h_k \dot{\omega}_k W_k - \frac{A}{\rho V}\sum_{k=1}^{K_g+M} h_k \dot{s}_k W_k, \tag{4.7}$$

respectively, with τ, Q, A, and V the reactor residence time, heat loss, surface and volume, respectively. A single surface phase is considered in Equation (4.7), with $M = K_s + K_b$ denoting the total number of surface (K_s) and bulk (K_b) species [126]. The surface species coverage is obtained by simplifying Equation (4.3) for a conserved surface site density Γ:

$$\frac{\partial \theta_m}{\partial t} = \sigma_m \frac{\dot{s}_m}{\Gamma}, \ m = 1, \dots, K_s. \tag{4.8}$$

Finally, the ideal and caloric gas laws close the system of equations:

$$p = \rho R^\circ T/\bar{W} \ \text{and} \ h_k = h_k^0(T_0) + \int_{T_0}^{T} c_{p,k} dT, \ k = 1, \dots K_g. \tag{4.9}$$

In the constant pressure batch reactor models, the first terms on the right side of Equations (4.6) and (4.7) are identically zero.

Such simplified models (sometimes augmented with transport models for the gaseous species) have been used to investigate hetero-/homogeneous kinetic coupling [55] and kinetically driven dynamic oscillatory behavior [127, 128] as will be discussed in Section 4.8.4.3. Since the microreactors in Figures 4.1 and 4.2 are inhomogeneous in space and time, zero-dimensional computations can only provide qualitative results. Figure 4.6 provides adiabatic steady PSR simulations with detailed hetero-/homogeneous chemical reaction schemes, which have been validated for fuel-lean CH_4/air combustion over Pt [55]. The residence times and surface-to-volume ratios in Figure 4.6 are relevant to catalytic microreactors, while the modest pressures are relevant to microturbine-based small power-generation systems. The regimes of importance for the gaseous pathway are delineated in this parametric plot; to the left of each line of fixed pressure and temperature, the gaseous chemistry cannot be neglected, as it accounts for at least 5 percent of the total methane conversion. It is evident that gaseous chemistry cannot be safely ignored in numerical models, particularly at the modest pressures of microturbine-based power devices.

In addition to the aforementioned zero-dimensional PSR models, zero-dimensional thermodynamic models have also been used to assess maximum theoretical microreactor efficiencies [129].

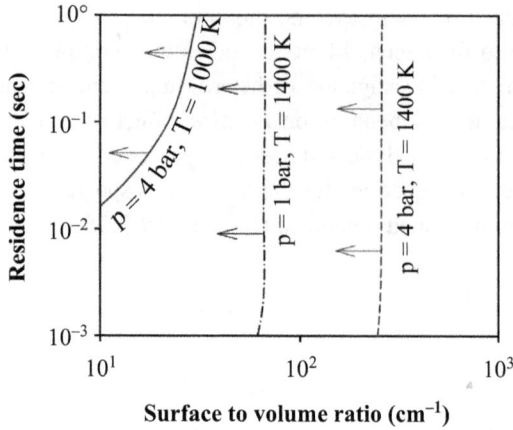

Figure 4.6. Regimes of significant gaseous chemistry contribution in fuel-lean methane/air hetero-/homogeneous combustion over Pt at modest pressures relevant to microreactors (reproduced by permission from Reinke et al. [55]). In the areas to the left of a line of constant pressure and temperature, homogeneous chemistry cannot be neglected.

4.6 ONE-DIMENSIONAL CHANNEL-FLOW MODELS

4.6.1 MATHEMATICAL FORMULATION

One-dimensional models with lumped heat and mass transport coefficients have been extensively used in the literature, initially with simplified chemistry and recently with detailed kinetics, for both catalytic and non-catalytic microreactors [60, 88–91,130–132]. The time-dependent 1D governing equations for a constant cross-sectional-area channel with heterogeneous and homogeneous chemical reactions (applicable to both catalytic and non-catalytic microreactors) are presented next in their most general form:
Continuity

$$\frac{\partial \rho}{\partial t} + \frac{\partial (\rho u)}{\partial x} = 0. \tag{4.10}$$

Momentum

$$\rho \frac{\partial u}{\partial t} + \rho u \frac{\partial u}{\partial x} + \frac{\partial p}{\partial x} + \frac{P \cdot f}{2A} \rho u^2 = 0. \tag{4.11}$$

Gas-phase species

$$\rho \frac{\partial Y_k}{\partial t} + \rho u \frac{\partial Y_k}{\partial x} + \frac{\partial}{\partial x}(\rho Y_k V_{x,k}) + \frac{P}{A}\alpha_k (Y_k - Y_{k,s}) - \dot{\omega}_k W_k = 0, \; k = 1,....K_g. \tag{4.12}$$

Energy

$$\rho\frac{\partial(c_pT)}{\partial t}+\rho u\frac{\partial(c_pT)}{\partial x}-\frac{\partial}{\partial x}\left(\lambda\frac{\partial T}{\partial x}\right)+\rho\sum_{k=1}^{Kg}Y_kV_{x,k}h_k+\frac{P}{A}\alpha_T(T-T_s)+\sum_{k=1}^{Kg}h_k\dot{\omega}_kW_k=0,\quad(4.13)$$

with P and A the perimeter and the constant cross-sectional area of the channel, respectively. The surface species coverage equations, Equations (4.8), are finally used along with the ideal and caloric gas laws, Equations (4.9). The interfacial boundary conditions for the gas-phase species are

$$\alpha_k(Y_k-Y_{k,s})+B\dot{s}_kW_k=0,\quad k=1,......K_g.\qquad(4.14)$$

Considering uniform properties for the solid wall over its cross-sectional area A_s, a 1D model for the solid phase can also be constructed:

$$A_s\frac{\partial(\rho_sc_sT_s)}{\partial t}+A_sk_s\frac{d^2T_s}{dx^2}-P\alpha_T(T-T_s)-P\dot{q}_{rad}+P\sum_{k=1}^{Kg+M}h_kB\dot{s}_kW_k=0.\qquad(4.15)$$

with α_k and α_T appropriate mass and heat transport lumped coefficients, respectively, and $\dot{q}_{rad}$ surface radiation exchange between the solid wall elements. Surface radiation has been shown [131] to reduce the steep axial solid temperature gradients created upon light-off of catalytic microreactors and to facilitate their ignition [21] by providing an additional heat recirculation route from the rear to the front of the channel. In the case of a prescribed wall temperature profile $T_p(x, t)$ (e.g., whenever temperature measurements along the wall are available, as in [105, 132]), Equation (4.15) is substituted by the boundary condition:

$$T_s(x,t)=T_p(x,t).\qquad(4.16)$$

For catalytic systems, the factor B in Equations (4.14) and (4.15) is the ratio of the catalytically active to the geometrical surface area, which is a parameter larger than unity for technical catalysts with good dispersion. It is clarified that B cannot account for pore (intraphase) diffusion in the catalyst washcoat but only for interfacial surface area increase. In the case of intraphase diffusion limitations, various modeling approaches (discussed briefly in Section 4.7.1) can be used. The formulation of this section can be applied to any type of microreactor (pure catalytic, pure gaseous, or hetero-/homogeneous) by keeping or dropping the catalytic ($\dot{s}_k$) or gaseous ($\dot{\omega}_k$) reaction terms in Equations (4.13)–(4.15).

For 1D models, a full multicomponent transport approach [15] for the diffusion velocities $V_{k,x}$ in Equations (4.12) and (4.13) is computationally manageable:

$$\frac{\partial X_k}{\partial x}=\sum_{\ell=1}^{Kg}\frac{X_kX_\ell}{D_{k\ell}}(V_{\ell,x}-V_{k,x})+(Y_k-X_k)\frac{1}{p}\frac{\partial p}{\partial x}+\sum_{\ell=1}^{Kg}\frac{X_kX_\ell}{\rho D_{k\ell}}\left(\frac{D_\ell^T}{Y_\ell}-\frac{D_k^T}{Y_k}\right)\frac{1}{T}\frac{\partial T}{\partial x}.\quad(4.17)$$

Simplified transport models typically consider mixture-average diffusion and may also include thermal diffusion for the light species [133]:

$$V_{x,k}=-D_{km}\frac{\partial}{\partial x}\left[ln\left(Y_k\bar{W}/W_k\right)\right]+\left[D_k^TW_k/(\rho Y_k\bar{W})\right]\frac{\partial(lnT)}{\partial x}.\qquad(4.18)$$

In many cases the flow is considered isobaric, such that Equation (4.11) reduces to $p(x, t) = p_0$. This is a key simplification that removes the strong pressure–velocity coupling and the time integration limitations associated with the compressible flow equations. The isobaric assumption is adequate considering the small pressure drop (less than 1 percent) in straight catalytic or non-catalytic channels with modest Reynolds number laminar flows and length-to-hydraulic diameter aspect ratios typically less than 50 (as in Figures 4.1 and 4.2). At both solid wall ends, front ($x = 0$) and rear ($x = L$), radiative boundary conditions can be used [9, 96]:

$$k_s \frac{\partial T_s}{\partial x}\bigg|_{x=0} = \varepsilon\sigma\left[T_s^4(x = 0) - T_{IN}^4\right], \quad -k_s \frac{\partial T_s}{\partial x}\bigg|_{x=L} = \varepsilon\sigma\left[T_s^4(x = L) - T_{OUT}^4\right], \quad (4.19)$$

with T_{IN} and T_{OUT} appropriate radiation exchange temperatures and ε the emissivity of the solid wall (Equation (4.19) further considers black body inlet and outlet radiation enclosures). The boundary conditions at the entry face can be further improved with the addition of convective heat losses due to flow impingement [60]. All gas-phase properties are prescribed at $x = 0$, while at $x = L$ zero Neumann conditions ($d\varphi/dx = 0$) are applied for the gas-phase variables φ. Finally, proper initial conditions for the solid and the gas phase have to be provided. A further simplification in 1D models is to neglect the interphase (gaseous) diffusion barrier in catalytic systems and to assume uniform properties at the gas and the solid–gas interface. This leads to the classical plug-flow reactor model. Plug-flow models are valid either for very low or very high (turbulent) Reynolds numbers [99]. For intermediate Reynolds numbers, the inclusion of lumped transfer coefficients as shown in Equations (4.12) and (4.13) is necessary.

One-dimensional models with lumped heat and mass transport coefficients (Nusselt and Sherwood numbers, $Nu = \alpha_T d_h/\lambda$, $Sh_k = \alpha_k d_h/(\rho D_{km})$, with d_h the channel hydraulic diameter) were the first used in microreactor research and are still a very common research tool. The models are appealing for their simplicity; however, unique lumped heat and mass transport coefficients for the entry channel-flow problem pertinent to microreactors (Figure 4.1c and 4.2c) are not always warranted in the presence of hetero-/homogeneous chemical reactions [91, 134]. In Groppi et al. [91], Nusselt correlations have been provided for catalytic channels, invoking a dependence on the catalytic reactivity:

$$\frac{Nu - Nu_H}{Nu_T - Nu_H} = \frac{Da_s Nu}{(Da_s + Nu)Nu_T}, \quad (4.20)$$

with Nu_T and Nu_H the constant temperature and constant heat flux solutions of the entry heat transfer problem, respectively [135], and Da_s a characteristic surface Damköhler number, defined as the ratio of the reaction rate to the diffusion rate. Similar correlations can be provided for the Sherwood numbers. Clearly, problem-specific transport correlations are needed for accurate 1D predictions at least for the more demanding hetero-/homogeneous systems where both heat and mass transport are important. This necessitates the comparison of 1D with 2D predictions (as in [136, 137]) so as to ensure the accuracy of the available correlations and to fine-tune them if necessary. Peak temperatures and hot spots such as those shown in Figure 4.3a are particularly sensitive to the employed transport correlations.

A key complication in 1D channel models with combined hetero-/homogeneous combustion systems is that a spatial dimensionality of at least two is required to correctly describe homogeneous combustion. This is because gas-phase combustion strongly depends on the limiting reactant boundary layer profile [14, 92] (see the Y_F profile in Figure 4.1c and consider that gas-phase reactions occur mostly in the near-wall regions whereby the temperature is the highest). Moreover, even in the absence of vigorous homogeneous combustion, the contribution of the gaseous reaction pathway may not always be neglected in catalytic combustion of hydrocarbons at elevated pressures due to the incomplete oxidation of the hydrocarbon fuel to CO [55], thus still necessitating a 2D description for many practical catalytic combustion systems.

For pure homogeneous combustion systems, the absence of near-wall reactant depletion renders 1D models more realistic -provided that a nearly planar flame forms in the channel (an issue addressed in Section 4.8.4.1). On the other hand, for pure catalytic systems (whenever gas-phase chemistry can be safely ignored, see Figure 4.6), the absence of large gas-phase axial gradients allows for a simpler parabolic solution (via marching in the streamwise direction) by neglecting the axial diffusion terms in Equations (4.12) and (4.13). In such systems it has been shown that the parabolic approach for the gas is acceptable for inlet Reynolds numbers down to 20 [99]. Finally, while in pure homogeneous combustion axial diffusion in the gas is always important due to the presence of a strong flame, in hetero-/homogeneous combustion systems axial diffusion may be neglected under certain conditions [100], as will be further discussed in Section 4.7.2. This property of hetero-/homogeneous combustion systems stems from their generally weaker gaseous combustion, owing to the upstream catalytic fuel depletion; nonetheless this advantage is truly relevant only for higher dimensionality (at least 2D) models since, as discussed before, 1D models suffer in their homogeneous combustion treatment due to the lack of boundary layer description for the reactants and temperature.

Given the long time scale for the solid substrate heat-up compared to the relevant chemical, convective and diffusive time scales in a channel reactor, the quasi-steady assumption can be applied for the gas-phase [110], whereby the only transient term retained in Equations (4.10) to (4.15) is that of the solid energy equation (4.15). Since the large thermal inertial of the solid leads to long integration times of the order of seconds, the quasi-steady assumption alleviates the otherwise excessive computational cost dictated by the short time steps imposed by flow and chemistry. Although this assumption does not necessarily hold at all spatial locations and/or times during the reactor history, it is nevertheless employed in most transient studies involving solution of the solid [110, 111, 138]. The validity of the quasi-steady assumption deserves attention and will be elaborated in Section 4.8.1.2.

4.6.2   APPLICATION OF 1D MODELS TO CATALYTIC MICROREACTORS

Fully transient 1D simulations for both gas and solid have been reported in early studies [138], but only to simulate the initial response due to a step change in the inlet reactant concentration; for longer times, the quasi-steady assumption was invoked. Transient 1D models with the quasi-steady assumption and simplified chemistry have been extensively used in the past [110, 138, 139]. More recently, studies with detailed surface chemistry, without invoking the quasi-steady assumption for the gas have appeared [60, 137]. Steady 1D models are also routinely used for the study of microreactor performance [90, 91, 136, 140]. In all previous investigations, the isobaric assumption $p = p(x = 0, t)$ is employed.

The application of a 1D fully transient model (without invoking the quasi-steady approximation) is illustrated in Figure 4.7 for the light-off of a fuel-lean ($\varphi = 0.16$) methane/air mixture over a supported Pd catalyst at a pressure of 3 bar [137]. A single tubular channel, with a

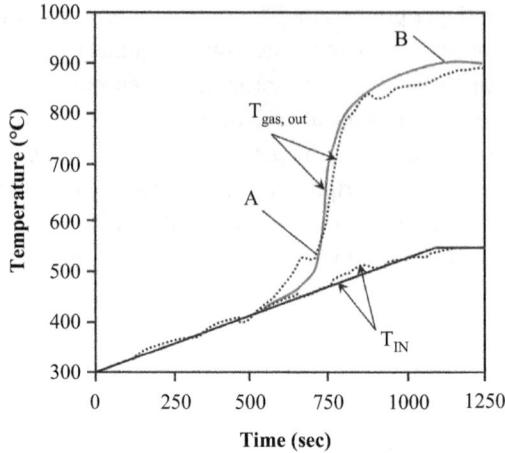

Figure 4.7. Transient 1D simulations (solid lines) and measurements (dotted lines) of temperatures during methane catalytic combustion in a single-channel Pd-coated catalytic reactor: equivalence ratio $\varphi =$ 0.16, $p = 3$ bar and constant mass flux of 0.4 g/cm²s. Point A denotes the light-off time and B the location of transport-limited fuel conversion (reproduced by permission from Robbins et al. [137]).

diameter of 1.2 mm and a length of 75 mm, is used to simulate the processes in a catalytic honeycomb structure. The detailed mechanism for the oxidation of CH_4 on Pd included 35 reactions for 10 surface species and two subsurface species, PdO (bulk) and Pd (bulk) [141]. Gas-phase chemistry played no role for this application. For interphase transfer, Nusselt and Sherwood empirical correlations were used, whose aptness has been established by selective comparisons with a 2D model. Predictions in Figure 4.7 reproduce the time history of the measured exit gas temperature, including the light-off time (point *A*) and the attainment of mass-transport-limited conversion (manifested by the temperature leveling at point *B*).

A steady 1D model has been used in [136] to investigate propane-fueled catalytic microreactors (gas-phase chemistry is not considered). The microreactor consists of two parallel 0.79 mm thick steel plates, with a length of 60 mm (out of which the central 50 mm is coated with platinum) and a channel gap of 0.3 mm; thermal spreaders are placed on top of the steel plates to achieve axially uniform temperature profiles (see the inset in Figure 4.8 and details in [142]). The 1D model is isobaric, with a simplified surface mechanism for the adsorption of propane and oxygen and a surface area factor $B = 1.7$ in Equations (4.14) and (4.15); in-channel or external radiation heat transfer has not been considered. The outer walls are non-adiabatic with an equivalent heat transfer coefficient $\alpha_T = 64$ W/(m²K), which accounts for both convective and radiation external heat losses. Pressure is atmospheric and the C_3H_8/air equivalence ratio varies between 0.60 and 0.95. Strategies are described [136] for devising appropriate *Nu* and *Sh* numbers, either using literature data or constructing custom-made correlations by comparing 1D to 2D simulations. The model predictions reproduce reasonably well the measured surface temperatures under a variety of reactor configurations (Figure 4.8). Stability diagrams are then constructed in terms of the inlet velocity and the maximum tolerable external heat loss (critical heat loss coefficient α_T) for various solid thermal conductivities (Figure 4.9). Such

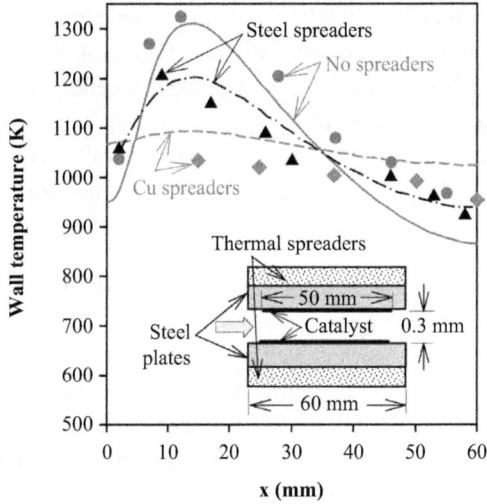

Figure 4.8. Comparison between 1D model predictions (solid lines) and measurements (symbols) of wall temperatures in a parallel plate Pt-coated microreactor (shown in the inset). Propane/air mixture with $\varphi = 0.95$, $p = 1$ bar, inlet flow rate of 2 slpm (reproduced by permission from Kaisare et al. [136]).

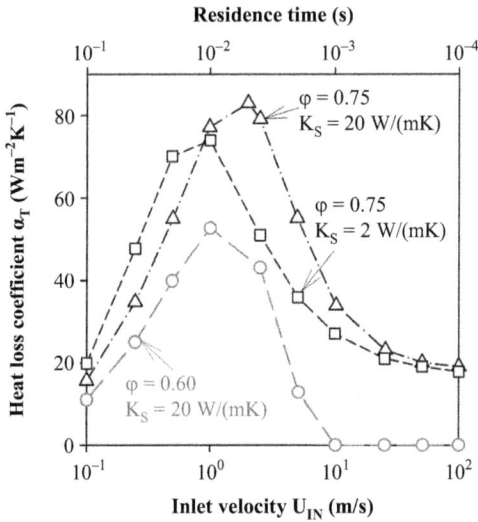

Figure 4.9. Computed stability maps for propane-fueled, Pt-coated catalytic microreactors (reactor geometry shown in Figure 4.8). Critical heat loss coefficient a_T at equivalence ratios of 0.60 and 0.75 and solid thermal conductivities k_s of 2 and 20 W/(mK), as a function of inlet velocity or reactor residence time. The area below each curve corresponds to stable combustion (reproduced by permission from Kaisare et al. [136]).

information is valuable as it defines practical operational envelopes of catalytic microreactors. Comparisons between experiments and simulations in Figures 4.7 and 4.8 indicate that 1D pure catalytic combustion models with suitable heterogeneous kinetics are capable of capturing crucial performance characteristics (light-off time and wall temperatures) in realistic microreactor configurations. The prime care is to use *Nu* and *Sh* numbers validated for each particular application and to further ensure that the contribution of gaseous chemistry is insignificant.

4.6.3 APPLICATION OF 1D MODELS TO NON-CATALYTIC MICROREACTORS

Flames established in microreactors can be either stationary or dynamic [89, 109, 143] depending on the flow velocity and wall temperature. At sufficiently low inflow velocities, stationary *weak flames* with very small heat release (of the order of a few degrees K) can be established [144]. The regime of weak flames allows for the validation of chemical reaction mechanisms of hydrocarbons at low-to-medium temperatures, by using 1D models with detailed chemistry and transport together with measurements of temperature and species concentrations. In Yamamoto et al. [89], the three-stage oxidation of stoichiometric gaseous *n*-heptane/air mixtures was investigated at pressures of 1–6 bar in a quartz tube with an inner diameter of 2 mm and length of 6 cm (see the inset in Figure 4.10). The wall temperature was controlled via external heating from a flat flame burner and monitored with thermocouples. At $p = 1$ bar, which is herein elaborated, gas sampling provided concentrations of major species and CH_2O, while CH chemiluminescence collected by a digital camera through the transparent quartz tube walls identified the heat release rate (HRR) locations.

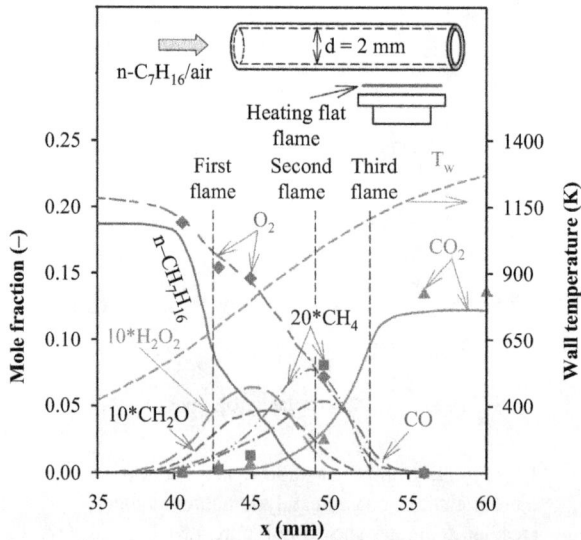

Figure 4.10. Computed (lines) and measured (symbols) species profiles in a microflow tubular reactor (see configuration in the inset) fed with stoichiometric *n*-heptane/air mixture, and a prescribed wall temperature measured by thermocouples (dashed line). The vertical dashed lines indicate the measured locations of three heat release rate (HRR) peaks, as assessed via CH-chemiluminescence measurements (reproduced by permission from Yamamoto et al. [89]).

Numerical simulations [89] comprise an isobaric steady 1D model adopted from the Premix code of Chemkin [145] with a species transport model according to Equation (4.18). The steady-state continuity and momentum Equations (4.10) and (4.11) reduce to $\rho(x)\, u(x) = \dot{M}$ with $\dot{M}$ the constant and unknown mass flux, and $p(x) = p_0$, with p_0 the constant and known pressure. The steady version of the energy Equation (4.13) is employed with a heat transfer coefficient $\alpha_T = \lambda Nu/d$ and a Nusselt number $Nu = 4$, while the prescribed wall energy boundary condition of Equation (4.16) is used in its time-independent form. Most importantly, for the associated low flow velocities, simulations were shown to be insensitive [89] to the particular value of the Nusselt number in the range $Nu_T = 3.66$ to $Nu_H = 4.36$. The steady versions of the species Equations (4.12) were used, without mass transport to the wall ($\alpha_k = 0$), as it was found that wall recombination reactions did not appreciably affect the results. Finally, a reduced gaseous mechanism for n-heptane [146] was included in the model. The resulting boundary value problem is solved using a Newton–Raphson iterative method [145].

Figure 4.10 provides predicted and measured species axial profiles, with the three vertical dashed lines denoting the chemiluminescence-assessed locations of three HRR peaks. The n-heptane and O_2 species are partially consumed at the first HRR peak. CO_2 starts rising at the first HRR peak and further increases through the second and third HRR peaks, reaching a nearly constant value after the third HRR peak. CO increases at the first HRR peak through the second HRR peak and it is finally consumed at the third HRR peak; CH_2O rises at the first HRR peak, then attains a nearly constant value, and finally drops at the second HRR peak. Measurements and simulations of major species and selected minor species profiles agree with each other, indicating that the model captures the three-stage oxidation process of n-heptane. Moreover, it is shown [89] that the chemiluminescence-measured HRR peaks are in good agreement with the predicted HRR peaks in terms of their corresponding temperatures. It is emphasized that the temperature at each HRR peak is of importance for the validation of low-temperature gaseous chemical reaction mechanisms, and the microreactor in Figure 4.10 has proven a very amenable platform for such studies.

In addition to the foregoing 1D model with detailed chemistry and transport, simplified models have been extensively used in microchannel combustion [108, 147–153]. The aforementioned works can be categorized in theoretical investigations (using activation energy asymptotics or bifurcation analyses) that identify the controlling parameters in either steady or transient microchannel combustion and in simplified (in terms of chemistry and transport) modeling works that address important engineering design issues [152, 153]. Two theoretical approaches are outlined next for steady and unsteady microreactor operation. The first example considers the steady flame propagation in a channel with inert walls (Figure 4.2c) having width d_i, flow speed u, flame propagation speed u_f (in laboratory coordinates), finite solid wall thickness, and convective heat-transfer coefficients for the inner and outer wall surfaces of h_i and h_o, respectively. Considering one-step chemistry controlled by the deficient reactant, a planar flame in the thin reaction zone regime, and constant properties for the gas-phase, the non-dimensional steady governing equations for the gas and solid become [147]

$$(U - U_f)\frac{d\theta}{d\xi} = \frac{d^2\theta}{d\xi^2} - H(\theta - \theta_w) + \exp\left[\beta(\theta_f - 1)/2\right]\delta(\xi - \xi_f), \qquad (4.21)$$

$$(U - U_f)\frac{dY}{d\xi} = \frac{1}{Le}\frac{d^2Y}{d\xi^2} - H(\theta - \theta_w) + \exp\left[\beta(\theta_f - 1)/2\right]\delta(\xi - \xi_f), \qquad (4.22)$$

$$-U_f \frac{d\theta_w}{d\xi} = \frac{\alpha_w}{\alpha_g} \frac{d^2\theta_w}{d\xi^2} + HC(\theta - \theta_w) - H_0 C\theta_w/\beta, \tag{4.23}$$

with all variables identified in the Nomenclature. A constant Nusselt number $Nu = 4.36$ has been used. The boundary conditions are

$$\xi \to -\infty, \quad \theta = \theta_w = 0, \quad Y = 1 \quad \text{and} \quad \xi \to +\infty, \quad \frac{d\theta}{d\xi} = \frac{d\theta_w}{d\xi} = 0, \quad Y = 0. \tag{4.24}$$

In the limit of a large activation energy reaction (large Zeldovich number β), up to first-order asymptotic expansions are carried out, while applying jump conditions across the flame surface [15]:

$$[\theta] = [Y] = \left[\frac{d\theta}{d\xi} + \frac{1}{Le} \frac{dY}{d\xi}\right] = 0 \quad \text{and} \quad -\frac{1}{Le} \frac{dY}{d\xi}\Big|_{\xi_f} = \exp\left[\frac{\beta}{2} \frac{\theta_f - 1}{1 + \gamma(\theta_f - 1)}\right]_{\xi_f}, \tag{4.25}$$

where the brackets denote differences across the flame, ξ_f the flame position, and γ the heat release parameter $(T_{ad} - T_\infty)/T_{ad}$.

The solution for the flame speed $m = U - U_f$ can then be computed for various gas and solid parameters [147]. Figure 4.11 provides the flame speed as a function of normalized heat transfer rate H with parameter the solid-to-gas heat capacity ratio C, for a case with $u = 0.1$, gas-to-solid thermal diffusivity ratio of 0.1, unity Lewis number, and external heat transfer rate $H_o = 0.02\, \beta H$. Two steady solutions for the flame speed are evident. For $C = 0.01$ (large wall thermal inertia) the flame extinguishes at point A due to wall cooling, which is conventionally used to determine the quenching diameter. However, at very low flame speeds there exists a new flame branch caused by exhaust gas heat recirculation via solid wall heat conduction. Owing to the large

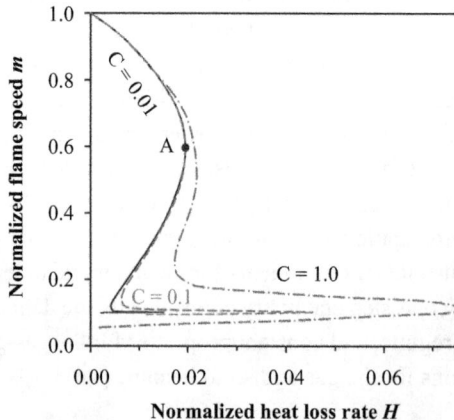

Figure 4.11. The analytical solution of microchannel normalized flame speed m as a function of the normalized heat transfer H, for various ratios of the gas-to-solid heat capacities C. Other parameters are: normalized flow velocity $U = 0.1$, ratio of gas-to-solid thermal diffusivity 0.1, gas thermal diffusivity 2.22×10^5 m²/s, chemical HRR 40 MJ/kgFuel, activation energy 124.65 kJ/mol, specific heat at constant pressure $c_p = 1.005$ kJ/mol, and unity Lewis number for the deficient fuel reactant (reproduced by permission from Ju et al. [147]).

thermal inertia of the wall, this flame exists only at a very small heat loss. When increasing the wall thermal inertia (e.g., see $C = 0.1$ or 1.0 in Figure 4.11), the new flame branch broadens such that this flame can be sustained even at considerable heat losses. This in turn shows that when the wall heat capacity is small, microscale gaseous combustion can be sustained at scales smaller than the classical quenching diameter. Additional parametric studies in terms of flow velocity, gas-to-solid thermal diffusivity ratio, equivalence ratio, and so forth, have been carried out in [147]. Results such as those in Figure 4.11, although approximate and having limited applicability to practical reactor design, have provided valuable fundamental understanding of the parameters controlling flame stability at the microscale. Asymptotic approaches based on higher dimensionality 2D models will be further discussed in Section 4.8.3.

Instead of seeking approximate solutions with activation energy asymptotics, full numerical integration of Equations (4.21) and (4.22) in their augmented transient form has been carried out in [108]. Therein, the flame—sheet model was again adopted and the wall temperature was constant, such that Equation (4.23) was substituted by $\theta = \theta_w$. The transient 1D simulation allowed for the identification of dynamic flame modes (oscillatory modes). Moreover, linear stability analysis of the steady solution θ_0 and Y_0 (performed by applying harmonic perturbations of the type $\theta = \theta_0(x) + \varepsilon \theta_1(x) \exp(\lambda t)$, $Y = Y_0(x) + \varepsilon Y_1(x) \exp(\lambda t)$, $x_f = x_{f,0} + \varepsilon \exp(\lambda t)$, with x_f the flame position, and ε a small parameter, which is inversely proportional to the Zeldovich number β in Equations (4.21) and (4.22)) delineated the regimes of steady and unsteady flame operation. Figure 4.12 provides a flame stability diagram in terms of normalized flame speed and wall temperature, for a heat loss coefficient $H = 0.4$. Eigenvalue analysis identifies regimes of stable flames, no flames, and unstable flames. The suppression of unstable combustion with increasing θ_w in Figure 4.12 is in qualitative agreement with 2D model simulations, also carried out in [108]. Microreactor flame stability will be discussed in more detail in the forthcoming Section 4.8.2 in the context of higher dimensionality models.

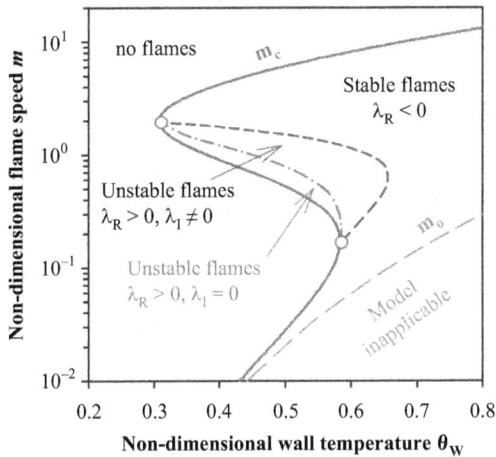

Figure 4.12. Flame stability diagram in a constant wall temperature microchannel, constructed using a 1D flame sheet model. Stable and unstable states are shown in the m versus θ_w parameter space. The solid line marked m_c denotes critical blow-off values and the dashed line m_0 the applicability limit of the flame sheet model; λ_R and λ_I are the real (growth rate) and imaginary parts, respectively, of the corresponding eigenvalues (reproduced by permission from Kurdyumov et al. [108]).

4.7 MULTIDIMENSIONAL MICROREACTOR MODELS

4.7.1 MATHEMATICAL FORMULATION

The governing equations in their vectorial form, appropriate for 2D and 3D geometries in any coordinate system and for either catalytic or non-catalytic reactors are given below. Although valid for any complex reactor geometry, the high computational cost makes them attractive for single-channel simulations. These models resolve all relevant spatiotemporal scales and are also referred to as direct numerical simulation (DNS).

Continuity equation:

$$\frac{\partial \rho}{\partial t} + \nabla \cdot (\rho \, \vec{u}) = 0 . \tag{4.26}$$

Momentum equations:

$$\frac{\partial (\rho \vec{u})}{\partial t} + \nabla \cdot (\rho \vec{u} \vec{u}) + \nabla p - \nabla \cdot \mu \left[\nabla \vec{u} + (\nabla \vec{u})^T - \frac{2}{3} (\nabla \cdot \vec{u}) \underline{\underline{I}} \right] = 0 . \tag{4.27}$$

Total enthalpy equation:

$$\frac{\partial (\rho h)}{\partial t} + \nabla \cdot (\rho \vec{u} h) + \nabla \cdot \left(\sum_{k=1}^{K_g} \rho Y_k h_k \vec{V}_k - \lambda \nabla T \right) = 0. \tag{4.28}$$

Gas-phase species equations:

$$\frac{\partial (\rho Y_k)}{\partial t} + \nabla \cdot \rho Y_k (\vec{u} + \vec{V}_k) - \dot{\omega}_k W_k = 0, k = 1, ..., K_g. \tag{4.29}$$

The surface coverage is provided by Equations (4.8) and the gas laws by Equations (4.9). Buoyancy is usually neglected in Equation (4.27), given the small reactor dimensions.

The diffusion velocities $\vec{V}_k$ in Equations (4.29) can be computed using the multidimensional analog of Equation (4.17) for multicomponent transport:

$$\nabla X_k = \sum_{l=1}^{K_g} \frac{X_k X_l}{D_{kl}} (\vec{V}_l - \vec{V}_k) + (Y_k - X_k) \frac{\nabla p}{p} + \sum_{l=1}^{K_g} \frac{X_k X_l}{\rho D_{kl}} \left(\frac{D_l^T}{Y_l} - \frac{D_k^T}{Y_k} \right) \frac{\nabla T}{T}, \tag{4.30}$$

or the multidimensional analog of Equation (4.18) for mixture-average plus thermal diffusion:

$$\vec{V}_k = -D_{km} \nabla \left[ln \left(Y_k \bar{W} / W_k \right) \right] + \left[D_k^T W_k / (\rho Y_k \bar{W}) \right] \nabla (lnT). \tag{4.31}$$

The interfacial gas-phase species boundary conditions become

$$\left[\rho Y_k (\vec{V}_k + \vec{u}_{st}) \right]_+ \cdot \vec{n}_+ = B W_k \dot{s}_k, \tag{4.32}$$

with $\vec{n}_+$ the outward-pointing normal to the catalytic walls, $\vec{u}_{st}$ the Stefan velocity, with magnitude:

$$\left| \vec{u}_{st} \right| \equiv (1/\rho)B \sum_{k=1}^{K_g} W_k \dot{s}_k .$$

(4.33)

The symbol "+" denotes gas properties just above the gas–wall interface. A multidimensional heat transfer for the solid wall heat transfer is also considered, given the rise of significant temperature gradients in the normal to the surface direction during transient microreactor operation [111]:

$$\frac{\partial(\rho_s c_s T_s)}{\partial t} - \nabla \cdot (k_s \nabla T_s) = 0,$$

(4.34)

with interfacial energy boundary condition:

$$\left[-\lambda \nabla T + \vec{q}_{rad} \right]_+ \cdot \vec{n}_+ + \left(k_s \nabla T_s \right)_- \cdot \vec{n}_+ + \sum_{k=1}^{K_g+M} B\, h_k\, \dot{s}_k W_k = 0 .$$

(4.35)

For non-catalytic microreactors, the terms $\dot{s}_k$ in Equations (4.32), (4.33) and (4.35) are zero, except in cases where radical wall recombination reactions are still considered. The Stefan velocity in Equation (4.33) is identically zero for catalytic systems at steady-state operation, since at such conditions there is no net mass deposition or etching of the surface. At transient catalytic operation, whereby the Stefan flux is nonzero, there is some confusion regarding its significance. Since the Stefan flux involves the deposition of a single monolayer at the cata-lyst surface, it is usually neglected. However, its importance changes as a function of time, depending on the associated time scales of surface coverage change. A specific example is provided next. Considering the deposition of atomic oxygen on a bare Pt surface with surface site density $\Gamma = 2.7 \times 10^{-9}$ mol/cm^2, the mass required to achieve a full O(s) coverage is $M_O = \Gamma W_O = 4.3 \times 10^{-8}$ g/cm^2. The Stefan flux needed to attain the deposition of this mass onto the surface is $2\rho_w Y_{w,O2} u_{st} \approx d\theta_o/dt\, M_O$, where the factor of two accounts for the two O(s) in an oxygen molecule and $d\theta_o/dt$ is the rate of change of the oxygen coverage, which depends on the particular transient process (i.e., light-off, oscillatory reactor operation, etc.) and can be monitored experimentally. Considering a gas density above the surface $\rho_w = 10^{-3}$ g/cm^3, an oxygen content $Y_{w,O2} = 0.1$ and an H$_2$/O$_2$ system in oscillatory dynamic operation over a Pt surface (whereby $d\theta_o/dt \sim 100/\text{s}^{-1}$, see [154]), the resulting Stefan velocity is $u_{st} \approx 0.02$ cm/s. Such velocities can be more than one order of magnitude smaller compared to the diffusion velocities $V_{k,y}$ in practical systems, but this largely depends on the specific application; during light-off, Stefan fluxes can reach magnitudes of a few cm/s. The difficulty in solving for the Stefan velocity lies in the fact that most multidimensional models employ the mixture average diffusion transport of Equation (4.31), which cannot conserve total mass as the full multicom-ponent model of Equation (4.30), thus requiring the introduction of a correction velocity [155]. Whenever the correction velocity is of the same order of magnitude as the Stefan velocity, the latter cannot be accurately computed in a CFD code. The use of the multicomponent transport approach removes this difficulty but at a substantial computational cost. Simple calculations of the type presented above should determine whether it is realistic to include the Stefan velocity in a transient catalytic simulation.

Intraphase transport may also be needed in microreactors coated with technical catalysts and having sufficiently thick washcoats and well-dispersed loading within the washcoat volume. High temperatures, in particular, enhance diffusion limitations through the porous catalyst structure. For certain combinations of fuels/catalysts, it has been shown [95, 130] that intraphase diffusion limitations may become important in washcoats as thin as 50 μm and temperatures as low as 700–800 K. There are a few approaches for modeling washcoat processes [94], including general reaction–diffusion models for multidimensional porous structures, 1D (normal to the wall) reaction diffusion models, and the effectiveness factor methodology. By neglecting convection, a set of transient transport reaction–diffusion equations for the surface and gaseous species inside the washcoat can be developed [141]. Spatial discretization of the resulting equations leads to a significant increase in computational cost, such that this approach is not used for transient multidimensional flow models. In these cases, a simpler effectiveness factor model is preferred.

Finally, the quasi-steady assumption for the gas-phase phase and surface chemistry can be invoked, similar to the foregoing discussion of 1D models in Section 4.6.1. In this case the only transient term kept is that in the solid heat conduction Equation (4.34).

4.7.2 BOUNDARY LAYER MODELS

A complete flow model considers the full NS (i.e., elliptic description, with axial diffusion included in Equations (4.27) to (4.29)). As stated in Section 4.6.1, boundary layer (parabolic) models are an option for channel flows with hetero-/homogeneous reactions, at least under certain operating conditions. The presence of heat conduction in the solid (Equation (4.34)) and of surface radiation heat transfer (Equation (4.35)) limits the full advantages of a computationally fast parabolic forward marching solution, since in these processes heat can also flow upstream. It is nonetheless possible to combine a parabolic solver for the gas with an elliptic description for the solid; such an approach has been used for single-channel pure homogeneous combustion simulations [156] and is also commonly used for the "*continuum*" reactor models discussed in Section 4.7.3. The parabolic solver retains its full computational advantages when the surface temperature is prescribed, thus eliminating the need for solid heat conduction and surface radiation modeling. The parabolic equations for the gas will be presented in the application of Section 4.8.3.1. The range of validity for the parabolic solver in catalytic and non-catalytic reactors is discussed next.

For pure catalytic systems (whenever gas-phase chemistry can be safely ignored, as discussed in the context of Figure 4.6), the absence of large gas-phase axial gradients renders the boundary layer approximation suitable, excluding the initial channel section with strong entry effects [99]. For the entry channel-flow problem of Figure 4.1c, investigations—in a fashion similar to the pure hydrodynamic problem—indicate that the key controlling parameter is the magnitude of the inlet Reynolds number or appropriately defined Peclet numbers. Figure 4.13 provides comparisons between 2D elliptic (NS) and parabolic (boundary layer) axial profiles of major and minor species during catalytic combustion of a lean ($\varphi = 0.285$) CH_4/air mixture in a Pt-coated tubular channel having diameter 2 mm, length of 100 mm (out of which the first 10 mm are catalytically inert) [99]. The wall temperature is prescribed $T_w = 1290$ K, the pressure is atmospheric, the inlet temperature is $T_{IN} = 600$ K and the inlet Reynolds number is $Re_{IN} = 200$. In both codes, a detailed heterogeneous mechanism for the oxidation of CH_4 on Pt [99]

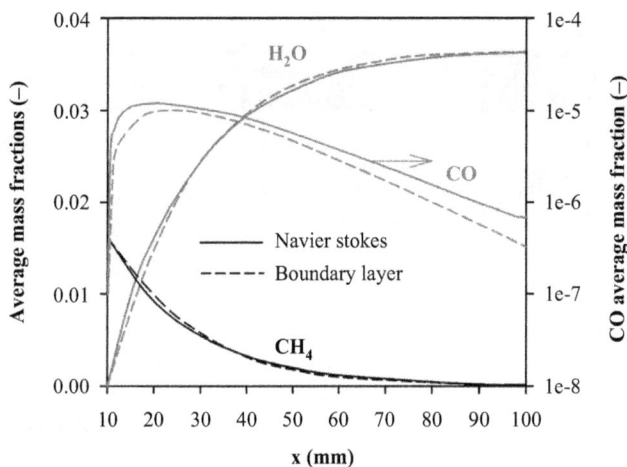

Figure 4.13. Catalytic combustion of a lean ($\varphi = 0.285$) CH_4/air mixture in a Pt-coated tubular catalytic channel (diameter of 2 mm, length of 100 mm out of which the first 10 mm are inert), inlet Reynolds number $Re_{IN} = 200$, constant wall temperature $T_w = 1290$ K and inlet temperature $T_{IN} = 600$ K. Comparison between axial profiles (averaged over the radial direction) of CH_4, H_2O and CO mass fractions, computed with a Navier–Stokes code (solid lines) and a boundary layer code (dashed lines) (reproduced by permission from Raja et al. [99]).

was used, along with the mixture-average transport model (Equation (4.31)). The agreement between both models is very good for the major species (CH_4, H_2O) and adequate for the minor species (CO), and this behavior holds for Re_{IN} as low as 20 for the examined 50:1 length-to-diameter geometry [99]. Entry effects invalidate the boundary layer approximation at very low Reynolds (LR) numbers due to appreciable rise of axial gas-phase gradients. While in catalytic reactors of large gas turbines Re_{IN} typically exceeds 200 [1], thus rendering the parabolic approach valid, this may not be the case for the lower mass throughput microreactors. Considering for example the planar catalytic microreactor in Figure 4.1b with a channel gap of 0.25 mm and an inlet velocity $U_{IN} = 10$ cm/s, the resulting inlet Reynolds number for $T_{IN} = 300$ K is $Re_{IN} \approx 3$. Therefore, care must be exercised when applying parabolic models in catalytic microreactors, as the boundary layer approximation may be invalidated either over the full length or a least over a substantial length of the reactor.

On the other hand, homogeneous combustion poses more severe restrictions for the parabolic model. The large axial gradients across the flame and the rise of appreciable transverse velocities due to the volumetric gaseous heat release can invalidate the boundary layer approximation. Simulations with elliptic and parabolic predictions have delineated the regimes of applicability of the parabolic approach in the presence of a flame [100].

Figure 4.14 compares 2D elliptic and parabolic predictions of the OH radical for CH_4/air atmospheric-pressure catalytic combustion over Pt in a planar mesoscale channel with a height of $2b = 6$ mm and $Re_{IN} = 1450$ (based on the height). In both codes the detailed heterogeneous scheme of Deutschmann et al. [46] and the elementary homogeneous scheme of Warnatz et al. [28] are used, together with a mixture-average diffusion model. The shorter ignition distance of the leaner mixture in Figure 4.14b compared to the richer mixture in Figure 4.14a is due to the well-known self-inhibition characteristics of methane gaseous ignition [157]. The elliptic

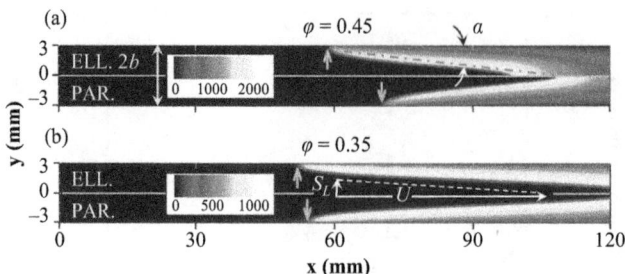

Figure 4.14. Computed two-dimensional distributions of the OH radical in a Pt-coated plane channel with half-height of 3 mm. Combustion of methane/air, $U_{IN} = 5$ m/s, $T_{IN} = 623$ K and $T_w = 1500$ K, $p = 1$ bar. Comparisons of elliptic (Navier–Stokes) and parabolic (boundary layer) predictions for two equivalence ratios: (a) $\varphi = 0.45$ and (b) $\varphi = 0.35$. The arrows indicate the onset of homogeneous ignition and the color bars the OH in ppmv [100].

approach always yields shorter ignition distances due to upstream diffusion of heat and species (Figure 4.14). As the inlet velocity U_{IN} increases, upstream diffusion is suppressed and the differences between elliptic and parabolic predictions are diminished. Finally, for U_{IN} greater than a minimum value $U_{m,IN}$ both computations yield the same homogeneous ignition distance. $U_{m,IN}$ depends mainly on the laminar flame speed S_L of the incoming combustible mixture, since the flow velocity and the propagation speed determine the flame sweep angle α (see Figure 4.14a). Small sweep angles result in flames mainly aligned parallel to the axial direction, thus minimizing axial diffusion. On the other hand, large sweep angles induced by flames with sufficiently large mixture strength and/or low flow velocities invalidate the boundary layer approximation. For atmospheric pressure methane combustion over Pt and for a wide range of operating conditions (1380 K $\leq T_w \leq$ 1600 K, 623 K $\leq T_{IN} \leq$ 743 K) and channel sizes (1.5 mm $\leq b \leq$ 15 mm) the minimum inflow velocity required to satisfy the boundary layer approximation ranges from $U_{m,IN} = 8$ m/s for $\varphi = 0.35$ to $U_{m,IN} = 15.5$ m/s for $\varphi = 0.50$ [100]. It is clarified that in the presence of gaseous combustion, and contrary to the pure catalytic combustion discussed before, the magnitude of the Reynolds number itself is not controlling the applicability of the parabolic model. While Re_{IN} is the key parameter for heat and mass transfer in channels without volumetric combustion [135], the channel height $2b$ does not impact the mechanism of aerodynamic flame stabilization depicted in Figure 4.14.

In natural-gas-fired turbine catalytic reactors, it is fortunate that the flow velocities are large and the laminar flame speeds are low due to the high pressure operation and the negative pressure dependence of the methane laminar flame speed ($S_{L,CH4} \sim p^{-0.5}$). In microreactors, however, the reduced flow rates (resulting to inflow velocities down to a few cm/s) and the moderate pressures (mostly atmospheric and for specific microturbine-based systems up to 5 bar [3]) render the applicability of the boundary layer approximation very questionable in the presence of a flame. For pure homogeneous combustion microreactors, the methodology established for the hetero-/homogeneous systems in Figure 4.14 still holds, the main difference being the lack of upstream fuel depletion due to the catalytic pathway. For the same wall temperature, the required $U_{m,IN}$ increases and thus further limits the applicability of the parabolic model.

In conclusion, although parabolic models are very popular due to their high computational efficiency, even when including large chemical reaction mechanisms, careful consideration is needed before applying them under microreactor-relevant operating conditions.

4.7.3 CONTINUUM REACTOR MODELS

The continuum approach is used for an entire honeycomb structure (Figure 4.1a), whenever a single channel cannot provide a realistic representation of the entire reactor (due to non-uniform inlet properties, non-adiabaticity at the reactor external surfaces, etc.). The approach has been laid down in [115, 116, 158] for transient reactor simulations. The equivalent continuum model requires that the entire structure exhibits geometric similarity and that the individual channels have diameters much smaller than the reactor diameter. The size disparity allows for a formal asymptotic pass from the microscopic to the macroscopic description [116]. Two transient coupled partial differential equations describe the heat transfer in the solid-phase structure and gas-phase, respectively. The quasi-steady assumption for the gas-phase is usually invoked in a manner analogous to that of single-channel modeling discussed in Section 4.6.1. The resulting equation for the solid phase becomes

$$\frac{\partial(\rho_s c_s T)}{\partial t} - \nabla \cdot (k_s \nabla T_s) - \dot{q} = 0 . \tag{4.36}$$

Coupling to the reactive flow fields of the catalytic channels in the monolith is achieved with the use of the heat source term $\dot{q}$, calculated for a representative number of channels, which accounts for the heat transfer between gas-phase and solid by the change in the integral enthalpy flux of the gas:

$$\dot{q} = -\sigma \frac{\partial \dot{H}_{gas}}{\partial x} , \tag{4.37}$$

with σ the channel density. In Equation (4.36) c_s, k_s, and ρ_s are the effective heat capacity, thermal conductivity, and density of the solid, respectively, accounting for the porosity of the material. Variable thermal conductivity in the axial and radial directions has to be considered, depending on the cross-section geometry of the individual channels [159]. Appropriate boundary conditions are provided at the reactor boundaries, for example prescribed temperature or heat flux. The representative channel calculations are typically carried out with a parabolic solver. Nonetheless, the limitations of the boundary layer approximation discussed in Section 4.7.2 must be carefully considered. To allow for computationally efficient 2D channel simulations, each channel is considered to be a tube with diameter equivalent to the hydraulic diameter of the actual channel cross section; however, the solid heat conduction Equation (4.36) can still be solved in 3D. Continuum models are very useful in technical applications with non-uniform reactor conditions; however, they do not resolve the temperature gradients inside the solid walls of each individual channel (and thus the resulting thermal stresses during reactor operation) and require careful assessment of the effective thermal conductivities in each spatial direction of the reactor. The application of the continuum model will be presented in Section 4.8.5.

4.8 APPLICATIONS OF MULTIDIMENSIONAL MODELS

Applications of the multidimensional models described in Section 4.7 are presented next for both catalytic and non-catalytic microreactors. Transient performance during the light-off of fuel-lean methane microreactors coated with Pt will be investigated in Section 4.8.1; a 2D quasi-steady (in the gas) transient model with detailed hetero-/homogeneous chemical reaction schemes is used in this application. Stability maps of catalytic and non-catalytic microreactors are then discussed in Section 4.8.2, using a 2D steady model. Analytical criteria for homogeneous ignition in catalytic and non-catalytic microchannels are presented in Section 4.8.3, based on matched activation energy asymptotics and 2D model description. Flame dynamics in non-catalytic and catalytic channels with prescribed wall temperature are discussed in Section 4.8.4, using fully transient 2D or 3D simulations with detailed hetero-/homogeneous chemistry. Finally, continuum 2D models are applied in Section 4.8.5 for a propane-fueled mesoscale catalytic honeycomb reactor.

4.8.1 LIGHT-OFF OF METHANE-FUELED CATALYTIC CHANNEL MICROREACTORS

The specific requirements for applying the quasi-steady assumption for the flow and chemistry in microchannels with catalytic and gas-phase chemical reactions will be laid down. Subsequently, the dependence of the catalytic ignition (light-off) characteristics on solid material properties, geometry, and in-channel heat transfer processes (see Figure 4.1c) will be elaborated. Computations will be carried out with either heterogeneous chemistry alone or with combined hetero-/homogeneous chemistry.

Hydrocarbon-fueled catalytic microreactors are investigated for portable power generation [160] (electric powers in the range 10–100 W). Applications range from scaled-down thermal engines [161, 162], to catalytic microthrusters for space applications [163] and to microreactors for fuel reforming in micro-SOFCs [164]. Most of the experimental and numerical studies have focused on the steady-state performance of catalytic microburners and microreformers [136, 165], with only limited investigations addressing the transient response and, in particular, the ignition process. Ignition at problem-specific operating conditions is a key design process in all microreactors.

4.8.1.1 NUMERICAL MODEL

A two-dimensional elliptic CFD code [21] has been used to simulate the laminar flow domain in a plane channel configuration, having length $L = 10$ mm, height $2b = 1$ mm, and wall thickness $\delta_s = 50$ μm (see Figure 4.15a). The first 1 mm length is catalytically inert, while the remaining $L_a = 9$ mm is coated with platinum. This geometry mimics single-channel catalytic microreactor geometries with large cross-flow aspect ratios (see Figure 4.1b). Due to symmetry, only half of the channel domain is modeled ($0 \leq y \leq b$, Figure 4.15a). The inlet temperature is $T_{IN} = 850$ K, a value practically achievable in recuperated microreactor thermal cycles driving microturbines [166]. The initial solid wall temperature is uniform and equal to the incoming mixture temperature, $T_s(x, t = 0) = 850$ K. Simulations have been performed at pressures up to 5 bar, inflow velocities U_{IN} of 0.3 to 1.5 m/s and methane/air equivalence ratios of 0.4 and 0.6 [21]. Two types of solid materials have been examined, cordierite (ceramic) and FeCr-alloy (metallic).

Under the quasi-steady assumption for the flow and hetero-/homogeneous chemistry, the general governing equations of Section 4.7.1 in 2D Cartesian coordinates are

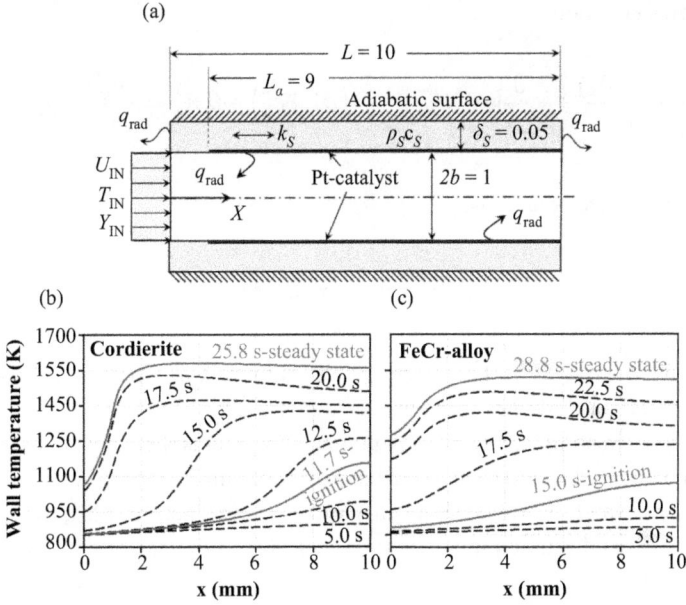

Figure 4.15. Transient simulations in a 2D planar catalytic microchannel, fed with a fuel-lean ($\varphi = 0.4$) methane/air mixture. (a) channel geometry (all dimensions in mm). Axial profiles of wall temperatures at various times during start-up for: (b) cordierite microreactor, and (c) FeCr-alloy microreactor (material properties are given in Table 4.2). Inlet velocity $U_{IN} = 0.3$ m/s, inlet temperature $T_{IN} = 850$ K and pressure $p = 5$ bar (reproduced by permission from Karagiannidis et al. [21]).

Continuity equation:

$$\frac{\partial(\rho u)}{\partial x} + \frac{\partial(\rho v)}{\partial y} = 0. \tag{4.38}$$

Momentum equations:

$$\frac{\partial(\rho uu)}{\partial x} + \frac{\partial(\rho vu)}{\partial y} + \frac{\partial p}{\partial x} - \frac{\partial}{\partial x}\left[2\mu\frac{\partial u}{\partial x} - \frac{2}{3}\mu\left(\frac{\partial u}{\partial x} + \frac{\partial v}{\partial y}\right)\right] - \frac{\partial}{\partial y}\left[\mu\left(\frac{\partial u}{\partial y} + \frac{\partial v}{\partial x}\right)\right] = 0, \tag{4.39}$$

$$\frac{\partial(\rho uv)}{\partial x} + \frac{\partial(\rho vv)}{\partial y} + \frac{\partial p}{\partial y} - \frac{\partial}{\partial x}\left[\mu\left(\frac{\partial v}{\partial x} + \frac{\partial u}{\partial y}\right)\right] - \frac{\partial}{\partial y}\left[2\mu\frac{\partial v}{\partial y} - \frac{2}{3}\mu\left(\frac{\partial u}{\partial x} + \frac{\partial v}{\partial y}\right)\right] = 0. \tag{4.40}$$

Energy equation:

$$\frac{\partial(\rho uh)}{\partial x} + \frac{\partial(\rho vh)}{\partial y} + \frac{\partial}{\partial x}\left(\rho\sum_{k=1}^{Kg}Y_k h_k V_{k,x} - \lambda_g\frac{\partial T}{\partial x}\right) + \frac{\partial}{\partial y}\left(\rho\sum_{k=1}^{Kg}Y_k h_k V_{k,y} - \lambda_g\frac{\partial T}{\partial y}\right) = 0. \tag{4.41}$$

Gas-phase species equations:

$$\frac{\partial(\rho u Y_k)}{\partial x} + \frac{\partial(\rho v Y_k)}{\partial y} + + \frac{\partial}{\partial x}\left(\rho Y_k V_{k,y}\right) + \frac{\partial}{\partial y}\left(\rho Y_k V_{k,y}\right) - \dot{\omega}_k W_k = 0, \ k = 1,..., K_g. \tag{4.42}$$

Surface species coverage equations:

$$\sigma_m \frac{\dot{s}_m}{\Gamma} = 0, \ m = 1,...., M_s. \tag{4.43}$$

The species diffusion velocities $\vec{V}_k$ are computed using mixture average diffusion (Equation (4.31)), and the ideal gas laws are given by Equations (4.9).

As discussed in Section 4.3, key to any numerical model is the use of appropriate catalytic and gas-phase kinetics. The oxidation of methane on platinum is modeled with a detailed heterogeneous scheme [27] and a $C_1/H/O$ elementary gas-phase mechanism [28]; the aptness of these chemical schemes was discussed in Section 4.3. A surface site density $\Gamma = 2.7 \times 10^{-9}$ mol/cm^2 has been used for Pt.

The interfacial boundary conditions for the gas-phase species Equations (4.32), under the quasi-steady assumption (zero Stefan flux) become

$$\left(\rho Y_k V_k\right)_{y=b} + \dot{s}_k W_k = 0, \ k = 1,..., K_g. \tag{4.44}$$

In Equation (4.44) the ratio of the catalytically active area to the geometrical area (factor B in Equation (4.32)) is considered unity. Radiative boundary conditions are finally applied to the vertical front and rear solid wall faces:

$$k_s \frac{\partial T_s}{\partial x}\bigg|_{x=0} = \varepsilon_{\text{eff, IN}}\sigma\left[T_s^4(x=0) - T_{\text{IN}}^4\right], \text{ and}$$

$$-k_s \frac{\partial T_s}{\partial x}\bigg|_{x=L} = \varepsilon_{\text{eff, OUT}}\sigma\left[T_s^4(x=L) - T_{\text{OUT}}^4\right]. \tag{4.45}$$

The effective emissivity ε_{eff} towards the inlet and outlet channel enclosures is

$$\varepsilon_{\text{eff,IN}} = \left[\left(\frac{1}{\varepsilon}\right) + \left(\frac{r_{\text{IN}}}{\varepsilon_{\text{IN}}}\right) - r_{\text{IN}}\right]^{-1} \text{ and } \varepsilon_{\text{eff,OUT}} = \left[\left(\frac{1}{\varepsilon}\right) + \left(\frac{r_{\text{OUT}}}{\varepsilon_{\text{OUT}}}\right) - r_{\text{OUT}}\right]^{-1}, \tag{4.46}$$

with $r_{\text{IN}}, r_{\text{OUT}}$ the ratios of the channel vertical front and rear areas over the areas they subtend at the inlet and outlet, respectively, and ε the emissivity of the catalytic surface. Since the vertical wall faces have a thickness of only $\delta_s = 50$ μm (Figure 4.15a), even a small inlet/outlet channel enclosure area justifies the approximation $\varepsilon_{\text{eff, IN}} = \varepsilon_{\text{eff, OUT}} = \varepsilon$. The formulation in Equations (4.45) and (4.46) is more general than that in Equations (4.19), allowing for gray-body radiation treatment of the inlet and outlet enclosures and accounting also for their specific geometrical arrangements.

The net radiation method for diffuse-gray areas [167] is employed to model radiation exchange between the discretized channel wall elements themselves and between each wall element and the inlet and outlet enclosures; gas radiative emission and adsorption are not considered, given the small optical paths and large nitrogen content. For the k-th channel surface element, the radiation balance becomes

$$\frac{1}{\varepsilon_k} q_k = \sum_{j=1}^{N+2} F_{k-j} \sigma \left(T_k^4 - T_j^4 \right) + \sum_{j=1}^{N+2} \left(\frac{1-\varepsilon_j}{\varepsilon_j} \right) F_{k-j} q_j, \qquad (4.47)$$

with $\varepsilon_k = \varepsilon = 0.6$ for each channel wall element, j running over the N surface elements as well as the inlet ($j = N + 1$) and outlet ($j = N + 2$), and F_{k-j} denoting the geometrical configuration factors between elements k and j.

The outer horizontal wall surfaces are considered adiabatic (see Figure 4.15a); however, the reactor is non-adiabatic due to radiation heat losses originating from the channel inner surface as well as from the vertical front solid wall face toward the colder inlet enclosure. Finally, the dimensionality of the solid wall heat conduction model is dictated by spatiotemporal scale considerations so as to satisfy the quasi-steady assumption and will be discussed next.

4.8.1.2 REQUIREMENTS FOR THE QUASI-STEADY ASSUMPTION

Under the quasi-steady formulation of Section 4.8.1.1, the characteristic time scales for heat conduction in the solid substrate must be longer than the characteristic convective, diffusive, and chemical time scales of the flow inside the channel. This ensures that the gaseous flow and chemistry equilibrate to the imposed, at every time step, solid wall temperature. To facilitate the discussion on characteristic time scales, the properties of the two investigated solid materials, FeCr alloy and cordierite, are provided in Table 4.2.

Characteristic times for axial convection, $t_{g,x} \approx L/U_{IN}$ (not accounting for flow acceleration due to combustion), range between 6 and 33 ms. Transverse gas diffusion time scales, $t_{g,y} \approx b^2/a_g$, computed using gas thermal diffusivities $\alpha_g = \lambda /(\rho c_p)$ at pressures $p = 1$ and 5 bar and temperatures of 850–1850 K (minimum and maximum gas temperatures attained in the microreactor), are in the range $t_{g,y} \sim 0.5$–2.0 ms for $p = 1$ bar and ~ 2.6–10.2 ms for $p = 5$ bar. The corresponding times for solid heat conduction in both directions are $t_{s,x} \sim L^2/a_s$ and $t_{s,y} \sim \delta_s^2/a_s$, with $a_s = k_s/(\rho_s c_s)$ the solid thermal diffusivity. For cordierite, $t_{s,x} \approx 190$ s and $t_{s,y} \approx 4.7$ ms, while for the FeCr alloy $t_{s,x} \approx 27.6$ s and $t_{s,y} \approx 0.7$ ms. Since $t_{s,x} \gg t_{g,x}$ and $t_{s,x} \gg t_{g,y}$ for both materials, the quasi-steady assumption could be safely invoked, as far as the gas-phase transport is concerned. However, since for the FeCr alloy, the transverse solid heat conduction

Table 4.2. Properties for cordierite and FeCr-alloy materials[a]

Material	k_s	ρ_s	c_s	$\rho_s c_s$
Cordierite	2	2600	1464	3806
FeCr alloy	16	7200	615	4428

[a]Thermal conductivity k_s [W/(mK)], density ρ_s [kg/m^3], specific heat capacity c_s [J/(kgK)] and heat capacity $\rho_s c_s$ [kJ/(m^{3}K)].

times $t_{s,y}$ are in most cases much shorter than the calculated times $t_{g,y}$, the time evolution of the temperatures across the 50 μm thick solid wall cannot be resolved. Therefore, a 1D axial energy balance has been employed for the solid wall:

$$\left(\rho_s c_s \frac{\partial T_s}{\partial t} - k_s \frac{\partial^2 T_s}{\partial x^2}\right)\delta_s - \left(\dot{q}_{rad} - \lambda\frac{\partial T}{\partial y}\Big|_{y=b} + \sum_{k=1}^{K_g+M} h_k \dot{s}_k W_k\right) = 0. \tag{4.48}$$

For consistency, and since for both employed materials $t_{s,x} \gg t_{s,y}$, the same 1D approach has been adopted for the cordierite wall, even though for this material $t_{s,y} > t_{g,y}$, at least for $p = 1$ bar. It is thus evident that the mathematical treatment of the solid and the capability to resolve transverse gradients inside the wall depends on the properties of the wall material, the channel geometry, and the flow conditions (residence time and pressure).

Characteristic chemical times for lean CH_4/air catalytic combustion on Pt must also be considered, since they can be quite long before and during light-off, especially if the initial surface and gas temperatures are low. The quasi-steady assumption, although widely used in many catalytic applications, may be invalidated at specific times and/or spatial locations of the reactor, whereby the characteristic catalytic chemical times become longer than the solid heat conduction times. For the examined microturbine-based application, however, the high initial solid temperature (850 K) largely removes such concerns, as shown next. A constant-pressure batch reactor model, with both catalytic and gas-phase reactions, is used to assess characteristic chemical time scales for all reacting gaseous species. The numerical code employed a steady batch reactor model (see Section 4.5). Operating conditions at $p = 1$ bar and $\varphi = 0.40$ pose the most stringent limitations (the catalytic reactivity increases with increasing methane concentration and rising pressure [55]). Moreover, calculations with an initial batch reactor temperature $T_{IN} = 850$ K provide strict upper estimates for the chemical times, as they mimic conditions at $t = 0$ in the channel reactor; at later times, the heat up of the channel wall and gas results in even higher effective inlet temperatures for the batch reactor simulations. For a more complete picture of the temperature effects, the initial batch reactor temperature is varied between 750 and 850 K. A surface-to-volume ratio of $A/V = 20$/cm is used for the batch reactor, which equals the A/V of the channel geometry in Figure 4.15a. With reactor ignition defined as the elapsed time until 50 percent of the incoming fuel is consumed, characteristic chemical time scales for each gas-phase species are defined as

$$\tau_{ch,k} = \frac{1}{t_{ig}}\int_0^{t_{ig}} \frac{[X_k]}{\dot{s}_k(A/V)+\dot{\omega}_k}\,dt. \tag{4.49}$$

For all reacting gas-phase species, computed chemical times are presented in Figure 4.16 (using the batch reactor model with both catalytic and gaseous reactions) for various inlet temperatures. With the characteristic chemical times ranging from 6×10^{-3} ms for minor species to 565 ms (O_2) for an initial reactor temperature of 750 K, a time step $\Delta t \sim 600$ ms is required for the solid in Equation (4.48) to ensure equilibration of all gaseous species to the given surface temperature. Due to the relatively short characteristic axial heat conduction time of the FeCr alloy ($t_{s,x} \sim 27$ s), such a high Δt will poorly describe the time evolution of the wall temperature during the start-up phase. A good choice for microreactor inlet temperature and time step is $T_{IN} = 850$ K and $\Delta t = 50$ ms, respectively; this choice is consistent with the quasi-steady

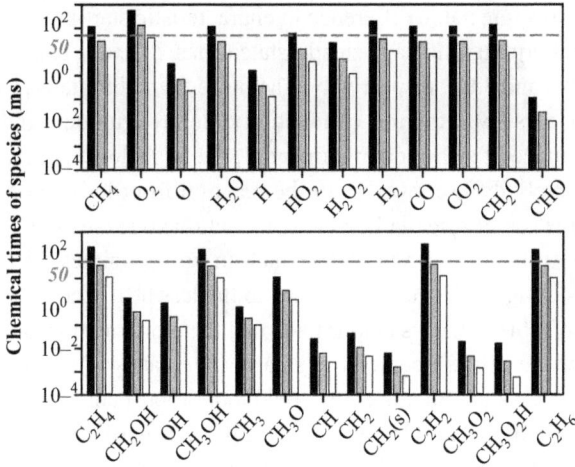

Figure 4.16. Chemical times for gaseous species in fuel-lean ($\varphi = 0.40$) methane/air combustion over Pt, computed in a batch reactor with detailed heterogeneous and homogeneous chemistry. Conditions: $p = 1$ bar, surface-to-volume ratio A/V $= 20$/cm, and initial temperatures: 750 K (black bars), 800 K (gray bars) and 850 K (white bars). The horizontal dashed line at 50 ms defines the time step used for the integration of the solid heat conduction Equation (4.48) (reproduced by permission from Karagiannidis et al. [21]).

assumption and adequately resolves the solid temperature time evolution. Characteristic chemical times for surface species should also be assessed. However, reactions involving only surface species are faster than the adsorption/desorption reactions for most power and chemical processing applications, leading to rapid equilibration of all surface intermediate species. It is emphasized that the analysis in Figure 4.16 is quite strict. As the solid starts heating above the initial temperature $T_s (x, t = 0) = 850$ K, the chemical time scales drop substantially: at 900 K they are already a factor of ~2.2–3.5 shorter than those at 850 K.

4.8.1.3 CATALYTIC MICROREACTOR START-UP

The effects of solid wall properties (Table 4.2), surface radiation heat transfer, and gas-phase chemistry on the ignition characteristics are herein investigated. The last two mechanisms, in particular, are typically not included in microreactor models.

At steady-state operation, the dominant heat transfer mechanism is conduction in the solid walls, directly impacting combustor stability by preheating the incoming fresh mixture. During the heat-up process, the solid heat capacity is also important, since materials with a higher $\rho_s c_s$ need larger energy input to raise their temperature. In Figure 4.15(b, c) simulations with only catalytic chemistry are provided, illustrating the key differences between cordierite and FeCr-alloy channel walls. For a cordierite material, rear-end ignition is attained (Figure 4.15b, $t = 11.7$ s), with the catalytic ignition (light-off) distance decreasing with increasing time until steady state is reached. In contrast, FeCr-alloy reactors (Figure 4.15c) display a spatially more uniform temperature profile. An appreciable difference in characteristic start-up times is observed, with cordierite exhibiting shorter ignition and steady-state times, t_{ig} and t_{st} respectively, compared to the FeCr alloy, by 3.3 s and 3.0 s, respectively. The more favorable start-up times for cordierite are mainly attributed to its lower thermal conductivity. Before ignition, axial heat conduction in the solid is less pronounced for the ceramic material, due to its lower k_s. Heat generated on the surface cannot effectively diffuse away from the reaction front located near the channel exit. This leads to the formation of a spatially confined reaction zone (see in Figure 4.15b the more pronounced hot spot at the reactor rear end), which in turn promotes faster fuel consumption and light-off. The faster light-off is also attributed to the less heat accumulated in the cordierite (see heat capacities in Table 4.2). It is emphasized that the aforementioned observations regarding material behavior are in stark contrast to steady-state catalytic microreactor performance [10]. At steady state, FeCr-alloy microreactors yield more robust combustion (extended stability limits) against imposed external heat losses when compared to cordierite ones. On the other hand, the transient results clearly show the advantages of low-thermal-conductivity ceramic materials in terms of start-up. However, in either steady or transient operation, low thermal conductivity ceramic materials lead to higher wall temperatures (Figure 4.15b). Both factors must be considered when selecting appropriate reactor materials for specific applications. Finally, models that can accurately predict the ignition times are essential for assessing the cumulative microreactor emissions; most emissions originate during the start-up phase, as shown in Figure 4.17.

The significance of in-channel surface radiation heat transfer, which is typically neglected in numerical models of catalytic channel reactors, is investigated next. The importance of radiation exchange in computational models when steep temperature gradients are present inside catalytic channels has been shown in [10, 12, 131]. During ignition and heat-up of microreactors with low-thermal-conductivity (mainly ceramic) walls, significant spatial temperature gradients can be created between the front- and rear-end sections (see, e.g., Figure 4.15b). To isolate the effect of in-channel surface radiation heat transfer between reactor elements on the

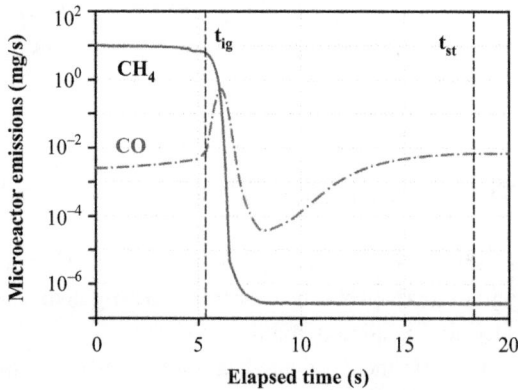

Figure 4.17. Unburned CH_4 and CO emissions at reactor outlet versus elapsed time. Catalytic microreactor configuration of Figure 4.15a (cordierite material), lean ($\varphi = 0.6$) CH_4/air mixture, $U_{IN} = 0.3$ m/s, $T_{IN} = 850$ K, p = 5 bar. The vertical dashed lines denote the ignition (t_{ig}) and steady-state (t_{st}) times (reproduced by permission from Karagiannidis et al. [21]).

ignition and steady-state times, t_{ig} and t_{st}, the case in Figure 4.15b has been recomputed under-adiabatic reactor conditions by suppressing both direct and reflective ($\varepsilon_{IN} = \varepsilon_{OUT} = 0.0$) radiation exchange with the inlet/outlet enclosures and allowing only for in-channel radiation exchange. Surface radiation of the channel walls was subsequently turned on or off by setting the channel surface emissivity either to its nominal value ($\varepsilon = 0.6$) or to zero ($\varepsilon = 0.0$). The corresponding wall temperature profiles at various times during the start-up phase are shown in Figure 4.18. By comparing t_{ig} and t_{st} in Figure 4.18, it is evident that surface radiation plays a dual role during start-up -at least for microreactors with low wall thermal conductivity. At ignition, radiation transfers heat away from the reaction zone (hot spot), reducing the wall temperature and thus elongating t_{ig}. Following ignition, however, radiation facilitates the faster redistribution and upstream transfer of heat inside the channel thus reducing t_{st}. Thus, for $t_{ig} < t < t_{st}$, radiation becomes a major upstream heat transfer mechanism in the channel, enhancing the upstream propagation of the reaction front. Characteristically, it was shown [21] that the ignition and steady-state times of a cordierite microreactor ($k_s = 2.0$ W/(mK)) with surface emissivity of $\varepsilon = 0.6$ were roughly the same with those of a $k_s = 5.5$ W/(mK) microreactor without radiation exchange ($\varepsilon = 0$).

Gas-phase reactions are usually neglected in numerical investigations—steady or transient—of catalytic microreactors. Moreover, the premise is that gas-phase chemistry does not affect the start-up of catalytic reactors. Recent studies have pointed out to the importance of homogeneous chemistry in enhancing steady-state combustion stability limits, especially at moderate pressures ($p = 5$ bar) [10], an issue that will be further addressed in Section 4.8.2. This enhancement of combustion stability is not always a result of total oxidation gaseous reactions. An important coupled hetero-/homogeneous reaction route is the incomplete oxidation of methane to CO via gas-phase reactions, followed by the main exothermic oxidation of the formed CO to CO_2 not via homogeneous but via catalytic reactions. To assess the impact of gaseous chemistry, the condition in Figure 4.15b has been recomputed with inclusion of

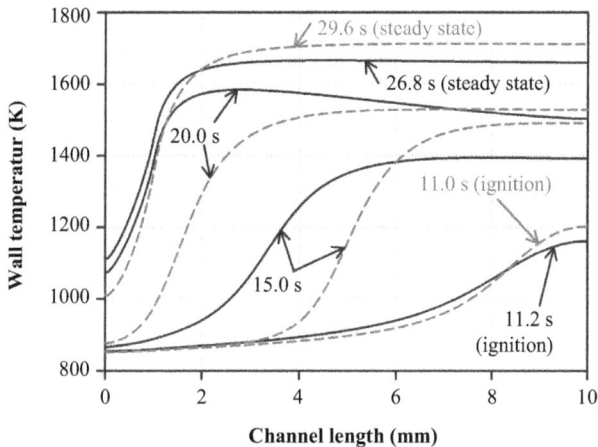

Figure 4.18. Computed wall temperature profiles at different times for the conditions in Figure 4.15b and cordierite wall material, but with different radiation treatment: $\varepsilon = \varepsilon_{IN} = \varepsilon_{OUT} = 0.0$ (dashed lines), and $\varepsilon = 0.6$, $\varepsilon_{IN} = \varepsilon_{OUT} = 0.0$ (solid lines) (reproduced by permission from Karagiannidis et al. [21]).

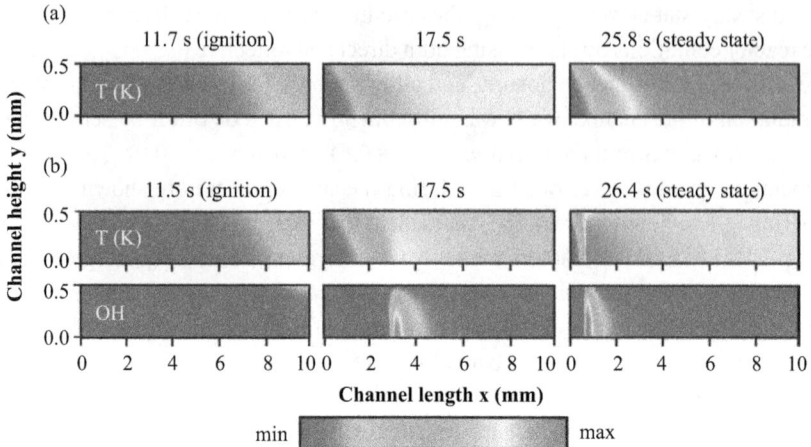

Figure 4.19. (a) Computed 2D distributions of temperature at three time instances during reactor start-up, for the geometry in Figure 4.15a and the conditions in Figure 4.15b (catalytic chemistry only); $y = 0$ is the symmetry plane and $y = 0.5$ mm the gas–wall interface. (b) Computed 2D distributions of temperature and OH mass fraction at three time instances during reactor start-up, same conditions as in (a) but with inclusion of combined hetero-/homogeneous chemistry. The scaling in the color bar corresponds to the ranges: for temperature 850–1708 K; for OH mass fraction $0.0–5.9 \times 10^{-7}$ ($t = 11.5$ s), $0.0–2.6 \times 10^{-4}$ ($t = 17.5$ s) and $0.0–7.6 \times 10^{-4}$ ($t = 26.4$ s) [21].

combined hetero-/homogeneous chemistry. In Figure 4.19a, 2D distributions of the temperature are shown at three times (catalytic chemistry only); in Figure 4.19b the temperature and OH mass fractions are shown for the same case when both catalytic and gas-phase reactions are included. Gas-phase chemistry appreciably affects the steady-state times by contributing to the heat generation inside the channel and altering the spatial extend of the reaction zone over which fuel is consumed. On the other hand, gas-phase reactions have only a small impact on t_{ig}, as shown by comparing the ignition times in Figure 4.19(a, b).

Following catalytic ignition, the contribution of the homogeneous reaction pathway increases substantially, with flames sustained in the channel by the exothermicity of the catalytic reactions, as illustrated in Figure 4.19b by the OH maps at the latest two times. With a fraction of the heat release zone now shifted from the surface to the gas, the solid substrate is partly heated via convection and partly via direct heat generation on the surface. At $t \sim 17.5$ s the maximum temperature location shifts from the channel wall to the gas phase. This shift is manifested by a high-temperature zone in the gas, as shown in Figure 4.19b at $t = 17.5$ s. The maximum reactor temperatures are subsequently maintained in the gas phase till steady state is reached (Figure 4.19b at $t = 26.4$ s). As the thermal conductivity of the hot gases is two orders of magnitude lower than the thermal conductivity of cordierite, both heat accumulation in the wall and upstream propagation of the high-temperature front zone of the solid are hindered by the additional thermal resistance of the gas phase, resulting in elongated characteristic steady-state times. The lower wall temperatures at $t = 17.5$ s in Figure 4.19b compared to those in Figure 4.19a attest to this phenomenon. Regardless of reactor material and operating conditions, the role of homogeneous chemistry in the start-up of methane-fueled catalytic microreactors is to elongate the steady-state times.

4.8.2 STABILITY MAPS OF CATALYTIC AND NON-CATALYTIC MICROREACTORS

The hetero-/homogeneous combustion and the stability limits of methane and propane-fueled Pt-coated catalytic microreactors have been investigated numerically in a plane channel with 1 mm height, 10 mm length, and 0.1 mm solid wall thickness [10, 22], in the same basic configuration as Figure 4.15a. Simulations were carried out using the 2D steady gas-phase model of Section 4.8.1.1 with $\varepsilon = \varepsilon_{IN} = \varepsilon_{OUT} = 0.6$. Moreover, a 2D steady model was used for solid heat conduction (see Equations (4.34) and (4.35)); the limitation for 1D transient solid heat conduction modeling discussed in the foregoing quasi-steady description of Section 4.8.1.2 is removed in the present steady formulation. Validated detailed heterogeneous and elementary homogeneous chemical reaction schemes were used for methane [27, 28] (see Section 4.3.1). For propane a pressure-corrected, one-step catalytic reaction was employed (see Section 4.3.1 and Figure 4.4) along with the elementary homogeneous mechanism of Qin et al. [168], both validated in [52]. It is noted that the combination of a properly defined single-step catalytic chemistry with an elementary gaseous reaction mechanism suffices for many applications. The reason is that major hetero-/homogeneous chemistry interactions arise from the near-wall catalytic fuel depletion and the presence of heterogeneously produced major combustion species (see Figure 4.1c), with a weak radical coupling between the two chemical pathways [17, 54, 55]. Thus, a single catalytic step when coupled to a detailed gas-phase reaction scheme can capture the catalytic reactant depletion and the onset of homogeneous ignition; this was successfully demonstrated by comparing *in situ* spatially resolved measurements of major (C_3H_8 and H_2O) and minor (OH) species with predictions in lean propane/air hetero-/homogeneous combustion [52].

External heat losses $\alpha_T(T_w - T_\infty)$ were imposed at the outer wall surfaces with a heat-transfer coefficient α_T and $T_\infty = 298$ K (see Figure 4.1c). Parametric studies were carried out by varying the solid thermal conductivity k_s, α_T, and the inlet velocity U_{IN}. Pressures of 1 and 5 bar were examined (the latter being of interest in microturbine-based microreactors [3]) and finally the inlet temperature and equivalence ratio were $T_{IN} = 700$ K and $\varphi = 0.40$, respectively, for both examined fuels. Figure 4.20 depicts stability maps for $p = 1$ and 5 bar and $k_s = 2$ W/(mK). The critical external heat transfer coefficient α_T is plotted as a function of inlet velocity; in the areas outside the bell-shaped curves, combustion cannot be sustained. Stability limits at large U_{IN} define blowouts due to insufficient residence times, whereas at low U_{IN} they pertain to heat-loss-induced extinction. Stability limits at $p = 1$ and 5 bar should be compared at the same mass throughput; thus two different velocity scales are provided in Figure 4.20. Irrespective of fuel, for the same mass throughput the stability limits at $p = 5$ bar are much wider than those at $p = 1$ bar. The reason is that both catalytic and gas-phase reactivities increase with rising pressure (the catalytic reactivity scales as $\sim p^{+0.47}$ for methane [55] and $\sim p^{+0.75}$ for propane [52], while the homogeneous reactivities scale roughly as $\sim p^{+1.0}$ for methane and $\sim p^{+1.75}$ for propane [169]).

Despite the higher catalytic and gas-phase reactivities of propane, the blow-out stability limits of methane are wider, particularly at 5 bar. This counterintuitive behavior is due to the slower transport of propane. At high pressures and sufficiently high mass throughputs, the catalytic fuel conversion is not kinetically-controlled but transport-controlled, leading to slower heterogeneous conversion for the larger Lewis number propane fuel [22]. On the other hand, at low mass throughputs the external heat losses become important for combustion stability and the low inlet velocity limits are dictated by catalytic kinetics. These stability characteristics

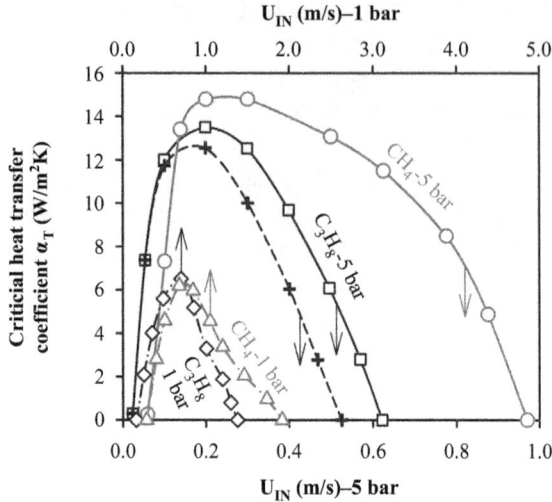

Figure 4.20. Stability maps in terms of inlet velocity U_{IN} versus critical heat transfer coefficient α_T for the catalytic microchannel geometry in Figure 4.15a, fuel-to-air equivalence ratio $\varphi = 0.40$, $T_{IN} = 700$ K and solid wall thermal conductivity $k_s = 2$W/mK. C_3H_8: squares (5 bar) and diamonds (1 bar). CH_4: circles (5 bar) and triangles (1 bar). Crosses/dashed line: C_3H_8 stability curve at 5 bar without including gaseous chemistry. Mass throughput ($\rho_{IN}U_{IN}$) is constant along the y-axis (reproduced by permission from Karagiannidis et al. [22]).

have to be carefully considered when using in catalytic microreactors heavier hydrocarbons (diesel and benzene) with even lower transport coefficients.

Computations with and without inclusion of gaseous chemistry indicate that the latter cannot be ignored at microreactor-relevant confinements and operating conditions (compare the open squares and crosses for C_3H_8 at 5 bar in Figure 4.20). The gaseous pathway extends appreciably the blowout stability limits (high inlet velocity branch) and moderately the extinction limits (low inlet velocity branch). This extension is mainly attributed not to the presence of a vigorous flame but to the incomplete gas-phase oxidation of the fuel to CO, as also discussed in Section 4.8.1.3. The impact of gaseous chemistry is qualitatively the same for both fuels; however, its effect is more pronounced for propane. The stability maps constructed in Figure 4.20 have a clear quantitative trait; the use of validated (detailed in most cases) chemical reaction mechanisms along with full elliptic 2D model for the solid and gas renders results such as those in Figure 4.20 valuable for assessing steady reactor performance and thus greatly facilitating reactor design.

Stability studies in methane and propane-fueled non-catalytic (pure homogeneous) microreactors have been performed in [88, 170], with a similar 2D elliptic model as the foregoing catalytic studies, however, using a one-step gaseous chemical reactions and without inclusion of surface radiation. The critical heat transfer coefficient is plotted in Figure 4.21 as a function of the wall thermal conductivity k_s, for a plane channel with height $2b = 0.6$ mm, wall thickness $\delta_s = 0.2$ mm and length $L = 10$ mm. The pressure is atmospheric, the inlet velocity $U_{IN} = 0.5$ m/s, the inlet and ambient temperatures $T_{IN} = T_\infty = 300$ K, and the fuel/air mixtures are stoichiometric. Contrary to the hetero-/homogenous stability maps in Figure 4.20, in pure homogeneous combustion, propane has much wider stability limits compared to methane.

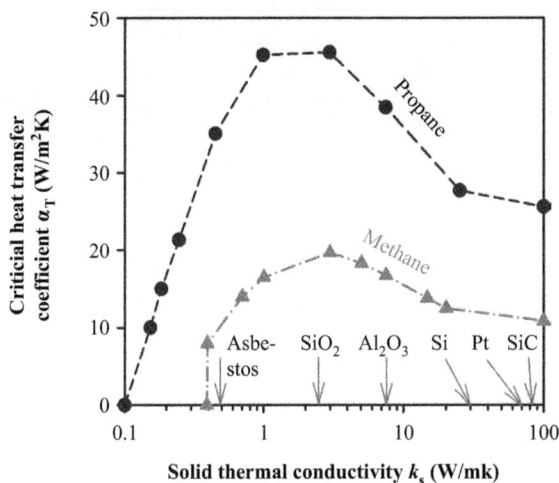

Figure 4.21. Stability maps in terms of solid wall thermal conductivity k_s versus critical external heat loss coefficient a_T for a non-catalytic microchannel, with stoichiometric fuel-to-air mixtures. Other operating parameters are $p = 1$ bar, $U_{IN} = 0.5$ m/s and $T_{IN} = 300$ K. Ceramics allow maximum external heat loss coefficients. Materials with lower k_s limit the upstream heat transfer while materials with higher k_s result in enhanced heat transfer to the surroundings (reproduced by permission from Norton et al. [170]).

Moreover, there is an optimum $k_s \approx 2$ W/(mK) for which the stability limits are wider for both fuels at the examined $U_{IN} = 0.5$ m/s. Lower k_s reduce upstream heat transfer while higher k_s lead to a spread of the high-temperature zone over larger axial extent; both effects narrow down the stability envelope, thus favoring intermediate values of k_s. Although direct comparison between Figures 4.20 and 4.21 cannot be made at $U_{IN} = 0.5$ m/s, $p = 1$ bar and $k_s = 2$ W/(mK) (due to the very different equivalence ratios, inlet temperatures, and geometry), it appears that the non-catalytic microreactors, particularly the propane-fueled, tolerate higher external heat losses. For a catalytic system, an external heat loss can be more detrimental to combustion stability due to the distributed heterogeneous reaction over the entire channel length. However, in homogeneous combustion systems, the reaction is concentrated in a narrow axial extent and the flame can be more resilient to surface heat losses [10]. In any case, the comparison of catalytic and non-catalytic microreactors under similar operating conditions and with detailed chemistry models is required to assess the operating advantages of each particular system. Moreover, transient start-up ignition considerations have also to be addressed as in Section 4.8.1.3.

4.8.3. ANALYTICAL STUDIES IN CATALYTIC AND NON-CATALYTIC MICROREACTORS

Matched activation energy asymptotics are powerful methods in providing approximate solutions that identify the key parameters controlling the onset of catalytic and gas-phase combustion in microreactors. Contrary to 1D analytical models described in Section 4.6.3, the 2D formulation presented here allows the proper description of hetero-/homogeneous combustion phenomena that exhibit strong boundary layer effects.

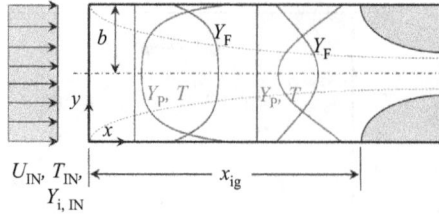

Figure 4.22. Planar channel geometry for analytical studies with hetero-/homogeneous combustion. Transverse profiles of the limiting reactant Y_F, the product Y_P, and the temperature T. The gas-phase ignition distance is denoted by x_{ig}.

4.8.3.1 Mathematical Formulation

The catalytic and gas-phase ignition of a fuel-lean premixed combustible gas is investigated [14, 92] in a developing laminar plane channel-flow established between two horizontal catalytic plates placed at a distance $2b$ apart (Figure 4.22). The incoming properties are uniform and the catalytic plates are maintained at a fixed temperature T_w. A single-step irreversible Arrhenius reaction is considered for both catalytic and gaseous pathways:

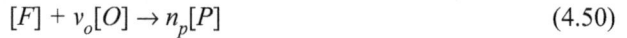

$$[F] + v_o[O] \rightarrow n_p[P] \tag{4.50}$$

To facilitate the ensuing analysis, a catalytic reaction first order with respect to the fuel and zero-order with respect to the oxidizer is considered. This is realistic, as most hydrocarbon fuels at fuel-lean stoichiometries exhibit first-order global kinetics on noble metals [101]. The mass-based catalytic surface reaction rate for the fuel, $\dot{s}_F$, can be written as

$$\dot{s}_F = W_F B_s \left(\frac{\rho Y_F}{W_F} \right)_w \exp\left(-E_s/RT_w \right). \tag{4.51}$$

A large activation energy gaseous reaction is considered with no limitations on its specific reaction orders. The mass-based gaseous fuel reaction rate, $\dot{\omega}_F$, becomes

$$\dot{\omega}_F = W_F B_g T^\gamma \left(\frac{\rho Y_F}{W_F} \right)^{n_F} \left(\frac{\rho Y_O}{W_O} \right)^{n_O} \exp(-E_g/RT). \tag{4.52}$$

Under the boundary layer approximation, the governing equations for the conservation of mass, momentum, energy, and species, become

$$\frac{\partial(\rho u)}{\partial x} + \frac{\partial(\rho v)}{\partial y} = 0, \tag{4.53}$$

$$\rho u \frac{\partial u}{\partial x} + \rho v \frac{\partial u}{\partial y} - \frac{\partial}{\partial y}\left(\mu \frac{\partial u}{\partial y} \right) = -\frac{dp}{dx}, \tag{4.54}$$

$$\rho u c_p \frac{\partial T}{\partial x} + \rho v c_p \frac{\partial T}{\partial y} - \frac{\partial}{\partial y}\left(\lambda \frac{\partial T}{\partial y} \right) = q\dot{\omega}_F, \tag{4.55}$$

$$\rho u \frac{\partial Y_i}{\partial x} + \rho v \frac{\partial Y_i}{\partial y} - \frac{\partial}{\partial y}\left(\rho D_i \frac{\partial Y_i}{\partial y}\right) = -\dot{\omega}_i, \tag{4.56}$$

respectively, with $\partial p/\partial y = 0$. Finally, the ideal gas law is used:

$$p = \rho \frac{R^o}{\overline{W}} T. \tag{4.57}$$

Equations (4.53)–(4.56) constitute a parabolic system with initial conditions ($x = 0$):

$$u = u_{IN}, v = 0, \quad T = T_{IN}, p = p_{IN}, \text{ and } Y_i = Y_{i,IN}. \tag{4.58}$$

The boundary conditions at the gas–wall interface ($x > 0, y = 0$) are

$$u = v = 0, T = T_w \quad \text{and} \quad \left(\rho D_i \frac{\partial Y_i}{\partial y}\right)_w = \dot{s}_i. \tag{4.59}$$

At the plane of symmetry ($x > 0, y = b$):

$$\partial u/\partial y = v = 0 \text{ and } \partial T/\partial y = \partial Y_i/\partial y = 0. \tag{4.60}$$

Non-dimensional streamwise and transverse coordinates are defined as

$$\zeta = x/(b\, Re_{IN}\, Pr) \text{ and } \eta = \frac{1}{b\rho_{IN}\sqrt{2Pr\zeta}} \int_0^y \rho\, dy, \tag{4.61}$$

with Re_{IN} the inlet Reynolds number based on the channel half-height:

$$Re_{IN} = \frac{\rho_{IN} U_{IN}\, b}{\mu_{IN}}. \tag{4.62}$$

Introducing the stream function $\psi (x, y)$ and the non-dimensional dependent variables

$$f = \frac{\psi}{b\rho_{IN} U_{IN}\sqrt{2\,Pr\zeta}}, \; p^* = \frac{p}{\rho_{IN}U_{IN}^2}, \; \theta = \frac{T_W - T}{T_W - T_{IN}}, \; \tilde{Y}_i = \left(\frac{W_F}{v_i W_i}\right)Y_i, \tag{4.63}$$

and further assuming equal species diffusivities $D_i = D$, constant values for $\rho\mu$, $\rho\lambda$, $\rho^2 D$ and c_p (resulting in constant Prandtl and Schmidt numbers), Equations (4.53)–(4.57) become

$$\frac{\partial^3 f}{\partial\eta^3} + f\frac{\partial^2 f}{\partial\eta^2} + 2\zeta\left(\frac{\partial f}{\partial\zeta}\right)\frac{\partial^2 f}{\partial\eta^2} - 2\zeta\left(\frac{\partial f}{\partial\eta}\right)\frac{\partial^2 f}{\partial\eta\partial\zeta} = 2\zeta\left(\frac{\rho_{IN}}{\rho}\right)\left(\frac{dp^*}{d\zeta}\right) \tag{4.64}$$

$$Pr^{-1}\frac{\partial^2\theta}{\partial\eta^2} + f\frac{\partial\theta}{\partial\eta} + 2\zeta\left(\frac{\partial f}{\partial\zeta}\right)\frac{\partial\theta}{\partial\eta} - 2\zeta\left(\frac{\partial f}{\partial\eta}\right)\frac{\partial\theta}{\partial\zeta} = 2Pr\zeta\frac{Re_{IN}bq}{U_{IN}c_p(T_W - T_{IN})}\left(\frac{\dot{\omega}_F}{\rho}\right) \tag{4.65}$$

$$Sc^{-1}\frac{\partial^2 \tilde{Y}_i}{\partial \eta^2} + f\frac{\partial \tilde{Y}_i}{\partial \eta} + 2\zeta\left(\frac{\partial f}{\partial \zeta}\right)\frac{\partial \tilde{Y}_i}{\partial \eta} - 2\zeta\left(\frac{\partial f}{\partial \eta}\right)\frac{\partial \tilde{Y}_i}{\partial \zeta} = 2Pr\zeta\frac{Re_{IN} b}{U_{IN}}\left(\frac{\dot{\omega}_F}{\rho}\right) \tag{4.66}$$

$$p = \rho\frac{R^o}{\overline{W}}\left[T_W - \theta(T_w - T_{IN})\right]. \tag{4.67}$$

The initial conditions ($\zeta = 0$) are

$$f' = 1,\ \theta = 1,\ p^* = p^*_{IN},\ \frac{1}{\tilde{Y}_{i,IN}}\tilde{Y}_i = 1. \tag{4.68}$$

The interfacial ($\zeta > 0$, $\eta = 0$) boundary conditions in Equations (4.59) become

$$f = 0,\ f' = 0,\ \theta = 0,\ \text{and}\ \frac{1}{\tilde{Y}_{F,IN}}\left(\frac{\partial \tilde{Y}_F}{\partial \eta}\right)_W = Le\sqrt{2Pr\zeta}\left(\frac{\rho_W}{\rho_{IN}}\right)\left(\frac{\tilde{Y}_{F,w}}{\tilde{Y}_{F,IN}}\right)Da_s, \tag{4.69}$$

with Da_s a characteristic surface (catalytic) Damköhler number,

$$Da_s = \frac{bB_s \exp(-E_s/RT_w)}{\alpha_{th,IN}}. \tag{4.70}$$

Given the equal diffusivity assumption, the interfacial gas-wall boundary conditions for species other than fuel are

$$\left(\frac{\partial \tilde{Y}_i}{\partial \eta}\right)_w = (-1)^r\left(\frac{\partial \tilde{Y}_F}{\partial \eta}\right)_w \tag{4.71}$$

with $r = 0$ for the oxidizer and $r = 1$ for the products. Finally, at the plane of symmetry ($\zeta > 0$, $\eta = \eta^*$):

$$f'' = 0,\ \theta' = 0,\ \tilde{Y}_i' = 0. \tag{4.72}$$

Equations (4.64)–(4.72) are used to obtain a closed-formed parametric solution for the gas-phase chemically frozen state (pure catalytic combustion), to formulate the weakly gas-phase reactive state and, finally, to determine via asymptotic matching of the inner and outer solutions the criticality equation for the onset of homogeneous ignition. The details are provided in [14, 92] and the physical significance of the derived solutions is discussed next.

4.8.3.2 Homogeneous Ignition Criteria

Homogeneous ignition criteria have been derived for infinitely fast and finite-rate surface chemistry. In the former case, the catalytic reactions are mass-transport-limited ($Da_s \rightarrow \infty$) and the closed-form homogeneous ignition criterion becomes [92]

$$\zeta_{ig} \frac{F(\zeta_{ig})}{\Delta_{cr}^*(\zeta_{ig})} = \frac{1}{2Pr} A \frac{1}{Da_g},$$ (4.73)

with ζ_{ig} the non-dimensional homogeneous ignition distance, Da_g the gas-phase Damköhler number defined as the ratio of the characteristic transverse diffusion time scale, τ_d, to the gas-phase chemical time $\tau_{ch,g}$:

$$Da_g = \tau_d / \tau_{ch,g}$$ (4.74)

$$\tau_d = b^2 / \alpha_{th,IN}$$ (4.75)

$\tau_{ch,g} =$

$$\left[B_g (p\bar{W}/R)^{n-1} W_F^{1-n_F} W_O^{-n_O} T_w^{1+\gamma-n} \left(Y_{O,IN} - \frac{W_O v_O}{W_F} Y_{F,IN}\right)^{n_O} Y_{F,IN}^{n_F} \exp(-E_g/RT_w) \right]^{-1}.$$ (4.76)

with $n = n_F + n_O$. The parameter A in Equation (4.73) is

$$A = \frac{(T_w - T_{IN})^{2+n_F}}{T_w^{2(1+n_F)}(R/E_g)^{1+n_F}(q/c_p)}.$$ (4.77)

$F(\zeta)$ is a monotonically increasing function of ζ, proportional to the inverse of the non-dimensional transverse gradient of the temperature at the wall, and Δ_{cr}^* is the critical ignition Damköhler number, obtained from the solution of an appropriate differential equation for the temperature perturbation [92].

The homogeneous ignition criterion of Equation (4.73) contains all the relevant chemical, flow, transport, and geometrical parameters of the channel-flow configuration. The left side of Equation (4.73) is a monotonically increasing function of ζ, given the corresponding monotonicity of $F(\zeta)$ and the particularly weak dependence of Δ_{cr}^* on ζ [92]. The competition between catalytic and gaseous pathways is clearly demonstrated through the increase of the non-dimensional ignition distance ζ_{ig} with decreasing Da_g. Short transverse diffusion times (narrow channels with small half-heights b, or equivalently channels with large surface-to-volume ratio) lead to increased transverse transport rates and hence to enhanced catalytic fuel depletion rates. The catalytic pathway converts fuel at the expense of the gaseous pathway, leading to inhibition of homogeneous ignition. In the limit $\zeta_{ig} \to 0$, it can be shown [92] that Equation (4.73) reduces to the homogeneous ignition criterion developed for flow over a flat catalytic plate [171].

In the case of finite rate surface chemistry, the homogeneous ignition criterion in Equation (4.73) becomes [14]

$$\zeta_{ig} \frac{F(\zeta_{ig})}{\Delta_{cr}^*(\zeta_{ig})} = \frac{1}{2Pr} A \frac{1}{Da_g} K(\zeta_{ig}, Da_s).$$ (4.78)

It can be shown [14] that $K(\zeta_{ig}, Da_s) \leq 1$ and $K(\zeta_{ig}, Da_s) \rightarrow 1$ as $Da_s \rightarrow \infty$, such that the criterion of Equation (4.73) is recovered at the mass-transport-limit. In addition, $K(\zeta_{ig}, Da_s)$ is a monotonically increasing function of Da_s. For $Da_s = 0$, a homogeneous ignition criterion for non-catalytic channels is derived. For direct comparisons with the mass-transport-limited solution of Equation (4.73), an effective transverse diffusion time scale is defined:

$$\tau_{d,\text{eff}} = \tau_d / K(\zeta_{ig}, Da_s), \qquad (4.79)$$

such that $\tau_{d,\text{eff}} \geq \tau_d$. The homogeneous ignition criterion in Equation (4.78) can be recast to

$$\zeta_{ig} \frac{F(\zeta_{ig})}{\Delta_{cr}^*(\zeta_{ig})} = \frac{1}{2Pr} A \frac{1}{Da_{g,\text{eff}}}, \qquad (4.80)$$

with $Da_{g,\text{eff}} = \tau_{d,\text{eff}}/\tau_{ch,g}$. The inspection of Equations (4.73) and (4.80) indicates that finite rate surface kinetics reduce the transverse transport rates by increasing the effective transverse diffusion time scale and, as a consequence, promote the onset of homogeneous ignition. An application of the ignition criterion of Equation (4.80) for fuel-lean ($\varphi = 0.32$) propane/air combustion with a gaseous reaction rate $B_g[C_3H_8] \exp(-E_g/RT)$, $E_g = 167.5$ kJ/mol, $B_g = 3.4 \times 10^{10}$/sec and varying surface kinetics in order to control Da_s is illustrated in Figure 4.23. The dependence of the homogeneous ignition distance ζ_{ig} on Da_g is presented parametrically for three Da_s and two wall temperatures, for various channel half-heights b. In addition, the catalytic ignition Damköhler numbers $Da_{s,ig}$ [14] are also provided on the horizontal axis. The promotion of homogeneous ignition with decreasing catalytic reactivity (smaller Da_s) and decreasing flow confinement (larger channel half-heights b) is evident. Homogeneous ignition is also strongly favored by higher wall temperatures due to the Arrhenius exponential dependence of the chemical time $\tau_{ch,g}$ (Equation (4.76)). The analytical predictions from Equation (4.80) are in good agreement with CFD numerical predictions (see Figure 4.23) that employ the boundary layer approximation and the same one-step chemistry description as the analytical solutions.

The sensitivity of the homogeneous ignition distance on both reaction pathways is demonstrated in Figure 4.24, which provides lines of constant ζ_{ig}. There are infinite combinations of gaseous and catalytic reactivities, Da_g and Da_s, which yield the same $Da_{g,\text{eff}}$ and hence the same ζ_{ig}. When a slow surface reaction (i.e., more pronounced near-wall fuel excess) is coupled to a slow gaseous reaction, it can produce the same ζ_{ig} to that of a faster surface reaction (reduced near-wall fuel excess) combined with a faster gaseous reaction. The existence of multiple combinations of gaseous and surface reactivities producing the same ζ_{ig} complicates the experimental validation of hetero-/homogeneous reaction schemes, which was discussed in Sections 4.3.1 and 4.3.2; to avoid such complications, simultaneous monitoring of the catalytic and gaseous reactivities with Raman and LIF is necessary.

4.8.3.3 Flame Propagation in Non-Catalytic Channels

The criteria in Section 4.8.3.2 provided analytical solutions based on 2D matched activation energy asymptotics, for the ignition distances in catalytic and non-catalytic channels with isothermal walls. In terms of flame propagation speeds, extension of the 1D steady simplified flame sheet models of Section 4.6.3 to 2D steady models with the addition of 2D steady heat

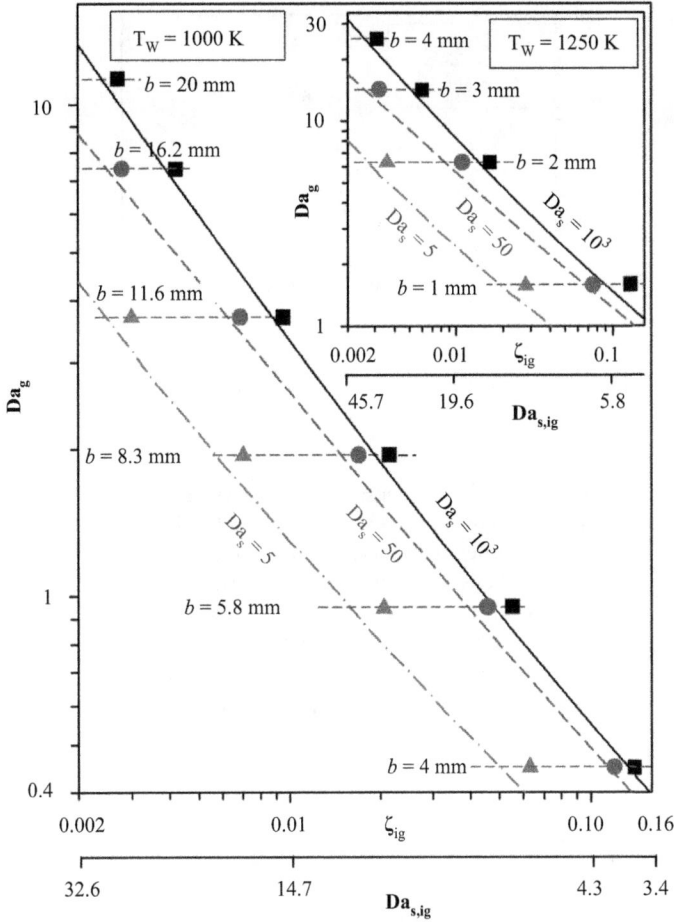

Figure 4.23. Homogeneous ignition distances ζ_{ig} versus gaseous Damköhler numbers Da_g for propane catalytic combustion. The solid lines are analytical solutions from Equation (4.80) and the symbols are numerical predictions. Parametric plots are shown for three surface Damköhler numbers Da_s. The inlet temperature is $T_{IN} = 600$ K and the wall temperature $T_w = 1000$ K ($T_w = 1250$ K in the inset). The catalytic ignition Damköhler numbers $Da_{s,ig}$ are also shown on the horizontal axes (reproduced by permission from Mantzaras et al. [14]).

conduction in the solid wall (conjugate heat transfer) has been performed in [172]. Key assumption is the consideration of planar and vertical flames in the channel (see the inset in Figure 4.25) and either a plug flow or fully developed velocity channel profile. The advantage of the 2D model is the removal of the ad-hoc Nusselt number considered in Section 4.6.3. The consideration of planar flame fronts allows for direct implementation of the jump conditions across the flame, in a fashion similar to that discussed in Section 4.6.3 (see Equations (4.25)). The burning velocity eigenvalue is then computed as a function of the external heat loss, defined in terms of a Nusselt number $Nu_E = \alpha_T d/\lambda$ and the ratio of the solid-to-gas thermal conductivity $\kappa = k_s/\lambda$. The nondimensional propagation speed S_L^* becomes [172]

$$S_L^* = \left[1 + \gamma(\theta_f - 1)\right]^2 \exp\frac{\beta(\theta_f - 1)}{2\left[1 + \gamma(\theta_f - 1)\right]}, \tag{4.81}$$

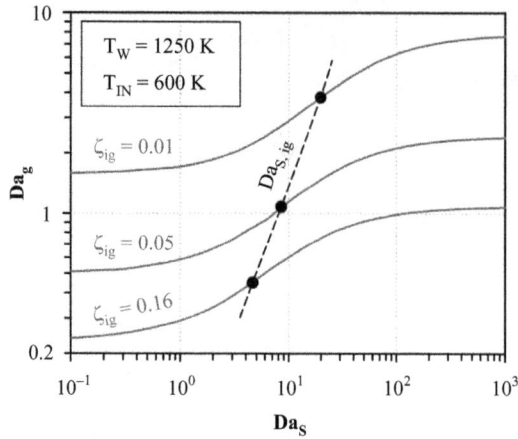

Figure 4.24. Lines of constant nondimensional homogeneous ignition distances (ζ_{ig}) calculated from the ignition criterion in Equation (4.80), for different combinations of gas-phase (Da_g) and catalytic (Da_s) Damköhler numbers. The symbols mark the catalytic ignition Damköhler numbers ($Da_{s,ig}$). There are infinite combinations of Da_g and Da_s yielding the same ζ_{ig} (reproduced by permission from Mantzaras et al. [14]).

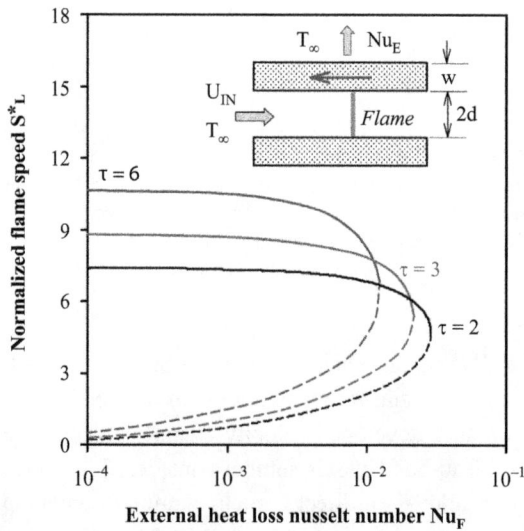

Figure 4.25. Computed dimensionless flame speed S_L^* (Equation (4.81)) in a plane channel (see the inset) as a function of the external heat loss Nusselt number Nu_E, for three normalized wall thickness ratios $\tau = (w+d)/d$ and a constant solid-to-gas thermal conductivity ratio of 100 (reproduced by permission from Veeraragavan et al. [172]).

with the non-dimensional temperature θ, heat release parameter γ, and Zeldovich number β defined as in Section 4.6.3 (see Nomenclature), and the subscript f denoting flame properties. In the case of no heat recirculation in the solid and no external heat loss ($k = Nu_E = 0$), $\theta_f = 1$ and Equation (4.81) reduces to $S_L^* = 1$, that is the flame speed equals the freely propagating flame speed. Figure 4.25 provides S_L^* as a function of external heat loss (Nu_E), for three different normalized wall thicknesses $\tau = (w + d)/d$ and a thermal conductivity ratio $\kappa = 100$. The maximum of Nu_E at the turning point, defines extinction. Moreover, only the upper branches of the curves (solid lines) correspond to stable solutions, although it is understood that some parts of these stable solutions may still lose stability to oscillatory and/or pulsating modes (see forthcoming Section 4.8.4), effects that cannot be captured by a steady model. As seen in Figure 4.25, by increasing the normalized wall thickness S_L^* also increases, as more heat is transferred upstream via the solid wall.

In conclusion, analytical studies based on steady 2D models have revealed the parameters controlling homogeneous ignition and flame propagation in catalytic and non-catalytic channels. For catalytic channels, in particular, the use of 2D models is essential to capture the near-wall fuel depletion that controls homogeneous ignition. These models have certain limitations, for example in the matched asymptotic analysis of Section 4.8.3.1 the required weak self-similarity of the perturbation analysis necessitates the boundary layer approximation with the resulting restrictions described in Section 4.7.2, while in the 2D flame propagation analysis the vertical planar front is an idealization (see Section 4.8.4). Moreover, as both analytical models are steady, they cannot account for the presence of a multitude of flame dynamics inherent in small channels. This effect is particularly important in non-catalytic microreactors, since nonstationary solutions are mostly suppressed in catalytic microreactors (see discussion in Section 4.8.4.2). Despite these limitations, the obtained analytical solutions are rich enough to clarify the underlying physics in microreactor combustion and can be used to guide computationally intensive detailed CFD simulations, since extensive parametric studies are currently not feasible.

4.8.4 FLAME DYNAMICS IN CATALYTIC AND NON-CATALYTIC MICROREACTORS

In channel-flow reactors, a flame experiences developing boundary layers, heat transfer to and from the walls, radical quenching, and/or surface reactivity. The nonlinear interaction between these effects gives rise to a wealth of combustion dynamics. Experimental and numerical investigations have shown different combustion modes for flames, predominantly in non-catalytic channels at small confinements (channel hydraulic diameters less than 10 mm, that is falling in the microscale and mesoscale combustion regimes). Dogwiler et al. [80] measured asymmetric flames during fuel-lean methane/air combustion in a rectangular channel with a height of 7 mm. Maruta et al. [173] studied experimentally hydrocarbon premixed flames in cylindrical tubes of 2 mm inner diameter; they observed stationary flames, oscillatory flames, and repetitive ignition and extinction, at different values of the inflow velocity.

4.8.4.1 Simulation of Flame Dynamics in Non-Catalytic Channels

Pizza et al. [143, 144] investigated numerically the stabilization and dynamics of lean ($\varphi = 0.50$) premixed hydrogen/air flames in planar non-catalytic channels with heights ranging between

$h = 0.3$ and 7.0 mm (micro- and mesoscale range) and lengths $L = 10\,h$. They used 2D transient elliptic simulations (referred to also as DNS) with an elementary chemical reaction mechanism for hydrogen [174]. A prescribed wall temperature profile was used in the simulation: the wall temperature was ramped, via a hyperbolic tangent function, over the initial 1/20 channel length from the inlet temperature $T_{IN} = 300$ K to a constant wall temperature $T_w = 960$ K. Flux boundary conditions were applied at the channel entry for the temperature and species, allowing for upstream diffusion at low Reynolds numbers. Non-catalytic boundary conditions were applied for the species at the wall (Equations (4.32) with $u_{st} = \dot{s}_k = 0$ or equivalently $B = 0$). Computations were performed over the entire channel height, without invoking symmetry at the channel midplane, so as to allow for the presence of asymmetric solutions, such as those reported in [80]. For a given channel height, the inlet velocity U_{IN} was increased and the resulting flame bifurcations were studied. Richer flame dynamics were reported for larger channel heights. In the mesoscale range ($2 \leq h \leq 7$ mm) [144], in particular, six different burning modes (mild combustion, ignition/extinction, closed steady symmetric flames, open steady symmetric flames, oscillating, and asymmetric flames) were established depending on the inflow velocity. Figure 4.26a summarizes the different flame modes and their bifurcations for the $h = 2$ mm non-catalytic channel ($B = 0$ signifies a non-catalytic channel according to Equations (4.32) and (4.33)). To facilitate the graphical presentation of all flame solutions, the parameter h_{Tmax} has been defined in the ordinate of Figure 4.26; it denotes the transverse distance from the lower channel wall ($y = 0$) to the point of the maximum temperature, normalized by the channel height h.

The different flame modes in Figure 4.26a are discussed next. At the lowest flow velocities, a stationary mode appears, termed "*mild combustion*", with a temperature rise of only a few degrees. For mild combustion at $U_{IN} = 0.5$ cm/s, Figure 4.27a provides axial profiles of temperature and species along the channel midplane ($y = 1$ mm) and Figure 4.27b the 2D map of the OH radical. Changes in intermediate species axial profiles are gradual and do not exhibit the localized peaks typical of vigorous combustion. However, even at such low velocities an appreciable boundary layer profile is evident in Figure 4.27b.

When U_{IN} increases from 0.8 to 0.9 cm/s, mild combustion loses stability to a nonstationary combustion mode termed *flame repetitive extinction* and *ignition* (FREI). This new mode, which persists for inflow velocities in the range $0.9 \leq U_{IN} \leq 5.0$ cm/s, is illustrated in Figure 4.28 for $U_{IN} = 1$ cm/s. The temporal history of the maximum temperature inside the channel (Figure 4.28a) reveals long periods of apparent inactivity with the maximum temperature being close to the wall temperature $T_w = 960$ K, associated with the flow of fresh mixture and the buildup of the radical pool. This period is followed by a sudden increase to a maximum temperature of 1312 K, associated with radical and thermal runway and subsequent fast flame propagation. The flame structure is shown with the 2D maps of Y_{OH} and Y_{H2} in Figure 4.28(c, d, e) and (f, g, h), respectively, at the three time instants marked in Figure 4.28b. The cold mixture ignites at a certain location inside the channel, leading to the formation of a nearly cylindrical flame, Figure 4.28(c, f), which expands until it approaches the tube walls. The confined flame can then propagate only horizontally, Figure 4.28(d, g), both upstream and downstream. The upstream propagating front changes its curvature from convex to concave towards the fresh mixture, Figure 4.28(e, h), and extinguishes at $x \approx 1.0$ mm due to heat losses to the cold inflow. On the other side, the downstream propagating flame preserves the initial convex shape toward the fresh mixture and extinguishes at $x \approx 8.0$ mm, wherein the local hydrogen mass fraction drops below the lean flammability limit (Figure 4.28h). After the flame extinguishes, the fuel fills the channel again until it reignites; the entire process repeats itself periodically with a constant

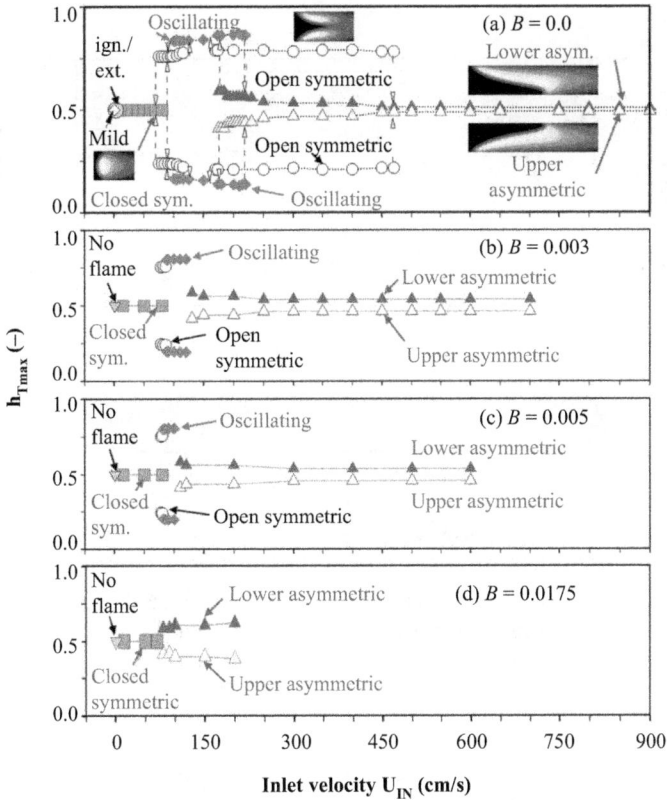

Figure 4.26. Computed combustion modes for H_2/air combustion ($\varphi = 0.5$) for a 2 mm plane channel, as a function of the inlet velocity U_{IN} and the ratio of the catalytically active area to geometrical surface area, B: (a) non-catalytic channel ($B = 0$), (b)–(d) catalytic Pt-coated channels with increasing parameter B. The vertical axis $h_{T,max}$ denotes the transverse distance from the lower channel wall ($y = 0$) of the point of the maximum temperature, normalized by the channel height h. For symmetric combustion modes, $h_{T,max} = 0.5$, while for asymmetric flames $h_{T,max}$ is either smaller or greater than 0.5. The vertical arrows in (a) indicate combustion mode transitions as U_{IN} is altered. The images in (a) provide typical flame shapes of various combustion modes via the distributions of the OH radical mass fraction at the flame vicinity (reproduced by permission from Pizza et al. [25]).

frequency of 9.9 Hz, defined as the inverse time between two consecutive bursts in the plot of Figure 4.28a. The behavior of the upstream propagating flame can be qualitatively regarded as flame acceleration in tubes [175] with the subsequent inversion of the flame curvature leading to a "tulip flame" [176]. This combustion mode exists for inlet velocities up to 5 cm/s, with both the amplitude and frequency of the temperature oscillations rising with increasing inlet velocity. The maximum frequency (for $U_{IN} = 5$ cm/s) is 33.3 Hz, a value comparable to the 50 Hz reported by Maruta et al. [173] in their experiments with propane/air flames at $\varphi = 0.5$ in tubes with internal diameter of 2 mm. The strong 2D nature of the fronts in Figure 4.28 (cylindrical shapes, with subsequent curvature changing gradually from convex to concave) points to the need of a 2D model to accurately capture the flame shape and propagation characteristics.

For $U_{IN} \geq 4$ cm/s, different combustion modes can coexist over certain velocity ranges (see Figure 4.26a). In this higher velocity range, stationary symmetric flames are established,

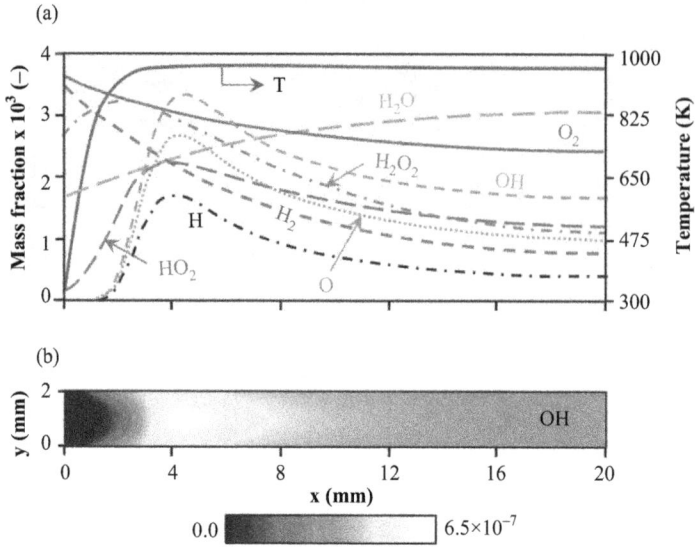

Figure 4.27. Computed mild combustion for H_2/air combustion ($\varphi = 0.5$) in a plane channel with height $h = 2$ mm and $U_{IN} = 0.5$ cm/s: (a) species mass fractions (4×10^4 Y_H, 6×10^3 Y_O, 6×10^3 Y_{OH}, 2×10^{-2} Y_{O2}, $2.5 \times 10^{-2}Y_{H2O}$, $30 \times Y_{HO2}$, $10 \times Y_{H2O2}$) and temperature along the channel midplane; (b) 2D map of Y_{OH} [144].

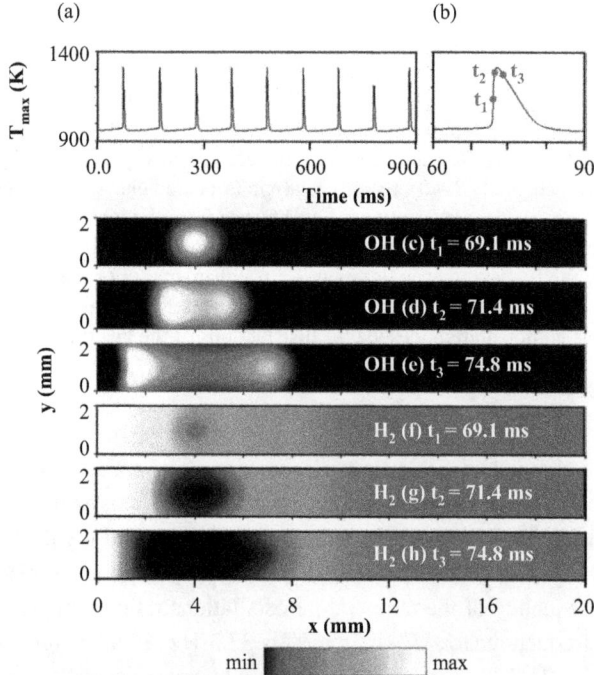

Figure 4.28. Computed repetitive ignition/extinction mode for H_2/air combustion ($\varphi = 0.5$) in a plane channel with height $h = 2$ mm and $U_{IN} = 1$ cm/s: (a) temporal evolution of the maximum temperature inside the channel; (b) detail of (a); (c)–(e) and (f)–(h) 2D maps of Y_{OH} and Y_{H2}, respectively, at the three times t_1 to t_3 marked in (b). The color bar provides values for species mass fractions: minimum corresponds to zero for both Y_{OH} and Y_{H2} while the maximum for Y_{OH} is 7.1×10^{-4}, 10^{-3}, and 6.0×10^{-4}, in (c)–(e), respectively, and for Y_{H2} is 5.1×10^{-3} [144].

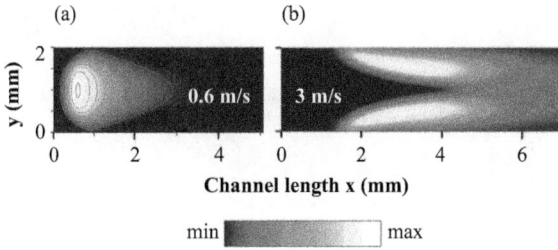

Figure 4.29. Computed steady symmetric flames for H_2/air combustion ($\varphi = 0.5$) in a plane channel with height $h = 2$ mm. Mass fraction of OH for: (a) closed symmetric flame at $U_{IN} = 0.6$ m/s, and (b) open symmetric flame at $U_{IN} = 3$ m/s. The minimum in the color bar is 0.0 and the maximum is 3.5×10^{-3} in (a) and 5.0×10^{-3} in (b) [144].

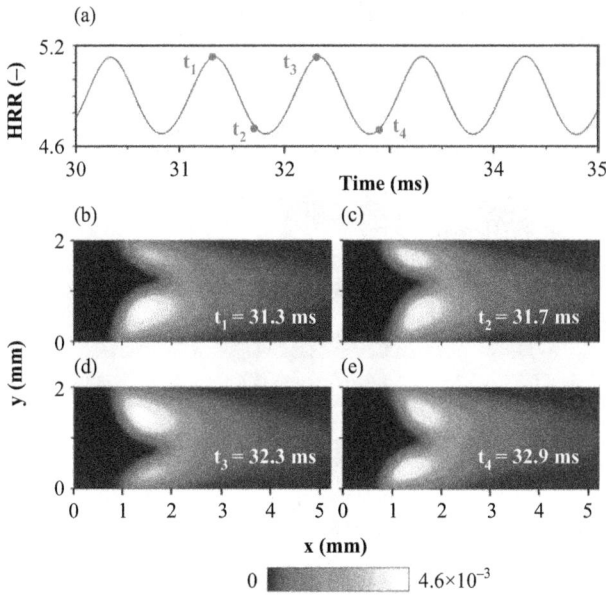

Figure 4.30. Computed oscillating flame for H_2/air combustion ($\varphi = 0.5$) in a plane channel with height $h = 2$ mm and $U_{IN} = 103.5$ cm/s: (a) temporal variation of the integral HRR, (b)–(e) maps of Y_{OH} at the four time instants t_1 to t_4 marked in (a) [144].

which are of closed or open shape (see Figure 4.29); the former appear over $4 \leq U_{IN} \leq 95$ cm/s and the latter over the intervals $75 \leq U_{IN} \leq 115$ cm/s and $170 \leq U_{IN} \leq 470$ cm/s. Oscillating flames form in the range $100 \leq U_{IN} \leq 220$ cm/s, coexisting over substantial velocity ranges with the open symmetric flames (see Figure 4.26a). Oscillating flames at the lowest velocities are periodic (Figure 4.30), as manifested by the normalized integrated HRR over the entire channel (Figure 4.30a) and the corresponding power spectra in Figure 4.31a. HRR is normalized by using a reference temperature $T_{ref} = T_{IN} = 300$ K, a reference heat capacity $c_{p,ref} = 1.2$ kJ/(kgK) and a reference time $t_{ref} = h/U_{IN}$. In the oscillatory mode of Figure 4.30, two flame branches are

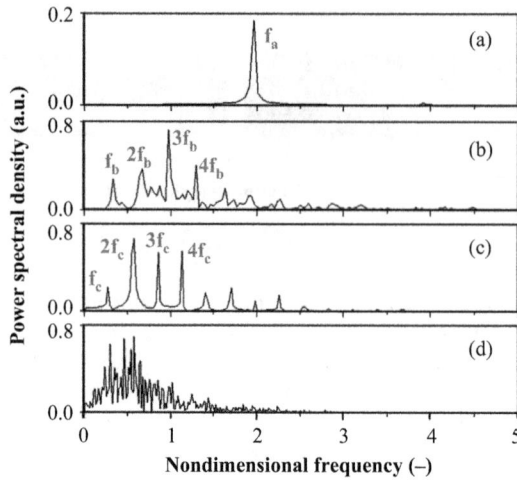

Figure 4.31. Power spectra of the integrated (HRR), in a plane channel with height $h = 2$ mm: (a) periodic oscillations of Figure 4.30 ($U_{IN} = 103.5$ cm/s), (b) aperiodic oscillations ($U_{IN} = 140$ cm/s), (c) mixed mode oscillations ($U_{IN} = 181$ cm/s), and (d) chaotic oscillations ($U_{IN} = 215$ cm/s) (reproduced by permission from Pizza et al. [144]).

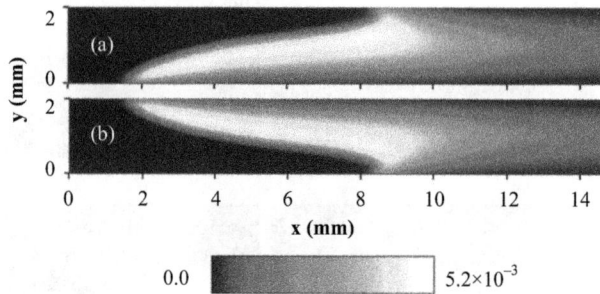

Figure 4.32. Computed asymmetric flames for H_2/air combustion ($\varphi = 0.5$) in a plane channel with height $h = 2$ mm and $U_{IN} = 400$ cm/s. Maps of OH mass fraction for (a) upper asymmetric flame, and (b) lower asymmetric flame [144].

propagating transversely and the burning intensity in one branch varies periodically and out of phase with respect to the other branch. At higher velocities aperiodic oscillations (Figure 4.31b), mixed mode oscillations (Figure 4.31c) and chaotic oscillations (Figure 4.31d) are present. The oscillatory modes have been investigated with nonlinear dynamics analysis (phase-space portraits and next amplitude maps) in [177], identifying a period-doubling cascade to chaos.

At the highest velocity range $174 \leq U_{IN} \leq 900$ cm/s, stationary asymmetric flames are computed, coexisting over an extended range with the open symmetric flames (Figure 4.26a). They form an obtuse or an acute angle with respect to the streamwise direction and are referred to as lower and upper asymmetric flames, respectively (Figure 4.32). Dogwiler et al. [80] measured upper and lower asymmetric flames in lean methane/air channel-flow non-catalytic combustion; however, the resulting flames were sensitive to small experimental perturbations, which could randomly trigger transition from one shape to the other. In general, channels with larger heights

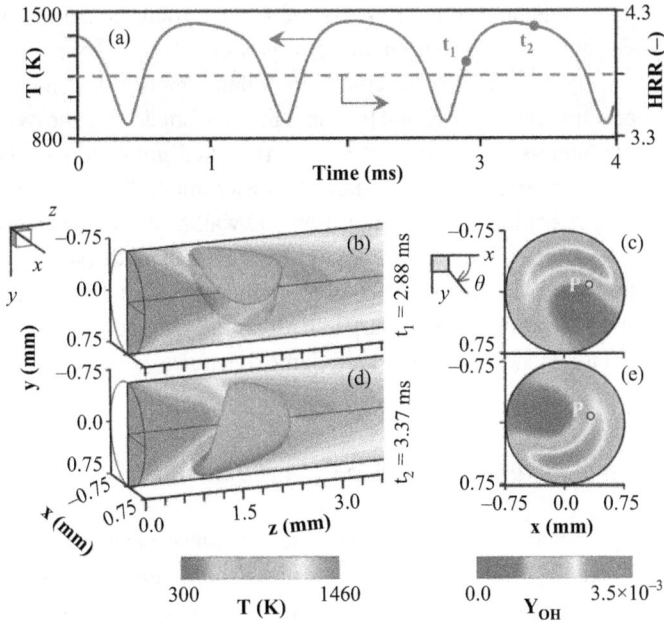

Figure 4.33. Computed clockwise spinning flame for an H_2/air mixture ($\varphi = 0.5$), at $U_{IN} = 150$ cm/s in a $d = 1.5$ mm tube: (a) temporal evolution of the temperature (solid line) at the reference point P with coordinates $(x, y, z) = (0.3, 0.1, 2.06)$ mm, and integral heat release HRR (dashed line); (b) and (d): iso-surfaces of $Y_{OH} = 1.7 \times 10^{-3}$, and temperature iso-contours on the y–z plane at the times t_1 and t_2 marked in (a); (c) and (e): iso-contours of Y_{OH} on the $z = 1.5$ mm plane, at time instances t_1 and t_2 [179].

exhibit a wider range of oscillatory combustion modes [143, 144]. Finally, tubular geometries exhibit similar dynamics with the difference that oscillatory flames appear mostly in the form of spinning modes [178]. Full 3D DNS in channels with internal diameter of 1.0 and 1.5 mm have been reported in [179]. The fuel, stoichiometry, inlet and wall temperatures were the same with the foregoing 2D plane channel simulations. For the 1.5 mm diameter channel, the spinning combustion mode is illustrated in Figure 4.33. The temperature at any position in the channel varies periodically, while the normalized integrated HRR is constant (Figure 4.33a), indicating a solid body rotation of the flame. The flame shape is shown in Figure 4.33(b–e), at two time instances of the spinning period, indicating a clockwise rotation (the rotation direction is random). Spinning modes appear over the range $100 \leq U_{IN} \leq 220$ cm/s for the 1.5 mm diameter channel, with rotational frequencies ranging from 300 to 880 Hz. The flame dynamics in non-catalytic channels originate mostly from flame–wall interactions, although the effect of transport and chemistry may also play a role, depending on the fuel.

The wealth of flame dynamics in non-catalytic channels implies certain restrictions for many simplified models. Steady models (e.g., see the 1D or 2D analytical studies in Sections 4.6.3 and 4.8.3.3) lead to a wider combustion stability envelope, since part of the computed steady solutions in reality loses stability to nonstationary combustion modes. Moreover, 1D models (steady or transient) necessitate planar and vertical flame fronts, a condition only satisfied by parts of the stable "closed flames" (Figures 4.26a and 4.29a). Transient thermodiffusive models (constant density assumption, one-step chemistry), particularly 2D, have been successful in qualitatively reproducing most of the aforementioned flame modes in non-catalytic channels [108, 178, 180], supporting the premise that the main mechanism driving the dynamics is flame/wall interactions. Nonetheless, for microreactor design detailed simulations of the type reported in this section are necessary. Irrespective of chemistry description, thermodiffusive models cannot account for flow acceleration due to combustion and wall heat transfer and the correct flow velocity is pivotal for capturing the precise ranges of each flame mode (see Figure 4.26a).

Extension of the herein described DNS tool to include conjugate heat transfer (see model equations in Section 4.7.1) would be necessary to address the stabilizing effects of heat recirculation in the solid and its impact to the observed flame dynamics.

4.8.4.2 Simulation of Flame Dynamics in Catalytic Channels

Flame dynamics have been investigated numerically in plane channels coated with Pt, for channel heights of 1 mm [24] and 2 mm [25]. The numerical platform is the same as in the non-catalytic DNS studies (Section 4.8.4.1). Fuel-lean ($\varphi = 0.5$) hydrogen/air combustion is investigated, with inlet temperature $T_{IN} = 300$ K and, similarly to the simulations in Section 4.8.4.1, the wall temperature ramps smoothly over the initial 1/20 channel length via a hyperbolic tangent function from the incoming mixture temperature, $T_{IN} = 300$ K, to a final value of 960 K. The elementary gas-phase mechanism by Kim et al. [174] is used, coupled to the detailed surface reaction scheme for the oxidation of H_2 on Pt by Deutschmann et al. [46].

The computed flame modes are mapped as a function of the inflow velocity, U_{IN}, and the catalytic reactivity, B (see Equations (4.32) and (4.33)). The summary of flame dynamics for the $h = 2$ mm channel height are presented in Figure 4.26(b–d) for various values of the catalytic reactivity parameter B. In technical systems with a well-dispersed catalyst, B is larger than unity such that the catalytic conversion rates of the typically less reactive hydrocarbon fuels are enhanced. However, because of the very high reactivity of hydrogen on platinum, a mass-transport-limited catalytic conversion is already achieved at $B = 1$. Therefore, the effects of finite-rate surface chemistry on the flame dynamics are investigated by reducing B to suitably low values ($3.0 \times 10^{-3} \leq B \leq 1.75 \times 10^{-2}$). Such values can be realized in practice by a very low loading of Pt on the catalyst washcoat.

Even the smallest examined catalytic reactivity ($B = 0.003$) suppresses the mild combustion and oscillatory ignition/extinction modes. This is because at the very low velocities (and hence long reactor residence times) of these modes, the catalyst effectively consumes all the fuel thus depriving it form the homogeneous pathway; the symbols marked with "no flame" in Figure 4.26b indicate the absence of homogeneous combustion at the lowest velocities. The effect of catalytic activity on the asymmetric flames is discussed next. As B increases, the flame first shifts downstream due to the upstream catalytic depletion of hydrogen and at the same time it becomes progressively more symmetric. For the narrower 1 mm channel height, in particular, complete flame symmetry can be restored for $B = 0.0115$ as shown in [24]. For the 2 mm channel height, examined here complete symmetry cannot be achieved. Finally, the flame extinguishes at $B = 0.018$ leading to pure heterogeneous conversion of hydrogen.

An increase in B has two effects on the oscillatory flames. First, the range of inflow velocities over which oscillating flames persist becomes narrower (Figure 4.26(a–c)). In addition, the

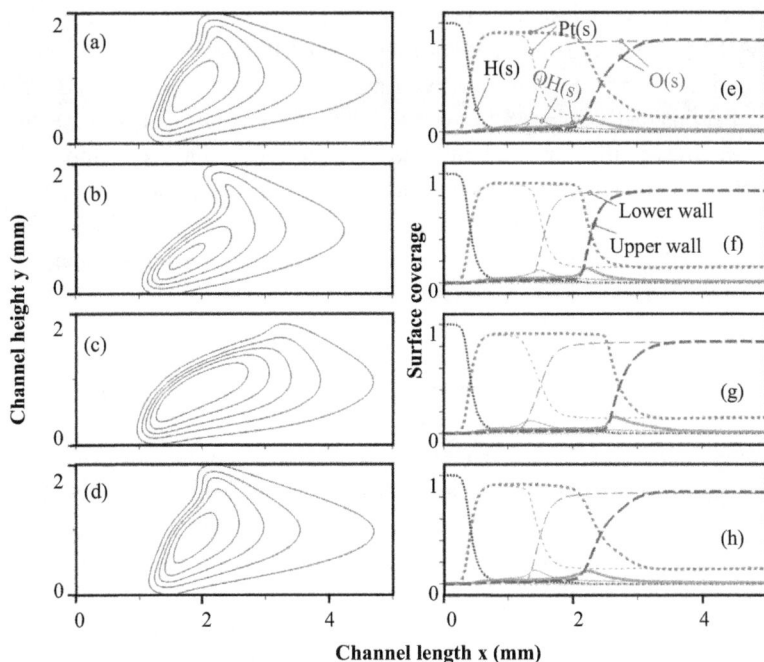

Figure 4.34. Computed pulsating flame for a 2-mm height Pt-coated channel at $U_{IN} = 100$ cm/s and catalytically active to the geometrical area ratio of $B = 0.005$: (a)–(d) Y_{OH} iso-lines (in the range $0.0007 \leq Y_{OH} \leq 0.0034$) at four time instants during the pulsating period; (e)–(h) the corresponding surface coverage on the lower (thin lines) and upper (thick lines) channel wall. The H(s) coverages for both upper and lower walls practically coincide (reproduced by permission from Pizza et al. [25]).

oscillations qualitatively change, and two new modes arise, the pulsating and the combined mode [25]. At $B = 0.003$, periodic and chaotic oscillations survive, but over a much narrower range of inflow velocities. At $B = 0.005$ and for the lowest U_{IN} (i.e., close to the lower limit of the stability range of oscillating flames), periodic oscillations are observed, qualitatively similar to the ones of the non-catalytic channel presented in Figure 4.30. When the inflow velocity approaches the upper limit of the oscillatory stability range, pulsating flames are observed. Figure 4.34(a–d) provides the Y_{OH} iso-lines and Figure 4.34(e–h) the surface coverage at four time instants during the pulsating period, for $B = 0.005$ and $U_{IN} = 100$ cm/s. One of the two flame anchoring points (either on the lower or on the upper wall) remains practically fixed while the other one moves back and forth in the streamwise direction in a periodic manner. The surface coverage in Figure 4.34(c–h) indicates temporal variation of the main surface species. Finally, at intermediate values of U_{IN} a combination of the harmonic oscillation and the pulsation is observed, giving rise to a new oscillatory mode, which is referred to as the combined mode. For this mode, the temporal evolution of the flame shape is such that during certain periods the flame oscillates as in Figure 4.30 with two separate branches alternatively expanding and contracting, while during other periods it pulsates as the flame in Figure 4.34.

In conclusion, the presence of catalytic surface reactions tends to suppress or to limit the ranges of appearance of oscillatory flame modes, although these modes persist for certain values of the catalytic reactivity. These findings are of prime interest in microreactors, as most dynamic phenomena are undesirable (they can compromise the reactor integrity and performance). A catalyst with a predetermined loading can thus be applied at the channel walls, so as to give the minimum heterogeneous reactivity necessary for suppressing the dynamic modes. It is stressed that this methodology still maintains a flame in the reactor, at least for a subrange of the inlet velocities of the non-catalytic case. For very high catalytic reactivities, gas-phase combustion may altogether be suppressed. Moreover, in the range of catalytic reactivities where flame dynamics are present, it is likely to have a coupling with potential catalytically driven dynamics (described next). The discovery of such interacting hetero-/homogeneous dynamics would be of great interest in catalytic microreactor applications.

4.8.4.3   *Heterogeneous-Driven Dynamics in Catalytic Microreactors*

Oscillatory dynamics driven by catalytic reactions have been the focus of intense research during the last years. Spatiotemporal oscillations during the oxidation of CO [181] and H_2 [182] over Pt were reported in the early seventies. The oscillatory behavior in CO oxidation has been subsequently studied by Ertl and co-workers [183] and was attributed to adsorbate-induced surface reconstruction. In the case of H_2 oxidation on Pt, oscillations were attributed to a combination of kinetics and transport, with a periodic transition between surface-kinetic-controlled and mass-transport controlled conditions. Oscillations can be periodic (typical periods up to kHz) but also chaotic [184] as in the flame dynamics of Section 4.8.4.1. Theoretical models have been constructed to understand the physicochemical processes leading to oscillations and also to simulate experimentally observed spatiotemporal variations of surface and gas-phase species above the catalyst [107, 185, 186]. The models were mainly based on either MC approaches (see Section 4.3.1) to understand the nature of surface reconstructions, or on simplified low-dimensionality macroscopic models (see Section 4.5) aiming at identifying the effect of mean-field kinetics (alone on in conjunction with transport) on the oscillatory behavior. Reactor-scale analysis (such as the one given in Section 4.8 with multidimensional models) is mostly lacking due to the problem complexity. These models once tested in ideal configurations can be implemented in advanced multidimensional reactor models such as those described in the foregoing sections.

Early investigations attributed all oscillatory behavior to intrinsic kinetics [186]. However, it is understood that the combination of kinetic bistability and mass transport limitations can cause oscillations in a PSR [107]. The following application [128] illustrates how this coupling leads to oscillations. The system under consideration is a spherical catalyst, located at radius $r = R$. On this catalyst, the surface reaction $2A + B_2 \rightarrow 2AB$ is considered (mimicking the oxidation of CO or H_2 on Pt), which proceeds with a reversible adsorption of A, an irreversible dissociative desorption of B_2, and a reaction between the surface species A and B to form the product AB. Under the assumption that the product AB desorbs rapidly such that its surface coverage is negligible, the mean field rate equations for the adsorbate coverages become

$$\frac{d\theta_A}{dt} = k_1 P_A (1-\theta_A) - k_2 \theta_A - k_3 \theta_A \theta_B, \tag{4.82}$$

$$\frac{d\theta_B}{dt} = k_4 P_{B2} (1-\theta_A - \theta_B)^2 - k_3 \theta_A \theta_B, \tag{4.83}$$

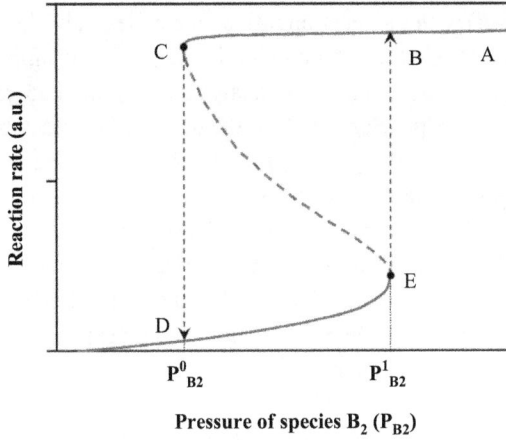

Figure 4.35. Surface reaction rate as a function of species B_2 pressure. The solid lines are stable states and the dashed line CE is an unstable state (reproduced by permission from Kulginov et al. [128]).

where P_A and P_{B2} are the partial pressures of the reactants, k_1 and k_4 their corresponding adsorption rate constants, k_2 is the rate constant for A desorption, and k_3 the rate constant for AB formation (considered irreversible). The rate expression in Equation (4.83) indicates that pre-adsorbed A surface species inhibit the dissociative adsorption of B_2 molecules via the quadratic term. In contrast, such site-blocking effect is not exerted by adsorbed B species on adsorbing A due to the absence of θ_B on the first right-hand term in Equation (4.82). For a given pressure P_A, there exists a unique steady-state solution of Equations (4.82) and (4.83) at sufficiently high or low values of P_{B2} (see Figure 4.35). On the other hand, at intermediate pressures $P_{B2}^0 \leq P_{B2} \leq P_{B2}^1$, there are three solutions in Equations (4.82) and (4.83) out of which the intermediate (dashed line region in Figure 4.35) is unstable and the other two are stable. The two stable solutions refer to fast (vigorous) and slow (weak) conversion rates. It is clarified that stability is examined within the employed mean-field approximation (see Section 4.3.1), which does not consider fluctuations in adsorbate coverages due to lateral interactions. When such effects are included, some of the stable solutions in Figure 4.35 may become metastable.

The impact of gas-phase transport limitations is treated by adding the boundary conditions at the catalyst surface ($r = R$) and considering a one-dimensional (spherically symmetric) problem:

$$D_A \frac{\partial n_A}{\partial r} = \dot{s} \text{ and } D_{B2} \frac{\partial n_{B2}}{\partial r} = \dot{s}/2, \tag{4.84}$$

with $\dot{s}$ the catalytic reaction rate calculated using the steady-state approximation for Equations (4.82) and (4.83) and D_A, D_{B2} are the species diffusivities. At $r \to \infty$, the reactant concentrations are $n_A = n_A^*$ and $n_{B2} = n_{B2}^*$. The steady-state solution of the diffusion equation in a spherically symmetric geometry is $n(r) = n(R) + \left(n^* - n(R)\right)(r - R)/r$, leading to the following diffusion flux at the catalyst surface:

$$J = D\left(n^* - n(R)\right)/R. \tag{4.85}$$

Comparing J_{max} $(= D\, n^*/R)$ with the reaction rate $\dot{s}$, it can be established whether the system is kinetically or diffusion controlled. For simplicity, in the present example diffusion limitations are considered only for species B_2. Then, the analysis of Figure 4.35 can give the conditions for oscillatory solutions. If the pressure of B_2 in the unperturbed reactant mixture corresponds to point A in Figure 4.35 and assuming that at $t = 0$ there are no spatial concentration gradients, the system lies in the vigorous reacting state. As time progresses, the local B_2 pressure will drop due to fast consumption and limited transport of B_2. The system then moves along line AC to the left (Figure 4.35). To achieve oscillations, the system must reach point C, whereby a transition to the weakly reacting state D will occur. This implies that the diffusion limitations for B_2 must be strong enough such that steady-state diffusion flux cannot support the high-reactive state on line AC. This requirement is expressed as [128]

$$D_{B2}\left(n^*_{B2} - n^0_{B2} \right) / R < \dot{s}_{max} / 2, \tag{4.86}$$

where n^*_{B2} and n^0_{B2} are the gas-phase concentrations corresponding to pressures P_{B2} and P^0_{B2} while $\dot{s}_{max}$ is the nearly constant reaction rate along line AC of the vigorously reacting branch. Once the system transitions to point D, the consumption of B_2 is reduced and if the diffusivity D_{B2} is not too small, the local B_2 concentration will rise again. The system moves along line DE and may then jump to the highly reacting state B, if the diffusion limitation of B_2 is not too strong:

$$D_{B2}\left(n^*_{B2} - n^1_{B2} \right) / R > \dot{s}_E / 2. \tag{4.87}$$

Equations (4.86) and (4.87) define the necessary conditions for an oscillatory solution (repeated oscillations along the path $B \rightarrow C \rightarrow D \rightarrow E \rightarrow B$). Applications are illustrated in Figure 4.36,

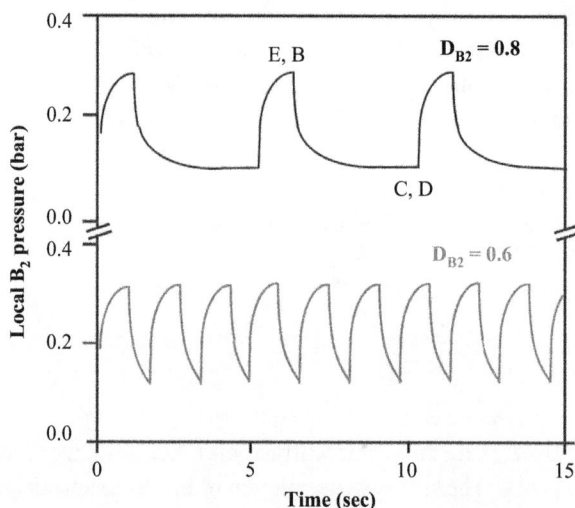

Figure 4.36. Predicted time history of the local pressure of reactant B_2 for two different diffusivities: $D_{B2} = 0.8$ and 0.6 cm^2/s. Points B, C, D, and E refer to the states shown in Figure 4.35 (reproduced by permission from Kulginov et al. [128]).

providing the time response of the local pressure of species B_2 for two different diffusivities: 0.8 and 0.6 cm²/s. These results are produced by numerical simulations [128], at reaction/diffusion conditions leading to oscillatory solutions. It is evident that the period of oscillation is a strong function of the diffusivity.

The application of kinetic schemes, such as those in Equations (4.82) and (4.83), to multidimensional transient models relevant to microreactors is of key interest to microreactor design and has not been addressed in the literature. It is plausible that for certain operating regimes oscillatory phenomena may appear in catalytic microreactors. Computer power is nowadays sufficient, at least for microchannel geometries, to apply the multidimensional models of Section 4.7.1 to investigate catalytically driven dynamics and their potential coupling to flame-driven dynamics. Of course, such a coupling requires the coexistence of catalytic and gaseous chemistries in microreactor confinements. To this direction, recent experimental studies, along the methodology outlined in Section 4.3.1, have validated hetero-/homogeneous chemical reaction schemes for hydrogen combustion over Pt at elevated pressures [187]. Using these validated elementary reactions schemes [27, 188] and steady 2D simulations, threshold wall temperatures were assessed for which the gaseous pathway contributed to 5 percent of the combined catalytic and gaseous hydrogen conversion. Results for Pt-coated channels with heights (or diameters) of 0.5 and 1 mm are shown in Figure 4.37, for microreactor-relevant

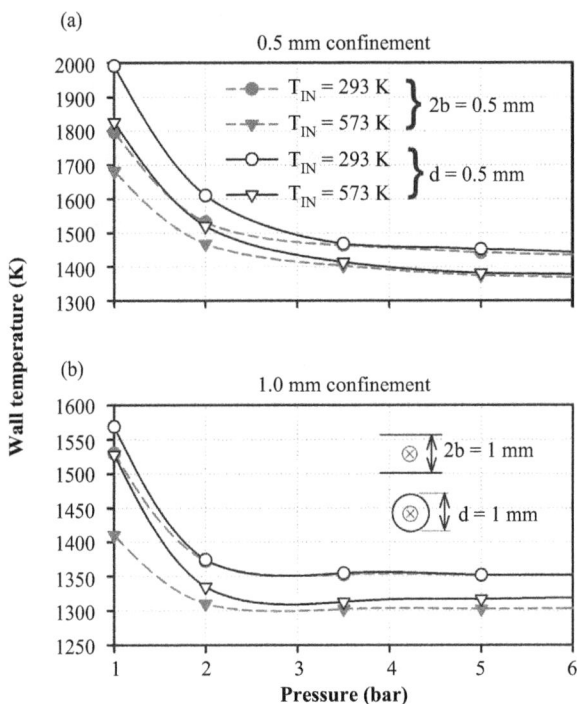

Figure 4.37. Computed threshold wall temperatures, T_w, corresponding to 5 percent gaseous hydrogen conversion, in hetero-/homogeneous combustion of H_2/air ($\varphi = 0.30$) over Pt-coated ducts as a function of pressure for two different preheats, two geometry types (tubular: open symbols, planar: filled symbols) and a constant mass flux $\dot{m} = 42.4$ kg/m²s. (a) 0.5 mm tubular channel diameter or planar channel full height, and (b) 1.0 mm tubular channel diameter or planar channel full height (reproduced by permission from Ghermay et al. [187]).

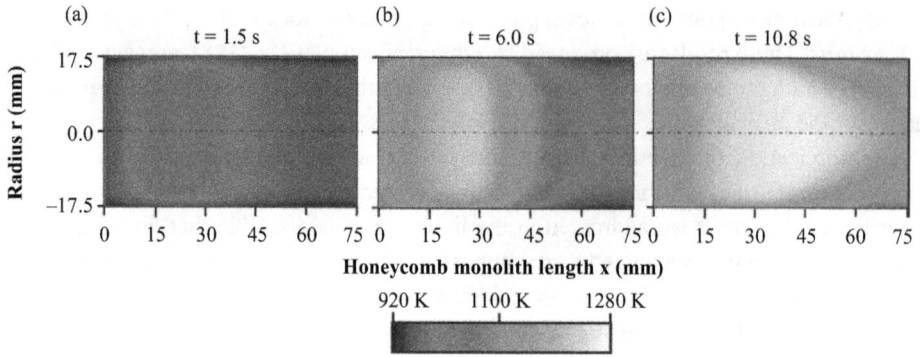

Figure 4.38. Transient simulations of honeycomb catalytic reactor start-up. The honeycomb cylindrical structure (35 mm diameter and 75 mm length) is made of FeCr-alloy foils coated with Pt. The operating conditions are: $p = 3$ bar, propane/air equivalence ratio $\varphi = 0.20$, inlet temperature $T_{IN} = 713$ K, inlet velocity $U_{IN} = 3.2$ m/s, and heat losses along the outer cylindrical surface $a_T(T_w - T_\infty)$ with $a_T = 12$ W/m²K and $T_\infty = 300$ K. Two-dimensional temperature maps are provided at three time instances ($t = 10.8$ s corresponds to steady state). The steady performance of this reactor has been studied in Karagiannidis et al. [3].

pressures up to 6 bar; it is evident that even at such small sizes, gas-phase chemistry cannot be ignored at realistic wall temperatures, particularly for $p \geq 2$ bar. DNS with hetero-/homogeneous chemistry and conjugate heat transfer is a feasible route to carry out detailed studies of potential hetero-/homogeneous dynamic interactions [189].

4.8.5 CONTINUUM MODES FOR CATALYTIC HONEYCOMBS

Since honeycomb structures mainly pertain to catalytic reactors, this model is of interest to catalytic mirco- or meso-scale reactors. Continuum modeling with elementary catalytic reaction schemes has been applied to study the transient 2D catalytic ignition of CH_4 over Pt in a cylindrical monolith [117], the transient 3D hydrogen-assisted catalytic combustion of methane over Pt [118], and the transient 2D partial catalytic oxidation of methane on Rh [119].

Figure 4.38 provides transient light-off simulations in a propane-fueled honeycomb structure coated with Pt, having a diameter of 35 mm, length of 75 mm, individual channel hydraulic diameters of 1.2 mm, and heat losses from the outside surfaces. The structure is made of FeCr-alloy (similar to that shown in Figure 4.1a) and was tested at a pressure of 3 bar for a mesoscale power-generation system based on a small turbine [3]. Axisymmetric 2D simulation was used for the heat conduction in the solid, Equation (4.36). The time history of the solid temperature is shown in Figure 4.38 and indicates that external heat losses lead to very different performances for channels close to the periphery of the cylindrical structure. Care must be exercised, however, when applying such models to small-sized honeycombs. While the structure in Figure 4.38 has over 600 individual catalytic channels, there exist microscale structures with only a few channels. In such cases the key requirement to have individual channel diameters much smaller than the reactor diameter is not met (see Section 4.7.3) and the continuum approach is invalidated.

4.9 TURBULENT MICROREACTOR COMBUSTION

In catalytic reactors, turbulent combustion within the individual honeycomb channels (Figure 4.1a) mainly occurs in gas turbine applications, whereby at full load the inlet Reynolds numbers can reach 30,000 [2]. Although turbulent catalytic combustion is mainly applicable to large-scale power generation, the fundamental turbulent combustion processes are of general interest since each channel falls in the microscale regime. For non-catalytic microreactors, turbulent flow may be created due to the injection of high-velocity fuel and air streams and their subsequent mixing [190]. A fundamental investigation of turbulent catalytic channel-flow combustion is first presented and then an application of turbulent models to non-catalytic microreactors with complex geometry is briefly discussed.

In entry channel flows with strong wall heat transfer, which is relevant to catalytic systems due to the heat release at the surface, the controlling parameters are the magnitude of the incoming turbulence and the laminarization of the flow due to the intense heating of the gas [191]. DNS in the entry region of a heated pipe air-flow at a moderate Re_{IN} of 4300 [192], supplemented with heat transfer experiments, has led to the development of advanced LR number models with suitable damping functions for strongly heated and developing channel flows. Appropriate models for turbulent flows in channels with hetero-/homogeneous reactions were subsequently developed and validated in [6, 103, 104]. Therein, fuel-lean hydrogen/air turbulent catalytic combustion was investigated experimentally and numerically at $p = 1$ bar.

Raman and OH-LIF measurements along with thermocouple-based surface temperature measurements were carried out in an optically accessible, Pt-coated mesoscale rectangular channel with height of 7 mm, in a fashion similar to the laminar studies of Section 4.3 (see also Figures 4.4 and 4.5). Moreover, particle image velocimetry (PIV) was used for the assessment of the instantaneous 2D velocity field [6]. Experiments have been performed with H$_2$/air equivalence ratios $\varphi = 0.18$–0.24 and inlet Reynolds numbers $Re_{IN} = 15,080$–$30,150$. Additional tests were carried out under nonreacting conditions (using air flows), to assess the impact of hetero-/homogeneous combustion on turbulent transport. All tests have been performed at statistically steady operating conditions.

In the modeling, a Reynolds Average Navier–Stokes (RANS) moment-closure approach was adopted [6, 103], whereby the Favre-averaged transport equation of any gas-phase variable φ is given by

$$\frac{\partial(\bar{\rho}\tilde{u}\tilde{\phi})}{\partial x} + \frac{\partial(\bar{\rho}\tilde{v}\tilde{\phi})}{\partial y} - \frac{\partial}{\partial x}\left(\Gamma_{eff}\frac{\partial \tilde{\phi}}{\partial x}\right) - \frac{\partial}{\partial y}\left(\Gamma_{eff}\frac{\partial \tilde{\phi}}{\partial y}\right) = \tilde{S}_\phi, \tag{4.88}$$

$$\Gamma_{eff} = \Gamma_\ell + \mu_t/\sigma_\phi \text{ and } \mu_t = c_\mu f_\mu \bar{\rho}\tilde{k}^2/\tilde{\varepsilon}. \tag{4.89}$$

The energy boundary condition at the gas–wall interface is a prescribed temperature profile (measured) according to Equation (4.16). For the catalytic reaction rates, the large thermal inertia of the solid wall suppresses the surface temperature fluctuations and eliminates the major reaction non-linearity [6], thus greatly simplifying the surface chemistry treatment under turbulent flow conditions. A laminar-like closure is therefore adopted for the catalytic reaction rates [6], for example, for adsorption reactions $\tilde{\dot{s}}_k = \dot{s}_k(T_w, \tilde{Y}_k, \theta_{Pt})$, and for reactions involving only surface species, $\tilde{\dot{s}}_m = \dot{s}_m(T_w, \theta_i)$ with i the participating surface species and T_w the measured

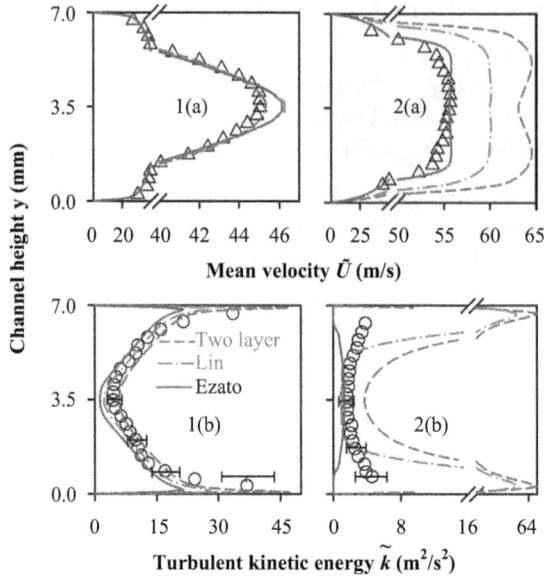

Figure 4.39. Computed and measured transverse profiles of mean velocities and turbulent kinetic energies at axial position $\xi = 195$ mm, in a Pt-coated planar channel with a height of 7 mm and a length of 300 mm. (a) Favre-average axial velocities $\tilde{U}$, and (b) Favre-average turbulent kinetic energies $\tilde{k}$ for two cases: the non-reacting Case 1 (isothermal flow of air at 296 K, $\tilde{U}_{IN}$ = 40 m/s, Re_{IN} = 35240) and the reacting Case 2 (H_2/air, φ = 0.24, $\tilde{U}_{IN}$ = 40 m/s, T_{IN} = 300 K, Re_{IN} = 30150, with maximum attainable wall temperature of 1220 K). Symbols are PIV measurements and lines are predictions with three different near-wall turbulence models. Dashed lines: two-layer model by Chen and Patel [193]; dashed-dotted lines: low Reynolds number (LR) model by Hwang and Lin [194]; solid lines: LR model by Ezato et al. [191] (reproduced by permission from Appel et al. [6]).

wall temperature. On the other hand, for gas-phase chemistry a presumed probability density function (PDF) reaction rate closure still has to be used to account for the large nonlinearity of these terms.

Three different LR near-wall turbulence models have been tested in [6, 103]. The first is the two-layer model of Chen and Patel [193] developed for non-heated flows. The other approaches are the LR model of Ezato et al. [191], which has been validated in strongly laminarizing heated channel flows, and the LR model of Hwang and Lin [194] that was developed for channel flows with and without heat transfer. The gas-phase scheme of Warnatz et al. [28] and the catalytic of Deutschmann et al. [27] were used in the simulations (both validated under laminar flow conditions in [17]).

Predicted and measured transverse profiles of the mean velocity $\tilde{U}$ and the turbulent kinetic energy $\tilde{k}$ re shown in Figure 4.39 for two cases with comparable Re_{IN}; one non-reacting (Case 1, flow consisting of air) and one reacting (Case 2). In the former, the two-layer model [193] provides very good agreement with the measurements, while the heat-transfer models of Hwang and Lin [194] and particularly Ezato et al. [191], underpredict $\tilde{k}$ and overpredict $\tilde{U}$ noticeably, as seen in Figure 4.39(1a, 1b). The last two models are thus overdissipative, clearly demonstrating the aptness of the two-layer model and the inapplicability of the heat transfer LR models in isothermal channel-flows. On the other hand, the LR model of Ezato provides good agreement to the measured $\tilde{U}$ for the reacting case in Figure 4.39(2a), while the Hwang and Lin model

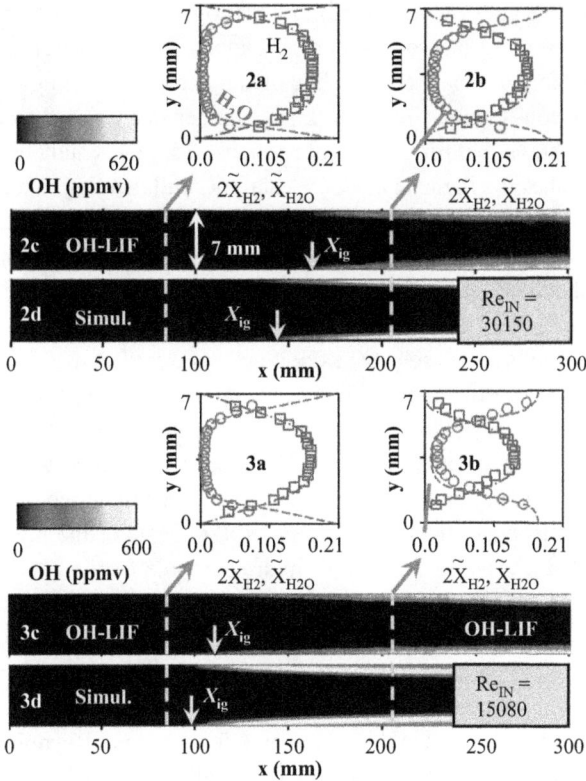

Figure 4.40. Measurements and simulations of turbulent catalytic channel-flow combustion in a Pt-coated planar channel with a height of 7 mm and a length of 300 mm, for two cases: Case 2 (plates 2a–2d): $\varphi = 0.24$, $T_{IN} = 300$ K, $\tilde{U}_{IN} = 40$ m/s, $Re_{IN} = 30150$; Case 3 (plates 3a–3d): $\varphi = 0.24$, $T_{IN} = 300$ K, $\tilde{U}_{IN} = 20$ m/s, $Re_{IN} = 15080$. (a, b) Raman-measured (symbols) and predicted (lines) transverse profiles of mean mole fractions for Cases 2 and 3 at two axial positions: (a) $x = 85$ mm and (b) $x = 205$ mm (H_2: squares and dashed-dotted lines; H_2O: circles and dashed lines). The thick solid lines through $y = 0$ in (b) define the predicted transverse gradients of the hydrogen mole fraction at the wall. (c) LIF-measured and (d) predicted 2D OH maps for the same two cases. The color bars provide OH in ppmv and the vertical arrows in (c) and (d) define the location of homogeneous ignition (x_{ig}) (reproduced by permission from Appel et al. [6]).

[194], and to a greater extent the two-layer model [193], substantially overpredict $\tilde{U}$. The last two models yield considerably higher near-wall $\tilde{k}$ (Figure 4.39(2b)) that in turn enhances the transport of heat from the hot wall to the channel core, leading to an unrealistic acceleration of the mean flow $\tilde{U}$ (Figure 4.39(2a)). In conclusion, from the tested heat-transfer LR models only the Ezato et al. [191] model provides a realistically strong turbulence damping for catalytic combustion applications.

Measured and predicted 2D maps of the OH radical have shown [6] that only the model of Ezato reproduces the measured location of homogeneous ignition and the flame shape. This is illustrated in Figure 4.40, providing comparison of OH maps for two reacting conditions (Cases 2 and 3, in Figure 4.40(2c, 2d) and Figure 4.40(3c, 3d), respectively), pertaining to two different Reynolds numbers. The two-layer and Hwang and Lin models, with their enhanced turbulence transport, lead to increased upstream catalytic conversion that deprives hydrogen from the homogeneous pathway thus inhibiting homogeneous ignition and failing to correctly capture the gaseous combustion processes in the channel. The increased transport under turbulent conditions has a profound impact on wall-bounded flames, by reducing the available residence times across the gaseous combustion zone. Early laminar stagnation-flow catalytic combustion studies [195] have shown that for a deficient reactant with Lewis number less than unity, a rise in the strain rate pushes the flame against the catalytic wall, leading to incomplete combustion through the gaseous reaction zone and to a subsequent catalytic conversion of the leaked fuel. A further rise in the strain rate extinguishes the flame. Turbulent transport in channel-flows has a role analogous to that of the strain rate in stagnation flows. This is illustrated in Figure 4.40(a, b) providing Raman-measured and predicted (using Ezato's model) transverse profiles of H_2 and H_2O mean mole fractions at two selected axial positions ($x = 85$ and 205 mm) for Cases 2 and 3. At $x = 85$ mm (position upstream of homogeneous ignition), Figure 4.40(2a, 3a) indicates mass transport limited catalytic conversion of hydrogen for both cases.

In the model of Ezato, the continuous flow laminarization allows for longer residence times of the fuel in the near-wall flame zones, leading to sustainable gaseous combustion, in agreement with the OH-LIF experiments in Figure 4.40(2c, 3c). For the higher Reynolds number Case 2, the predicted transverse gradient of hydrogen mole fraction at the wall is non-zero at the post-ignition location $x = 205$ mm (shown with the thick solid line passing through $y = 0$ in Figure 4.40(2b)), clearly indicating fuel leakage through the flame zone. This is also supported by the measurements, despite the near-wall limitations of the Raman experiments. Therefore, even well-downstream of homogeneous ignition, the catalytic pathway converts hydrogen in parallel to the gaseous pathway. On the other hand, for the lower Reynolds number Case 3, the corresponding hydrogen gradient at the wall is practically zero (see Figure 4.40(3b)), indicating that catalytic conversion ceases shortly after the onset of homogeneous ignition. The aforementioned suppression of gaseous combustion in favor of catalytic conversion at higher Re_{IN} is actually detrimental to the catalyst integrity (as discussed in Section 4.2 and Figure 4.3a for fuels with Lewis number less than unity).

Using the validated LR model of Ezato et al. [191], additional computations of hydrogen/air catalytic combustion over platinum have been carried out in microchannel geometries [6]. Uniform turbulent kinetic energies and turbulent dissipation rates are used as inlet boundary conditions, according to

$$\tilde{k}_{IN} = \alpha_1 \tilde{U}_{IN}^2 \text{ and } \tilde{\varepsilon}_{IN} = \left(c_\mu / \alpha_2\right) k_{IN}^{3/2} / r_h, \qquad (4.90)$$

with r_h the channel hydraulic radius and $\alpha_2 = 0.03$. The computed flow laminarization domains are depicted in Figure 4.41. In the shaded zones, the flow laminarizes and the turbulent hydrogen conversion differs by less than 10 percent from the corresponding conversion computed by a simpler laminar model. The inlet turbulence (determined by the parameter α_1 in Equation (4.90)) is crucial in determining the extent of the flow laminarization domain. It is emphasized that turbulent mass and heat transport in channel flows has not received proper attention. The DNS tool described in Section 4.8.4 has been recently used for catalytic mesoscale channel simulations at realistic Reynolds numbers [196]; such studies can provide valuable input for the development of appropriate mass and heat transfer near-wall submodels for RANS or large eddy simulation (LES) codes.

For non-catalytic systems, 3D LES modeling has been used to simulate complex-flows in microreactors [190]. The reactor volume is cylindrical with a diameter and length of 6 mm

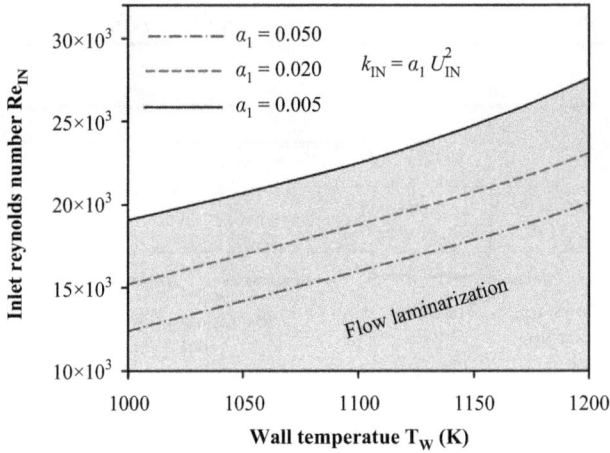

Figure 4.41. Domains of flow laminarization (defined by the shaded areas below each line of constant inlet turbulent kinetic energy) in turbulent channel catalytic combustion as a function of inlet Reynolds number and wall temperature (reproduced by permission from Appel et al. [6]).

and 9 mm, respectively. Two 2-mm-diameter inlet nozzles, at 90° with respect to each other, supply methane and air at jet exit Reynolds numbers up to 4,000 thus creating a swirling flow with high turbulence. Measurements of exit temperature and species concentrations are used to validate the model. A commercial flow solver is used with a steady flamelet combustion model, although better predictions are reported with an eddy dissipation concept (EDC). A WALE sub-grid model [197], suitable for near wall flows, is employed for turbulence modeling. Predictions are encouraging when compared to measurements in terms of reactor efficiency [190]. Simulation of complex microreactor geometries with commercial software is expected to greatly aid microreactor design. Still, fundamental studies of the type reported earlier in this section are necessary for developing appropriate combustion and turbulence submodels.

4.10 NON-CONTINUUM FLOWS IN MICROREACTORS

The governing equations in Section 4.7.1 have been derived under the continuum flow assumption. However, as the Knudsen number Kn ($= \lambda/L$, ratio of the mean free path λ of the gas to a characteristic hydrodynamic length L) increases, the continuum postulation breaks down and the use of kinetic-theory-based models is required. Figure 4.42 depicts different flow regimes in terms of the Knudsen number; the correspondence between Kn and flow lengths (L) in this figure has been established for air at atmospheric pressure and room temperature. Continuum modeling and hence NS are valid for $Kn < 0.01$, free molecular flows appear for $Kn > 2$, while the intermediate range $0.1 < Kn < 2$ covers the transitional and slip regimes. Non-continuum (i.e., finite Knudsen number) effects may be present in certain microreactor flows. The microchannels investigated in Sections 4.3 to 4.8 with gaps or diameters down to 0.25 mm (see Figure 4.1b) still fall in the continuous flow regime under atmospheric pressure and realistic combustion temperatures. At such scales, however, microthrusters for space applications (non-catalytic on catalytic

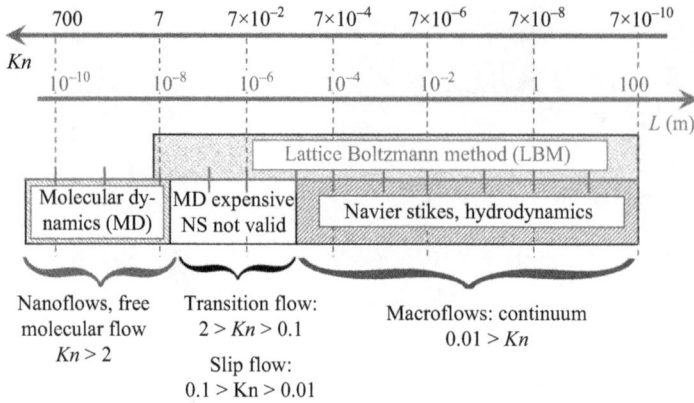

Figure 4.42. Flow regimes at standard conditions as a function of length scale. Flows in microreactors fall mostly in the continuum regime, but may also enter the slip regime.

[163, 198]) can fall in the slip flow regime due to the low operating pressures (~0.01 bar). At atmospheric pressure it is likely that, while the microreactors themselves fall in the continuum regime, certain microreactor components may operate in the slip regime. Such examples encompass the flow inside the porous washcoat of catalytic microreactors and inside the porous electrodes of micro-SOFCs, where pore sizes of a few microns can be encountered. Finite Knudsen number effects manifest themselves with the appearance of velocity, concentration, and temperature slips (the latter also referred to as temperature jumps) at the gas–wall interface. Slip effects were found negligible in microchannels [87], while other studies indicated that near-wall Knudsen layers may create noticeable slips (including concentration slips) even though the bulk flow in the microchannel could still obey the NS equations [199, 200].

A fundamental investigation of potential slip effects at the microscale entails the use of *ab initio* computations such as molecular dynamics (MD) or direct simulation Monte Carlo (DSMC). DSMC of velocity slips and temperature jumps in a Couette flow was carried out in Marques et al. [201], yielding temperature jumps of only ~0.9 percent for a relatively high Knudsen number of $Kn = 0.01$. For practical microreactors at atmospheric pressure, even lower Knudsen numbers are expected ($Kn \sim 0.001$), suggesting even smaller (practically insignificant) temperature jumps. DSMC and kinetic modeling becomes computationally expensive at the slip regime (see Figure 4.42) and new methods have been established for microflow modeling. The LB method is based on a discrete-velocity description of the continuous Boltzmann equation. For a single component mixture, the discrete Boltzmann equation becomes [202]:

$$\frac{\partial f_i}{\partial t} + c_{ij}\frac{\partial f_i}{\partial x_j} = \Omega_i, \quad i = 1, 2, ..., N, \tag{4.91}$$

with f_i the discrete velocity population distribution, N is the total number of discrete velocities, c_{ij} are the molecular velocities of populations f_i, and Ω_i is the collision integral. A BGK (Bhatnagar, Gross, and Krook [203]) model for the collision integral is typically used, leading to

$$\frac{\partial f_i}{\partial t} + c_{ij}\frac{\partial f_i}{\partial x_j} = \frac{1}{\tau}(f_i - f_i^{eq}), \quad i = 1, 2, ..., N, \tag{4.92}$$

with f_i^{eq} the equilibrium distribution and τ a relaxation time for equilibration, which is linked to specific transport processes (viscosity for flow, thermal conductivity for energy, diffusivity for species). The number of discrete velocities N depends on the spatial dimensionality of the system and the model requirements (only flow solution, capacity to solve flow and energy, etc.). For 2D problems, a popular LB model is the D2Q9, using nine discrete velocities arranged in a square lattice (Figure 4.43a), while for 3D systems a D3Q27 lattice (27 velocities) is a common approach (Figure 4.43b). The equilibrium functions f_i^{eq} are tuned so as to recover the macroscopic NS equations at the limit $Kn \to 0$. The LB methodology is thus suited for the entire continuum flow range but due to its kinetic origin can also be applied to the mesoscopic slip flow regime.

Slip effects in single-component flows have been studied with LB methods [204]. To further investigate slip effects in microreactors, LB models with appropriate boundary conditions at the gas–wall interface have been developed for isothermal binary and multicomponent flows [205, 206]. Therein, each species k has its own velocity distribution f_{ki} contrary to most common approaches that use a passive scalar approach to decouple the flow and species conservation equations. LB simulations in a Couette flow (flow between two parallel plates moving at equal and opposite velocities) have been carried out for binary He–Xe mixtures [206]. Figure 4.44 compares LB simulations to DSMC [207] and linearized kinetic theory results [208] for $Kn = 0.02$ in terms of the binary slip coefficient:

$$U_o - U_w = \sigma_{12}\lambda \left(\frac{\partial U}{\partial y}\right)_{y=0},$$
(4.93)

where U_o and U_w are the fluid velocity at the wall ($y = 0$) and the wall velocity, respectively, and λ is the mean free path of the gas. The LB results are in good agreement with the kinetic predictions and, in particular, capture the strong dependence of the slip coefficient on mixture composition and the resulting peak at $X_{He} \approx 0.9$. LB methods for non-isothermal flows have also

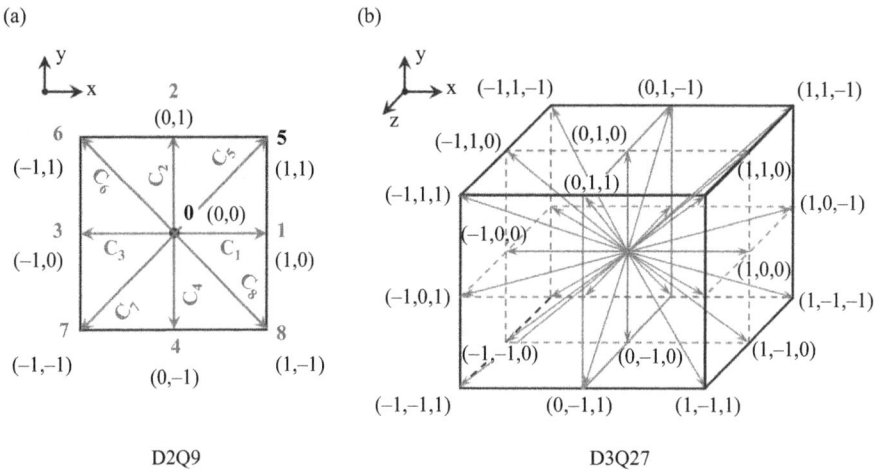

Figure 4.43. Standard lattices used in lattice Boltzmann (LB) methods: (a) 2D lattice with 9 velocities (D2Q9), (b) 3D lattice with 27 velocities (D3Q27).

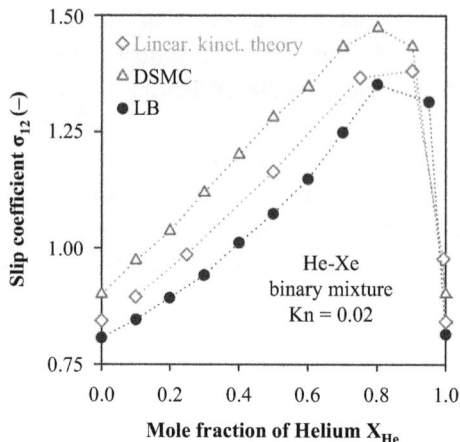

Figure 4.44. Binary slip coefficient σ_{12} (see Equation (4.93)) in Couette flow of a Helium / Xenon mixture, as a function of the helium molar fraction X_{He}, for a Knudsen number $Kn = 0.02$. Diamonds: linearized kinetic theory results [208]; triangles: direct simulation Monte Carlo (DSMC) [207]; circles: lattice Boltzmann (LB) method [206] (reproduced by permission from Arcidiacono et al. [206]).

been established [209] and the development of combined thermal, reactive, multicomponent LB models [210] facilitates the investigation of slip effects (in terms of velocity, temperature, and concentration) in practical systems.

Applications of LB models were reported in [211] for mesoscale catalytic channels, a range where no slip effects are present. Moreover, simulations (2D and 3D) of a nonreactive flow of CH_4 and H_2O at 800 K and $p = 1$ bar inside the anode electrode of a SOFC have been performed in [212], whereby the complex porous geometry was acquired with X-ray tomography (see the inset in Figure 4.45, providing an 8 mm × 8 mm 2D slice of the SOFC electrode). Two-dimensional LB simulations of a force-driven CH_4/H_2O flow (70 percent H_2O per volume) for the geometry shown in the inset of Figure 4.45 (flow from left to right, with periodic boundary conditions in the y-direction) have provided the velocity magnitude $U = (u_x^2 + u_y^2)^{0.5}$ in the electrode domain. Moreover, simulations have been carried out with both LB and standard continuum NS codes. The size of the square electrode domain was varied, so as to control the Knudsen number, while still maintaining $Re_d < 1$ based on the average pore size d. The NS simulations result in the same normalized maximum velocity U_{max}/U_{IN} irrespective of Kn, whereas the LB predictions yield a substantial drop of U_{max}/U_{IN} for $Kn > 10^{-3}$ due to velocity slip effects. The dashed zone in Figure 4.45 indicates ranges applicable for typical SOFC electrodes. It is thus evident that for SOFCs (either micro-SOFCs or large-scale SOFCs), the flow inside the electrodes can be non-continuum. Slip effects, even if absent at the microreactor scale, may be present in many microreactor subsystems, thus requiring proper non-continuum description.

LB simulations of entire subsystems (such as the porous structures in Figure 4.45) can be coupled via multiscale modeling to NS simulations of the main macroscopic system (i.e., the anode and cathode channel flows in a SOFC). Since such methods are very computationally intensive, alternate approaches are to use LB simulation results of subsystems in the slip regime in order to develop appropriately corrected submodels for a full NS continuum treatment of the entire microreactor device.

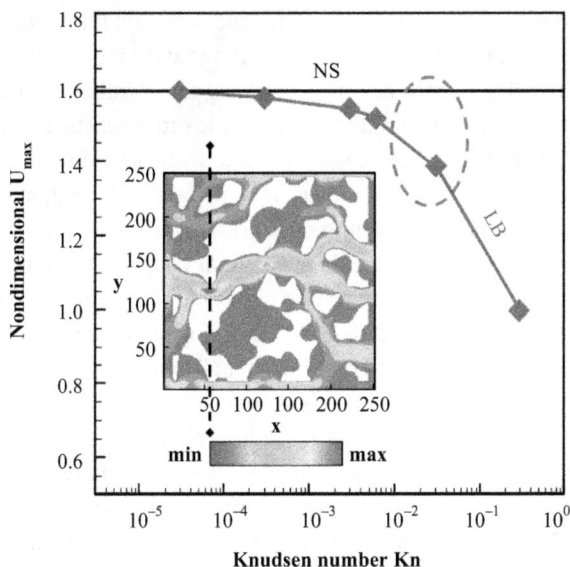

Figure 4.45. Lattice Boltzmann (LB) simulations of CH_4/H_2O nonreacting flow in the anode electrode of a solid oxide fuel cell (SOFC). The 2D domain (8 mm × 8 mm) is discretized with 250 lattices in each direction (see the inset, white-colored regions denote solid) and a D2Q9 LB model is used (see Figure 4.43a). The flow is from left to right and the color map provides the computed velocity magnitude. LB simulations are compared to standard continuum Navier–Stokes (NS) predictions by varying the Knudsen number (accomplished by artificially changing the physical size of the simulated domain). LB results capture strong slip effects at $Kn > 10^{-3}$. The dashed area signifies the operating regimes of SOFC anodes (reproduced by permission from Kang et al. [212])

4.11 CONCLUSION AND FUTURE WORK

Simulation tools for the investigation of steady and dynamic behavior of non-catalytic and catalytic microreactors are reviewed. Numerical models are laid down, with emphasis on multi-dimensional approaches using elementary homogeneous (gas-phase) reactions, detailed heterogeneous (catalytic) mean-field chemical description, and detailed gas-phase transport. Realistic microchannel geometries are employed to address the complex interactions between kinetics, fluid mechanical transport, heat conduction in the solid wall, and surface radiation heat transfer. The requirements for invoking the quasi-steady-state assumption for the gas in dynamic micro-reactor modeling are outlined. Dynamic response due to reactor varying operating conditions is presented for the ignition (light-off) of catalytic microchannels. The impact of solid material properties, catalytic reactivity, gas-phase chemistry, and in-channel heat transfer mechanisms on the light-off behavior is discussed.

Intrinsic dynamic behavior driven by interactions between gas-phase chemistry and walls is subsequently demonstrated using full DNS in 2D and 3D microchannel geometries. Geometrical confinements at the microscale, and especially at the mesoscale, lead to a wealth of flame dynamics ranging from periodic to intermittent, mixed mode, and finally to chaotic oscillations. Moreover, it is illustrated that the presence of a modest catalytic reactivity on the wall can suppress or at least moderate the flame/wall-driven instabilities. The origin of heterogeneously driven dynamic oscillations is subsequently addressed using low-dimensionality reactive/diffusive systems. Heterogeneous kinetics and diffusion limitations are responsible for many of the observed dynamics. Monte Carlo simulations at the nanometer scale have improved the understanding of oscillatory behavior by accounting for lateral interactions between species and for surface reconstruction. Such understanding leads to sophisticated mean-field catalytic chemical reaction schemes, which can be used together with gas-phase kinetics in realistic multidimensional reactor models. Computer power is nowadays sufficient, at least for microchannel geometries, to apply such advanced multidimensional models in order to gain much needed insight on catalytically driven dynamics, their potential coupling with homogeneously driven dynamics, and finally their response to time-varying reactor operating conditions (ignition, load change, etc.). This approach could greatly aid in devising start-up procedures and operational envelopes so as to circumvent unstable combustion modes in microreactor systems.

The applicability of the boundary layer approximation (parabolic model) for the gaseous flow under microreactor-relevant conditions is studied. Narrower applicability is illustrated for microreactors compared to their large-scale counterparts. For pure catalytic microreactors (no gas-phase chemistry) these stronger restrictions originate from the typically low flow rates and modest length-to-height microchannel aspect ratios. Gas-phase combustion in non-catalytic, or even catalytic, microchannels poses even more severe limitations for the parabolic model, the reason being the low inflow velocities and the high laminar flame speeds of hydrocarbon fuels due to the typically low operating microreactor pressures. It is finally shown that gas-phase chemistry may not be always neglected in catalytic microreactors, despite their tight geometrical confinements. This is especially the case for moderate-pressure applications (up to five bar in microturbine-based concepts for portable power generation).

For catalytic honeycomb reactors, *continuum* models for the entire structure are presented and their applicability to microreactors is discussed. While for mesoscale reactors the *continuum* model is largely valid, care must be exercised at the microscale due to the insufficient number of channels in the entire honeycomb structure.

Turbulent combustion is investigated (numerically and experimentally) in mesoscale channels with heterogeneous and homogeneous reactions. For limiting reactants having Lewis numbers less than unity, increased turbulent transport allows for combined hetero-/homogeneous conversion via reactant leakage through the gaseous reaction zone toward the catalytic wall. Even higher turbulent transport may eventually lead to flame extinction, leading to pure catalytic conversion. Key for the aptness of the near-wall turbulence models is the capture of the flow laminarization induced by heat transfer from the hot catalytic walls. The application of multidimensional LES to complex 3D turbulent flows in microreactors exemplifies the need for fundamental understanding of the near-wall thermoscalar transport.

Analytical studies based either on activation energy asymptotics or on simplified description of the flow (thermodiffusive models) expose the controlling processes in microreactor combustion and provide closed-form criteria for microreactor ignition or extinction. Although such results may not be directly used for microreactor design, they can facilitate elaborate CFD simulations by reducing the parameter space under investigation. Finally, finite Knudsen number effects, (i.e., non-continuum flows, manifested by the presence of velocity, concentration and temperature slips at the gas-wall interface) are elaborated. Although for atmospheric pressure applications most microreactors fall in the continuum regime, certain microreactor components, such as electrodes in micro-SOFCs or porous washcoats in catalytic microreactors, may operate in the slip regime. It is shown that LB models are particularly suited for the slip regime and are able to reproduce key slip dependences of kinetic-based models such as DSMC. While multiscale modeling with combination of LB and NS codes is a computationally intensive methodology for microreactors, a simpler approach is the creation of empirical submodels for the slip regime-based on LB results.

ACKNOWLEDGEMENTS

Support was provided by the European Union project HRC-Power and the Swiss Competence Centers for Energy Research (SCCER)

REFERENCES

[1]   R. Carroni, T. Griffin, J. Mantzaras, M. Reinke, High-pressure experiments and modeling of methane/air catalytic combustion for power-generation applications, Catalysis Today 83 (2003) 157–170. doi: http://dx.doi.org/10.1016/S0920-5861(03)00226-8.

[2]   R. A. Dalla Betta, T. Rostrup-Nielsen, Application of catalytic combustion to a 1.5 MW industrial gas turbine, Catalysis Today 47 (1999) 369–375. doi: http://dx.doi.org/10.1016/S0920-5861(98)00319-8.

[3]   S. Karagiannidis, K. Marketos, J. Mantzaras, R. Schaeren, K. Boulouchos, Experimental and numerical investigation of a propane-fueled, catalytic mesoscale combustor, Catalysis Today 155 (2010) 108–115. doi: http://dx.doi.org/10.1016/j.cattod.2010.04.030.

[4]   D. G. Norton, E. D. Wetzel, D. G. Vlachos, Fabrication of single-channel catalytic microburners: Effect of confinement on the oxidation of hydrogen/air mixtures, Industrial and Engineering Chemical Research 43 (2004) 4833–4840. doi: http://dx.doi.org/10.1021/ie049798b.

[5]   N. I. Kim, S. Kato, T. Kataoka, T. Yokomori, S. Maruyama, T. Fujimori, K. Maruta, Flame stabilization and emission of small Swiss-roll combustors as heaters, Combustion and Flame 141 (2005) 229–240. doi: http://dx.doi.org/10.1016/j.combustflame.2005.01.006.

[6]   C. Appel, J. Mantzaras, R. Schaeren, R. Bombach, A. Inauen, Turbulent catalytically stabilized combustion of hydrogen/air mixtures in entry channel flows, Combustion and Flame 140 (2005) 70–92. doi: http://dx.doi.org/10.1016/j.combustflame.2004.10.006.

[7]   J. Mantzaras, Understanding and modeling of thermofluidic processes in catalytic combustion, Catalysis Today 117 (2006) 394–406. doi: http://dx.doi.org/10.1016/j.cattod.2006.06.047.

[8]   W. C. Pfefferle, L. D. Pfefferle, Catalytically stabilized combustion, Progress in Energy and Combustion Science 12 (1986) 25–41. doi: http://dx.doi.org/10.1016/0360-1285(86)90012-2.

[9]   U. Dogwiler, P. Benz, J. Mantzaras, Two-dimensional modelling for catalytically stabilized combustion of a lean methane-air mixture with elementary homogeneous and heterogeneous chemical reactions, Combustion and Flame 116 (1999) 243–258. doi: http://dx.doi.org/10.1016/S0010-2180(98)00036-4.

[10] S. Karagiannidis, J. Mantzaras, G. Jackson, K. Boulouchos, Hetero-/homogeneous combustion and stability maps in methane-fueled microreactors, Proceedings Combustion Institute 31 (2007) 3309–3317. doi: http://dx.doi.org/10.1016/j.proci.2006.07.121.

[11] S. T. Kolaczkowski, Modelling catalytic combustion in monolith reactors—challenges faced, Catalysis Today 47 (1999) 209–218. doi: http://dx.doi.org/10.1016/S0920-5861(98)00301-0.

[12] A. L. Boehman, Radiation heat transfer in catalytic monoliths, AIChE Journal 44 (1998) 2745–2755. doi: http://dx.doi.org/10.1002/aic.690441215.

[13] R. E. Hayes, S. T. Kolaczkowski, Introduction to Catalytic Combustion, Gordon and Breach, New York, NY, 1997.

[14] J. Mantzaras, C. Appel, Effects of finite rate heterogeneous kinetics on homogeneous ignition in catalytically stabilized channel flow combustion, Combustion and Flame 130 (2002) 336–351. doi: http://dx.doi.org/10.1016/S0010-2180(02)00384-X.

[15] F. A. Williams, Combustion Theory, Benjamin/Cummings, Menlo Park, CA, 1985.

[16] X. Song, W. R. Williams, L. D. Schmidt, R. Aris, Ignition and extinction of homogeneous-heterogeneous combustion: CH_4 and C_3H_8 oxidation on Pt, Proceedings Combustion Institute 23 (1990) 1129–1137. doi: http://dx.doi.org/10.1016/S0082-0784(06)80372-3.

[17] C. Appel, J. Mantzaras, R. Schaeren, R. Bombach, A. Inauen, B. Kaeppeli, B. Hemmerling, A. Stampanoni, An experimental and numerical investigation of homogeneous ignition in catalytically stabilized combustion of hydrogen/air mixtures over platinum, Combustion and Flame 128 (2002) 340–368. doi: http://dx.doi.org/10.1016/S0010-2180(01)00363-7.

[18] M. Reinke, J. Mantzaras, R. Bombach, S. Schenker, N. Tylli, K. Boulouchos, Effects of H_2O and CO_2 dilution on the catalytic and gas-phase combustion of methane over platinum at elevated pressures, Combustion Science and Technology 179 (2006) 553–600. doi: http://dx.doi.org/10.1080/00102200600671930.

[19] J. C. G. Andrae, P. H. Björnbom, Wall effects of laminar hydrogen flames over platinum and inert surfaces, AIChE Journal 46 (2000) 1454–1460. doi: http://dx.doi.org/10.1002/aic.690460718.

[20] P. A. Bui, D. G. Vlachos, P. R. Westmoreland, Homogeneous ignition of hydrogen/air mixtures over platinum, Proceedings Combustion Institute 26 (1996) 1763–1770. doi: http://dx.doi.org/10.1016/S0082-0784(96)80402-4.

[21] S. Karagiannidis, J. Mantzaras, Numerical investigation on the start-up of methane-fueled catalytic microreactors, Combustion and Flame 157 (2010) 1400–1413. doi: http://dx.doi.org/10.1016/j.combustflame.2010.01.008.

[22] S. Karagiannidis, J. Mantzaras, K. Boulouchos, Stability of hetero-/homogeneous combustion in propane- and methane-fueled catalytic microreactors: channel confinement and molecular transport effects, Proceedings Combustion Institute 33 (2011) 3241–3249. doi: http://dx.doi.org/10.1016/j.proci.2010.05.107.

[23] J. Mantzaras, Catalytic combustion of syngas, Combustion Science and Technology 180 (2008) 1137–1168. doi: http://dx.doi.org/10.1080/00102200801963342.

[24] G. Pizza, J. Mantzaras, C. E. Frouzakis, A. G. Tomboulides, K. Boulouchos, Suppression of combustion instabilities of premixed hydrogen/air flames in microchannels using heterogeneous reactions, Proceedings Combustion Institute 32 (2009) 3051–3058. doi: http://dx.doi.org/10.1016/j.proci.2008.05.055.

[25] G. Pizza, J. Mantzaras, C. E. Frouzakis, Flame dynamics in catalytic and non-catalytic mesoscale microreactors, Catalysis Today 155 (2010) 123–130. doi: http://dx.doi.org/10.1016/j.cattod.2010.04.029.

[26] N. I. Kim, K. Maruta, A numerical study on propagation of premixed flames in small tubes, Combustion and Flame 146 (2006) 283–301. doi: http://dx.doi.org/10.1016/j.combustflame.2006.04.001.

[27] O. Deutschmann, L. I. Maier, U. Riedel, A. H. Stroemman, R. W. Dibble, Hydrogen assisted catalytic combustion of methane on platinum, Catalysis Today 59 (2000) 141–150. doi: http://dx.doi.org/10.1016/S0920-5861(00)00279-0.

[28] J. Warnatz, R. W. Dibble, U. Maas, Combustion, Physical and Chemical Fundamentals, Modeling and Simulation, Springer-Verlag, New York, NY, 1996.

[29] Y. Ghermay, J. Mantzaras, R. Bombach, Effects of hydrogen preconversion on the homogeneous ignition of fuel lean $H_2/O_2/N_2/CO_2$ mixtures over platinum at moderate pressures, Combustion and Flame 157 (2010) 1942–1958. doi: http://dx.doi.org/10.1016/j.combustflame.2010.02.016.

[30] R. W. Schefer, Catalyzed combustion of H_2/air mixtures in a flat boundary layer: II. Numerical model, Combustion and Flame 45 (1982) 171–190. doi: http://dx.doi.org/10.1016/0010-2180(82)90043-8.

[31] G. C. Snow, W. V. Krill, E. K. Chu, R. K. Kendall, Mechanisms and kinetics in catalytic combustion, Report No. EPA-600/9-84-002, 1984.

[32] O. A. Hougen, K. M. Watson, Solid catalysts and reaction rates—general principles, Journal of Industrial and Engineering Chemistry 35 (1943) 529–541. doi: http://dx.doi.org/10.1021/ie50401a005.

[33] R. J. Kee, M. E. Coltrin, P. Glaborg, Chemically Reacting Flow, Theory and Practice, Wiley, Hoboken, NJ, 2003. doi: http://dx.doi.org/10.1002/0471461296.

[34] H. Motz, H. Wise, Diffusion and heterogeneous reaction. III. Atom recombination at a catalytic boundary, Journal of Chemical Physics 32 (1960) 1893–1894. doi: http://dx.doi.org/10.1063/1.1731060.

[35] R. Kissel-Osterrieder, F. Behrendt, J. Warnatz, U. Metka, H. R. Volpp, J. Wolfrum, Experimental and theoretical investigation of CO oxidation on platinum: bridging the pressure and materials gap, Proceedings Combustion Institute 28 (2000) 1341–1348. doi: http://dx.doi.org/10.1016/S0082-0784(00)80348-3.

[36] R. Kissel-Osterrieder, F. Behrendt, J. Warnatz, Detailed modeling of the oxidation of CO on platinum: a Monte Carlo model, Proceedings Combustion Institute 27 (1998) 2267–2274. doi: http://dx.doi.org/10.1016/S0082-0784(98)80076-3.

[37] V. Zhdanov, B. Kasemo, Monte Carlo simulation of oscillations in the NO-H_2 reaction on Pt(100), Applied Catalysis A: General 187 (1999) 61–71. doi: http://dx.doi.org/10.1016/S0926-860X(99)00183-0.

[38] V. Zhdanov, Kinetic oscillations on nm-sized catalyst particles: NO reduction by CO on Pt, Catalysis Letters 93 (2004) 135–138. doi: http://dx.doi.org/10.1023/B:CATL.0000017066.40546.98.

[39] V. Zhdanov, Pattern formation in catalytic reactions due to lateral adsorbate-adsorbate interactions, Langmuir 17 (2001) 1793–1799. doi: http://dx.doi.org/10.1021/la0104222.

[40] J. Frenken, B. Hendriksen, The reactor-STM: A real-space probe for operando nanocatalysis, Materials Research Society 32 (2007) 1015–1021.

[41] H. Harle, A. Lehnert, U. Metka, H. R. Volpp, L. Willms, J. Wolfrum, In-situ detection of chemisorbed CO on a polycrystalline platinum foil using infrared-visible sum-frequency generation, Chemical Physics Letters 293 (1998) 26–32. doi: http://dx.doi.org/10.1016/S0009-2614(98)00754-4.

[42] N. McMillan, C. Snively, J. Lauterbach, Spatio-temporal dynamics of the NO + NH_3 reaction on polycrystalline platinum, Surface Science 601 (2007) 772–780. doi: http://dx.doi.org/10.1016/j.susc.2006.11.005.

[43] C. Perego, S. Peratello, Experimental methods in catalytic kinetics, Catalysis Today 52 (1999) 133–145. doi: http://dx.doi.org/10.1016/S0920-5861(99)00071-1.

[44] J. G. McCarty, Kinetics of PdO combustion catalysis, Catalysis Today 26 (1995) 283–293. doi: http://dx.doi.org/10.1016/0920-5861(95)00150-7.

[45] M. Lyubovsky, L. Pfefferle, Complete methane oxidation over Pd catalyst supported on alpha-alumina. Influence of temperature and oxygen pressure on the catalyst activity, Catalysis Today 47 (1999) 29–44. doi: http://dx.doi.org/10.1016/S0920-5861(98)00281-8.

[46] O. Deutschmann, R. Schmidt, F. Behrendt, J. Warnatz, Numerical modeling of catalytic ignition, Proceedings Combustion Institute 26 (1996) 1747–1754. doi: http://dx.doi.org/10.1016/S0082-0784(96)80400-0.

[47] H. Ikeda, J. Sato, F. A. Williams, Surface kinetics for catalytic combustion of hydrogen-air mixtures on platinum at atmospheric-pressure in stagnation flows, Surface Science 326 (1995) 11–26. doi: http://dx.doi.org/10.1016/0039-6028(95)00765-2.

[48] Y. K. Park, P. Aghalayam, D. G. Vlachos, A generalized approach for predicting coverage-dependent reaction parameters of complex surface reactions: Application to H_2 oxidation over platinum, Journal Physical Chemistry A 103 (1999) 8101–8107. doi: http://dx.doi.org/10.1021/jp9916485.

[49] V. Dupont, S. H. Zhang, A. Williams, Experiments and simulations of methane oxidation on a platinum surface, Chemical Engineering Science 56 (2001) 2659–2670. doi: http://dx.doi.org/10.1016/S0009-2509(00)00536-4.

[50] M. E. Coltrin, R. J. Kee, G. H. Evans, E. Meeks, F. M. Rupley, J. M. Grcar, Spin: A Fortran program for modeling one-dimensional rotating-disk/stagnation-flow chemical vapor deposition reactors, Report No. SAND 91-8003, Sandia National Laboratories, Livermore, CA, 1996.

[51] A. Schneider, J. Mantzaras, R. Bombach, S. Schenker, N. Tylli, P. Jansohn, Laser induced fluorescence of formaldehyde and Raman measurements of major species during partial catalytic oxidation of methane with large H_2O and CO_2 dilution at pressures up to 10 bar, Proceedings Combustion Institute 31 (2007) 1973–1981. doi: http://dx.doi.org/10.1016/j.proci.2006.08.076.

[52] S. Karagiannidis, J. Mantzaras, S. Schenker, K. Boulouchos, Experimental and numerical investigation of the hetero-/homogeneous combustion of lean propane/air mixtures over platinum, Proceedings Combustion Institute 32 (2009) 1947–1955. doi: http://dx.doi.org/10.1016/j.proci.2008.06.063.

[53] J. Mantzaras, R. Bombach, R. Schaeren, Hetero-/homogeneous combustion of hydrogen/air mixtures over platinum at pressures up to 10 bar, Proceedings Combustion Institute 32 (2009) 1937–1945. doi: http://dx.doi.org/10.1016/j.proci.2008.06.067.

[54] M. Reinke, J. Mantzaras, R. Schaeren, R. Bombach, A. Inauen, S. Schenker, Homogeneous ignition of CH_4/air and H_2O- and CO_2-diluted CH_4/O_2 mixtures over platinum; an experimental and numerical investigation at pressures up to 16 bar, Proceedings Combustion Institute 30 (2005) 2519–2527. doi: http://dx.doi.org/10.1016/j.proci.2004.08.054.

[55] M. Reinke, J. Mantzaras, R. Schaeren, R. Bombach, A. Inauen, S. Schenker, High-pressure catalytic combustion of methane over platinum: In situ experiments and detailed numerical predictions, Combustion and Flame 136 (2004) 217–240. doi: http://dx.doi.org/10.1016/j.combustflame.2003.10.003.

[56] M. Reinke, J. Mantzaras, R. Bombach, S. Schenker, A. Inauen, Gas phase chemistry in catalytic combustion of methane/air mixtures over platinum at pressures of 1 bar to 16 bar, Combustion and Flame 141 (2005) 448–468. doi: http://dx.doi.org/10.1016/j.combustflame.2005.01.016.

[57] C. Appel, J. Mantzaras, R. Schaeren, R. Bombach, A. Inauen, N. Tylli, M. Wolf, T. Griffin, D. Winkler, R. Carroni, Partial catalytic oxidation of methane to synthesis gas over rhodium: In situ Raman measurements and detailed simulations, Proceedings Combustion Institute 30 (2005) 2509–2517. doi: http://dx.doi.org/10.1016/j.proci.2004.08.055.

[58] J. D. Taylor, M. D. Allendorf, A. H. McDaniel, S. F. Rice, In situ diagnostics and modeling of methane catalytic partial oxidation on Pt in a stagnation-flow reactor, Industrial and Engineering Chemical Research 42 (2003) 6559–6566. doi: http://dx.doi.org/10.1021/ie020934r.

[59] T. F. Garetto, E. Rincon, C. R. Apesteguia, Deep oxidation of propane on Pt-supported catalysts: Drastic turnover rate enhancement using zeolite supports, Applied Catalysis B: Environmental 48 (2004) 167–174. doi: http://dx.doi.org/10.1016/j.apcatb.2003.10.004.

[60] J. F. Kramer, S. A. S. Reihani, G. S. Jackson, Low-temperature combustion of hydrogen on supported Pd catalysts, Proceedings Combustion Institute 29 (2002) 989–996. doi: http://dx.doi.org/10.1016/S1540-7489(02)80125-0.

[61] B. Hellsing, B. Kasemo, V. P. Zhdanov, Kinetics of the hydrogen oxygen reaction on platinum, Journal of Catalysis 132 (1991) 210–228. doi: http://dx.doi.org/10.1016/0021-9517(91)90258-6.

[62] E. Fridell, A. Rosen, B. Kasemo, A laser-induced fluorescence study of OH desorption from Pt in H_2O/O_2 and H_2O/H_2 mixtures, Langmuir 10 (1994) 699–708. doi: http://dx.doi.org/10.1021/la00015a018.

[63] W. R. Williams, C. M. Marks, L. D. Schmidt, Steps in the reaction $H_2 + O_2 = H_2O$ on Pt—OH desorption at high-temperatures, Journal of Physical Chemistry 96 (1992) 5922–5931. doi: http://dx.doi.org/10.1021/j100193a051.

[64] P. Aghalayam, Y. K. Park, D. G. Vlachos, Construction and optimization of complex surface-reaction mechanisms, AIChE Journal 46 (2000) 2017–2029. doi: http://dx.doi.org/10.1002/aic.690461013.

[65] M. Maestri, A. Beretta, T. Faravelli, G. Groppi, E. Tronconi, D. G. Vlachos, Two-dimensional modeling of fuel rich H_2 combustion over Rh/Al_2O_3 catalyst, Chemical Engineering Science 63 (2008) 2657–2669. doi: http://dx.doi.org/10.1016/j.ces.2008.02.024.

[66] M. Rinnemo, D. Kulginov, S. Johansson, K. L. Wong, V. P. Zhdanov, B. Kasemo, Catalytic ignition in the $CO-O_2$ reaction on platinum: Experiment and simulations, Surface Science 376 (1997) 297–309. doi: http://dx.doi.org/10.1016/S0039-6028(96)01572-5.

[67] P. Aghalayam, Y. K. Park, D. G. Vlachos, A detailed surface reaction mechanism for CO oxidation on Pt, Proceedings Combustion Institute 28 (2000) 1331–1339. doi: http://dx.doi.org/10.1016/S0082-0784(00)80347-1.

[68] A. B. Mhadeshwar, D. G. Vlachos, A thermodynamically consistent surface reaction mechanism for CO oxidation on Pt, Combustion and Flame 142 (2005) 289–298. doi: http://dx.doi.org/10.1016/j.combustflame.2005.01.019.

[69] A. B. Mhadeshwar, D. G. Vlachos, Hierarchical, multiscale surface reaction mechanism development: CO and H_2 oxidation, water-gas shift, and preferential oxidation of CO on Rh, Journal of Catalysis 234 (2005) 48–63. doi: http://dx.doi.org/10.1016/j.jcat.2005.05.016.

[70] A. B. Mhadeshwar, P. Aghalayam, V. Papavassiliou, D. G. Vlachos, Surface reaction mechanism development for platinum-catalyzed oxidation of methane, Proceedings Combustion Institute 29 (2002) 997–1004. doi: http://dx.doi.org/10.1016/S1540-7489(02)80126-2.

[71] P. A. Bui, D. G. Vlachos, P. R. Westmoreland, Catalytic ignition of methane/oxygen mixtures over platinum surfaces: comparison of detailed simulations and experiments, Surface Science 385 (1997) L1029–L1034. doi: http://dx.doi.org/10.1016/S0039-6028(97)00438-X.

[72] P. Aghalayam, Y. K. Park, N. Fernandes, V. Papavassiliou, A. B. Mhadeshwar, D. G. Vlachos, A C1 mechanism for methane oxidation on platinum, Journal of Catalysis 213 (2003) 23–38. doi: http://dx.doi.org/10.1016/S0021-9517(02)00045-3.

[73] A. B. Mhadeshwar, D. G. Vlachos, Hierarchical multiscale mechanism development for methane partial oxidation and reforming and for thermal decomposition of oxygenates on Rh, Journal of Physical Chemistry B 109 (2005) 16819–16835. doi: http://dx.doi.org/10.1021/jp052479t.

[74] D. A. Hickman, L. D. Schmidt, Steps in CH_4 oxidation on Pt and Rh surfaces—high temperature reactor simulations, AIChE Journal 39 (1993) 1164–1177. doi: http://dx.doi.org/10.1002/aic.690390708.

[75] O. Deutschmann, L. D. Schmidt, Modeling the partial oxidation of methane in a short-contact-time reactor, AIChE Journal 44 (1998) 2465–2477. doi: http://dx.doi.org/10.1002/aic.690441114.

[76] M. Reinke, J. Mantzaras, R. Schaeren, R. Bombach, W. Kreutner, A. Inauen, Homogeneous ignition in high-pressure combustion of methane/air over platinum: Comparison of measurements and detailed numerical predictions, Proceedings Combustion Institute 29 (2002) 1021–1029. doi: http://dx.doi.org/10.1016/S1540-7489(02)80129-8.

[77] T. A. Griffin, L. D. Pfefferle, M. J. Dyer, D. R. Crosley, The ignition of methane-ethane boundary-layer flows by heated catalytic surface, Combustion Science and Technology 65 (1989) 19–37. doi: http://dx.doi.org/10.1080/00102208908924040.

[78] C. M. Marks, L. D. Schmidt, Hydroxyl radical desorption in catalytic combustion, Chemical Physics Letters 178 (1991) 358–362. doi: http://dx.doi.org/10.1016/0009-2614(91)90265-B.

[79] S. Buser, P. Benz, A. Schlegel, H. Bockhorn, Measurements of OH by LIF and temperatures by holographic interferometry in catalytically stabilized combustion of hydrogen – comparison of measurements and numerical simulation, Physical Chemistry Chemical Physics 97 (1993) 1719–1723.

[80] U. Dogwiler, J. Mantzaras, P. Benz, B. Kaeppeli, R. Bombach, A. Arnold, Homogeneous ignition of methane/air mixtures over platinum: Comparison of measurements and detailed numerical predictions, Proceedings Combustion Institute 27 (1998) 2275–2282. doi: http://dx.doi.org/10.1016/S0082-0784(98)80077-5.

[81] M. Forsth, F. Gudmundson, J. L. Persson, A. Rosen, The influence of a catalytic surface on the gas-phase combustion of $H_2 + O_2$, Combustion and Flame 119 (1999) 144–153. doi: http://dx.doi.org/10.1016/S0010-2180(99)00045-0.

[82] N. Khadiya, N. G. Glumac, Destruction of NO during catalytic combustion on platinum and palladium, Combustion Science and Technology 165 (2001) 249–266. doi: http://dx.doi.org/10.1080/00102200108935834.

[83] G. P. Smith, D. M. Golden, M. Frenklach, N. W. Moriarty, B. Eiteneer, M. Goldenberg, C. T. Bowman, R. K. Hanson, S. Song, W. C. Gardiner, V. Lissianski, Z. Qin, An Optimized Detailed Chemical Reaction Mechanism for Methane Combustion, 2000, http://www.me.berkeley.edu/gri_mech

[84] K. J. Hughes, T. Turanyi, A. Clague, M. J. Pilling, Development and testing of a comprehensive chemical mechanism for the oxidation of methane, International Journal of Chemical Kinetics 33 (2001) 513–538. doi: http://dx.doi.org/10.1002/kin.1048.

[85] P. A. Bui, E. A. Wilder, D. G. Vlachos, P. R. Westmoreland, Hierarchical reduced models for catalytic combustion: H_2/air mixtures near platinum surfaces, Combustion Science and Technology 129 (1997) 243–275. doi: http://dx.doi.org/10.1080/00102209708935728.

[86] X. Yan, U. Maas, Intrinsic low-dimensional manifolds of heterogeneous combustion processes, Proceedings Combustion Institute 28 (2000) 1615–1621. doi: http://dx.doi.org/10.1016/S0082-0784(00)80559-7.

[87] S. Raimondeau, D. Norton, D. G. Vlachos, R. I. Masel, Modeling of high-temperature microburners, Proceedings of the Combustion Institute 29 (2003) 901–907. doi: http://dx.doi.org/10.1016/S1540-7489(02)80114-6.

[88] D. G. Norton, D. G. Vlachos, Combustion characteristics and flame stability at the microscale: A CFD study of premixed methane/air mixtures, Chemical Engineering Science 58 (2003) 4871–4882. doi: http://dx.doi.org/10.1016/j.ces.2002.12.005.

[89] A. Yamamoto, H. Oschibe, H. Nakamura, T. Tezuka, S. Hasegawa, K. Maruta, Stabilized three-stage oxidation of gaseous n-heptane/air mixture in a micro flow reactor with a controlled temperature profile, Proceedings Combustion Institute 33 (2011) 3259–3266. doi: http://dx.doi.org/10.1016/j.proci.2010.06.087.

[90] A. E. Cerkanowicz, R. B. Cole, J. G. Stevens, Catalytic combustion modeling—comparisons with experimental data, Journal of Engineering Power Transactions 99 (1977) 593–600. doi: http://dx.doi.org/10.1115/1.3446556.

[91] G. Groppi, A. Belloli, E. Tronconi, P. Forzatti, A comparison of lumped and distributed models of monolith catalytic combustors, Chemical Engineering Science 50 (1995) 2705–2715. doi: http://dx.doi.org/10.1016/0009-2509(95)00099-Q.

[92] J. Mantzaras, P. Benz, An asymptotic and numerical investigation of homogeneous ignition in catalytically stabilized channel flow combustion, Combustion and Flame 119 (1999) 455–472. doi: http://dx.doi.org/10.1016/S0010-2180(99)00071-1.

[93] P. Canu, S. Vecchi, CFD simulation of reactive flows: Catalytic combustion in a monolith, AIChE Journal 48 (2002) 2921–2935. doi: http://dx.doi.org/10.1002/aic.690481219.

[94] N. Mladenov, J. Koop, S. Tischer, O. Deutschmann, Modeling of transport and chemistry in channel flows of automotive catalytic converters, Chemical Engineering Science 65 (2010) 812–826. doi: http://dx.doi.org/10.1016/j.ces.2009.09.034.

[95] R. E. Hayes, S. T. Kolaczkowski, Mass and heat transfer effects in catalytic monolith reactors, Chemical Engineering Science 49 (1994) 3587–3599. doi: http://dx.doi.org/10.1016/0009-2509(94)00164-2.

[96] C. Bruno, P. M. Walsh, D. A. Santavicca, N. Sinha, Y. Yaw, F. V. Bracco, Catalytic combustion of propane-air mixtures on platinum, Combustion Science and Technology 31 (1983) 43–74. doi: http://dx.doi.org/10.1080/00102208308923630.

[97] R. E. Hayes, S. T. Kolaczkowski, W. J. Thomas, Finite-element model for a catalytic monolith reactor, Computers and Chemical Engineering 16 (1992) 645–657. doi: http://dx.doi.org/10.1016/0098-1354(92)80014-Z.

[98] R. Wanker, H. Raupenstrauch, G. Staudinger, A fully distributed model for the simulation of a catalytic combustor, Chemical Engineering Science 55 (2000) 4709–4718. doi: http://dx.doi.org/10.1016/S0009-2509(00)00060-9.

[99] L. L. Raja, R. J. Kee, O. Deutschmann, J. Warnatz, L. D. Schmidt, A critical evaluation of Navier–Stokes, boundary-layer, and plug-flow models of the flow and chemistry in a catalytic

combustion monolith, Catalysis Today 59 (2000) 47–60. doi: http://dx.doi.org/10.1016/S0920-5861(00)00271-6.

[100] J. Mantzaras, C. Appel, P. Benz, Catalytic combustion of methane/air mixtures over platinum: homogeneous ignition distances in channel flow configurations, Proceedings Combustion Institute 28 (2000) 1349–1357. doi: http://dx.doi.org/10.1016/S0082-0784(00)80349-5.

[101] P. Markatou, L. D. Pfefferle, M. D. Smooke, A computational study of methane air combustion over heated catalytic and non-catalytic surfaces, Combustion and Flame 93 (1993) 185–201. doi: http://dx.doi.org/10.1016/0010-2180(93)90146-T.

[102] M. E. Coltrin, H. K. Moffat, R. J. Kee, F. M. Rupley, Creslaf: A Fortran program for modeling laminar, chemically reacting, boundary-layer flow in cylindrical or planar channels, Report No. SAND 93-0478, Sandia National Laboratories, Livermore, CA, 1996.

[103] C. Appel, J. Mantzaras, R. Schaeren, R. Bombach, B. Kaeppeli, A. Inauen, An experimental and numerical investigation of turbulent catalytically stabilized channel flow combustion of hydrogen/air mixtures over platinum, Proceedings Combustion Institute 29 (2002) 1031–1038. doi: http://dx.doi.org/10.1016/S1540-7489(02)80130-4.

[104] J. Mantzaras, C. Appel, P. Benz, U. Dogwiler, Numerical modelling of turbulent catalytically stabilized channel flow combustion, Catalysis Today 59 (2000) 3–17. doi: http://dx.doi.org/10.1016/S0920-5861(00)00268-6.

[105] S. Y. Yamamoto, C. M. Surko, M. B. Maple, R. K. Pina, Spatio-temporal dynamics of oscillatory heterogeneous catalysis: CO oxidation on platinum, Journal of Chemical Physics 102 (1995) 8614–8625. doi: http://dx.doi.org/10.1063/1.468963.

[106] P. A. Carlsson, V. Zhdanov, M. Skoglundh, Self-sustained kinetic oscillations in CO oxidation over silica-supported Pt, Physical Chemistry Chemical Physics 8 (2006) 2703–2706. doi: http://dx.doi.org/10.1039/b608486a.

[107] R. Imbihl, G. Ertl, Oscillatory kinetics in heterogeneous catalysis, Chemical Reviews 95 (1995) 697–733. doi: http://dx.doi.org/10.1021/cr00035a012.

[108] V. N. Kurdyumov, G. Pizza, C. E. Frouzakis, J. Mantzaras, Dynamics of premixed flames in a narrow channel with a step-wise wall temperature, Combustion and Flame 156 (2009) 2190–2200. doi: http://dx.doi.org/10.1016/j.combustflame.2009.08.001.

[109] F. Richecoeur, D. C. Kyritsis, Experimental study of flame stabilization in low Reynolds and Dean number flows in curved mesoscale ducts, Proceedings Combustion Institute 30 (2005) 2419–2427. doi: http://dx.doi.org/10.1016/j.proci.2004.08.015.

[110] J. S. Tien, Transient catalytic combustor model, Combustion Science and Technology 26 (1981) 65–75. doi: http://dx.doi.org/10.1080/00102208108946947.

[111] N. Sinha, C. Bruno, F. V. Bracco, Two-dimensional, transient catalytic combustion of CO-air on platinum, Physicochemical Hydrodynamics 6 (1985) 373–391.

[112] R. E. Hayes, S. T. Kolaczkowski, W. J. Thomas, J. Titiloye, Transient experiments and modeling of the catalytic combustion of methane in a monolith reactor, Industrial and Engineering Chemical Research 35 (1996) 406–414. doi: http://dx.doi.org/10.1021/ie950308c.

[113] O. Deutschmann, F. Behrendt, J. Warnatz, Formal treatment of catalytic combustion and catalytic conversion of methane, Catalysis Today 46 (1998) 155–163. doi: http://dx.doi.org/10.1016/S0920-5861(98)00337-X.

[114] L. L. Raja, R. J. Kee, L. R. Petzold, Simulation of the transient, compressible, gas-dynamic behavior of catalytic combustion ignition in stagnation flows, Proceedings Combustion Institute 27 (1998) 2249–2257. doi: http://dx.doi.org/10.1016/S0082-0784(98)80074-X.

[115] R. Aris, Models of the catalytic monolith, Proceedings, 1st Levich Conference, Oxford, 1977.

[116] Y. H. Liu, Equivalent continuum models for nonadiabatic monolith catalytic reactors, Ph.D. Dissertation, Rice University, Houston, TX, 1986.

[117] R. Schwiedernoch, S. Tischer, O. Deutschmann, J. Warnatz, Experimental and numerical investigation of the ignition of methane combustion in a platinum-coated honeycomb monolith,

Proceedings Combustion Institute 29 (2002) 1005–1011. doi: http://dx.doi.org/10.1016/S1540-7489(02)80127-4.

[118] S. Tischer, C. Correa, O. Deutschmann, Transient three-dimensional simulations of a catalytic combustion monolith using detailed models for heterogeneous and homogeneous reactions and transport phenomena, Catalysis Today 69 (2001) 57–62. doi: http://dx.doi.org/10.1016/S0920-5861(01)00355-8.

[119] R. Schwiedernoch, S. Tischer, C. Correa, O. Deutschmann, Experimental and numerical study on the transient behavior of partial oxidation of methane in a catalytic, monolith, Chemical Engineering Science 58 (2003) 633–642. doi: http://dx.doi.org/10.1016/S0009-2509(02)00589-4.

[120] S. Kalamatianos, Y. K. Park, D. G. Vlachos, Two-parameter continuation algorithms for sensitivity analysis, parametric dependence, reduced mechanisms, and stability criteria of ignition and extinction, Combustion and Flame 112 (1998) 45–61. doi: http://dx.doi.org/10.1016/S0010-2180(97)81756-7.

[121] S. Kalamatianos, D. G. Vlachos, Bifurcation behavior of premixed hydrogen/air mixtures in a continuous stirred tank reactor, Combustion Science and Technology 109 (1995) 347–371. doi: http://dx.doi.org/10.1080/00102209508951909.

[122] Y. K. Park, D. G. Vlachos, Kinetically driven instabilities and selectivities in methane oxidation, AIChE Journal 43 (1997) 2083–2095. doi: http://dx.doi.org/10.1002/aic.690430911.

[123] O. N. Temkin, A. V. Zeigarnik, R. E. Valdes-Perez, L. G. Bruk, Critical phenomena as a discriminating factor in the studies of reaction mechanisms, Kinetics and Catalysis 41 (2000) 298–299. doi: http://dx.doi.org/10.1007/BF02755364.

[124] A. Schlegel, P. Benz, T. Griffin, W. Weisenstein, H. Bockhorn, Catalytic stabilization of lean premixed combustion: Method for improving NO_x emissions, Combustion and Flame 105 (1996) 332–340. doi: http://dx.doi.org/10.1016/0010-2180(95)00211-1.

[125] H. K. Moffat, P. Glaborg, R. J. Kee, J. F. Grcar, J. A. Miller, Surface PSR: A Fortran program for modeling well-stirred reactors with gas and surface reactions, SAND91-8001 Report, Sandia National Laboratories, Livermore, CA, 1993.

[126] M. E. Coltrin, R. J. Kee, F. M. Rupley, Surface Chemkin: A Fortran package for analyzing heterogeneous chemical kinetics at a solid-surface–gas-phase interphase, Report No. SAND90-8003C, Sandia National Laboratories, Livermore, CA, 1991.

[127] M. Fassihi, Catalytic oxidation of H_2 on Pt, Licenciate report, Department of Physics, Goteborg, 1991.

[128] D. Kulginov, V. Zhdanov, B. Kasemo, Oscillatory surface reaction kinetics due to coupling of bistability and diffusion limitations, Journal of Chemical Physics 106 (1997) 3117–3128. doi: http://dx.doi.org/10.1063/1.473086.

[129] F. J. Weinberg, D. M. Rowe, G. Min, P. D. Ronney, On thermoelectric power conversion from heat recirculating combustion systems, Proceedings Combustion Institute 29 (2002) 941–947. doi: http://dx.doi.org/10.1016/S1540-7489(02)80119-5.

[130] S. T. Kolaczkowski, S. Serbetcioglu, Development of combustion catalysts for monolith reactors: a consideration of transport limitations, Applied Catalysis A: General 138 (1996) 199–214. doi: http://dx.doi.org/10.1016/0926-860X(95)00296-0.

[131] S.-T. Lee, R. Aris, On the effects of radiative heat transfer in monoliths, Chemical Engineering Science 37 (1977) 827–837. doi: http://dx.doi.org/10.1016/0009-2509(77)80068-7.

[132] H. Oschibe, H. Nakamura, T. Tezuka, S. Hasegawa, K. Maruta, Stabilized three-stage oxidation of DME/air mixture in a micro flow reactor with a controlled temperature profile, Combustion and Flame 157 (2010) 1572–1580. doi: http://dx.doi.org/10.1016/j.combustflame.2010.03.004.

[133] R. J. Kee, G. Dixon-Lewis, J. Warnatz, M. E. Coltrin, J. A. Miller, A Fortran computer code package for the evaluation of gas-phase multicomponent transport properties, Report No. SAND86-8246, Sandia National Laboratories, Livermore, CA, 1996.

[134] R. E. Hayes, S. T. Kolaczkowski, A study of Nusselt and Sherwood numbers in a monolith reactor, Catalysis Today 47 (1999) 295–303. doi: http://dx.doi.org/10.1016/S0920-5861(98)00310-1.

[135] R. K. Shah, A. L. London, Laminar Flow Forced Convection in Ducts, Academic Press, New York, NY, 1978.

[136] N. S. Kaisare, S. R. Deshmukh, D. G. Vlachos, Stability and performance of catalytic microreactors: Simulations of propane catalytic combustion on Pt, Chemical Engineering Science 63 (2008) 1098–1116. doi: http://dx.doi.org/10.1016/j.ces.2007.11.014.

[137] F. A. Robbins, H. Y. Zhu, G. S. Jackson, Transient modeling of combined catalytic combustion/ CH_4 steam reforming, Catalysis Today 83 (2003) 141–156. doi: http://dx.doi.org/10.1016/S0920-5861(03)00225-6.

[138] R. Prasad, L. A. Kennedy, E. Ruckenstein, A model for the transient behavior of catalytic combustors, Combustion Science and Technology 30 (1983) 59–88. doi: http://dx.doi.org/10.1080/00102208308923612.

[139] T. Ahn, W. V. Pinczewski, D. L. Trimm, Transient performance of catalytic combustors for gas-turbine applications, Chemical Engineering Science 41 (1986) 55–64. doi: http://dx.doi.org/10.1016/0009-2509(86)85197-1.

[140] C. Trevino, M. Sen, Catalytic combustion in monolith reactors, Chemical Engineering Science 41 (1986) 2253–2260. doi: http://dx.doi.org/10.1016/0009-2509(86)85076-X.

[141] H. Zhu, G. S. Jackson, Transient modeling for assessing catalytic combustor performance in small gas turbine applications, ASME paper No. 2001-GT-0520, ASME Turbo Expo 2001, New Orleans, LA, 2001.

[142] D. G. Norton, E. D. Wetzel, D. G. Vlachos, Thermal management in catalytic microreactors, Industrial and Engineering Chemical Research 45 (2006) 76–84. doi: http://dx.doi.org/10.1021/ie050674o.

[143] G. Pizza, C. E. Frouzakis, J. Mantzaras, A. G. Tomboulides, K. Boulouchos, Dynamics of premixed hydrogen/air flames in microchannels, Combustion and Flame 152 (2008) 433–450. doi: http://dx.doi.org/10.1016/j.combustflame.2007.07.013.

[144] G. Pizza, C. E. Frouzakis, J. Mantzaras, A. G. Tomboulides, K. Boulouchos, Dynamics of premixed hydrogen/air flames in mesoscale channels, Combustion and Flame 155 (2008) 2–20. doi: http://dx.doi.org/10.1016/j.combustflame.2008.08.006.

[145] F. M. Rupley, R. J. Kee, J. A. Miller, Premix: A Fortran program for modeling steady laminar one-dimensional premixed flames, Report No. SAND85-8240, Sandia National Laboratories, Livermore, CA, 1995.

[146] R. Seiser, H. Pitsch, K. Seshadri, W. J. Pitz, H. J. Curran, Extinction and autoignition of n-heptane in counterflow configuration, Proceedings Combustion Institute 28 (2000) 2029–2036. doi: http://dx.doi.org/10.1016/S0082-0784(00)80610-4.

[147] Y. Ju, X. Bo, Theoretical and experimental studies on mesoscale flame propagation and extinction, Proceedings Combustion Institute 30 (2005) 2445–2453. doi: http://dx.doi.org/10.1016/j.proci.2004.08.234.

[148] S. Minaev, K. Maruta, R. Fursenko, Nonlinear dynamics in a narrow channel with a temperature gradient, Combustion Theory and Modelling 11 (2007) 187–203. doi: http://dx.doi.org/10.1080/13647830600649364.

[149] T. T. Leach, C. P. Cadou, G. S. Jackson, Effect of structural conduction and heat loss on combustion in micro-channels, Combustion Theory and Modelling 10 (2006) 85–103. doi: http://dx.doi.org/10.1080/13647830500277332.

[150] Y. Ju, C. W. Choi, An analysis of sub-limit flame dynamics using opposite propagating flames in mesoscale channels, Combustion and Flame 133 (2003) 483–493. doi: http://dx.doi.org/10.1016/S0010-2180(03)00058-0.

[151] P. D. Ronney, Analysis of non-adiabatic heat-recirculating combustors, Combustion and Flame 135 (2003) 421–439. doi: http://dx.doi.org/10.1016/j.combustflame.2003.07.003.

[152] N. S. Kaisare, D. G. Vlachos, Optimal reactor dimensions for homogeneous combustion in small channels, Catalysis Today 120 (2007) 96–106. doi: http://dx.doi.org/10.1016/j.cattod.2006.07.036.

[153] N. S. Kaisare, D. G. Vlachos, Extending the region of homogeneous micro-combustion through forced unsteady operation, Proceedings Combustion Institute 31 (2007) 3293–3300. doi: http://dx.doi.org/10.1016/j.proci.2006.07.031.

[154] B. Kasemo, K. E. Keck, T. Högberg, A fast response, local gas-sampling system for studies of catalytic reactions at $1-10^3$ Torr; application to the H_2-D_2 exchange and H_2 oxidation reactions on Pt, Journal of Catalysis 66 (1980) 441–450. doi: http://dx.doi.org/10.1016/0021-9517(80)90046-9.

[155] T. Poinsot, D. Veynante, Theoretical and Numerical Combustion, Edwards, Philadelphia, PA, 2005.

[156] J. Mantzaras, Gas phase combustion of ZEP mixtures in inert channel flows, Internal Report, EU project Advanced Zero Emissions Power, 2004.

[157] L. J. Spadaccini, M. B. Colket, Ignition delay characteristics of methane fuels, Progress in Energy and Combustion Science 20 (1994) 431–460. doi: http://dx.doi.org/10.1016/0360-1285(94)90011-6.

[158] K. Zygourakis, Transient operation of monolith catalytic converters: A two-dimensional reactor model and the effects of radially nonuniform flow distributions, Chemical Engineering Science 44 (1989) 2075–2086. doi: http://dx.doi.org/10.1016/0009-2509(89)85143-7.

[159] G. Groppi, E. Tronconi, Continuous vs. discrete models of nonadiabatic monolith catalysts, AIChE Journal 42 (1996) 2382–2387. doi: http://dx.doi.org/10.1002/aic.690420829.

[160] A. C. Fernandez-Pello, Micropower generation using combustion: Issues and approaches, Proceedings Combustion Institute 29 (2002) 883–899. doi: http://dx.doi.org/10.1016/S1540-7489(02)80113-4.

[161] A. Gomez, J. J. Berry, S. Roychoudhury, B. Coriton, J. Huth, From jet fuel to electric power using a mesoscale, efficient stirling cycle, Proceedings Combustion Institute 31 (2007) 3251–3259. doi: http://dx.doi.org/10.1016/j.proci.2006.07.203.

[162] A. M. Karim, J. A. Federici, D. G. Vlachos, Portable power production from methanol in an integrated thermoelectric/microreactor system, Journal of Power Sources 179 (2008) 113–120. doi: http://dx.doi.org/10.1016/j.jpowsour.2007.12.119.

[163] S. J. Volchko, C. J. Sung, Y. Huang, S. J. Schneider, Catalytic combustion of rich methane/oxygen mixtures for micropropulsion applications, Journal of Propulsion and Power 22 (2006) 684–693. doi: http://dx.doi.org/10.2514/1.19809.

[164] P. K. Cheekatamarla, C. M. Finnerty, C. R. Robinson, S. M. Andrews, J. A. Brodie, Y. Lu, P. G. Dewald, Design, integration and demonstration of a 50 W JP8/kerosene fueled portable SOFC power generator, Journal Power Sources 193 (2009) 797–803. doi: http://dx.doi.org/10.1016/j.jpowsour.2009.04.060.

[165] M. J. Stutz, D. Poulikakos, Optimum washcoat thickness of a monolith reactor for syngas production by partial oxidation of methane, Chemical Engineering Science 63 (2008) 1761–1770. doi: http://dx.doi.org/10.1016/j.ces.2007.11.032.

[166] B. Schneider, S. Karagiannidis, M. Bruderer, D. Dyntar, C. Zwyssig, Q. Guangchun, M. Diener, K. Boulouchos, R. S. Abhari, L. Guzzella, J. W. Kolar, Ultra-high-energy-density converter for portable power, Power-MEMS 2005, Tokyo, Japan, November 28–30, 2005.

[167] R. Siegel, J. R. Howell, Thermal Radiation Heat Transfer, Hemisphere, New York, NY, p. 271, 1981.

[168] Z. Qin, V. V. Lissianski, H. Yang, W. C. Gardiner, S. G. Davis, H. Wang, Combustion chemistry of propane: A case study of detailed reaction mechanism optimization, Proceedings Combustion Institute 28 (2000) 1663–1669. doi: http://dx.doi.org/10.1016/S0082-0784(00)80565-2.

[169] C. K. Westbrook, F. L. Dryer, Simplified reaction mechanisms for the oxidation of hydrocarbon fuels in flames, Combustion Science and Technology 27 (1981) 31–43. doi: http://dx.doi.org/10.1080/00102208108946970.

[170] D. G. Norton, D. G. Vlachos, A CFD study of propane/air microflame stability, Combustion and Flame 138 (2004) 97–107. doi: http://dx.doi.org/10.1016/j.combustflame.2004.04.004.

[171] C. Trevino, A. C. Fernandez-Pello, Catalytic flat plate boundary layer ignition, Combustion Science and Technology 26 (1981) 245–251. doi: http://dx.doi.org/10.1080/00102208108946966.

[172] A. Veeraragavan, C. P. Cadou, Flame speed predictions in planar micro/mesoscale combustors with conjugate heat transfer, Combustion and Flame 158 (2011) 2178–2187. doi: http://dx.doi.org/10.1016/j.combustflame.2011.04.006.

[173] K. Maruta, T. Kataoka, N. I. Kim, S. Minaev, R. Fursenko, Characteristics of microscale combustion in a narrow channel with a temperature gradient, Proceedings Combustion Institute 30 (2005) 2429–2436. doi: http://dx.doi.org/10.1016/j.proci.2004.08.245.

[174] T. J. Kim, R. A. Yetter, F. L. Dryer, New results on moist CO oxidation: High pressure, high temperature experiments and comprehensive kinetic modeling, Proceedings Combustion Institute 25 (1994) 759–766. doi: http://dx.doi.org/10.1016/S0082-0784(06)80708-3.

[175] V. Bychkov, V. Akkerman, G. Fru, A. Petchenko, L. E. Eriksson, Flame acceleration in the early stages of burning in tubes, Combustion and Flame 150 (2007) 263–276. doi: http://dx.doi.org/10.1016/j.combustflame.2007.01.004.

[176] F. S. Marra, G. Continillo, Numerical study of premixed laminar flame propagation in a closed tube with a full Navier–Stokes approach, Proceedings Combustion Institute 26 (1996) 907–913. doi: http://dx.doi.org/10.1016/S0082-0784(96)80301-8.

[177] G. Pizza, C. E. Frouzakis, J. Mantzaras, Chaotic dynamics in premixed hydrogen/air channel flow combustion, Combustion Theory and Modelling 16 (2011) 275–299. doi: http://dx.doi.org/10.1080/13647830.2011.620174.

[178] A. Fan, S. Minaev, E. V. Sereshchenko, Y. Tsuboi, H. Nakamura, K. Maruta, Dynamic behavior of splitting flames in a heated channel, Combustion Explosion and Shock Waves 45 (2009) 245–250. doi: http://dx.doi.org/10.1007/s10573-009-0032-6.

[179] G. Pizza, C. E. Frouzakis, J. Mantzaras, A. G. Tomboulides, K. Boulouchos, Three-dimensional simulations of premixed hydrogen/air flames in micro tubes, Journal of Fluid Mechanics 658 (2010) 463–491. doi: http://dx.doi.org/10.1017/S0022112010001837.

[180] V. Kurdyumov, J. M. Truffaut, J. Quinard, A. Wangher, G. Shearby, Oscillations of premixed flames in tubes near the flashback condition, Combustion Science and Technology 180 (2008) 731–742. doi: http://dx.doi.org/10.1080/00102200801893689.

[181] P. Hugo, Stabilität und zeitverhalten von durchfluss-kreislauf-reaktoren, Berichte der Bunsen-Gesellschaft für Physikalische Chemie 74 (1970) 121–127.

[182] H. Beusch, P. Fieguth, E. Wicke, Thermisch und kinetisch verursachte Instabilitaeten im reaktionsverhalten einzelner Katalysatorkoerner, Chemie Ingenieur Technik 44 (1972) 445–451. doi: http://dx.doi.org/10.1002/cite.330440702.

[183] M. Eiswirth, G. Ertl, Kinetic oscillations in the catalytic CO oxidation on a Pt(110) surface, Surface Science 177 (1986) 90–100. doi: http://dx.doi.org/10.1016/0039-6028(86)90259-1.

[184] S. L. Lane, M. D. Graham, D. Luss, Spatiotemporal temperature patterns during hydrogen oxidation on a nickel disk, AIChE Journal 39 (1993) 1497–1508. doi: http://dx.doi.org/10.1002/aic.690390909.

[185] M. M. Slinko, N. I. Jaeger, Oscillatory heterogeneous catalytic systems, Studies in Surface Science and Catalysis, Elsevier, Amsterdam, 1994.

[186] L. F. Razon, R. A. Schmitz, Intrinsically unstable behavior during the oxidation of carbon monoxide on platinum, Catalysis Reviews, Science and Engineering 28 (1986) 89–164. doi: http://dx.doi.org/10.1080/03602458608068086.

[187] Y. Ghermay, J. Mantzaras, R. Bombach, K. Boulouchos, Homogeneous combustion of fuel lean $H_2/O_2/N_2$ mixtures over platinum at elevated pressures and preheats, Combustion and Flame 158 (2011) 1491–1506. doi: http://dx.doi.org/10.1016/j.combustflame.2010.12.025.

[188] J. Li, Z. Zhao, A. Kazakov, F. L. Dryer, An updated comprehensive kinetic model of hydrogen combustion, International Journal of Chemical Kinetics 36 (2004) 566–575. doi: http://dx.doi.org/10.1002/kin.20026.

[189] A. Brambilla, C. E. Frouzakis, J. Mantzaras, A. Tomboulides, S. Kerkemeier, K. Boulouchos, Detailed transient numerical simulation of H_2/air hetero-/homogeneous combustion in platinum-coated channels with conjugate heat transfer, Combustion and Flame, doi.org/10.1016/j.combustflame.2014.04.003, in press, 2014.

[190] A. Minotti, C. Bruno, F. Cozzi, A LES simulation of a CH_4/air microcombustor with detailed chemistry, Combustion Science and Technology 183 (2011) 554–574. doi: http://dx.doi.org/10.1080/00102202.2010.523031.

[191] K. Ezato, A. M. Shehata, T. Kunugi, D. M. McEligot, Numerical prediction of transitional features of turbulent forced gas flows in circular tubes with strong heating, Journal of Heat Transfer, Transactions ASME 121 (1999) 546–555. doi: http://dx.doi.org/10.1115/1.2826015.

[192] S. Satake, T. Kunugi, A. M. Shehata, D. M. McEligot, Direct numerical simulation for laminarization of turbulent forced gas flows in circular tubes with strong heating, International Journal of Heat and Fluid Flow 21 (2000) 526–534. doi: http://dx.doi.org/10.1016/S0142-727X(00)00041-2.

[193] H. C. Chen, V. C. Patel, Near-wall turbulence models for complex flows including separation, AIAA Journal 26 (1988) 641–648. doi: http://dx.doi.org/10.2514/3.9960.

[194] C. B. Hwang, C. A. Lin, Improved low-Reynolds-number k-e model based on direct numerical simulation data, AIAA Journal 36 (1998) 38–43. doi: http://dx.doi.org/10.2514/2.349.

[195] C. K. Law, G. I. Sivashinsky, Catalytic extension of extinction limits of stretched premixed flames, Combustion Science and Technology 29 (1982) 277–286. doi: http://dx.doi.org/10.1080/00102208208923594.

[196] F. Lucci, C. E. Frouzakis, J. Mantzaras, Three-dimensional direct numerical simulation of turbulent channel flow catalytic combustion of hydrogen over platinum, Proceedings Combustion Institute 34 (2013) 2295–2302. doi: http://dx.doi.org/10.1016/j.proci.2012.06.110.

[197] F. Nicoud, F. Ducros, Subgrid-scale stress modelling based on the square of the velocity gradient tensor, Flow Turbulence and Combustion 62 (1999) 183–200. doi: http://dx.doi.org/10.1023/A:1009995426001.

[198] G. A. Boyarko, C. J. Sung, S. J. Schneider, Catalyzed combustion of hydrogen-oxygen in platinum tubes for micro-propulsion applications, Proceedings Combustion Institute 30 (2005) 2481–2488. doi: http://dx.doi.org/10.1016/j.proci.2004.08.203.

[199] B. Xu, Y. Ju, Concentration slip and its impact on heterogeneous combustion in a microscale chemical reactor, Chemical Engineering Science 60 (2005) 3561–3572. doi: http://dx.doi.org/10.1016/j.ces.2005.02.040.

[200] B. Xu, Y. Ju, Theoretical and numerical studies of non-equilibrium slip effects on a catalytic surface, Combustion Theory and Modelling 10 (2006) 961–979. doi: http://dx.doi.org/10.1080/13647830600792313.

[201] W. Marques Jr., G. M. Cremer, F. M. Shapirov, Couette flow with slip and jump boundary conditions, Continuum Mechanics and Thermodynamics 12 (2000) 379–386. doi: http://dx.doi.org/10.1007/s001610050143.

[202] S. Succi, The Lattice Boltzmann Equation, Oxford University Press, New York, NY, 2001.

[203] P. L. Bhatnagar, E. P. Gross, M. A. Krook, A model for collisional processes in gases I: Small amplitude processes in charged and in neutral one-component systems, Physics Review (1954) 511–525. doi: http://dx.doi.org/10.1103/PhysRev.94.511.

[204] S. Succi, Mesoscopic modeling of slip motion at fluid-solid interface interfaces with heterogeneous catalysis, Physical Review Letters 89 (2002) 064502. doi: http://dx.doi.org/10.1103/PhysRevLett.89.064502.

[205] S. Arcidiacono, J. Mantzaras, S. Ansumali, I. V. Karlin, C. Frouzakis, K. Boulouchos, A discrete velocity model for binary mixtures, Physical Review E 74 (2006) 056707. doi: http://dx.doi.org/10.1103/PhysRevE.74.056707.

[206] S. Arcidiacono, I. Karlin, J. Mantzaras, C. Frouzakis, Lattice Boltzmann model for the simulation of multicomponent mixtures, Physical Review E 76 (2007) 046703. doi: http://dx.doi.org/10.1103/PhysRevE.76.046703.

[207] T. Hyakutake, K. Yamamoto, H. Takeuchi, Flow of gas mixtures through micro channel. In: Capitelli M, editor. Rarefied Gas Dynamics: 24th International Symposium. AIP Press, New York, NY, USA: 2005. pp. 780–788.

[208] F. Sharipov, D. Kalempa, Velocity slip and temperature jump coefficients for gaseous mixtures. I. Viscous slip coefficient, Physics of Fluids 15 (2003) 1800–1806. doi: http://dx.doi.org/10.1063/1.1574815.

[209] N. Prasianakis, I. Karlin, Lattice Boltzmann method for simulation of compressible flows on standard lattices, Physical Review E 78 (2008) 016704. doi: http://dx.doi.org/10.1103/PhysRevE.78.016704.

[210] J. Kang, N. Prasianakis, J. Mantzaras, Lattice Boltzmann model for thermal binary-mixture gas flows, Physical Review E 87 (2013) 053304. doi: http://dx.doi.org/10.1103/PhysRevE.87.053304.

[211] S. Arcidiacono, J. Mantzaras, I. Karlin, Lattice Boltzmann simulation of catalytic reactions, Physical Review E 78 (2008) 046711. doi: http://dx.doi.org/10.1103/PhysRevE.78.046711.

[212] J. Kang, N. Prasianakis, J. Mantzaras, I. Zinovik, D. Poulikakos, L. Holtzer, Simulation of flow in the anode of a solid oxide fuel-cell: Continuum versus finite Knudsen number effects, (2014) in preparation.

NOMENCLATURE

A	Cross-sectional flow area, Equations (4.12) and (4.13)
A_s	Cross-sectional area of solid wall, Equation (4.15)
b	Channel half-height
B	Ratio of active to geometrical surface area, Equations (4.14) and (4.32)
B_g, B_s	Arrhenius pre-exponentials, Equations (4.51) and (4.52)
c_p, c_s	Specific heat of gas at constant pressure, specific heat of solid
c_μ	Turbulence constant, Equation (4.89)
d_i	Channel gap
$D_{k\ell}$	Binary diffusion coefficient between species k and ℓ, Equation (4.17)
D_k^T	Thermal diffusion coefficient of k-th species, Equations (4.17) and (4.30)
D_{km}	Mixture average diffusion coefficient of k-th species, Equation (4.18)
Da_g	Gas-phase Damköhler number, Equation (4.74)
Da_s	Catalytic Damköhler number, Equation (4.70)
E	Activation energy
f	Fanning friction factor, Equation (4.11)
f	Non-dimensional stream function, Equation (4.63)
f_μ	Turbulent damping function, Equation (4.89)
f_i	i-th velocity population in discrete Boltzmann equation, Equation (4.91)
$F_{k\text{-}j}$	Radiation view factor between surface elements k and j, Equation (4.47)
h, h_k	Total enthalpy, enthalpy of gaseous species k, Equation (4.9)
h_i, h_0	Inner and outer heat transfer coefficient on the channel wall
H	Normalized rate of heat transfer on the inner wall surface, $2h_i \lambda / d_i (\rho\, c_p U_{ad})^2$, Equation (4.23)
H_0	Normalized rate of heat transfer on the outer wall surface, $2\beta\, h_0 \lambda / d_i (\rho\, c_p U_{ad})^2$, Equation (4.23)

$\underset{\sim}{I}$	Unity matrix, Equation (4.27)
$\bar{\tilde{k}}$	Turbulent kinetic energy
k_s	Thermal conductivity of solid, Equations (4.15) and (4.34)
K_g	Total number of gas-phase species
K_s	Total number of surface species
K_b	Total number of bulk species
L	Channel length
Le_k	Lewis number of k-th gaseous species, $\lambda/(\rho c_p D_{km})$
m	Normalized flame speed $U - U_f$, Equations (4.21) and (4.22)
M	Sum of K_s and K_b for a given surface phase, Equation (4.7)
Nu	Nusselt number
p	Pressure
P	Channel wetted perimeter, Equation (4.11)
q_{rad}, q_k	Radiative flux, radiative flux of kth element, Equations (4.35) and (4.47)
R^o	Universal gas constant
Re	Reynolds number
r	radial coordinate
$\dot{s}_k$	Heterogeneous molar production rate of kth species
$\tilde{S}_\phi$	Source term for variable φ in turbulent transport equations, Equation (4.88)
T	Temperature
t_{ig}, t_{st}	Ignition time, steady-state time
u, v	Streamwise and transverse velocity
u_{st}	Stefan velocity, Equation (4.33)
U_{IN}	Inlet streamwise velocity
U	Normalized flow velocity u/U_{ad}, Equations (4.21) and (4.22)
U_{ad}	Adiabatic flame speed
U_f	Normalized flame propagation speed u_f/U_{ad}, Equations (4.21)–(4.23)
V	Reactor volume, Equation (4.6)
$\vec{V}_k$	Diffusion velocity vector for k-th species, Equations (4.17) and (4.30)
W_k	Molecular weight of k-th species
$\bar{W}$	Average mixture molecular weight
x, y, z	Streamwise, transverse and lateral coordinate
X_k, $[X_k]$	Mole fraction and molar concentration of k-th gaseous species
Y_k	Mass fraction of k-th gaseous species
Y	Normalized fuel concentration $Y_F/Y_{F,\infty}$, Equation (4.22)

GREEK SYMBOLS

α_k	Mass transfer coefficient of k-th gaseous species, Equation (4.12)
α_T	Heat transfer coefficient, Equation (4.13)
β	Zeldovich number, $E(T_{ad} - T_\infty)/R^o T_{ad}^2$, Equations (4.21)–(4.23)
χ_k	Sticking coefficient of k-th gas-phase species, Equation (4.5)
γ	Heat release parameter $(T_{ad} - T_\infty)/T_{ad}$, Equation (4.25)

γ	Pre-exponential temperature dependence of reaction rate, Equation (4.52)
Γ	Surface site density, Equation (4.3)
$\Gamma_\lambda, \Gamma_{\text{eff}}$	Laminar and effective turbulent transport coefficients, Equation (4.89)
δ_{s}	Channel wall thickness
δ	Delta function, Equations (4.21)–(4.23)
$\tilde{\varepsilon}$	Dissipation rate of the turbulent kinetic energy, Equation (4.89)
ε	Surface emissivity, Equations (4.19) and (4.47)
ζ	Non-dimensional axial distance, Equation (4.61)
η	Non-dimensional transverse distance, Equation (4.61)
θ	Normalized temperature, $(T - T_\infty)/(T_{\text{ad}} - T_\infty)$, Equation (4.21)
θ_{m}	Coverage of m-th surface species, Equation (4.3)
λ	Thermal conductivity of the gas, Equation (4.13) and (4.28)
λ	Mean free path of gas, Equation (4.93)
μ, μ_{t}	Laminar viscosity, turbulent viscosity, Equation (4.89)
$v''_{k,i}, v'_{k,i}$	Stoichiometric coefficients of k-th species in i-th reaction, Equation (4.2)
ξ	Streamwise coordinate normalized by $\lambda/\rho\, c_p U_{\text{ad}}$, Equations (4.21)–(4.24)
ρ	Gas density
ρ_{s}	Solid density, Equation (4.15) and (4.34)
σ	Stefan-Boltzmann constant
σ_ϕ	Turbulent Schmidt number, Equation (4.89)
σ_{12}	Binary velocity slip coefficient, Equation (4.93)
σ_m	Surface species site occupancy, Equation (4.3)
τ	Reactor residence time, Equations (4.6) and (4.7)
φ	Equivalence ratio
$\dot{\omega}_k$	Gas-phase species molar production rate, Equations (4.12) and (4.29)
Ω_{i}	Collision integral in discrete Boltzmann equation, Equation (4.91)

SUBSCRIPTS

ad	Adiabatic
f	Flame front
ig	Ignition
IN	Inlet
k, m	Index for gas-phase species, index for surface species
s	Solid, surface
st	Steady state
w	Wall
x, y	Streamwise and transverse components
∞	Unburned mixture

TURBULENCE AVERAGING

~, -	Favre- and Reynolds-averaging, Equation (4.88)

ACRONYMS

DSMC	Direct simulation Monte Carlo
HRR	Heat release rate
LB	Lattice Boltzmann
LIF	Laser induced fluorescence
LR	Low Reynolds number
MC	Monte Carlo
PSR	Perfectly stirred reactor
SOFC	Solid oxide fuel cell

CHAPTER 5

NON-PREMIXED MICRO COMBUSTION

Yei-Chin Chao and Chih-Peng Chen

The advancement of human civilization is closely related to the ability to use and control flames. Traditionally, flames have been used for heating, cooking, illumination, and power generation for industrial and civil applications. Modern applications of flames include power generation for electricity and transportation, and propulsion for both atmospheric aviation and space exploration.

Recently, with increasing demands of micro devices such as microsatellites and micro aerial vehicles, needs for a micro power source to activate these systems have significantly increased [1]. These microsystems require high-density power sources for long periods of operation. The energy density of typical hydrocarbon fuels is about 100 times higher than that of the most advanced batteries. Even with heat losses in the process of extracting power from burning the fuel, a micro-scale combustion system has been considered as a viable alternative to batteries. To develop such combustion systems, an understanding of the physics of laminar microflames is of vital importance.

For microscale applications, premixing fuel and air is not favored as the process requires additional volume and weight. The current investigation concentrates specifically on the flame shape, length, structure, and stabilization mechanism of microjet diffusion flames. Numerical and advanced non-intrusive diagnostic tools are used in the study. A comprehensive literature survey including studies of typical diffusion flames and microjet diffusion flames is given in the following sections.

5.1   MICROJET DIFFUSION FLAMES

Diffusion flame is the most common type of flame in practical combustion devices. The ability to predict the coupled effects of complex transport phenomena with detailed chemical kinetics in these systems is critical for modeling reacting flows, improving engine efficiency, and understanding the phenomena such as flame extinction and pollutant formation. Laminar jet diffusion flames are fundamental to combustion and have received much attention as a model system due to their relatively simple geometry and fluid mechanical characteristics. Efforts have been devoted to the detailed measurements of thermophysical properties and structures of laminar flames along with computational efforts [2–12].

Laminar jet diffusion flames fall into three categories: (1) the Burke–Schumann flame [13] controlled by diffusion, (2) the Roper flame [14] controlled by momentum or buoyancy, and (3) the microflame [15] controlled by diffusion or momentum. The first two types of flames have long been extensively investigated [16]. The third type of flame was only recently investigated [15, 17–23]. However, these studies were concerned with bulk flame geometrical and thermal characteristics. Due to its small size and negligible buoyancy force, the microjet flame can be used as a point heat source with no preference to the orientation in future micro power devices. Therefore, it is necessary to determine the smallest stable flame that can be maintained and to investigate its characteristic structure near extinction.

Laminar microjet diffusion flames were theoretically investigated [15] and numerically studied [17]. The importance of axial diffusion in the theoretical model and the insignificance of buoyancy effect on the height and shape of the microjet diffusion flame have been identified [15]. More recent investigations of microjet diffusion flames include those of various fuels: methane [19, 20, 23, 24], propane [21, 22], and hydrogen [25] flames. These studies found that the microjet flame is sufficiently smaller in size (few millimeters) and is nearly spherical in shape independent of orientation, indicating its buoyancy insensitivity. Because the spherical shape is similar to that of a microgravity flame, it may serve as a model of microgravity flames [26]. Also, due to its small size, the heat loss to the burner is relatively large. Thus, the flame might always operate in a severe, near-extinction condition. It is also known that different fuels produce different visible flame shapes near the burner port. For example, the C_2 class of hydrocarbon [15] and hydrogen [25] flames extend upstream of the burner port, while the propane [21, 22] and methane yield standoff flames [19, 20, 23, 24]. Experimental results of flame shapes of microjet methane diffusion flames are different using the direct photograph and laser shadowgraph [20]. This implies that an intensively hot region may exist just ahead of the flame zone.

5.2 BASIC MICROFLAME STRUCTURE

The laminar microjet methane diffusion flames investigated here are stabilized on vertical straight stainless-steel tubes with the inner diameter (d) ranging from 186 to 778 μm. The tube wall thickness is 79, 88, 122, 147, 148, and 124 μm with d = 186, 212, 324, 382, 529, and 778 μm, respectively. Methane fuel is introduced through the tube into the quiescent atmospheric air. The flame shapes are recorded using a color charge-coupled device (CCD) camera with a macro lens. Figures 5.1 and 5.2 respectively show photographs and numerically predicted longitudinal distributions of OH radicals of the microjet flames. Photographical

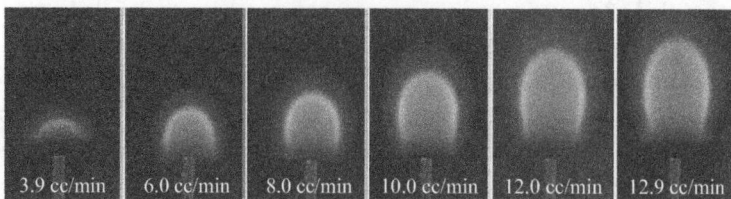

| 3.9 cc/min | 6.0 cc/min | 8.0 cc/min | 10.0 cc/min | 12.0 cc/min | 12.9 cc/min |

Figure 5.1. Photographs of flames supported on the 186 μm tube.

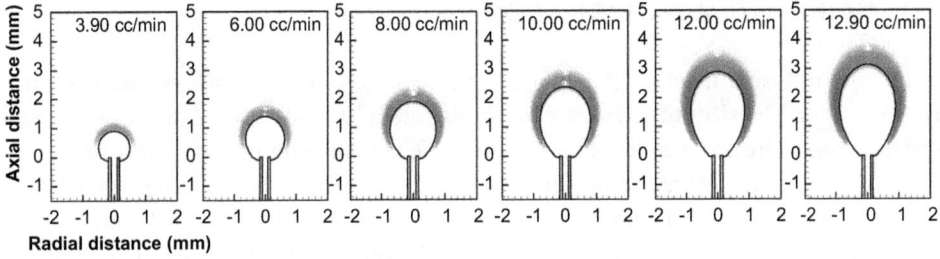

Figure 5.2. Computed OH mass fraction isopleths for the $d = 186$ μm flames in contrast to the stoichiometric mixture fraction contour.

images and predicted results clearly indicate the characteristic flame structure of the microjet diffusion flame, which severely quenches locally at the flame base and the flame stands off significantly from the tube.

5.3 METHODOLOGY

5.3.1 EXPERIMENTAL SETUP

Because of the small flame sizes, intrusive techniques are unsuitable for the study of microjet diffusion flames. Non-intrusive optical methods are used for the present study. Observations and measurements are described in the following.

5.3.1.1 Burner and Fuel System

The laminar microjet diffusion flames investigated here are stabilized on vertical straight stainless-steel (AISI 304) tubes. The fuel flow is fully developed when it exits the burner. The fuel flow rate is measured with an electrical mass flowmeter (Brooks 5850E), which is well calibrated using a soap-bubble flowmeter. The fuel Reynolds numbers range in the order of 10 to 100.

5.3.1.2 Image Capture System

The traditional way to define qualitative flame properties is usually based on photographs. A high-quality recording system is beneficial for later digital image processing. Microflame images are obtained by a highly sensitive 3-chip color CCD camera (Sony DXC-9000) with external triggering and recorded in the NTSC format. The camera shutter time can be adjusted for best images of various flame conditions. The captured images are then digitized for further analysis. The visible flame heights have previously been measured using an identical imaging system by Cheng et al. [25]. Moreover, the flame CH^* chemiluminescence image is recorded by a cooled CCD camera (Cooke SensiCam) with a macro lens and a narrowband interference filter centered at $\lambda = 431$ nm with a full width at half-maximum of 10 ± 2 nm.

5.3.2 MATHEMATICAL FORMULATION

For microjet diffusion flames, detailed measurements are difficult to achieve because of the tiny flame sizes. The sizes of microjet flame in this study are only several millimeters and the flame thicknesses are on the order of micrometers. Neither intrusive nor non-intrusive measurements can obtain sufficient resolution for species distribution by the existing experimental technologies. Therefore, detailed numerical simulations [27] with sufficiently small grid resolution are used to comprehend the flame structures and characteristics. Theoretical models are also applied to examine the measured flame shapes and flame lengths in this work.

5.3.2.1 Numerical Simulation

To numerically model the laminar microjet diffusion flames, the governing equations of mass, momentum, energy, and chemical species for a steady axisymmetric reacting flow can be written in the cylindrical (r, x) coordinate system as

$$\nabla \cdot (\rho v) = 0 \tag{5.1}$$

$$\nabla \cdot (\rho v v) = -\nabla p + \nabla \cdot (\mu \nabla v) + \rho g_x \tag{5.2}$$

$$\nabla \cdot (\rho v T) = \frac{1}{c_p} \nabla \cdot (\lambda \nabla T) - \frac{1}{c_p} \sum_i h_i \left\{ w_i + \nabla \cdot [\rho D_i \nabla Y_i + \rho D_i^T \nabla (\ln T)] \right\} \tag{5.3}$$

$$\nabla \cdot (\rho v Y_i) = \nabla \cdot [\rho D_i \nabla Y_i + \rho D_i^T \nabla (\ln T)] + w_i \tag{5.4}$$

and the state equation

$$p = \rho R_0 T \sum_i \frac{Y_i}{M_i} \tag{5.5}$$

where ρ, p, T, Y, c_p, h, w, R_0, M, g_x, and $v = (u, v)$ are the density, pressure, temperature, mass fraction, specific heat capacity of the mixture, enthalpy, species production rate, universal gas constant, molecular weight, gravitational acceleration in the x-direction, and velocity vector, respectively. μ, λ, and D are the viscosity, thermal conductivity, and mass diffusivity, respectively. The subscript i in Equations (5.3)–(5.5) stands for the i-th chemical species. The second term in the bracket of Equations (5.3) and (5.4) is the thermo-diffusion or Soret diffusion due to the effect of temperature gradient. The concentration-driven diffusion coefficient is calculated as

$$D_i = \frac{1 - x_i}{\left[\sum_{j=1}^{N} \frac{x_j}{D_{ij}} \right]_{j \neq i}} \tag{5.6}$$

where D_{ij} is the binary diffusion coefficient. The binary mass diffusivity is determined by the Chapman–Enskog kinetic theory using Lennard-Jones parameters. The thermo-diffusion coefficient is calculated as

$$D_i^T = \left[\sum_{j=1}^{N} \frac{M_i M_j}{M^2} k_{ij} D_{ij} \right]_{j \neq i} \tag{5.7}$$

where M is the mixture molecular weight and k_{ij} is the thermo-diffusion ratio.

A schematic illustration of the computational domain coupled with boundary conditions is shown in Figure 5.3. The governing equations are solved using a commercial flow package. An orthogonal, non-uniform staggered-grid system is used for solving the discretized equations with a control volume formulation in accordance with the Semi-Implicit Method for Pressure Linked Equations–Consistent (SIMPLEC) [28] algorithm. The momentum equations are solved using the second-order upwind scheme while the central difference method is used for the energy and species equations. The above equations are solved along the mesh lines in the computational domain using iterative ADI and TDMA techniques. The input of the molecular transport data is obtained from the CHEMKIN package [29] and then the code calculates the thermal conductivity and viscosity of the mixture using Wilke's formula. Thermal diffusion and buoyancy are included in this analysis but radiation heat loss is neglected. A skeletal chemical kinetic mechanism [30] is coupled with the computational fluid dynamic program to predict the shape, length, and structure of the micro methane flame. This skeletal chemical mechanism consists of ten reversible and fifteen irreversible reactions. It has been demonstrated to be sufficiently accurate for predicting flame speeds, extinction limits, and the thermo-chemical

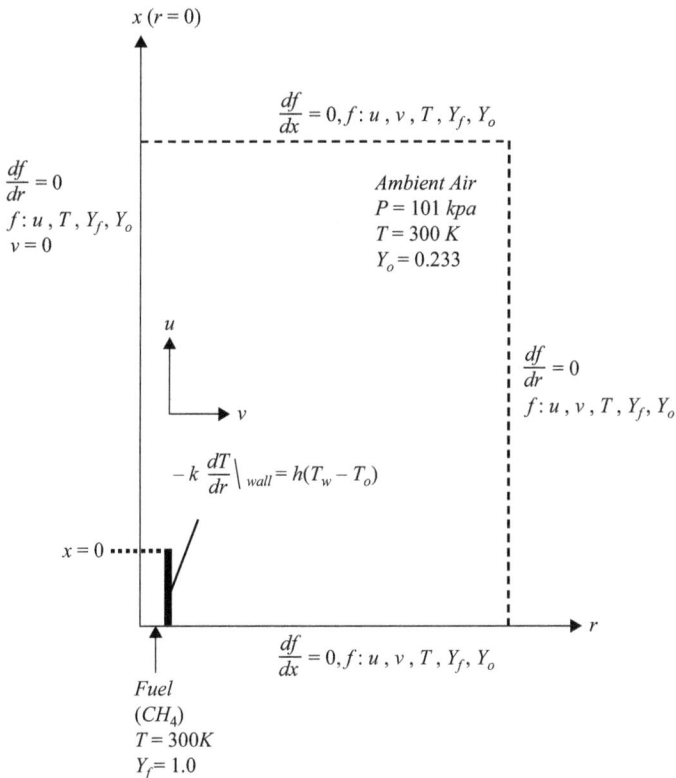

Figure 5.3. Schematic illustration of computational domain coupled with boundary conditions.

properties [30]. Furthermore, full CH_4/Air chemical kinetic mechanism, GRI-Mech 3.0 [31], is used to investigate the detailed flame structure and stabilization mechanism.

5.3.2.2 Theoretical Models

It has been shown [15, 21, 32] that simple models derived from similarity analysis can adequately predict flame height and flame shape for laminar jet diffusion flames. It is imperative to see the suitability of these models as applied to microjet methane diffusion flames. For a steady, axisymmetric, vertical laminar jet at low Mach numbers, uniform pressure, negligible buoyancy, and negligible mass diffusion, heat conduction as well as viscous action in the axial direction but fast chemical reaction rates, the governing Equations (5.1) to (5.4) can be simplified and represented as the following differential equation [15]:

$$\zeta F'' + F'(F-1) = 0. \tag{5.8}$$

where ζ is the similarity variable and F is the dependent variable of the similarity equation and prime superscript denotes derivative. The dimensionless axial velocity component ($u*$) and mixture fraction (f) become

$$u^* = f = \frac{\gamma^2}{Re\,x^*}\frac{F'(\zeta)}{\zeta}. \tag{5.9}$$

where γ is a constant in the similarity variable. The flame shape can be determined from Equation (5.9) by assuming that the flame exists at the stoichiometric value of f:

$$x^* = \frac{\gamma^2}{Re\,f_{st}}\frac{F'(\zeta)}{\zeta} \tag{5.10}$$

and the flame length is at the axial position as $\zeta \to 0$,

$$x_{fl}^* = \frac{\gamma^2}{Re\,f_{st}}F''(0) \tag{5.11}$$

where x_{fl}^* is the dimensionless flame height, f_{st} is the stoichiometric value of the mixture fraction (0.055 for CH_4—air), Re is the Reynolds number based on the jet diameter, $\gamma = (\sqrt{3}/8)\,Re$, and $F''(0) = 2$.

Equations (5.10) and (5.11) are derived by assuming negligible axial diffusion in the flame. The flame shapes calculated including the axial diffusion term were shown to better agree with those measured in experiment for low Reynolds number C_2 class hydrocarbon microflames [15]. If the axial diffusion term is taken into account, then Equation (5.8) becomes

$$\left(\frac{\zeta^3}{\gamma^2} + \zeta\right)F'' + F'(F-1) = 0 \tag{5.12}$$

with the far-field boundary conditions $F = 0$ and $F' = 0$ as $\zeta \to \infty$. The flame shape is then determined by solving Equation (5.12) numerically and substituting F' into Equation (5.10).

In the study of microjet propane flames, Matta et al. [21] postulated that flame behaviors are primarily controlled by diffusion and that buoyancy becomes less important, at least for cases near the quenching limit. Therefore, they adopted the jet diffusion flame model of Turns [33] that does not include buoyancy effects for the flame length predictions. The flame length depends only upon the volumetric fuel flow rate, Q_f, and not on the Reynolds number:

$$L_f = \frac{3}{8\pi} \frac{Q_f}{D Y_{F,stoic}} \tag{5.13}$$

where D is the mass diffusivity and $Y_{F,stoic}$ is the stoichiometric fuel mass fraction. In addition, Matta et al. [21] hypothesized that the critical fuel flow rate at quenching is that at which the predicted flame length equals the measured standoff distance. Equating the flame length in Equation (5.13) to the measured standoff distance yields a fuel flow rate at quenching. Comparisons of the predictions with experimental measurements indicated that the behavior of miniature diffusion flames can be adequately modeled by this simple jet diffusion flame model.

For the estimation of jet diffusion flame lengths, Roper [14] modified the Burke—Schumann theory to allow the mass velocity of fuel gas to vary with axial distance as affected by buoyancy and in accordance with continuity. For the circular burner port, the following expression can be used to calculate the flame length:

$$L_f = \frac{Q_f}{4\pi D_O \ln(1+1/S)} \left(\frac{T_O}{T_f}\right)^{0.67} \tag{5.14}$$

where S is the molar stoichiometric oxidizer—fuel ratio, D_O is a mean diffusion coefficient evaluated for the oxidizer at the oxidizer stream temperature, T_O, and T_f is the mean flame temperature. Equation (5.14) can be applied to estimate the flame length regardless of whether or not buoyancy is important, and is applicable for fuel jets emerging into either a quiescent oxidizer or a co-flowing stream, as long as the flames are over-ventilated.

The solutions of jet diffusion flame theory for the flame length estimations are mainly dependent upon the volumetric flow rate and hence on the Reynolds number. Chung and Law [34] extended the Burke—Schumann theory to include the effects of both streamwise (axial) and preferential diffusion and derived a solution for the flame length prediction. The solution that is independent of the Reynolds number but depends on the Peclet number (Pe) can be expressed as

$$\left[c(1+Y_{O0}) - Y_{O0}\right] + 2(1+Y_{O0})\sum_1 \sin(n\pi c)\exp\left[(Pe-\alpha_n)Z_f^*/2\right]/(n\pi) = 0 \tag{5.15}$$

where c is the normalized half-width of the inner wall, Y_{O0} is the mass fraction of the oxidizer at the oxidizer stream, $\alpha_n = (Pe^2 + 4\pi^2 n^2)^{1/2}$, and Z_f^* is the normalized flame height. Equation (5.15) suggests that with the limit of $Pe \rightarrow 0$ the flame becomes independent of Pe, while the streamwise diffusion can be substantially more significant than streamwise convection. This, however, has not been experimentally validated.

Equations (5.10), (5.13), (5.14), and (5.15) are all derived by assuming unity Lewis (Le) and Schmidt (Sc) numbers. If Sc is not equal to unity, 0.704 for methane, then the flame shape can be represented as [35]

$$x^* = \frac{(2\,Sc+1)\,Re}{32\,f_{st}} \frac{1}{\left[1+0.25\,\zeta^2\right]^{2Sc}}, \zeta^2 = \frac{3}{64}Re^2\left(\frac{r}{x}\right)^2,\tag{5.16}$$

where x^* is the flame shape normalized by the nozzle diameter and the flame length is at the axial position as $\zeta \to 0$.

In the present study, Equations (5.8) (without axial diffusion) and (5.12) (with axial diffusion) as well as Equation (5.16) are solved numerically to predict the flame shapes and flame lengths. Equations (5.13), (5.14), and (5.15) are used only for the flame length calculations. Comparisons, in the following section, of the measurement and theoretical results are made to assess the applicability of the simple jet flame models for microjet methane diffusion flames.

5.4 CHARACTERISTICS OF MICROJET METHANE DIFFUSION FLAMES

5.4.1 FLAME SHAPE

Equation (5.16) as well as Equations (5.8) and (5.12) coupled with Equation (5.10) are solved numerically to obtain the flame shape. The calculated flame shapes with the measured and predicted results of Ban et al. [15] for the $d = 186$ μm flames operated at several fuel flow rates are compared in Figure 5.4. The measured flame shapes are indicated by symbols, the solid and dashed lines denote those from the calculation with and without axial diffusion, respectively, and the dashed-dot-dashed lines depict the calculated results of Equation (5.16). Results of Figure 5.4 indicate that flame shapes calculated with and without axial diffusion are in poor agreement with the experimental data and the flame heights are also significantly over-predicted by nearly an order of magnitude for all ranges of fuel flow rates. With the consideration of the non-unity Schmidt number effect, the predictions of flame heights are slightly improved. The comparison shown in Figure 5.4 clearly demonstrates that the simple jet flame theory fails completely in the prediction of the flame shapes of microjet methane flames. For the $d = 324$ μm flames (not shown here), similar discrepancies between the predicted and measured flame

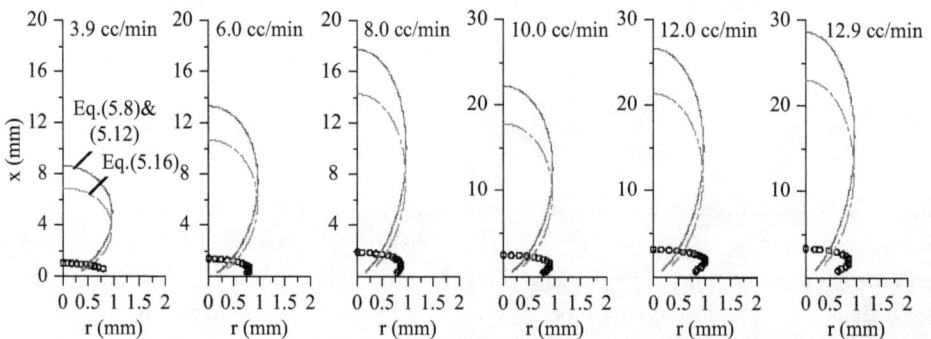

Figure 5.4. Comparison of the calculated and measured flame shapes for the $d = 186$ μm tube. Symbols are the measured data and lines denote those from calculations.

shapes are also observed. The failure of the simple jet flame model may be attributed to the failure of the similarity assumption, which is used in the derivation of the simple model, as applied to microjet diffusion flames.

5.4.2 FLAME LENGTH

Equations (5.11) and (5.16) indicate that the flame length depends only upon the Reynolds number. It means that different diameter tubes with the same fuel exit Reynolds number will produce the same flame length. Although the Reynolds number does not explicitly appear in Equations (5.13) and (5.14), the flame length can still be related to the Reynolds number through the dependence on the flow rate (Q_f). Comparison of the measured and calculated non-dimensional flame lengths as functions of Reynolds and Peclet numbers for $d = 186$ and $324\ \mu m$ flames is shown in Figure 5.5. Figure 5.5(a) shows that the measured nondimensional flame length data collapse into a straight line and can be scaled with the Reynolds number. Comparison of the measured and calculated data indicates that the models of Ban et al. [15], Turns [33], and Lee and Chung [35] all result in over-prediction of the flame lengths. Figure 5.5(a) indicates that the flame length predicted by Roper's model [14] agrees very well with the measured

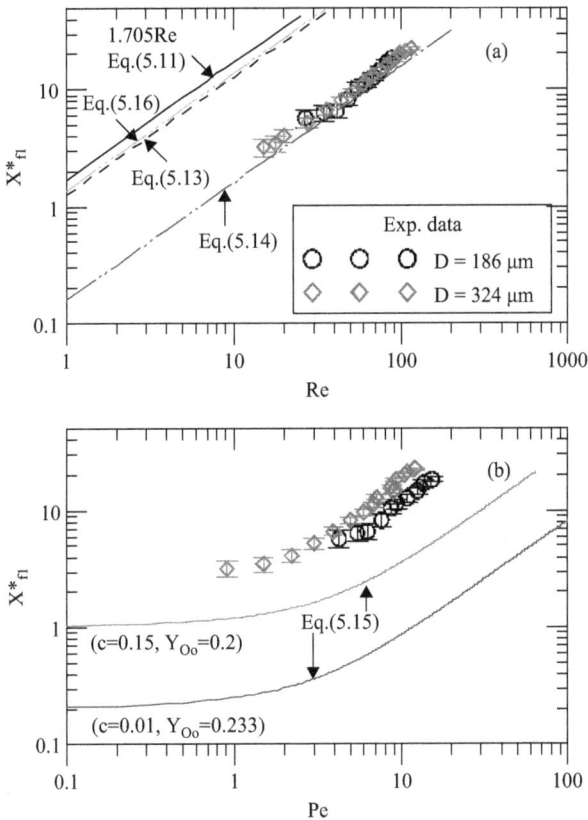

Figure 5.5. Comparison of the calculated and measured non-dimensional flame lengths as a function of (a) Reynolds and (b) Peclet numbers for the $d = 186$ and $324\ \mu m$ tubes.

data by assuming a mean flame temperature of 1500 K in Equation (5.14). The slight difference between the predictions and measurements could be due to minor effects like buoyancy.

Figure 5.5(b) illustrates the nondimensional flame length as a function of Peclet number. It should be noted that the model of Chung and Law [34] was derived for a coflowing Burke—Schumann flame and the definition of Peclet number ($Pe = u_e b /D$) is slightly different from that used in the present study ($Pe = u_e l_D /D$), where b is the half-width of the outer wall and l_D is the radial position of the flame edge. The purpose of this comparison is not to search for a theoretical representation of the experimental data. Instead, it is to experimentally assess the model prediction when accounting for the streamwise (axial) diffusion in Burke—Schumann flames [34]. Both measurement and theoretical data show that in the limit of $Pe \rightarrow 0$ the flame length becomes independent of Pe. This suggests that the streamwise diffusion can be substantially more significant than streamwise convection at low Peclet numbers.

From these comparisons, one may expect that buoyancy acceleration would be more important than molecular diffusion in the microjet flames. The diffusion—buoyancy and diffusion—momentum parameters of Baker et al. [22] are examined. The diffusion-to-buoyancy parameter is defined as

$$N_{DB} = \frac{D^2}{a L_f^3} \tag{5.17}$$

where D is the fuel gas mass diffusivity estimated at the mean flame temperature and $a \cong 0.6g[(T_f / T_0)]$ is the mean buoyant acceleration [14]. If $N_{DB} << 1$ then the flame is buoyancy controlled. If $N_{DB} >> 1$ then the flame is diffusion controlled. Similarly, N_{DM} is the diffusion-to-momentum parameter defined as

$$N_{DM} = \frac{D}{u_{z0} L_f} \tag{5.18}$$

where $u_{z0} = Q_f I Y_{F,stoic}/A$ is the ratio of the actual initial momentum flux to that for uniform flow (for parabolic exit velocity $I = 1.5$ and for uniform flow $I = 1.0$) and A is the cross-section area of the burner port. If $N_{DM} << 1$, then the flame is momentum controlled. If $N_{DM} >> 1$, then the flame is diffusion controlled. It is noted that the Froude number, based on the flame length instead of burner port diameter as defined by Roper [14], can be recast as $Fr = N_{DB}/N_{DM}^2$. An examination of the Peclet number, the Froude number, the diffusion-to-buoyancy parameter, and the diffusion-to-momentum parameter is sufficient to determine the relative importance of the three transport mechanisms (buoyancy, momentum, or diffusion) for microjet diffusion flames. The calculated parameters, based upon experimental data and baseline data ($T_f = 1500$ K, $T_0 = 300$ K, $I = 1.5$, $Y_{F,stoic} = 0.055$, and $D = 2.8195$ cm^2/s) are listed in Table 5.1 for the $d = 186$ and 324 μm flames. It is clear that in the analysis considering only diffusion and buoyancy, the flames are more in the diffusion-dominated regime for $Q_f \leq 5$ cc/min because N_{DB} is greater than 1. For $Q_f \leq 5$ cc/min the microjet flames are in spherical shape. When $Q_f > 6$ cc/min, the buoyancy acceleration would be more important than molecular diffusion in the microjet flames because of $N_{DB} < 1$. The Froude number for the flames studied here ranges from $O(10^{-1})$ to $O(1)$. This indicates that the flames fall within the transitional to buoyancy-controlled regime. In addition, the Peclet number ranges from $O(1)$ to $O(10)$ indicating that the molecular-diffusion velocity is comparable to the convective velocity only for $Q_f \leq 5$ cc/min.

This is consistent with the N_{DM} parameter analysis. From the above parameter analysis, we can conclude that the microjet diffusion flames studied here are not completely buoyancy-free and the molecular diffusion is effective only for $N_{DB} > 2$ and $Pe < 2$.

5.4.3 QUENCHING VELOCITY

To investigate the effect of tube size on extinction behavior, different tube diameters are used. Figure 5.6 shows photographs of flames operated at fuel exit velocity near extinction for tube diameters varying from 186 to 778 μm. The most notable feature of Figure 5.6 is that the stand-off distance is essentially the same, approximately 0.78 mm, for all the tubes. In addition, the flame shapes are remarkably similar over the range of tube diameters. It has been demonstrated in the previous section and shown in Figure 5.5 that the measured and predicted flame length is a linear function of Reynolds number. Accordingly, a flame cannot be sustained if the predicted flame length for a fuel exit velocity is smaller than the measured standoff distance. Thus, it

Table 5.1. Dimensionless parameters

	d = 186 μm					d = 324 μm			
Q_f(cc/min)	N_{DB}	N_{DM}	Fr	Pe	Q_f(cc/min)	N_{DB}	N_{DM}	Fr	Pe
3.9	2.83	1.35	1.56	4.25	3.8	3.10	4.32	0.17	0.91
5	2.05	0.94	2.30	5.45	4.5	2.44	3.37	0.22	1.49
6	1.84	0.76	3.20	6.26	5	1.49	2.57	0.23	2.20
7	0.95	0.52	3.49	7.63	9	0.34	0.88	0.45	3.92
8	0.46	0.36	3.59	8.72	13	0.11	0.42	0.64	6.01
9	0.36	0.29	4.18	9.39	17	0.05	0.24	0.82	7.16
10	0.26	0.24	4.62	10.90	19	0.03	0.19	0.91	8.79
11	0.17	0.19	4.85	12.51	21	0.02	0.15	1.00	8.99
12	0.11	0.15	4.97	13.65	25	0.01	0.10	1.13	10.47
12.9	0.09	0.13	5.34	15.28	29	0.01	0.08	1.37	12.14

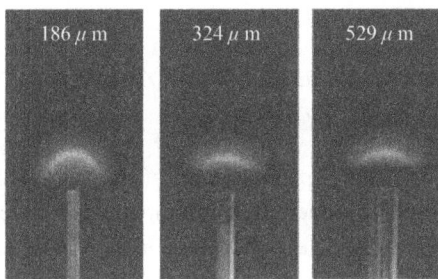

Figure 5.6. Photographs of flames just above quenching limit for different tube diameters.

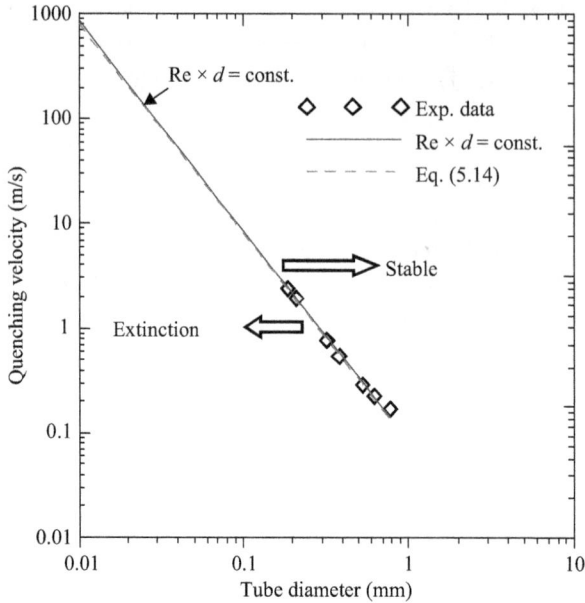

Figure 5.7. Comparison of the predicted and measured quenching velocities as a function of tube diameter.

is hypothesized that the critical fuel exit velocity at quenching is that at which the predicted flame length equals the measured standoff distance [21]. Since Roper's model [14] is shown above to well predict the flame lengths, the model is further used for quenching velocity predictions. The quenching velocity for different tube sizes is determined by equating the flame length in Equation (5.14) to the measured standoff distance. Comparison of the measured and calculated quenching velocities is depicted in Figure 5.7. The predicted quenching velocities (dashed line) are in excellent agreement with the measured data. The good agreement between the calculated and measured quenching velocities is due that near the quenching limit; the microjet flames are mainly diffusion controlled and buoyancy has a minor effect on the flame length. It is also found that the measured quenching velocities follow $Re \times d =$ constant ($u \times d^2 =$ constant) curve. This finding is in agreement with the relationship proposed [20] but different from $Re \times d^2 =$ constant that was obtained numerically for the adiabatic wall conditions [19].

5.4.4 EFFECT OF BURNER WALL BOUNDARY CONDITION ON THE STANDOFF DISTANCE

In order to clarify the effect of tube materials on the standoff distance, the computed OH isopleths with different wall thermal conductivity are compared in Figure 5.8 for flames of $d = 186$, 324, and 529 µm. The standoff distance is measured along the jet centerline from the burner exit to the bottom of blue flame cap. For the same tube diameter, for example $d = 186$ µm, the image of Figures 5.8(a), (d), and (g) are calculated with thermal conductivity $k = 16.2$ W/m.K (AISI 304), $k = 8$ W/m.K (quartz tube), and $k = 1$ W/m.K, respectively. It can be seen from

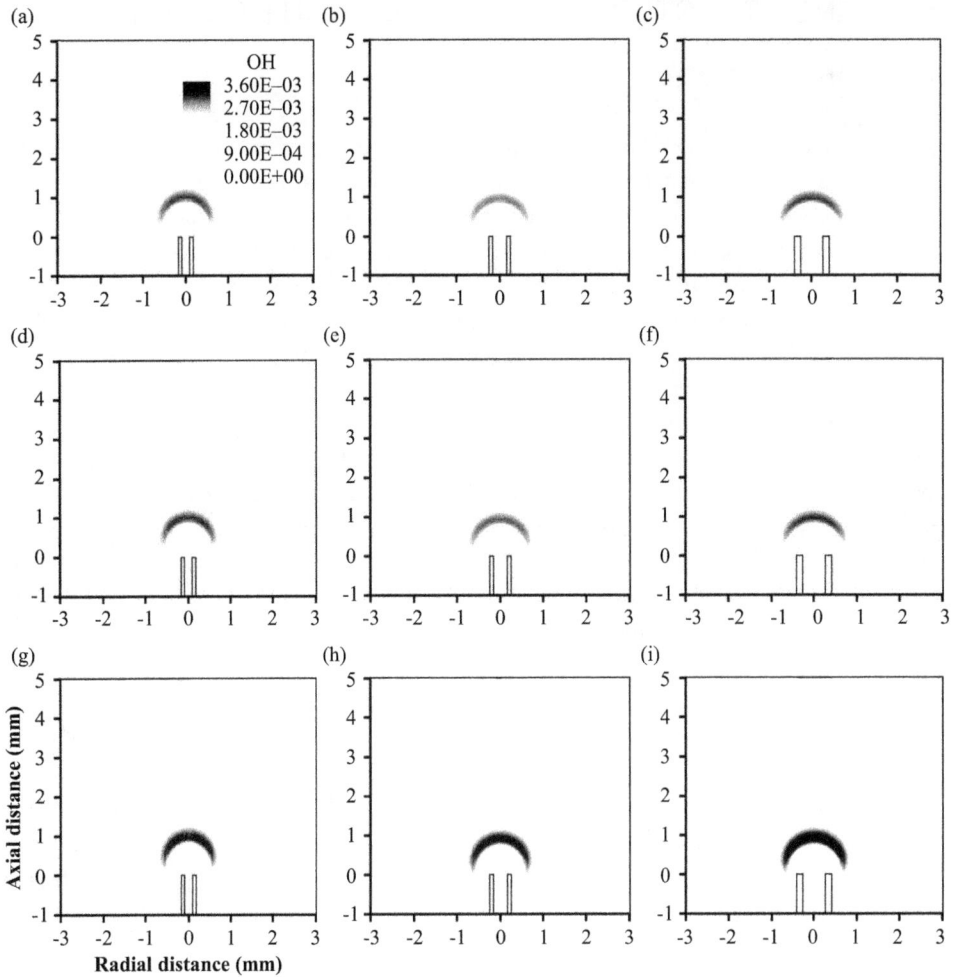

Figure 5.8. Computed OH mass fraction isopleths for the flames with d = 186, 324, and 529 μm near the extinction limit with different wall thermal conductivities. (a)–(c): k = 16.2 W/m.K (AISI 304), (d)–(f): k = 8 W/m.K (quartz tube), and (g)–(i): k = 1 W/m.K.

Figures 5.8(a)–(c) that the computed standoff distance (0.8 mm) is in good agreement with the measured data (0.78 mm) for the stainless-steel tube. The maximum variation of the calculated standoff distance is less than 3 percent for the same tube material with different tube diameters and that is less than 5 percent for a fixed tube diameter with different tube materials. Therefore, it can be concluded that the change of tube materials only has a negligible effect on the predicted standoff distance, but does influence the quenching gap between the flame and the tube. It is noted that the detachment of the flame base from the burner rim is reduced as the thermal conductivity of the tube is reduced. This is due that reduced thermal conductivity results in less conduction of heat from the flame edge to the tube, and hence reduces the quenching gap. If an adiabatic wall condition is imposed, the flame would attach to the burner and no quenching gap is produced.

5.5 FLAME STRUCTURE AND STABILIZATION MECHANISM

5.5.1 FLAME STRUCTURE

In order to have a further look into the detailed flame structure of the microjet methane diffusion flame, the case of a stainless-steel (AISI 304) tube with an inner diameter (d) of 186 µm and wall thickness of 79 µm at a volumetric flow rate of 3.9 cc/min corresponding to a bulk velocity of 2.39 m/s is adopted. This flow rate is just slightly above the extinction limit and the corresponding Reynolds and Froude numbers are 27 and 1.56, respectively.

The computed CH species mass fraction isopleths are compared with the measured images of the excited CH* chemiluminescence and flame in Figure 5.9. Note that the CH* image is Abel inverted and it is performed only for the flame height and flame shape measurements. The CH* intensity cannot be correlated to the CH concentration because the CH* reaction mechanisms are not included in the present numerical calculation. It can be seen that the comparisons in terms of flame height, flame shape, and standoff distance are in very good agreement. Both experimental and numerical results indicate that the flame is quenched by the tube wall and creates a gap to allow oxidizer entrainment. As a result, the flame stands off from the tube. The characteristics of semi-spherical flame shape, non-buoyancy flame configuration, and large quenching gap between the flame base and the tube are very similar to those of a microgravity flame, but the heat-release rate is about two orders of magnitude larger than that of a microgravity flame [36]. In view of the success of predicting the overall flame characteristics of flame heights and shapes of microjet flames by numerical simulation, computations using GRI-Mech 3.0 are extended to investigate the flame structure and reaction characteristics [37].

The computed results of the 2D temperature, CH, O_2, CH_4, CO_2, H_2O, H_2, and CO species mass fraction contours as compared with mixture fraction contours of lean ($\xi = 0.029$), stoichiometric ($\xi = 0.055$), and rich ($\xi = 0.089$) limits of the methane flame are shown in Figure 5.10. The mixture fraction is defined as [38]

$$\xi = \frac{\left\{ \left(2Y_C / W_C \right) + 0.5Y_H / W_H + \left[\left(Y_{O,0} - Y_O \right) / W_O \right] \right\}}{\left[\left(2Y_{C,f} / W_C \right) + \left(0.5Y_{H,f} / W_H \right) + \left(Y_{O,0} / W_O \right) \right]} \qquad (5.19)$$

where Y_C, Y_H, and Y_O are the mass fraction for carbon, hydrogen, and oxygen atoms, W_i is the atomic weight of species i, and subscripts f and O refer to fuel jet and ambient air, respectively.

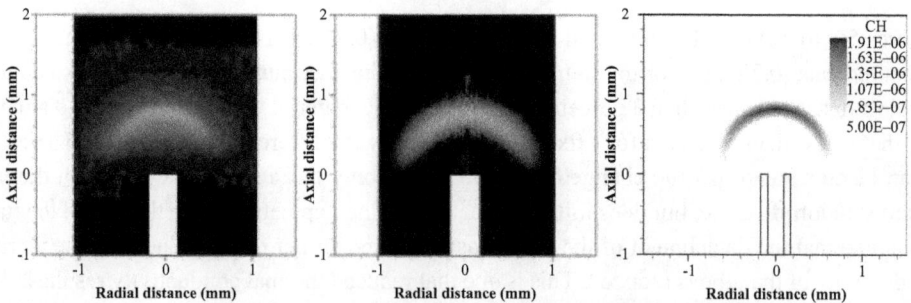

Figure 5.9. Comparison of the computed CH mass fraction isopleths (right) with measured photograph (left) and CH* image (middle) for the $d = 186$ µm flame.

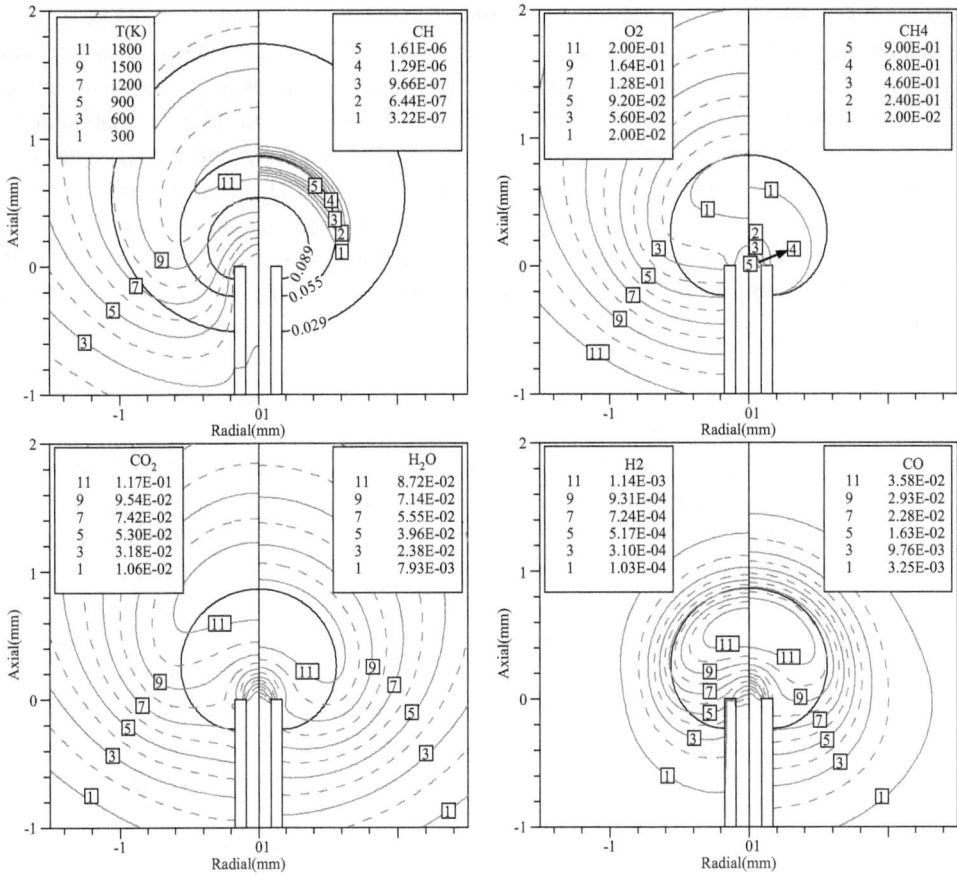

Figure 5.10. Computed 2D temperature and species mass fraction contours for the $d = 186$ μm flame near the extinction. The contours denote the lean ($\xi = 0.029$), stoichiometric ($\xi = 0.055$), and rich ($\xi = 0.089$) mixture fractions.

It can be seen that a small amount of O_2 is entrained into the standoff region from the gap between the burner wall and flame base and a small amount of CH_4 has diffused upstream of the burner port. The entrainment could result in partial premixing of fuel and oxygen over the standoff distance. The computed temperature contour shows that the maximum flame temperature locates at the jet centerline near the stoichiometric mixture fraction contour and the maximum of the CH isopleths. The unburned mixtures, the burner wall, and the fuel stream are heated to a temperature higher than 700 K. This suggests that the standoff and the flame stabilization are strongly related to the characteristic hot zone and heating of the fuel stream and unburned mixture through the tube wall. The appearance of more stable species of CO and H_2 and products of CO_2 and H_2O within the standoff region may further suggest that the high temperature in the standoff region is due to heat release from the final-product formation reactions ($CO + OH \rightarrow CO_2 + H$, $OH + H_2 \rightarrow H_2O + H$) or due to formation of double reaction zones in a partially premixed flame. However, the axial distributions of reaction rates (not shown here) and species mass fractions along the jet centerline indicate that the final-product formation reactions occur in the downstream region of the maximum temperature location and no double flame structure exists in this region. Therefore, this high temperature in the standoff region is most likely due to molecular heat conduction from the flame or other key radical reactions in the upstream region.

The computed axial profiles of temperature and selected species along the centerline of the jet are shown in Figure 5.11. The vertical dashed line denotes the location of stoichiometric mixture fraction. It can be seen that oxygen is entrained and diffused into the jet center near the burner exit and its concentration decreases with decreasing CH_4 concentration as it flows downstream. The consumption of CH_4 and O_2 intersects at a point slightly upstream of where the maximum temperature ($T = 1870$ K) is located. This is usually seen in a normal laminar diffusion flame except that no O_2 is present upstream of the intersection point in a diffusion flame. The distribution of CO, H_2, H_2O, and CO_2 is different from a typical diffusion flame structure [33], indicating appreciable amount of reactions in the standoff region. Moreover, there is only one CH peak occurring at the maximum temperature position (Figure 5.11(b)). This is also consistent with the computed pure CH_4 diffusion flame structure [39].

To further examine the role of a tribrachial flame in the microjet flame stabilization, the computed velocity vectors coupled with CH and stoichiometric mixture fraction contours are depicted in Figure 5.12 for a small region near the burner exit (0.8 mm × 1.2 mm). Note that the computational domain is 4 mm × 16 mm in the radial and axial direction, respectively. The maximum fuel velocity is 9.7 m/s at the centerline of the burner exit. The velocity vectors show lateral expansion and longitudinal acceleration as the fuel emerges from the burner exit and approaches the hot zone around the flame base due to thermal expansion. The flow velocity in the ambient air as well as in the downstream of flame base remains relatively constant and parallel to the axial direction, indicating a negligible buoyancy effect. Also, a small amount of air is entrained into the standoff region from the gap between the burner wall and the flame base. Figure 5.12 indicates that the flow velocity (0.2 m/s) near the flame edge is less than the laminar burning velocity (0.4 m/s) and no increase of velocity is observed after passing the flame edge similar to the observation by Puri et al. [40] (2001). In addition, the flame edge is not located on the stoichiometric mixture fraction point [41]. All of these outstanding features suggest that a tribrachial flame may not exist in a microjet flame near extinction and the flame stabilization must be due to other unrevealed mechanisms.

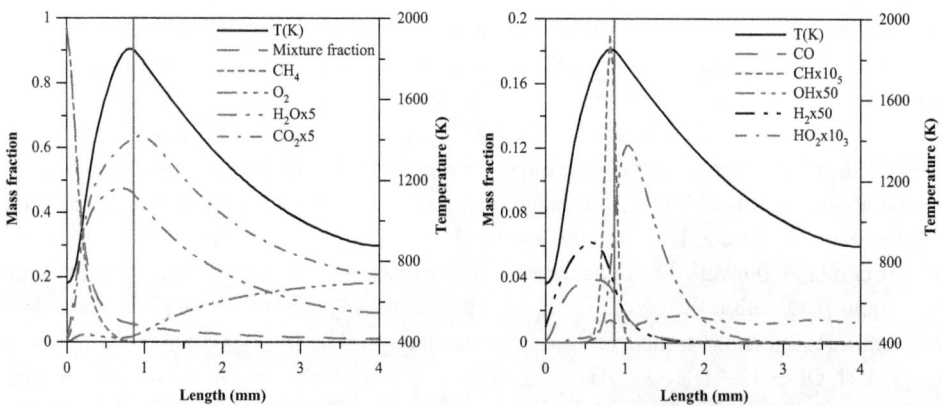

Figure 5.11. Computed axial distribution of temperature and scalars along the centerline of the jet flame. Vertical line denotes the location of stoichiometric mixture fraction.

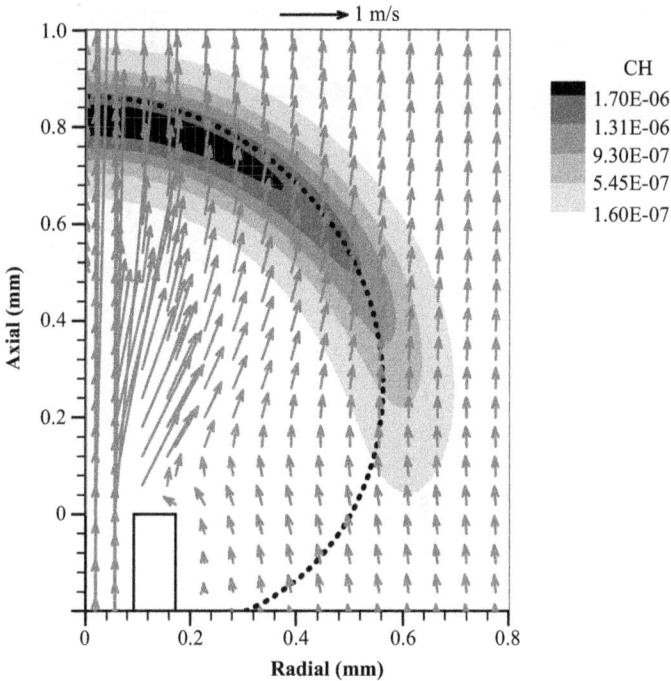

Figure 5.12. Computed 2-D velocity vector coupled with CH isopleths in the standoff region. The maximum fuel exit velocity is 9.7 m/s.

5.5.2 STABILIZATION MECHANISM

From the above discussion one might conjecture that the stabilization of a standoff microjet flame is due to the interaction of flame quenching, preheating of partially premixed mixture and sustained preflame reaction. The characteristics of flame quenching are illustrated in Figure 5.13 by comparing two computed CH contours under adiabatic and variable wall temperature conditions. With the adiabatic wall, the flame zone extends upstream of the port and connects to the tube as indicated by CH contours, while with the variable wall temperature condition (a realistic situation), the flame is quenched on the tube wall and creates a gap to allow oxidizer entrainment. As the flame is quenched by the tube wall, heat is transferred through the wall to accelerate fuel decomposition, initiate further reaction, and produce intermediate radicals in the vicinity of the exit. As usual, the HO_2 radicals are brisk in the chain-terminating reactions when the flame is quenched on the wall. Detailed examination of the intermediates indicates that the HO_2 radicals play an important role in connecting with the stabilization of the flame, as described in the following.

Figure 5.14 shows the calculated results of distribution contours of HO_2, isopleths of total heat-release rate and CH mass fraction, and mass flux vectors of H, O, OH, and HO_2 radicals in contrast to the stoichiometric mixture fraction location. The CH mass fraction isopleths and mass flux vectors of O and H radicals are shown on the left hand side of the figure, while the total heat-release rate isopleths, HO_2 contours, and mass flux vectors of OH and HO_2 on the right hand side. It can be seen that the maximum HO_2 appears in the gap region then decreases to connect with heat-release rate isopleths. The heat-release rate isopleth depicts a peak reactivity

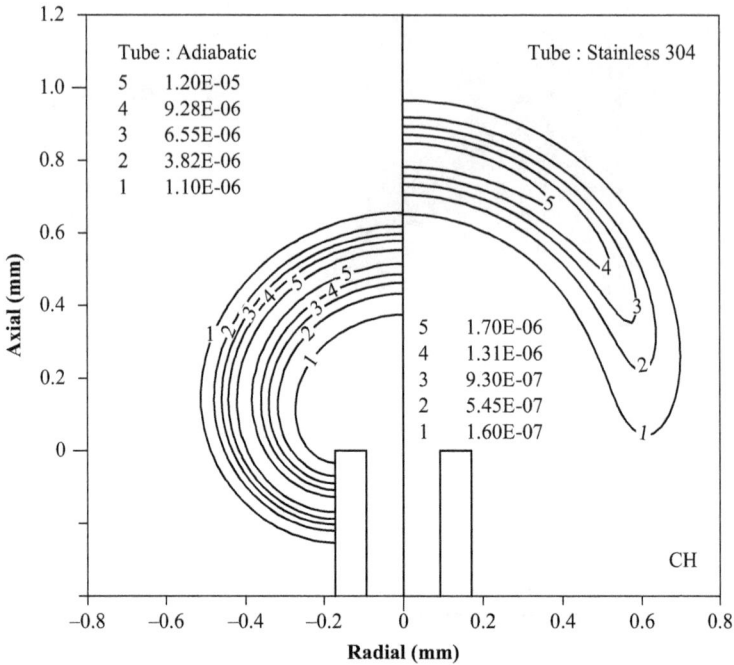

Figure 5.13. Comparison of CH contours calculated with adiabatic and variable wall temperature conditions.

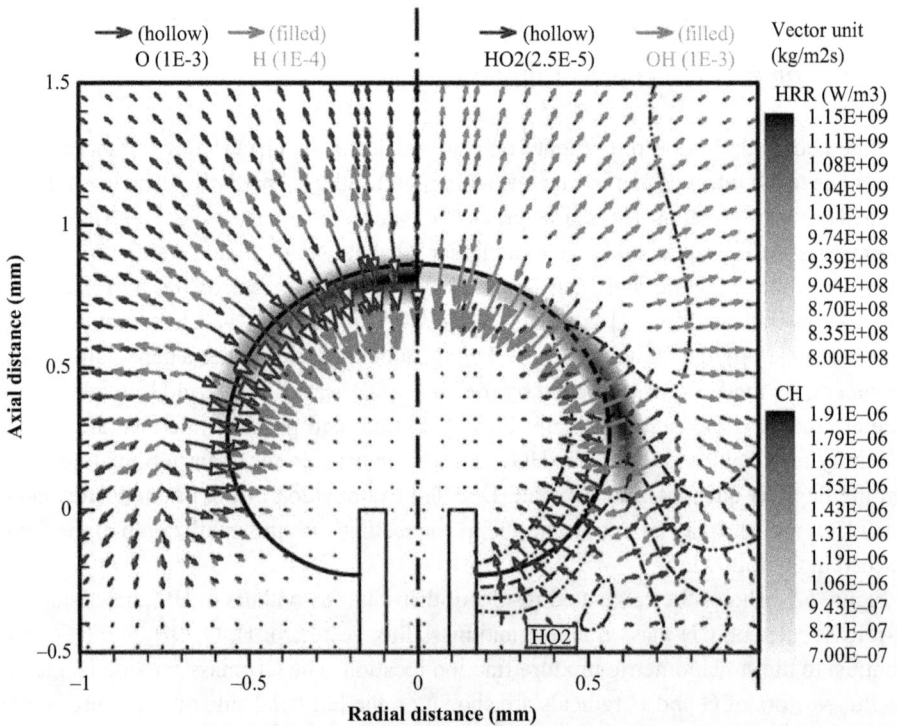

Figure 5.14. Computed stoichiometric mixture fraction and HO_2 contours, total heat-release rate and CH mass fraction isopleths, and the mass flux vectors of H, O, OH, and HO_2 for the $d = 186$ μm flame.

spot (called reaction kernel [36]) located on the lean side of the stoichiometric mixture fraction with the position of $x \approx 0.33$ mm and $r \approx 0.6$ mm. Hydrogen and oxygen atoms as well as hydroxyl radicals diffuse to both sides of the flame zone and upstream against the incoming fuel-rich flow, while the HO_2 radicals predominantly diffuse and displace in the inward direction to the fuel stream near the tube exit.

To further delineate the stabilization process of the microjet flame, the detailed flame structures across the standoff region ($x = -0.1$ mm) and the reaction kernel ($x = 0.33$ mm) are examined. The radial profiles of the temperature, species mass fractions, formation rates, net reaction rates, and net heat-release rates of the major elementary steps at these two characteristic heights are illustrated in Figures 5.15 and 5.16. In the figures, the radial location of the tube wall and the stoichiometric mixture fraction ($\xi = 0.055$) are indicated by the solid and dashed lines, respectively. For the case in Figure 5.15 ($x = -0.1$ mm), the distributions of temperature and major species mass fractions are typical of diffusion flames, except for the appreciable amount of penetration of O_2 to the fuel-side and the efflux of CH_4 to the air-side through the gap (Figure 5.15(a)). The peak temperature and the production rates of minor species are primarily formed on the lean-side of stoichiometric mixture fraction. The peak CH_4 consumption rate and the production rates of H_2O and CO_2 are also on the air-side (Figure 5.15(b)). These facts suggest that in the quenching gap region lean chemical reactions take place. Figures 5.15(c) and 5.15(d)

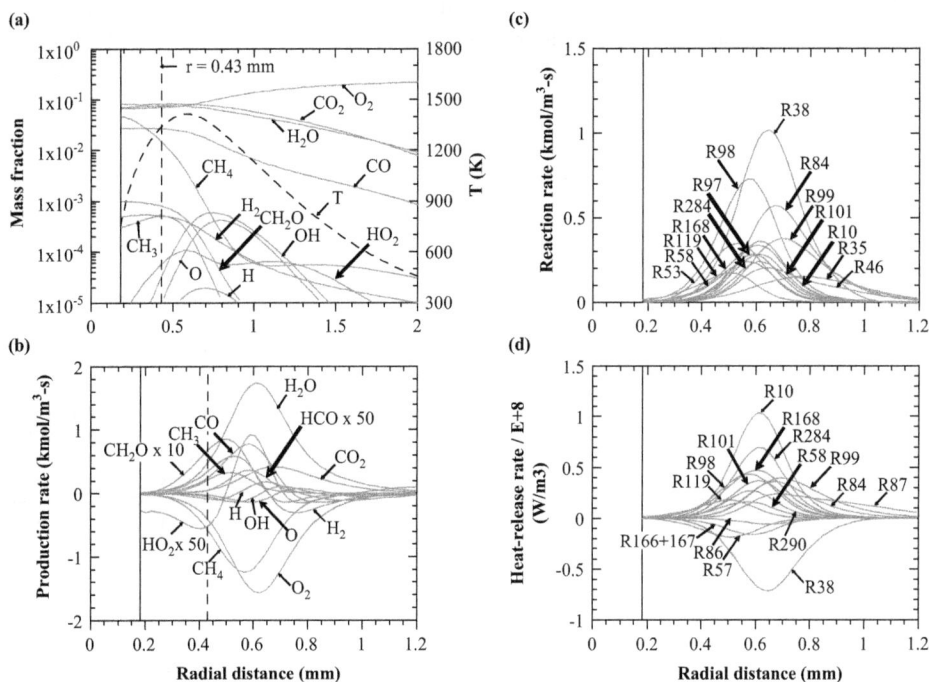

Figure 5.15. Computed radial profiles of temperature, species mass fraction, production rate, net reaction rate, and heat-release rate at x = −0.1 mm. The major reaction steps are: H + O_2 ↔ OH + O (R38), OH + CH_4 ↔ CH_3 + H_2O (R98), OH + H_2 ↔ H + H_2O (R84), OH + CO ↔ H + CO_2 (R99), O + CH_3 ↔ H + CH_2O (R10), HO_2 + H ↔ OH + OH (R46), H + CH_4 ↔ H_2 + CH_3 (R53), O + CH_3 → H + H_2 + CO (R284), O_2 + HCO ↔ HO_2 + CO (R168), OH + CH_2O ↔ HCO + H_2O (R101), O_2 + CH_2 → 2H + CO_2 (R290), OH + HO_2 ↔ O_2 + H_2O (R87), HO_2 + CH_3 ↔ OH + CH_3O (R119).

show the net reaction and heat-release rates of major elementary steps. It is obvious that the most significant reaction is the chain-branching reaction $H + O_2 \leftrightarrow OH + O$ (R38), followed by the dehydrogenation of methane primary by $OH + CH_4 \leftrightarrow CH_3 + H_2O$ (R98) and $OH + H_2 \leftrightarrow H + H_2O$ (R84) reactions to form the product H_2O and to further build up the radical pool. Although the rate for H production through $OH + H_2 \leftrightarrow H + H_2O$ (R84), $OH + CO \leftrightarrow H + CO_2$ (R99), and $O + CH_3 \leftrightarrow H + CH_2O$ (R10) reactions is more significant than that for OH formation through $HO_2 + CH_3 \leftrightarrow OH + CH_3O$ (R119) and $HO_2 + H \leftrightarrow OH + OH$ (R46) reactions, the rate of methyl (CH_3) radical production by OH radical attacking on CH_4 (R98) exceeds that by H atoms (R53) in this relatively low temperature region. In addition, the formation (R168) and destruction (R119) of HO_2 reactions are believed to play an important role for the hot zone and reaction kernel formation in the quenching gap region. Figure 5.15(d) indicates that the major contributors to the positive heat release are the reactions, including $O + CH_3 \leftrightarrow H + CH_2O$ (R10), $O + CH_3 \rightarrow H + H_2 + CO$ (R284), $O_2 + HCO \leftrightarrow HO_2 + CO$ (R168), $OH + CH_4 \leftrightarrow CH_3 + H_2O$ (R98), $OH + CO \leftrightarrow H + CO_2$ (R99), $OH + H_2 \leftrightarrow H + H_2O$ (R84), $OH + CH_2O \leftrightarrow HCO + H_2O$ (R101), $O_2 + CH_2 \rightarrow 2H + CO_2$ (R290), $OH + HO_2 \leftrightarrow O_2 + H_2O$ (R87), and $HO_2 + CH_3 \leftrightarrow OH + CH_3O$ (R119). The negative contributors are R38, R166 + R167, R57, and R86.

Figure 5.16 shows the radial distributions of calculated variables across the reaction kernel ($x = 0.33$ mm). In general, the distributions of variables are similar to those at $x = -0.1$ mm, except for the increased levels of species production rates, net reaction rates, and heat-release rates resulting in higher species concentrations and peak temperature ($T = 1746$ K) in the

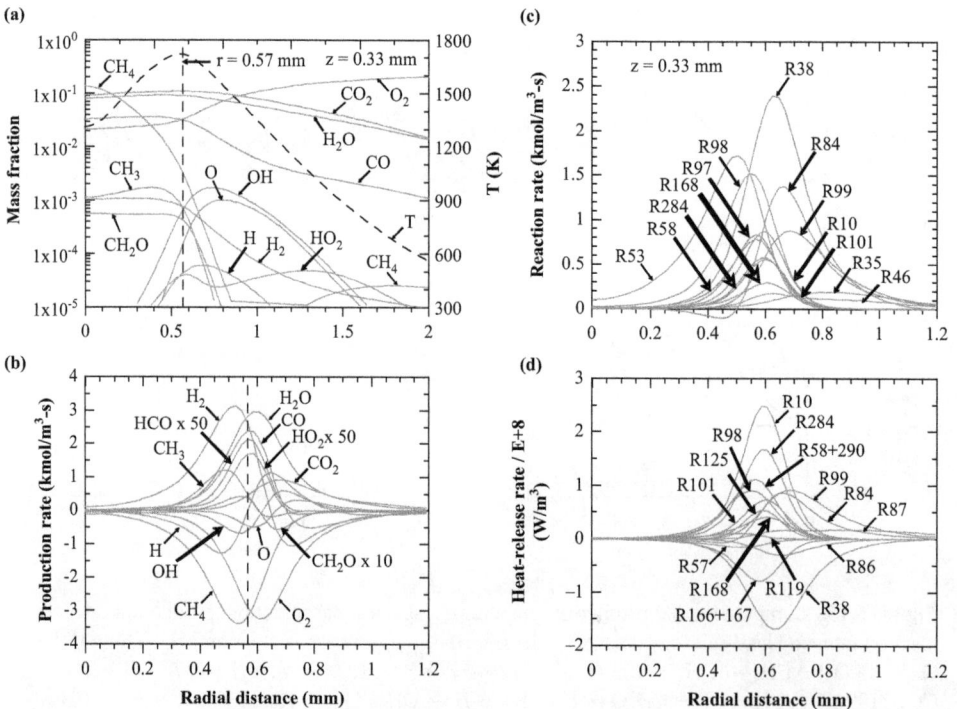

Figure 5.16. Computed radial profiles of temperature, species mass fraction, production rate, net reaction rate, and heat-release rate at $x = 0.33$ mm.

form of a flame. In addition, the production rates (Figure 5.16(b)) show that unlike those at $x = -0.1$ mm, the H_2, CO, CH_3, and CH_2O species are primarily formed on the fuel-side and consumed on the air-side of the stoichiometric mixture fraction; and the H, O, and OH radicals are produced on the air-side and consumed on the fuel-side. The peak consumption rate of CH_4 is on the fuel-side and that of O_2 and the production rates of the H_2O and CO_2 are on the air-side. Figures 5.16(c) and (d) show that the chain-branching reaction (R38) is still the most significant reaction and the methyl oxidation reaction (R10) is the major contributor to the heat-release rate. It should be noted that the methyl (CH_3) radical is produced more by the H atom through (R53) than by the OH radical through (R98) as compared to those reactions at $x = -0.1$ mm. Also, the formation (R168) and destruction (R119) of HO_2 reactions, that play an important role for the hot zone formation in the gap region ($x = -0.1$ mm), become insignificant at this location. The dominant exothermic reactions, including R10, R284, R58, R290, R98, R99, R84, R101, R125, and R168, contribute up to 81.5 percent of the total heat-release rate at the reaction kernel (1.15×10^9 W/m^3), thereby stabilizing a standoff microjet flame.

From the analysis of chemical kinetic structures, the sequence of flame stabilization can be summarized schematically in Figure 5.17 and explained as follows. As the flame is quenched on the tube wall, heat is transferred through the tube wall to accelerate fuel pyrolysis near the tube exit, produce CH_3 and H intermediate radicals in the standoff region, and initiate further reaction in the vicinity of the flame base. Consequently, the chain-branching reaction (R38), the methane consumption reactions (R98) and (R53), and the final product formation (R84) and (R99) are enhanced to further build up the radical pool of H, O, OH, and CH_3 in the gap region.

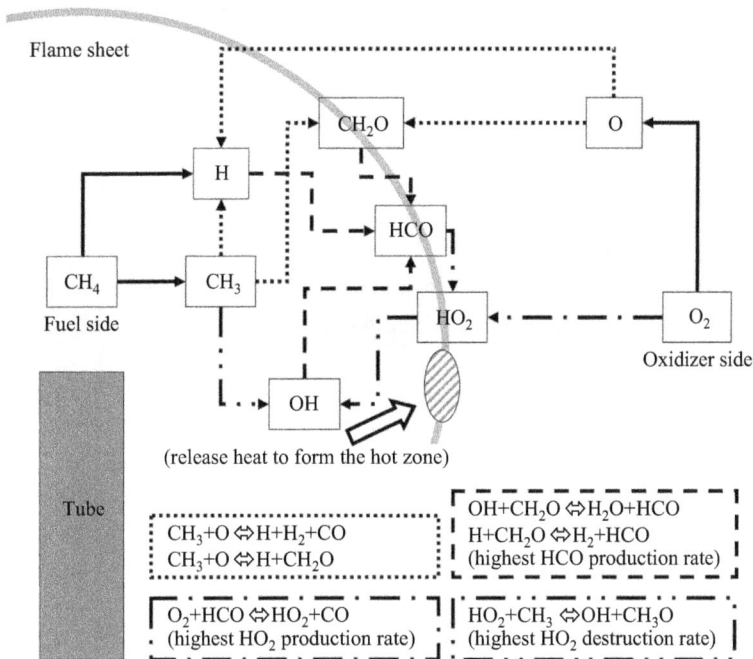

Figure 5.17. Schematic diagram of key reactions leading to microjet diffusion flame stabilization.

Then the methyl (CH_3) radical oxidations (R10) and (R284) not only release significant heat to enhance further reactions but also produce the formaldehyde (CH_2O) radical. The CH_2O radical is attacked by the OH radical (R101) and H atom (R58) to form the formyl (HCO) radical and followed by the HCO oxidation (R168) to produce the HO_2 radical. Finally the HO_2 radical predominantly diffuses and moves inward to the fuel stream near the tube exit and reacts with CH_3 (R119) to release heat near the tube wall region. These key reactions are believed to form the hot zone ($T = 800$–1450 K) that connects the visible flame base and provides heat for sustaining and enhancing further H_2–O_2 chain reactions and CH_3 formation (R53 and R98) and oxidation (R10 and R284) at the flame base, which in turn result in the formation of the reaction kernel responsible for flame stabilization. It should be noted that although the reaction rate and heat-release rate of the HO_2 formation (R168) and consumption (R119) reactions are not as significant as those of the other reactions, they indeed play an important role in the formation of the hot zone and high heat-release reaction kernel [42].

5.6 CONCLUSION AND FUTURE WORK

Flame characteristics, in terms of flame shape, flame length, flame structure, reaction rate profiles, quenching limit, and stabilization mechanism, of microjet methane diffusion flames operated at various fuel exit velocities ranging from just above quenching to below blowoff are investigated. Comparisons of the measured primary flame parameters, such as flame heights, flame shapes, and quenching velocities with theoretical predictions indicate that Roper's model [14] can satisfactorily predict the characteristics of microjet methane flames. Comparisons of the predicted quenching velocity with measured results indicate that quenching occurs when the flame length equals the standoff distance. It is also found that the quenching velocity for different tube diameters collapses to the curve $Re \times d =$ constant. An order of magnitude analysis of the dimensionless parameters suggests that the microjet diffusion flames studied here are not buoyancy-free and the molecular diffusion is effective only for $N_{DB} > 2$ and $Pe < 2$. Numerical simulations of the flames stabilized at the tip of a 186 μm tube indicate that the computed flame shape and flame length are in excellent agreement with experimental results. Furthermore, the calculated flame structures show that the diffusion process dominates over the premixing process in the standoff region, suggesting that the flame burns in a diffusion mode. Besides, the calculated OH mass fraction isopleths indicate that the change of tube materials has a minor effect on the standoff distance near extinction, but does influence the quenching gap between the flame and the tube.

The computed flame heights as a function of fuel exit velocities indicate that buoyancy may play a role for flames with larger tube diameters and higher exit velocities. However, for the case of $d = 186$ μm flame with $u_e < 3.68$ m/s, the buoyancy effect becomes minor and the flame is diffusion controlled. Although partial premixing may occur in the standoff region, its mixing intensity is not strong enough to generate reaction and hence no double flame structure is observed. Further examination of the computed flame structure reveals that the flame is stabilized by a consequence of flame quenching on the tube wall, formation of HO_2 layer, forming the hot zone through reaction of HO_2 with CH_3, yielding subsequent key radical reactions, and finally forming the reaction kernel. The major reaction path near the flame base is also summarized schematically.

REFERENCES

[1] Kovacs, G. T. A., 1998, *Micromachined Transducers-Source Book.* McGraw-Hill, New York.

[2] Mitchell, R. E., Sarofim, A. F., and Clomburg, L. A., 1980, "Experimental and Numerical Investigation of Confined Laminar Diffusion Flames," *Combustion and. Flame*, vol. 37, pp. 227–244. doi: http://dx.doi.org/10.1016/0010-2180(80)90092-9.

[3] Smooke, M. D., Mitchell, R. E., and Grcar, J. F., 1984, in: *Elliptic Problem Solvers II*, edited by Birkhoff, G. and Schoenstadt, A., Academic, New York.

[4] Smooke, M. D., Lin, P., Lam, J., and Long, M. B., 1990, "Computational and Experimental Study of a Laminar Axisymmetric Methane/Air Diffusion Flame," *Proceeding of Combustion Institute*, vol. 23, pp. 575–582. doi: http://dx.doi.org/10.1016/S0082-0784(06)80305-X.

[5] Smooke, M. D., Xu, Y., Zurn, R., Lin, P., Frank, J., and Long, M. B., 1992, "Computational and Experimental Study of OH and CH Radicals in Axisymmetric Laminar Diffusion Flames," *Proceedings of Combustion Institute*, vol. 24, pp. 813–821. doi: http://dx.doi.org/10.1016/S0082-0784(06)80099-8.

[6] Smooke, M. D., Ern, A., Tanoff, M. A., Valdati, B. A., Mohammed, R. K., Marran, D. F., and Long, M. B., 1996, "Computational and Experimental Study of NO in an Axisymmetric Laminar Diffusion Flame," *Proceedings of Combustion Institute*, vol. 26, pp. 2161–2170. doi: http://dx.doi.org/10.1016/S0082-0784(96)80042-7.

[7] Norton, T. S., Smyth, K. C., Miller, J. H. and Smooke, M. D., 1993, "Comparison of Experimental and Computed Species Concentration and Temperature Profiles in Laminar, Two-dimensional Methane/Air Diffusion Flames," *Combustion Science and Technology*, vol. 90, pp. 1–34. doi: http://dx.doi.org/10.1080/00102209308907601.

[8] Takagi, T. and Xu, Z., 1994, "Numerical Analysis of Laminar Diffusion Flames—Effects of Preferential Diffusion of Heat and Species," *Combustion and Flame*, vol. 96, pp. 50–59. doi: http://dx.doi.org/10.1016/0010-2180(94)90157-0.

[9] Walsh, K. T., Long, M. B., Tanoff, M. A., and Smooke, M. D., 1998, "Experimental and Computational Study of CH, CH*, and OH* in an Axisymmetric Laminar Diffusion Flame," *Proceedings of Combustion Institute*, vol. 27, pp. 615–623. doi: http://dx.doi.org/10.1016/S0082-0784(98)80453-0.

[10] McEnally, C. S., Pfefferle, L. D., Schaffer, A. M., Long, M. B., Mohammed, R. K., Smooke, M. D., and Colket, M. B., 2000, "Characterization of a Coflowing Methane/Air Non-premixed Flame with Computer Modeling, Rayleigh-Raman Imaging, and On-line Mass Spectrometry," *Proceedings of Combustion Institute*, vol. 28, pp. 2063–2070. doi: http://dx.doi.org/10.1016/S0082-0784(00)80614-1.

[11] Nguyen, Q. V., Dibble, R. W., Carter, C. D., Fiechtner, G. J., and Barlow, R. S., 1996, "Raman-LIF Measurements of Temperature, Major Species, OH, and NO in a Methane-Air Bunsen Flame," *Combustion and Flame*, vol. 105, pp. 499–510. doi: http://dx.doi.org/10.1016/0010-2180(96)00226-X.

[12] Chou, C.-P., Chen, J.-Y., Yam, C. G., and Marx, K. D., 1998, "Numerical Modeling of NO Formation in Laminar Bunsen Flames—A Flamelet Approach," *Combustion and Flame*, vol. 114, pp. 420–435. doi: http://dx.doi.org/10.1016/S0010-2180(97)00317-9.

[13] Burke, S. P. and Schumann, T. E. W., 1928, "Diffusion Flames," *Industrial Engineering Chemistry*, vol. 20, pp. 998–1004. doi: http://dx.doi.org/10.1021/ie50226a005.

[14] Roper, F. G., 1977, "The Prediction of Laminar Jet Diffusion Flame Sizes: Part I. Theoretical Model and Part II. Experimental Verification," *Combustion and Flame*, vol. 29, pp. 219–234. doi: http://dx.doi.org/10.1016/0010-2180(77)90112-2.

[15] Ban, H., Venkatesh, S., and Saito, K., 1994, "Convection-Diffusion Controlled Laminar Micro Flames," *Transactions of the ASME, Journal of Heat Transfer*, vol. 116, pp. 954–959.

[16] Williams, F. A., 1985, *Combustion Theory.* Addison-Wesley, New York.

[17] Nakamura, Y., Ban, H., Saito, K., and Takeno, T., 1997, "Micro Diffusion Flames in a Cold Boundary," *Proceeding of the Central State Section Meeting*, pp. 160–163.

[18] Nakamura, Y. and Saito, K. 2001, "Thermal and Fluid Dynamic Structures of Micro-Diffusion Flames," *Nagare* (in Japanese), vol. 20, pp. 74–82.

[19] Nakamura, Y., Kubota, A., Yamashita, H., and Saito, K., November 30, 2003, "Near Extinction Flame Structure of Micro-Diffusion Flames," *The International Symposium on Micro-Mechanical Engineering*, Paper No. ISMME2003-111, pp. 163–170.

[20] Ida, T., Fuchihata, M., and Mizutani, Y., 2000, "Microscopic Diffusion Structures with Micro Flames," *Proceedings of the Third International Symposium on Scale Modeling*, ISSM3-E7, Nagoya, Japan.

[21] Matta, L. M., Neumeier, Y., Lemon, B., and Zinn, B. T., 2002, "Characteristics of Microscale Diffusion Flames," *Proceedings of Combustion Institute*, vol. 29, pp. 933–939. doi: http://dx.doi.org/10.1016/S1540-7489(02)80118-3.

[22] Baker, J., Calvert, M. E., and Murphy, D. W., 2002, "Structure and Dynamics of Laminar Jet Micro-Slot Diffusion Flames," *Transactions of the ASME, Journal of Heat Transfer*, vol. 124, pp. 783–790.

[23] Cheng, T. S., Chao, Y.-C., Wu, C.-Y., Li, Y.-H., Nakamura, Y., Lee, K.-Y., Yuan, T., and Leu, T. S., 2005, "Experimental and Numerical Investigation of Microscale Hydrogen Diffusion Flames," *Proceedings of Combustion Institute*, vol. 30, pp. 2489–2497. doi: http://dx.doi.org/10.1016/j.proci.2004.07.025.

[24] Nakamura, Y., Ban, H., Saito, K., and Takeno, T., 2000, "Structures of Micro (Millimeter Size) Diffusion Flames," *Proceedings of the Third International Symposium on Scale Modeling*, ISSM3-E7, Nagoya, Japan.

[25] Cheng, T. S., Wu, C.-Y., Chen, C.-P., Li, Y.-H., Chao, Y.-C., Yuan, T., and Leu, T. S., 2005, "Detailed Measurement and Assessment of Laminar Hydrogen Jet Diffusion Flames," *Combustion and Flame*, vol. 146, pp. 268–282. doi: http://dx.doi.org/10.1016/j.combustflame.2006.03.005.

[26] Yoo, S. W., Law, C. K., and Tse, S. D., 2002, "Chemiluminesent OH* and CH* Flame Structure and Aerodynamic Scaling of Weakly Buoyant, Nearly Spherical Diffusion Flames," *Proceedings of Combustion Institute*, vol. 29, pp. 1663–1670. doi: http://dx.doi.org/10.1016/S1540-7489(02)80204-8.

[27] Cheng, T. S., Chen, C.-P., Chen, C.-S., Li, Y.-H., Wu, C.-Y., and Chao, Y.-C., 2006, "Characteristics of microjet methane diffusion flames," *Combustion Theory and Modeling*, vol. 10, pp. 861–881. doi: http://dx.doi.org/10.1080/13647830600551917a.

[28] Van Doormal, J. P. and Raithby, G. D., 1984, "Enhancements of the SIMPLE Method for Predicting Incompressible Fluid Flows," *Numerical Heat Transfer*, vol. 7, pp. 147–163.

[29] Kee, R. J., Rupley, F., Miller, J., Coltrin, M., Grcar, J., Meeks, E., Moffat, H., Lutz, A., Dixon-Lewis, G., Smooke, M. D., Warnatz, J., Evans, G., Larson, R., Mitchell, R., Petzold, L., Reynolds, L., Caracotsios, M., Stewart, W., and Glarborg, P., 1999, *User Manual, The CHEMKIN Collection Release 3.5*, Reaction Design, Inc., San Diego.

[30] Smooke, M. D., 1991, "Reduced Kinetic Mechanisms and Asymptotic Approximations for Methane-Air Flames," *Lecture Notes in Physics*, Springer-Verlag, Berlin, vol. 384, pp. 1–28. doi: http://dx.doi.org/10.1007/BFb0035362.

[31] Smith, G. P., Golden, D. M., Frenklach, M., Moriarty, N. W., Eiteneer, B., Goldenberg, M., Bowman, C. T., Hanson, R. K., Song, S., Gardiner, W. C., Lissianski, V. V., and Qin, Z., 1999, GRI-Mech Homepage, Gas Research Institute, Chicago, www.me.berkeley.edu/gri_mech/.

[32] Spalding, D. B., 1979, *Combustion and Mass Transfer.* Pergamon Press, New York.

[33] Turns, S. R., 2000, *An Introduction to Combustion: Concepts and Applications*, McGraw-Hill, New York, p. 352.

[34] Chung, S. H. and Law, C. K., 1984, "Burke-Schumann Flame With Streamwise and Preferential Diffusion," *Combustion Science and Technology*, vol. 37, pp. 21–46. doi: http://dx.doi.org/10.1080/00102208408923744.

[35] Lee, B. J. and Chung, S. H., 1997, "Stabilization of Lifted Tribrachial Flames in a Laminar Nonpre-mixed Jet," *Combustion and Flame*, vol. 109, pp. 163–172. doi: http://dx.doi.org/10.1016/S0010-2180(96)00145-9.

[36] Takahashi, F. and Katta, V. R., 2002, "Reaction Kernel Structure and Stabilization Mechanisms of Jet Diffusion Flames in Microgravity," *Proceedings of Combustion Institute*, vol. 29, pp. 2509–2518. doi: http://dx.doi.org/10.1016/S1540-7489(02)80306-6.

[37] Chen, C.-P., Chao, Y.-C., Cheng, T. S., Chen, G.-B., and Wu, C.-Y., 2007, "Structure and Stabili-zation Mechanism of a Microjet Methane Diffusion Flame Near Extinction," *Proceedings of the Combustion Institute*, vol. 31, pp. 3301–3308. doi: http://dx.doi.org/10.1016/j.proci.2006.08.069.

[38] Bilger, R. W., Stårner, S. H., Kee, and R. J., 1990, "On Reduced Mechanisms for Methane – Air Combustion in Nonpremixed Flames," *Combustion and Flame*, vol. 80, pp. 135–149. doi: http://dx.doi.org/10.1016/0010-2180(90)90122-8.

[39] Blevins, L. G. and Gore, J. P., 1999, "Computed Structure of Low Strain Rate Partially Premixed CH/Air Counterflow Flames: Implications for NO Formation," *Combustion and Flame*, vol. 116, pp. 546–566. doi: http://dx.doi.org/10.1016/S0010-2180(98)00059-5.

[40] Puri, I. K., Aggarwal, S. K., Ratti, S., and Azzoni R., 2001, "On the Similitude Between Lifted and Burner-Stabilized Triple Flames: A Numerical and Experimental Investigation," *Combustion and Flame*, vol. 124, pp. 311–325. doi: http://dx.doi.org/10.1016/S0010-2180(00)00201-7.

[41] Echekki, T. and Chen, J. H., 1998, "Structure and Propagation of Methanol–Air Triple Flames," *Combustion and Flame*, vol. 114, pp. 231–245. doi: http://dx.doi.org/10.1016/S0010-2180(97)00287-3.

[42] Cheng, T. S., Chao, Y.-C., Chen, C.-P., and Wu, C.-Y., 2008, "Further Analysis of Chemical Kinetic Structure of a Standoff Microjet Methane Diffusion Flame Near Extinction," *Combustion and Flame*, vol. 152, pp. 461–467. doi: http://dx.doi.org/10.1016/j.combustflame.2007.10.007.

NOMENCLATURE

b half-width of the outer wall
c normalized half-width of the inner wall
c_p specific heat capacity
d inner diameter
D mass diffusivity
D_i concentration-driven diffusion coefficient
D_{ij} binary diffusion coefficient
D_i^T thermal-diffusion coefficient
D_O mean diffusion coefficient evaluated for the oxidizer at the oxidizer stream temperature
F dependent variable in the similarity equation
f focal length; mixture fraction
Fr Froude number
f_{st} stoichiometric value of mixture fraction
g_x gravitational acceleration in the x-direction
h enthalpy
I ratio of the actual initial momentum to that for uniform flow
k thermal conductivity
k_{ij} thermal-diffusion ratio
l_D radial position of the flame edge
Le Lewis number
L_f flame length

M	molecular weight
N_{DB}	diffusion to buoyancy parameter
N_{DM}	diffusion to momentum parameter
p	pressure
Pe	Peclet number
P_i	power of incident laser
Q_f	volumetric fuel flow rate
Re	Reynolds number
R_0	universal gas constant
S	molar stoichiometric oxidizer—fuel ratio
Sc	Schmidt number
T	temperature
u^*	dimensionless axial velocity component
u_d	molecular-diffusion velocity
u_e	fuel exit velocity
w	species production rate
W_i	atomic weight of species i
x^*	flame shape
x_{fl}^*	dimensionless flame height
$Y_{F,stoic}$	stoichiometric fuel mass fraction
Y_i	mass fraction of species i
Y_{O0}	mass fraction of the oxidizer at the oxidizer stream
Z_f^*	normalized flame height
γ	constant
λ	wave length; thermal conductivity
μ	viscosity
ν	(u, v) velocity vector; kinetic viscosity
ζ	similarity variable
ξ	mixture fraction
ρ	density
Φ	equivalence ratio

CHAPTER 6

DIFFUSION FLAME INSTABILITY AND CELL FORMATION IN MESO- AND MICROSCALE COMBUSTION

Yiguang Ju and Sang Hee Won

In most practical applications for microscale combustion, fuel and oxidizer are often injected separately into the combustion chamber and then mixed in a laminar or turbulent flow. As the combustor size decreases, the flow Reynolds number becomes very low (1–100) [1]. As a result, the molecular diffusion is the dominant mechanism for viscous flow and fuel–air mixing [2–4]. Due to the short flow residence time, reactants and oxidizers may not be able to be fully mixed before combustion. As such, non-premixed combustion plays an important role in these flow conditions. When the wall and structure heat loss increases, the limited fuel–air mixing due to molecular diffusion will dramatically change the flame dynamics, leading to diffusion flame instability, flame cells [5], flame streets, and unsteady flame propagation [6]. In liquid fuel combustion, the vaporization of liquid fuel and mixing of the fuel with air create new challenges to microscale combustion [7]. In this chapter, the diffusion flame instability and cell formation in gas-phase meso- and microscale combustion will be described.

6.1 CELL FLAME FORMATION IN A MICROSCALE DIFFUSION FLAME REACTOR

The joint effect of fuel–oxidizer mixing and wall heat loss in a non-premixed microscale combustor creates new flame regimes. Cellular instabilities of laminar non-premixed diffusion flames were observed in a polycrystalline alumina microburner with a channel of 0.75 mm by Miesse et al. [5]. The burner consists of an inverted Y-shaped geometry having fuel and oxidizer supply legs (Figure 6.1, left), with a rectangular combustion channel of 30 mm in length and 0.75 mm × 5 mm in cross-section situated above the splitter plate. The top of the burner is open to atmospheric air so that the exhaust gas and unreacted material may exit. Changes in the flame structure were observed as a function of the fuel type (H_2, CH_4, and C_3H_8) and diluents. As shown in Figure 6.1 (right), multiple isolated reaction zones or flame cells were observed. In

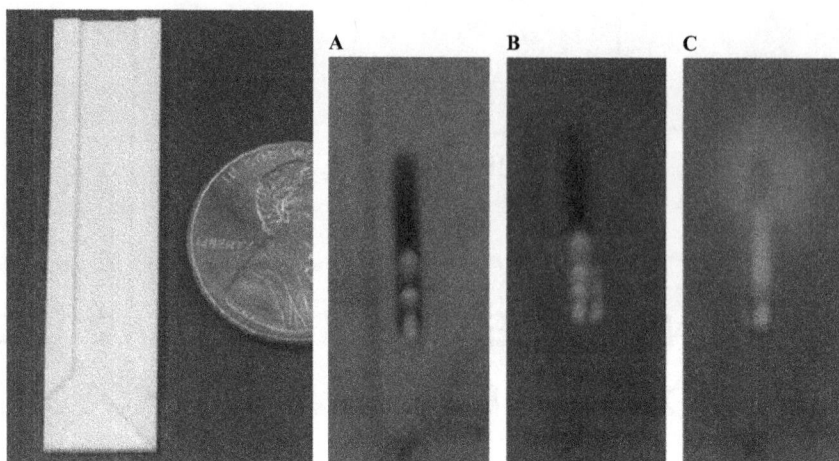

Figure 6.1. Left: A photograph of a section of the Y-shaped alumina burner. Right: chemiluminescent images—(a) three flame cells formed with 100 sccm CH_4/200 sccm O_2. The flame cell closest to the top of the burner is the one at the top of the image. The bottom cell is slightly elongated due to the strained flow from the angled burner inlets. (b) Four flame cells for 100 sccm CH_4/130 sccm O_2. (c) A laminar diffusion flame a top a single cell (the exclamation mark flame) for 65 sccm CH_4/150 sccm O_2. A methane/air diffusion flame can also be observed at the exit to the burner (reproduced by permission from Miesse et al. [5]).

addition, the number of flame cells observed inside the burner was dependent on the initial supply flow rates and stoichiometry and varied between one and four for the length of the burners considered here. By analyzing the effects of mixture Lewis numbers (at low Lewis numbers), it was concluded that these cells were resulted from a cellular instability of the underlying diffusion flame.

6.2 FLAME STREETS AND UNSTEADY FLAME PROPAGATION IN MESOSCALE DIFFUSION FLAMES

To understand the combined effect of molecular diffusion, wall heat loss, and wall skin friction on the non-premixed flame dynamics, flame structures in a fuel–air mixing layer of a mesoscale channel were experimentally and analytically studied by Xu and Ju [6] by using methane and propane–air mixtures.

6.2.1 EXPERIMENTAL METHOD

A mesoscale combustor with temperature-controlled heating surfaces was constructed to study non-premixed combustion with enhanced heat losses, increased diffusion timescale, and skin friction. The combustor (Figure 6.2 (a)) consisted of two horizontal silicon carbide plates, which were 241 mm long (L), 95 mm wide, and 6 mm thick. Two rectangular quartz windows were positioned between the two ceramic plates to allow the optical access and keep a constant

(a)

(b)

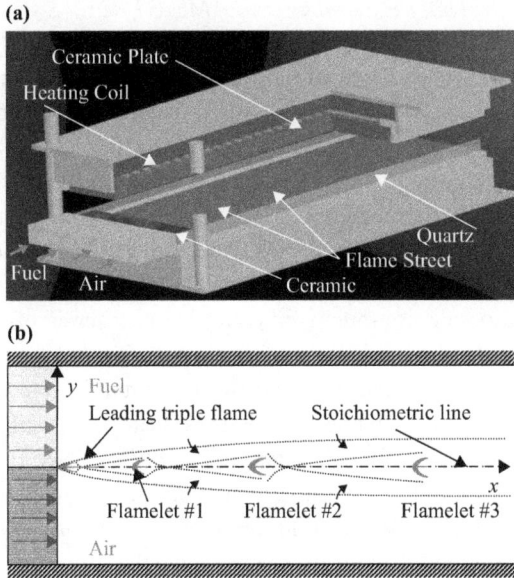

Figure 6.2. (a) Schematic of the channel combustor and (b) non-premixed flamelets formed in the mixing layer (Flame street, reproduced by permission from Xu and Ju [6]).

channel height of 6 mm (h). The resulting mesoscale channel combustor size was delineated by the 241 × 95 × 6 mm³ enclosure. The silicon carbide plates had a high thermal conductivity (~120 W/m·K) and low thermal expansion coefficient and thus provided a uniform wall temperature distribution in a broad temperature range. The two plates are separated from the aluminum frame by using two ceramic support rims, which were made of alumina silicate with a thermal conductivity of 1.3 W/m·K. The support rims were placed on a 241 × 114 × 50 mm³ silica foam block with a thermal conductivity of 0.187 W/m·K. Between the silicon carbide plate and the silica foam, a metal heating coil was placed to control the temperature of the silicon carbide plates. The heating coil controlled by an AC power supply had a resistance of 11 Ω which gave a maximum heating power of 1 kW. The temperature of the silicon carbide plates could be raised up to 1000 K by using multiple heating coils. The surface temperatures of the plates were measured using a K-type thermal couple with the quartz plate taken off and the heating power fixed. Due to high heat conductivity of the silicon carbide and the insulations surrounding the plates, the maximum surface temperature difference at full heating power was less than 15°C.

At the inlet, a 400 cell honeycomb was mounted into the inlet connector. The ceramic honeycomb was divided into two separate parts, so that it could discharge uniform flows of fuel and oxidizer between the two parallel plates and generate a mixing layer between the fuel and air streams. The flow rates of fuel (methane and propane) and air were controlled by chocked sonic nozzles. In order to visualize the flames in the mesoscale mixing layer, the top half burner was replaced with a quartz plate. A Canon Digital Rebel Xt 350D camera was used to capture static images and a high-speed video camera (PHOTRON 120K) to record transient flame images.

6.2.2 FORMATION OF NON-PREMIXED FLAME STREETS

The schematic top view of the formation of fuel–air mixing layer and the flame street (multiple flamelets) is shown in Figure 6.2(b). A mixing layer is formed between the fuel and air streams along the flow direction from left to right. At the leading edge of the mixing layer, fuel and air mix quickly and form a partially premixed region with the stoichiometric line centered in the mixing layer. Therefore, experimentally, there is an anchored leading triple flame due to the lower flow speed at the splitter than the propagation speed of the triple flame. It is interesting to note that, different from the single triple flame structure observed in the conventional non-premixed flames, in mesoscale non-premixed combustion, a series of triple flame structures emerged in the mixing layer after the first anchored triple flame. Due to the similarity to the *vortex street* in a laminar flow shear layer, we call these multi-triple flamelet structures the *flame street*. The occurrence of the flame street in mesoscale combustion can be explained by the triple flame extinction, fuel diffusion, reignition, and triple flame propagation and stabilization processes. At first, the diffusion flame branch of each triple flamelet extinguishes due to the heat loss to the channel walls, fuel dilution by the burned products, and the insufficient diffusion of reactants. Secondly, after the flame quenching the reactants continue to diffuse into the extinguished high temperature mixing layer. After a certain delay time, re-ignition occurs in the mixing layer and a new triple flame is formed. Thirdly, the new triple flame propagates upstream at a decreasing flame speed and is stabilized at the location where the local flame speed is balanced by the flow speed. This process repeats and forms a series of triple flamelets until the reactant gradients across the mixing layer become so small that it cannot support a triple flame beyond the flammability limit. Therefore, the flame street structure will be dominantly affected by the fuel diffusion, flow residence time (diffusion time), the heat loss to the wall, and the temperature dependence of chemical kinetics. In order to examine the effects of fuel Lewis number and temperature dependence of chemical kinetics, methane and propane, which have very distinct Lewis numbers and activation energies are investigated in the experiments.

By controlling the flow velocity and wall temperature, it was found that there exist two flame regimes: the stable non-premixed flame street regime and the unsteady bimodal flame propagation regime. The boundaries of different flame regimes are shown in Figure 6.3 for methane and propane. The unsteady bimodal flame regime exists at low flow rates where only

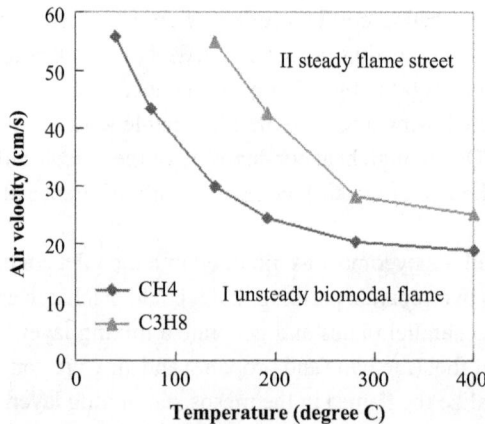

Figure 6.3. Flame street regimes of methane and propane flames.

repetitively unsteady propagating and extinction triple flames can be observed in the mixing layer. The stable flame street regime existed at high flow speeds. With the increase of wall temperature, the flame street boundary became broader due to the extended flammability limit of the triple flame. At a higher wall temperature, the flame was observed for much leaner mixtures and stabilized at a lower flow speeds. The non-premixed flame street was comprised of one or multiple triple flamelets inside the mixing layer depending on the wall temperature and the flow velocity. The onset mechanism of the triple flamelets is explained schematically in Figure 6.2(b). Along the stoichiometric line, flame speed decreases due to both the dilution of the products from the previous flamelet, the heat loss to the wall, and the diffusion rates of reactants. Meanwhile, the flame curvature also increases and flame size decreases. If the flow velocity is too low, the flame will be extinguished due to the increased product dilution and heat loss as well as decreased reactant diffusion, leading to the unsteady bimodal flame regime. On the other hand, if the flow velocity is high enough, the flame can stabilize where the flow speed balances with the flame speed, leading to the stable flame street regime.

Figure 6.4(a–c) shows, respectively, the top and side views of a flame street with four stable flamelets (including the leading triple flame) and the flame structure of these flamelets (a_1–a_4). It is seen that each flamelet has a triple flame structure. The two "wings" correspond to the two partially premixed flames, a rich flame on the fuel stream side and a lean flame on the air side. The luminescence intensity of the rich flame branch on the fuel side is stronger than that of the lean flame branch on the air side due to the higher concentrations of CH* and C_2 radicals on the rich side. The unburned fuel and oxidizer behind the two partially premixed flame branches diffuse toward each other and form a diffusion flame branch along the stoichiometric contour. The non-premixed flame branch is very weak due to the heat loss. The distance between the ith and the $(i+1)$th flamelets is defined as d_i. Therefore, d_0 is the length of the leading triple flame and d_1 is the tail–head distance between the leading flame and the first flamelet.

In the experiments, in order to avoid excessive unburned fuel, the flow speed of fuel stream was kept lower than that of the air stream. Due to the viscous stress on the wall, a higher flow velocity corresponded to a higher pressure drop in the channel ($dp/dx \sim U_{max}$). Since the exit pressure was the same, the stream with a higher flow speed had a higher pressure in the channel. Due to the pressure effect, the centerline of the mixing layer was pushed towards the fuel side when the air flow velocity was higher. Since oxygen in the air stream was only 21 percent,

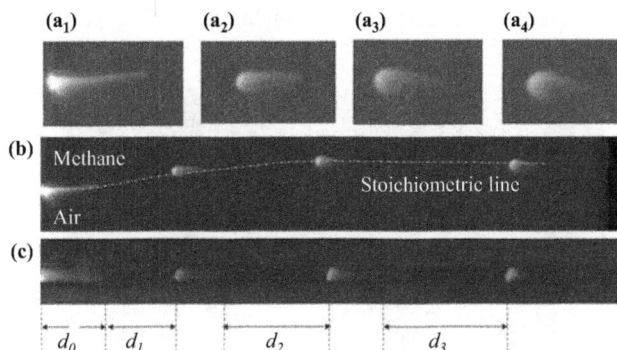

Figure 6.4. The methane Flame Street at heating power 670 W, U_{air} = 45 cm/s, U_{CH4} = 12 cm/s. (a) Top view of the four flamelets; (b) top view of the Flame Street; and (c) side view of the Flame Street (reproduced by permission from Xu and Ju [6]).

the stoichiometric line for methane–air mixture was on the air side of the centerline of the mixing layer. As a result, the flamelets were more sensitive to the air flow speed than the fuel flow speed. It was also observed experimentally that an increase of the fuel flow rate only slightly changed the flame separation distances.

6.2.3 EFFECTS OF WALL TEMPERATURE AND AIR FLOW VELOCITY

The top views of the flamelets at different wall temperatures in the flame street regime are shown in Figure 6.5. The change of wall temperature affects the flame distance in two different ways. First, at a higher wall temperature, the heat loss from the flame to the wall is reduced and the flame temperature increases. As such, the flamelets can stabilize at more diluted conditions and the flame separation distance becomes shorter. Second, as the wall temperature increases, the mean flow velocity in the channel increases due to thermal expansion. As such, the fuel and oxidizer requires longer distance to diffuse into each other. As a result, the flame distance increases linearly with the increase of flow speed. Since methane has a higher activation energy (48.4 kcal/mole) than propane (30 kcal/mole), due to the exponential dependence of flame speed on flame temperature, the methane flame speed is more sensitive to the increase of mixture temperature than propane. The experiments further showed that, for methane flames, as the wall temperature increased, the first effect (activation energy) dominated (Figure 6.5). However, for propane flames, the thermal expansion effect dominated, and the flame distance slightly increased. Figure 6.5 also shows that the length of the diffusion branch increases as the wall temperature increases. This increase implies that the heat loss to the wall plays an important role to the extinction of the triple flame. At a very high wall temperature around 500 K, the diffusion flame branch does not quench and all flamelets merge, leading to a single triple flame structure, which is the case reported in the conventional triple flame studies.

Figure 6.6 shows that the distances between the flamelets increases nonlinearly with the increase of air flow speed. In addition, at all wall temperature and flow rate conditions, the flame separation distance of the downstream flamelet is larger than that of the upstream one. This increase of the tail–head distance is a result of the dilution effect of the burned product from the neighboring flamelet at upstream and the increase of diffusion length scale. After each flamelet, the reactants are consumed and the wake of the triple flame has a very high concentration of the burned products. Due to the growth of the mixing layer and products dilution, it takes longer distance for diffusion before the reactants at the stoichiometric line reach the flammable limit.

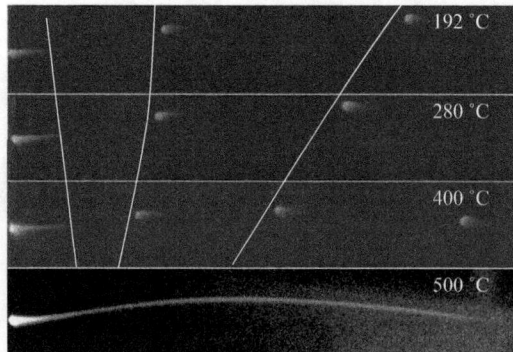

Figure 6.5. Top view of Flame Street at different flow and wall conditions (U_{air} = 26 cm/s, U_{CH4} = 12 cm/s) (reproduced by permission from Xu and Ju [6]).

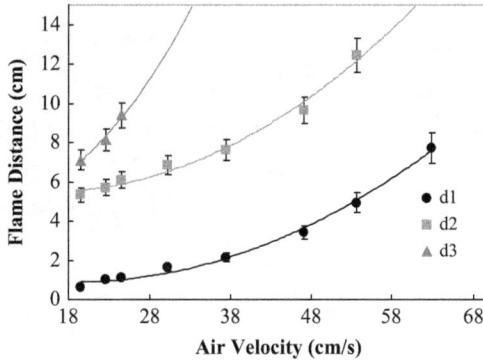

Figure 6.6. Measured flame distances at $P = 670$ W, $U_{CH4} = 12$ cm/s.

6.2.4   UNSTEADY BIMODAL FLAME REGIME

In the unsteady bimodal flame regime where the flow rate is below the speed of the triple flame stabilization limit, a triple flame will propagate close to the anchored leading triple flame. Since the convection velocity is too low to transport the reactants into and the products out of the mixing layer, the propagating triple flamelet will be quenched by a small external heat loss via both radiation and heat conduction [7].

The unsteady triple flame propagation history is shown in Figure 6.7(a). It is seen that both the flame size and the flame speed decrease as the flame propagate upstream. The flame size in the transverse direction is roughly proportional to the mixing layer thickness generated by the previous triple flamelet. The onset mechanism of the Flame Street and the triple flame interaction in mesoscale non-premixed combustion can be explained by the evolution of the unsteady bimodal flame regime shown in Figure 6.7(b). Initially, there was only a single stable triple flamelet "A" formed in the upstream of the mixing layer. Then, a new triple flamelet "B" was formed at the downstream and propagated upstream. Both its size and flame speed decreased as it moved upstream. Due to the triple flame interaction between "A" and "B," the triple flamelet "B" decelerated and finally stabilized in the product wake of the triple flame "A." At $t = 0.16$ s, triple flamelet "A" extinguished and caused a rapid decrease in the product dilution in the upstream of triple flamelet "B." Therefore, at $t = 0.48$ s, triple flamelet "B" grew in size and upstream propagation. Finally, it stabilized at the same location as flamelet "A" and extinguished due to the heat loss. These extinction, re-ignition, and unsteady propagation processes repeated, resulting in the unsteady bimodal propagating triple flame regime.

The wall temperature also had a significant effect on the flame street. At a heating power of $P = 360$ W, there existed three triple flamelets, leading to the stable Flame Street regime. When the heating power was reduced to 160 W with the same flow speed, no stable flamelet existed, leading to the unsteady bimodal flame regime. Figure 6.8 shows the flame size as a function of the flame position at an average flow speed of 20 cm/s. Position "0" corresponds to the end position of the anchored leading triple flame tail. It is seen that at the transient state when the flamelet propagated along the stoichiometric line in the mixing layer, it had a larger flame size compared to the stabilized flamelet inside the flame street. This is because the size of the flamelet is determined by the width of the mixing layer in the wake of the previous triple flamelet. The flame size will increase as the distance between two neighboring flamelets increases.

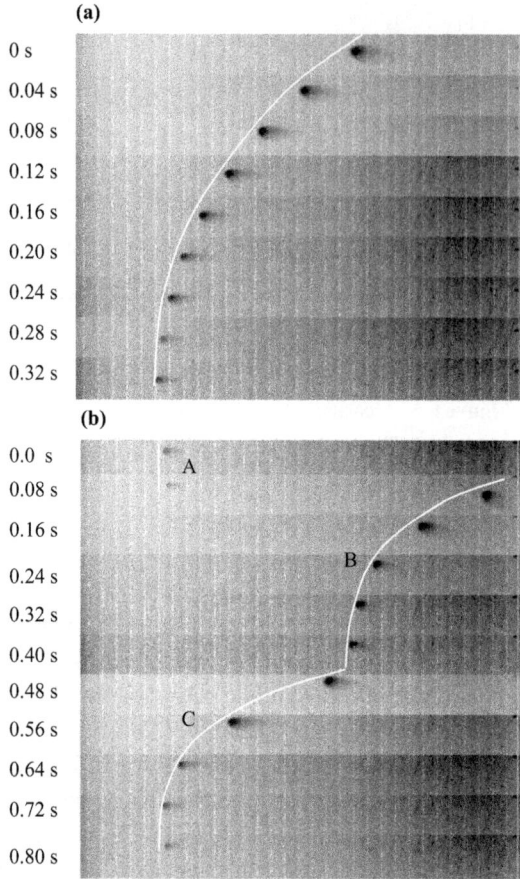

Figure 6.7. Unsteady bimodal triple flamelet propagation history at heating power 160 W, $U_{CH4} = 11$ cm/s, $U_{air} = 29$ cm/s: (a) one flamelet; (b) two flamelets (reproduced by permission from Xu and Ju [6]).

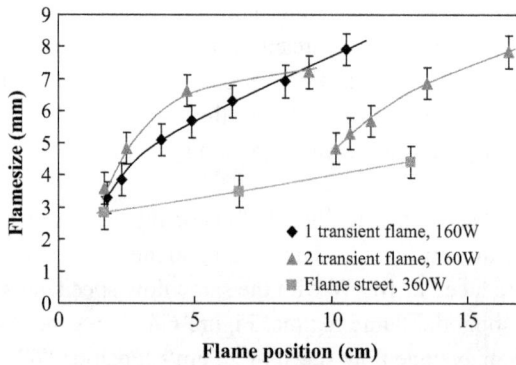

Figure 6.8. Vertical flame size for flame streets and transient triple flames ($U_{CH4} = 11$ cm/s, $U_{air} = 20$ cm/s).

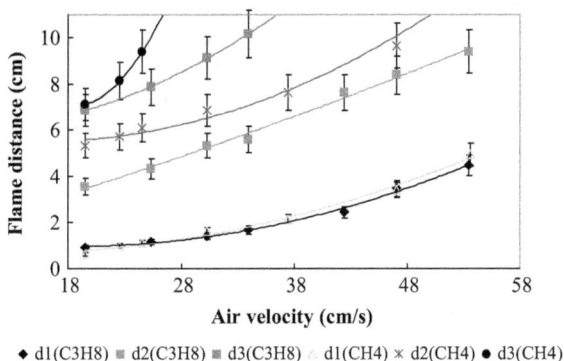

Figure 6.9. Comparison of methane and propane tail-head distances at different air flow rates (U_{C3H8} = 10 cm/s, U_{CH4} = 12 cm/s, P = 700 W).

The sensitivity of flame position to the air flow velocity depends strongly on the location of stoichiometric line. Higher fuel diffusivity and stoichiometric air-fuel ratio shift the stoichiometric line to the air side. The measured distances of methane and propane flames are shown in Figure 6.9 at the same wall temperature. It is seen that with the increase of the air flow speed, the flame distances for both methane and propane flames increase. Compared to propane, the methane flame distance is affected more with the increase of the air flow speed. This larger increase implies that the fuel diffusivity affects the dependence of the triple flame distance on the air flow speed. The larger the fuel diffusion velocity, the stronger the flame distance is affected by the increase of air flow speed.

6.2.5 THE EFFECT OF WALL SHEAR STRESS AND THE PRESSURE-DRIVEN TRIPLE FLAME PROPAGATION REGIME

In addition to the effects of thermal and concentration stratifications on flame street, the wall shear stress will also change the triple flame speed via the change in pressure distribution in mesoscale combustion. Unlike the unconfined triple flame propagation, flame propagation in a confined space is strongly affected by the pressure change via thermal expansion. It is well known that in a mesoscale channel [9], the pressure gradient is proportional to the product of viscosity and the maximum velocity of the flow. Since both the viscosity and the flow velocity are strongly dependent on flow velocity, for a flame in a confined space, the pressure gradients before and after the flame front have a large difference. Figure 6.10 shows a schematic of the pressure distributions of a premixed flame in a free space and in a confined channel with and without considering the dependence of flow viscosity on temperature. For flame propagation in free space, the pressure increase slightly ahead of the flame front is due to viscous forces and the pressure decrease in the burned gas is due to thermal expansion (case a). However, in a confined space, due to wall surface skin friction, the pressure decreases linearly along the flow direction (case b) when the flow viscosity is constant. Since viscosity is a strong function of temperature and the burned gas velocity is also much greater than that of unburned gas, the

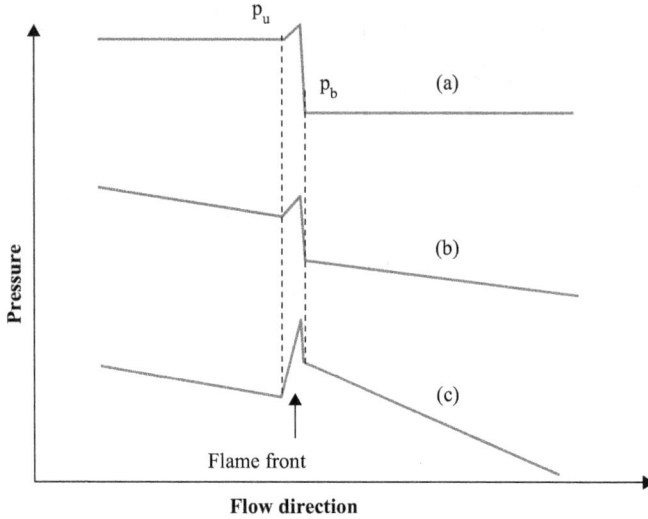

Figure 6.10. Schematic pressure distributions across the flame front (a), in a free space (b), and (c) in a confined space with and without viscosity dependence on temperature, respectively.

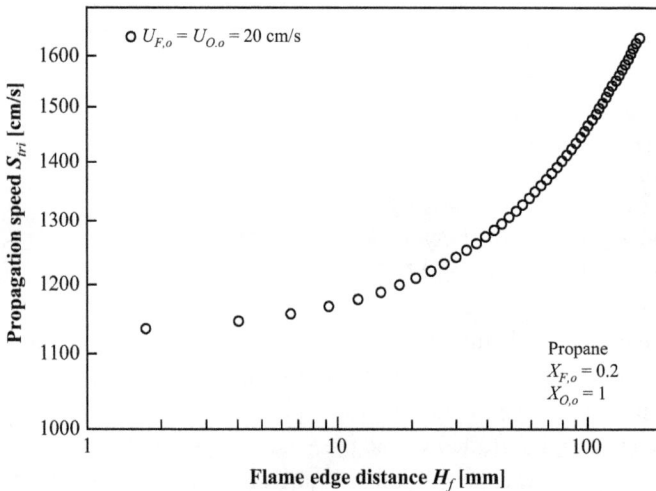

Figure 6.11. Triple flame propagation speed in a confined mesoscale combustor (methane–nitrogen diluted oxygen flame).

pressure gradient in the burned gas is much larger than that on the side of unburned gas (case c). Therefore, when the thermal expansion is strong, the large pressure gradient in the burned gas will result in a local higher pressure than that in the unburned side, leading to a new flame regime known "*fast pressure-driven flame propagation*" in a confined space.

This new flame regime was observed in our preliminary experiments. Figure 6.11 shows the dependence of the triple flame propagation speed of a nitrogen-diluted methane-oxygen system in a confined channel on the distance the flame travels. The corresponding triple flame speed is 194 cm/s, which is at least seven times faster than the corresponding adiabatic laminar flame speed. The rapid increase of the propagation speed cannot be explained by the existing triple flame or edge flame theory [10, 11]. This experimental result revealed an interesting combustion mode in mesoscale combustion. Further studies of this new flame regime are needed to understand quantitatively the fundamental mechanism.

6.2.6 SCALING ANALYSIS

The schematic of the flame street configuration is shown in Figure 6.2 (b). Following Ghosal and Vervisch [12], by assuming a constant density, zero thermal expansion, unity Lewis numbers, and using the characteristic diffusion length $l_D = k/U_\infty$ for normalization, the governing equation of the fuel–air mixture fraction can be written as

$$\frac{\partial Z}{\partial x^*} = \frac{\partial^2 Z}{\partial y^{*2}} \tag{6.1}$$

where the superscript "*" represents nondimensional variables, and Z is the mass fractions of fuel and oxidizer or the fuel mixture fraction. The analytical solution of the pure diffusion symmetric mixing layer problem can be easily solved. The mixing layer thickness is derived as $\delta(x^*) = 4\sqrt{x^*}$.

For a triple flame to be stabilized in the mixing layer, its local flame speed must be equal to the local flow velocity at the stoichiometric surface. The triple flame speed depends on many parameters such as the curvature of the partially premixed front, thermal expansion, heat loss, dilution, and preheating by the burning products. Following Ghosal and Vervisch [12], the normalized triple flame speed is given as a function of the dimensionless mixture fraction gradient

$$S_t / S_L(\alpha) = f\left(\frac{1}{Z}\frac{\partial Z}{\partial y^*}, \alpha, \beta\right) \tag{6.2}$$

where $S_L(\alpha)$ is the speed of a stoichiometric planar flame without curvature, and S_t is the triple flame speed; β is the Zeldovich number; α is the thermal expansion rate.

The thermal expansion effect increases the flame speed by a constant factor and significantly modifies the triple flame speed [11, 12]. The laminar flame speed with the thermal expansion effect is obtained by Ghosal and Vervisch [12]

$$S_L(\alpha) = (1-\alpha)^\nu S_L(0) \tag{6.3}$$

For a planar flame, it is well known that the normalized planar flame speed ($m = S_0/S_a$) depends on the normalized heat loss (H) as [24]

$$\ln m = -2H / m^2 \tag{6.4}$$

H is the normalized rate of heat loss rate on the inner wall surface defined as $H = 2h_i D_T / d(\rho c_p S_{L,ad})^2$; d is the characteristic channel width, D_T is the thermal conductivity, and h_i is the heat transfer coefficient on the wall.

The triple flame speed can also be modified by the dilution and preheat of the products from the previous flamelet. If we consider a simple one-step combustion process with v_F molecules of a fuel react and v_O molecules of an oxidizer to form v_P product molecules. Without dilution, the mixture fractions of the reactants have to sum up to one: $Y_O + Y_F = 1$. As such, the reaction rate of adiabatic flame without dilution is as follows:

$$\varpi_0 = \left(\rho \frac{1}{1+r}\right)^{v_F} \left(\rho \frac{r}{1+r}\right)^{v_O} k(T) \tag{6.5}$$

where r is the stoichiometric ratio. However, at the stoichiometric line behind a flamelet, the sum of the mixture fractions of the reactants is less than unity, and the laminar flame speed is reduced. Since the laminar flame speed is proportional to the square root of the reaction rate, the flame speed of a diluted flame is written as

$$\frac{S_{L,d}}{S_{L,ad}} = \left(\frac{\varpi}{\varpi_0}\right)^{1/2} = \left((1+r)Y_F\right)^{v_F/2} \left(\frac{1+r}{r}Y_O\right)^{v_O/2} \tag{6.6}$$

Therefore, the triple flame speed modified due to dilution of the products, heat loss, and curvature can be expressed as follows:

$$\frac{S_t}{S_{L,ad}} = \left((1+r)Y_F\right)^{v_F/2} \left(\frac{1+r}{r}Y_O\right)^{v_O/2} m(H) f\left(\frac{1}{Z_{F,s}}\frac{\partial Z_{F,s}}{\partial y^*}, \alpha, \beta\right) \tag{6.7}$$

where the subscript "s" represents the stoichiometric condition. The preheating effect from the previous flamelet might increase the laminar triple flame speed. However, a further analysis shows that this preheating effect can be ignored. In our experiment, the distance between the two ceramic plates is 6 mm. The flow velocity is in a range from 10 cm/s to 40 cm/s, which gives a Reynolds number of around 100. The thermal boundary layer thickness of a laminar flow is

$$\delta_t = \frac{\delta}{1.026\sqrt[3]{Pr}} = \frac{4.64\sqrt{\mu x/(\rho V)}}{1.026\sqrt[3]{Pr}} \tag{6.8}$$

A thickness of 3 mm gives a characteristic inlet length of 1 cm at $U_\infty = 40$ cm/s. A simple analysis showed that the hot gas would be significantly cooled down close to the wall temperature within 2 cm, which is typically smaller than the average flamelet distance. Therefore, this preheating effect from the previous flamelet can be neglected.

For a stable triple flamelet, its triple flame speed is equal to the flow speed at the infinity, that is, $S_t = U_\infty$. Using equations (6.7 and 6.8), the flame position can be obtained. For a single propagating triple flame, the mixture fraction distribution is obtained using Equation (6.1) and the flame curvature at the stoichiometric line can be derived as a function of the flame position

$$\frac{1}{r^*_{cur}} = \kappa = \frac{\beta}{\sqrt{4v-2}}\frac{1}{Z_s}\frac{\partial Z}{\partial y^*}\Big|_{y^*=0} = \frac{\beta}{\sqrt{4v-2}}\frac{1}{\sqrt{\pi x^*}} \tag{6.9}$$

$$r^*_{cur} = \frac{\sqrt{\pi x^*}\sqrt{4v-2}}{\beta} = \frac{\sqrt{\pi}\sqrt{4v-2}}{4\beta}\delta(x^*) \tag{6.10}$$

As such, the flame size is roughly proportional to the size of the mixing layer at a transient state. The numerical procedure is as follows: (1) a fully explicit method is used to solve Equation (6.1) to obtain the mixture fraction distribution; (2) the first flame position is found by evaluating Equation (6.7) and comparing the triple flame speed with the flow speed; (3) after that, Equation (6.9) is used to get the flame surface profile. The flame surface is cut off when the gradient is larger than 10 ($dx^*/dy^*>10$) such that the transverse flame size does not increase. The fractions of reactants are set to zero after the flame surface (infinite fast reaction rate). Based on the boundary conditions obtained after each flamelets, the mixture fraction distributions of further downstream can then be calculated and another flamelet is obtained.

In the current scaling analysis, the structure of the diffusion flame branch is not included. Nevertheless, the current model can qualitatively capture most of the phenomena identified in the experiments. The heating power is taken into account by changing the heat loss parameter H. Figure 6.12 shows the simulated fuel fraction distribution and the configuration of the flame street. S^* is the normalized flow speed by the adiabatic planar flame speed. It is seen that the distance between two neighboring flamelets increases as the flamelets move further away from the leading edge. Although the results are not shown here, the analysis also showed that an increase in either air flow speed or heat loss would increase the distance between two neighboring flamelets. All these results agree qualitatively well with our experiments.

The calculated flame sizes at a transient process and inside a flame street are shown in Figure 6.13. The scales of x and y axes are normalized by the characteristic diffusion length. In a flame street, a flamelet that is stabilized further away from the leading edge is also slightly bigger than the previous flamelet. However, the size of flamelets in a flame street is much smaller than a triple flame in a transient state. Those results are also consistent with our experimental observation shown in Figure 6.4. In the current case, using the mass diffusivity of methane (0.254 cm^2/s) and the average inlet velocity (25 cm/s) of methane and air, the calculated characteristic diffusion length is about 0.1 mm. The calculated maximum flame sizes in a flame street and in a transient state are about 3 mm and 8 mm, respectively. The corresponding values in Figure 6.8 are 4.4 mm and 7.9 mm, respectively. Considering the simplicity of the scaling analysis, the results agree qualitatively well with the experimental results. Therefore, the fuel–air mixing, wall heat loss, and triple flame–triple flame interaction dominate the dynamics of the flame street in mesoscale non-premixed combustion.

A numerical simulation was also carried by using a 3D model of a rectangular-shaped mesoscale burner ($241 \times 95 \times 6$ mm^3) to assess quantitatively the effect of wall temperature and

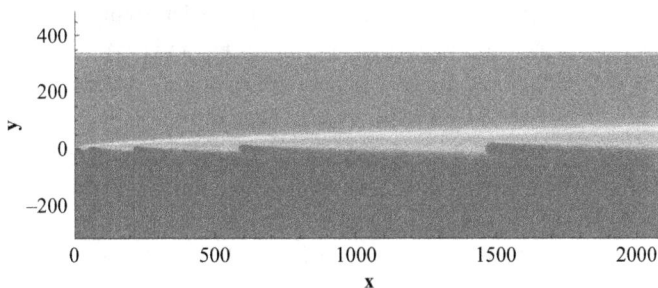

Figure 6.12. The calculated fuel fraction distribution and Flame Street configuration (x and y are normalized by the diffusion length, $\alpha = 0$, $S^* = 0.5$, $H = 0.05$, $\beta = 7.5$).

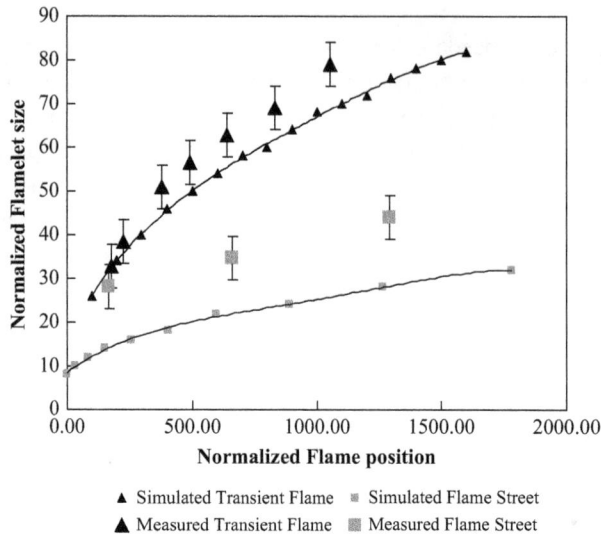

Figure 6.13. The calculated vertical flame sizes of a Flame Street and transient flamelet (position is normalized by the diffusion length, $S^* = 0.375$, $H = 0$).

flow velocity on the development of a non-premixed flame in a simple mesoscale channel [13]. In order to compare with the experimental results (Figure 6.4), different factors were analyzed in the 3D model. The inlet length, inlet widths, and mesh sizes are modified throughout the creation of a mesh to accurately represent the geometry to be studied. Additionally, four main parameters affecting the non-premixed flame street in the mesoscale channel are investigated: the turbulence model employed, the Schmidt number, chemistry model, and simulation solution parameters to get the optimal configuration of the mesoscale combustor and accurate boundary conditions.

Four different wall temperatures, three flow velocities and two different types of fuels are studied to see the effect of wall heat transfer and diffusion time in the behavior of the triple flame street. Figure 6.14 illustrates the behavior of the triple flame street if the pressure output and the pressure–velocity coupling are modified. As it can be seen the effect of the wall heat transfer and diffusion time remains the same as in the previous case but now the cold zone in the middle of the combustor chamber does not longer exist giving us a very accurate and realistic representation of the experimental data (Figures 6.4 and 6.5). Non-premixed flamelets of a mixing layer within the constrained channel are captured. A flame street is formed by several steady triple flames and its behavior clearly depends on the flow rate and surface temperature. The distance between two close flamelets increases with increasing the flow speed. Also as the temperature of the wall is increasing the triple flames tend to merge into a unique anchored flame that propagates along the channel. It can be said that the phenomenon presence is expected regardless of the kind of fuel that is used and that the behavior of the flamelet street is very similar. It is concluded then that the two main parameters affecting the development of a non-premixed flame street in a mesoscale channel being the heat transfer to and from the wall and the diffusion time of the flow.

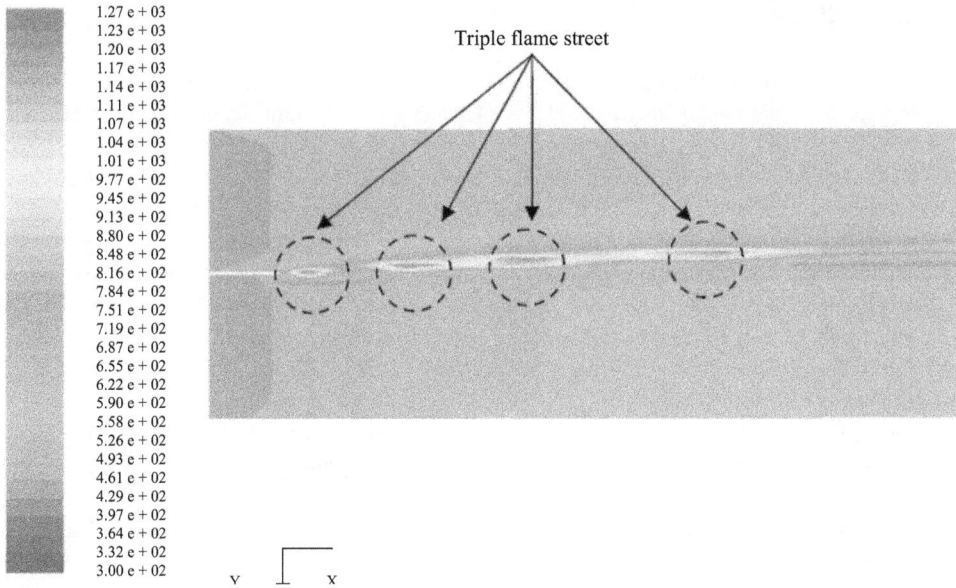

Figure 6.14. Temperature contour of the mesoscale channel simulated under the effect of the modification of all the parameters studied.

6.3 CONCLUSION AND FUTURE WORK

Non-premixed combustion in a meso- and microscale combustor induces new phenomena such as flame cells, flame street, unsteady bimodal triple flame propagation, and pressure-driven flame propagation due to the effect of limited mixing, enhanced heat loss, reignition, and increased surface shear stress. The formation of flame cells was observed in a microscale non-premixed microcombustion. The flame street, unsteady bimodal triple flame propagation, and pressure-driven fast flame propagation were observed experimentally for methane and propane fuels in a mesoscale reactor. It was found that the stable flame street occurred at high air flow speeds and the unsteady bimodal flame regime appeared at low air flow speeds. For a stable flame street, the number of flamelets and the flame distance in the mixing layer depends on the flow rate, wall surface temperature, and fuel diffusivity. The distance between two neighboring flamelets increases with increasing flow speed. The onset of a flame street also highly depends on the wall heat loss, production dilution effect, mixing layer thickness, and re-ignition. The size of the flamelets is constrained by the width of the mixing layer. The unsteady bimodal flame regime is dominated by the triple flame–triple flame interaction and heat loss-induced triple flame extinction. The pressure-driven fast flame propagation was identified for non-premixed flames with increased oxygen concentration. It was demonstrated that the increased wall shear force in mesoscale combustion can lead to a flame propagation speed much higher than the adiabatic triple flame speed. A scaling analysis model and numerical simulations were conducted to describe the mechanism of the multi-flamelet flame street. The predicted flame cell distance and occurrence of flame street agree qualitatively well with the experimental results. These results will contribute to the understanding of combustion phenomena in micro-engines.

REFERENCES

[1]   Fernandez-Pello A C, Micropower generation using combustion: Issues and approaches. Proceedings of the Combustion Institute 2002; 29(1): 883–899. doi: http://dx.doi.org/10.1016/S1540-7489(02)80113-4.

[2]   Dellimore K and Cadou C, Fuel-air mixing challenges in micro-power systems. 42nd AIAA Aerospace Sciences Meeting and Exhibit, AIAA-2004-301, Reno, Nevada, 5–8 January 2004.

[3]   Cheng T S, Chao Y C, Wu C Y, Li Y H, Nakamura Y, Lee K Y, Yuan T, and Leu T S, Experimental and numerical investigation of microscale hydrogen diffusion flames. Proceedings of the Combustion Institute 2005; 30(2): 2489–2497. doi: http://dx.doi.org/10.1016/j.proci.2004.07.025.

[4]   Matta L M, Neumeier Y, Lemon B, and Zinn B T, Characteristics of micro-scale diffusion flames. Proceedings of the Combustion Institute 2002; 29(1): 933–939. doi: http://dx.doi.org/10.1016/S1540-7489(02)80118-3.

[5]   Miesse C, Masel R I, Short M, and Shannon M A, Diffusion flame instabilities in a 0.75 mm non-premixed microburner. Proceedings of the Combustion Institute 2005; 30(2): 2499–2507. doi: http://dx.doi.org/10.1016/j.proci.2004.08.140.

[6]   Xu B and Ju Y, Studies on non-premixed flame streets in a mesoscale channel. Proceedings of the Combustion Institute 2009; 32(1): 1375–1382. doi: http://dx.doi.org/10.1016/j.proci.2008.07.027.

[7]   Sirignano W A, Pham T K, and Dunn-Rankin D, Miniature-scale liquid-fuel-film combustor. Proceedings of the Combustion Institute 2002; 29(1): 925–931. doi: http://dx.doi.org/10.1016/S1540-7489(02)80117-1.

[8]   Maruta K, Ju Y, Honda A, Niioka T, Lewis number effect on extinction characteristics of radiative counterflow CH_4–O_2–N_2–He flames. Proceedings of the Combustion Institute 1998; 27(2): 2611–2617. doi: http://dx.doi.org/10.1016/S0082-0784(98)80115-X.

[9]   Schlichting H, Boundary-layer Theory. New York, NY: McGraw-Hill, 1955.

[10] Buckmaster J and Weber R, Edge-flame holding. Proceedings of the Combustion Institute 1996; 26(1): 1143–1149. doi: http://dx.doi.org/10.1016/S0082-0784(96)80330-4.

[11] Chung S H, Stabilization, propagation and instability of tribrachial triple flames. Proceedings of the Combustion Institute 2007; 31(1): 877–892. doi: http://dx.doi.org/10.1016/j.proci.2006.08.117.

[12] Ghosal S and Vervisch L, Theoretical and numerical study of a symmetrical triple flame using the parabolic flame path approximation. Journal of Fluid Mechanics 2000; 415(1): 227–260. doi: http://dx.doi.org/10.1017/S0022112000008685.

[13] Contreras M A G, Ju Y, and Ng H D, Numerical investigation of the dynamics of a non-premixed flame in a mesoscale channel. Proceedings of CFD Society of Canada 18th Annual Conference, London, Ontario, Canada, May 17, 2010.

CHAPTER 7

MICRO-COMBUSTION IN NON-CATALYTIC NARROW DUCTS

David P. Tse and Dimitrios C. Kyritsis

One of the groundbreaking contributions of micro-combustion-related research is the challenge that it posed to the classical notion of the quenching diameter. The classical picture of ducted flames in large-scale hardware is well established: the non-adiabatic flame can stabilize at a location in the duct against the opposing flow or it can flashback. If the diameter is smaller than a particular value (defined as the quenching diameter), neither stabilization in the duct nor flashback is possible and the flame sits outside the tube. This is because in classical, large-scale combustion, the solid hardware surrounding the flame can only act as a sink of radicals and heat. However, in micro-combustion, that is, in combustion in hardware whose size is on the order of the laminar flame thickness, the solid hardware can act in a more complicated way. This is because the Biot and Fourier numbers are such that the tube wall can actually provide heat to the flame. This generates very rich phenomenology that is briefly reviewed in this chapter for the case of non-catalytically coated ducts.

7.1 OVERVIEW OF THE STATE OF THE ART

The emergence of accurate micro-manufacturing technologies at the beginning of the century gave rise to the general idea of putting the high-energy density of combustion at work in compact power-generation devices that would operate as "liquid-fuel" batteries [1–3]. The experience of the last decade or so has proven that the task is substantially tougher than what initially thought. In fact, the only compact power source that is based on micro-combustion and has reported an acceptable overall efficiency on the order of 20 percent is the one described in [4], which employs a Stirling engine along with a combination of a small-scale catalytic reactor and electrosprays that were utilized for fuel dispersion. The initially envisioned coupling with direct energy conversion modules such as thermoelectrics and thermophotovolatics has not realized in practical devices and in any case can only yield very low overall efficiencies. The small-scale direct adaptation of large-scale engines (such as micro-turbines and micro-IC-engines) has proven to be even more difficult.

At the heart of the matter lies the difficulty in sustaining clean and efficient combustion in small-size enclosures. Combustion in a narrow duct is the paradigm for such reactive flows, and the whole idea of micro-combustion basically amounts to a challenge to the classical notion of the quenching diameter as described for example in [5]. At a fundamental level, this was highlighted by Ju and Xu [6], who defined a normalized heat loss as the ratio of heat loss to the wall to the total chemical heat release and showed that this was proportional to the square of the ratio of the flame thickness to the inner diameter of the tube [2, 6]. Also, it has been pointed out [7–9] that the elevated wall temperatures of micro-combustors increase the effects of chemical quenching. In particular, Miesse et al. [7] reported that at low wall temperatures (~773 K), the quenching distance was independent of wall material. However, at high wall temperatures (~1273 K), quenching distance varied based on wall material. Kim et al. [8] also concluded that thermal quenching played a dominant role at low temperatures while radical quenching was significant at high temperatures, whereas Evans and Kyritsis [9] observed that wall material properties, especially thermal conductivity, played a big role in wall temperature profile and flame phenomenology.

Catalysis is certainly a plausible way forward. The chemistry and physics of reactive flow in narrow channels with catalytic walls have been researched extensively, especially in the context of catalytic reactors for gas turbines and catalytic converters for exhaust gas after treatment. In an impressive line of work, Mantzaras and his collaborators have recently studied micro-combustion in catalytically coated channels of elongated rectangular cross-section and a hydraulic diameter of a few mm [10–13] utilizing a variety of computational and experimental methodologies. Lean hydrogen and propane flows were mainly studied and particular emphasis was placed on the interaction of gaseous and catalytic chemistry. The results revealed rich flame dynamics that included oscillatory and occasionally chaotic phenomena [10]. The computational results of [13] suggest that the application of catalyst at the walls can stabilize flame behavior. Catalytic and catalytically stabilized combustion in narrow channels can facilitate low-emission gas turbine operation by allowing lower combustion temperatures, especially if the onset of catalytic combustion is facilitated by hydrogen content in the mixture of reactants. However, experience from the field of fuel cells shows that the combination of hydrogen management with precious metal loading can be problematic when it comes to compact power generation. For this reason, we will limit ourselves in this chapter to micro-combustion in tubes with non-catalytic walls and point to several phenomena that challenge the notion of the quenching diameter and may point to novel modes of power generation. It has been argued that because of the elevated temperatures, the wall can act catalytically in *any* microchannel, independently of its particular chemical composition, but this is not our point here; we will simply refer to tubes that have not been coated catalytically with precious metals.

An interesting category of such micro-combustors is heat-recirculation-type burners that employ what is referred to as "excess enthalpy" combustion. The idea was first introduced by Weinberg and co-workers [14] who proposed the so-called swiss-roll burners that use the elevated enthalpy of the reactants in order to preheat the products such that heat loss (which is a major limiting factor in micro-combustion) is minimized. Experimental studies of these burner have been conducted with and without a catalyst [15, 16]. Ronney and co-workers [15] observed an "out-of-center reaction zone" where flames stabilized near the burner inlet and could only be re-centered to the burner center in the presence of a catalyst. For the particular configuration, low-Reynolds-number flows yielded lower peak flame temperatures and sustained combustion in the gaseous phase, facilitated by catalysis.

For combustion in straight and curved narrow ducts, a phenomenon observed by Maruta et al. [17, 18], Fan et al. [19], and Kyritsis and co-workers [9, 20–22] is that of flame "oscillations" in which a premixed flame undergoes a succession of ignition, propagation, extinction, and reignition events. This periodic phenomenon is stable and can be sustained for several hours provided that the configuration does not change [20]. Maruta et al. [17] studied the combustion of a premixed methane and air mixture in a cylindrical quartz tube with an inner diameter of 2 mm. A uniform wall temperature of 1273 K was maintained by employing two heating plates at the top and bottom of the tube. Four regimes were reported, corresponding to a stable flame, a "flame with repetitive extinction and ignition" (FREI), a pulsating flame, and a flame with characteristics of both the pulsating flame and FREI [17]. Fan et al. [19] studied methane and air combustion in quartz combustors heated by infrared lamps and found that similar flame oscillations occurred as the air flow rate increased and equivalence ratio decreased. Richecoeur and Kyritsis [20] observed oscillations in rich methane-oxygen flames in curved mesoscale quartz tubes.

In terms of determining the salient physical characteristics that would differentiate micro-combustion in narrow ducts from combustion in large-scale ducts, Ju and Xu [6] theoretically and experimentally investigated mesoscale flame propagation and extinction and obtained the flame dynamics due to the flame–wall interaction. They found that a decrease in channel width and equivalence ratio created both a fast (order of cm/s) and a slow (order of mm/s) flame propagation regime. A flame in a mesoscale channel could propagate faster than an adiabatic flame depending on wall thermal properties and heat capacity [6]. Kessler and Short [21] used a numerical simulation to show that upon ignition, the progression of the reaction down the channel proceeded as a cyclic sequence of ignition, propagation, and quenching events. Also, the presence of a temperature gradient in the wall was important in determining combustion evolution. Maruta et al. [17, 18] performed a stability analysis to understand the mechanism of the flames using a conventional one-dimensional, constant density, localized reaction-zone model. They concluded that the flame oscillations occurred at transitions between stable and unstable branches of the solution. Taking a similar approach but a different perspective, Jackson et al. [22] proposed a model that captured successfully oscillation dynamics and transitions in the parameter space between steady solutions, unsteady solutions, and a return to steady solutions, by treating the phenomenon as an edge-flame propagation driven thermally, without any contribution from hydrodynamics. In the introduction to that paper, it was pointed out that the problem of flame propagation/stabilization in narrow ducts has similarities with the one of the propagation/stabilization of an edge flame. This was highlighted further by Matalon and collaborators in [23]. The same group has also studied the hydrodynamics of heat-recirculating micro-combustors [24], thus adding to the thermal analyses of [14–16]. Numerical coupling between the gaseous flow and the conjugate heat transfer problem in the gaseous phase has been recently demonstrated in the work of Veeraragavan and Cadou [25].

As a final note on the state of the art, one should remark that micro-combustion in small enclosures was perhaps inspired by the quest for portable power generation, but it seems to be making faster progress toward practical applications in the field of analytic chemistry and miniaturization of flame-ionization detectors that employ either premixed [26, 27] or (more recently) diffusion flames [28]. This field proceeds in a manner opposite to micro-combustion for power generation, because practical devices have progressed more than the fundamentals.

In what follows, we focus on non-catalytic ducts, because we feel that if micro-combustion is to make a breakthrough in the context of compact power generation, the impact will be greatest if it does not involve precious metal loading. We focus on what we see as the main characteristics of energy conversion in these narrow ducts, that is the relatively simple hydro-dynamics, acoustic emission, and the generation of almost uniform high-temperature surfaces on refractory materials. Nevertheless, the route that will convert these features to compact, efficient, and clean power sources remains elusive.

7.2 STRUCTURE OF THE TUBULAR MICRO-REACTIVE FLOW

A major characteristic of micro-combustion flows in narrow ducts that has been established by the early experimental studies of Maruta et al. [17, 18] and Richecoeur and Kyritsis [20] is their capability to develop oscillatory behavior. Notably, this is a new mode of flame behavior that does not exist in large-scale ducts. There, the situation is fairly simple: either flame speed is higher than the flow speed and the flame flashes back, or the flow speed is higher and the flame is stabilized outside the duct. For a non-adiabatic tube, a location exists where the speed of the non-adiabatic reaction front is equal to the local flow speed and the flame stabilizes at a location in the tube.

However, as pointed out in detail by Evans and Kyritsis [9, 29, 30], the interaction of the wall with the propagating flame can generate much richer phenomenology in micro-combustion flows. In particular, for combustion at atmospheric pressure in tubes with inner diameters on the order of a few mm (<4 mm) and wall thickness on the order of 1 mm, the wall can act in a double role: at the early stages of propagation, the flame loses heat to the tube wall. The radial temperature gradients are substantial, heat loss proceeds fast and extinction occurs, pretty much along the lines of the analysis that generated the classical notion of the quenching diameter. In a micro-combustion flow though, the situation is different in that the thermal inertia of the hardware itself is small, so before extinction the temperature of the wall has risen to a point that the wall can act as a source of re-ignition. As a result, high-frequency extinction re-ignition phe-nomena occur that cause acoustic emission and can give the oscillating flame the macroscopic appearance of Figure 7.1.

This interaction can generate oscillations that either span the whole full length or cover the tube only partially, in addition to the "classical" externally stabilized flame, stationary flame

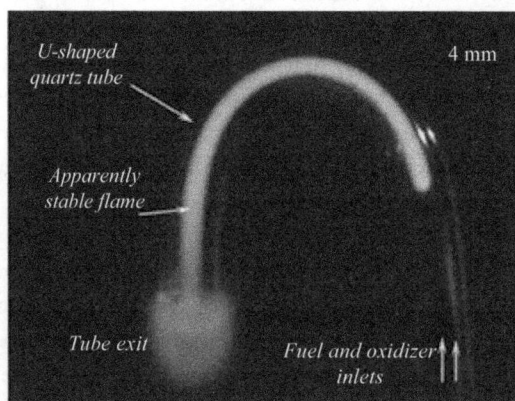

Figure 7.1. Rapidly oscillating flame in a quartz tube (reproduced by permission from Richecoeur and Kyritsis [20]).

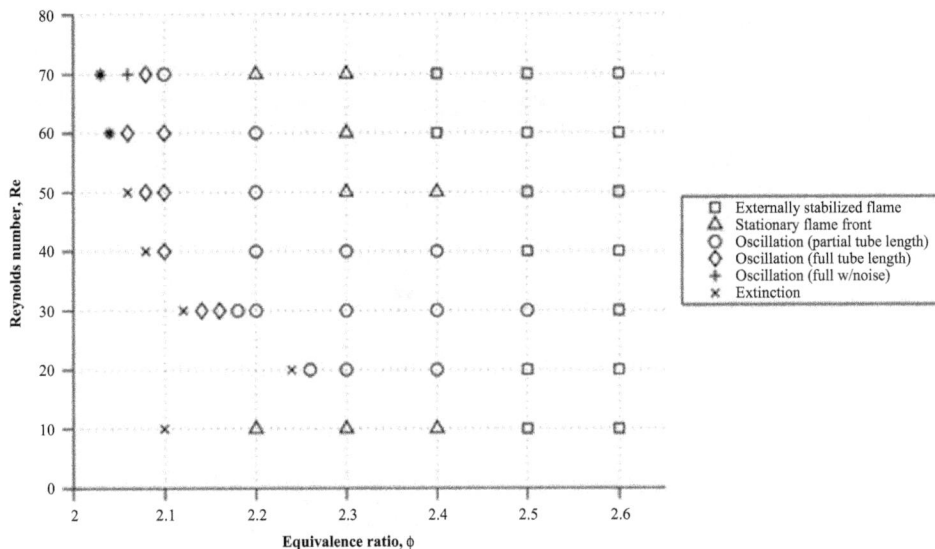

Figure 7.2. Flame phenomenology map for combustion in narrow tubes.

fronts, and flashbacks. Notably these oscillatory phenomena have been observed only for rich mixtures (in methane–oxygen flames) for reasons that are not currently clear. The mode of flame behavior that is ultimately established depends on both the Reynolds number of the cold flow of the reactants and the equivalence ratio of the mixture. Typical phenomenology maps are shown in Figure 7.2. For very rich mixtures ($\phi > 2.5$), an externally stabilized flame is established, independently of Reynolds number. Oscillations covering the full tube length were generally not observed at low Reynolds numbers (Re < 20); instead, stationary flame fronts were prevalent. Oscillatory flows cease to occur after the equivalence ratio decreases past a certain threshold, usually around $\phi = 2$. Acoustic phenomena typically occur at the lowest possible equivalence ratios that sustain oscillatory flows. Note that oscillations with sound emission were observed at flow conditions very close to extinction. Figure 7.2 shows that both full-tube oscillations with sound emission and extinction were possible at $\varphi \approx 2.05$ for Re = 60 and Re = 70. This denoted that full-tube oscillations with acoustic emissions could transition to extinction under steady flow conditions. In fact, it can be claimed that acoustic emissionis useful as a predictor of extinction during equivalence ratio variations.

An interesting question is whether these oscillations are driven thermally or hydrodynamically. The success of the theoretical/computational analysis of Jackson et al. [22] in capturing the oscillatory behavior with a model that did not include hydrodynamics but rather imposed a Poiseuille profile of flow velocity undoubtedly points to the importance of thermal considerations, as does the thermal modeling of Ju and collaborators [6] and Cadou and collaborators [25]. However, at the same time Richecoeur and Kyritsis [20] calculated the Strouhal numbers based on the frequency of the observed oscillations and showed that these corresponded to the Strouhal numbers of the instability of the jet emerging from the micro-tube (St ≈ 0.4) as they were determined in the early work of Crowe and Champagne [31]. The matter has been conclusively decided through the Particle-Streak-Velocimetry (PSV) measurements of the flow field by Tse [32].

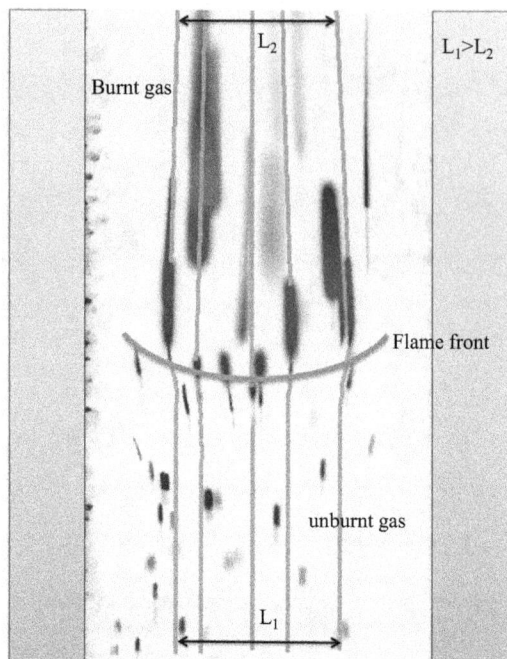

Figure 7.3. Flow structure of flame oscillations. The shorter streaks correspond to unburnt gas and the longer streaks to burnt gas.

A typical flow structure as revealed by PSV data is shown in Figure 7.3. This is a long-exposure image of the flow around the oscillating flame that was seeded with TiO_2 particles and the length of each streak is proportional to the flow speed at the particular location. The longer streaks near the top of the figure, representing a higher particle velocity, showed the flow of burnt gas after the flame front had passed. The shorter streaks near the bottom of the figure, representing lower particle velocity, showed the flow of the unburnt methane–oxygen pre-mixture. The flame front propagated downwards and could be easily visualized as a curved boundary between the burnt and unburnt gases. Particles glowed intensely once they passed the flame front; as a result, these particles appeared to be thicker than particles in the unburnt gas that were illuminated with a continuous wave Nd-YAG laser. The depiction in Figure 7.3 was constructed using one frame of a high-speed video of the flow. The tube wall, flame front, and path lines were added to better illustrate the structure of the flow. By looking at a sequence of frames as the flame front propagated down the tube, individual particles could be traced and path lines could be established. Unburnt gas traveled in a straight up-and-down trajectory and only deflected away from the tube centerline very near the flame front. In contrast, burnt gas traveled downstream of the tube while deflecting toward the tube centerline. The flow tube got larger as it approached the flame and smaller as it moved away from the flame; this result stemmed from the fact that the flame sheet was convex toward the unburnt gas [33]. As a result, the width of the flow tube L_1 in front of the flame front was slightly larger than the corresponding width L_2 behind the flame.

The quantitative analysis of the velocity PSV measurements is presented in detail by Tse [32]; however the main conclusion of the measurement is that this is indeed an essentially one-dimensional flow with a Poiseuille velocity profile. Small deviations from one-dimensionality are only present in the immediate vicinity of the slightly curved flame front, and their effect is inconsequential. The assumption underlying the work of Jackson et al. [22], Ju et al. [6], and Veeraragavan and Cadou [25] that the oscillatory behavior can be modeled as a thermal phenomenon without complicated hydrodynamics was thus confirmed.

In fact, using the PSV data and reasonable assumptions about the chemical reaction one can compare the entropy generated in this phenomenon through the generation of vorticity with the one that is due to the chemical reaction. Specifically, if we are to consider separately each of the entropy-generation terms of the Crocco theorem:

$$(\bar{w} \times \bar{u}) = T\nabla S - \nabla h_{tot}$$

we can estimate the entropy generation term by assuming the adiabatic reaction (although this is definitely a very weak assumption), which for a methane–oxygen mixture of at $\phi = 2$ would give a value of $\Delta S = 7874.20$ J/kgK obtained from GASEQ [34]. Taking reasonable estimates for the flame thickness and determining flame orientation from the PSV data, we can estimate the radial component of the entropy generation term as

$$(T\nabla S)_{\hat{r}} = T\frac{\partial S}{\partial r}$$

Similarly, the vorticity term can be estimated as

$$(\bar{w} \times \bar{u})_{\hat{r}} = -\left(u_z \frac{\partial u_z}{\partial r}\right)$$

from the PSV data. Detailed processing of flow field data has shown that the vorticity-related term is at least three orders of magnitude smaller than the chemical term. Reactant conversion and heat release dominate entropy generation and drive the reactive flow and the oscillatory phenomena that accompany it.

Utilizing PSV measurements in the vicinity of the tube axis, where flame curvature is minimal, one can formulate arguments about the flame speed in these micro-combustion flows. Figure 7.4 shows the axial velocity for burnt and unburnt gas as a function of Reynolds number (for φ approximately equal to 2), as measured with PSV. Mass conservation in a reference frame that follows the flame front (i.e., a reference frame in which the flame is steady) yields

$$\rho_{unburnt}(u_{unburnt} + u_F) = \rho_{burnt}(u_{burnt} + u_F)$$

The velocity of the flame front, u_F, was determined by high-speed visualization. Table 7.1 shows velocity and density data corresponding to several Reynolds numbers and equivalence ratios for which flame oscillations were observed. The inlet conditions were obtained from the mass flow rates used in the experiment, with properties of the mixture at 298K. The laminar flame speed, S_L, was numerically calculated using a premixed flat flame model with the GRI-3.0 mechanism using Cantera. The theoretical burnt gas density, ρ_{burnt}*, for adiabatic conditions was obtained with GASEQ.

A first notable result is that the inlet velocity, u_{inlet}, was substantially smaller than the unburnt gas velocity $u_{unburnt}$. This suggested that the temperature of the unburnt gas was elevated near the flame front. Careful analysis of PSV data shows that the unburnt gas was not preheated by the flame front directly, as the velocities remained fairly constant ahead of the flame front. Rather, heat transfer from the flame to the duct wall elevated the wall temperatures and

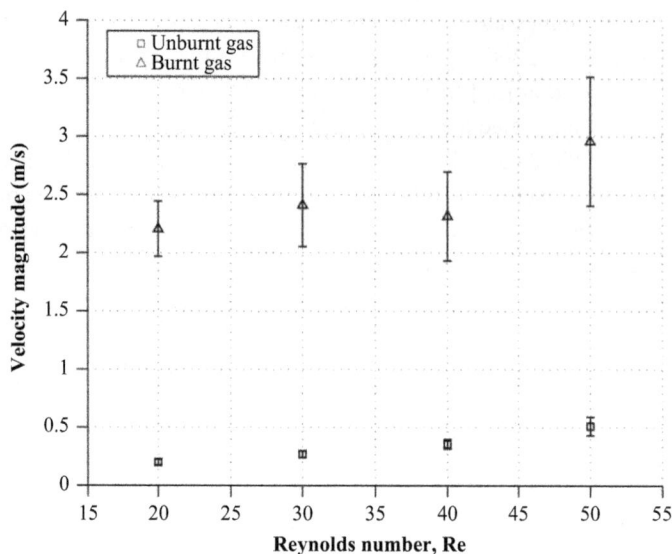

Figure 7.4. Axial velocity for unburnt and burnt gases for varying Reynolds numbers. The error bar corresponds to one standard deviation. Specific data on the flow conditions can be seen in Table 7.2.

Table 7.1. Velocity and density values for flame oscillations

Re	ϕ	ρ_{inlet} (kg/m³)	u_{inlet} (m/s)	$\rho_{unburnt}$ (kg/m³)	$u_{unburnt}$ (m/s)	u_{burnt} (m/s)	u_F (m/s)	S_L (m/s)	ρ_{burnt} (kg/m³)	ρ_{burnt}^* (kg/m³)
20	1.9	0.99	0.08	0.40	0.20	2.21	0.63	1.28	0.12	0.07
30	2.0	0.98	0.12	0.44	0.27	2.41	0.72	1.28	0.14	0.07
40	2.0	0.98	0.16	0.44	0.36	2.32	0.67	1.28	0.15	0.07
50	2.0	0.98	0.20	0.38	0.51	2.96	0.89	1.28	0.14	0.07

preheated the incoming gases. This is in agreement with the computational findings of Kessler and Short [21] who found that heating of incoming reactants was typically achieved by streamwise conduction in the walls. The density of the unburnt gas actually corresponded to the density of the methane–oxygen mixture at a temperature of ~370 K. This was consistent with infrared thermography studies by Evans and Kyritsis [9], indicating that the quartz tube wall maintained a steady temperature of around 500 K during flame oscillations.

Second, the effect of intense heat loss can be seen on the flame speed. The laminar flame speed, S_L, as calculated for adiabatic combustion can be as much as 40 percent greater than the sum of $u_F + u_{unburnt}$, a good estimate of the flame speed in the particular configuration. Notably, the "measured" flame speed ($u_F + u_{unburnt}$) increased as a function of Reynolds number and for Re=50 the flame speed is slightly higher and within the error of the PSV measurements equal to the adiabatic flame speed corresponding to the cold reactants. This is because the reactants are preheated more effectively by the tube wall in the higher-Reynolds-number flow. Furthermore, the theoretical density of the products from adiabatic combustion was smaller than the measured density of the burnt gas; this resulted from heat loss from the tube that reduced the gaseous temperature compared to the one of adiabatic flames.

7.3 THERMOACOUSTICS

There seems to be experimental consensus on the fact that micro-combustion in narrow ducts is accompanied by sound emission [9, 17–20, 29, 30]. The reasons behind these acoustic phenomena have not yet been conclusively determined. The connection between sound emission and combustion flow in ducts dates back to the seminal work of Rijke [35]. The theoretical analysis of this coupling has been revisited relatively recently by Nicoud and Poinsot [36]. However, as we showed in the previous section, this micro-combustion flow is driven almost exclusively by thermal mechanisms, and hydrodynamics that would cause the pressure fluctuations discussed in [35, 36] does not seem to play an important role. As a result, sound generation cannot be attributed to flow hydrodynamics, but rather to the contraction/expansion that happens during the extinction/re-ignition events that are characteristic of such flows. The importance of thermal phenomena is exacerbated by the experimental finding that sound emission or lack thereof is basically determined by the wall material [29, 30, 32]. The more conductive the wall material, the stronger the sound emission.

Typical results of a microphone output measuring sound emission are presented in Figure 7.5. The data was taken for a methane–oxygen mixture of equivalence ratio equal to 2.1 and cold-flow Reynolds number equal to 130. The flow took place in a quartz tube with a 4 mm ID and a 1 mm wall thickness. Data acquisition there was synchronized with high-speed visualization from the camera and the correlation with the extinction/re-ignition events is evident. Tse [32] has pointed out that the phenomenon is apparatus-dependent and has observed that it is caused by homogeneous explosions that happen at the tube entrance during re-ignition.

Assuming a weak discontinuity in velocity and density, the acoustic approximations can be used to calculate the pressure difference across the wave (it is, of course, realized that this

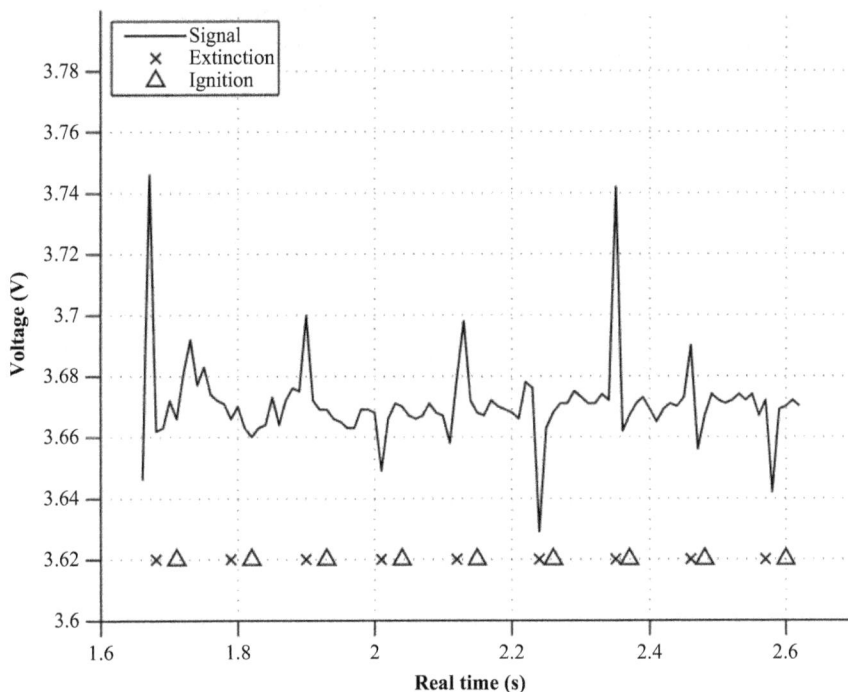

Figure 7.5. Acoustic emissions from flame oscillations covering the full-tube length at Re = 130 and corresponding extinction and ignition events.

is a very weak approximation). Using a control volume as shown in Figure 7.6, the flame front with control surface, A*, can be assumed to be a thin discontinuity traveling with a velocity u_F.

The integral forms of the balance of mass and momentum conservation:

$$\int_{A*} \rho(u+u_F) \cdot n \, dA = 0$$

$$\int_{A*} \rho u(u+u_F) \cdot n \, dA = \int_{A*} -Pn \, dA$$

can be reduced to yield, respectively,

$$\rho_1(u_1 + u_F) = \rho_2(u_2 + u_F)$$

$$\rho_1 u_1 (u_1 + u_F) + P_1 = \rho_2 u_2 (u_2 + u_F) + P_2$$

Combining the last two equations and substituting the relevant variables with $\rho_{unburnt}$, $u_{unburnt}$, u_{burnt}, we obtain

$$\delta P = \rho_{unburnt} (u_{unburnt} + u_F)(u_{burnt} - u_{unburnt})$$

The sound pressure level (SPL) can then be obtained in decibels (dB) as

$$SPL = 20 \log_{10} \frac{\delta P}{P_{ref}}$$

where $P_{ref} = 2.2 \times 10^{-5} \frac{N}{m^2}$

Table 7.2 shows the results for and, calculated using PSV velocity data and density calculations similar to the ones of Table 7.2 for Re = 20, 30, 40, and 50 and an equivalence ratio equal to ~2. (SPL: sound pressure level; PSV: particle-streak-velocimetry.)

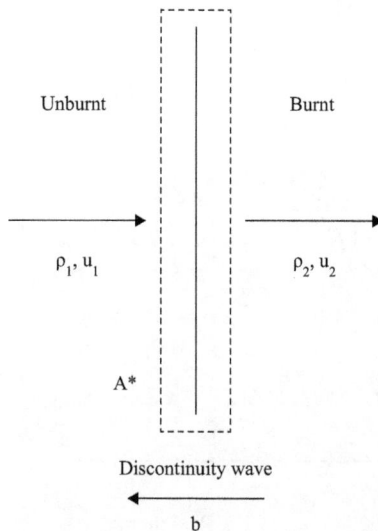

Figure 7.6. Control volume around acoustic discontinuity.

Table 7.2. Calculations of δP and SPL using measurements from PSV

Re	$\rho_{unburnt}$ (kg/m^3)	$u_{unburnt}$ (m/s)	u_{burnt} (m/s)	u_F (m/s)	δP (Pa)	SPL (dB)
20	0.40	0.20	2.21	0.63	0.66	90
30	0.44	0.27	2.41	0.72	0.93	93
40	0.44	0.36	2.32	0.67	0.89	92
50	0.38	0.51	2.96	0.89	1.32	96

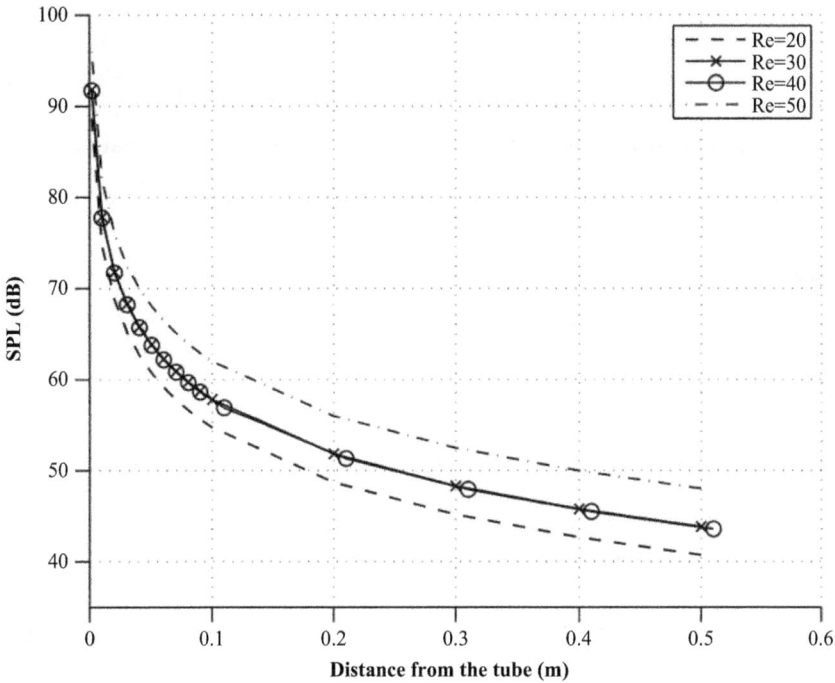

Figure 7.7. Sound pressure levels (dB) from flame oscillations estimated based on PSV data.

Assuming further spherically symmetric propagation, the acoustic energy flux attenuates as $1/r^2$, which means that the pressure level scales as $1/r$. So, if we assume that

$$\delta P_A \cdot r_A = \delta P_B \cdot r_B$$

and use as r_A the tube radius and r_B the location of the sound measurement, we can get SPL as a function of distance from the tube as shown in Figure 7.7. SPL values in the vicinity of 40–50 dB are equivalent to the sound emission from usual conversations and compatible with the perception of sound emitted from these phenomena for distances 40 to 50 cm from the tube. This analysis proves that the recorded levels of sound emission are indeed compatible with the expansion caused by the re-ignition event, but does not conclusively prove the mechanism of sound generation. The results also point to the thermoacoustics as a potential mode of energy recovery, if micro-combustion is to be used for practical power generation.

7.4   OUTER WALL TEMPERATURE

Another important parameter in the context of coupling micro-combustion with compact power generation is the temperature that is established on the outer wall of small-scale ducts in which micro-combustion takes place. This is because several direct energy conversion modules (e.g., thermoelectrics, thermophotovoltaics, etc.) are currently only (somewhat) efficient in relatively narrow temperature ranges. In this context, the finding that flame oscillations neither generate nor pre-require oscillations of the wall temperature is particularly important. This finding was first strongly suggested by the early work of Maruta and collaborators [17, 18], and then it was con-clusively shown by the experimental work of Evans and Kyritsis [9, 29, 30] who measured wall temperature for a series of flame phenomoenologies for both metal and refractory tube materials.

Typical results of such measurements are shown in Figure 7.8. The main conclusions in terms of the temperature field can be summarized as follows. When a partially premixed flame is stabilized at the tube exit, the wall temperature is maximum at the tube exit and decreases with distance from the tube end. Similarly, when a steady, non-adiabatic flame is stabilized in the tube, the temperature decreases with distance from the flame. Expectedly, a less conductive material (quartz) can sustain substantially larger temperature gradients than a more conductive one (steel). As shown in panels (c) and (e) of Figure 7.8, this has the interesting consequence that flame oscillations in a tube of a refractory material can lead to regions of practically uni-form temperature on the external tube wall, something that can be very useful for coupling with direct energy conversion modules.

Rationalizing these results in terms of the relevant Biot and Fourier numbers, Evans and Kyritsis [9, 30] offer an insight as to how the heat transfer to the tube works during micro-combustion. The Biot numbers based on either the length ($Bi_L=hL/k$) or the thickness of the tube wall ($Bi_t=ht/k$) compare the thermal resistance due to convection with the one due to conduction (h: heat transfer coefficient, k: tube material conductivity). By making a reasonable assumption of $h =10$ W/m^{2}K for the heat transfer coefficient of the natural convection in the outer surface of the tube and approximating the Nusselt number as approximately equal to 4 for the tube flow (4.36 is the accurate value for low-Reynolds-number, steady flow), they were able to calculate an area-weighted average of the heat transfer coefficient and use it in order to calculate the Biot numbers based on tube length (Bi_L) and tube thickness (Bi_t) that are shown on Table 7.3. The small values of Bi_t indicate that the thermal resistance due to conduction in the radial sense is negligible independent of material, whereas the close-to-unity values of Bi_L (especially) in the case of quartz establish the capability of the tubes to sustain substantial tem-perature gradients in the axial sense.

Since the radial thermal resistance is negligible, whereas the one in the axial direction is substantial, this means that the mesoscale duct acts in terms of heat transfer as a fin. The spatial variation of temperature along the fin was characterized by the fin parameter βL, where β is calculated as

$$\beta^2 = \frac{hP}{kA_c}$$

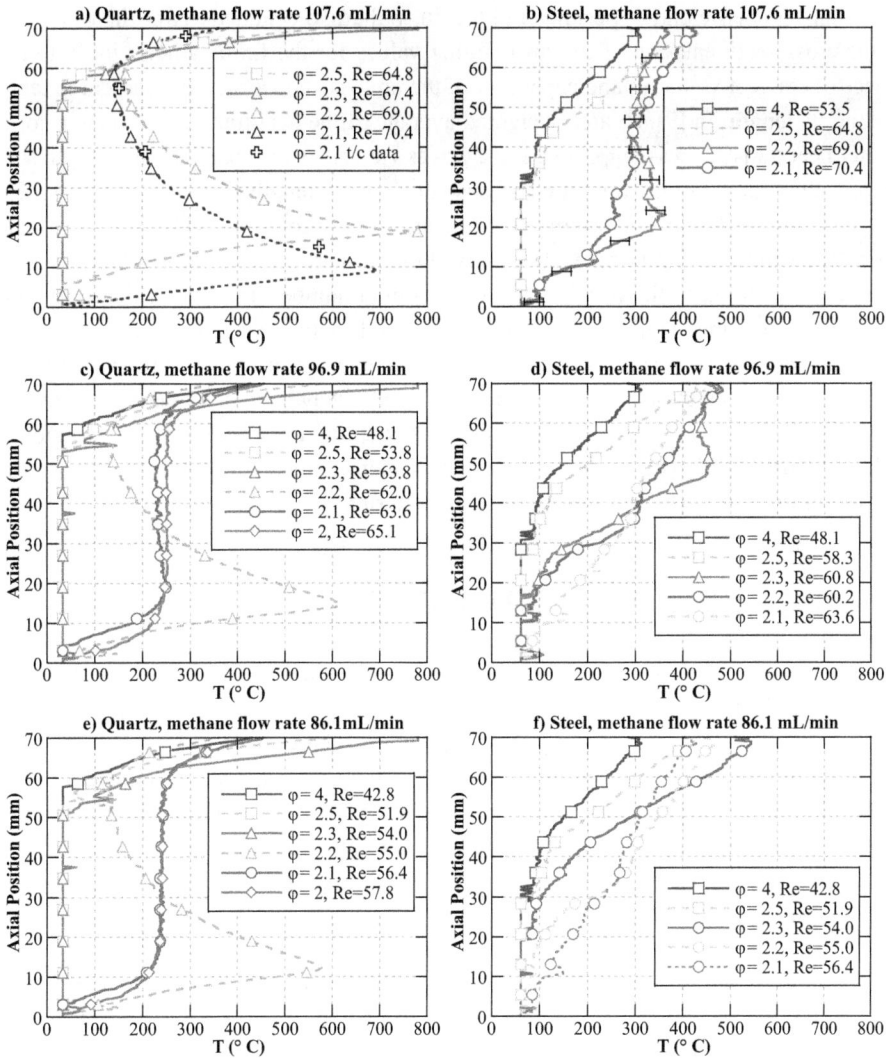

Figure 7.8. Temperature distributions for two duct materials and several fuel flow rate. Each curve represents the steady-state temperature distribution for a single equivalence ratio. The fuel flow rate is uniform for all curves within each subplot. The flame phenomenology for each equivalence ratio is indicated using the symbols defined in Figure 7.2. The symbols in panel (a) indicate the locations of thermocouple measurements whereas in panel (b) indicative error bars were inserted (reproduced by permission from Evans and Kyritsis [9]).

Table 7.3. Physical parameters and non-dimensional numbers relating to the heat transfer mode in the duct

Material	K (W/m-K)	α (m²/s)	Bi_L	Bi_t	β (m⁻¹)	βL	Fo
Quartz	1.38	8.34e-07	8.86E-01	1.26E-02	159	11.14	1.13E-05
Steel	15.1	3.91e-06	8.10E-02	1.16E-03	48	3.37	5.32E-05

(h: heat transfer coefficient, P: perimeter of the fin, k: thermal conductivity of the wall material, A_c is the cross-sectional area of the fin.) Higher values for the fin parameter indicate greater influence of convective losses and capability of the fin to sustain a sharper gradient in temperature along its length. In [9], an area-weighted average of the heat transfer coefficient for inner and outer heat transfers was employed and $1/\beta$ was viewed as the characteristic length scale of the temperature gradient that the fin could sustain along its length. The data of Table 7.3 indicate a characteristic length of ~6 mm for quartz and 21 mm for steel, which justifies the much sharper temperature gradients that were observed in the quartz tubes. The steadiness of the wall temperature is justified by the very low values of Fourier numbers that were calculated for these flows, even for the smallest frequencies of the observed oscillations (~15 Hz).

In closure, exactly because of the oscillatory phenomenon, micro-combustion in small-scale ducts can be used in order to generate solids of relatively high, homogeneous temperature. This can have favorable consequences in terms of coupling to direct energy conversion modules. However, this will necessitate multiplexing of micro-combustion elements into reasonably sized heat sources through accurate micro-manufacturing.

7.5   CONCLUSION AND FUTURE WORK

Micro-combustion in narrow, non-catalytic ducts has challenged the classical notion of the quenching diameter. When the duct is small enough, it does not only act as a heat sink that causes extinction, but also as a high-temperature source that can re-ignite the mixture. This causes, under certain conditions, a stable oscillatory behavior that involves repetitive extinction/re-ignition and sound emission. The phenomenon is driven thermally, resembling the propagation of an edge front without the need for complicated hydrodynamics and is thus qualitatively different from the thermoacoustics of a Rijke tube. The oscillations in the gas are not accompanied by oscillations of the temperature on the outer wall of the tube, which is steady. In refractory materials, it is possible to establish relatively long areas on the outer wall that have an elevated, uniform, and steady temperature. This is theoretically very favorable for coupling with direct energy conversion modules; however this coupling has not yet been demonstrated in an efficient manner.

REFERENCES

[1] Ju, Y., and Maruta, K. (2011). Microscale combustion: Technology development and fundamental research. Progress in Energy and Combustion Science, 37(6), 669–715. doi: http://dx.doi.org/10.1016/j.pecs.2011.03.001.

[2] Fernandez-Pello, A.C. (2002). Micropower generation using combustion: Issues and approaches. Proceedings of the Combustion Institute, 29(1), 883–899. doi: http://dx.doi.org/10.1016/S1540-7489(02)80113-4.

[3] Kyritsis, D.C., Roychoudhury, S., McEnally, C.S., Pfefferle, L., and Gomez, A. (2004). Mesoscale combustion: A first step towards liquid fueled batteries. Experimental Fluid and Thermal Science, 28(7), 763–770. doi: http://dx.doi.org/10.1016/j.expthermflusci.2003.12.014.

[4] Gomez, A., Berry, J.J., Roychudgury, S., Coriton, B., and Huth, J. (2007). From fuel jet to electric power, using a mesoscale, efficient stirling cycle. Proceedings of the Combustion Institute, 31(2), 3251–3259. doi: http://dx.doi.org/10.1016/j.proci.2006.07.203.

[5] Glassman, I., and Yetter, R.A. (2008). Combustion (4th ed.). Burlington, MA: Academic Press.

[6] Ju, Y., and Xu, B. (2005). Theoretical and experimental studies on mesoscale flame propagation and extinction. Proceedings of the Combustion Institute, 30(2), 2445–2453. doi: http://dx.doi.org/10.1016/j.proci.2004.08.234.

[7] Miesse, C., Masel, R., Short, M., and Shannon, M. (2005). Experimental observations of methane–oxygen diffusion flame structure in a sub-millimeter micro-burner. Combustion Theory and Modelling, 9(1), 77–92. doi: http://dx.doi.org/10.1080/13647830500051661.

[8] Kim, K.T., Lee, D.H., and Kwon, S. (2006). Effects of thermal and chemical surface-flame interaction on flame quenching. Combustion and Flame, 146(1–2), 19–28. doi: http://dx.doi.org/10.1016/j.combustflame.2006.04.012.

[9] Evans, C.J., and Kyritsis, D.C. (2011). Experimental investigation of the effects of flame phenomenology on the wall temperature distribution of mesoscale nonadiabatic ducts. Combustion Science and Technology, 183(9), 847–867. doi: http://dx.doi.org/10.1080/00102202.2011.567199.

[10] Pizza, G., Frouzakis, C.E., Mantzaras, J., Tomboulides, A.G., and Boulouchos, K. (2010). Three-dimensional simulations of premixed hydrogen/air flames in microtubes. Journal of Fluid Mechanics, 658, 463–491. doi: http://dx.doi.org/10.1017/S0022112010001837.

[11] Ghermay, Y., Mantzaras, J., Bombach, R., and Boulouchos, K. (2011). Homogeneous combustion of fuel-lean $H_2/O_2/N_2$ mixtures over platinum at elevated pressures and preheats. Combustion and Flame 158(8), 1491–1506. doi: http://dx.doi.org/10.1016/j.combustflame.2010.12.025.

[12] Karagiannidis, S., Marketos, K., Mantzaras, J., Schaeren, R., and Boulouchos, K. (2010). Experimental and numerical investigation of a propane-fueled, catalytic mesoscale combustor. Catalysis Today, 155(1–2), 108–115. doi: http://dx.doi.org/10.1016/j.cattod.2010.04.030.

[13] Pizza, G., Frouzakis, C.E., and Mantzaras, J. (2010). Flame dynamics in catalytic and non-catalytic mesoscale microreactors. Catalysis Today, 155(1–2), 123–130. doi: http://dx.doi.org/10.1016/j.cattod.2010.04.029.

[14] Lloyd, S.A., and Weinberg, F.J. (1974). A burner for mixtures of very low heat content. Nature, 251, 47–49. doi: http://dx.doi.org/10.1038/251047a0.

[15] Ahn, J., Eastwood, C., Sitzki, L., and Ronney, P.D. (2005). Gas-phase and catalytic combustion in heat-recirculating burners. Proceedings of the Combustion Institute, 30(2), 2463–2472. doi: http://dx.doi.org/10.1016/j.proci.2004.08.265.

[16] Shirsat, V., and Gupta, A.K. (2011). A review of progress in heat recirculating meso-scale combustors. Applied Energy, 88(12), 4294–4309. doi: http://dx.doi.org/10.1016/j.apenergy.2011.07.019.

[17] Maruta, K., Kataoka, T., Kim, N.I., Minaev, S.S., and Fursenko, R.V. (2005). Characteristics of combustion in a narrow channel with a temperature gradient. Proceedings of the Combustion Institute, 30(2), 2429–2436. doi: http://dx.doi.org/10.1016/j.proci.2004.08.245.

[18] Maruta, K., Parc, J.K., Oh, K.C., Fujimori, T., Minaev, S.S., and Fursenko, R.V. (2004). Characteristics of combustion in a narrow heated channel. Combustion, Explosion, and Shock Waves, 40(5), 516–523. doi: http://dx.doi.org/10.1023/B:CESW.0000041403.16095.a8.

[19] Fan, Y., Suzuki, Y., and Kasagi, N. (2009). Experimental study of micro-scale premixed flame in quartz channels. Proceedings of the Combustion Institute, 32(2), 3083–3090. doi: http://dx.doi.org/10.1016/j.proci.2008.06.219.

[20] Richecoeur, F., and Kyritsis, D.C. (2005). Experimental study of flame stabilization in low Reynolds and Dean number flows in curved mesoscale ducts. Proceedings of the Combustion Institute, 30(2), 2419–2427. doi: http://dx.doi.org/10.1016/j.proci.2004.08.015.

[21] Kessler, D.A., and Short, M. (2008). Ignition and transient dynamics of sub-limit premixed flames in microchannels. Combustion Theory and Modelling, 12(5), 809–829. doi: http://dx.doi.org/10.1080/13647830801956295.

[22] Jackson, T.L., Buckmaster, J., Lu, Z., Kyritsis, D.C., and Massa, L. (2007). Flames in narrow circular tubes. Proceedings of the Combustion Institute, 31(1), 955–962. doi: http://dx.doi.org/10.1016/j.proci.2006.07.032.

[23] Bieri, J.A., Kurdyumov, V.N., and Matalon, M. (2011). The effect of gas expansion on edge flames stabilized in narrow channels. Proceedings of the Combustion Institute, 33(1), 1227–1234. doi: http://dx.doi.org/10.1016/j.proci.2010.06.161.

[24] Kurdyumov, V.N., and Matalon, M. (2011). Analysis of an idealized, heat-recirculating microcombustor. Proceedings of the Combustion Institute, 33(2), 3275–3284. doi: http://dx.doi.org/10.1016/j.proci.2010.07.041.

[25] Veeraragavan, A., and Cadou C.P. (2011). Flame predictions in planarmicro/meso-scale combustors with conjugate heat transfer. Combustion and Flame, 158(11), 2178–2187. doi: http://dx.doi.org/10.1016/j.combustflame.2011.04.006.

[26] Zimmermann, S., Krippner, P., Vogel, A., and Müller, J. (2002). Miniaturized flame ionization detector for gas chromatography. Sensors and Actuators, B: Chemical, 83(1–3), 285–289. doi: http://dx.doi.org/10.1016/S0925-4005(01)01060-7.

[27] Kuipers, W., and Müller, J. (2010). Sensitivity of a planar micro-flame ionization detector. Talanta, 82(5), 1674–1679. doi: http://dx.doi.org/10.1016/j.talanta.2010.07.042.

[28] Kim, J., Yeom, J., Shannon, M.A., Chen, Q., and Bae, B. (July 15–19, 2012). Development of a portable gas analyzer using a micro-gas chromatograph/flame ionization detector (micro-gc/fid) for nasa's environmental missions. 42nd International Conference on Environmental Systems:eISBN: 978-1-60086-934-1, San Diego, California.

[29] Evans, C.J., and Kyritsis, D.C. (2009). Operational regimes of rich methane and propane/oxygen flames in mesoscale non-adiabatic ducts. Proceedings of the Combustion Institute, 32(2), 3107–3114. doi: http://dx.doi.org/10.1016/j.proci.2008.06.089.

[30] Evans, C.J. (2010). Experimental and computational investigation of combustion phenomena in mesoscale ducts. Ph.D. Dissertation, University of Illinois at Urbana-Champaign Department of Mechanical Sci. & Engineering.

[31] Crow, S.C., and Champagne, F.H. (1971). Orderly structure in jet turbulence. Journal of Fluid Mechanics, 48(3), 547–591. doi: http://dx.doi.org/10.1017/S0022112071001745.

[32] Tse, D.P. (2012). Experimental investigation of the flow-field during combustion in narrow circular ducts, MS Thesis, University of Illinois at Urbana-Champaign, Department of Mechanical Sci. & Engineering.

[33] Williams, F.A. (1985). Combustion Theory (2nd Ed.). Menlo Park, CA: The Benjamin/Cummings Publishing Company, Inc.

[34] Morley, C. (2005). Gaseq Chemical Equilibrium Program. http://www.gaseq.co.uk/

[35] Rijke, P.L. (1859). On the vibration of the air in a tube open at both ends. Philosophical Magazine, 17, 419–422.

[36] Nicoud, F., and Poinsot, T. (2005). Thermoacoustic instabilities: Should the Rayleigh criterion be extended to include entropy changes? Combustion and Flame, 142(1–2), 153–159. doi: http://dx.doi.org/10.1016/j.combustflame.2005.02.013.

CHAPTER 8

FUNDAMENTALS OF MICROSCALE CATALYTIC COMBUSTION

Yei-Chin Chao and Guan-Bang Chen

In this chapter, the fundamental aspects and outstanding combustion phenomena in a micro-catalytic combustor are studied and discussed through detailed numerical simulations of hydrogen–air catalytic reactions in a catalytic platinum microtube. Emphasis is placed on the complicated interactions between the homogeneous and heterogeneous reactions leading to the characteristic combustion modes in the microscale catalytic combustor. The parameters studied include inlet flow velocity (residence time), tube diameter, wall thermal conductivity, and catalyst segmentation. The characteristic combustion modes and the role of the catalytic wall in the reaction characteristics in terms of its sustaining and competing effects on homogeneous combustion in the catalytic microtube are identified and discussed.

8.1 INTRODUCTION

8.1.1 BACKGROUND

With the prevailing microelectromechanical systems (MEMS) devices, micropower generation systems and microscale energy converters with characteristic length scales of few millimeters to sub-millimeters are receiving intensive research interest recently [1, 2]. For micropower generation systems, the energy and power densities become the key issues of its performance. Combustion converts chemical energy of fuels into thermal energy, which can then be converted into electricity via mechanical or thermal-electric devices. Since the energy density of typical hydrocarbon fuels is about 100 times higher than that of the most advanced lithium batteries, the power output of hydrocarbon combustion will still be much higher than that of current batteries even with very poor conversion efficiency. In addition, it can provide longer operational periods and reduce mass and volume fractions of the power system. Micro-combustors can be applied in many specific applications, including micro-reformers for fuel cells and micro-thrusters for micro- and pico-satellite missions, and so forth.

In practice, maintaining a stable combustion inside a microreactor still remains a major challenge for the micro-combustor design. High surface-to-volume ratio (S/V) of a microscale device causes enhanced heat loss through the walls. Flame extinction in a micro-combustor is mainly due to thermal and radical quenching on the wall. Thermal quenching occurs when too much heat is removed from the flame to sustain the flame, and radical quenching occurs when excessive reaction radicals are adsorbed or recombined on the walls, resulting in flame extinction. It has been demonstrated [2–4] that with suitable thermal management and fine balance between flow residence and chemical times, stable combustion can be maintained in a micro-combustor. Several useful strategies have been proposed to solve these problems, such as the "Swiss roll" and heat-regeneration combustors to reduce heat losses [4, 5] and the catalytic combustors to enhance reaction and to suppress radical quenching on the wall [3, 6, 7].

8.1.2   CATALYTIC COMBUSTION

It has been well known that the platinum surface can promote surface reaction and cause combustion of flammable mixtures "without flames." In general, catalytic combustion can be defined as the complete oxidation of a combustible compound on the surface of a catalyst. Relying upon the highly active surface, catalyst enhances the reaction of a low equivalence ratio mixture providing a highly stable combustion process at a rather low temperature. Catalyst is defined as a substance that helps speed up a chemical reaction without being consumed in the reaction.

Catalytic combustion is achieved through a sequence of five distinctive steps that are fundamental to all heterogeneous reactions. The first step is the diffusion of the reactants to the catalyst surface. Subsequently, some of the reactant molecules are adsorbed on the surface to form the respective molecule–catalyst complexes. The chemical reaction may occur on the surface between either two adsorbed molecules or an adsorbed molecule and a molecule in the gas phase. The products formed are then desorbed from the surface and diffuse back into the gas stream. Out of these five individual steps the first and the last are purely physical processes of mass diffusion through which the reactants and products are, respectively, brought to and from the catalyst surface. The adsorption, reaction, and desorption are all chemical processes with a unique activation energy associated with each step. For practical combustion applications, catalysts may be supported on a variety of materials and with various configurations, for example, pellets and honeycomb monoliths. Catalytic combustion involves the coupling of the reactive flow and the processes on the catalyst leading to complex interactions between the gas phase and the surface. It is necessary to consider chemical reactions as well as heat transport both on the surface and in the gas phase. In a catalytic combustion monolith, the mechanism of catalytic combustion can be described as a complex interaction among axial convection, diffusion to and from the catalytic channel wall and inside the washcoat, several heat transport modes, and chemical reactions on the catalyst and possibly in the gas phase. The very fast adsorption and desorption processes are balanced by species mass diffusion at the gas–surface interface.

8.1.3   MICROSCALE CATALYTIC COMBUSTION

Catalyst with highly active surface elements enhances the reaction of a low equivalence ratio and low heating value mixture by providing a highly stable combustion process at a rather low temperature. The main obstacles encountered in combustor miniaturization are inadequate residence time for complete combustion and flame stability and quenching due to the high S/V ratio. Using the catalyst wall, on the other hand, can reduce radical depletion and can ensure an intensive reaction on the wall. Therefore, catalytic combustion is particularly useful for energy conversion in a miniaturized reactor. Boyarko et al. [8] and Volchko et al. [9] studied catalytic combustion of hydrogen–oxygen and rich methane–oxygen mixtures in sub-millimeter platinum tubes. Their results showed that the catalytic reaction of methane can be self-sustained in a platinum tube with 0.4 mm in inner diameter.

Being limited by the size, it is very difficult to obtain information inside a microreactor by experimental studies. In fact, most experimental studies acquired only outer wall surface and outlet information. On the other hand, numerical simulation provides a convenient and cost-effective approach to study the micro-combustion phenomena and mechanism. Through numerical studies with a multi-step reaction mechanism, Maruta et al. [10] showed that the equivalence ratio at the extinction limit in a micro-catalytic channel decreased monotonically with increasing Reynolds number for adiabatic wall conditions, while for non-adiabatic conditions, the extinction curve exhibited a U-shaped dual-limit behavior due to heat loss and insufficient residence time as compared to chemical time. Methane–air flame propagation in a straight tubular microchannel was numerically simulated with detailed gas-phase chemistry by Raimondeau et al. [11]. They found that heat loss in the near-entrance region and radical quenching on the wall were key factors that control flame propagation inside the microchannels.

Most previous computational studies for micro-combustors dealt with the heterogeneous or homogeneous reaction separately and concentrated on the extinction limits and flame stabilities. As the combustor size is reduced, the interaction between heterogeneous and homogeneous reactions becomes very intensive and complicated in the micro-catalytic combustors. Integrated catalytic surface reaction and homogeneous gas-phase reaction mechanisms should be employed in the computational studies to reveal this complicated catalytic reaction interaction in a micro-combustor. The complicated catalytic combustion of hydrogen–air in a microtube was studied by Chen et al. [12]. However, the effects of catalytic walls on combustion characteristics inside microchannels are still not fully understood nor well documented.

8.2 METHODOLOGY

In principle, the complex phenomena occurring in the catalytic monolith channels include the heterogeneous reactions on the catalyst wall and the homogeneous reactions in the gas phase, the transports phenomena of the mass, momentum, and heat in the gas phase and at the gas–solid interface, heat transfer in the solid and so forth. In modeling the catalytic combustion one should include the fluid mechanics in the flow field, chemical reactions in the gas phase and on the catalytic surface, the relating energy and species transports, and the complicated coupling interaction between the heterogeneous and homogeneous reactions. Hydrogen–air combustion is employed here for illustration since it has a fast reaction rate, reliable detailed chemical mechanisms, and is also an important and clean fuel in the future. Governing equations and related mathematical setup for numerical simulation of catalytic combustion in a microchannel are described in the following.

8.2.1 GOVERNING EQUATIONS

The numerical model used to simulate the catalytic combustion consists of the coupling of the flow field and the chemical reaction in the gas phase and on the catalyst surface. In practical applications monolithic catalyst is usually employed to meet the low-pressure drop requirement, typically few percent of the inlet pressure. In the literature, one-dimensional models of single monolith channel, lumping the cross-sectional and peripheral distributed variables into average values, are usually used for the analysis and design of a catalytic combustor. For these 1D models, the evaluation of transverse diffusion of heat and mass is critical. It has been showed that 1D models may be inaccurate in predicting the light-off behavior due to the difficulty in estimating local interphase coefficients in the presence of very fast variations of the heat flux on the wall in the light-off region. In particular, incorrect results have been obtained when heat conduction in the solid is considered in the model. In addition, 1D models failed in predicting the ignition of the homogeneous combustion in the boundary layer adjacent to the catalyst wall.

The governing equations for axisymmetric, reacting channel flows can be described as:
Mixture Continuity equation

$$\frac{\partial \rho}{\partial t} + \frac{1}{r}\frac{\partial}{\partial r}(r\rho v_{\mathrm{r}}) + \frac{1}{r}\frac{\partial}{\partial z}(r\rho v_{z}) = 0 \tag{8.1}$$

Navier–Stokes Momentum equation

$$\frac{\partial(\rho\varphi)}{\partial t} + \frac{1}{r}\frac{\partial}{\partial r}\left(r\rho v_{\mathrm{r}}\varphi - r\mu\frac{\partial\varphi}{\partial r}\right) + \frac{1}{r}\frac{\partial}{\partial z}\left(r\rho v_{z}\varphi - r\mu\frac{\partial\varphi}{\partial z}\right)$$
$$= S_{\varphi}, \tag{8.2}$$

Species continuity equation

$$\frac{\partial(\rho Y_{k})}{\partial t} + \frac{1}{r}\frac{\partial}{\partial r}\left(r\rho v_{r}Y_{k} - r\frac{\rho D_{\mathrm{km}}}{M}\frac{\partial(Y_{k}M)}{\partial r} - \frac{rD_{k}^{T}}{T}\frac{\partial T}{\partial r}\right) +$$
$$\frac{1}{r}\frac{\partial}{\partial z}\left(r\rho v_{z}Y_{k} - r\frac{\rho D_{\mathrm{km}}}{M}\frac{\partial(Y_{k}M)}{\partial z} - \frac{rD_{k}^{T}}{T}\frac{\partial T}{\partial z}\right) = \dot{\omega}_{k}, \qquad k = 1, ..k_{g} - 1. \tag{8.3}$$

Energy equation

$$\frac{\partial(\rho T)}{\partial t} + \frac{1}{r}\frac{\partial}{\partial r}(r\rho v_{r}T) + \frac{1}{r}\frac{\partial}{\partial z}(r\rho v_{z}T) = -\frac{1}{c_{p}}[\frac{\partial}{\partial z}\left(\lambda\frac{\partial T}{\partial z}\right) +$$
$$\frac{1}{r}\frac{\partial}{\partial r}\left(r\lambda\frac{\partial T}{\partial r}\right).] - \frac{1}{c_{p}}\sum_{k=1}^{k_{g}} h_{k}\{\dot{\omega}_{k} + \frac{1}{r}\frac{\partial}{\partial r}(r\frac{\rho D_{\mathrm{km}}}{M}\frac{\partial(Y_{k}M)}{\partial r} + \frac{rD_{k}^{T}}{T}\frac{\partial T}{\partial r})$$
$$+ \frac{1}{r}\frac{\partial}{\partial z}\left(r\frac{\rho D_{\mathrm{km}}}{M}\frac{\partial(Y_{k}M)}{\partial z} + \frac{rD_{k}^{T}}{T}\frac{\partial T}{\partial z}\right)\} \tag{8.4}$$

Equation of state

$$p = \rho RT \sum_{k=1}^{k_{g}}\frac{Y_{k}}{M_{k}} \tag{8.5}$$

where ρ represents density; the variable φ in Equation (8.2) represents v_r, v_z the radial and axial velocities, respectively; Y_k, $\dot{\omega}_k$, and h_k are the mass fraction, reaction rate, and enthalpy of species k, respectively; λ is the mixture thermal conductivity, and M_k is the molecular weight of species k.

In the species continuity equation, the Fickian diffusion law is used for diffusive mass flux and is given by

$$\vec{j}_k^c = -\frac{\rho D_{km}}{M} \nabla (Y_k M) \tag{8.6}$$

The mixture-averaged diffusion coefficient is described as

$$D_{km} = \frac{1 - Y_k}{\displaystyle\sum_{\substack{i=1 \\ i \neq k}}^{K} X_i / D_{ik}} \tag{8.7}$$

where X_i is the mole fraction of species i, M is the molecular weight of mixture and D_{ik} are the binary diffusion coefficient. The mass diffusion flux caused by temperature gradient (thermal diffusion) is also included as it becomes important for lighter species, such as H_2. It is given by

$$\vec{j}_k^T = -D_k^T \nabla (\ln T) \tag{8.8}$$

The thermal diffusion coefficient D_k^T, the binary diffusion coefficient D_{ik}, and the viscosities and the thermal conductivity of individual gas species are obtained from kinetic theory using Lennard-Jones and Stockmayer potentials [13, 14]. In the energy equation, viscous dissipation and diffusion-thermo energy flux (the Dufour effect) are neglected. The diffusive energy flux includes the conductive energy, $-\lambda \nabla T$, and the energy flux due to species interdiffusion in a multicomponent mixture, $\displaystyle\sum_{k=1}^{K_g} h_k \left(\vec{j}_k^c + \vec{j}_k^T \right)$.

The viscosity, μ, and the thermal conductivity, λ, of the mixture are determined from

$$\mu = \sum_{k=1}^{K} \frac{X_k \mu_k}{\displaystyle\sum_{j=1}^{K} X_j \varphi_{kj}} \tag{8.9}$$

$$\lambda = \frac{1}{2} \left(\sum_{k=1}^{K} X_k \lambda_k + \frac{1}{\displaystyle\sum_{k=1}^{K} X_k / \lambda_k} \right) \tag{8.10}$$

where

$$\varphi_{kj} = \frac{1}{\sqrt{8}} \left(1 + \frac{M_k}{M_j} \right)^{\frac{1}{2}} \left[1 + \left(\frac{\mu_k}{\mu_j} \right)^{\frac{1}{2}} \left(\frac{M_j}{M_k} \right)^{\frac{1}{4}} \right]^2 \tag{8.11}$$

The specific heat of the mixture is given by

$$c_{\mathrm{p}} = \sum_{k=1}^{K_g} c_{\mathrm{p}k} Y_k \tag{8.12}$$

with the specific heat of the individual species $c_{\mathrm{p}k}$ determined from polynomial fits over specific temperature ranges in the CHEMKIN package [15].

An orthogonal, non-uniform staggered-grid system is used to solve the discretized equations with a control volume formulation in accordance with the SIMPLEC (Semi-Implicit Method for Pressure Linked Equations—Consistent) algorithm. The second-order upwind scheme is used to solve the momentum equations, while the central difference scheme is used for the energy and species equations. Iterative alternating direction implicit (ADI) and the tridiagonal matrix algorithm (TDMA) techniques are used to solve the discretized equations along the mesh lines in the computational domain.

8.2.2 COMPUTATIONAL DOMAIN

The numerical model consists of axisymmetric Navier–Stokes equations, mass, and energy conservation equations and one species equation for each chemical species. The simplified geometry of a typical catalytic microtube for numerical simulation is shown in Figure 8.1. In the simulation, the channel wall is assumed thermally thin so that the axial heat conduction along the wall is negligible. Axial heat conduction within the wall may have a significant effect and will be considered in a later study case of catalyst segmentation.

Figure 8.2 also shows the schematic diagram of the computational domain for the catalytic microreactor modeled for the case of multiple catalyst segments. The computational domain contains both the gas phase and the surrounding channel walls. Due to symmetry, simulations

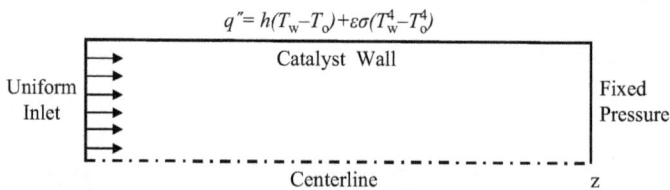

$$q'' = h(T_{\mathrm{w}} - T_o) + \varepsilon\sigma(T_{\mathrm{w}}^4 - T_o^4)$$

Catalyst Wall

Uniform Inlet

Fixed Pressure

Centerline

z

Figure 8.1. Schematic of computational domain for a single channel.

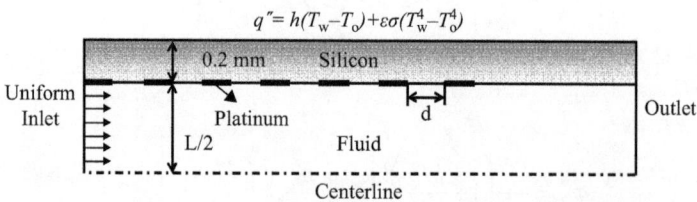

$$q'' = h(T_{\mathrm{w}} - T_o) + \varepsilon\sigma(T_{\mathrm{w}}^4 - T_o^4)$$

0.2 mm   Silicon

Uniform Inlet

Platinum

d

Outlet

L/2   Fluid

Centerline

Figure 8.2. Schematic of computational domain for catalyst segmentation.

are performed in one half of the microchannel. The reactor is 3 cm in length and the wall thickness is 0.2 mm. The heat conduction equation for the wall is also solved simultaneously for a better description of the wall temperature distribution. To further delineate the complicated interactions between heterogeneous and homogeneous reactions in a micro-catalytic combustor, in addition to the single section of uniform catalyst bed disposition used in most previous researches, effects of catalyst segmentation are studied. The inner wall is coated with platinum catalyst and the total catalyst length is 1 cm for cases both with and without catalyst segmentation. The 1 cm catalyst length is divided into multiple segments. The segment length and the separation distance (d) between two adjacent segments are variable. The boundary conditions are described as follows.

8.2.3 BOUNDARY CONDITIONS AT THE CATALYST SURFACE

The boundary conditions on the catalyst surface are complicated due to the heterogeneous reactions at the gas–surface interface. Boundary conditions for species and energy equations near the catalyst surface are derived by balancing the flux and source/sink of an interface control volume with an infinitesimal thickness. The convective and diffusive mass fluxes of gas-phase species on the surface are balanced by the production (or depletion) rates of gas-phase species by surface reactions. The relationship is

$$\left[\rho Y_k \left(\vec{V}_k + \vec{u}\right)\right] \cdot \vec{n} = \dot{s}_k W_k \qquad k = 1,....K_g \qquad (8.13)$$

where ρ is the gas density, $\vec{n}$ is the unit outward-pointing normal vector to the surface, Y_k is the gas-phase mass fraction, $\dot{s}_k$ is the net production rate $[\textit{mole} / cm^2 - s]$ of gas-phase species k by surface reactions, V_k is the gas-phase diffusion velocity, and u is the mass averaged Stefan velocity given by

$$\vec{u} \cdot \vec{n} = \frac{1}{\rho}\sum_{k=1}^{K_g} \dot{s}_k W_k \qquad (8.14)$$

Exothermicity (or endothermicity) of surface reactions contributes to the energy balance at the interface. The temperature boundary condition is derived by balancing the diffusive and convective fluxes in the gas phase with thermal radiation and chemical heat release on the surface. It is stated as

$$\left(\vec{n} \cdot k\nabla T - \sum_{k=1}^{K_g} \vec{n} \cdot \rho Y_k \left(\vec{V}_k + \vec{u}\right) h_k\right)_g = \sigma\varepsilon(T^4 - T_0^4) + \left(\vec{n} \cdot k\nabla T + \sum_{k=K_g+1}^{K_g+K_s} \dot{s}_k W_k h_k\right)_s \qquad (8.15)$$

By substituting Equation (8.14) into the flux term on the left-hand side, Equation (8.15) can be written in a more compact form as

$$\vec{n} \cdot k\nabla T\big|_g = \sigma\varepsilon(T^4 - T_0^4) + \vec{n} \cdot k\nabla T\big|_s + \sum_{k=1}^{K_g+K_s} \dot{s}_k W_k h_k \qquad (8.16)$$

where h_k is the enthalpy of species k and k_g and k_s are the total number of gas phase and surface species, respectively. Subscripts g and s denote gas phase and solid phase, respectively.

8.2.4 INLET AND BOUNDARY CONDITIONS FOR THE MICRO-CATALYTIC REACTOR

The concentration of the hydrogen–air mixture is specified at the inlet. The inlet temperature for the mixture is 300 K. A uniform velocity profile is specified at the inlet and the flow is laminar for all cases studied. The thermal boundary condition at the wall is the heat loss to the ambient air at the room temperature 300 K. The heat loss to the ambience includes the heat convection by air and thermal radiation, described by

$$q'' = h(T_w - 300) + \varepsilon\sigma(T_w^4 - 300^4) \tag{8.17}$$

where the heat transfer coefficient h equals to 20 W/m^2/K in the study. T_w is the wall temperature and the value of 300 K is specified for ambient air temperature. The wall emissivity ε is 0.5 and σ is the Stefan–Boltzmann constant. At the combustor exit, a constant ambient pressure of 101 kPa is specified and an extrapolation scheme is used for species and temperature.

8.2.5 CHEMICAL KINETICS

Chemical reaction mechanisms are used in the gas phase as well as on the catalyst surface of the combustor inner wall. The reaction rate is represented by the modified Arrhenius expression and for the heterogeneous reactions all temperature exponents are set to zero. The homogeneous reaction mechanism of hydrogen–air combustion used here composes of 9 species and 19 reaction steps; these are adopted from that proposed by Miller and Bowman [16]. The surface reaction mechanism used in the study is compiled from that proposed by Deutschmann et al. [17] and shown in Table 8.1. These reaction mechanisms have been used previously and comparisons with experimental results were found satisfactory [18–20]. For hydrogen fuel, five surface species, H(s), O(s), OH(s), H$_2$O(s), Pt(s), are used to describe the coverage of the surface with adsorbed species. Pt(s) indicates free surface sites, which are available for adsorption and further catalytic reaction. The gas-phase chemical kinetics with the CHEMKIN format and heterogeneous chemical kinetics with Surface CHEMKIN are imported into the code. Details of the rate formulation of the chemical reaction and CHEMKIN format can be found in the user's manual [15, 21].

8.3 CHARACTERISTICS OF MICROSCALE CATALYTIC COMBUSTION

8.3.1 MODEL VALIDATION

In order to validate the present numerical model and chemical mechanisms used in the study, simulations of catalytic combustion in a single platinum microtube were performed first and the results are compared with experiments of the previous study [3]. Figure 8.3 shows the predicted

Table 8.1. Surface reactions of hydrogen/air on platinum

Reaction	A	b	Ea (kJ/mole)	
$H_2 + 2PT(S) => 2H(S)$	0.046	0.0	0	STICK FORD/PT(S) 1/
$2H(S) => H_2 + 2PT(S)$	3.7E + 21	0.0	67.4 − 6.0H(S)	
$O_2 + 2PT(S) => 2O(S)$	21	−1.0	0	STICK
$2O(S) => O_2 + 2PT(S)$	3.7E + 21	0.0	213.2 − 60O(S)	
$H + PT(*) => H(S)$	1.00	0.0	0	STICK
$O + PT(*) => O(S)$	1.00	0.0	0	STICK
$OH + PT(*) => OH(*)$	1.00	0.0	0	STICK
$H_2O + PT(*) => H_2O(S)$	0.75	0.0	0	STICK
$H(*) + O(*) = OH(*) + PT(S)$	3.7E + 21	0.0	11.5	
$H(*) + OH(*) = H_2O(*) + PT(S)$	3.7E + 21	0.0	17.4	
$2OH(*) = H_2O(*) + O(*)$	3.7E + 21	0.0	48.2	
$OH(*) => OH + PT(S)$	1.0E + 13	0.0	192.8	
$H_2O(*) => H_2O + PT(S)$	1.0E + 13	0.0	40.3	

A: pre-exponential factor in the Arrhenius expression; b: temperature exponent; Ea: activation energy.

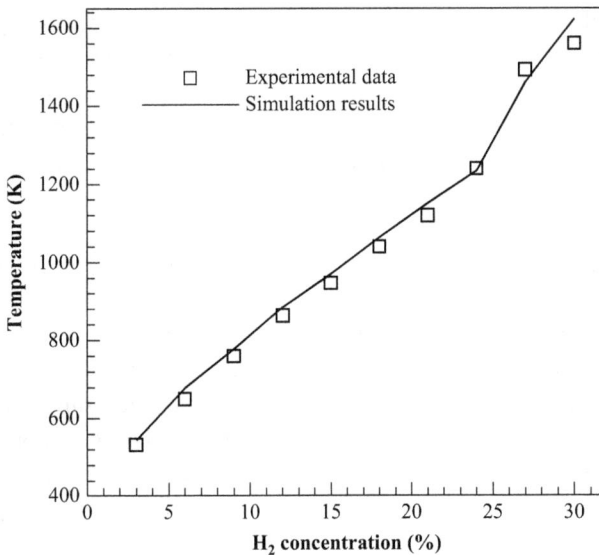

Figure 8.3. Comparisons of predicted and measured outlet temperature (K) for a 1 mm platinum tube ($V = 15$ m/s).

outlet temperature of a typical case of velocity 15 m/s in a 1 mm microtube and they agree well with the experimental data for the range of hydrogen concentration tested. In particular, the simulation also successfully predicted gas-phase combustion near the tube exit at higher hydrogen-concentration cases, indicated by the sudden increase of the tube outlet temperature in Figure 8.3.

The agreement suggests the feasibility of further application of the present numerical models in the investigations of reaction characteristics inside a micro-catalytic reactor where experimental measurements are very difficult.

8.3.2 COMBUSTION CHARACTERISTICS FOR DIFFERENT REACTION MODELS

To delineate the complicated hydrogen–air catalytic combustion characteristics in terms of interaction of heterogeneous and homogeneous reactions inside a catalytic microscale reactor is the main issue of this study. First, three different reaction modes by combinations of the homogeneous and heterogeneous reaction mechanisms are studied to identify and highlight the effects of heterogeneous reactions; they are both homogeneous and heterogeneous, homogeneous-alone and heterogeneous-alone mechanisms respectively. Figure 8.4 shows the computed temperature and OH concentration contours and Figure 8.5 shows the fuel and H_2O concentration contours for these three reaction modes. For these cases, the stoichiometric mixture flows into a 1 mm tube and the inlet velocity is 2 m/s, which is close to the laminar burning velocity of the hydrogen–air mixture. There are obvious differences in these figures. In the homogeneous-alone case, the highest temperature and highest OH concentration regions are located in the central region of the tube and the flame structure shows conical in shape. Temperature decreases radially and heat is dissipated through the wall. By assuming an inert wall in the homogeneous-alone case, the OH concentration near the wall reaches a certain value. High-OH concentration contours can be used to identify the high-temperature and flame regions inside the microtube. For the heterogeneous-alone case by switching off the homogeneous gas-phase reaction mechanism, reactions can only occur on the wall. Therefore, the highest temperature is found on the wall near the inlet and heat is transported downstream by convection. The highest temperature reached is lower than that for cases of the other reaction modes. The highest OH concentration appears on the wall and its strength is weak since OH species has high-absorption ability by the catalytic wall. The difference in OH concentration profiles also reveals the variation of reaction characteristics of heterogeneous and homogeneous reactions. For the complete case employing both gas-phase and surface reaction mechanisms, the high-temperature regions exist both on the wall near the entrance and in the central region of the tube. The formation of the high-temperature zone is not simply due to heat convection from the wall. Both homogeneous and heterogeneous reactions exist inside the catalyst tube. The homogeneous reaction takes place in the central region of the tube but the peak location shifts slightly downstream as compared with the homogeneous-alone case (see the OH concentration). To some extent, the homogeneous reaction is indeed affected by the presence of the heterogeneous reaction and the fuel concentration distribution in Figure 8.5 also exhibits the same phenomenon. The fuel concentration in the axial direction decreases dramatically once there is homogeneous combustion. When we employ both homogeneous and heterogeneous reaction mechanisms in the simulation, complicated results characterizing the interaction between these two reaction modes can clearly be identified as they are compared with the two "alone" models. In this complete case, the fuel concentration along the tube centerline is consumed in the upstream region as compared with the heterogeneous-alone case, but it also shows that homogeneous gas-phase combustion is delayed by the presence of the heterogeneous reactions when compared with the homogeneous-alone case. Since fuel is only consumed on the wall for the heterogeneous-alone

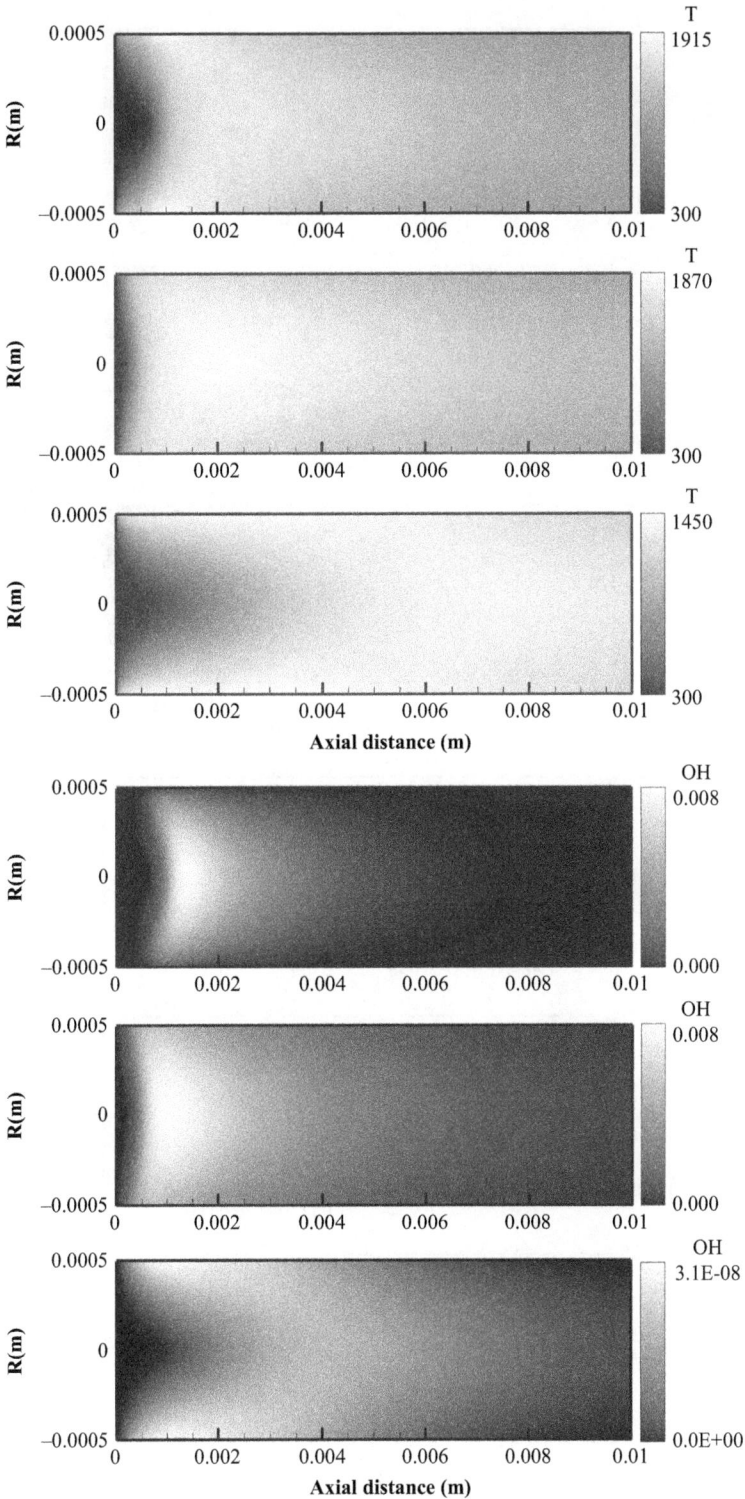

Figure 8.4. The computed temperature and OH concentration contours for three different reactions: both homogeneous and heterogeneous, homogeneous-alone, and heterogeneous-alone.

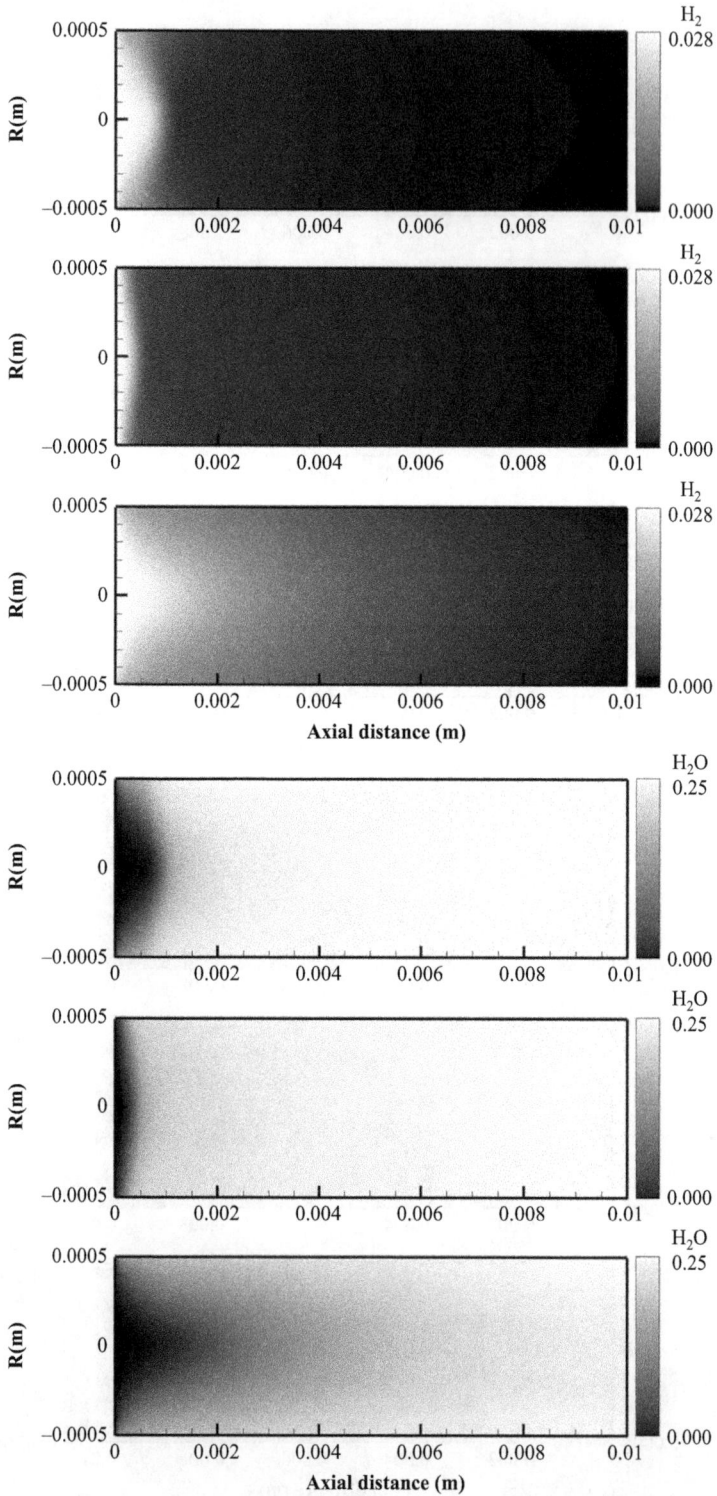

Figure 8.5. The computed H_2 and H_2O concentration contours for three different reactions: both homogeneous and heterogeneous, homogeneous-alone, and heterogeneous-alone reactions.

case, it must diffuse to and adsorbed by the wall before it can be consumed. The heterogeneous reaction on the wall can be seen to interfere with the reactions and delay the homogeneous combustion in the tube, if one carefully examines the fuel consumption along the axial direction.

It has been identified and proposed that the homogeneous reaction can be inhibited by the catalytic surface reaction, especially in fuel lean conditions. The surface reactions will compete with the homogeneous gas-phase reaction for fuel and oxidizer as well as reaction radicals. From the results of Figures 8.4 and 8.5, even in the stoichiometric $(\Phi = 1)$ and high-fuel concentration conditions, the effect of catalytic inhibition still exists. Consequent differences among these reaction models can also be found in the combustion product concentration contours. In particular, the simulation results show that the homogeneous reaction cannot be ignored inside a micro-catalytic reactor for most operation conditions.

8.3.3 EFFECT OF FLOW VELOCITY

The inlet velocity and the equivalence ratio are the major operational parameters in a catalytic micro-combustor. Figures 8.6–8.8 show the temperature, OH mass fraction, and fuel mass fraction profiles along the central axis for different inlet velocities. The inlet velocities are varied from 2 to 20 m/s and some cases are shown here. The tube diameter is 1 mm. In Figure 8.6, for all cases the temperature increases along the central axis due to heat release by reactions. For smaller inlet velocities, the temperature increases sharply and the fuel is consumed abruptly close to the entrance due to the occurrence of homogeneous combustion. As the inlet velocity is increased above 6 m/s, the curves for temperature increase and the curves for fuel consumption show two different slopes. These two slopes imply two different reaction modes. The first slope of the curve is smoother denoting the region dominated by heterogeneous reactions. Fuel consumption for heterogeneous reactions is dominated by diffusion and the generated heat is transported from the wall back to the tube center by convection. Therefore, the temperature

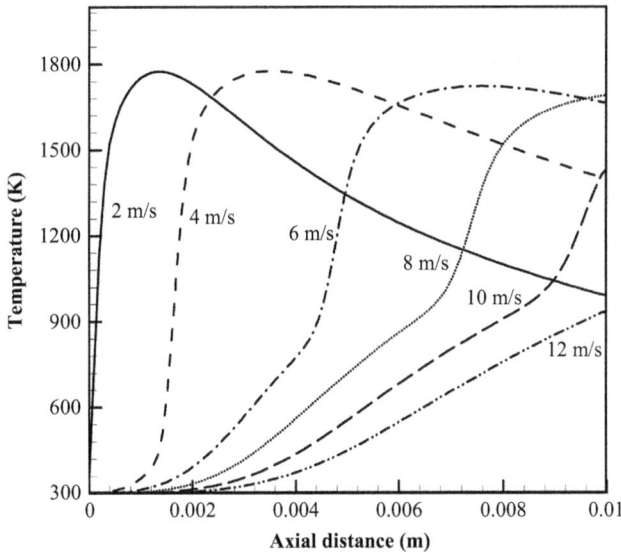

Figure 8.6. Temperature profile along the central axis for different inlet velocities.

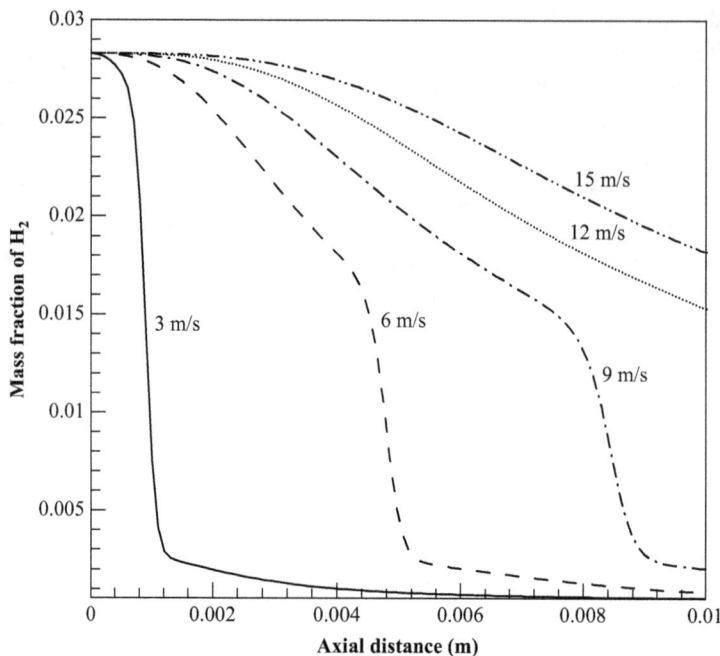

Figure 8.7. H_2 mass fraction profile along the central axis for different inlet velocities.

increase becomes smooth. On the other hand, the second slope is due to homogeneous combustion. Homogeneous combustion occurs in the tube center and the chemical time scale is much smaller than the flow time scale. Therefore, the temperature increase and the fuel consumption are very intense in the tube. From these results, the interaction and competition between the heterogeneous and homogeneous reactions can clearly be identified in the catalytic microtube. It is obvious that the reaction inside the catalyst tube can be divided into two sections. The upstream region is dominated by the heterogeneous reaction and the downstream one is dominated by the homogeneous reaction for high inlet velocities. For lower inlet velocities, the heterogeneous reaction becomes weak and homogeneous combustion occurs near the flow entrance. The temperature first sharply increases due to combustion and then decreases along the axial direction as the product gases is cooled due to heat loss through the wall. The heat loss effect is relatively obvious in low flow-rate cases. When the inlet velocity increases, the mixture residence time decreases and the axial heat transport increases. In Figure 8.7, when the inlet velocity increases to 10 m/s, the homogeneous combustion can maintain near the tube exit. This inlet velocity much exceeds the stoichiometric hydrogen flame speed (~2 m/s). Complete combustion inside a micro-combustor depends on the ratio between the flow residence time and the chemical reaction time (the Damköhler number). For a fixed combustor scale, the increasing flow rate represents a smaller residence time, which can be detrimental for a complete reaction. The above results show that the existence of a catalyst wall in the combustor will extend the homogeneous combustion region and will help stabilize the homogeneous combustion and reach the complete reaction inside the microtube. When the inlet velocity is further increased to exceed 10 m/s, the temperature along the central axis increases smoothly in general and the fuel consumption does not show a sharp decrease. At this condition, homogeneous combustion cannot occur inside

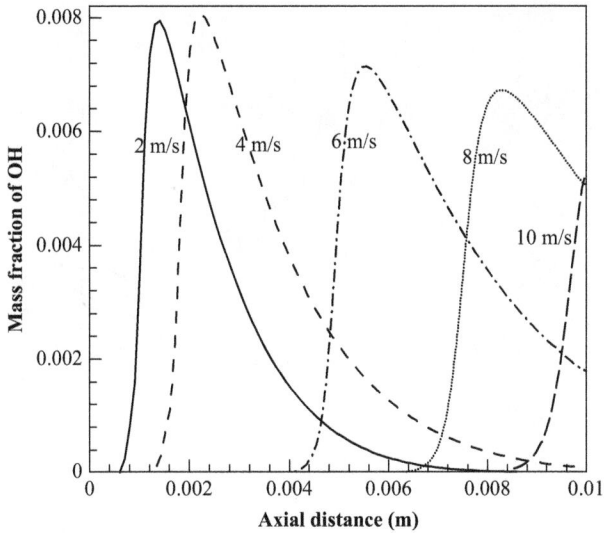

Figure 8.8. OH mass fraction profile along the central axis for different inlet velocities.

the catalyst reactor. The reaction in the microtube is dominated by heterogeneous reactions. In Figure 8.8, the peak OH concentration along the central axis decreases and the peak location shifts downstream with the increase of the inlet velocity. When the inlet velocity exceeds 10 m/s, OH concentration will decrease substantially.

The existence of the heterogeneous reactions has two effects on homogeneous combustion. On the one hand, the heterogeneous reactions will consume part of the fuel in the upstream region. The reduced fuel concentration and the reaction product will slow down the homogeneous reactions. On the other hand, high temperature and radicals generated on the wall by heterogeneous reactions will enhance the gas-phase reaction and reduce heat losses from the gas in the micro-combustor. Therefore, the homogeneous combustion can be sustained inside the micro-catalytic reactor with the size smaller than the quenching distance of the mixture. These competing effects determine whether gas-phase combustion can occur inside the micro-catalytic combustor. To help further examine the effects of inlet velocity on the performance, Figure 8.9 shows the average temperature and fuel conversion at the outlet. When homogeneous combustion occurs near the entrance for smaller inlet velocities, the outlet temperature is low due to the intense heat loss of the microtube. As the inlet velocity increases, homogeneous combustion moves downstream and the outlet temperature increases. As shown in the figure, if homogeneous combustion occurs inside the catalyst tube, complete combustion can be attained for most cases. When the inlet velocity exceeds 10 m/s and homogeneous combustion cannot be maintained inside the microtube, the outlet temperature and the fuel conversion ratio at the tube exit decrease obviously. Nevertheless, the residual fuel at the exit and the high temperature from the catalytic reaction will help stabilize a flame at the tube exit [3].

8.3.4 EFFECT OF TUBE DIAMETER

To look further into effects of the tube diameter on reaction characteristics in a catalytic microtube, the tube diameter is reduced from 1 mm to 0.2 mm. Figure 8.10 shows the OH

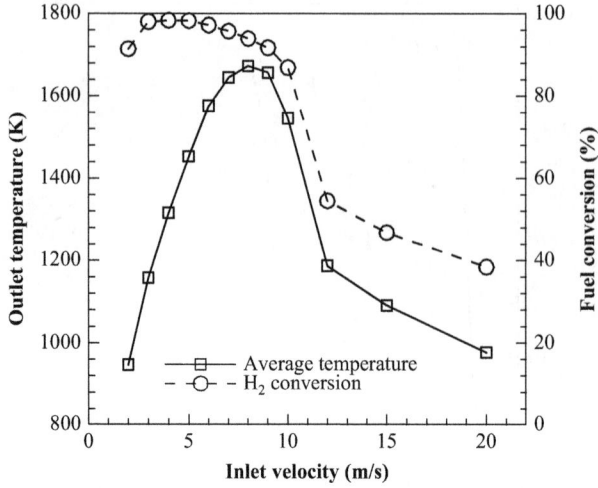

Figure 8.9. Average temperature and fuel conversion ratio at the outlet for different inlet velocities.

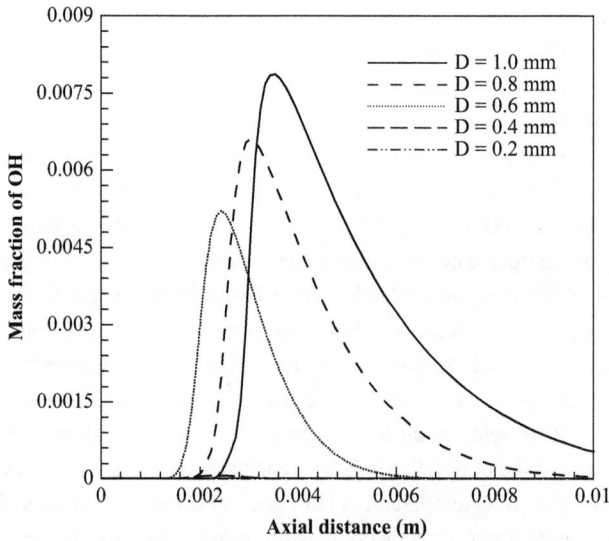

Figure 8.10. OH concentration profile along the tube center for different tube diameters at a fixed inlet velocity of V = 5 m/s.

concentration contour along the tube center for five different tube diameters. The inlet velocity is fixed at 5 m/s. In the figure, the peak and the OH concentration distributions are found to shift upstream and become weaker when the tube diameter is reduced from 1 mm down to 0.6 mm. For diameters smaller than 0.4 mm, the OH concentration along the central axis is almost indiscernible, indicating quenching of the homogeneous gas-phase combustion. However, the heterogeneous reaction on the catalyst surface becomes enhanced and dominant.

The heterogeneous reaction is affected by the mass diffusion. The mass diffusion time can be estimated by the following equation:

$$\tau_D = \frac{R^2}{D_{AB}} \tag{8.18}$$

where D_{AB} and R are the mass diffusivity and the diffusive length, respectively. For smaller diameter tubes, the fuel and the oxidizer in the bulk gas mixture need less diffusive time to reach the catalytic surface. The surface reaction takes the dominant role in the reactions and consumes most of the fuel–air mixture leading to quenching of the homogeneous combustion. For large diameter tubes, the heat and radicals produced by heterogeneous reactions also help maintain homogeneous combustion inside the microtube with higher inlet velocities. Figure 8.11 shows the maximum allowable inlet velocity for maintaining homogeneous combustion inside the microreactor with different tube sizes. With the decrease of the tube diameter, the maximum allowable inlet velocity will increase first to 18 m/s, a velocity far exceeding the hydrogen–air flame speed, when the diameter is 0.6 mm and then decrease drastically and quench at a diameter of 0.2 mm. These results clearly indicate that the inhibition effect of the heterogeneous reaction surpasses the enhancement effects of surface-generated heat and radicals on the homogeneous combustion. For the 0.2 mm tube diameter case, homogeneous combustion simply quenches for all inlet velocities in our simulation. This can also be identified by the substantial decrease of OH concentration in the centerline.

From above results in a micro-catalytic reactor, the interaction between heterogeneous and homogeneous reactions can be divided into three typical modes. In the first mode, the homogeneous combustion is weakened by the catalyst but it survives over a large range of

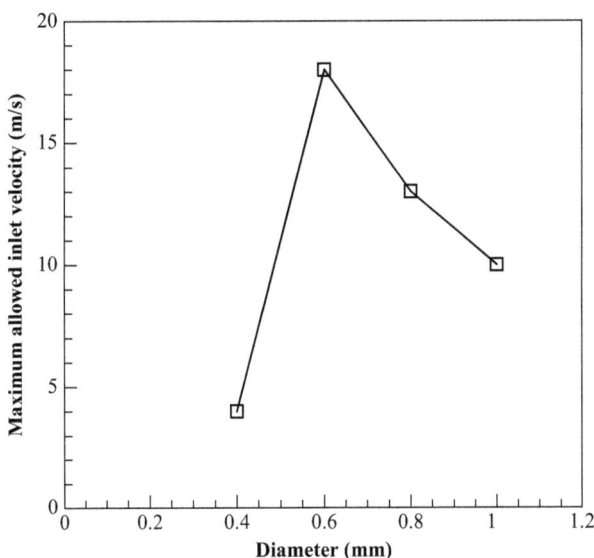

Figure 8.11. The maximum allowed inlet velocity for homogeneous combustion inside the micro reactor with different tube sizes.

inlet velocities, as mentioned in the previous section. The resultant temperature, major species, and radicals profiles are apparently different whether homogeneous reactions occur or not. The second mode of interaction is shown in Figures 8.12 and 8.13. Figure 8.12 shows the temperature, H_2 mass fraction, and H_2O mass fraction contours, and Figure 8.13 includes the OH mass fraction contours for the tube diameter of 0.4 mm and the inlet velocity 5 m/s. In these figures, the results of computation using both heterogeneous and homogeneous reaction mechanisms are shown on top and the results using only heterogeneous reaction mechanism at the bottom. The results are almost identical for temperature and major species contours whether homogeneous reactions are included or not. Obviously, homogeneous reactions are significantly inhibited by the heterogeneous catalytic reaction. Homogeneous

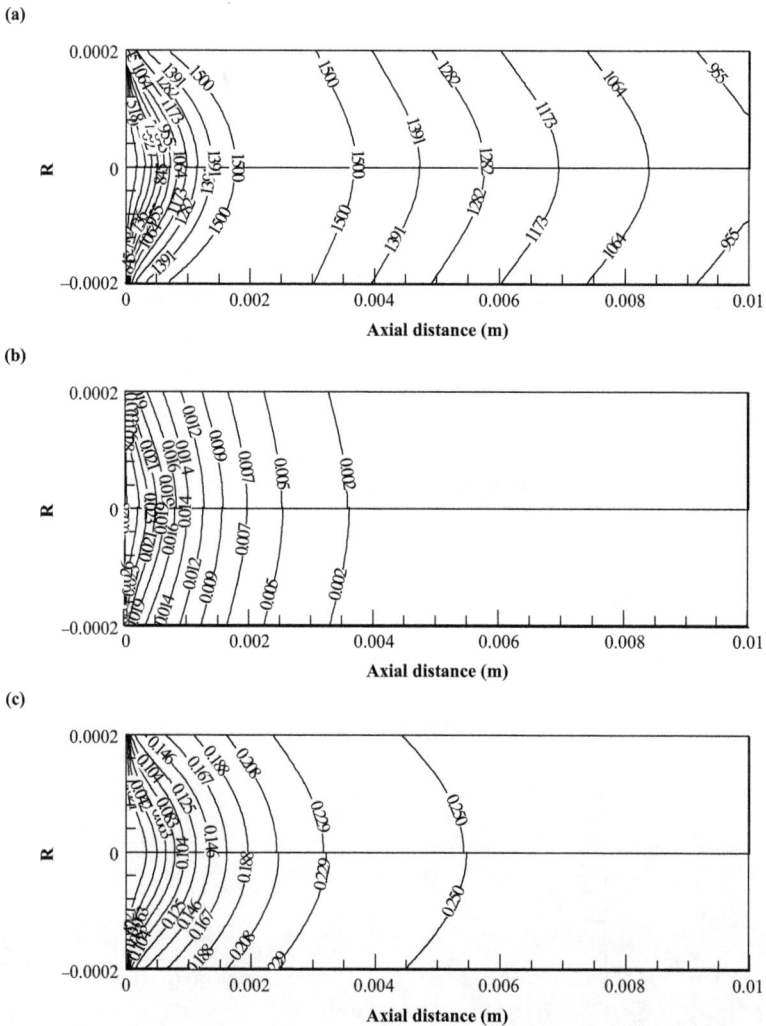

Figure 8.12. The computed (a) temperature, (b) H_2 mass fraction, and (c) H_2O mass fraction contours for both homogeneous and heterogeneous (top) and heterogeneous-alone (bottom) reactions with D = 0.4 mm and V = 5 m/s.

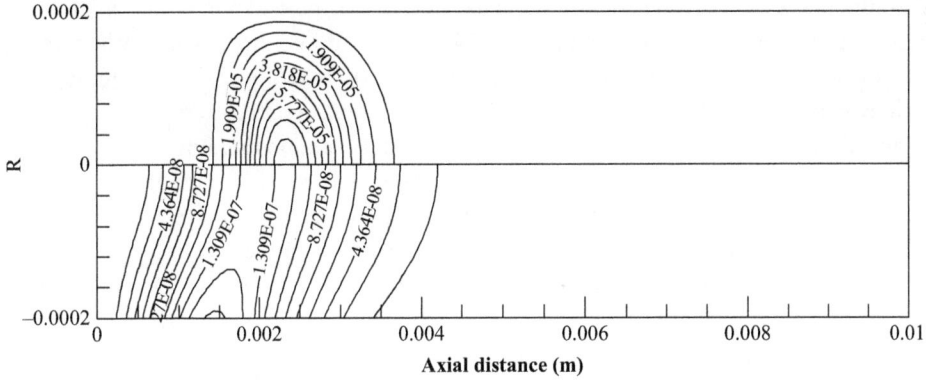

Figure 8.13. The computed OH mass fraction contours for both homogeneous and heterogeneous (top) and heterogeneous-alone (bottom) reactions with D = 0.4 mm and V = 5 m/s.

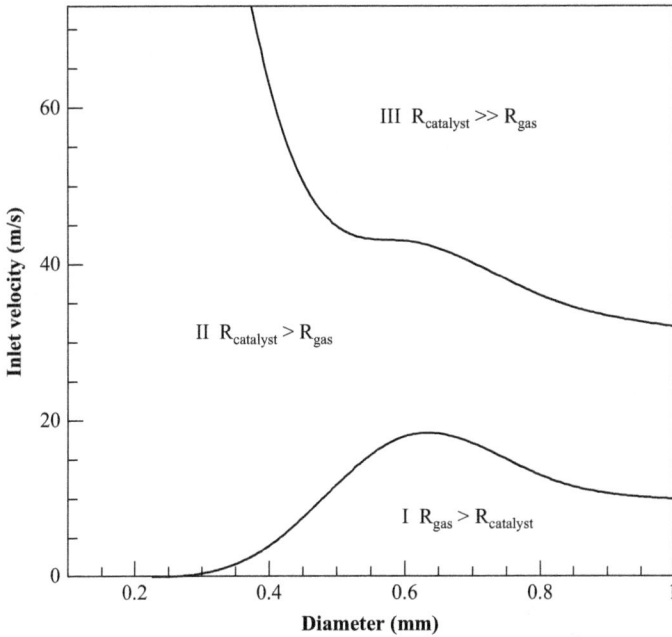

Figure 8.14. Three reaction regions for the effect of heterogeneous reaction on the homogeneous reaction.

reactions contribute very little to the final spatial distributions of temperature and major species. However, the mismatched peak and pattern of OH concentration distributions in Figure 8.13 implying that the central high-temperature region (see Figure 8.12) supports the homogeneous reaction to generate OH radicals to some extent but not enough to induce gas-phase combustion. The homogeneous reaction mechanism should be included in the simulation to catch the intrinsic reaction characteristics in the catalytic microtube. The third mode of interaction is that homogeneous reactions can be completely neglected. The whole tube is dominated by heterogeneous reactions. In summary, these three characteristic reaction modes can be schematically depicted in terms of tube diameter and inlet velocity in Figure 8.14. Reaction regions in Figure 8.14 are divided into three characteristic regions. In region I where the velocity is low and the diameter is relatively large, homogeneous reactions can sustain inside the tube but the homogeneous combustion is affected by the presence of heterogeneous reactions. The homogeneous reaction rate obviously exceeds the heterogeneous reaction rate. When the velocity is increased or the diameter is reduced, the characteristic reaction shifts to region II where the surface catalytic reaction is enhanced and surpasses the homogeneous reaction, and the surface catalytic reaction becomes dominant in the tube. Nevertheless, homogeneous gas-phase reactions still affect the expression of intermediate species. When the operation condition locates in region III, with very high inlet velocity and relatively large tube diameter, the homogeneous reaction is inhibited and almost has no effects in the tube.

8.3.5 EFFECT OF WALL MATERIAL

The heat conduction through the tube wall is also a key element of the combustion characteristics in a catalytic microreactor and is also investigated here in terms of different wall materials. In the study, the case of 1 mm tube diameter, 0.1 mm wall thickness, and 2 m/s inlet velocity is employed. The inner surface of the wall is coated with platinum. Three kinds of characteristic wall materials are selected from practical considerations, platinum, silicon, and alumina with thermal conductivities of 78, 23, and 3.3 W/m/K, respectively. These materials are often used in the catalyst reactors. For the bare wall microreactors, it is shown in the literature that the wall material is critical for flame ignition and stability inside the channel. There are two competing effects of the wall thermal conductivity. Heat conduction in the wall will provide a convenient route for heat transfer from the flame zone and post-combustion region to preheat the unburned mixture in the upstream region. On the other hand, the radial heat conduction through the wall to the environment will delay flame ignition or even quench the flame.

For the catalytic microreactors, Figure 8.15 shows the temperature distributions along the inner wall surface and the central axis for different wall materials. For the lowest thermal conductivity case, the resultant high temperature of the hot spot on the wall is able to melt the material or degrade the catalyst and the temperature gradient is high, indicating intense heat transfer. The wall temperature distribution becomes uniform as the wall thermal conductivity increases. In Figure 8.15(b), the temperature distribution along the central axis shows a higher temperature in the low wall thermal conductivity. Figure 8.16 shows the OH distribution along the central axis. A lower peak OH concentration is found and the location shifts slightly downstream for the highest thermal wall conductivity case. A high-temperature gradient on the wall (low thermal conductivity) will shift the homogeneous combustion upstream and the system will have a higher peak temperature. This behavior is different from that in the bare-wall microreactor. Recall that with bare wall, the flame core will shift downstream when the wall thermal conductivity decreases. The current results with catalytic wall indicate that the effect of wall thermal conductivity is not as pronounced as that in the bare-wall microreactor. Since the heat to ignite the homogeneous reaction is primarily from the catalytic reaction on the surface, not from the heat conduction from downstream, the effect of wall thermal conductivity is then not so pronounced.

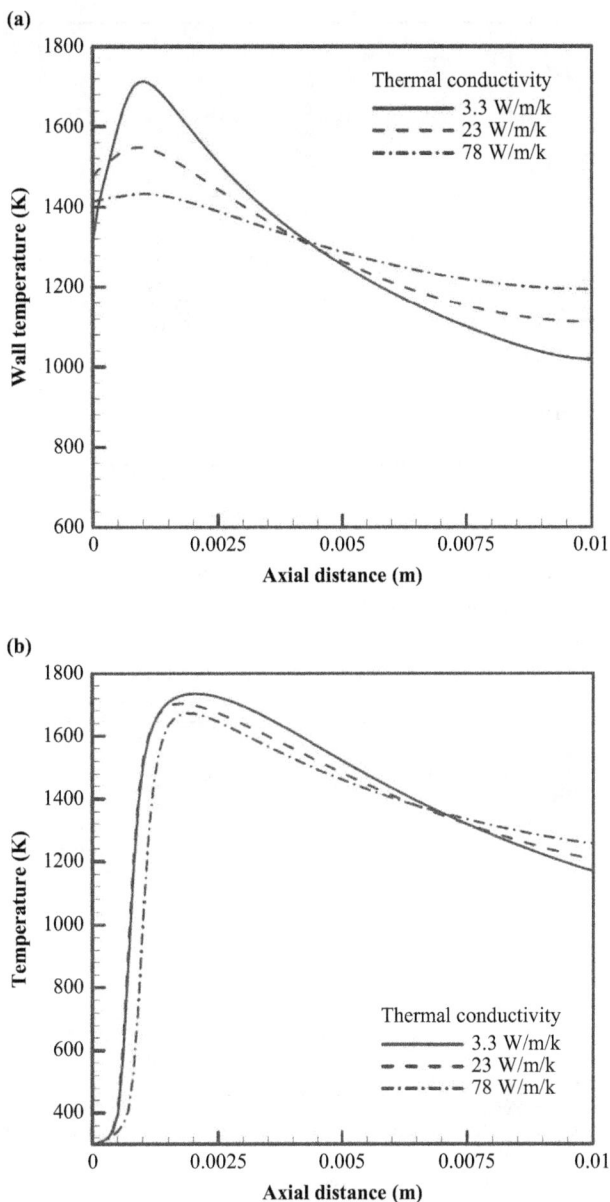

Figure 8.15. Temperature profiles along (a) the wall and (b) the central axis for different wall thermal conductivity values.

8.4 ENHANCEMENT OF MICROSCALE COMBUSTION BY CATALYST SEGMENTATION

8.4.1 COMBUSTION CHARACTERISTICS FOR CATALYSTS OF MULTIPLE SEGMENTS

To further delineate the other type of mutual supporting interaction between heterogeneous and homogeneous reactions, the combustion characteristics of multi-segment catalyst dispositions

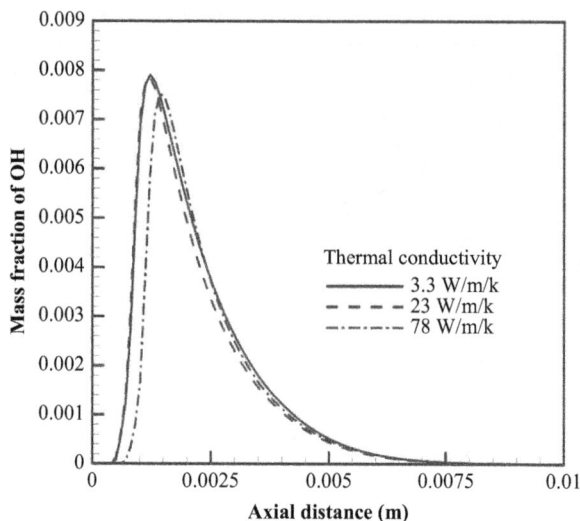

Figure 8.16. OH mass fraction profile along the central axis for different wall thermal conductivity values.

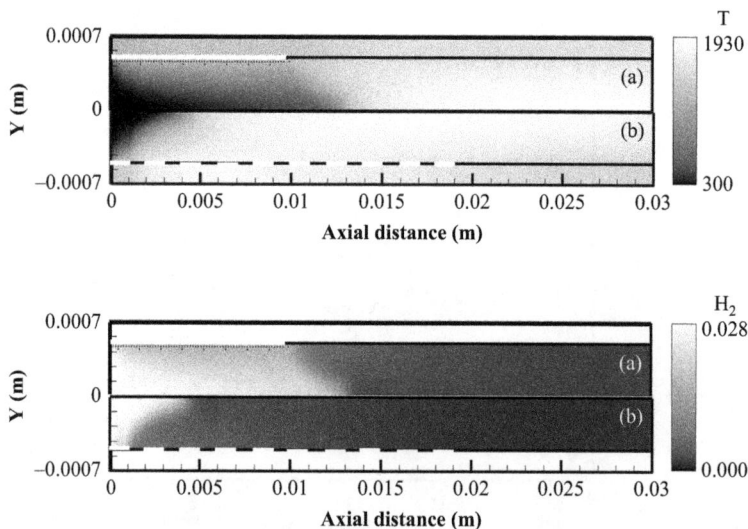

Figure 8.17. The computed temperature and H_2 mass fraction contours for different catalyst dispositions: single catalyst (top) and multiple catalyst segments (bottom).

for hydrogen–air reactions inside a microreactor are investigated. First, results of 1 cm single catalyst channel and the channel with ten 1-mm catalyst segments are compared for illustration. The separation distance between segments is 1 mm in the multi-segment case. Figure 8.17 shows the computed temperature and fuel mass fraction contours in the microreactors with the single catalyst case on top and the multi-segment case at the bottom. Figure 8.18 shows major reaction radicals OH and H mass fraction contours. For these cases, stoichiometric fuel/air mixture flows into the channel reactor of 1 mm in height and the inlet velocity is 20 m/s.

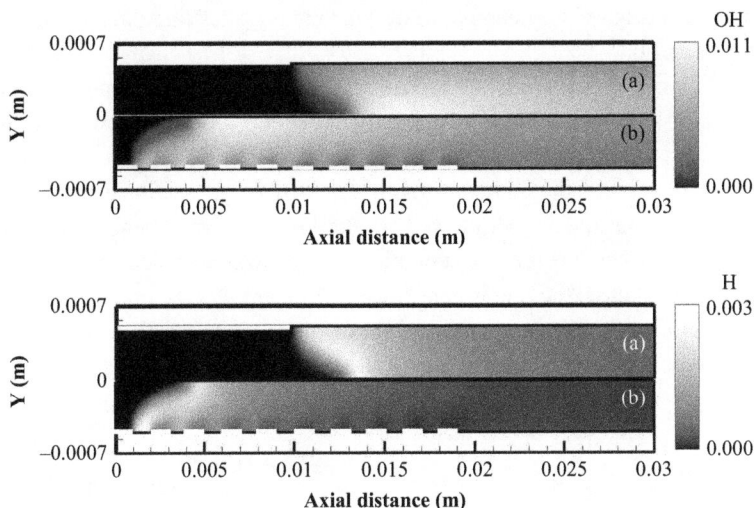

Figure 8.18. The computed OH mass fraction and H mass fraction contours for different catalyst dispositions: single catalyst (top) and multiple catalyst segments (bottom).

High-velocity case is shown here to emphasize the significant differences between these two catalyst dispositions. This velocity far exceeds the stoichiometric flame speed of hydrogen–air mixture and homogeneous combustion cannot sustain in the microreactor without catalyst. As shown in the figures, homogeneous combustion do occur in the center line for both cases but the flame location, as indicated by high-temperature regions (Figure 8.17) and high-OH concentration regions (Figure 8.18), has large variance. In the single catalyst case, some of the fuel is consumed by heterogeneous reactions inside the catalyst section and there is unburned fuel at the end of the catalyst section. The high-temperature fuel and products leaving the catalyst section stabilize the homogeneous combustion downstream of the catalyst section. The highest temperature and highest fuel consumption gradients locate near the outlet of the catalyst section. Complete combustion is achieved at the axial distance ~1.38 cm from the inlet. This is a typical case of catalytically stabilized homogeneous combustion. On the other hand, in the multi-segment case, homogeneous combustion obviously shifts upstream and complete combustion occurs near the channel inlet. The radical distributions shown in Figure 8.18 further display the difference between these two cases. H radical is the first product in the hydrogen decomposition reaction and it is often used to identify the start of reaction. OH concentration contours can be used to mark the high-temperature and flame locations. For the single catalyst case, the OH concentration inside the catalyst section is very sparse since OH species is produced on the surface and adsorbed by the wall with its high absorption ability. The highest OH and H concentrations are found behind the catalyst section near the homogeneous reaction region. High-H concentration layer can be found to extend from the wall near the outlet of the catalyst section to the centerline to connect the high-OH concentration region in the center. For the multi-segment case, the high-H concentration layer is seen to extend from the first catalyst segment gap to the centerline to connect the high-OH concentration region. This radical behavior clearly displays the phenomenon of catalytically stabilized homogeneous combustion in the microchannel. Apparently, different wall catalyst disposition shifts the homogeneous combustion upstream. Furthermore, the multi-segment catalyst disposition can help maintain the wall temperature and reduce heat loss to the surroundings resulting in a higher and uniform temperature distribution.

Figure 8.19 shows the fuel and OH mass fraction distributions along an axial direction near the inner wall. For the single catalyst case, the fuel consumption shows two distinct characteristics. The first smooth-decrease region is caused by catalytic combustion and is kinetic controlled. The reaction and the fuel consumption in the catalyst section are slow, and the wall temperature is low. This region is followed by the homogeneous combustion in the second characteristic region, where the fuel concentration near the wall then decreases abruptly. For the multi-segment catalyst case, the fuel is consumed significantly at the first catalyst gap and then decreases smoothly due to the occurrence of the homogeneous reaction between the catalyst gaps. At these gaps, high-OH concentration peaks are found. The radicals produced can move

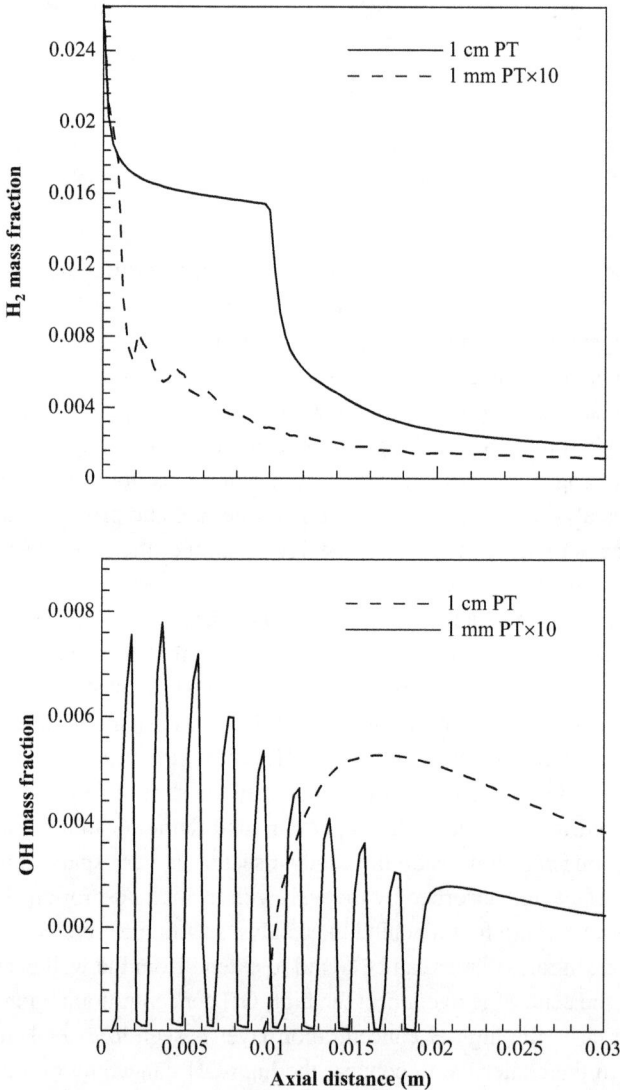

Figure 8.19. H_2 mass fraction and OH mass fraction profiles along the wall for different catalyst dispositions.

downstream with the hot gases and induce homogeneous combustion in the centerline of the channel; see Figure 8.18.

The existence of heterogeneous reaction has two effects on homogeneous combustion. First, the heterogeneous reaction consumes part of the fuel in the upstream region. The reduced fuel concentration and the reaction product will degenerate the homogeneous reaction. Second, high temperature and radicals generated by heterogeneous reactions can enhance the gas-phase reaction in the volume and help overcome heat loss from the gas. These two competing effects will determine whether the homogeneous combustion can occur inside the micro-catalytic combustor. The results show that even in the high-velocity condition (20 m/s in these cases), homogeneous combustion can exist near the entrance of the microchannel with superior performance by disposition of catalyst segmentation.

8.4.2 COMBUSTION CHARACTERISTICS UNDER DIFFERENT CATALYST DISPOSITIONS

Variations of multi-segment dispositions are also tested and three representative cases are shown here. These three types are 1 mm catalyst with 10 segments, 2 mm catalyst with 5 segments, and 5 mm catalyst with 2 segments. The channel height (L) is 1 mm and the catalyst gap (d) is 1 mm. The effect of gap distance (d) will be discussed in the next paragraph. The single-catalyst result is also shown here for comparison. Figure 8.20 shows the fuel mass fraction and OH mass fraction distributions along the centerline for different catalyst dispositions. In Figure 8.20, 1 mm catalyst case has the best performance and the homogeneous combustion shifts downstream with the increase of the catalyst segment length. As the increase of the catalyst segment length, the inhibition effect on the homogeneous reaction increases. Moreover, the increase of the catalyst segment length shifts the intensive homogeneous reaction in the first gap downstream. Figure 8.21 shows the fuel mass fraction near the inner wall. The location of the highest fuel consumption always occurs in the first catalyst gap for all catalyst dispositions, implying the strict homogeneous reaction in the gap region. In the catalyst gaps, the produced high-temperature radicals can move downstream with the hot gases and it is the homogeneous reaction in the catalyst gap that promotes global homogeneous combustion in the central region of the channel.

The effect of the catalyst separation distance (d) is also investigated. Figure 8.22 shows fuel mass fraction and OH mass fraction distributions along the centerline for different separation distances. Ten segments of 1 mm catalyst with separation distances of 1 mm, 2 mm, and 3 mm in the channel are used in the study. In Figure 8.22, the fuel mass fraction distributions almost overlap with each other and the OH mass fraction distributions deviate insignificantly for these three cases. The gap distance does not affect the global reaction. The catalyst gap can maintain high temperature due to the heterogeneous reactions on the neighboring catalytic surface to light-off the fast reaction of hydrogen combustion in the gap and support the global homogeneous combustion in the channel.

8.5 CONCLUSION AND FUTURE WORK

The fundamental and outstanding combustion phenomena in a micro-catalytic combustor are studied. Numerical simulations of hydrogen–air reaction in a catalytic platinum microtube are performed to identify the characteristic combustion modes and the interaction between the

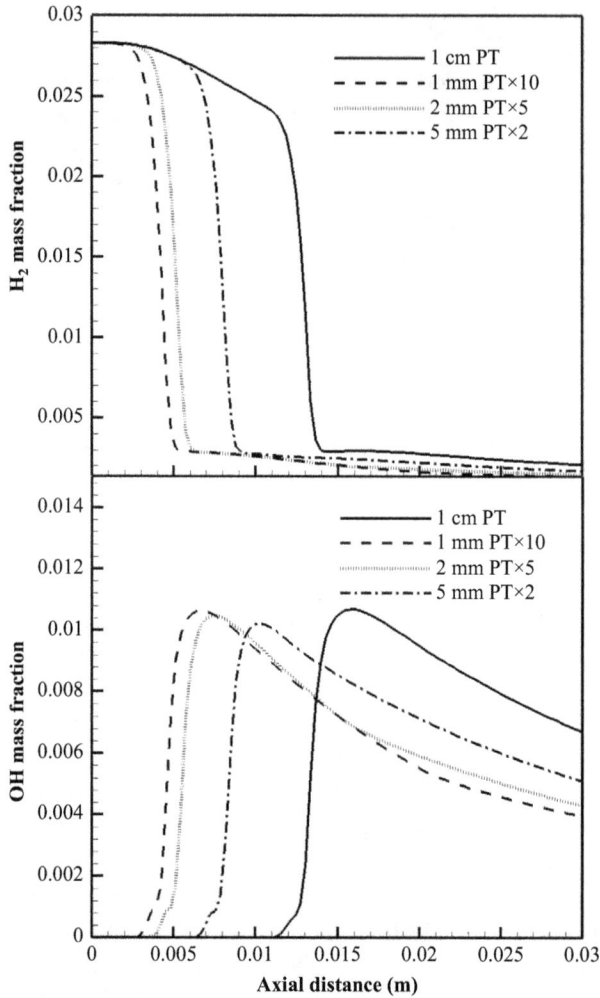

Figure 8.20. H_2 mass fraction and OH mass fraction profiles along the centerline for different catalyst dispositions.

homogeneous and heterogeneous reactions and to verify the role of the catalytic wall in the reaction characteristics in terms of its sustaining and competing effects on homogeneous combustion in the catalytic microtube. The parameters studied include inlet flow velocity (residence time), tube diameter, wall thermal conductivity, and catalyst segmentation. The following conclusions on catalytic micro-combustion are obtained from this study:

The existence of heterogeneous catalytic surface reactions has two effects on homogeneous combustion. On the one hand, the heterogeneous reaction competes with the homogeneous reaction for fuel and oxidizer as well as reaction radicals in the microtube. As the heterogeneous reaction will consume part of the fuel (hydrogen) in the upstream region, the reduced fuel concentration and the generated reaction products will slow down the homogeneous reaction. On the other hand, high temperature and radicals generated by heterogeneous reactions will enhance the homogeneous reaction to overcome heat loss from the gas phase and help sustain the homogeneous combustion.

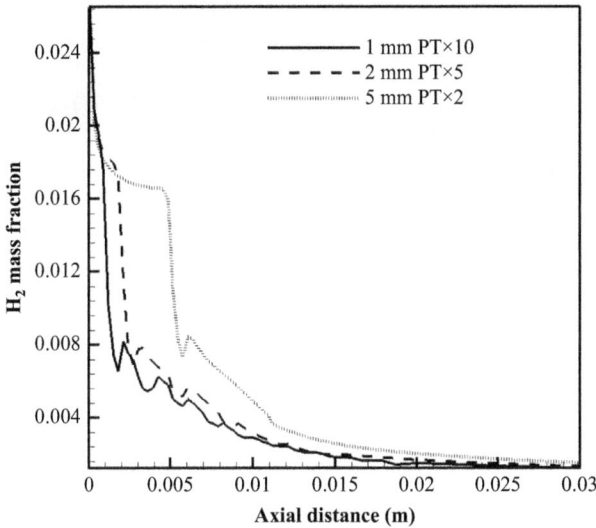

Figure 8.21. H_2 mass fraction along the wall for different catalyst dispositions.

In a catalytic microtube, decreasing the tube diameter will enhance the heterogeneous reaction, and the increased heat release on the catalytic surface will benefit homogeneous combustion. With the decrease of the tube diameter, the maximum allowable inlet velocity will increase first to 18 m/s, a velocity far exceeding the hydrogen–air flame speed, when the diameter is 0.6 mm, and then decrease drastically and quench at a diameter of 0.2 mm.

The effects of heterogeneous reaction on the homogeneous reaction can be divided into three characteristic reaction modes expressed in terms of the tube diameter and inlet velocity. In the first mode, where the velocity is low and the diameter is relatively large, homogeneous reactions can sustain inside the tube but the homogeneous combustion is affected by the presence of heterogeneous reactions. The homogeneous reaction rate apparently exceeds the heterogeneous reaction rate. When the velocity is increased or the diameter is reduced, the characteristic reaction shifts to the second mode where the surface catalytic reaction is enhanced and surpasses the homogeneous reaction, and the surface catalytic reaction becomes dominant in the tube. Nevertheless, homogeneous gas-phase reactions still affect the expression of intermediate species. Finally, with very high inlet velocity and relatively large tube diameter, the characteristic combustion turns into the third mode where the homogeneous reaction is inhibited and almost has no effects in the tube.

Since the heat to ignite the homogeneous reaction is primarily from the catalytic reaction on the surface, not from the heat conduction from downstream, the effect of wall thermal conductivity is then not so pronounced.

With a fixed total catalyst length (1 cm), the multi-segment catalyst reveals better performance than the single catalyst. Heterogeneous reactions help maintain a high wall temperature and the gap space between catalyst segments reduces the inhibition effect of the catalyst and promotes homogeneous reactions in the gap region. Therefore, homogeneous combustion shifts upstream. The catalyst separation distance has no obvious influence on homogeneous combustion due to the fast reaction rate of hydrogen.

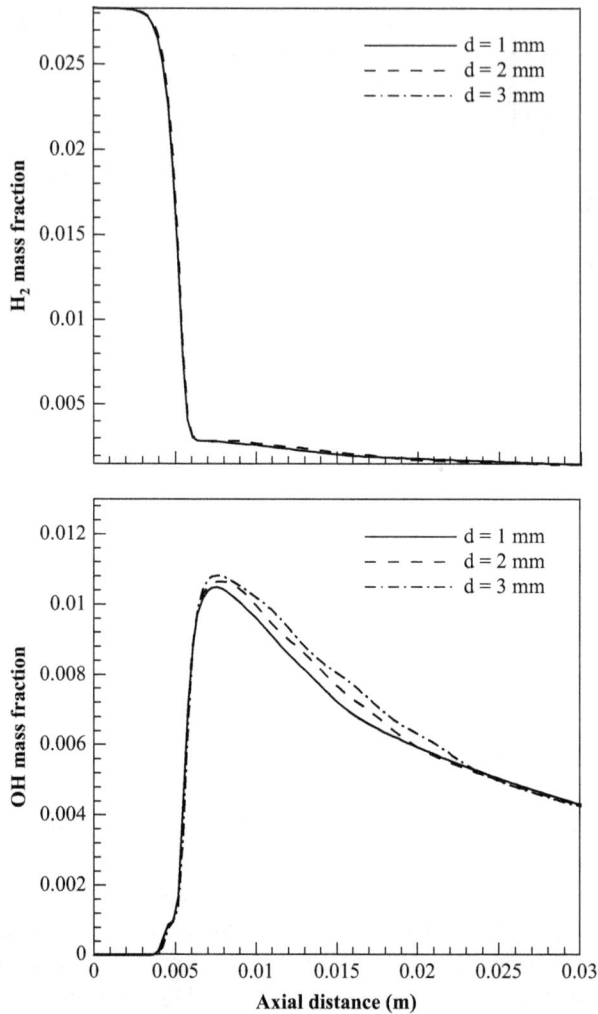

Figure 8.22. H_2 mass fraction and OH mass fraction profiles along the centerline for different catalyst space distances.

REFERENCES

[1] Sirignano, W.A., Pham, T.K., and Dunn-Rankin, D., 2002, "Miniature liquid-fuel-film combustor," Proceedings of the 29th Symposium (International) on Combustion, The Combustion Institute, Pittsburgh, PA, pp. 925–931.

[2] Fernandez-Pello, A.C., 2002, "Micro-power generation using combustion: issues and approaches," Proceedings of the 29th Symposium (International) on Combustion, The Combustion Institute, Pittsburgh, PA, pp. 883–899.

[3] Chao, Y.C., Chen, G.B., Hsu, C.J., Leu, T.S., Wu, C.Y., and Cheng, T.S., 2004, "Operational characteristics of catalytic micro-combustion in a platinum micro-tube," Combustion Science and Technology Vol. 176, No. 10, 1755–1777. doi: http://dx.doi.org/10.1080/00102200490487599.

[4] Sitzki, L., Borer, K., Wussow, S., Schuster, E., Maruta, K., Ronney, P., and Cohen, A., 2001, "Combustion in microscale heat-recirculating burners," AIAA-2001-1078, 38th AIAA Space Sciences & Exhibit, Reno, NV.

[5] Ronney, P.D., 2003, "Analysis of non-adiabatic heat-recirculating combustors," Combustion and Flame, Vol. 135, No. 4, 421–439. doi: http://dx.doi.org/10.1016/j.combustflame.2003.07.003.

[6] Veser, G., 2001, "Experimental and theoretical investigation of H_2 oxidation in a high-temperature catalytic microreactor," Chemical Engineering Science, Vol. 56, No. 4, 1265–1273. doi: http://dx.doi.org/10.1016/S0009-2509(00)00348-1.

[7] Norton, D.G., Wetzel, E.D., and Vlachos, D.G., 2004, "Fabrication of single-channel catalytic micro-burners: Effect of confinement on the oxidation of hydrogen/air mixtures," Industrial and Engineering Chemistry Research, Vol. 43, No. 16, 4833–4840. doi: http://dx.doi.org/10.1021/ie049798b.

[8] Boyarko, G.A., Sung, C.J., and Schneider, S.J., 2005, "Catalyzed combustion of hydrogen–oxygen in platinum tubes for micro-propulsion applications," Proceedings of the 30th Symposium (International) on Combustion, The Combustion Institute, Pittsburgh, PA, pp. 2481–2488.

[9] Volchko, S.J., Sung, C.J., Huang, Y.M., and Schneider, S.J., 2006, "Catalytic combustion of rich methane/oxygen mixtures for micropropulsion applications," Journal of Propulsion and Power, Vol. 22, No. 3, 684–693. doi: http://dx.doi.org/10.2514/1.19809.

[10] Maruta, K., Takeda, K., Ahn, J., Borer, K., Sitzki, L, Ronney, P.D., and Deutchman, O., 2002, "Extinction limits of catalytic combustion in microchannels," Proceedings of the 29th Symposium (International) on Combustion, The Combustion Institute, Pittsburgh, PA, pp. 957–963.

[11] Raimondeau, S., Vlachos, D.G., and Masel, R.I., 2002, "Modeling of high- temperature microburners," Proceedings of the 29th Symposium (International) on Combustion, The Combustion Institute, Pittsburgh, PA, pp. 901–907.

[12] Chen, G.B., Chen, C.P., Wu, C.Y., and Chao, Y.C., 2007, "Effects of Catalytic Wall on Hydrogen/Air Combustion Inside a Micro Tube," Applied Catalysis A: General, Vol. 332, No. 1, 89–97. doi: http://dx.doi.org/10.1016/j.apcata.2007.08.011.

[13] Kee, R.J., Dixon-Lewis, G., Warnatz, J., Coltrin, M.E., and Miller, J.A., 1991, "A fortran computer code package for the evaluation of gas-phase multicomponent transport properties," Sandia Report SAND86-8246, Sandia National Laboratories, Livermore, CA.

[14] Dixon-Lewis, G., 1968, "Flame structure and flame reaction kinetics ii. Transport phenomena in multicomponent systems," Proceedings of the Royal Society A., Vol. 307, No. 1488, 111–135. doi: http://dx.doi.org/10.1098/rspa.1968.0178.

[15] Kee, R.J., Rupley, F.M., Meeks, E., and Miller, J.A., 1996, "Chemkin-III: A fortran chemical kinetics package for the analysis of gas phase chemical and plasma kinetics," Sandia Report SAND96-8216, Sandia National Laboratories, Livermore, CA.

[16] Miller, J.A., and Bowman, C.T., 1989, "Mechanism and modeling of nitrogen chemistry in combustion," Progress in Energy and Combustion Science, Vol. 15, No. 4, 287–338. doi: http://dx.doi.org/10.1016/0360-1285(89)90017-8.

[17] Deutschmann, O., Schmidt, R., Behrendt, F., and Warnatz, J., 1996, "Numerical modeling of catalytic ignition," Proceedings of the 26th Symposium (International) on Combustion, The Combustion Institute, Pittsburgh, PA, pp. 1747–1754.

[18] Chen, C.P., Chao, Y.C., Wu, C.Y., Lee, J.C., and Chen, G.B., 2006, "Development of a catalytic hydrogen micro-propulsion system," Combustion Science and Technology, Vol. 178, No. 10–11, 2039–2060. doi: http://dx.doi.org/10.1080/00102200600793395.

[19] Deutschmann, O., Maier, L.I., Riedel, U., Stroemman, A.H., and Dibble, R.W., 2000, "Hydrogen asssisted catalytic combustion of methane on platinum," Catalysis Today, Vol. 59, No. 1–2, 141–150. doi: http://dx.doi.org/10.1016/S0920-5861(00)00279-0.

[20] Cheng, T.S., Wu, C.Y., Chen, C.P., Li, Y.H., Chao, Y.C., Yuan, T., and Leu, T.S., 2006, "Detailed measurement and assessment of laminar hydrogen jet diffusion flames," Combustion and Flame, Vol. 146, No. 1–2, 268–282. doi: http://dx.doi.org/10.1016/j.combustflame.2006.03.005.

[21] Coltrin, M.E., Kee, R.J., Rupley, F.M., and Meeks, E., 1996, "A fortran package for analyzing heterogeneous chemical kinetics at a solid-surface–gas-phase interface," Sandia National Laboratories Report SAND96-8217, Livermore, CA.

NOMENCLATURE

c_p	constant pressure specific heat per unit mass of the mixture
c_{pk}	constant pressure specific heat per unit mass of species k
D_{AB}	the mass diffusivity
D_{km}	mixture averaged diffusion coefficient
D_{ik}	binary diffusion coefficient
D_K^T	thermal diffusion coefficient
d	separation distance between catalyst segments
h	(1) Planck's constant
	(2) convection heat transfer coefficient
h_k	enthalpy per unit mass of species k
$\vec{j}_k^c$	diffusion mass flux vector of species k due to concentration gradients
$\vec{j}_k^T$	diffusion mass flux vector of species k due to temperature gradients
k	(1) reaction rate constant
	(2) thermal conductivity
K_g	total number of gas-phase species
K_s	total number of surface species
L	channel height
M	molecular weight of mixture
M_k	molecular weight of species k
$\vec{n}$	outward normal vector from surface
p	pressure
q''	heat loss to the ambience
R	(1) universal gas constant
	(2) the diffusive length
r	radial coordinate
S_φ	source term
$\dot{s}_k$	production rate of species k due to surface reactions
t	time
T	temperature
T_o	ambient temperature
T_w	wall temperature
$\vec{u}$	velocity vector
V	scaled radial velocity
$\vec{V}_k$	diffusion velocity of species k
v_r	radial component of velocity
v_z	axial component of velocity
W_k	molecular weight of species k
$\dot{\omega}_k$	production rate of species k due to gas phase reactions
X_k	mole fraction of species k
Y_k	mass fraction of species k
z	(1) axial coordinate
	(2) number of nearest-neighboring sites of one surface site
λ	thermal conductivity of mixture
λ_k	thermal conductivity of species k

ρ	mixture density
μ	mixture viscosity
μ_k	viscosity of species k
ε	emissivity
Φ	Equivalence ratio
φ	variable in Equation (8.2)
τ_D	the mass diffusion time

CHAPTER 9

MINIATURE LIQUID FUEL COMBUSTION

Derek Dunn-Rankin, William A. Sirignano, Yueh-Heng Li, and Yei-Chin Chao

The remarkable specific energy content of liquid hydrocarbon fuels encourages the notion of combustion-based compact power sources. Miniature devices, however, allow a very limited residence time for evaporation, mixing, and chemical reaction of the fuel, making miniature liquid fueled combustion a significant challenge. This chapter describes and demonstrates the viable use of liquid film combustors to meet this challenge. The design and performance of devices that burn fuel off the wall of cylindrical chambers and combustors that burn fuel from the surface of a central porous plug are presented. Electrical power generation is demonstrated in the latter case after coupling the combustor with a thermophotovoltaic converter.

9.1 OVERVIEW

As seen elsewhere in this volume, microcombustion research has largely been motivated by the promise of high-volumetric and mass-specific energy provided by liquid hydrocarbon fuels typical in macroscale systems (e.g., automobile engines and aircraft engines).The interesting practical applications for microcombustion include power for small unmanned ground and air vehicles, small robots, and personal (i.e., mobile) power sources for the delivery of work or electricity generation. The mobility theme leads immediately to a high power-to-mass demand, and liquid fuel combustion is the only practical candidate to meet that demand.

Liquid hydrocarbon fuels have specific energy, on the order of 13 kWh/kg, which with 25 percent conversion efficiency would provide close to 3.25 kWh/kg. This value is near the upper end of the desired personal power range. Further encouraging combustion, a very common comparison [1, 2] has been made between the specific energy of hydrocarbons and that of batteries, with hydrocarbons favored by factors ranging from 100 to 500 times (e.g., hydrocarbons have specific energy on the order of 13 kWh/kg or 45 MJ/kg and batteries are in the range of 100 Wh/kg or 0.4 MJ/kg—in fact, a typical 9V alkaline battery has mass of ~40 g and contains 17 kJ of electrical energy). This comparison, however, does not incorporate any balance-of-plant considerations, and the favorability of combustion diminishes if the conversion of thermochemical energy to electricity is included in the comparison.

Another potentially misleading element of combustion-derived power is that the energy of hydrocarbons cannot be released without an oxidizer, yet the oxidizer mass is not included in the specific energy calculation. This incomplete comparison leads, for example, to the non-intuitive result that butter contains nearly 10 times more energy per unit mass than does TNT (naturally, TNT has the far higher energy release rate but a comparatively anemic-specific energy). Also note that lithium metal has a specific energy of 41 MJ/kg, which is very much comparable to gasoline with a specific energy of 43 MJ/kg. It is only that we do not burn lithium in a battery but rather reversibly reconfigure a lithium compound. In a lithium/air primary battery, we get 9 MJ/kg, and for a rechargeable lithium/ion battery, we are down to 0.72 MJ/kg. It is not, therefore, the energy properties of the pure fuel that we should focus on as being responsible for the advantages of combustion for portable power. Instead, it is important, to make performance comparisons based on relevant factors for the entire power system. While some publications [3] have tried to point out that such comparisons are fraught with misleading elements, the seductiveness of high specific energy fuel persists.

With this issue in mind, the logical progression of concepts that led to the mesoscale liquid fueled combustion devices highlighted in this chapter are as follows:

1. As shown in Figure 9.1, the clear victor for portable power is combustion (i.e., thermo-chemical conversion) when high specific power and long operating duration are required. Even the most advanced (seemingly) lithium–sulfur battery [4], which has reached the amazing battery performance of 350 W/kg and 130 W/kg, still falls short of ideal portable power systems.
2. Long-operating duration necessitates high specific energy fuel with the oxidizer coming from the environment; the most logical candidate for such fuel is liquid hydrocarbon. It can be shown quite clearly that even in the case of hydrogen-based fuel systems, the chemical storage of hydrogen in hydrocarbons is extremely effective [5].

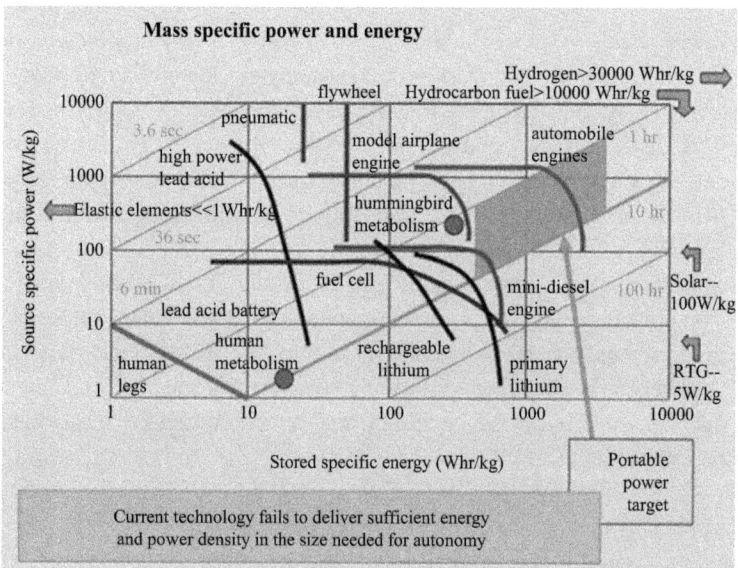

Figure 9.1. Energy and power densities.

3. Substantial power delivery (hundreds of Watts, as is appropriate for the thermochemical approach) and an operating duration on the order of hours requires a fuel tank that will be on the order of tens to hundreds of cubic centimeters, even with ideal systems. Since the fuel volume is already relatively large, there is little advantage to shrinking arbitrarily the scale of the combustion chamber alone. Hence, we focus on the mesoscale range of devices, where at least one dimension is on the order of a centimeter or less (i.e., 5 to 25 percent of the fuel volume).

9.2   BRIEF REVIEW OF MESOSCALE LIQUID FUEL COMBUSTORS

There is relatively little research and development in mesoscale power systems that use liquid fuels, in spite of the fact that, for portable and mobile applications, gaseous fuels will attract limited interest outside of the laboratory. A recent review [2] highlights this lack of activity by showing the focus in gaseous fuel systems. Nevertheless, there have been a few studies examining liquid-fueled engines, including external combustion engines like those based on the Stirling cycle. One advantage of external combustion engines, as well as turbomachines, is that the combustor operates continuously. It has proven to be a challenge, however, to run a small-scale (on the order of centimeters) continuous combustor on liquid fuel. While the typical approach is to spray the fuel as tiny droplets, this method is problematic at small sizes where the wall surfaces are so close together that substantial impingement and minimal air entrainment occurs except in the very special case of miniature hybrid rocket motors, where the oxidizer is the liquid in contact with the solid fuel wall [6]. One alternative for the air-breathing case that has been explored was to burn the fuel after it was delivered via electrospray onto a catalytic mesh [7–10]. See the review paper by Dunn-Rankin et al. [3] for a description of the historical approaches to 2006. These are not reviewed again here as we concentrate on more recent developments.

Internal combustion piston engines at small scale have also been developed. The smallest stock internal combustion engine is often recognized to be the Cox .049 (0.8 cc) two-stroke, glowplug ignited airplane engine. It is truly a mesoscale liquid fueled engine but it has extremely low efficiency because the unburned fuel mist is used intentionally as an internal coolant. A slightly larger but more reliable engine with high power density is the commercially available model airplane O.S. FS-30-S engine, which was studied as an archetypical mesoscale combustion system [11, 12]. The O.S. Engine FS-30-S was chosen for this study because it is an example of a modern mass-produced model engine design. The engine is a single cylinder, 4.89 cc displacement four-stroke design, with single intake and exhaust valves driven by pushrods. The piston has a single piston ring. The weight of the engine is 278 g, and the maximum power output is 0.5 brake horsepower (373 W), which occurs at 10,000 rpm. Based on these specifications, the specific power is 1300 W/kg (neglecting the weight of the fuel). The operating efficiency is below 3 percent, however, which limits its specific energy performance. Cadou and co-workers have shown clearly the dramatic degradation in performance of piston engines as their size decreases [13, 14]. The small-scale engine aspect of miniature combustion is described more fully in another chapter of this volume.

Besides the liquid film combustors that are the focus of this chapter, there have been few recent mesoscale liquid fuel combustion studies. Agrawal and co-workers examined a double-wall combustor design where liquid fuel was sprayed into an outer annulus where it evaporated and was carried through a porous base plug into the main combustion chamber [15].

The combustion process was stabilized by the porous medium. One of the key aspects of this design was a novel spray injector [16], but the design still relied on wall evaporation of the liquid fuel. Further details on the role played by the spray and porous medium in this system appear in the literature [17]. Gupta and co-workers describe a liquid fueled "Swiss-roll" burner, which is a dual channel heat recirculating combustor design [18, 19]. Air and fuel enter one channel and burn at the center of the system. The exhaust gas leaves through another channel, but heat transfer between the hot exhaust channel and the incoming fuel/air channel provides the opportunity for prevaporizing the liquid fuel. Generally, devices of this kind are meant to include thermoelectric elements along the channel to harvest the energy available. A more fundamental version of the heat recirculation concept applied to liquid fuel combustion involves two concentric tubes forming an inner channel for incoming air, and the outer annulus provides the escape path for the combustion products [20]. A tiny capillary is centered in the inner tube and liquid fuel is supplied through that capillary by a syringe pump. The diffusion flame at the end of the capillary then provides heating that recirculates and preheats the incoming air. The study shows substantially improved stability when the heat recirculation is permitted.

Considering the potential importance of liquid fuel combustion at the mesoscale, it is somewhat surprising that so little work focuses on these systems. The following sections help explain this circumstance by describing in detail one such system, the liquid film combustor, to show the challenges and potential associated with mesoscale combustion for portable power.

9.3 LIQUID FILM COMBUSTOR FUNDAMENTALS

A miniature combustor is defined as the one where the critical dimensions are less than 1 cm. At these sizes, in the millimeter range, the surface-to-volume ratio of the combustor becomes so large that substantial heat losses occur as heat transfers from the hot combustion products to the solid internal combustor surfaces. These heat losses reduce the thermal efficiency and can result in quenching (i.e., a flame cannot be sustained). As the name implies, liquid-fuel film combustors involve the creation of liquid films on surfaces where large heat transfer occurs. The film evaporates, thereby providing gaseous fuel and simultaneously keeping the surfaces cool. Current technology for larger liquid-fueled combustors uses the formation of a fuel spray to maintain a sufficiently large ratio of liquid surface area to liquid volume in order to sustain the required fuel vaporization rate. At this larger scale, the intention is to vaporize the liquid before liquid accumulates on the walls or solid surfaces of the combustor. If the fuel were filmed on the walls in these larger engines, the surface area of the liquid would be insufficient to sustain the needed vaporization rate. In fact, the resulting fuel pooling is one of the major challenges for direct injection gasoline engines. Fortunately, with the smaller portable power systems, the surface-to-volume ratio of the combustor and therefore that of any wall film will grow as the combustor size decreases. At some characteristic dimension, the wall film can provide a sufficient surface area to evaporate fuel at the rate needed to sustain combustion. It is this miniature fuel film combustor condition that we are interested in exploiting for portable power.

9.3.1 LIQUID FUEL FILM COMBUSTION CONCEPT

At the heart of the liquid fuel film combustor is the realization that, in the sub-centimeter range under discussion here, liquid films can offer as high a liquid surface area for vaporization as a vaporizing spray can, particularly when considering the realistic size of droplets achievable by a microspray device. In addition, it is difficult to atomize liquid fuel without imparting high momentum to the liquid (as in typical simplex or air-blast approaches), and this momentum creates wall wetting (i.e., a wall film) in the confined spaces of miniature combustors anyway. Furthermore, the liquid film can offer protection from heat losses and/or quenching that a vaporizing spray does not. With the liquid film on the solid surface, the wall temperature will not exceed the boiling point of the liquid, which is substantially lower than current wall cooling technology can achieve, and, as mentioned above, this is the essential reason that model airplane engines run so much excess fuel.

When burning liquid fuels in miniature combustors, a logical choice is then to inject all or a portion of the fuel directly as a film on the solid surfaces where high heat transfer from the combustion products occurs. When non-liquid fuels are involved in miniature combustors, it can still be sensible to augment the combustion with liquid fuel that is filmed on internal solid surfaces. It can also be sensible to use inert liquids for filming in some cases. In order to enhance the likelihood of combusting in a confined chamber at high flow rates, this new technology can be combined with other well-known technologies, such as the use of swirl generators and vortex generators. Based on the above discussion and some preliminary experiments, the miniature liquid fuel film combustor concept was patented [21] by the University of California in 2005. This section describes the foundation theory of that patent and the realization of fuel film combustors that have followed.

Cylindrical liquid fuel film combustors provide a convenient geometry for both theory and practice. The main idea is to inject the liquid fuel into the combustion chamber in a way that allows the fuel to spread over the inside face of the chamber's wall. The film will become the fuel source of the combustion as the evaporated liquid mixes and burns with air inside the chamber. The flame should then be located between the fuel film and the exhaust gas region in the center of the combustion chamber, as shown pictorially in Figure 9.2.

Having a flame inside a small cylinder, which is the case of high surface-to-volume ratio combustors, means that the wall can get very hot as heat from the flame is transferred to the confining chamber. As mentioned earlier, however, the fuel film provides thermal protection

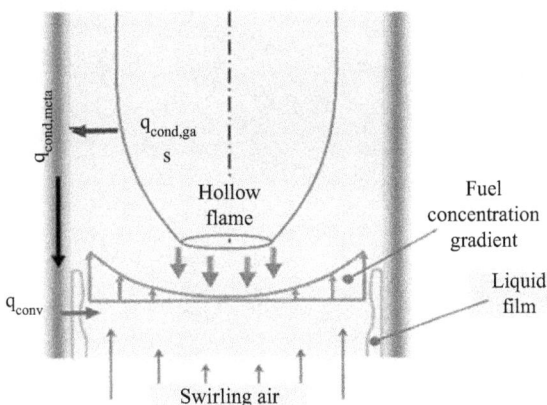

Figure 9.2. Schematic view of film combustor. Presumed flame/flow structure at the anchor point. The flame ring at the base is a triple flame point.

to the wall since the heat is absorbed by the phase change of the fuel itself. This evaporative enthalpy is then returned to the combustion process since the gaseous fuel reaction provides the heat release. Hence, the overall heat loss from the combustor walls and their temperature can be minimal. This cooling effect is one advantage of the fuel film device over droplet spray systems. Despite this potential advantage, it is important to confirm that there is sufficient surface area in the wall film to provide the fuel evaporation rate needed to sustain a chemical reaction at a level that is possible with a spray.

Figure 9.3 confirms that a liquid film covering a cylindrical chamber of diameter d can have as much surface area as an equivalent spray of a given droplet radius R, if the combustor diameter is sufficiently small and/or the pressure is sufficiently low [22].

The analysis shown graphically in Figure 9.3 is based on the following general arguments and assumptions. The global vaporization rate depends on two factors: the total liquid surface area and the local vaporization rate per unit area. Also, the combustion rate depends on the vaporization rate. Assuming that the laminar flow hypothesis is valid, we make some rough quantitative estimates for the design and operation of a miniature combustor. The ratio of air mass flow rate to the fuel mass flow rate at stoichiometric proportion is O(10) for typical liquid hydrocarbon fuels. For example, the ratio is 14.71 for C_nH_{2n} and 6.435 for methyl alcohol. We will proceed by considering stoichiometric or near-stoichiometric conditions; however, the concept and the analysis are extendable to rich or lean operations (and, in fact, we see in practice that overall fuel-rich conditions are needed). If we consider a tube of internal diameter d between 5 mm and 10 mm and an axial air velocity u_g of 1–10 m/s, the air volumetric flow rate V_g is 2.5×10^{-5} to 10^{-3} m^3/s. The density ratio of liquid fuel to air will vary from O(10^3)

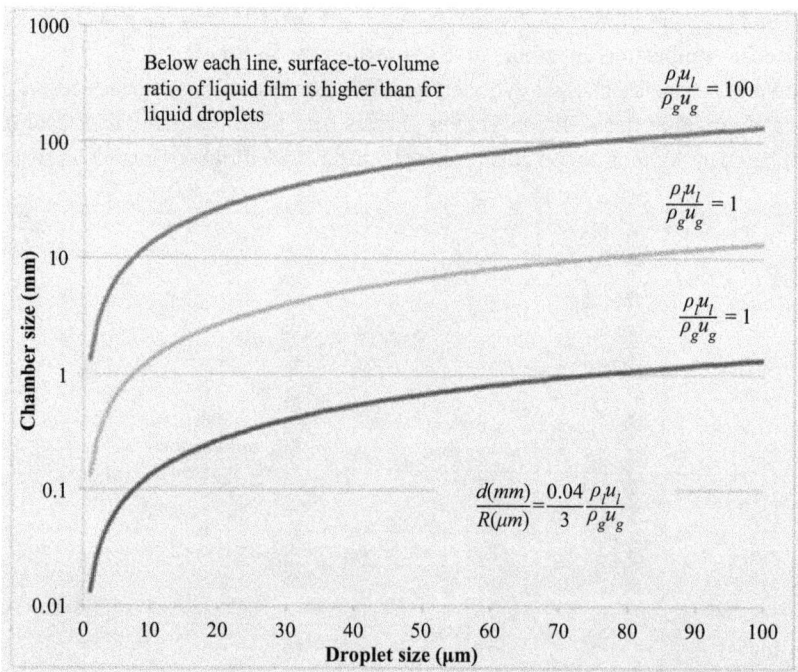

Figure 9.3. Plot showing the conditions under which a liquid film can provide higher surface area than droplets.

at atmospheric pressure to $O(10^2)$ at ten atmospheres; so with stoichiometric proportions, the volumetric flow rate ratio (air to liquid) is between $O(10^3)$ and $O(10^4)$. Therefore, the liquid volume flow rate V_l must have a value between $O(10^{-9})$ and $O(10^{-6})$ m³/s. With a liquid density ρ_l of $O(1$ g/cm³), this flow rate implies a variation from about one milligram of fuel per second (for a 5-mm-diameter combustor, one meter per second for air velocity, and one-atmosphere-pressure operation) to about one gram per second (with 10 mm diameter, 10 m/s axial air velocity, and 10 atmospheres of pressure). The power range for these fuel flow rates can be significant; chemical energy release rates with typical hydrocarbon fuels will vary between 10 and 10^4 calories per second. Even with an overall power-system efficiency of 10 percent, the power produced will be between 4 W and 4 kW for the range of parameters considered here.

The liquid axial and swirl velocity components will be primarily determined by the action of viscous shear from the gas flow. We can expect the liquid velocity u_l on average through the film to be several-fold smaller than the gas velocity. We originally estimated that the reduction factor is $O(10^{-1})$. Recent work by Gupta and co-workers using a slip velocity correlation has shown, however, that the ratio is closer [19] to $O(10^{-2})$, which ultimately leads to substantially larger film thicknesses. Despite this difference, however, the essential point remains the same, and for historical consistency we continue the order-of-magnitude analysis with the original ratio. The thickness t of the liquid film can be estimated from a continuity equation that yields $t = V_l/(\pi d u_l)$ if all of the liquid enters the chamber at the upstream end. If the liquid enters at N positions along the axial coordinate, the value calculated above should be divided by N to estimate the film thickness. This indicates that, for stoichiometric proportions, the film thickness t might vary from about a micron (at one atmosphere and the lower diameter value) to about 25 microns at ten atmospheres and the larger diameter. With the larger slip velocity determined by Vijayan and Gupta, the film thickness is an order of magnitude larger. In either case, film thickness will increase (decrease) as we go to richer (leaner) mixtures, larger (smaller) diameter, and/or higher (lower) pressures.

The Reynolds number Re_d for the air flow based on average velocity and combustor diameter d can vary between $O(10^2)$ and $O(10^5)$ for our range of parameters. Over these three orders of magnitude of difference in Re_d, only two relate to configuration change and one depends on the temperature that we use to evaluate viscosity and density. The range extends from the laminar to the turbulent regime, though there are entrance length and development issues that complicate the characterization of the flow. The Reynolds number Re_l for the liquid flow based upon its density, viscosity, and average velocity, and film thickness will vary from $O(10^{-1})$ to $O(10^2)$. Furthermore, an estimate of the gas-phase laminar sublayer thickness in the turbulent case indicates that it is greater than the estimated film thickness. So, the liquid film is expected to always be in the laminar range, although not necessarily a Stokes flow. This expectation remains true even with the increased film thickness proposed above.

The global mass vaporization rate must equal the fuel mass flow rate into the chamber for the device to operate properly. The volume of combustion chamber required for this match should not be too large if the device is to be miniature. The global vaporization rate depends on two factors: the total liquid surface area in the combustor and the local vaporization rate per unit area. For a spray, the ratio of total liquid surface area to total liquid volume is the surface-to-volume ratio for the average droplet:

$$(S/V)_{\text{drop}} = 4\pi R^2 / \left(4\pi R^3/3\right) = 3/R$$

where R is the radius of an average droplet. In typical fuel sprays, an average droplet size will be in the range of 10 μm to 100 μm. For a thin cylindrical wall film of axial length L, the exposed surface area to volume ratio is approximately

$$(S/V)_{film} = \pi dL / (\pi dLt) = 1/t$$

From the continuity relations for gas and liquid, the stoichiometric proportion for the mass flows, and the above results for surface-to-volume ratios, it follows that

$$(S/V)_{film} / (S/V)_{drop} \sim (40/3)(\rho_l u_l / \rho_g u_g) R/d$$

From this result, it is seen that the film will have as much or more surface area if the combustor diameter is sufficiently small and/or the pressure is sufficiently low. In particular, a linear relation of d versus R whose slope decreases as pressure increases gives the boundary where the two ratios are equal. The qualitative analysis outlined above is unchanged by the assumption of air-to-film velocity ratio, but, as described earlier, the quantitative analysis of film thickness depends on this ratio directly. Our early estimate was a ratio of 10:1 for air: film velocity. The more recent studies (including our own experiments described later in this chapter) suggest that a ratio of 100:1 is more likely, which corresponds to a thicker film than we projected originally. Despite this difference, the general conclusion remains that for sufficiently narrow tube diameters there can be higher surface area available for evaporation than would be possible with typical spray droplet sizes.

In addition to providing surface area for fuel evaporation, the wall film's prevention of wall heat losses maintains efficiency and reduces potential flame quenching. Estimates of the radial acceleration show that swirl velocities that are an order of magnitude less than the axial velocity can be enough to maintain the liquid on the wall. However, a stronger swirl might enhance vaporization and mixing, as well as affecting the residence time.

Different configurations of fuel film combustor are presented later, but typically fuel and air are introduced via tangential injectors located at one end of the combustion chamber, usually the bottom end considering a vertical position. Injecting the air tangentially creates a swirling flow field, which as explained in previous research [22] helps to spread the fuel and stabilize the film on the combustor surface. Although conceptually straightforward, the fuel film combustion process has been shown to be deceptively complex. For example, it has been discovered that the fuel film provides fuel-rich and fuel-lean zones useful for flame anchoring and establishment. It was further determined [23, 24] that flame stabilization requires a balance of air strain on the flame base that does not exceed the maximum flame propagation speed, a swirling flow to generate fuel concentration gradients and enhance heat transfer to the film, a liquid film that remains upstream of the flame anchor point, and a thermally conductive chamber material that permits heat transfer from the flame through the walls and to the film. These discoveries are described more fully in the following sections.

9.3.2 EARLY FILM COMBUSTOR DESIGN AND PERFORMANCE

Several promising configurations of fuel-film combustors have been developed since the concept of the fuel film was introduced as an alternative to spray combustion systems at small scales. In particular, the previous work has shown the possibility of creating combustors with diameter of one centimeter with the capability to burn liquid fuel from the inside wall of a cylindrical combustion chamber. The earliest version of the miniature liquid film combustor was presented [22] in 2001. In order to provide complete optical access (and because it was presumed that the evaporating wall film would protect the wall from excessive thermal load), the design comprised a transparent combustion chamber made of Pyrex with eight tangential fuel inlets and a mechanical vane to create swirling air flow moving axially through the tube. The volume was 6.3 cm^3 (see Figure 9.4). With the transparent combustor, we confirmed that the swirling air flow does spread the liquid film over the inside cylindrical surface. This film was produced despite the fact that due to some difficulties in controlling the liquid flow rate between the fuel tubes, only the bottom inlets were used on the device in the first proof-of-concept studies.

The original miniature film combustor design was based on the expectation that the flame would behave as a standard continuous combustor, with the flame stabilized at a recirculation zone produced just downstream of the swirler. An example image of the first successful fuel film combustor demonstration is shown in Figure 9.5. The results were promising, as the flame appears to be anchored just downstream of the swirl vanes and is nearly confined within the combustor. However, we were unable to sustain the less volatile liquid methanol fuel flame without the addition of some gaseous methane. Reducing the methane fuel flow until the flame

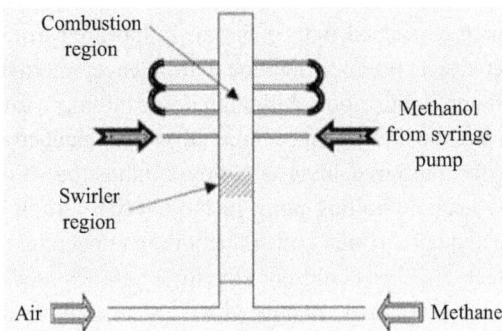

Figure 9.4. Pyrex fuel film prototype.

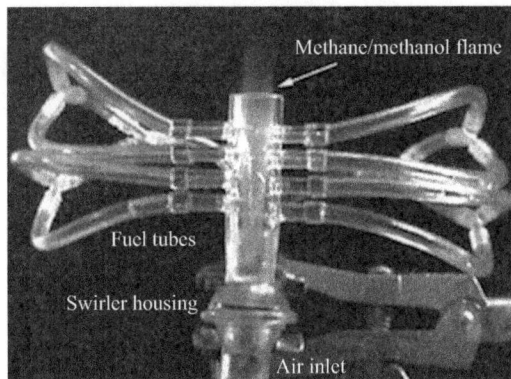

Figure 9.5. Original 1-cm diameter pyrex liquid film combustor in operation.

extinguished produced a minimum operating boundary characterized by 8.5 L/min airflow, 4.8 mg/s methanol, and 0.25 L/min methane, which corresponds to an overall equivalence ratio (ER) of 0.5. The heptane flame was later readily stabilized in the cylindrical chamber. An additional interesting discovery from these early tests was that the flame was not stabilized in the tube with a purely gaseous fuel but remained above the exit external to the chamber, burning like a Bunsen flame. Surprisingly, we found that to operate successfully with methanol the film combustor required both liquid and gaseous fuels.

The next advance in the film combustor design and operation was a change of fuel to heptane with its lower latent heat of vaporization (as compared to methanol), and the use of a metal-walled combustion chamber with tangential air injection rather than a mechanical swirler, and only two fuel injection ports. The device, shown in Figure 9.6, comprises a steel combustion chamber tubing 70 mm long and 9.5 mm in diameter with stainless steel capillary side tubes silver soldered to the main chamber. Two flats were machined to support a sapphire window for optical access.

The metal chamber more evenly heats the wall film and can better withstand the mechanical stresses associated with non-uniform heat loads. Thermocouples at three locations measured the wall temperature. The heptane fuel was fed via two syringe pumps, each providing a flow rate up to 9.7 mg/s. Air was injected tangentially below the fuel port at average bulk rates from 1.5 to 2.4 m/s, resulting in cold-flow Reynolds numbers ranging from 1070 to 1500 and an overall ER from 1.4 to 2.2. High-speed videos of the cold flow were taken using copper vapor laser illumination and air flow seeded with glycerin. It appeared from these videos that the tangentially oriented inlet air jets produce high local turbulence, improving mixing locally and possibly contributing to flame stabilization. Although it was initially assumed (based on the fuel film combustor concept) that the liquid film covered the entire chamber wall, wall temperatures suggested otherwise. In fact, infrared images during combustion showed that temperatures remained well below the heptane boiling point in the bottom portion of the chamber where the fuel and air were injected but are well above the boiling temperature near the exit. Like the quartz combustor, the steel chamber could not contain a gaseous fuel/air flame. However, in contrast with the quartz combustor, experiments with the steel chamber showed that an entirely liquid fueled flame could be sustained. These results suggested that wall heat transfer plays an important role in the miniature film combustor design. It is this dominant wall interaction that distinguishes mesoscale combustion devices from their macroscale counterparts.

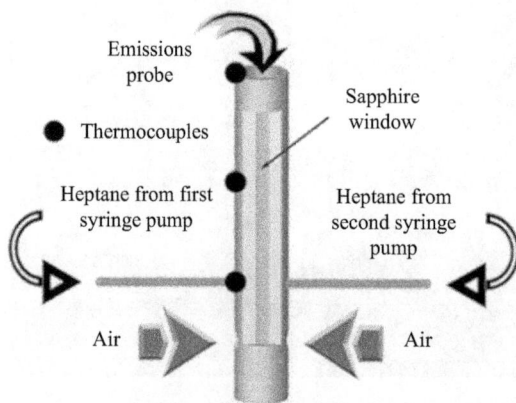

Figure 9.6. Sketch of a metal chamber.

To explore the role of chamber thermal properties, tubes of other material such as stainless steel, chromoly, aluminum, copper, and sapphire were tested with heptane fuel. Figure 9.7 shows stable operational envelopes of combustors with diverse thermal conductivities. As thermal conductivity is increased, the stable operational range is broadened because heat transfer occurs more rapidly and the vaporization rate of the liquid fuel increases.

The use of a transparent sapphire tube permitted the identification by direct observation of flame stabilization factors such as the flame shapes and film locations relative to the flame. From these observations, it became clear that one of the interesting elements of the tangential air injection film combustor regards the role of wall heat transfer and film evaporation on stable combustion. The operational map of the combustor showed clearly that combustion inside the chamber required sufficient heat flow from the flame to the combustor walls, which was then conducted down the wall to the base region where the liquid film resided. There was then a thermal gradient from the exit of the tube to the base where the film kept the walls cool. Insufficiently volatile fuel or insufficient thermal conductivity of the wall material prevented internal combustion. The second interesting element was the flame anchoring mechanism. Figures 9.2 and 9.8 show schematically the flame structure proposed. Three postulated mechanisms for flame anchoring were: (1) a recirculation zone near the base of the combustor generated by the tangential injection of air, (2) the external diffusion flame at the lip of the combustor acts as an ignition source for the internal flame, and (3) a triple flame point near the flame base is created by the rich mixture near the wall where fuel is evaporating and the lean mixture in the center of the tube where the concentration of oxygen is high. Mechanism (1) was the original intent of the film combustor and may be important in the case where mechanical swirl vanes are included. For the purely tangential air injection case, however, there does not seem to be any evidence of a recirculation zone that could account for a flame anchor point. The swirl number is significantly less than unity in the film combustor. Computations for the conditions of the experiment suggest that any pressure gradient would be too weak to produce a recirculation.

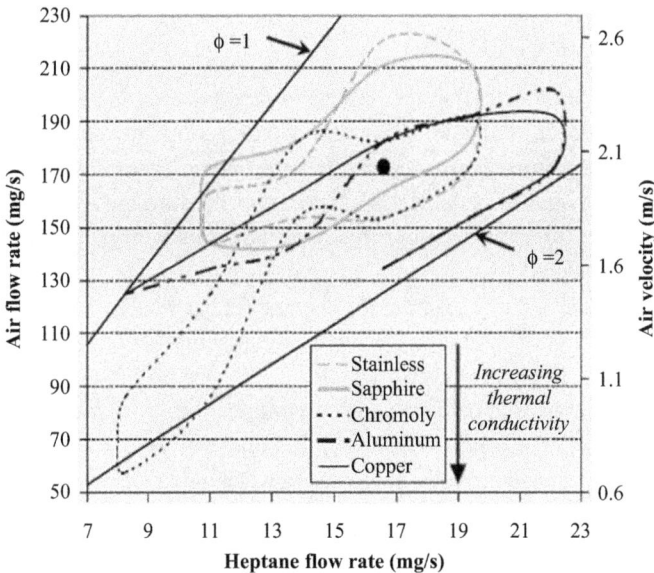

Figure 9.7. Operational envelope for five different combustor materials.

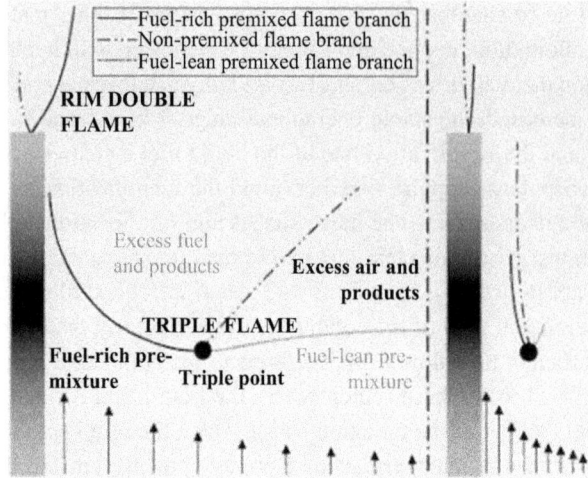

Figure 9.8. Triple flame.

Finally, we find that the axial position of the anchor point follows closely the downstream edge of the fuel film (that is, the point on the wall where the film evaporates) independently of the tangential air injection rate. If a recirculation zone were responsible, we would have expected the anchor point to remain fairly fixed in space. Mechanism (2) is definitely not responsible for the flame anchoring because we have injected nitrogen below the external lip of the combustor and extinguished the secondary flame without any effect on the internal flame being observed. This leaves the triple flame mechanism (3) as the most likely candidate for flame stabilization in the miniature film combustor. Visual observation, CARS measurements, chemiluminescence measurements, and the literature [23, 24] support this theory.

9.3.3 LINEAR ANALYSIS OF FILM COMBUSTOR

In parallel with the experiments described above, a theoretical effort was performed [25, 26] using a linearized approach to the advection/diffusion/reaction equations. The purpose of the analysis was to develop important size scaling information that can guide film combustor design. The model does not consider the application of the film but assumes that the wall is lined with a liquid fuel film. Air flowing into a cylindrical chamber with a thin film of fuel on the walls was considered. Vaporization, mass diffusion, heat diffusion, and fast oxidation kinetics were modeled. The flame position for heptane fuel at a bulk temperature of 298 K was calculated at three different Peclet numbers under the assumption of infinite chemical kinetic rate so that a flame of zero thickness results.

Steady-state axisymmetric combustion in a cylindrical chamber with the liquid fuel introduced through a wall film was analyzed [25, 26]. Laminar flow with a swirling core flow of air and diffusion of fuel vapor and products at unitary Lewis number were considered. The analysis of the scalar fields took advantage of the existence of a linear combination of the scalar properties, known as the Super Scalar, which is spatially uniform. The base flow was assumed to have constant density and to be either a fully developed flow or a plug flow. Perturbations accounted separately for swirl effects, Stefan flow, and density variations. The effects of each perturbation type on the velocity fields and scalar fields were then determined. The motion in the liquid film was coupled to the core gas motion. Then, the diffusion flame character was portrayed, vaporization rates and burning rates were determined, and the influences of radial velocity perturbations, due to swirl and Stefan flow, and of the density variations on the transport, vaporization, and burning rates were predicted.

The analysis [25, 26] yielded closed-form solutions in terms of eigenfunction expansions. Series solutions were found to certain ordinary differential equations describing radial variations in the dependent variables. Large Damköhler number solutions were calculated as functions of Reynolds number and Peclet number, showing strong dependencies. Methanol and heptane fuels were considered. Limits of high Peclet number were also examined. The results allowed for the prediction of required chamber length for given chamber diameter, inlet flow conditions, liquid fuel physico-chemical properties, initial fuel temperature, and overall mixture ratio.

As expected, the flame moved further downstream with increasing Peclet number. At $Pe = 500$, the combustion was completed at ~12.5 diameters length as indicated by the flame collapsing to a point on the axis. If relatively high mass flows and relatively short chamber lengths are desired, it is clear from these findings that the effective diffusivity must be increased through the generation of vortices or turbulence.

Temperature and mass fraction profiles for heptane are presented in Figures 9.9 and 9.10. Peak temperature occurs at the flame with fuel vapor and oxidizer (air) each existing on only one side of the flame with both mass fractions decreasing with radial distance from the flame and going to zero values at the flame. The heptane case leads to faster vaporization and shorter flame length, while a change in the initial fuel temperature has little effect. Figure 9.11 shows an important implication of the integrated vaporization rate. L_{flame} is the downstream position

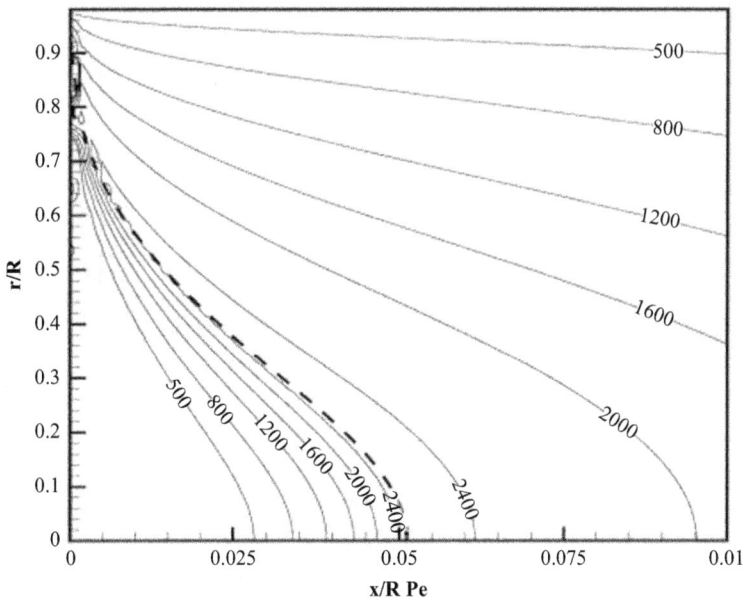

Figure 9.9. Calculated temperature contour in axisymmetric heptane/air liquid-film combustor. Air flows from left to right. Peclet number = 500. Parabolic axial-velocity profile.

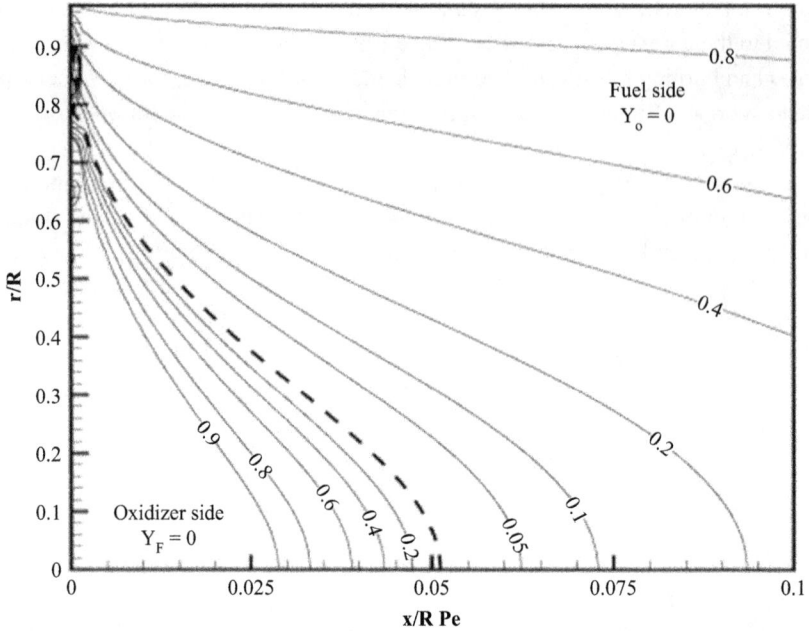

Figure 9.10. Calculated heptane fuel-vapor and oxygen mass fractions contours in axisymmetric liquid-film combustor. Air flows from left to right. Lengths are normalized by cylinder radius. Peclet number = 500. Parabolic axial-velocity profile.

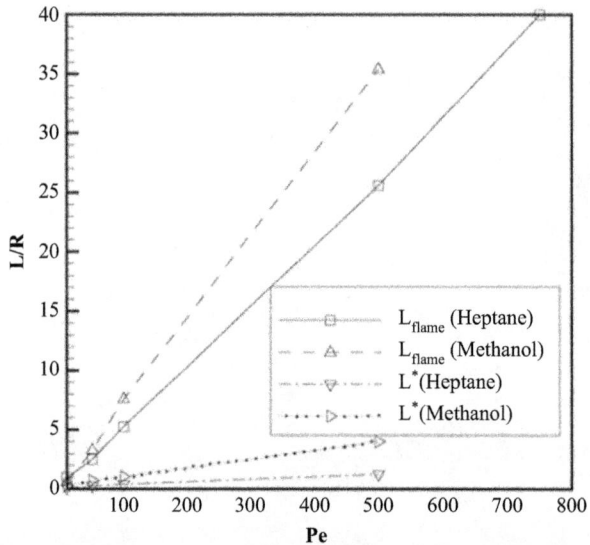

Figure 9.11. Film length for stoichiometric flux proportion and flame length versus Peclet number for methanol and heptane fuels. Plug-flow axial-velocity profile.

at which combustion is completed; that is, it is value of the x-position where the flame crosses the axis of symmetry. It is seen to increase roughly linearly with increasing Peclet number. Essentially, an increase in air velocity or a decrease in diffusivity will elongate the flame. Furthermore, a greater distance is required for the less volatile fuel, methanol, than for heptane. L^* is the fuel-film length required to vaporize at an integrated rate that matches, in stoichiometric proportion, the inflow of air. It is seen in Figure 9.11 that $L^* < L_{\text{flame}}$ for $Pe = 10$ or greater.

A careful examination of the data indicates that the dependence of L^* on Pe is sublinear; that result is qualitatively in agreement with boundary layer theory, which predicts a square-root dependence. The implication of the linear and sublinear dependencies is that a crossing of the curves will occur but, apparently, it happens for a very small Pe value. The result implies that, in order to protect the chamber wall for the full length needed to complete combustion, the fuel must be supplied in excess of stoichiometric mixtures. An alternative option not yet explored is to mix the fuel with a less volatile inert liquid (e.g., alcohol and water) to get the wall coverage for heat protection without exceeding stoichiometric proportions significantly. Alternatively, the inert liquid could be applied to the chamber walls downstream of the end of the fuel film. Also, the addition of turbulence or vorticity in the gas can create a larger value of the effective diffusivity. Qualitatively, this is similar to decreasing the value of Pe implying that the increased transport rates will shorten the required lengths for the completion of vaporization and combustion.

Generally, the analysis supports the feasibility of the liquid-film concept, which has been demonstrated in the laboratory. A solution with a standing diffusion flame and cool walls was obtained. The flame length at higher Peclet number values is many combustor radii and is also substantially longer than the film length required for vaporizing a stoichiometric amount of fuel. This situation can result in dry walls exposed to high heat fluxes. This means that, for higher mass flows, it is desirable to increase significantly the effective diffusivity (so, as to decrease the effective Peclet number and flame length) via vortex generation. The perturbations to the velocity field due to vaporization and gas expansion have significant effects on the combustion process, extending the flame length and modifying the vaporization rate as a function of downstream position.

This linearized analysis gives some indication of the important physics and of some length scales of the flow downstream of the air inlet. It cannot represent well the inflow and flame holding details. The swirling inflow would have a three-dimensional character and would typically involve jets from ports, wakes of flows through vanes, and/or recirculation zones. Computational-fluid-dynamical methods are required to give better resolution of all of the physics in the inflow and flame-holding region.

9.4 COMBUSTOR DESIGN EVALUATION

9.4.1 REVIEW OF BASIC DESIGN

Based on the extensive research with the basic tangential air injection film combustor design, the following key findings are important to reiterate—(1) chamber diameters of ~1 cm are the most reliable, (2) the combustor only operates steadily under overall fuel-rich combustion conditions, (3) wall heat transfer from the exit to the base of the combustor is needed to help vaporize the liquid fuel, and (4) an ignition/stabilization that involves a triple flame is needed near the base of the combustor. Unfortunately, these features are not desirable for the ideal miniature combustor design. Rich combustion means that some fuel burns outside the combustion chamber, reducing the system performance. The fact that wall heat transfer is needed to accelerate fuel evaporation means that part of the combustor is not covered with the wall film and is therefore exposed to higher temperatures and higher heat losses. The need for a triple flame stabilization mechanism reduces the range of operation and makes the system more sensitive to changes in fuel flow rates since we must maintain a partial premixing condition that is lean near the core and rich near the walls of the chamber. To ameliorate these disadvantageous features of the basic design, several alternative miniature film combustors were examined.

9.4.2 DOUBLE-WALL FUEL FILM COMBUSTOR

To prevent overheating of the exit section of the tube, a double-wall design was developed and realized [27, 28]. Increasing the residence time for evaporation and mixing inside the chamber and uniform-fuel-film distribution are the points of interest of this design. The characteristic of this combustor, shown in Figure 9.12, is the presence of a pre-chamber surrounding the main combustion chamber where both air and fuel are injected tangentially. Air and fuel move across the pre-chamber from the top inlets toward the bottom, premixing and vaporizing the fuel by the heat of the combustion chamber and doubling the total residence time. Two air inlets were displaced at the bottom of the chamber to increase the swirl and helping stabilize the flame. A porous surface was applied inside the pre-chamber to create capillary forces large enough to contrast the centrifugal strengths on the film and help the fuel film maintain contact with the wall. The chamber length was 75 mm, overall diameter was 2 cm, and the inside diameter of the stainless steel combustion chamber was 8 mm.

The outer surface porosity of the inner wall was needed to enhance capillary spreading of the liquid fuel. The porosity was realized as a knurled surface that was supported by a stainless steel wire mesh wrapped around the cylinder to reinforce the capillary effect. Experiments with heptane fuel proved that stable self-sustained combustion was possible within the double-wall device, but only in rich conditions (ER between 1.5 and 2). As expected at these rich conditions, the presence of a plume suggested that a large part of the combustion continued to occur

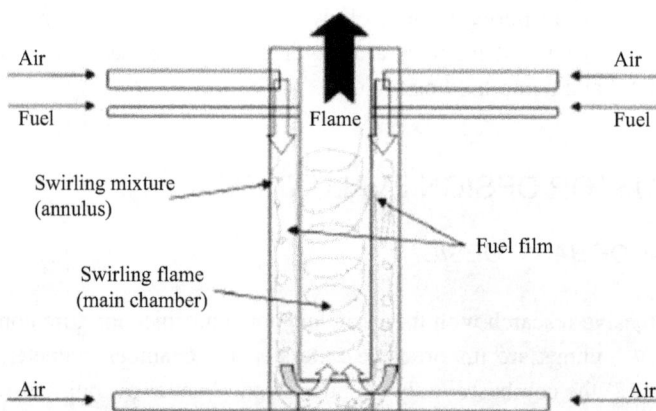

Figure 9.12. Diagram of the double-wall combustor design.

outside the chamber. The plume's length increased with the ER. Hence, the double-wall design had only limited combustion performance improvement over the single-wall system, though the outer wall was kept cool by fuel evaporation as desired. As mentioned earlier, a similar double-wall design with a more complex fuel injection system and porous plug stabilization [15].

9.4.3 SECONDARY AIR INJECTION

As is documented many times [23–30], the first generation of miniature fuel film combustors had significant reaction and heat release taking place outside the combustion chamber. For this reason, we consider a two-part flame, one that exists in an inside domain, the combustion chamber, and one burning in an outside domain, which is effectively a plume. The volume of the outside flame is basically proportional to the ER and measurements show that the length of the plume outside varies between 2 cm to more than 10 cm [29] for a fuel film combustor of 1 cm diameter and 8 cm in height running at ER2; the higher the ER the longer the plume. This behavior indicates a lack of oxygen in the chamber that forces chemical reactions outside where there is more fresh oxygen. This two-stage combustion concept is used intentionally in some large-scale systems, such as in the RQL gas turbine [31], where secondary air injection had the function to minimize the NO_x.

Previous film combustors commonly had, at the exit, a higher axial velocity component compared to the tangential component. In fact, the air inlets are usually placed at the bottom of the chamber to enhance mixing of air and fuel. The effect of using this position is that the swirl motion rapidly decreases as the flow moves toward the exit of the chamber. With these considerations, a combustor was designed to increase the swirl in the upper section of the chamber together with increasing mixing and residence time. A first prototype was built based on the original combustor described earlier by adding a top section to the chamber with four tangential air inlets, as shown in Figure 9.13.

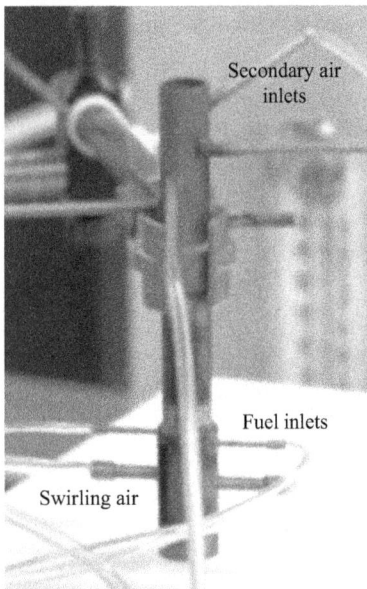

Figure 9.13. Fuel film combustor prototype with secondary air injection in the top four inlet tubes.

The top chamber has a constant diameter of 9.8 mm with a total length of almost 11 cm subdivided in three main sections, bottom, middle, and top of respectively 25.7 mm, 40 mm, and 44 mm. The diameter of the air inlets is estimated by requiring a double total airflow of a burning condition characterized by the long plume at ER2 where air was injected only through the base inlets. The airflow increase is nearly 8 L/min and splitting the airflow into four injectors gives a volumetric flow of 3.3×10^{-5} cubic meters per second. Considering a flow velocity range between 5 m/s and 10 m/s, the inside air injector's radius is assumed to be 1.1 mm. Each air inlet is silver soldered to the main chamber and both chamber and inlets are made of stainless steel AISI type 304, which has a thermal conductivity value of 16 W/mK.

Early burning tests with the combustor revealed weaknesses; so, a tougher, heat-resistant design was developed. This newly designed chamber included a different middle section with four pieces of different lengths that could be used singly or in combination. In addition, a new top section was added that, like the previous combustor, contained four tangential air injectors. Each secondary air injector had a diameter of 2.15 mm and they were threaded to the main chamber. The bottom section of the chamber was the same as was used in the previous combustor in order to ensure the same flame anchoring behavior. The base had two air inlets and two fuel inlets with an inside diameter of respectively 2.7 and 1 millimeters. Threaded rings were designed to connect the bottom section to the middle and top sections, all made of stainless steel type 316. In this configuration, the length of the chamber from the fuel injectors to the exit tip was 65 mm and the overall length, including the bottom mixing section, was 88.9 mm, with a constant inside diameter of 9.7 mm. To explore the effect of pressure confinement on stability and generic combustion behavior in the secondary air injection combustor, a set of 10 nozzles, 6 with conical and 4 with hemispherical profiles, were included in the design. The results with the nozzles are too preliminary to include at this stage, but they are the basis for our upcoming future work. A transparent sapphire tube was occasionally used as the middle section to allow the visualization of the flame anchor point and combustion behaviors during the regime operation. This version of the combustor allowed rapid changes in configuration and a better resistance to high temperature than previous designs.

Tests [32, 33] with the secondary air injection film combustor demonstrated robust performance over a wide range of conditions. Figure 9.14, for example, shows the film combustor

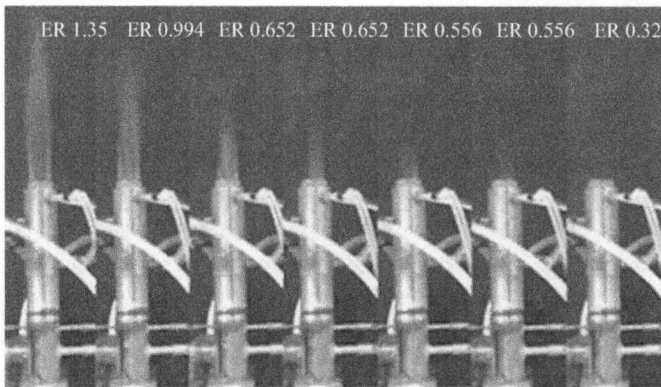

Figure 9.14. Flame confinement as secondary air injection is increased. ER refers to the overall equivalence ratio, including the secondary air injected.

operating on heptane as the secondary air injection is increased to produce flame confinement. Figure 9.15 shows a detail of the flame anchor point through the sapphire window in the base, which demonstrates that the secondary air injection does not affect the flame anchoring mechanism of the film combustor. Figure 9.16 is the overall operational map for the film combustor with secondary air injection. It demonstrates the considerable improvement provided by air injection in terms of the operational potential of the film combustor. Note that the operating range increases with the addition of secondary air injection. Under the conditions tested, these combustors were able to operate stably at thermal power output of between 380 W and 660 W.

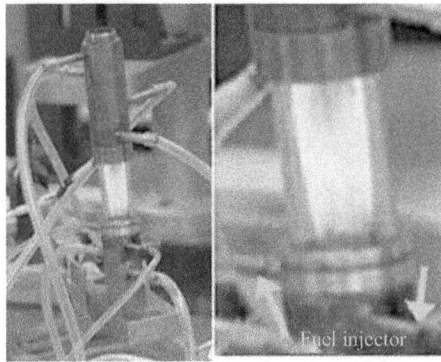

Figure 9.15. Anchoring of flame with secondary injection.

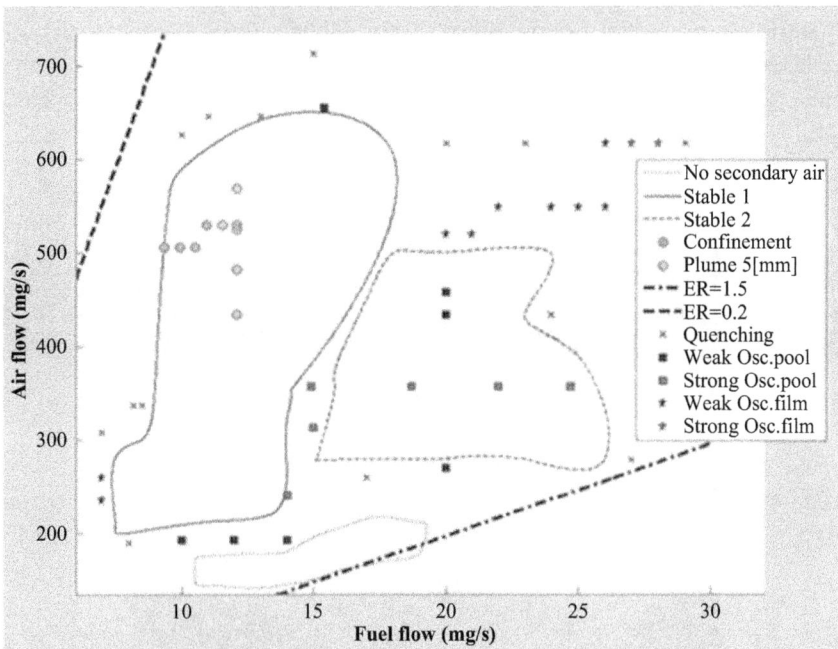

Figure 9.16. Stable operating map and the type of processes for a range of operating conditions.

9.4.4   SUB-CENTIMETER MINIATURIZATION

Another recent development in film combustor performance concerns the dimension of the combustion chamber. All prior studies used combustors on the order of 1 cm in diameter, and a very important concept underlined in the use of these combustors has been the stabilization of the flame. The overall length and diameter of the combustor can affect the capacity to completely burn the fuel inside the chamber by changing the fuel film coverage and thickness. Furthermore, it is clear that stability, oscillations and the overall equilibrium between heat exchange, fuel vaporization, kinetic reactions, and flame holding are fundamental for the correct functioning of the combustor. According to the results obtained with the secondary air injection combustor described above, we assumed that any reduced size film combustor would also need secondary air injection, as shown in the left part of Figure 9.17.

As investigated by Sirignano and co-workers [22, 30] plausible diameters of liquid fuel film combustors go from a few millimeters to one centimeter. Thus, a 5 mm diameter was selected for testing. Expectation on air velocity is that with further size reduction the velocity range should be reduced proportionally in order to allow fluid dynamic stabilization and anchoring of the flame. Due to a volume reduction of the chamber down to 1.73 cm^3 a reduction of the air flow rate is obtained by decreasing the number of secondary injectors from four to two, while two other injectors, whose diameter is 2.2 mm, were located at the base of the combustor.

To prevent overheating, a cooling system was set up that consisted of four plastic tubes with a pliable metallic covering to allow adjustable orientation. The cooling air supply was separate from combustion-air supplies so that they could be controlled separately in pressure and flow rate. To retain simplicity and minimize cost, the combustor was a one-piece chamber with no possibility of changing configurations or length. Specifically, this prototype had a constant inside diameter of 5.3 mm, wall thickness of 2.2 mm; length was 78 mm, with a total of four air inlets and two fuel inlets of 1 mm inside diameter. All injectors were tangentially displaced and threaded onto the main chamber to prevent meltdown during the regime operation at high temperature. The material was stainless steel AISI type 316 whose conductivity is the same as that of AISI type 304.

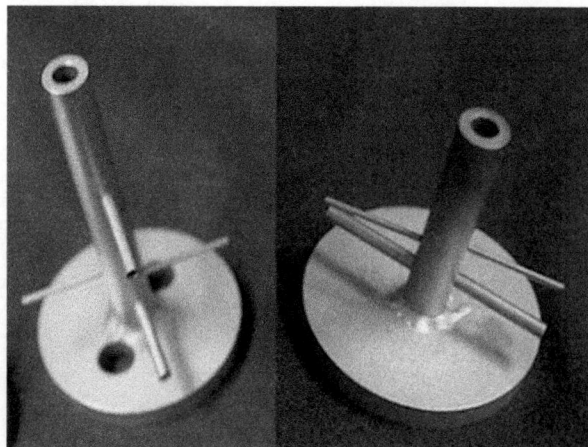

Figure 9.17. Sub-centimeter prototype with secondary air injection (left) and without (right).

With this combustor, as in the previous efforts, the first combustion test revealed some difficulties in burning and igniting the fuel, and the flame was barely able to anchor inside the chamber. In response to these difficulties, a second prototype of the chamber (see right part of Figure 9.17) was constructed. This second version of the sub-centimeter combustor has the same diameter and bottom air and fuel injectors as the first. The difference was in the top part of the chamber, since the tube was 3 cm shorter than the previous one and there was no secondary air injection. The total length was 4.8 cm and the chamber volume was 1.06 cm^3. This combustor was able to operate stably but only under rich conditions with a large external plume, as shown in Figure 9.18. The maximum output of this sub-centimeter combustor was ~170 W thermal.

9.4.5 SWIRL VANE STABILIZATION

A third combustor prototype was designed to work by axially introducing the bottom air flow instead of using tangential injectors. The new base section of the chamber was open and a plastic swirl vane section (see Figure 9.19) created the swirl that was previously generated by two tangential air inlets. The middle and top sections were the same used to explore the secondary air injection combustor so that the only difference between the two resided in the base section. The goal was to explore how swirl vanes act as a flame holder and to determine the consequences on the anchor mechanism and combustion stability when the secondary air

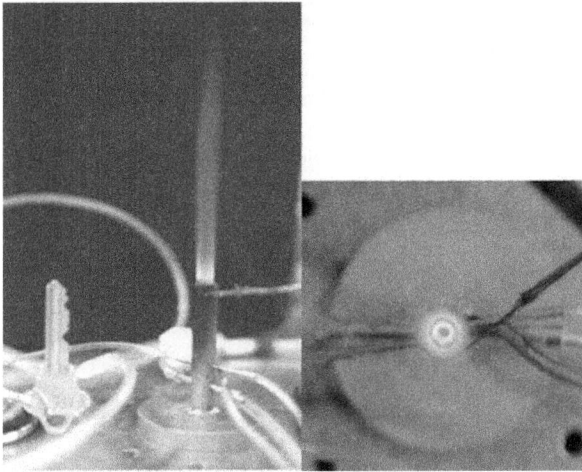

Figure 9.18. Sub-centimeter film combustor in operation with heptane. Note the external plume.

Figure 9.19. Bottom section, swirl-vanes, and tangential fuel injectors.

injection was applied. The swirl vane was attached to the stainless steel support, where the fuel injectors were located, with thermally resistant glue. The chamber length varied from 65 mm to 105 mm; internal diameter was 9.7 mm and inside volume was 5.5 cm^3. The total number of air tangential injectors was four, with two tangential fuel injectors. The assembled combustor is shown in Figure 9.20, and the combustor operating is shown in Figure 9.21. Somewhat surprisingly, the results of these experiments demonstrated that the swirl vanes were inferior to the tangential injection method for the film combustor [33]. We assumed that the inferior performance reflected an inferior swirl vane design and so the next step explored a range of swirlers.

Variations [34] of the new prototype based on the combustor described above were made using two different materials, stainless steel, and aluminum, at three different chamber lengths (from 25 mm to 75 mm). The highlights are repeated in the following. As before, this provided several combinations to help identify the features leading to the best combustion stability

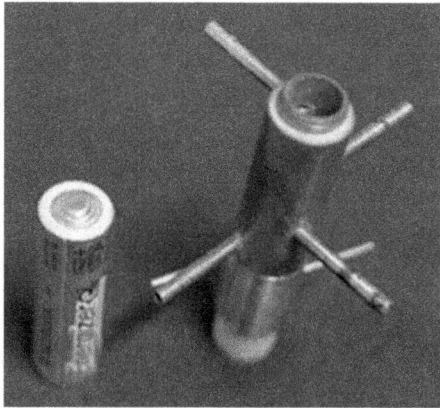

Figure 9.20. Assembled combustor with secondary air inlets.

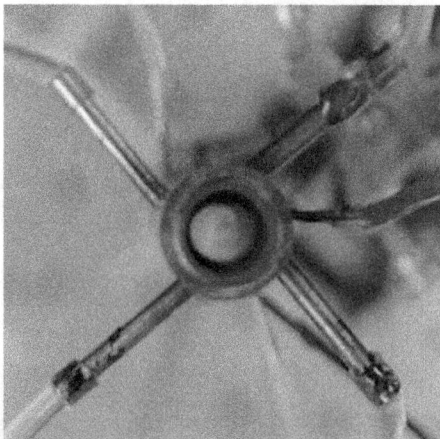

Figure 9.21. Top view of swirl vane design in operation on heptane. Light region at the center is the flame. Performance was satisfactory but not superior to tangential injection design.

and flame properties. Also, as before, we found that the temperature variation using stainless steel was drastically higher than when using aluminum (see Figure 9.22) and the operational stability for the stainless steel was correspondingly poor. Hence, we concentrated on the aluminum chambers, all with a constant inside diameter of 9.7 mm. In contrast with previous works [32, 33], however, the middle part design, which connects the swirler with the combustion chamber, was changed, to allow the easy introduction of different swirl vane designs. In this way, a top view allows a single uniform cylinder surface inside the combustor, simplifying the geometry and allowing a study of different swirl design effects. The swirl vanes were built from *Zcorp 131* powder, a non-transparent material with fairly high superficial roughness. The material is non-toxic, however, and relatively easy to work with in the 3D printing facility. Four different swirl vanes were designed, using *SolidWorks*, varying the geometry and the fluid dynamic features. Figure 9.23 shows the variety of swirl vanes developed and Figure 9.24

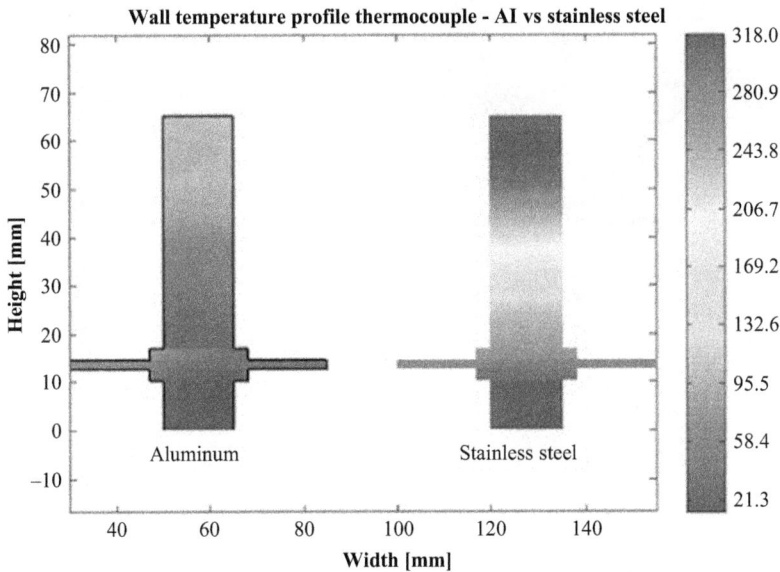

Figure 9.22. Surface temperature variation between aluminum and stainless steel film combustors. Temperatures were determined by thermocouple and confirmed using an infrared camera.

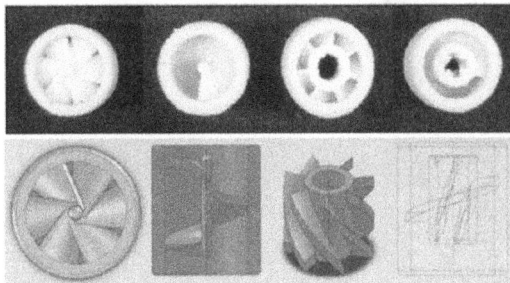

Figure 9.23. Drawings of swirl vane designs tested and the rapid prototype outcomes when 3D printed. All vanes have 1 cm diameter.

shows the various vanes and the different chambers used. Figure 9.25 shows how those components fit into the overall film combustor. During burning tests operating limits and combustion stability were analyzed, looking for a stable condition able to burn steadily for a long period (typical tests ran for at least 10 min to demonstrate a stable operation).

Figure 9.26 compares the burning behavior using three different swirl vane designs [34]. The results are surprisingly similar considering the substantial differences in swirler shape. A top view of the combustion process shows the characteristic tubular flame in the core (Figure 9.27), suggesting that the stabilization mechanism is still the triple flame described for the tangential air injection designs.

Figure 9.24. Sample chambers, swirl vanes, and fuel inlet housing for the variety of film combustors tested. The two chambers on the left are stainless steel and the one to the right is aluminum.

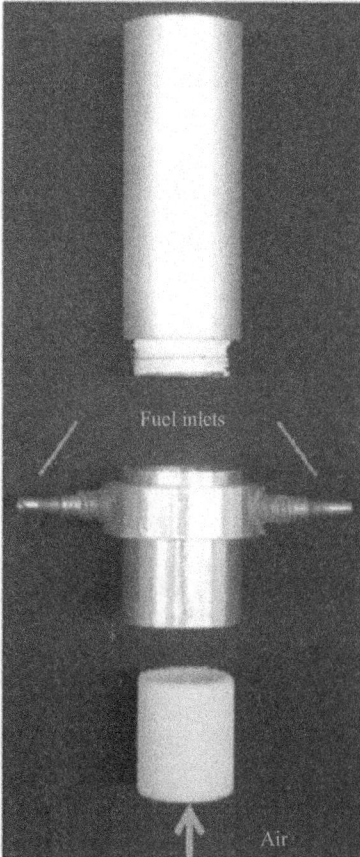

Figure 9.25. Exploded assembly showing the swirl vane at the base, the fuel inlet section above it, and the aluminum combustion chamber threaded in above the fuel inlets.

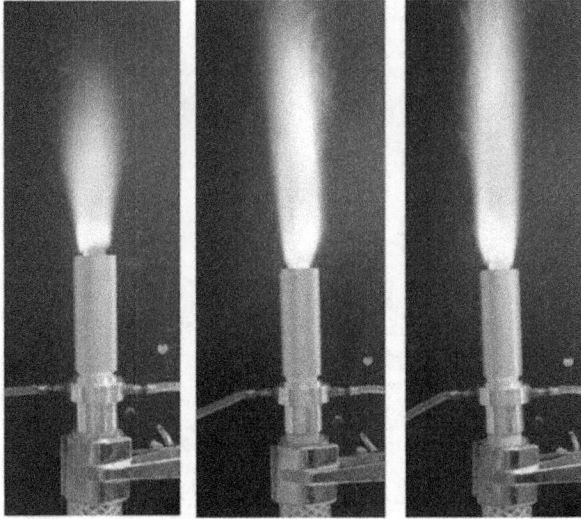

Figure 9.26. Aluminum film combustor operating with three different swirl vanes (the first three left-right from Figure 9.23).

Figure 9.27. Top view of film combustion process with swirl vanes. Flame shows the distinct features of a tubular flame.

9.4.6 CENTRAL POROUS MEDIA FILM COMBUSTOR

The continued desire for wide operating range of ERs (particularly lean conditions) and complete flame confinement led to a related design with a central porous media feeding the fuel film. Utilizing a metal-porous medium is an effective method to increase the contact surface area and conduction heat transfer for liquid fuel vaporization and flame stabilization. The major purpose of designing the combustor with a porous media is that liquid fuel can be spread over a chosen surface through the porous materials on account of its own momentum and surface tension, as well as friction forces from neighboring flowing gases, and gravity. Figure 9.28

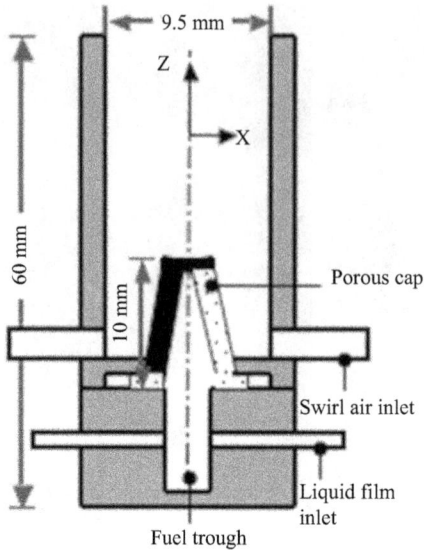

Figure 9.28. Central porous media combustor design.

shows the proposed fuel film combustor [29, 35–39] with a porous liquid fuel inlet, which has the shape of a cap.

Swirling air enters the chamber, draws the liquid fuel out, generates a liquid fuel film on the porous cap, and helps to create a recirculation mechanism (shown in Figure 9.29) that stabilizes the flame. As in the standard film combustor, a tri-brachial flame structure is proposed for the combustor, with a diffusion flame between the chamber wall and the porous cap, a rich premixed flame leaving the cap, and a lean premixed flame feeding the diffusion flame. The porous cap works like a bluff body and creates a recirculation zone behind it, characterized by a low-velocity flow field where the flame anchors. Three bead size of the bronze porous media were tested and the results indicated that the flame structure and anchor position do not change much with the air flow rate nor with the bead size.

Further studies with the porous media combustor have focused on thermo photo voltaic (TPV) applications, where the flame heats a radiant wall (the combustor itself) and surrounding photovoltaic (PV) cells convert the radiation directly to electricity. A recirculation system that carries the hot exhaust gas past the outside of the combustor wall improves the uniformity of the photothermal flux and improves the overall efficiency of the power system [35, 36]. The final section of this chapter describes the power system in some detail to show how balance-of-plant issues can be an important part of any combustion-driven portable power system. Air handling and fuel storage and delivery are common requirements for all systems. In addition, for traditional electrical power generation, the balance-of-plant would include compressors, turbines, and generators. For motive power, there would be thermochemical to mechanical energy conversion elements that might include shafts, gears, and the like, all of which add to system parasitic losses and complexity. Even without additional moving components, miniature power system design includes much more than miniature combustor design. The following section provides a specific example of how these elements can interact.

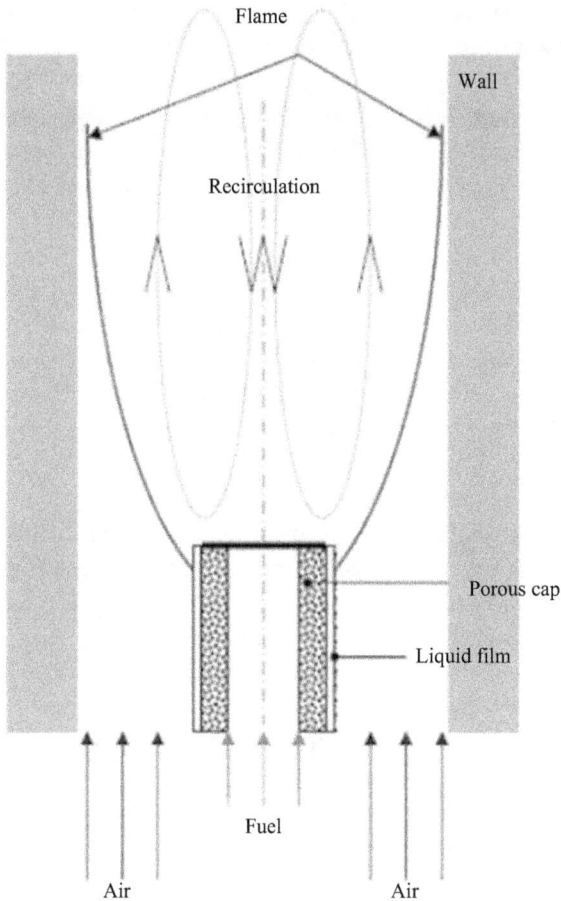

Figure 9.29. Sketch of the proposed porous plug combustor recirculation behavior.

9.5. LIQUID FILM COMBUSTION-DRIVEN TPV POWER SYSTEM

As one relatively simple example of the fundamental and system issues raised in the above sections, this section demonstrates how the combustor is only one part of the portable power challenge; additional effort is always needed to develop an effective overall system.

9.5.1 TPV SYSTEM DESIGN

Meso-scale combustion has been considered as one of the potential solutions for the increasing demand of higher-density power supplies for miniaturized devices. Simultaneous production of electric power and heat can be achieved by means of a TPV power generator in fuel-fired residential and commercial heating appliances. Therefore, research in cogeneration and combustion in meso-scale TPV systems [36, 37, 40–43] has been inspired. In general, the TPV power system mainly consists of a heat source, an emitter, and a PV cell array. The characteristics of emitters and PV cells have been broadly discussed over decades [44–47]. Especially, technological improvements in selective emitter [44, 45] and low band gap PV cells [46, 47] have evoked a renewed interest in TPV generation of electricity. Most TPV heat sources are generated by chemical combustion. A fully aerated flame burns close to the emitter surface, heating the surface to incandescence. Radiant-burner design encourages a high surface temperature, while conventional gas-burner design often requires minimizing the effect of high burner wall temperature. Moreover, the radiant power of a selective thermal radiator at specific wavelengths exhibits an exceptional dependence on wall temperature. It can therefore be expected that further increasing the combustion gas temperature will considerably improve useful radiation output. It is determined not only by the design of the burner/emitters but also by combustion operating conditions. As the size of the combustor is reduced, an increasing surface-to-volume ratio for a TPV power system enlarges the specific contact surface between the flame and the combustor/emitter wall, and further enhances heat transfer to the wall to raise the temperature of the emitter.

Nevertheless, there are three well-known approaches to liquid fuel burning in a small and confined chamber. One is using electrospray to produce micro-scale droplets for quick evaporation [19], but it is restricted by confined configurations. Another is using porous media [48–50], as a porous medium contributes to recuperate heat from burned gases to evaporate the liquid fuel via thermal conduction and radiation. However, porous media inherently possess low thermal conductivity and large heat capacity depending upon the material. Consequently, it takes a long time to ignite a porous medium combustor, unless gaseous burning and external thermal assistance are necessitated in the ignition process. The third approach is applying a fuel-film concept in a small combustor, in which liquid fuels are ejected onto the chamber wall and absorb heat for vaporization instead of heat loss to the wall. However, it is not feasible in a miniature TPV power system due to the fact that the chamber wall is often used as a TPV emitter. Alternatively, the concept of the central porous medium combustor becomes an attractive replacement in TPV power systems.

One of the key factors of an efficient meso-scale liquid-fuel-burning TPV system is to properly design an effective combustor, which not only can stabilize the flame inside the chamber but also can use the chamber wall as an effective emitter of thermal radiation. The fuel-film concept is considered to be a superior approach for stable, continuous burning of liquid fuels in a confined meso-scale space. To further apply fuel-film combustion in a TPV system, a porous medium is incorporated to provide the primary fuel–film surface. A porous medium facilitates a surface area large enough to produce the necessary fuel vaporization rates, and a convenient pathway for thermal recuperation between the liquid fuel and the flame. For these reasons, the fuel–film concept can be applied to enhance the efficiency in small-scale TPV systems. As shown photographically in Figure 9.30, the combustion chamber consists of three major parts: (1) a combustion chamber of 9.5 mm inside diameter (ID), with two tangential air inlet ports and the chamber wall as an emitter or a quartz wall for flame visualization experiments, (2) a chamber seat with a fuel trough of 10 mm ID and two fuel inlet ports, and (3) a porous medium. The results are replicated from Li et al. [36]. The combustion chamber is made of an infrared thermal tube, 60 mm in length and 3 mm in thickness. The bronze porous medium has a truncated conical shape, 4 mm in diameter on top, 6.5 mm in diameter in the bottom, 10 mm long, and with a pore size of ~20 μm. The system does not involve any moving parts. It is relatively easy to fabricate and ensemble. Convenience and low-cost in fabrication, reliable in operation, and easy for maintenance are the key elements of an effective TPV system. The

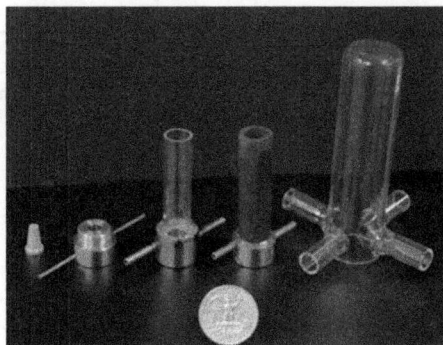

Figure 9.30. Photograph of the experimental combustor [36].

main chamber and the liquid-fuel trough are separated by a porous medium. Liquid hydrocarbon fuel is injected smoothly from two inlet ports into the fuel trough using a syringe pump, which provides continuous fuel inflow in the range of a few cubic-centimeters per hour. The air, metered by an electronic flowmeter, is injected tangentially from the air ports at the chamber seat. Swirling air enters the cylindrical chamber to mix with the fuel vapor from the fuel–film surface on the porous media for combustion. The swirl also provides a recirculation mechanism whereby the flame can be stabilized. The swirl system is used mainly to enhance fuel–air mixing and increase residence time in order to overcome the disadvantageous effects caused by the increasing surface-to-volume ratio in the miniature systems.

9.5.2   TPV SYSTEM TESTING AND EFFICIENCY DEMONSTRATION

9.5.2.1   Burner Combustion Performance

Understanding the combustion characteristics of a liquid fuel central porous combustor is of vital importance to the current design of the porous fuel–film TPV system. The porous medium and the fuel type strongly influence the combustion phenomena and operation range of the TPV combustor. The porous medium provides a large surface area and heat recuperation for liquid fuel evaporation. Theoretically, effective thermal conductivity of the porous medium filled with fluid can be expressed in terms of porosity of the porous medium and thermal conductivity of the solid and the fluid. Figure 9.31 shows the stable experimental operation envelopes for different fuel types and porosities. As shown in Figure 9.31(a), the porosity of the porous medium does not change the envelope region significantly as it is varied from 32 percent to 48 percent. The stable combustion envelope region for the small porosity combustor is slightly larger than that for large porosity ones. The reason is that the small-porosity porous medium has a greater portion of solid, which produces a larger effective thermal conductivity, so that it can transfer heat from flames more effectively. The latent heat of the liquid fuel is the other important factor affecting the porous burner behavior. Figure 9.31(b) indicates the stable operation envelopes for two different fuel types, pentane and heptane, respectively. The stable operation envelope for pentane is significantly larger than that for n-heptane. The results are replicated from Li et al. [36]. Owing to its low boiling temperature and latent heat, pentane evaporates easily

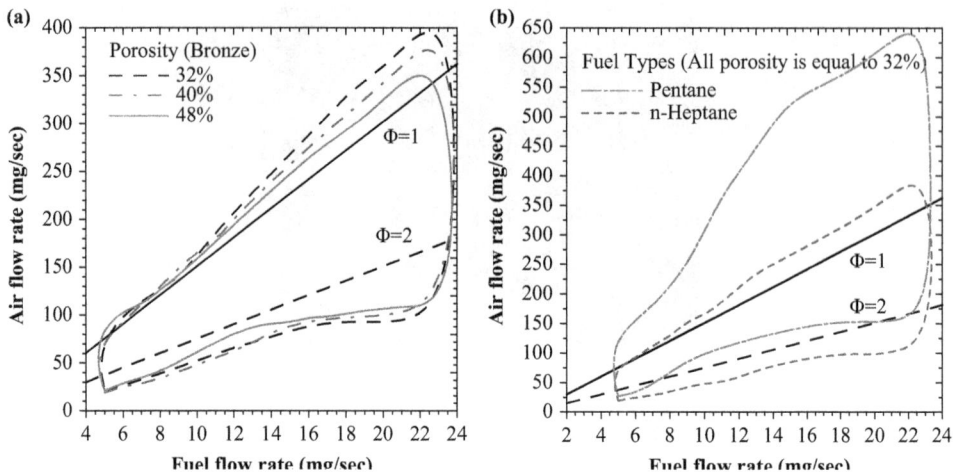

Figure 9.31. Stable operating envelopes for the combustor with (a) different porosities and (b) different fuel types [36].

so that it can operate into a much lower fuel-lean condition. Fuel adsorption into the porous medium is related to the capillary effect and surface tension. Fuel surface tension and fluid viscosity may influence the spread rate of liquid fuel-film on the porous surface. However, there is not much difference of these factors between the two fuels. Most importantly, the feeding pressure of the syringe pump dominates over these passive effects.

9.5.2.2 Emitter Radiant Characteristics

High and uniform illumination of an emitter is essential for a high radiant efficiency. Radiant efficiency of the emitter is strongly related to surface temperature of the emitter and flame-wall (emitter) interaction in the combustion chamber. The miniature combustor has inherent drawbacks for short residence time and potentially incomplete combustion. Incomplete combustion leads to insufficient heat release. An even more detrimental effect is that heat cannot effectively transfer to the emitter during short residence, producing an uneven and faded illumination of the emitter, as shown in Figure 9.32(a). The bright illumination region congregates at the bottom of the emitter and exposes the position of flame attachment. It appears that the illumination image has a spiral pattern along the emitter caused by strong swirling air flow, and a long flame plume near the chamber exit. To ameliorate this situation, a reverse tube is implemented to redirect the hot product gas to reheat the emitter. Figure 9.32(b) indicates the intensity and uniformity of illumination on the emitter after this improvement. In addition, flame burns completely inside the chamber. The results are replicated from Li et al. [36].

Figure 9.33 shows the temperature distribution along the emitter measured by an infrared thermometer for the same fuel input (11 mg/s) and varied ERs of n-heptane and air. The results are replicated from Li et al. [36]. The emitter has a high and uniform temperature distribution when the ER approaches the stoichiometric condition. Burning in the fuel-rich condition may be necessary to provide sufficient fuel evaporation, but this would sacrifice the opportunity of heating the emitter leading to a reduced emitter temperature. However, evaporation heat requirement is related to fuel characteristics. Figure 9.34 indicates the temperature measurement in a

(a)   (b)

Figure 9.32. Photographs of combustion chamber operation (a) without a reverse tube and (b) with a reverse tube [36].

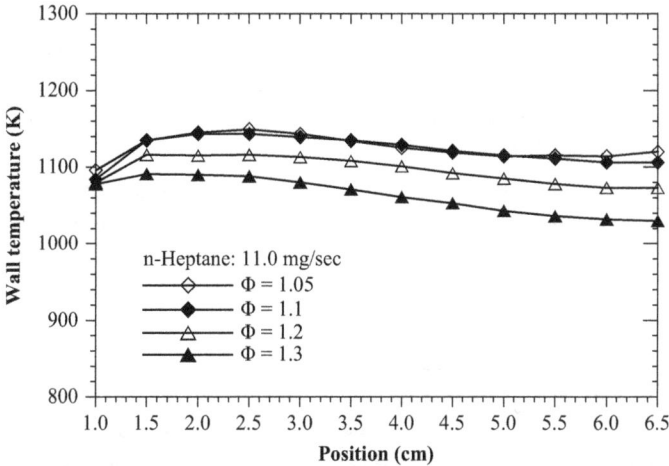

Figure 9.33. Measured wall temperature distribution along the axial direction for *n*-heptane fuel of 11 mg/s and varied equivalence ratio [36].

flow condition of air at 165 mg/s and fuel at 11.0 mg/s for pentane and *n*-heptane. The results are replicated from Li et al. [36]. No obvious differences, except for the no-reverse-tube cases, can be seen for different liquid hydrocarbon fuels. Pentane can evaporate prematurely due to its low boiling temperature, so that the flame attachment position locates further upstream than that of *n*-heptane. The temperature distribution along the surface of an emitter without a reverse tube has a steep decrease, while that with a reverse tube has only slight temperature variation. It appears that a reverse tube promotes illumination and uniformity of the emitter. Another option

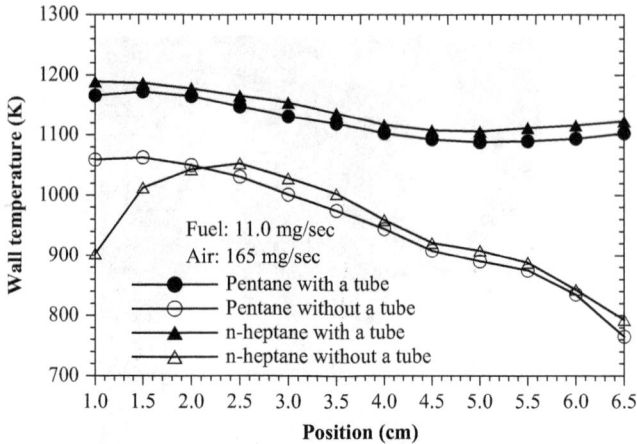

Figure 9.34. Measured wall temperature distribution along the axial direction for pentane and *n*-heptane with air flow rate of 165 mg/s and fuel flowrate of 11 mg/s [36].

might be to increase the number of tangential air inlet to enhance swirling intensity that can further improve the uniformity of illumination.

In order to monitor the chemical efficiency of the present combustion system, measurements are conducted using the gas analyzer to obtain CO_2, CO, O_2, and unburned hydrocarbons (UHC). Since no other carbon containing species except CO_2, CO, and UHC are observed in these conditions, the chemical efficiency of methane combustion is defined as the ratio of the measured volume percentage between carbon dioxide and all carbon-containing species. Figure 9.35 displays the CO and NO_x emissions for different thermal inputs of *n*-heptane with a fixed ER of 1.1. The results are replicated from Li et al. [36]. In the cases without a reverse tube, NO_x emission does not change significantly as the thermal input is increased, while CO emission rises with a fuel flow rate; note that CO emission is in the logarithmic scale in the figure. As regards chemical efficiency of the non-reverse-tube cases, all conditions tested are able to reach up to 95 percent. For the cases with a reverse tube, CO emission is relatively high for low thermal inputs and decreases sharply with the increase of thermal input to less than 100 ppm, and the corresponding chemical efficiency varies from 82 to 99 percent. Increasing thermal input enhances heat release and flame length and leads to a higher emitter temperature when a reverse tube is applied to enhance residence time and to reheat the emitter. UHC and CO passing through the emitter can react again since the emitter has high surface temperature and thermal radiation. Relatively speaking, NO_x emission increases only slightly since chemical efficiency also increases. The reverse tube reinforces thermal recuperation and prolongs residence time. It does not only enhance illumination of an emitter, but also promote chemical efficiency.

9.5.2.3 Systematic Efficiency Demonstration

In order to evaluate the feasibility of using the proposed miniature central porous combustor in the TPV system, GaSb PV cells are employed to collect emitter illumination emission and convert it into electricity. The electrical power output of the prototype meso-scale liquid fuel TPV

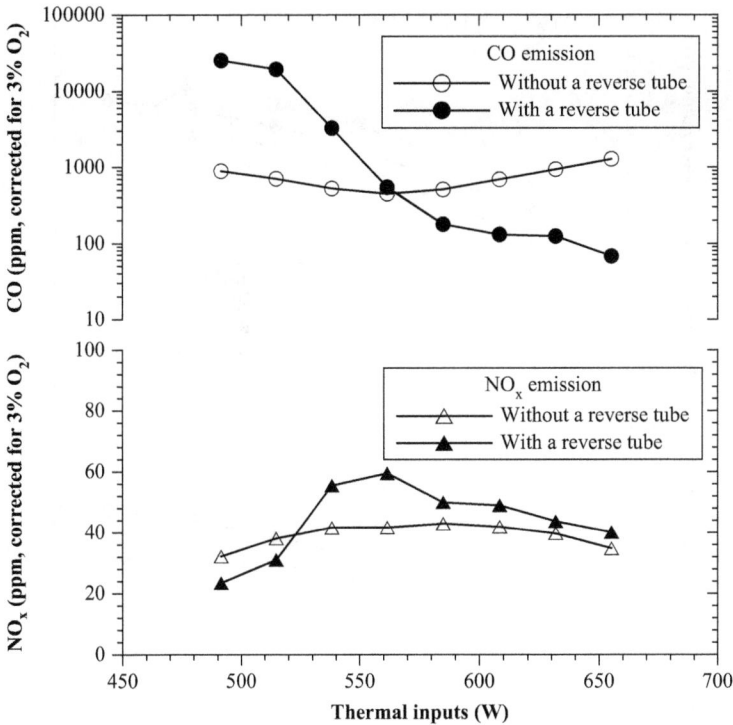

Figure 9.35. Effect of thermal input on CO and NO$_x$ emissions [36].

system incorporating GaSb cell modules is then measured for various ERs and different fuel types. The prototype TPV system consists of the central-porous medium combustor surrounded by four pieces of GaSb PV cell modules. Figure 9.36 shows the maximum electrical power output. For a specific fuel type, maximum electrical power output is found to depend upon thermal input and ER, as shown in Figure 9.36(a). The decrease of power output for the highest thermal input for the stoichiometric condition in Figure 9.36(a) may attribute to the variation of the flame stabilization and combustion characteristics. Since high thermal input of a fixed ER requires a high air flow rate, the flame attachment point moves to a downstream location that changes combustion characteristics and emitter illumination and ends up decreasing electrical power output. Figure 9.36(b) exhibits the electrical power output for two fuel types with a fixed ER. The results are replicated from Li et al. [36]. The TPV system burning *n*-heptane obtains higher power output than that burning pentane. The two fuels have similar energy density and so the difference in electrical power output is primarily due to the particular combustion mode. With a reverse tube, combustion modes inside the chamber are complicated and coupled with fuel parameters, such as latent heat, viscosity, and fuel density. Further observation on flow dynamics inside the tube will help to determine the relationship between fuel type and power output.

Based upon the power output shown in Figure 9.36, the average overall efficiency from fuel to electricity for all conditions tested is estimated to be ~1 percent. For the case of *n*-heptane with the flow rate at 12 mg/s (thermal input 561.6 W) and an ER of 1.0, an electrical power output of 8.3 W has been achieved for the small fuel-film TPV system, corresponding to an overall efficiency of

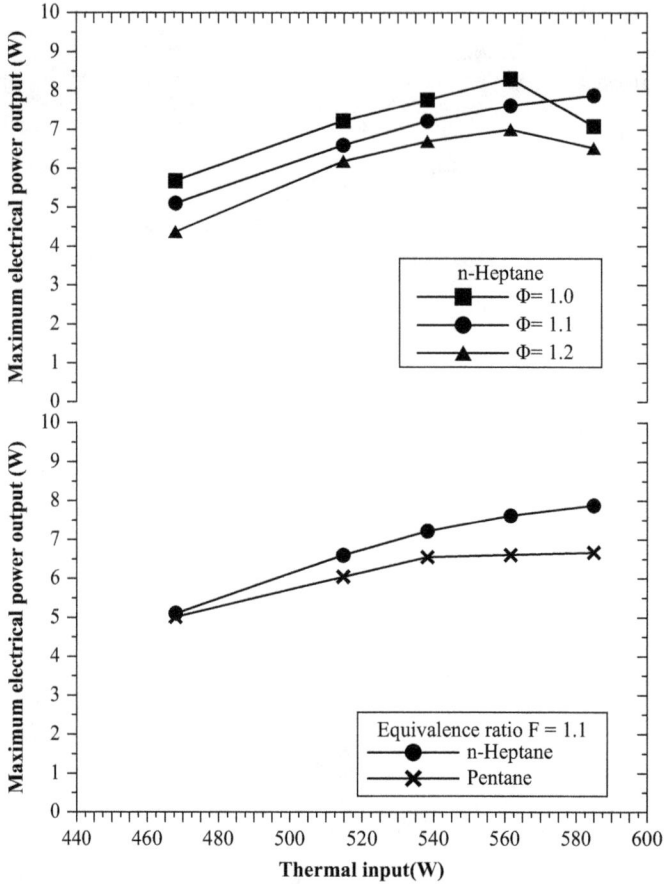

Figure 9.36. The maximum electrical power output under (a) different equivalence ratios of *n*-heptane and (b) different fuel types [36].

1.47 percent, which is twice of that reported in the literature [51]. The open-circuit voltage and short-circuit current are 1.81 V and 6.6 A, respectively. The fill factor reaches 0.695. The high efficiency region of the GaSb PV cell spans between 600 nm and 1700 nm in wavelength, but the output of the present emitter ranges between 1300 nm and 2100 nm. The non-matching photons are not converted to electricity by PV cells, but act to increase the cell temperature that reduces the conversion efficiency of the cell. In order to further improve the efficiency of the small fuel-film TPV system, it is necessary to employ a lower band gap GaInAsSb PV cells where high quantum efficiency is achieved in the matching ranges between 800 nm and 2400 nm. Further enhancement of the efficiency can be achieved by means of utilizing a selective emitter in future designs to replace the present infrared thermal broadband emitters, as most of the emitted photons can be located in the short wavelength range with energies greater than the band gap of PV cells.

9.5.3 Potential Improvements and Performance Expectations

The above example shows clearly the level of complexity increase when a power converter is included in the system design. The example still does not include a fuel or air delivery system, though with electrical power output it is feasible to consider a simple pump and blower to provide these flows. It is clear, however, that the overall system size and performance cannot be fairly assessed without including all components.

9.6   CONCLUSION AND FUTURE WORK

The relatively simple scaling assumptions for the film combustor concept have been reviewed. The concept has been shown to be valid. With proper design, fuel film combustors are capable of providing a thermal power that ranges from 380 W to 660 W for the secondary air combustor or 170 W for the sub-centimeter version. Total weight including the combustor and fuel with syringe is around 300 g and the combustion conversion efficiency is estimated at >90 percent. Tests show further that the secondary air combustor can operate in any orientation with respect to gravity. Using liquid film evaporation from the wall as the source of combustible fuel, it is possible to hold flames in volumes of a few cubic centimeters. The roles of swirls of the liquid film on the wall and the primary and secondary flow are important in maintaining film coverage of the wall, holding the flame, and mixing sufficiently rapidly to complete combustion within the small combustion chamber. In particular, secondary air injection increases dramatically the stable operating range of the film combustor and allows complete confinement of the reaction within the combustor. Tangential injection is found to be equivalent or superior to the use of swirl vanes. The flame has a tri-brachial character due to stratification of the mixture ratio in the gas phase, and this character is found to be critical for flame stability in the absence of recirculation zones for all film combustors. The tri-brachial stabilization is somewhat problematic since it requires rich combustion. Secondary air in some form is therefore necessary for complete combustion.

An analysis of the combustor operation has been performed using a linear model. The model provides useful information, but it does not recover some of the interesting details such as the triple-flame structure. In the future, nonlinear Navier–Stokes solutions should be obtained.

Variations to the basic configuration show some promise. Double-wall chambers and porous surfaces are two interesting design choices. The central porous design has been integrated into a complete power system, producing TPV electricity. System design, as with all miniature combustion power devices, remains a major aspect of the portable power challenge.

Practical development of miniature liquid-fueled combustion devices will require attention to all components of the power system. For example, a continuous miniature combustor must be coupled with a compressor and turbine to develop shaft power.

ACKNOWLEDGMENTS

The work reviewed in this chapter, as is clear by the included references, involved a number of graduate student researchers over the years, including, Ben Strayer, Trinh Pham, Nicola Sarzi-Amade, Simone Stanchi, Roberto Mattioli, and Claudio Giani. The work was supported in part by the National Science Foundation.

REFERENCES

[1]   A.C. Fernandez-Pello, "Micro-power generation using combustion: Issues and approaches," *Proceedings of the Combustion Institute* 29, 883–899, 2002. doi: http://dx.doi.org/10.1016/S1540-7489(02)80113-4.

[2] K. Maruta, "Micro and mesoscale combustion," *Proceedings of the Combustion Institute* 33, 125–150, 2011. doi: http://dx.doi.org/10.1016/j.proci.2010.09.005.

[3] D. Dunn-Rankin, E. Leal, and D. Walther, "Personal power systems," *Progress in Energy and Combustion Science* 31, 422–465, 2006. doi: http://dx.doi.org/10.1016/j.pecs.2005.04.001.

[4] Sion Power—The Rechargeable Battery Company, "Technology Overview," http://www.sionpower.com/technology (accessed March 30, 2014).

[5] L. Schlapbach, and A. Züttel, "Hydrogen-storage materials for mobile applications," *Nature* 414, 353–358, 2011. doi: http://dx.doi.org/10.1038/35104634.

[6] A.D. Pelosi, and A. Gany, "Combustion of a solid fuel tube with contained liquid oxidizer in a hot gas atmosphere," *Combustion Science and Technology* 179, 265–280, 2007. doi: http://dx.doi.org/10.1080/00102200600809522.

[7] D.C. Kyritsis, I. Guerrero-Arias, S. Roychoudhury, and A. Gomez, "Mesoscale power generation by a catalytic combustor using electrosprayed liquid hydrocarbons," *Proceedings of the Combustion Institute* 29, 965–972, 2002. doi: http://dx.doi.org/10.1016/S1540-7489(02)80122-5.

[8] D.C. Kyritsis, B. Coriton, F. Faure, S. Roychoudhury, and A. Gomez, "Optimization of a catalytic combustor using electrosprayed liquid hydrocarbons for mesoscale power generation," *Combustion and Flame* 139, 77–89, 2004. doi: http://dx.doi.org/10.1016/j.combustflame.2004.06.010.

[9] D.C. Kyritsis, S. Roychoudhury, C.S. McEnally, L.D. Pfefferle, and A. Gomez, "Mesoscale combustion: A first step towards liquid fueled batteries," *Experimental Thermal and Fluid Science,* 28, 763–770, 2004. doi: http://dx.doi.org/10.1016/j.expthermflusci.2003.12.014.

[10] A. Gomez, J.J. Berry, S. Roychoudhury, B. Coriton, and J. Huth. "From jet fuel to electric power using a mesoscale, efficient stirling cycle," *Proceedings of the Combustion Institute* 31, 3251–3259, 2007. doi: http://dx.doi.org/10.1016/j.proci.2006.07.203.

[11] J. Papac, and D. Dunn-Rankin, "Characteristics of combustion in a miniature four-stroke engine," *Journal of Aeronautics, Astronautics and Aviation, invited paper, Series A* 38, 77–88, 2006.

[12] J. Pompa, S. Karnani, and D. Dunn-Rankin, "Performance characterization and combustion analysis of a centimeter-scale internal combustion engine," *Journal of Aeronautics, Astronautics and Aviation, Series A* 40, 205–216, 2008.

[13] C.P. Cadou, T. Sookdeo, N. Moulton, and T. Leach, "Performance scaling and measurement for hydrocarbon-fueled engines with mass less than 1 kg," *Aerospace Sciences Meeting, paper AIAA* 2002–2825, Reno, NV; 2002.

[14] S. Heatwole, A. Veeraragavan, C.P. Cadou, and S.G. Buckley, "In situ species and temperature measurements in a millimeter-scale combustor," *Nanoscale and Microscale Thermophysical Engineering* 13, 4–76, 2009. doi: http://dx.doi.org/10.1080/15567260802662455.

[15] V. Sadasivuni, and A.K. Agrawal, "A novel meso-scale combustion system for operation with liquid fuels," *Proceedings of the Combustion Institute* 32, 3155–3162, 2009. doi: http://dx.doi.org/10.1016/j.proci.2008.06.039.

[16] H. Panchasara, D. Sequera, W. Schreiber, and A.K. Agrawal, "Combustion performance of a novel injector using flow-blurring for efficient fuel atomization," *Paper G-12-Spray*, 5th US Combustion Meeting, San Diego, CA, 2007.

[17] S. Vijaykant, *Meso-scale Combustion of Liquid Fuels using Porous Inert Media*, Ph.D. dissertation, Department of Mechanical Engineering, University of Alabama, 2008.

[18] V. Shirsat, and A.K. Gupta, "Performance characteristics of methanol and kerosene fuelled meso-scale heat-recirculating combustors," *Applied Energy* 88, 5069–5082, 2011. doi: http://dx.doi.org/10.1016/j.apenergy.2011.07.019.

[19] V. Vijayan, and A.K. Gupta, "Thermal performance of a meso-scale liquid-fuel combustor," *Applied Energy* 88, 2335–2343, 2011. doi: http://dx.doi.org/10.1016/j.apenergy.2011.01.012.

[20] J. Li, J. Huang, D. Zhao, J.Y. Zhao, M. Yan, and N. Wang, "Study on diffusion combustion characteristics of liquid heptane in a small tube with and without heat recirculating," *Combustion Science and Technology*, ICDERS special issue 184, 1591–1607, 2012.

[21] W.A. Sirignano, and D. Dunn-Rankin, "Miniature liquid-fueled combustion chamber," United States Patent No. US 6,877,978 B2, April 12, 2005.

[22] W.A. Sirignano, D. Dunn-Rankin, and B.A. Strayer, "Miniature combustor with liquid-fuel film," Paper 01F-36, *Western States Section/The Combustion Institute Fall Meeting*, Salt Lake City, UT, October 15–16, 2001. doi: http://dx.doi.org/10.2514/1.14156.

[23] T.K. Pham, *Flame Structure and Stabilization in Miniature Liquid Film Combustors*, Ph.D. Dissertation, University of California, Irvine, 2006.

[24] T.K. Pham, D. Dunn-Rankin, and W.A. Sirignano, "Flame structure in small-scale liquid film combustors," *Proceedings of the Combustion Institute* 31/32, 3269–3275, 2006.

[25] W.A. Sirignano, S. Stanchi, and R. Imaoka, "Linear analysis of liquid-film combustor," *Journal of Propulsion and Power* 21, 1075–1091, 2005.

[26] W.A. Sirignano, and S. Stanchi, *"Linearized analysis of liquid film combustor,"* Proceedings of the 3rd Joint Meeting of the US Sections of the Combustion Institute,Chicago, Illinois, March 2003.

[27] N. Sarzi-Amade, Y.-H. Li, T.K. Pham, D. Dunn-Rankin, and W.A. Sirignano, *"Miniature Liquid Film Combustors with Double Chamber or Central Porous Fuel Inlets,"* Paper F31, U.S. Combustion Meeting, San Diego, March 25–28, 2007.

[28] N. Sarzi Amadè, *Double-Wall Miniature Combustor With Liquid Fuel-Film*, M.S. Thesis, University of California, Irvine, and Politecnico di Milano, 2005.

[29] Y.-H. Li, Y.-C. Chao, N. Sarzi-Amade, and D. Dunn-Rankin, "Miniature liquid film combustors: Double chamber and central porous inlet," *Experimental Thermal and Fluid Science* 32, 1118–1131, 2008. doi: http://dx.doi.org/10.1016/j.expthermflusci.2008.01.005.

[30] W.A. Sirignano, T.K. Pham, and D. Dunn-Rankin, "Miniature scale liquid-fuel film combustor," *Proceedings of the Combustion Institute* 29, 925–931, 2002. doi: http://dx.doi.org/10.1016/S1540-7489(02)80117-1.

[31] D.L. Straub, K.H. Casleton, R.E. Lewis, T.G. Sidwell, D.J. Maloney, and G.A. Richards, "Assessment of rich-burn, quick-mix, lean-burn trapped vortex combustor for stationary gas turbines," *Journal of Engineering for Gas Turbines and Power–Transactions of the ASME* 127, 36–41, 2005.

[32] R. Mattioli, T.K. Pham, and D. Dunn-Rankin, "Secondary air injection in miniature liquid fuel film combustors," *Proceedings of the Combustion Institute,* 32, 3091–3098, 2009. doi: http://dx.doi.org/10.1016/j.proci.2008.06.174.

[33] R. Mattioli, *Liquid Fuel Film Technology: Combustion Confinement, Miniaturization, and Flame Stabilization*, M. Eng. Thesis, Politecnico di Milano, 2009.

[34] C. Giani, D. Dunn-Rankin, and J. Garman, "Swirl vane design for miniature fuel film combustor," *Paper 11F-27, Fall Technical Meeting of the Western States Section/The Combustion Institute*, Riverside, CA, October 17–18, 2011.

[35] Y.-H. Li, D. Dunn-Rankin, and Y.-C. Chao, "Combustion in a meso-scale liquid-fuel-film combustor with central-porous fuel inlet," *Combustion Science and Technology* 180, 1900–1919, 2008. doi: http://dx.doi.org/10.1080/00102200802261464.

[36] Y.H. Li, Y.S. Lien, Y.C. Chao, and D. Dunn-Rankin, "Performance of a mesoscale liquid fuel-film combustion driven TPV power system," *Progress in Photovoltaics: Research and Applications* 17, 327–336, 2009. doi: http://dx.doi.org/10.1002/pip.877.

[37] Y.H. Li, H.Y. Li, D. Dunn-Rankin, and Y.C. Chao, "Enhancing thermal, electrical efficiencies of a miniature combustion-driven the rmophotovoltaic system," *Progress in Photovoltaics: Research and Applications* 17, 502–512, 2009. doi: http://dx.doi.org/10.1002/pip.905.

[38] Y.-H. Li, G.-C.Chen, Y.-C. Chao, and D. Dunn-Rankin, *"A Meso-scale Liquid-Fuel-Film Combustor with Central-Porous Fuel Injection,"* 6th Asia-Pacific Conference on Combustion (ASPACC07), Nagoya, Japan, May 20–23, 2007.

[39] Y.-H. Li, Y.-C. Chao, and D. Dunn-Rankin, "Combustion in small-scale central-porous-media liquid film combustors," *21st International Colloquium on Dynamics of Explosions and Reactive Systems*, Poitiers, France, July 23–27, 2007.

[40] W. Durish, B. Bitnar, F. von Roth, and G. Palfinger, "Small thermophotovoltaic prototype systems," *Solar Energy* 75, 11–15, 2003. doi: http://dx.doi.org/10.1016/S0038-092X(03)00232-9.

[41] K. Qiu, and A.C.S. Hayden, "Development of a silicon concentrator solar cell based TPV power system," *Energy Conversion and Management* 47, 365–376, 2006. doi: http://dx.doi.org/10.1016/j.enconman.2005.04.008.

[42] Y.H. Li, C.Y. Wu, H.Y. Li, and Y.C. Chao, "Concept and combustion characteristics of the high-luminescence flame for the rmophotovoltaic systems," *Proceedings of the Combustion Institute* 33, 3447–3454, 2011. doi: http://dx.doi.org/10.1016/j.proci.2010.06.142.

[43] Y.H. Li, T.S. Cheng, Y.S. Lien, and Y.C. Chao, "Development of a tubular flame combustor for thermophotovoltaic power systems," *Proceedings of the Combustion Institute*, 33, 3439–3445, 2011. doi: http://dx.doi.org/10.1016/j.proci.2010.05.051.

[44] L.G. Ferguson, and F.A. Dogan, "Highly efficient NiO-doped MgO-matched emitter for thermophotovoltaic energy conversion," *Materials Science and Engineering B-Solid* 83, 35–41, 2001. doi: http://dx.doi.org/10.1016/S0921-5107(00)00795-9.

[45] B. Bitnar, W. Durisch, J.-C.Mayor, H. Sigg, and H.R. Tschudi, "Characterisation of rare earth selective emitters for thermophotovoltaic applications," *Solar Energy Materials and Solar Cells* 73, 221–234, 2002. doi: http://dx.doi.org/10.1016/S0927-0248(01)00127-1.

[46] C.A. Wang, H.K. Choi, S.L. Ransom, G.W. Charache, L.R. Danielson, and D.M. DePoy, "High-quantum-efficiency 0.5 eVGaInAsSb/GaSbthermophotovoltaic devices," *Applied Physics Letters* 75, 1305–1307, 1999. doi: http://dx.doi.org/10.1063/1.124676.

[47] M.W. Wanlass, J.S. Ward, K.A. Emery, M.M. Al-Jassim, K.M. Jones, and T.J. Coutts, "GaxIn1-x As thermophotovoltaic converters," *Solar Energy Materials and Solar Cells* 41, 405–417, 1996. doi: http://dx.doi.org/10.1016/0927-0248(95)00124-7.

[48] M.A. Mujeebu, M.Z. Abdullah, M.Z. Abu Bakar, A.A. Mohamad, M.K. Abdullah, "A review of investigation on liquid fuel combustion in porous inert media," *Progress in Energy and Combustion Science* 35, 216–230, 2009. doi: http://dx.doi.org/10.1016/j.pecs.2008.11.001.

[49] M.A. Mujeebu, M.Z. Abdullah, M.Z. Abu Bakar, A.A. Mohamad, and M.K. Abdullah, "Combustion in porous media and its applications – A comprehensive survey," *Journal of Environmental Management* 90, 2297–2312, 2009. doi: http://dx.doi.org/10.1016/j.jenvman.2008.10.009.

[50] S. Jugjai, and N. Polmart, "Enhancement of evaporation and combustion of liquid fuels through porous media," *Experimental Thermal and Fluid Science* 27, 901–909, 2003. doi: http://dx.doi.org/10.1016/S0894-1777(03)00062-1.

[51] W.M. Yang, S.K. Chou, C. Shu, H. Xue, and Z.W. Li, "Development of a prototype micro-thermophotovoltaic power generator," *Journal of Physics D: Applied Physics* 37, 1017–1020, 2004. http://dx.doi.org/10.1088/0022-3727/37/7/011.

CHAPTER 10

HEAT-RECIRCULATING COMBUSTORS

Paul D. Ronney

A major challenge for any small-scale combustion device is to avoid flame extinction via heat losses. These losses are more significant at small scales than larger ones due to the larger surface area to volume ratio and thus larger heat loss to heat generation ratio as the device scale decreases. For this reason, many researchers have considered the use of heat-recirculating combustors to minimize the detrimental effects of such losses. In this chapter, a simple analysis of linear and spiral counter-current heat-recirculating combustors is conducted to identify the dimensionless scaling parameters expected to quantify the performance of such heat-recirculating combustors. The predictions of this simple analysis are compared to 3D numerical models. By adjustment of property values, it is shown that four dimensionless parameters are sufficient to characterize combustor performance at all scales: the Reynolds number, a heat loss coefficient, a Damköhler number, and a radiative transfer number. The key (and detrimental) role of streamwise heat conduction along the heat exchange dividing wall is discussed and characterized through a Biot number. There are substantial differences between the performance of linear and spiral combustors which can be explained in terms of the effects of the area exposed to heat loss to ambient and the sometimes detrimental effect of increasing heat transfer to adjacent outlet turns of the spiral exchanger. Practical aspects of the performance of heat-recirculating combustors, including the effects of the number of turns, height, wall thermal conductivity, turbulence, and catalysis, are discussed.

10.1   INTRODUCTION

It is well known [1–5] that hydrocarbon fuels store at least 50 times more energy per unit mass than the state-of-the-art batteries. As a consequence, many researchers have attempted to develop devices that convert hydrocarbon fuels to electrical power at small scales in applications where traditionally batteries are employed. Although conversion of hydrocarbons to electricity at large scales using internal combustion engines is routine, because of issues associated with heat and friction losses at small scales it has proven difficult to employ the same technologies at small scales. An alternative approach frequently studied is that of minimizing the impact of heat losses and avoiding moving parts using heat-recirculating reactors and coupling these reactors to thermoelectric devices [6] or fuel cells [7] to generate electrical power. This chapter discusses the behavior of heat-recirculating combustors with an emphasis on scaling, particularly how size affects performance as characterized by reactor temperature and extinction limits, as a function of mass flow or Reynolds number (Re). Because of fabrication limitations it is difficult to build and test geometrically similar devices of widely varying scales; instead, it is often preferable to test scaled-up (i.e., laboratory-scale) devices that are easily built and instrumented and use appropriate dimensionless parameters to predict performance at smaller scales.

In this chapter, first an approximate mixing-cup analysis of a simple linear counter-current combustion is used to identify the governing dimensionless parameters and general performance characteristics of heat-recirculating combustors. These results are then extended to more detailed analyses and computations to understand the limitation of the simplified analyses. Finally, the practical limitations on the performance of heat-recirculating combustors due to manufacturing, materials, and catalysis are discussed. Emphasis is placed on spiral counter-current combustors because, as the results will show, for practical conditions the performance (in terms of the range of mass flow rates, extinction limits, excess enthalpy, etc.) attainable from spiral combustors greatly exceeds that of combustors in linear or other geometries.

10.2 SIMPLIFIED ANALYSIS

10.2.1 LINEAR EXCHANGER

An elementary model of a linear counter-current heat-recirculating combustor (Figure 10.1) is used to identify the relevant non-dimensional parameters, using approximations enabling simple closed-form solutions to be obtained. The reactants are preheated from ambient temperature T_1 to the preheat temperature T_2 in the inlet arm of the heat exchanger. The reactants are presumed to be inert in this region. The temperature then increases from T_2 to T_3 due to the chemical reaction in the combustor, which initially will be presumed to be a well-stirred reactor (WSR). Finally, the temperature of the products of combustion decreases from T_3 to T_4 in the outlet arm of the heat exchanger due to heat transfer to the reactants. The products are presumed to be inert in this region as well. Assuming equal heat capacity (C_P) of the reactant and product streams, the energy balances for the inlet and outlet arms of the heat exchanger and the WSR are, respectively,

$$\dot{m}C_P\left(T_2 - T_1\right) = Q_T - Q_{L,i} \tag{10.1a}$$

$$\dot{m}C_P\left(T_4 - T_3\right) = Q_T - Q_{L,e} \tag{10.1b}$$

$$\dot{m}C_P\left(T_3 - T_2\right) = \dot{m}Q\left(Y_3 - Y_2\right) \tag{10.1c}$$

where Q_T is the rate of heat transfer from the product to reactants across the heat exchanger and $Q_{L,i}$ and $Q_{L,e}$ are, respectively, the rates of heat loss the reactant (inlet) and product (exhaust) sides of the heat exchanger to ambient. It is presumed that the volume of the WSR is small compared with the overall size of the heat exchanger and thus heat loss from the WSR can be neglected. Initially it is presumed that complete conversion of reactants to products occurs in the WSR, that is, that the chemical reaction time scale is much shorter than the residence time in the WSR. In this case $Y_3 = 0$ and thus the temperature jump in the WSR (ΔT_R) is given by $T_3 - T_2 = Y_\infty Q/C_P = \Delta T_C$, where ΔT_C is the adiabatic temperature rise for complete combustion

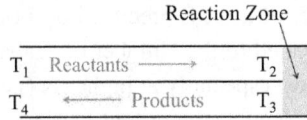

Figure 10.1. Schematic diagram
of the linear counter-current
heat-recirculating combustor.

and $Y_\infty = Y_1 = Y_2$ is the fuel mass fraction in the incoming fresh mixture. Finite-rate chemical reaction will be discussed in a subsequent section.

Equations (10.1a) and (10.1b) are the same as employed by Jones et al. [8]; these authors estimated the heat recirculation Q_T via the "mixing cup temperature" assumption, that is, the average temperatures for the purposes of heat transfer on the inlet and outlet arms are $(T_1 + T_2)/2$ and $(T_3 + T_4)/2$, respectively. Using this assumption here also and furthermore using mixing cup temperatures to estimate the terms for heat loss to ambient one obtains:

$$Q_T = U_T A_T \left(\frac{T_3 + T_4}{2} - \frac{T_1 + T_2}{2} \right) \tag{10.2a}$$

$$Q_{L,i} = U_L A_L \left(\frac{T_1 + T_2}{2} - T_1 \right) \tag{10.2b}$$

$$Q_{L,o} = U_L A_L \left(\frac{T_3 + T_4}{2} - T_1 \right) \tag{10.2c}$$

where U_T is the overall heat transfer coefficient between the product and reactant sides of the heat exchanger and U_L is the coefficient of heat loss to ambient. If the dividing wall between the inlet and outlet arms of the heat exchanger has negligible thermal resistance compared to the thermal resistance from the hot combustion products to this wall and from this wall to the cold reactants, $U_T = h_T/2$, where h_T is the usual convective heat transfer coefficient for channel flow, assumed to be the same for the inlet and outlet arms of the exchanger. For a linear exchanger the heat exchange area (A_T) and the area exposed to heat loss (A_L) will be nearly the same, though of course there are two surfaces exposed to heat loss, one on the reactant side and one on the product side, whereas there is only one surface for heat exchange between the product and reactant sides.

Equations (10.1) and (10.2) represent six equations for the unknown temperatures T_2, T_3, and T_4 and heat fluxes Q_T, $Q_{L,i}$, and $Q_{L,o}$ (T_1 and temperature rise due to combustion $\Delta T_R = \Delta T_C = T_3 - T_2$ are specified values). These equations can be solved to obtain the performance of the exchanger in terms of the excess enthalpy (E) (the temperature rise in the reactants due to heat recirculation non-dimensionalized by the temperature rise due to combustion) as a function of the Number of Transfer Units (N) and a dimensionless heat loss coefficient (α):

$$E = \frac{4N}{4 + \alpha N \left[4 + N(2 + \alpha) \right]} ; E \equiv \frac{T_2 - T_1}{\Delta T_C} ; N \equiv \frac{U_T A_T}{\dot{m} C_P} ; \alpha \equiv \frac{U_L A_L}{U_T A_T} \tag{10.3}$$

Figure 10.2a shows the effect of a on E predicted by Equation (10.3). Without heat loss ($\alpha = 0$), the excess enthalpy is related to the Number of Transfer Units by the simple relation E = N. As α increases, E decreases, especially at higher N (i.e., lower flow rate or more rapid heat transfer from products to reactants). According to this model, for any $\alpha > 0$, when N is sufficiently large, increasing N further actually yields lower E due to recycled thermal enthalpy being lost to ambient rather than increasing the enthalpy of the reactants. This partially explains why a low-velocity (large N) extinction limit always exists in heat-recirculating combustors despite the fact that N can be extremely large in some experiments [9] – as much as 10 for gas-phase combustion and 330 for catalytic combustion. Such large values of N would lead to extremely large values of E and thus very broad extinction limits in a linear device if it were truly adiabatic. On the other hand heat losses alone do not explain the low-velocity limit as will be discussed in Section 10.2.4.

Since Equation (10.3) uses N to represent different operating conditions (e.g., different flow rates and sizes), it requires knowledge of U_T and thus h_T, which in turn must be estimated from the Nusselt number $Nu \equiv h_T d / k_g$ which is not known *a priori*. Moreover, the scaling of h_T with flow rate changes between laminar and turbulent flows. For these reasons it is preferable to characterize the flow in terms of the Reynolds number $Re \equiv \rho v d / \mu$ since all of these properties are known *a priori* from the experimental conditions. Assuming fully-developed channel flow, Nu can usually be approximated by an expression of the form $Nu \sim Re^a Pr^b$, where Pr is the Prandtl number; since Pr is close to unity for most gases and usually $b < 1$, the effects of Pr will be neglected here. Hence, for geometrically similar devices ($A_T \sim d^2$, $A_X \sim d^2$):

$$N \equiv \frac{U_T A_T}{\dot{m} C_P} \sim \frac{\dfrac{1}{2} \dfrac{Nu \, k_g}{d} A_T}{\rho v A_X C_P} \sim Nu \frac{\kappa}{vd} \sim Nu \frac{\rho \kappa}{\mu} \frac{\mu}{\rho v d} \sim Re^a \frac{1}{Pr} \frac{1}{Re} \sim Re^{a-1} \qquad (10.4)$$

For laminar flow $\alpha = 0$ whereas for turbulent flow in straight channels, $\alpha \approx 0.8$ for straight channels. For the spiral heat exchangers discussed in Section 10.2.2, the curvature of the channels may affect the value of a due to the formation of Dean vortices. These effects were discussed in the context of heat-recirculating combustors in [10] and it was found that when the combined effects of turbulence and curvature are considered, for Re > 500, Nu is very nearly proportional to Re^1, thus the scaling relation $Nu \sim Re^a$ at high Re is considered valid. Whether $\alpha = 0.8$ or 1.0 does not affect the following discussion; it is only important that Nu scales with Re only for geometrically similar heat exchangers without separate influence of other parameters. Consequently, for both laminar and turbulent flows Re may be substituted for N as a scaling parameter. Also, since a is rarely if ever larger than unity, Equation (10.4) shows that N will not increase with increasing Re even for turbulent flows because the increase in U_T is offset by the increase in $\dot{m}$.

For geometrically similar devices a scales according to

$$\alpha \sim \frac{U_L}{\dfrac{1}{2} \dfrac{Nu \, k_g}{d}} \sim \frac{1}{Re^a} \frac{U_L d}{k_g} \qquad (10.5)$$

Typically the external heat loss is due to buoyant convection or radiative transfer, for both of which the loss per unit area is nearly independent of scale. If this is the case then for fixed Re, $a \sim d^1$, that is, a increases linearly with increasing scale. For extremely small combustors heat loss to the surroundings may be dominated by conduction instead of convection or radiation, in which case $U_L \sim d^{-1}$ and thus a is independent of scale.

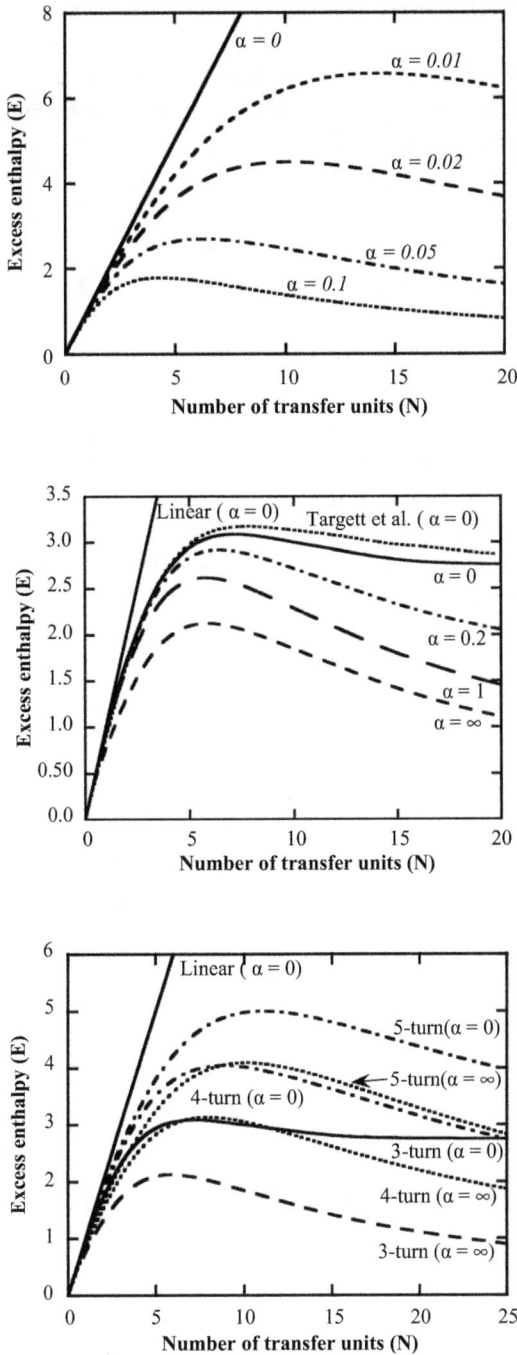

Figure 10.2. Dimensionless excess enthalpy (E) versus number of transfer units (N) for different heat loss coefficients (α): (a) linear exchanger (Equation (10.3)) (data from [17]), (b) simulated 3-turn spiral exchanger (Equation (10.7)), with comparison to results by Targett et al. [13] (data from [17]), (c) simulated 3, 4, and 5-turn spiral exchangers.

10.2.2 SPIRAL EXCHANGER

Linear exchangers suffer from a large ratio of heat loss area (A_L) to heat exchange area (A_T), generally about 2 as discussed earlier. One way of minimizing the impact of heat losses is to roll up the linear device into an n-turn spiral "Swiss roll" combustor [11, 12]. In this way for a heat exchanger of overall length L, the length exposed to heat loss to ambient decreases from 2L to L/n and thus A_L decreases by a factor of about 1/2n. Additionally, the remaining portion of what were the outer walls of the linear heat exchanger exposed to heat loss become the heat transfer area and thus A_T increases by a factor of (2L − L/n)/L= 2 − 1/n. Hence, rolling a linear exchanger into a spiral device increases N by a factor of about 2 − 1/n and decreases α by a factor of 2(2n−1). Consequently, the Swiss roll is one of the most thermally efficient types of heat exchangers, though it has limitations at high N as will be discussed in Section 10.2.2.

Analysis for spiral exchangers is more elaborate since the energy balances on individual turns are coupled and their contributions to A_T vary according to the distance from the center of the spiral. For the current purposes the device of Figure 10.1 can be cut into thirds and stacked to simulate (in the thermal sense) a spiral having three equal-length "turns," with the outlet of each turn being the inlet to the next (Figure 10.3). The WSR sits between the end of the last inlet arm of the exchanger and the beginning of the first outlet arm. For the inlet turns the energy conservation equations are (refer to Equations (10.1) and (10.2)):

$$\dot{m}C_P\left(T_2 - T_1\right) = U_T A_T \left(\frac{T_7 + T_8}{2} - \frac{T_1 + T_2}{2}\right) - U_L A_L \left(\frac{T_1 + T_2}{2} - T_1\right)$$

$$\dot{m}C_P\left(T_3 - T_2\right) = U_T A_T \left(\frac{T_6 + T_7}{2} - \frac{T_2 + T_3}{2}\right) + U_T A_T \left(\frac{T_7 + T_8}{2} - \frac{T_2 + T_3}{2}\right) - U_L A_L \left(\frac{T_2 + T_3}{2} - T_1\right) \quad (10.6a)$$

$$\dot{m}C_P\left(T_4 - T_3\right) = U_T A_T \left(\frac{T_5 + T_6}{2} - \frac{T_3 + T_4}{2}\right) + U_T A_T \left(\frac{T_6 + T_7}{2} - \frac{T_3 + T_4}{2}\right) - U_L A_L \left(\frac{T_3 + T_4}{2} - T_1\right)$$

For the outlet turns energy conservation requires:

$$\dot{m}C_P\left(T_8 - T_7\right) = U_T A_T \left(\frac{T_1 + T_2}{2} - \frac{T_7 + T_8}{2}\right) + U_T A_T \left(\frac{T_2 + T_3}{2} - \frac{T_7 + T_8}{2}\right) - U_L A_L \left(\frac{T_7 + T_8}{2} - T_1\right)$$

$$\dot{m}C_P\left(T_7 - T_6\right) = U_T A_T \left(\frac{T_2 + T_3}{2} - \frac{T_6 + T_7}{2}\right) + U_T A_T \left(\frac{T_3 + T_4}{2} - \frac{T_6 + T_7}{2}\right) - U_L A_L \left(\frac{T_6 + T_7}{2} - T_1\right) \quad (10.6b)$$

$$\dot{m}C_P\left(T_6 - T_5\right) = U_T A_T \left(\frac{T_3 + T_4}{2} - \frac{T_5 + T_6}{2}\right) - U_L A_L \left(\frac{T_5 + T_6}{2} - T_1\right)$$

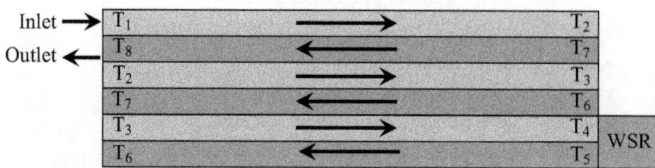

Figure 10.3. Stacked linear device used to simulate 3-turn spiral Swiss roll combustor.

Simplifying these equations and non-dimensionalizing temperatures with respect to T_1 (denoted by $\tilde{T}$) yields

$$(2+N+\alpha N)\tilde{T}_2 - N\tilde{T}_7 - N\tilde{T}_8 = 2 - N + \alpha N$$

$$(-2+2N+\alpha N)\tilde{T}_2 + (2+2N+\alpha N)\tilde{T}_3 - N\tilde{T}_6 - 2N\tilde{T}_7 - N\tilde{T}_8 = 2\alpha N$$

$$(-2+2N+\alpha N)\tilde{T}_3 + (2+N+\alpha N)\tilde{T}_4 - 2N\tilde{T}_6 - N\tilde{T}_7 = 2\alpha N + N\Delta\tilde{T}_C$$

$$-2N\tilde{T}_2 - N\tilde{T}_3 + (-2+2N+\alpha N)T_7 + (2+2N+\alpha N)T_8 = N + 2\alpha N \qquad (10.7)$$

$$-N\tilde{T}_2 - 2N\tilde{T}_3 - N\tilde{T}_4 + (-2+2N+\alpha N)\tilde{T}_6 + (2+2N+\alpha N)\tilde{T}_7 = 2\alpha N$$

$$-N\tilde{T}_3 + (-2+\alpha N)\tilde{T}_4 + (2+N+\alpha N)\tilde{T}_6 = 2\alpha N + (2 - N - \alpha N)\Delta\tilde{T}_C$$

This is a set of six linear equations for the unknowns $\tilde{T}_2$, $\tilde{T}_3$, $\tilde{T}_4$, $\tilde{T}_6$, $\tilde{T}_7$, and $\tilde{T}_8$ in terms of specified thermal parameters N, α, and $\Delta\tilde{T}_R = \Delta\tilde{T}_C = \tilde{T}_5 - \tilde{T}_4$. Results are expressed in Figure 10.2b in terms of the excess enthalpy $E = (\tilde{T}_4 - 1)/\Delta\tilde{T}_C = (T_4 - T_1)/\Delta T_C$ as a function of N for varying α.

The analysis results in the same dimensionless parameters as with the linear device but to compare the results of the linear and simulated spiral devices, three observations should be noted. First, in the above equations N is based on the area (A_T) of one heat exchange surface. Since there are $2n - 1 = 5$ heat transfer surfaces between products and reactants, A_T for the purposes of defining N must be based on the total area available for heat recirculation, meaning $N_{spiral} = N_{linear}/5$. Second, heat loss occurs only from the first inlet turn and thus to simulate a spiral exchanger α is set to zero for all but the first equation, but for comparison with the linear exchanger the same definition of α is retained, that is, A_L is the area of one exterior side of the linear exchanger; if instead α is defined based on the area exposed to heat loss to ambient, the values of α for the spiral exchanger reported here should be decreased by a factor of 3. Third, in this analysis heat loss in the third dimension (out of the plane of the spiral) is neglected; this is justified for a spiral heat exchanger that is sufficiently tall compared to its diameter or one that is wrapped in the third dimension to create a toroidal device. This assumption will be justified in Section 10.4.2; however, it will also be shown that for a short device this mode of heat loss cannot be neglected.

The results of the analysis of the simulated spiral device are shown in Figure 10.2b, along with results of a much more detailed analysis of an adiabatic 3-turn spiral exchanger by Targett et al. [13]. The close agreement with Targett et al.'s results indicates that the highly simplified analysis given here is satisfactory for the current purposes. Figure 10.2b reveals that at small N, E = N as with the linear exchanger, whereas at larger N, even for the adiabatic case E reaches a maximum value and then decreases. As discussed by Churchill and collaborators [13, 14], this occurs because if N is too large, heat transfer from one outlet channel to the adjoining inlet channel will be too rapid and the temperature of this inlet channel will become hotter than the next-cooler (farther toward the outside of the device) outlet channel and some heat transfer from this inlet channel to the cooler adjacent outlet channel will result, rather than the inlet channel receiving thermal enthalpy from both adjacent outlet channels. This cannot occur with linear device since heat transfer only occurs from one side of the outlet channel to the adjacent side of the inlet channel. As a result, for a truly adiabatic system the linear device provides larger excess enthalpy for a given N and thus will exhibit broader extinction limits; however, in the presence of heat losses, spiral exchangers provide substantially larger excess enthalpy. Comparing Figures 10.2a and 10.2b it can be seen that for a given N (thus Re), the spiral exchanger can provide the same value of excess enthalpy E at much larger value of the heat loss α. Even in the limit $\alpha \to \infty$, only the outermost inlet turn becomes ineffective in terms of heat recirculation; this turn insulates the inner turns from heat losses. Of course, as N increases the rate of heat transfer from the inner turns to the outermost one increases and more enthalpy is lost to ambient, resulting in significant decreases in E at large N even for the spiral device.

Figure 10.2c shows a comparison of the performance of simulated 3, 4, and 5-turn spiral exchangers, the latter two using sets of equations identical to Equation (10.7) extended to include the effects of the additional turns. Results are shown in the two extreme cases of zero and infinite heat loss coefficient α. It can be seen that, as expected, a larger number of turns results in better performance, even for the same N (i.e., the same mass flow rate, heat transfer coefficient, and overall exchanger length). Moreover, the performance of an adiabatic n-turn device is practically the same as an n+1 turn device with infinite heat loss coefficient, because in this limit the outermost inlet turn (the one exposed to heat loss) remains at ambient temperature but all other turns are nearly unaffected by this loss.

10.2.3 FINITE-RATE CHEMISTRY

While the above equations identify the dimensionless groups N and α needed to describe the heat exchanger performance, additionally the finite rate of thermal enthalpy release due to the chemical reaction must be considered in order to determine extinction limits. The coupling between the heat exchange and the chemical reaction exists because any factor that decreases the reactor temperature ($T_R = T_3$ for the linear device, T_5 for the simulated 3-turn device) will decrease the thermal enthalpy release rate; if this rate drops sufficiently some reactants will not be converted to products within the available residence time in the WSR and thus the temperature rise due to combustion ($\Delta T_R = T_3 - T_2$ for the linear device, $T_5 - T_4$ for the simulated 3-turn device) will decrease below the value for the complete reaction (ΔT_C), leading to less heat recirculation from products to reactants, leading to a further decrease in reactor temperature and eventually to extinction. Probably the simplest approach to modeling this thermal enthalpy release is a classical WSR with a single-step chemical reaction. The motivation for using a WSR model is that experimental [9] and numerical [10, 15] results show that, at least for sufficiently high Re, near extinction limits the reaction zone structures in Swiss roll combustors are more similar to WSRs than propagating premixed flames; compared to propagating flames the reaction zones in Swiss roll combustors have much smaller temperature gradients, lower peak temperatures, and longer residence times at high temperature. Combining a first-order reaction rate expression (mass per unit volume per unit time) of the form $Z\rho Y_R \exp\left(-E_a/\Re T_R\right)$ with the enthalpy balance across the reactor $Q\left(Y_\infty - Y_R\right) = C_P \Delta T_R$, a form of the usual WSR expression is obtained:

$$\Delta T_R = \frac{\Delta T_C}{1 + 1/\left[DaN \exp\left(-E_a/\Re T_R\right)\right]}; Da \equiv \frac{\rho C_P ZV}{U_T A_T} \qquad (10.8)$$

where the DaN = $\rho ZV/\dot{m}$ has been written to show the dependence on N explicitly. Written this way, the Damköhler number Da is a constant (independent of $\dot{m}$), except for turbulent flows in which case U_T will become approximately linearly proportional to $\dot{m}$. Note that in the limit of sufficiently high Da or N, the reaction is nearly complete and ΔT_R approaches ΔT_C. The scaling of Da is given by

$$Da \equiv \frac{\rho C_p Z V}{U_T A_T} \sim \frac{\rho C_p Z d^3}{\frac{1}{2} \frac{Nu}{d} k_g \, d^2} \sim \frac{Z \rho d^2 \, Pr}{Nu \; \mu} \sim \frac{Z \rho d^2}{Re^a \mu} \tag{10.9a}$$

and that of DaN by

$$DaN \sim \frac{Z \rho d^2}{Re^a \mu} Re^{a-1} \sim \frac{Z \rho d^2}{Re \; \mu} \tag{10.9b}$$

Since μ, ρ, and Z are molecular properties independent of scale, for fixed Re, both $Da \sim d^2$ and $DaN \sim d^2$.

By replacing ΔT_C (for complete combustion) with ΔT_R (for the finite-rate reaction) as given by Equation (10.3) (for the linear exchanger) or Equation (10.7) (for the simulated 3-turn exchanger) and combining these expressions with Equation (10.8), the effects of finite-rate chemistry on combustor performance can be assessed. For the linear exchanger a single relation for the reactor temperature $T_R = T_3$ is obtained, which in the dimensionless form is

$$\frac{\tilde{T}_3 - 1}{F + 1} = \frac{\Delta \tilde{T}_C}{1 + 1 / \left[DaN \exp \left(-\beta / \tilde{T}_3 \right) \right]} ; F \equiv \frac{4N}{4 + \alpha N \left[4 + N \left(2 + \alpha \right) \right]} \tag{10.10}$$

Figure 10.4 shows the response of dimensionless reactor temperature $\tilde{T}_R$ to N for adiabatic ($\alpha = 0$) and non-adiabatic ($\alpha = 0.1$) conditions for the linear and simulated 3-turn exchangers. As discussed in Section 10.2.4, the values of Da and β are chosen to match those employed in a

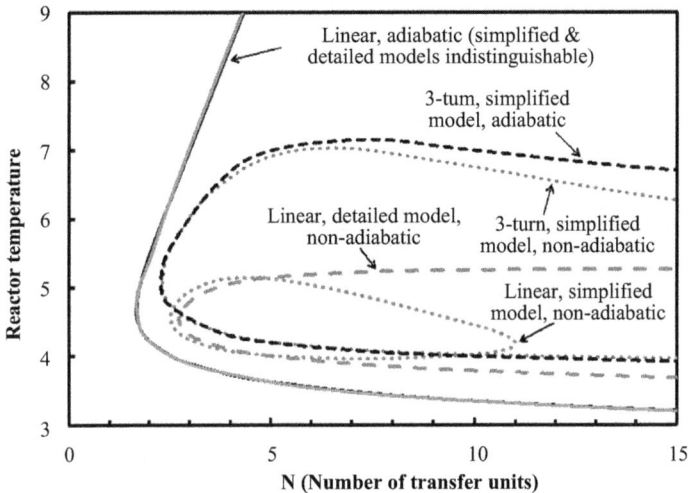

Figure 10.4. Reactor temperature (T_R) as a function of the number of transfer units (N) for linear and simulated 3-turn spiral heat exchangers with Damköhler number $Da = 2 \times 10^7$, non-dimensional activation energy $\beta = 70$ and non-dimensional heat loss coefficient $\alpha = 0.1$. Also shown are comparisons between simplified and detailed models for the linear device.

prior investigation using a more detailed analysis of linear exchangers, the results of which are also shown in Figure 10.4. First note that for both the linear and simulated spiral exchangers, without heat loss there is a low-N (high flow rate) extinction limit but no high-N (low flow rate) limit. The value of N at the low-N limit is nearly the same for the linear and simulated spiral exchangers, which is reasonable because for adiabatic conditions at low values of N, the excess enthalpy (E) is nearly the same for the two types of exchangers (Figure 10.2b). For large N, the temperature increases much more rapidly for the linear exchanger and unlike the 3-turn device does not reach a maximum value. This is consistent with the response of excess enthalpy (E) to N for the linear device as compared to the simulated 3-turn devices (Figure 10.2b). With heat loss, the low-N limits do not change significantly for either type of device but as N increases, the 3-turn device exhibits considerably better performance. This is also consistent with the results shown in Figure 10.2b.

Figure 10.4 shows a high-N (low flow rate) extinction limit with heat loss as a consequence of the isola response of $\tilde{T}_R$ to N. While dual (low and high-flow rate) limits are common in combustion systems, for heat-recirculating combustors this low flow rate limit shown in Figure 10.4 is actually an artifact of the use of the mixing-cup approximation which uses average temperature differences to compute heat recirculation and heat loss. The mixing-cup approximation is reasonable if the temperature profiles along the inlet and outlet arms of the heat exchanger are nearly linear, which is valid for moderate heat loss. If there are substantial heat losses then the entire inlet end of the exchanger is at near-ambient temperature with only a small region near the WSR end having temperatures above ambient, in which case the mixing-cup method overestimates the impact of heat loss. A more detailed model that does not suffer from this limitation is presented in the next section.

10.2.4   DETAILED ANALYTICAL MODEL

In this section the model developed in Section 10.2.2 is replaced by one using energy balances applied to infinitesimal elements of the heat exchanger rather than one applied globally for each arm of the exchanger. The model in this section also includes the effect of heat conduction along the wall dividing the inlet and outlet arms of the exchanger, which is critical to understanding the performance of heat-recirculating combustors [16]. Referring to Figure 10.5, energy balances on the wall, inlet arm, and outlet arm of the exchanger readily yield

$$k_w \tau \frac{d^2 T_w}{dx^2} + 2U_T(T_e - T_{w,e}) - 2U_T(T_i - T_{w,i}) = 0 \tag{10.11a}$$

$$\frac{\dot{m}C_P}{A_T / L} \frac{dT_i}{dx} - 2U_T(T_{w,i} - T_i) + U_L(T_i - T_\infty) = 0 \tag{10.11b}$$

$$\frac{\dot{m}C_P}{A_T / L} \frac{dT_e}{dx} - 2U_T(T_e - T_{w,e}) - U_L(T_e - T_\infty) = 0 \tag{10.11c}$$

In the above relations it has been assumed, as discussed in Section 10.2.1, that the heat transfer coefficients are constant and equal on the inlet and outlet arms of the exchanger with negligible thermal resistance across the dividing wall, in which case $U_T = h_T/2$. Using the mean wall temperature $T_w \approx (T_{w,e} + T_{w,i})/2$, invoking the thermally-thin assumption ($T_{w,e} - T_{i,w} \ll T_e - T_i$)

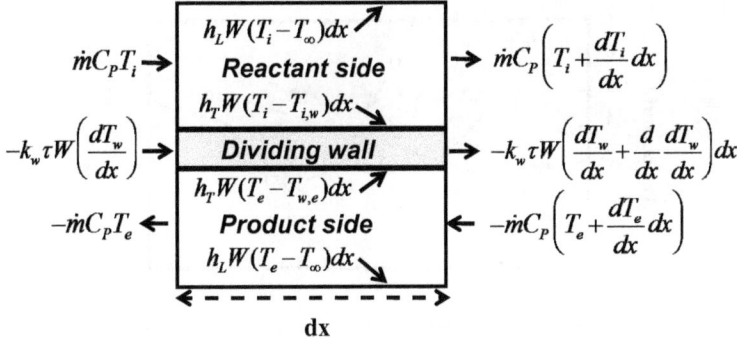

Figure 10.5. Schematic diagram of enthalpy fluxes assumed in detailed analysis of linear counter-current heat recirculating combustor.

and combining Equations (10.11a)–(10.11c) yield a fourth-order differential equation for wall temperature:

$$\frac{1}{N^2 Bi\,\alpha(2+\alpha)}\frac{d^4\tilde{T}_w}{d\tilde{x}^4} - \left[\frac{1}{N^2\alpha(2+\alpha)}+\frac{2+\alpha}{Bi\,\alpha}\right]\frac{d^2\tilde{T}_w}{d\tilde{x}^2}+\tilde{T}_w = 1; \quad \tilde{T} \equiv \frac{T}{T_1}; \tilde{x} \equiv \frac{x}{L}; Bi \equiv \frac{4U_T L^2}{k_w \tau}$$

(10.12)

where introducing the effects of wall thermal conductivity generates a new parameter, namely a Biot number (Bi). Note that for geometrically similar devices the scaling of Bi is given by

$$Bi \equiv \frac{4U_T L^2}{k_w \tau} \sim \frac{\dfrac{Nu\, k_g}{d}d^2}{k_w d} \sim Re^a \frac{k_g}{k_w}$$

(10.13)

and thus for a given Re, Bi is independent of scale and for laminar flow (a = 0) is simply a constant except to the extent that k_g and k_w vary with temperature.

The boundary conditions for Equation (10.12) are (1) the temperature at the inlet of the exchanger is ambient; (2) the temperature rise across the WSR is given by Equation (10.8); (3) the inlet end of the wall is adiabatic; and (4) the WSR end of the wall is adiabatic. The assumption of adiabatic wall ends does not affect the results substantially; using a convection boundary condition rather than an adiabatic one changes the results only slightly for realistic choices of property values [16]. Since the first two boundary conditions are in terms of $\tilde{T}_i$ and $\tilde{T}_e$ rather than $\tilde{T}_w$, to solve this system of equations first $\tilde{T}_w(\tilde{x})$ is found using Equation (10.12), then this result is applied to Equations (10.11a)–(10.11c) to find $\tilde{T}_i(\tilde{x})$ and $\tilde{T}_e(\tilde{x})$. The process is straightforward but tedious and not reproduced here; see [16] for a detailed derivation. Results of the analysis are shown in Figures 10.4, 10.6, and 10.7.

Examples of the heat exchanger temperature profiles $\tilde{T}_i(\tilde{x})$, $\tilde{T}_w(\tilde{x})$, and $\tilde{T}_e(\tilde{x})$ predicted by this detailed analysis are shown in Figure 10.6. For the infinite reaction rate (Da = ∞), adiabatic (α = 0) and wall-conduction-free (Bi = ∞) conditions, Figure 10.6 (top) shows that temperature profiles are linear. For this case, for any N the reactor temperature is $\tilde{T}_R = 1 + N\Delta\tilde{T}_C$, corresponding E = N as predicted by Equation (10.3), and the exhaust temperature $\tilde{T}_e(0)$ is simply the adiabatic flame temperature $1 + \Delta\tilde{T}_C$ as energy conservation requires. The response of $\tilde{T}_R$ to

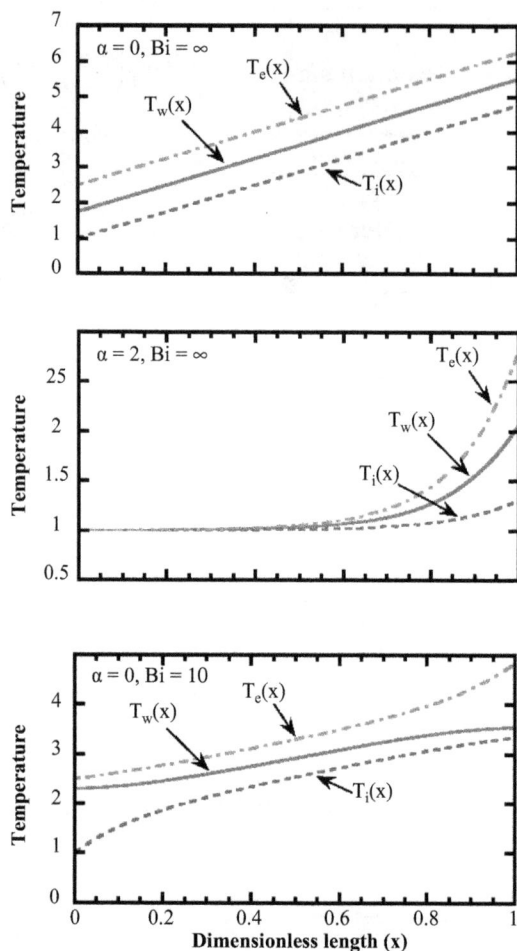

Figure 10.6. Temperature profiles in the linear heat exchanger for Da = ∞, N = 2.5, $\Delta \tilde{T}_C$ = 1.5. Top: α = 0, Bi = ∞; middle: α = 2, Bi = ∞; bottom: α = 0, Bi = 10. Adapted from [16].

N predicted by this analysis is shown in Figure 10.4, where it is compared to the results of the simple analysis discussed in Section 10.2.2. It can be seen that the results of the detailed model are identical to those of the simplified model, which is understandable since the temperature profiles are linear and thus the mixing-cup model provides an accurate estimate of the heat transfer across the heat exchanger. For α = 0 and Bi = ∞ but the finite-rate reaction (Da ≠ ∞), the temperature profiles will still be linear but the temperature jump at the WSR will decrease according to Equation (10.8) and thus the reactor temperature $\tilde{T}_R$ will be lower.

Figure 10.6 (middle) shows heat exchanger temperature profiles predicted by the detailed analytical model for the infinite reaction rate (Da = ∞) and wall-conduction-free (Bi = ∞) conditions with heat loss (α > 0). It can be seen that the temperatures are near ambient except near the WSR end of the exchanger ($\tilde{x}$ = 1) and thus there is heat loss only from this end. In contrast, the simplified mixing-cup model essentially assumes linear profiles and thus has no means to capture this behavior. It should also be noted that for the case shown in Figure 10.6 (middle), all of the thermal enthalpy generated can be lost to ambient rather than exhausted at the exchanger exit without extinction occurring. Thus, simply equating rates of heat generation and loss cannot always yield an extinction criterion.

The response of $\tilde{T}_R$ to N for this non-adiabatic case is also shown in Figure 10.4. It can be seen that for low N (large mass flow rates) where heat loss effects are minimal, the simplified and detailed models predict nearly the same extinction limit, whereas for large N the detailed model predicts no extinction limit whatsoever (as $N \rightarrow \infty$, $\tilde{T}_R$ asymptotes to a fixed value), whereas the simplified model predicts isola behavior indicating an extinction limit. This is because as $N \rightarrow \infty$, the detailed model predicts that heat recirculation is balanced by heat loss; above a certain value of N, increasing N further (*e.g., in an experiment by decreasing the mass flow rate or increasing the length of the heat exchanger*) has no effect other than to increase the fraction of the length of the exchanger where both the reactant and product streams remain at near-ambient temperatures. It can be shown that the reactor temperature in the limit $N \rightarrow \infty$ is

$$\tilde{T}_e(1) = \frac{(1+\Delta\tilde{T}_C)G-1}{G-1}; G \equiv \sqrt{\frac{1+\alpha+\sqrt{\alpha(2+\alpha)}}{1+\alpha-\sqrt{\alpha(2+\alpha)}}} \quad (Bi \rightarrow \infty, N \rightarrow \infty) \qquad (10.14)$$

This observation is crucial to understanding the high-N (low Re) extinction limits because it indicates that, even in the presence of heat losses, without thermal conduction along the wall there is no means to reduce the reactor temperature as N is increased. This is quite different from combustors without heat recirculation, where sufficient reduction in the mass flow rate or Re (thus increase in N) will nearly always lead to extinction due to heat losses.

While the detailed model predicts no extinction limit at high N/low flow rates, experiments [9–12] do show that high-N limits do indeed exist, indicating that an additional mechanism is required to predict such extinction limits. One readily identifiable mechanism is that of heat losses in the out-of-plane dimension, which will be discussed in Section 10.4.2. Even this mechanism can be essentially eliminated by extending the height of the exchanger in the third dimension or wrapping the exchanger in the third dimension to create a toroidal device. However, even if heat losses in the third dimension are completely eliminated, another mechanism, namely thermal conduction along the wall dividing the inlet and outlet arms of the exchanger, can lead to high-N extinction limits without an additional loss mechanism. Figure 10.6 (bottom), which shows temperature profiles in the case of an adiabatic exchanger with wall heat conduction effects (Bi $\neq \infty$, $\alpha = 0$), provides insight into this mechanism. The temperature profiles clearly show that even though there is no heat loss in the system whatsoever, wall heat conduction removes thermal enthalpy from the high-temperature gas near $\tilde{x} = 1$ and returns this enthalpy to the gas at lower temperatures (smaller $\tilde{x}$), resulting in a lower reactor temperature than the adiabatic exchanger without wall heat conduction (Bi = ∞). (Nevertheless, the exit temperature $\tilde{T}_e(0)$ is $1+\Delta\tilde{T}_C$ for both infinite (Figure 10.6, top) and finite Bi (Figure 10.6, bottom) because in both cases the system is adiabatic with respect to the surroundings ($\alpha = 0$)).

The significance of streamwise wall heat conduction is further elucidated in Figure 10.7, which shows the response of WSR temperature $\tilde{T}_R = \tilde{T}_e(1)$ to N for several values of the Biot number (Bi) under adiabatic and non-adiabatic conditions. As already shown in Figure 10.4, for the infinite reaction rate, adiabatic conditions and no streamwise wall thermal conduction response is monotically increasing corresponding to E = N and with the finite-rate reaction a C-shaped extinction curve is found. In contrast, with wall conduction (finite Bi), the C-shaped response of $\tilde{T}_R$ to N that occur for both adiabatic and non-adiabatic conditions becomes isolas with both lower and upper limits on N because conduction of thermal energy away from the WSR vicinity through the wall becomes significant at large N. Once conducted away from the WSR vicinity, some thermal energy is transferred back to the gas via convection and a portion of this energy is then lost to ambient. It is emphasized that this mechanism is important only at large N (low Re or mass flow rates), where wall conduction is competitive with gas-phase convection. Figure 10.7 also shows that the small-N extinction limit is extended slightly by wall conduction, since heat recirculation (thus WSR temperature) is low at small N (Equation (10.3)); thus the increase in heat recirculation provided by wall conduction increases the WSR temperature slightly.

Figure 10.8 shows the effect of N on the fuel concentration expressed in terms of the minimum adiabatic temperature rise due to combustion $\Delta \tilde{T}_C$ required to sustain combustion (corresponding to the minimum fuel concentration, thus the extinction limit). Without streamwise wall conduction (Bi = ∞) no small-M extinction limit exists for the reasons given in the

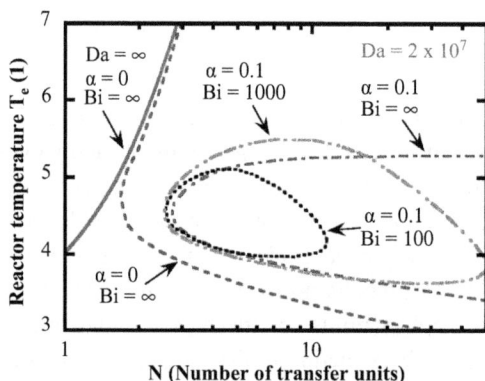

Figure 10.7. Effect of N on WSR temperature in the counter-current combustor for finite reaction rates $(Da = 2 \times 10^7)$ with $\Delta \tilde{T}_C = 1.5$. For reference, the adiabatic, wall-conduction-free curve from Figure 10.4 is also shown (data from [16]).

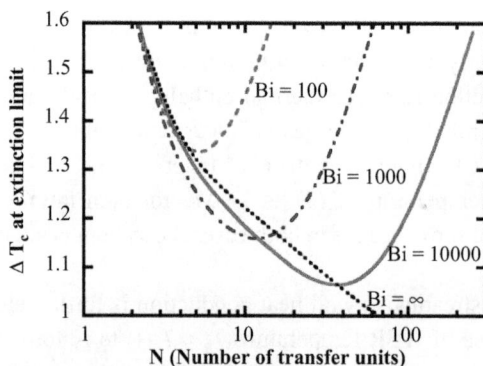

Figure 10.8. Effect of N on adiabatic temperature rise $\Delta \tilde{T}_C$ (i.e., fuel concentration) at extinction limit in the non-adiabatic ($\alpha = 0.1$) counter-current combustor for finite reaction rates $(Da = 2 \times 10^7)$ and varying Biot number (Bi) (data from [16]).

previous paragraph. For finite Bi, both small-N (high mass flow rate) and large-N (low mass flow rate) limits exist due to finite residence time and heat losses, respectively. Consistent with Figure 10.7, the large-N limit is slightly extended by decreasing Bi (thereby slightly increasing heat recirculation), whereas the small-N limit is drastically narrowed by decreasing Bi due to the mechanism described in the previous paragraph, that is, conduction removing thermal enthalpy from the WSR region and thus reducing the reactor temperature and the reaction rate.

It should be stressed that the value of Bi needed to affect extinction is much smaller than that which might be expected based on simplistic estimates. The overall ratio of streamwise convection to wall conduction is of the order $\dot{m}C_P/(k_w \tau A_T/L^2) = Bi/4N$. Even for the Bi = 10,000 case shown in Figure 10.8, where the effects of wall conduction might be thought to be negligible, the extinction limits are affected for all N > 50, thus Bi/4N > 50. Based on simplistic estimates, no effect wall conduction effects would be expected unless Bi/4N > 1. The powerful wall conduction effects result from the fact that the streamwise wall temperature gradients near the WSR are much larger when heat losses are present (Figure 10.6, middle), that is, much larger than the mean gradient under these conditions.

10.3   SCALING

10.3.1   OBJECTIVES

The analyses of Section 10.2 suggest that if the dimensionless groups N (or Re), α, Da, and Bi are constant, the extinction limits should be the same for a given value of the adiabatic temperature rise due to combustion $\Delta \tilde{T}_C$ regardless of the physical size of the combustor. In particular the following questions arise:

1. Can a highly simplified analysis (i.e., one-dimensional heat transfer, constant property values, simplified heat recirculation and heat loss models, WSR) be used to identify the dimensionless parameters describing the performance of heat-recirculating combustors?
2. Are Re, α, Da and Bi a complete set of parameters?
3. Are these parameters applicable to both laminar and turbulent flows?
4. Are these parameters applicable to both linear and spiral (Swiss roll) combustors?

One way of addressing these questions is via experimentation, that is, construct geometrically similar devices of different sizes, test them at the same values of Re, α, Da, and Bi, and determine if the operating temperatures for a given $\Delta \tilde{T}_C$ (or the equivalence ratio or the fuel mass fraction) and the values of $\Delta \tilde{T}_C$ at extinction are the same. In experiments it is very difficult to keep all dimensionless groups constant for combustors of different scales because, as will be shown in the next subsection, this requires adjustment of the heat loss coefficients, surface emissivities, and reaction rate parameters. Instead, these questions will be addressed via numerical simulation since the material properties and operating conditions can readily be adjusted in the numerical model.

The results of this section are presented in terms of the Reynolds number (based on the channel width (d), inlet flow velocity, and the viscosity of the incoming fuel-air mixture at ambient conditions) rather than the number of transfer units N because N is built with the overall heat transfer coefficient U_T which is a calculated quantity rather than an input parameter per se and because U_T will vary within the exchanger due to changes in mixture temperature and composition. Of course, Re also varies within the exchanger due to increases in temperature (which affects both the local flow velocity and mixture viscosity), but if the scaling analysis presented here is valid, the same temperature profiles in the exchanger and combustor will occur regardless of scale and thus temperature-dependent effects will be the same for all scales.

10.3.2   COMPUTATIONAL MODEL

A fully three-dimensional computational model using FLUENT 12.1 was used for the simulations. Details of the model and validation with experiments have been reported previously [10, 17]. For this study three 3.5 turn Swiss roll combustors were modeled. The nominal-scale combustor (Figure 10.9) was 5 cm tall with a channel width d = 3.5 mm and a wall thickness $\tau = 0.5$ mm; the other two combustors were geometrically identical but half and double the size of the nominal-scale device. The computational domain included the gaseous reactants and products, solid combustor walls and insulation on the top and bottom surfaces. As discussed in Section 10.4.2, computations with devices of different heights showed that the nominal 5 cm height chosen was sufficient to minimize the effects of heat loss in the direction out of the plane of the spiral. Both convective (nominally $h_L = 10$ W/m²K, typical of buoyant convection in ambient air) and radiative (nominally $\varepsilon_L = 0.8$ for exterior walls and $\varepsilon_L = 1$ for insulation) boundary conditions were used to simulate heat loss from the combustors. Symmetry was assumed at the midplane of the device (i.e., the bottom surface in Figure 10.3) and thus only half the device was modeled, which would be inaccurate if buoyancy effects were important; however, calculations made without the assumption of symmetry showed that buoyancy effects were negligible [17]. The Reynolds Stress Model (RSM) was employed to simulate the effects of turbulence on heat transfer. Propane-air combustion was simulated using single-step finite rate gas-phase chemistry with an activation energy $E_a = 40$ kcal/mole and the pre-exponential term adjusted to obtain agreement between the model and the experiment at Re = 1,000. No model of turbulence-chemistry interactions was used, which is consistent with the use of a WSR model in the analysis of Section 10.2.

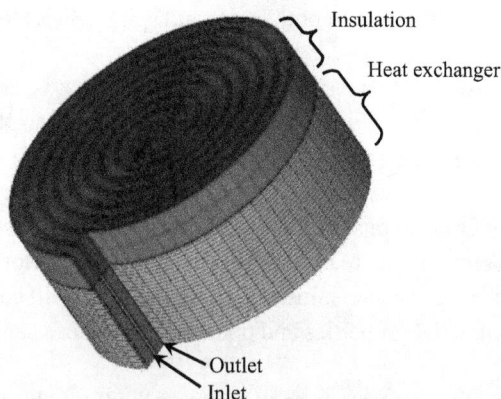

Figure 10.9. Wire-frame model of the spiral combustor used for scaling analyses showing grid resolution.

10.3.3   EXTINCTION LIMITS WITHOUT SCALING

Figure 10.10 (upper) shows the predicted extinction limits as a function of Re for the three combustors without any adjustment of property values to obtain constant α or Da. (Note that Bi is independent of scale according to Equation (10.13).) All three extinction limit curves exhibit the expected U-shaped behavior but the performance of the three combustors is clearly not identical even at the same Re (thus same N). In particular, at lower Re, smaller-scale combustors show better performance (lower lean extinction limits), whereas at higher Re, larger-scale combustors show better performance. These results can be explained as follows. As previously discussed, at low Re, extinction behavior is dominated by heat losses and according to Equation (10.5) the heat loss parameter $\alpha \sim d^1$; thus smaller combustors are subject to less impact of heat loss (specifically $\alpha_{double} : \alpha_{full} : \alpha_{half} = 4 : 2 : 1$) and consequently will have wider extinction limits. On the other hand at high Re, extinction limits are caused by insufficient residence time relative to the chemical reaction time, that is, the WSR blows out. The ratio of residence time to chemical reaction time is of course characterized by Da. According to Equation (10.9), Da $\sim d^2$; thus larger combustors have more residence time relative to the chemical reaction time (specifically $Da_{double} : Da_{full} : Da_{half} = 16 : 4 : 1$) and thus will have wider extinction limits.

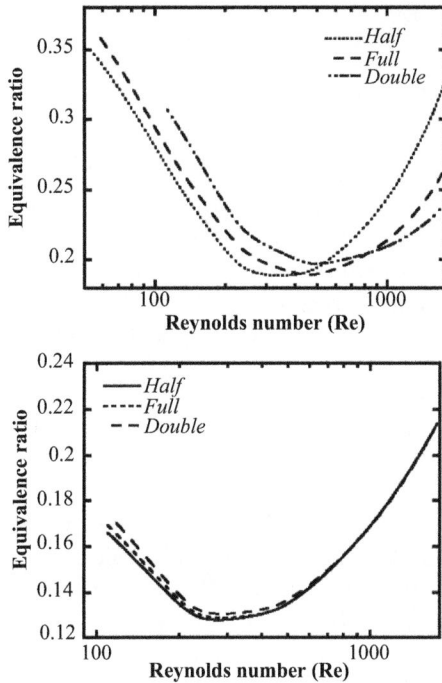

Figure 10.10. Computed propane-air equivalence ratios at the extinction limits for geometrically-similar Swiss roll combustors of three different overall sizes (data from [17]). Top: with baseline property values; bottom: with property values adjusted (Table 10.1) to obtain constant values of the dimensionless parameters describing heat loss (α), reaction rate (Da), and internal radiative transfer (R).

10.3.4 *EXTINCTION LIMITS WITH SCALING*

To determine whether Re, α, Da, and Bi are sufficient to characterize the performance of Swiss roll combustors, the simulations were repeated with property values adjusted so that these dimensionless parameters are the same for all three combustors. Specifically, the convective heat loss coefficient h_L and the emissivities for external radiative loss ε_L were artificially adjusted in proportion to d^{-1} and the pre-exponential term in the reaction rate expression was artificially adjusted in proportion to d^{-2} (see Table 10.1). The computed results (not shown) were practically identical at high Re but at low Re the larger combustors perform more poorly, implying that other scale-dependent process(es) are significant at low Re. The dominant process was identified [17] as wall-to-wall radiative transfer, which previous studies [15] had shown significantly affects the extinction limits at low Re. In particular, radiation between internal heat exchange surfaces is a means of transferring thermal enthalpy away from the high-temperature reaction zone without recycling the enthalpy to the incoming reactants in a manner very similar to that of streamwise wall heat conduction and thus is detrimental to combustor performance. (This mode of heat transfer does not exist in linear exchangers and thus could not be identified by the analysis in Section 10.2.) The radiative heat transfer (Q_R) between neighboring walls j and k of a spiral counter-current heat exchanger can be estimated assuming infinite parallel isothermal walls, that is, $Q_R = \dfrac{\varepsilon_i}{2-\varepsilon_i}\sigma A_T\left(T_j^4 - T_k^4\right)$, and thus a radiative heat transfer coefficient $U_R = Q_R/A_T(T_j - T_k)$ can be estimated as

$$U_R = \frac{\varepsilon_i}{2-\varepsilon_i}\sigma\left(T_j^2 + T_k^2\right)\left(T_j + T_k\right) \tag{10.15}.$$

Of course U_R is not a true coefficient since it depends on the temperature and thus will vary within the combustor, but if the scaling parameter for internal radiation is valid the temperature at a given location with the combustor and thus U_R will be the same regardless of scale. Consequently another dimensionless group, namely the internal radiation coefficient R, can be identified:

$$R \equiv \frac{U_R}{U_T} = \frac{\sigma}{U_T}\frac{\varepsilon_i}{2-\varepsilon_i}\left(T_i^2 + T_j^2\right)\left(T_i + T_j\right) \tag{10.16}.$$

Table 10.1. Values of the heat loss coefficients, emissivities, and pre-exponential term for combustors of varying scale. The values for the "Full" scale device are the nominal properties for the device that was tested experimentally [10] to verify the accuracy of the computational model, whereas the values for the "Half" and "Double" scale devices are artificially adjusted to obtain constant values of the dimensionless parameters α, Da, and R

Property	Half	Full	Double
h_L (W/m²K)	10	5	2.5
ε_L (external wall)	0.8	0.4	0.2
ε_L (insulation)	1	0.5	0.25
Z(m-sec-kmole units)	1.44×10^{11}	3.6×10^{10}	9.0×10^9
ε_i (internal wall)	0.8	0.5	0.2857

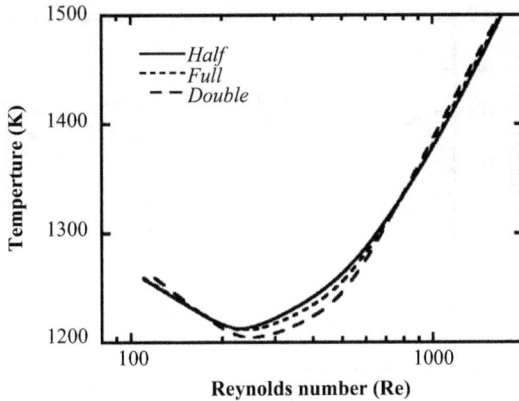

Figure 10.11. Computed maximum temperatures at the extinction limits of propane-air mixtures for geometrically similar Swiss roll combustors of three different overall sizes with property values adjusted to obtain constant α, Da, and R (data from [17]).

The scaling of R is given by

$$R = \frac{\sigma}{U_T} \frac{\varepsilon_i}{2-\varepsilon_i}\left(T_i^2 + T_j^2\right)\left(T_i + T_j\right) \sim \frac{\sigma d}{Nu \, k_g} \frac{\varepsilon_i}{2-\varepsilon_i} \sim \frac{\sigma d}{Re^a \, k_g} \frac{\varepsilon_i}{2-\varepsilon_i} \qquad (10.17).$$

Since $R \sim d^1$, the three combustors will have different R according to $R_{double} : R_{full} : R_{half} = 4 : 2 : 1$. The larger impact of internal radiative transfer at larger scales explains why the larger combustors perform more poorly at small Re even when α and Da are the same for all combustors. To keep R constant for combustors of varying scale, ε_i was adjusted as shown in Table 10.1. (Note that $\varepsilon_i/(2-\varepsilon_i)$ rather than ε_i itself must be adjusted in proportion to d^{-1}.) The extinction limits computed with the adjusted values of ε_i are shown in Figure 10.10 (lower). It can be see that the limits are nearly the same for all combustors at all Re, encompassing both laminar and turbulent flow regimes. It should be emphasized that the parameter values listed in Table 10.1 were adjusted based entirely on scaling considerations and not readjusted in an empirical fashion to obtain the favorable results shown in Figure 10.10 (lower). Additionally Figure 10.11 shows that the maximum temperatures at the extinction limits for the three combustors are nearly identical, which justifies the use of R as a scaling parameter even though U_R is not temperature-independent. Moreover, this similarity of temperatures validates the definition of Da employed here which does not include the temperature-dependent Arrhenius term $\left(\exp\left(-E_a/\Re T_3\right)\right)$. These results demonstrate that the dimensionless parameters Re, α, Da, Bi, and R are sufficient to characterize the performance of heat-recirculating combustors with similar geometry but widely varying scale.

10.3.5 *LINEAR VERSUS SWISS ROLL HEAT EXCHANGERS*

While the simple analyses of the linear and simulated Swiss roll devices in Section 10.2 identified the governing dimensionless parameters, it is still instructive to compare combustor performance in the two different geometries using detailed numerical simulations. With this

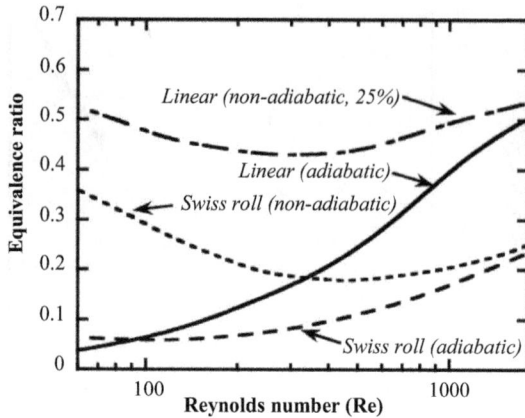

Figure 10.12. Comparison of extinction limits of Swiss roll and linear combustors with the same exchanger length, with and without heat losses (data from [17]). For the linear combustor, the heat loss coefficients are set to 25% of their nominal values (see the text.)

motivation an "unrolled" linear combustor model having the same channel width d, total heat exchanger length L and same central volume and shape as the nominal-scale Swiss roll combustor was created. A problem arose in the simulations in that the linear combustor could not sustain combustion at low Re (high N) even with a stoichiometric mixture. Such difficulties could be expected based on the poorer performance of non-adiabatic linear exchangers versus spiral ones as shown in Figure 10.2a versus Figure 10.2b. To rectify this situation, for the linear device the external heat loss coefficients h_L and ε_L were artificially reduced to 25% of their values for the Swiss roll device. (Even with this scheme, the effective α is larger in the linear device than the Swiss roll since rolling a linear exchanger into a 3.5 turn spiral reduces A_L by a factor of about 7.)

Figure 10.12 shows the predicted extinction limits for the Swiss roll combustor and the linear combustor (the latter with modified h_L and ε_L), along with the corresponding results for adiabatic conditions and no internal radiation ($h_L = \varepsilon_L = \varepsilon_i = 0$). First note that even at high Re, where a comparison of the adiabatic and non-adiabatic results shows that heat losses are insignificant, the Swiss roll combustor still shows broader extinction limits. This is probably because the inlet arm of the Swiss roll exchanger receives thermal enthalpy from both walls of the channel, whereas the linear exchanger receives enthalpy from only one wall, so the Swiss roll device has about twice the effective heat transfer area (A_T) and thus twice as large a value of N at the same Re. Consistent with this explanation, the equivalence ratios at the extinction limits for the Swiss roll combustor are approximately half that of the linear combustor; to obtain the same T_3 with half the N and thus half the E, twice as much temperature rise due to combustion ($T_3 - T_2$) is required and thus twice as high a fuel concentration in the fresh mixture ($Y_{f,\infty}$ or equivalence ratio) is required.

Figure 10.12 also shows that in the (unrealistic) adiabatic case, as Re decreases (and thus N increases) the benefit of the Swiss roll combustor over the linear device in terms of extinction limits decreases and at very low Re/high N, the extinction limits of linear device are leaner than those of the Swiss roll despite the linear device having half the N at the same Re. This result can be understood considering the properties of spiral heat exchangers discussed in Section 10.2. Since $N_{linear} \approx N_{spiral}/2$ and for adiabatic conditions $N_{linear} = E$, the cross-over to leaner extinction limits for the linear device should occur when $N_{spiral} \approx E/2$. For the 3-turn device modeled here, the condition for which $N_{spiral} = E/2$ occurs is at $N \approx 6.1$ for $\alpha = 0$, corresponding to Re $\approx 400/N = 66$. This is comparable to the cross-over value Re ≈ 90 seen in Figure 10.12, which suggests that the aforementioned behavior of spiral heat exchangers is responsible for the cross-over in extinction limits seen at low Re for adiabatic combustors.

10.4 PRACTICAL PERSPECTIVES

10.4.1 NUMBER OF TURNS

The scaling analyses presented in Sections 10.3.1–10.3.3 presume that there is no limitation on the ability to manufacture arbitrarily large or small devices while enforcing geometrical similarity. A more practical situation would be a constraint on the volume available for the device, with a fixed wall thickness limited by manufacturing capabilities. Given these constraints, for an n-turn device with average channel length per turn of L^* (which would be nearly constant for a device of fixed overall size), a scaling analysis similar to that leading to Equation (10.4) yields

$$N \equiv \frac{U_T A_T}{\dot{m} C_P} \sim \frac{\frac{1}{2} \frac{Nu}{d} k_g A_T}{\rho v A_X C_P} \sim Re^{a-1} \frac{L^*}{d} n \qquad (10.18)$$

For fixed overall size, the channel width (d) is inversely proportional to the number of turns and thus for fixed Re, $N \sim n^2$. This would seem to indicate a significant advantage to increasing the number of turns, thus increasing N. However, for spiral exchangers increasing N is beneficial only up to a certain limit as Figures 10.2(b) and 10.2(c) show. Moreover, for the stated constraints, more turns also imply more wall material and thus more heat transfer along the walls in the direction perpendicular to the plane of the spiral, which results in additional heat loss to ambient. Clearly if the number of turns is sufficient, the structure will be mostly wall material with massive heat losses in the third dimension (out of the plane of the spiral) and little space for reactant and product flow. (While the impact of heat loss in the third dimension was stated in Section 10.3.2 to be insignificant for the 3.5 turn exchanger, this is not necessarily true for an increased n and thus increased wall material.) Consequently, an optimum value of n must exist.

With this motivation, a set of two-dimensional calculations were performed using a model for heat loss in the third dimension [15] that has been shown [17] to closely match results of fully 3D calculations. Combustors of n = 3.5 to 12 having the same overall dimensions and wall thickness (Figure 10.13) were simulated at varying Re and the extinction limits determined. Again a one-step Arrhenius reaction rate model for propane was employed. Results are shown in Figure 10.14. It can be see that an optimum n does indeed exist though it is rather flat for the conditions examined. The optimum is somewhat sharper and shifted to smaller n for smaller Re. For the dimensions of the combustors shown in Figure 10.12, $N \approx (24n^2/Re)/(1 - 0.012n)$

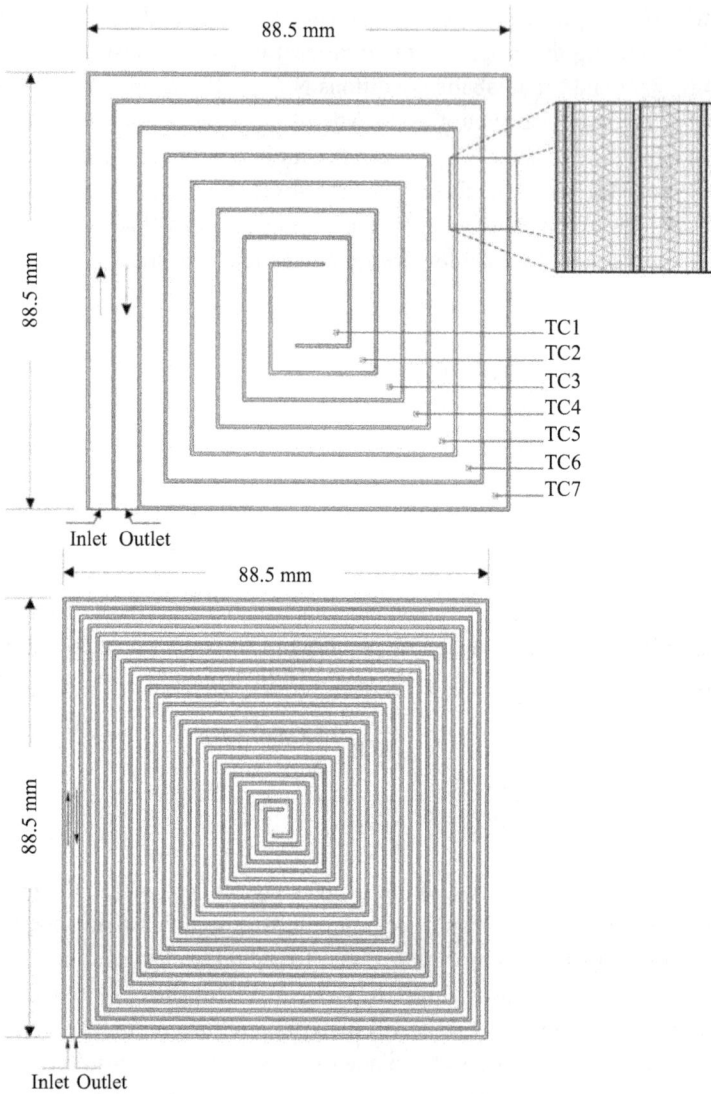

Figure 10.13. 3.5-turn and 12-turn spiral combustors used for the computational study of the effects of the number of turns on combustor performance when the overall device size and wall thickness are fixed (data from [18]). Not shown for brevity: 5.5, 7, and 10-turn devices.

which varies from about 6 to 81 for Re = 50 and proportionally less for the higher values of Re. Since the optimal N for spiral exchangers varies roughly in proportion to N (Figure 10.2c), due to the n^2 effect the larger values of n (more turns) correspond to values of N beyond the optimum. This, coupled to the additional heat loss at large n just mentioned, results in an optimum not only the number of transfer units (N) but also in the number of turns (n). At higher Re, the effect is not as pronounced since N is smaller (not far beyond the optimal value if at all) and the impact of heat loss is less; thus the adverse effect of too many turns is not as pronounced.

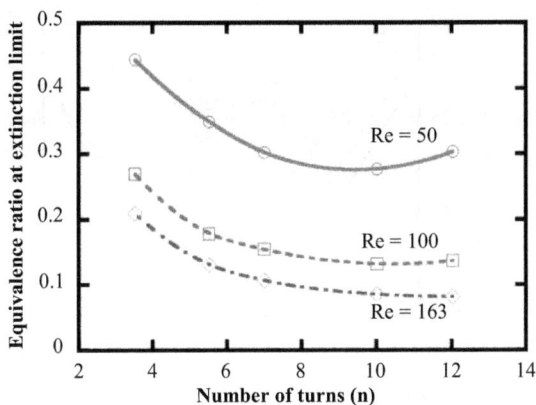

Figure 10.14. Computed propane-air equivalence ratios at the extinction limits for Swiss roll combustors with varying numbers of turns (n) but the same overall size and same wall thickness (data from [18]).

10.4.2 EFFECT OF HEAT EXCHANGER HEIGHT AND OUT-OF-PLANE HEAT LOSSES

The scaling analysis of Section 10.3 was performed without any consideration of heat loss in the third dimension, that is, perpendicular to the plane of the spiral. In contrast, the computations described in Section 10.3.2 were fully three-dimensional and did include these losses. Figure 10.15 shows a comparison of the extinction limits in the nominal-size (5 cm total height) combustor with identical devices extruded to 10 cm or cut to 2.85 cm height. There is very little difference in the extinction limits between the 5 and 10 cm devices, justifying the neglect of heat losses in the third dimension for the computations in Section 10.3; however, the shorter device does show significantly decreased performance due to an increased out-of-plane heat loss rate relative to the heat release rate, indicating that out-of-plane losses must be considered in such cases. Moreover, the limits are practically independent of height at high Re where extinction results from residence time limitations (low Da) and heat losses are unimportant, whereas the limits are very dependent on height at low Re where heat losses are responsible for the extinction limits. These results are consistent with experiments employing Swiss roll propane-air combustors of varying heights but otherwise identical geometries [19, 20] where it is found that the taller device has wider extinction limits (e.g., device "D" vs. "S" in [20]). On the other hand due to their small height and thick dividing walls, these burners suffer substantial heat losses in the third dimension and thus are not technically heat-recirculating combustors; comparing a 2-turn device "RD" with a 4-turn device "D" of equal channel width, height, and wall thickness it is found that the extinction limits are nearly identical. This is consistent with Figure 10.6 (middle), where the analysis of Section 10.2.4 shows that with substantial heat losses the part of the exchanger farthest away from the reaction region does not provide any preheating of the reactants because exhaust enthalpy is lost to ambient instead of being transferred to the reactants, and thus a longer device (more turns) would have the same performance as one with fewer turns. Additionally, because of these losses, the extinction limits in these devices are

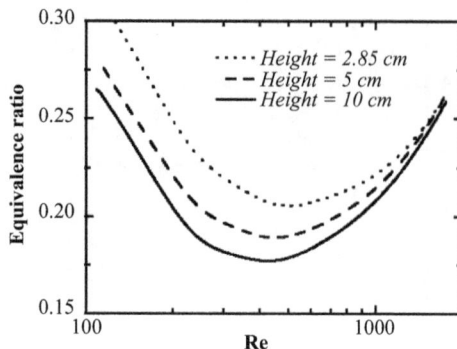

Figure 10.15. Computed propane-air equivalence ratios at the extinction limits for nominal-
scale Swiss roll combustors of different heights
(data from [17]).

within the conventional flammability limits, whereas with taller devices and thinner dividing walls, the extinction limits may be substantially outside the conventional flammability limits (e.g., Figure 10.17).

10.4.3 OPTIMAL THERMAL CONDUCTIVITY

The analytical presented in Section 10.2.4 show that streamwise wall heat conduction is detrimental to performance; however, for zero wall thermal conductivity (k_w), no heat recirculation is possible. Thus, an optimum k_w causing the widest possible extinction limits must exist. (This behavior was not predicted by the analysis in Section 10.2.4 because thermally-thin heat exchanger walls were assumed *a priori*, thus neglecting the thermal resistance across the wall.) Figure 10.16 shows the computed effect of k_w on extinction limits for varying Re for a 3.5-turn combustor with one-step propane-air chemistry. It can be seen that the optimum k_w is extremely small—in fact, smaller than the conductivity of air (0.026 W/m°C). The following estimate of this optimum is proposed. There is no disadvantage to lower k_w until the wall thermal resistance $\sim \tau/k_w$ is comparable to the thermal resistance between gas and wall $\sim 1/U_T$. At higher k_w, streamwise wall conduction reduces performance, whereas at lower k_w heat recirculation via conduction across the wall is diminished. Thus at the optimum condition $k_w \approx U_T\tau \approx ((10.6/2)$ W/m²K)(0.0005mm) = 0.0028W/mK (here U_T is the computed value averaged over all interior walls), which is comparable to the calculated optimal value seen in Figure 10.15. The optimum is more pronounced at lower Re (higher N) where the impact of both streamwise wall heat conduction and heat losses are greater. At high Re (lower N), where the impact of these effects is greatly reduced, there is no effect of the wall thermal conductivity except at very low values of k_w. These results demonstrate that while a theoretical optimal thermal conductivity exists, for any solid material from which heat exchanger walls might be constructed, lower thermal conductivity is always advantageous. (While one could reach such low values of the effective thermal conductivity using a Dewar-type wall construction, this would only suppress heat transfer across the wall, that is, in the direction for which heat transfer is desired, and would not decrease heat transfer in the streamwise direction along the wall, for which heat transfer is detrimental. Consequently Dewar-type walls would be of no value for this application.)

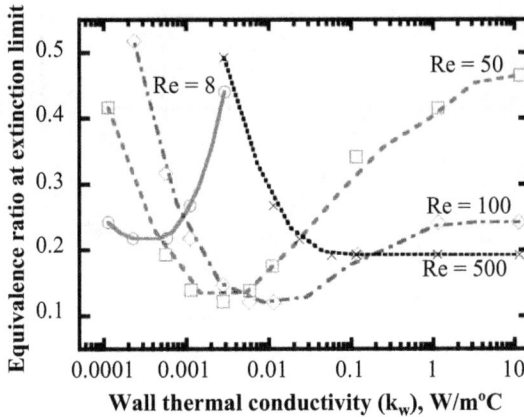

Figure 10.16. Computed propane-air equivalence ratios at the extinction limits for 3.5-turn Swiss roll combustors with varying wall thermal conductivity (k_w). While the values of k_w vary with temperature within the combustor, the values show on the horizontal axis are those corresponding to the reactants at ambient conditions (data from [18]).

10.4.4 TURBULENCE EFFECTS

While invocation of the word "microcombustion" suggests low Reynolds numbers for which the effects of turbulent flow would be absent, the possibility of transitional values of Re cannot be ruled out. The primary effect turbulence has on heat-recirculating combustors is of course the increase in U_T and thus increased N when Re is sufficiently high. This would increase the excess enthalpy E (except for large N for spiral reactors, as discussed in Section 10.2.2) and thus reduce the fuel concentration and equivalence ratio at the extinction limit compared to that which would occur without turbulence. Of course this supposition cannot be tested experimentally because one cannot arbitrarily suppress turbulence at will, but this can be tested computationally. Figure 10.17 (upper) shows a comparison of the experimentally-measured extinction limits of lean propane-air mixtures to those predicted with a 3D RSM of turbulent flow and heat transfer enabled and disabled. What is remarkable about this figure is that the predictions are practically the same with and without the turbulence model enabled. The reason for this is that without turbulence, Dean vortices form in the curved channels that enhance heat transport and thus heat recirculation by nearly the same amount as turbulence does [10]. Note also that the predictions of both 3D models are in good agreement with the experimental extinction limits. Figure 10.17 (lower) shows that, in contrast, for the 2D simulations the extinction limits are very different with and without the turbulence model activated because Dean vortices cannot form in 2D simulations. Without the turbulence model, the 2D simulation significantly underpredicts the amount of heat transfer and thus heat recirculation, leading to a requirement for a higher fuel concentration to reach a given flame temperature and thus a higher equivalence ratio at the lean extinction limit.

Figure 10.18 shows a comparison of temperatures predicted by 2D simulations [15] at the same locations as the thermocouples in the experiments [9]. (These computed temperatures are not the maxima over the entire computational domain but provide the most realistic

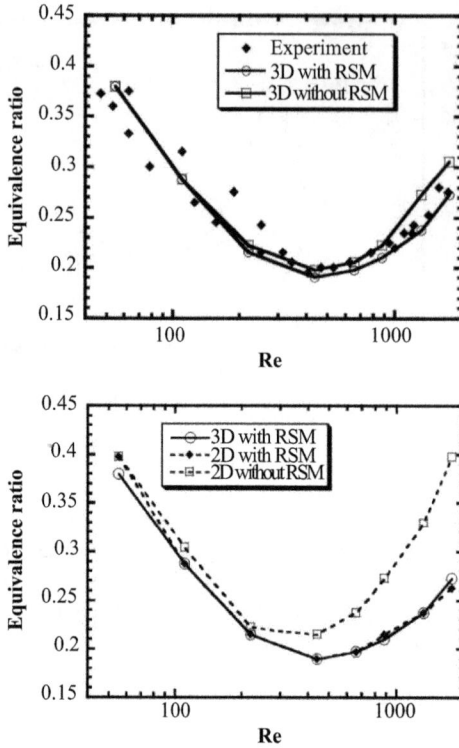

Figure 10.17. Upper: comparison of measured extinction limits in Swiss-roll combustors to predictions of the three-dimensional numerical model with and without the RSM turbulence model activated [10]. Lower: comparison of extinction limits predicted by the 2D simulations with and without the RSM turbulence model activated and the 3D simulations with the RSM activated.

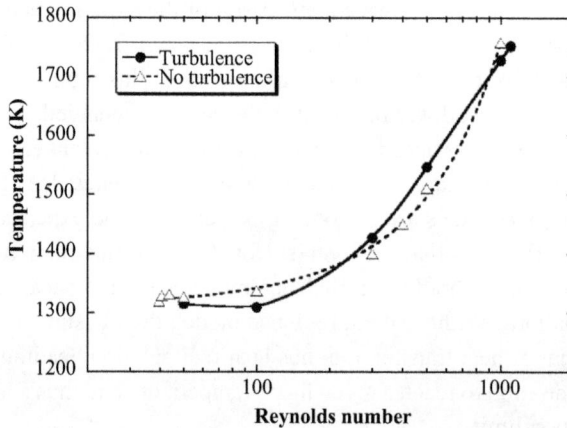

Figure 10.18. Maximum calculated [15] temperatures at extinction limits obtained using a 2D model with and without turbulent flow and transport suppressed.

comparison between the model and the experiment.) An interesting feature of Figure 10.18 is that the limit temperatures with and without turbulent transport are nearly identical even though the limit T_{ad} and mixture strengths (see for example Figure 10.17, lower) are very different. This is because at high-Re conditions where turbulent flow is present, extinction is caused by insufficient residence time compared to the reaction time scale rather than heat losses. This residence time is set by the flow velocity (thus Re). Consequently, Re sets the chemical reaction rate required to avoid extinction, which is far more sensitive to temperature than any other property. Thus, to a very close approximation it can be stated that a given Re requires a given reaction temperature to avoid extinction, regardless of the transport environment required to obtain this temperature.

10.4.5 CATALYSIS

The analysis and experiments on heat-recirculating combustors reported above correspond to conditions where only gas-phase reactions can occur. In practice, it is often advantageous to employ a catalyst to extend the extinction limits, particularly for conditions where high reaction temperatures are undesirable or unsustainable (e.g., due to heat losses). Catalytic combustion at small scales is discussed in more detail elsewhere in this book; however, some discussion of the effect of catalytic combustion on the performance of heat-recirculating combustors is given in this chapter. Figure 10.19 shows extinction limits measured in a 3.5-turn Swiss roll combustor, 5 cm in height with 7.5 cm overall width and depth, with and without the use of a catalyst. For experiments with catalyst, strips of bare platinum foil were placed along the walls in the central section of the combustor, with an exposed area of ≈ 30 cm^2. Four curves are shown in Figure 10.19 indicating extinction limits for catalytic combustion using untreated Pt, catalytic combustion using Pt treated via combustion in an ammonia-air mixture, gas-phase combustion (i.e., without catalyst installed), and the boundaries of the "out-of-center reaction zone" regime discussed below. The salient features of Figure 10.19 include:

- With the ammonia-treated catalyst, combustion can be sustained over a range of Re from about 1 to 2000 (higher Re are potentially attainable but corresponded to minimum temperatures required to sustain combustion in excess of limit for the Inconel from which the combustor was built). This range of Re corresponds to a 2000:1 range of mass flow; this "turn-down ratio" far exceeds that of any other type of combustor known to the author.
- Without catalyst, the minimum Re is much higher (about 40) and thus the attainable turn-down ratio smaller.
- The catalyst improves lean-limit performance only slightly at moderate Re and not at all at the highest Re tested. This is probably because at sufficiently high Re the ratio of the mass flux to the catalyst to the total mass flux becomes very small and only a small fraction of the fuel can be burned on the catalyst, in which case gas-phase combustion is the only means to the self-sustain reaction.
- The rich limits are extended drastically with catalyst; for example the equivalence ratio at the rich extinction limit is about 40 for Re = 15. Product analysis [9] shows that these rich mixtures are not "reformed" or partially oxidized; the products do not contain significant amounts CO or H_2 but instead only unburned fuel, CO_2 and H_2O.

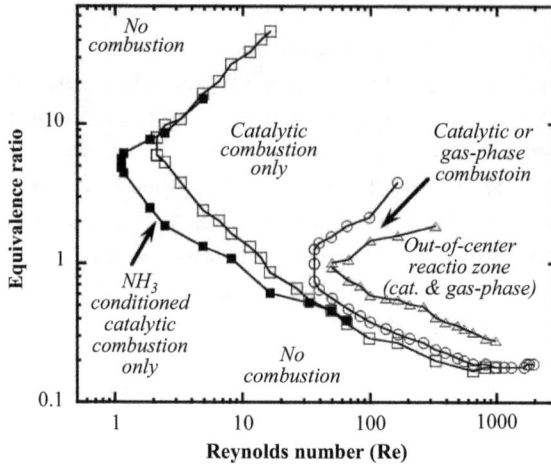

Figure 10.19. Experimental extinction limit map for
catalytic and gas-phase combustion in a 3.5-turn Swiss roll
using propane-air mixtures (data from [9]).

- Ammonia treatment of the catalyst extends the extinction limits, but only at low Re corresponding to low temperatures (see Figure 10.20).

- For Re < 15, the <u>lean</u> catalytic extinction limit is actually <u>rich</u> of stoichiometric. No similar trend was found without catalyst, for which the limits are nearly symmetric about stoichiometric. The asymmetry with catalytic combustion was attributed to (1) at low Re and thus low temperatures, desorption of O(s) or CO(s) is the rate-limiting process, and (2) there is a change in the Pt surface coverage from O(s) to CO(s) for sufficiently rich mixtures, with a corresponding increase in the net reaction rate due to the lower activation energy for desorption of CO(s) than O(s) [21].

- For sufficiently high reaction temperatures (i.e., near-stoichiometric mixtures and higher Re), the reaction is sufficiently strong to self-sustain without the benefit of heat recirculation and the reaction zone moves out of the center of the spiral towards the inlet. The out-of-center limit is the same for catalytic and gas-phase combustion, which is expected since the catalyst cannot affect the reaction when it is essentially completed upstream of the catalyst.

Figure 10.20 shows the maximum measured temperatures <u>at the extinction limit</u> as a function of Re for both gas-phase and catalytic combustion. Even without catalyst, temperatures required to support combustion (typically 1100 K) are lower than for propagating hydrocarbon-air flames (1500 K), indicating conditions more similar to plug-flow reactors than flames (the difference being that plug-flow reactors have prescribed temperature profiles adjusted using electrical heaters, whereas for the Swiss roll combustor the temperature profile is strongly coupled to heat transfer and heat release). For Re > 750, temperatures gas-phase and catalytic temperatures converge, corresponding to the convergence of lean extinction limits (Figure 10.19). Minimum temperatures required to support combustion with catalyst are typically 400K less than for the gas-phase reaction, though the difference in limit mixture compositions is small (Figure 10.19). For this reason an additional curve is given in Figure 10.20 that shows T_{max} at the limit with catalytic reaction *for the mixture at the gas-phase lean extinction limit*. For

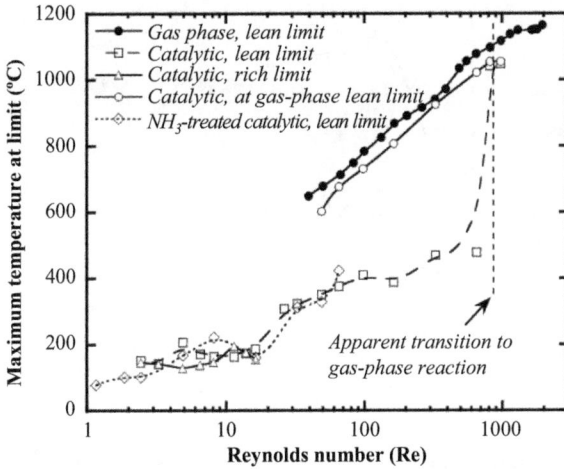

Figure 10.20. Maximum burner temperatures at the extinction limits (data from [9]).

this case it can be see that the temperatures with catalyst are actually slightly lower than with the gas-phase reaction only. This shows that the catalyst is actually slightly detrimental to the gas-phase reaction; the catalytic reaction is beneficial only when the gas-phase reaction is not possible.

Perhaps the most noteworthy aspect of Figure 10.20 is the low temperatures capable of sustaining combustion with catalyst. The minimum temperature observed is only 78°C at Re = 1.2, which to the author's knowledge is the lowest reported self-sustaining hydrocarbon flame temperature (as described in [9], care was taken to ensure that the maximum catalyst temperature was measured.) For both catalytic and gas-phase reactions, the minimum temperature required to support combustion increases as Re increases. This is expected since higher Re means higher velocities (since neither channel width nor viscosity was changed), thus shorter residence times and consequently faster reaction required to sustain combustion. The faster reaction in turn requires higher temperatures. In fact, Figure 10.20 can be redrawn as an Arrhenius plot (ln(Re) vs. 1/T); the resulting plot (not shown) is nearly linear with slopes corresponding to effective activation energies (E) of about 19 and 6.4 kcal/mole for gas-phase and catalytic limits, respectively. On first glance the latter seems to be rather low, but the dimensionless activation energy (E/RT) for the Re = 1.2 limit case is 9.2, indicating significant sensitivity to temperature even in this case.

Having low steady-state flame temperatures is of some merit because it results in smaller heat losses, fewer issues with thermal expansion and allows a wider choice of materials. However, in some applications higher temperatures are desired to supply thermal power to thermoelectric or pyroelectric generators, solid oxide fuel cells, and so forth. In contrast, having low ignition temperatures is always meritorious because it reduces the energy storage (e.g., from a battery or supplemental fuel) required for a self-contained micropower generation system. Figure 10.21 shows the ignition temperature (determined by slowly increasing the power delivered to an electrically-heated wire in the center of the spiral and measuring the temperature at which a sudden jump corresponding to ignition occurs) for a Swiss roll combustor using the ammonia-treated Pt catalyst. Also shown in Figure 10.21 is the corresponding combustion temperature measured after steady-state conditions are achieved. The lean and rich extinction

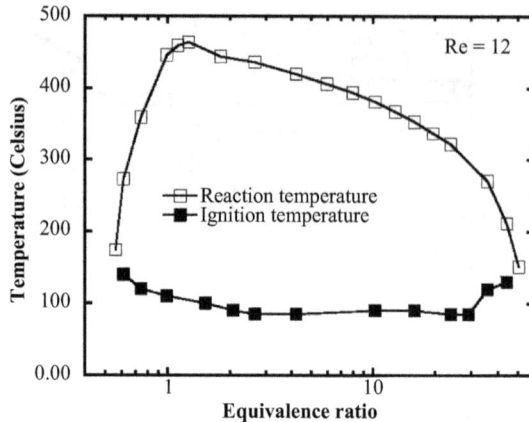

Figure 10.21. Ignition temperatures and steady-state flame temperatures in a Swiss-roll combustor using propane-air mixtures as a function of equivalence ratio for Re = 12.

limits correspond to conditions where the steady-state combustion temperature is nearly as low as the ignition temperature. It can be seen that due to the thermal management provided by the Swiss roll and the low-temperature reaction provided by the catalyst, ignition temperatures as low as 85°C are readily obtained. It is well known that hydrogen will ignite at room temperature on Pt catalyst; hydrocarbons do not, but to the author's knowledge 85°C is the lowest reported ignition temperature for self-sustaining hydrocarbon combustion.

10.5   CONCLUSION AND FUTURE WORK

Heat-recirculating combustors are an important component of many micropower generation systems because they enable the self-sustaining chemical reaction over a wider range of mixtures and flow rates than conventional combustors. However, at small scales heat-recirculating combustors suffer from many of the same issues as other types of combustors, namely heat losses, insufficient residence time, and undesirable redistribution of thermal energy via solid-phase heat conduction. This chapter discussed ways of assessing these issues particularly with regard to identification and use of scaling parameters. The key design parameters examined in this chapter are:

- Geometry—rolling a linear counter-current heat exchanger into a spiral (Swiss roll) yields two advantages: a significantly smaller heat loss parameter (α) and a doubling of the number of transfer units (N), both of which benefit performance substantially. While at low Re (large N) it is theoretically possible a linear exchanger to exhibit better performance than a spiral one, in practice it is unlikely that heat losses can be minimized to the point where this would be observable.
- Number of turns—for a given overall device size and wall thickness, there is an optimal choice of the number of turns of the spiral. A device with too few turns has a small

value of N and thus small excess enthalpy (E); a device with too many turns also has small E (Figures 10.2b, 10.2c) and additionally suffers from greater heat losses due to heat conduction through the walls in the direction out of the plane of the spiral (Figure 10.14).

- Device height—a short device will have substantial out-of-plane heat losses and thus a taller device is preferred; however, if total volume is constrained then an optimal height will exist which minimizes out-of-plane heat loss (Figure 10.15) yet provides space for a larger number of spiral turns and/or larger channel width (d).
- Dividing wall thermal conductivity—although a perfectly insulating wall between the inlet and outlet arms of the heat exchanger does not allow heat recirculation, for all practical purposes a wall material having lowest possible thermal conductivity provides the widest extinction limits. Among dense (non-porous) solid materials, polymers have the lowest thermal conductivities. Polymers generally cannot tolerate sufficiently high temperatures to sustain gas-phase hydrocarbon combustion, but with catalysis the combustion temperatures are well within the temperature limits of many polymers (particularly polyimides) and have been used to construct Swiss roll combustors [22].
- Wall emissivity—while difficult to modify, lower wall emissivity leads less turn-to-turn heat transfer and improves performance in a manner similar to lower wall thermal conductivity.
- Catalytic combustion—the use of appropriate catalysts results in greatly reduced flame temperatures and thus smaller heat losses, fewer issues with thermal expansion, and a wider choice of materials. Additionally, ignition temperatures are substantially lower for catalytic combustion than gas-phase combustion and room-temperature ignition of hydrocarbons may be possible. Considering that the lowest observed ignition temperature is about 85°C (Figure 10.21) and the apparent activation energy for low-temperature catalytic combustion is 6.4 kcal/mole, to obtain ignition at 25°C would require a factor of 6 increase in residence time within the combustor (i.e., a 6-fold increase in N) without a corresponding increase in heat losses (meaning a 6-fold decrease in the heat loss α, since at large N, Equation (10.3) shows that $E \sim 1/\alpha N$)—a challenging but possibly achievable goal.

Finally, it is noted that the scaling parameters identified in Section 10.2 and tested in Section 10.3 can be used not only to predict the behavior of combustors of varying scale, but also to extrapolate the performance of relatively large, easily tested laboratory-scale devices to predict the performance of small-scale devices. The scaling is sometimes unintuitive, for example, the radiation parameter R (Equation (10.17)) scales in proportion to d^1 for laminar flow and thus radiation may not be significant at small scales but if not considered in the extrapolation of large-scale devices to smaller scales, inaccurate predictions may result.

ACKNOWLEDGMENTS

The author wishes to thank graduate students Jeongmin Ahn, Chien-Hua Chen, and James Kuofor for their contributions to the study of heat-recirculating combustors. This research has been supported by the Defense Advanced Research Projects Agency, National Aeronautics and

Space Administration and the Defense Threat Reduction Agency. This chapter is dedicated to the memory of Prof. Felix J. Weinberg, a valued colleague and scholar.

REFERENCES

[1] Dunn-Rankin, D., E.M. Leal, and D.C. Walther. 2005. "Personal Power Systems." *Progress in Energy and Combustion Science* 31, no. 5, pp. 422–65. doi: http://dx.doi.org/10.1016/j.pecs.2005.04.001

[2] Walther, D.C., and J. Ahn. 2011. "Advances and Challenges in the Development of Power Generation Systems at Small Scales." *Progress in Energy and Combustion Science* 37, no. 5, pp. 583–610. doi: http://dx.doi.org/10.1016/j.pecs.2010.12.002

[3] Maruta, K. 2011. "Micro and Mesoscale Combustion: Technology Development and Fundamental Research." *Proceedings of the Combustion Institute* 33, no. 1, pp. 125–50. doi: http://dx.doi.org/10.1016/j.proci.2010.09.005

[4] Kaisare, S.N., and D. Vlachos. 2012. "A Review on Microcombustion: Fundamentals, Devices and Applications." *Progress in Energy and Combustion Science* 38, no. 3, pp. 321–59. doi: http://dx.doi.org/10.1016/j.pecs.2012.01.001

[5] Ju, Y., and K. Maruta. 2011. "Microscale Combustion: Technology Development and Fundamental Research." *Progress in Energy and Combustion Science* 37, no. 6, pp. 669–715. doi: http://dx.doi.org/10.1016/j.pecs.2011.03.001

[6] Karim, A.M., J.A. Federici, and D.G. Vlachos. 2008. "Portable Power Production from Methanol in an Integrated Thermoeletric/Microreactor System." *Journal of Power Sources* 179, no. 1, pp. 113–20. doi: http://dx.doi.org/10.1016/j.jpowsour.2007.12.119

[7] Ahn, J., Z. Shao, P.D. Ronney, and S.M. Haile. November 2009. "A Thermally Self-Sustaining Miniature Solid Oxide Fuel Cell." *Journal of Fuel Cell Science and Technology* 6, no. 4. doi: http://dx.doi.org/10.1115/1.3081425

[8] Jones, A.R., S.A. Lloyd, and F.J. Weinberg. 1978. "Combustion in Heat Exchangers." *Proceedings of the Royal Society of London Series A: Mathematical and Physical Science* 360, no. 1700, pp. 97–115. doi: http://dx.doi.org/10.1098/rspa.1978.0059

[9] Ahn, J., C. Eastwood, L. Sitzki, and P.D. Ronney. 2005. "Gas-phase and Catalytic Combustion in Heat-recirculating Burners." *Proceedings of the Combustion Institute* 30, no. 2, pp. 2463–72. doi: http://dx.doi.org/10.1016/j.proci.2004.08.265

[10] Chen, C.-H., and P.D. Ronney. 2011. "Three-dimensional Effects in Counterflow Heat-recirculating Combustors." *Proceedings of the Combustion Institute* 33, no. 2, pp. 3285–91. doi: http://dx.doi.org/10.1016/j.proci.2010.06.081

[11] Lloyd, S.A., and F.J. Weinberg. 1974. "A Burner for Mixtures of Very Low Heat Content." *Nature* 251, no. 5470, pp. 47–49. doi: http://dx.doi.org/10.1038/251047a0

[12] Lloyd, S.A., and F.J. Weinberg. 1975. "Limits to Energy Release Utilization from Chemical Fuels." *Nature* 257, no. 5525, pp. 367–70. doi: http://dx.doi.org/10.1038/257367a0

[13] Targett, M., W. Retallick, and S. Churchill. 1992. "Solutions in Closed form for a Double-spiral Heat Exchanger." *Industrial and Engineering Chemical Research* 31, no. 3, pp. 658–69. doi: http://dx.doi.org/10.1021/ie00003a003

[14] Strenger, M., S. Churchill, and W. Retallick. 1990. "Operational Characteristics of a Double-spiral Heat Exchanger for the Catalytic Incineration of Contaminated Air." *Industrial and Engineering Chemical Research* 29, no. 9, pp. 1977–84. doi: http://dx.doi.org/10.1021/ie00105a033

[15] Kuo, C.H, and P.D. Ronney. 2007. "Numerical Modeling of Non-Adiabatic Heat-Recirculating Combustors." *Proceedings of the Combustion Institute* 31, no. 2, pp. 3277–84. doi: http://dx.doi.org/10.1016/j.proci.2006.08.082

[16] Ronney, P.D. 2003. "Analysis of Non-Adiabatic Heat-recirculating Combustors." *Combustion and Flame* 135, no. 4, pp. 421–39. doi: http://dx.doi.org/10.1016/j.combustflame.2003.07.003

[17] Chen, C.-H., and P.D. Ronney. 2013. "Scale and Geometry Effects on Heat-Recirculating Combustors." *Combustion Theory and Modelling* 17, no. 5. doi: http://dx.doi.org/10.1080/13647830.2013.812807 to appear

[18] Kuo, C.H. 2006. "Numerical Modeling of Heat-Recirculating Combustors" [PhD. thesis]: University of Southern California, Los Angeles, USA.

[19] Kim, N.I., S. Kato, T. Kataoka, T. Yokomori, S. Maruyama, T. Fujimori, and K. Maruta. 2005. "Flame Stabilization and Emission of Small Swiss-roll Combustors as Heaters." *Combustion and Flame* 141, no. 3, pp. 229–40. doi: http://dx.doi.org/10.1016/j.combustflame.2005.01.006

[20] Kim, N.I., S. Aizumi, T. Yokomori, S. Kato, T. Fujimori·, and K. Maruta. 2007. "Development and Scale Effects of Small Swiss-roll Combustors." *Proceedings of the Combustion Institute* 31, no. 2, pp. 3243–50. doi: http://dx.doi.org/10.1016/j.proci.2006.08.077

[21] Maruta, K., K. Takeda, J. Ahn, K. Borer, L. Sitzki, P.D. Ronney, and O. Deutschman. 2002. "Extinction Limits of Catalytic Combustion in Microchannels." *Proceedings of the Combustion Institute* 29, no. 1, pp. 957–63. doi: http://dx.doi.org/10.1016/s1540-7489(02)80121-3

[22] Sanford, L.L., S.Y.J. Huang, C.S. Lin, J.M. Lee, J.M. Ahn, and P.D. Ronney. 2008. "Plastic Mesoscale Combustors/Heat Exchangers." *Proceedings of the ASME International Mechanical Engineering Congress and Exposition 2007, Vol. 6: Energy Systems: Analysis, Thermodynamics and Sustainability (2008)*, pp. 141–45.

NOMENCLATURE

A Exponent in scaling relation for turbulent flow ($Nu \sim Re^n$)

A_T Heat exchange area (m^2)

A_L Heat loss area (m^2)

A_X Cross-section area of heat exchanger channel (m^2)

Bi Biot number

C_p Gas heat capacity (J/kgK)

d Channel width (m)

D Mass diffusivity

Da Damköhler number

E Dimensionless excess enthalpy

E_a Activation energy (J/mole)

h_T Convective heat transfer coefficient inside channel (W/m^2K)

h_L Convective heat loss coefficient (W/m^2K)

k_g Gas thermal conductivity (W/mK)

k_w Wall thermal conductivity (W/mK)

L Channel length (m)

$\dot{m}$ Mass flow rate (kg/s)

M Fuel molecular weight (kg/mole)

n Number of turns of spiral heat exchanger

N	Number of transfer units
Nu	Nusselt number inside channel
Pr	Prandtl number = μ/rk
Q	Fuel heating value (J/kg)
$Q_{L,i}$	Heat loss from the inlet channel (W)
$Q_{L,o}$	Heat loss from the outlet channel (W)
Q_T	Heat exchange rate from products to reactants (W)
Q_w	Wall-to-wall radiative transfer rate (W)
R	Internal radiation coefficient
$\mathcal{R}$	Gas constant (J/moleK)
Re	Reynolds number of flow inside channel
T	Temperature (K)
U_T	Overall heat transfer coefficient inside the channel (W/m²K)
U_L	Overall heat transfer coefficient to environment (W/m²K)
U_R	Heat transfer coefficient for internal wall-to-wall radiation (W/m²K)
v	Gas flow velocity inside channel
V	Reaction zone volume (m^3)
Y	Fuel mass fraction
Y_∞	Fuel mass fraction in the fresh reactants ($= Y_1$)
Z	Pre-exponential factor in the Arrhenius reaction rate (1/s)
α	Heat loss coefficient
ε_i	Internal wall emissivity
ε_L	External wall emissivity
κ	Gas thermal diffusivity
μ	Gas dynamic viscosity
ρ	Gas density
σ	Stefan-Boltzmann constant (W/m²K⁴)
τ	Heat exchanger wall thickness

CHAPTER 11

CATALYTIC REACTORS: POWER GENERATION AND FUEL PROCESSING

Marco Schultze and John Mantzaras

The large surface-to-volume ratios of microreactors favor the use of catalytic reactions as opposed to volumetric gas-phase reactions. Latest designs of efficient catalytic microreactors for portable power generation and for on-board fuel reforming are outlined. The underlying physicochemical processes are presented and state-of-the-art numerical models for catalytic microreactors are reviewed. The key mechanisms affecting the thermal management of catalytic microreactors are elaborated and, finally, numerical and experimental investigations of steady and transient operation of catalytic microreactors and microreformers are presented.

11.1 INTRODUCTION TO CATALYTIC COMBUSTION

Complete or partial catalytic oxidation of low hydrocarbons over catalytically active surfaces is used in many technical processes, such as chemical synthesis, exhaust gas treatment, fuel reforming, fuel cells, microreactors, and gas turbines for power generation. These processes have been the subject of many research and development efforts over the last years. Heterogeneous (catalytic) combustion has been intensively investigated for gas turbines of power generation systems since the early 1970s, aiming at a significant reduction of NO_x emissions and improved combustion stability.

Initially, the catalytically stabilized thermal combustion (CST) concept was developed [1]. In the CST methodology, part of the fuel is converted heterogeneously in one or more sequential catalytic reactor modules, while the remaining fuel is combusted homogeneously in a post-catalyst flame. Since heterogeneous combustion is a flameless process and does not contribute to NO_x formation, CST provides an ultralow NO_x combustion technology with demonstrated NO_x emissions less than 3 ppm under turbine-relevant conditions [2, 3]. Compared to NO_x after-treatment techniques, this technology can lead to a significant cost reduction and offers a possibility to meet the stringent NO_x emission regulations in Europe and the United States. Furthermore, CST improves the flame stability and reduces combustion-induced pressure pulsations when compared to the conventional lean-premixed combustion approaches. By using heterogeneous combustion it is also possible to handle renewable and "dirty" fuels containing a high amount of NH_3 (fuel-bound nitrogen), such as biogases. Instead of contributing to additional NO_x emissions, NH_3 can be oxidized catalytically back to N_2.

A schematic of a CST-based natural gas turbine burner is given in Figure 11.1, including typical design temperatures. In such a burner, fuel is premixed with preheated air upstream of the catalytic reactor. In most cases, flow velocities (defined by the required mass flow rates) lead to laminar flows inside the catalytic channels, although transitional flows can also be attained [4]. The number of catalytic modules used, their length and cross-sectional area, as well as their cell density and the active material are determined by the specific operating conditions. For turbines operating with natural gas at fuel-lean stoichiometries, palladium-based catalysts are typically used, since they combine the highest activity for methane oxidation and a low volatilization rate of the active catalyst. In order to attain catalyst light-off (i.e., ignition of the catalytic reactions), it is sometimes necessary to increase the gas inlet temperature by a preburner upstream of the catalytic module (see Figure 11.1). Because of materials' limitations owing to the high temperatures of modern gas turbines (~1750 K), concepts featuring complete combustion of the fuel inside the catalytic reaction module [5] are not practical. Instead, a hybrid CST approach is favored, combining a fractional heterogeneous fuel conversion and a follow-up homogeneous fuel conversion. The catalyst is used to increase the gas temperature to a level at which homogeneous combustion can be initiated and stabilized aerodynamically downstream of the catalyst. The temperature at the reactor outlet depends on the catalytic reaction rate and the amount of fuel and air passed through the catalytic module; although in the approach shown in Figure 11.1 all the fuel and air flow through the catalytic reactor, in other CST variants part of the air and/or fuel bypass the catalytic module [6]. In doing so, the catalytic reactor products establish a pilot flame, which in turn stabilizes the homogeneous combustion of the bypassed reactants.

A recent alternative to CST for natural-gas-fueled turbine burners is the catalytic-rich/gaseous-lean combustion methodology, which entails catalytic partial oxidation (CPO) of the hydrocarbon fuel to synthesis gas. A schematic of this concept is shown in Figure 11.2 [7, 8]. Therein, fuel and part of the preheated air are mixed and driven, at an overall fuel-rich stoichiometry, in a CPO reactor (which is coated with a noble metal, with rhodium the preferred catalyst). Catalyst coating is applied to the outer surfaces of a bundle of tubes through which

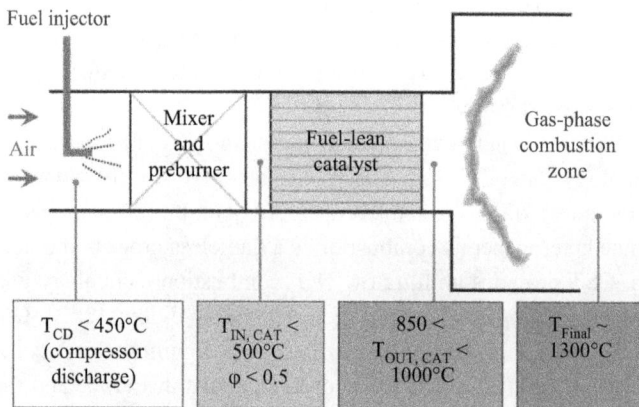

Figure 11.1. Catalytically stabilized thermal (CST) concept.

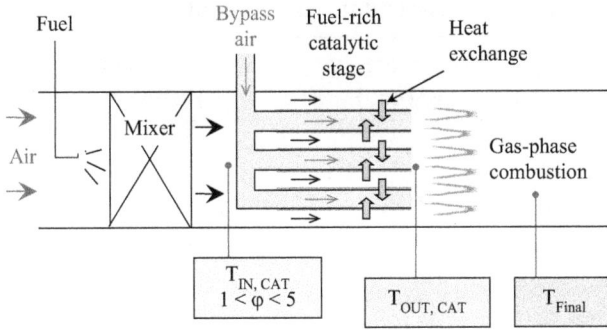

Figure 11.2. Catalytic-rich/gaseous lean combustion approach.

bypass air flows (Fig. 11.2). The catalyst surfaces are cooled by the bypass air, thus avoiding reactor overheating. The products of the CPO reactor, mainly hydrogen-rich synthesis gas and unconverted reactants, are mixed with the bypass air, forming overall fuel-lean flames downstream of the catalyst. A major advantage of this concept is the lower catalyst light-off temperature compared with that of the CST approach, the reason being that hydrocarbons ignite easily over noble metals at the conditions of oxygen deficiency [9]. Thus, the preburner shown in Figure 11.1, which is a potential source for NO$_x$ emissions, is no longer necessary. Furthermore, the CPO extinction limit is extended compared to lean catalytic combustion [7, 8, 10]; the catalytic combustion is controlled by the availability of oxygen (which negates complete fuel consumption inside the CPO even in the case of accidental homogeneous ignition) and finally the stability of the ensuing flame is enhanced because of the hydrogen content in the CPO products [11, 12]. The CPO concept also features single-digit NO$_x$ emissions [7].

The fuel-rich catalytic concept is applicable for turbines using not only natural gas but also syngas or pure hydrogen fuels [7, 13, 14]. In this case, the catalyst does not have a prime CPO function (at least for syngas fuels with low methane content) but rather acts as a preheater and stabilizer for the downstream homogeneous combustion zone. This concept can be used for a wide range of different syngas fuels, especially for those having low calorific value, whereby flame stability is an issue, and fuels with high hydrogen content (e.g., coal-derived syngas), where conventional lean-premixed combustion may lead to flame flashback. Finally, a variant of the fuel-rich catalytic concept for syngas fuels is the rich-quick-lean (RQL) approach [15]. This concept is based on the fact that the catalytic consumption of hydrogen is favored over that of carbon monoxide. Thus, the catalyst is used to primarily oxidize the hydrogen of the fuel mixture in the first stage. The exhaust stream of the catalytic module, containing mainly carbon monoxide, is mixed with air and burned homogeneously in a following combustion stage without the risk of flame flashback.

11.2 CATALYTIC MICROREACTORS

With advancing miniaturization of electromechanical devices in the past years, there has been a growing demand for a variety of microscale portable power generation systems. However, most of today's available portable devices are powered by batteries, which result in low energy density. Typical problems are short operation cycles between recharging or replacing the battery and a high contribution of the battery to the total weight and volume of the device.

Table 11.1. Specific energy densities of different fuels and battery types

Technology	Energy density (kWeh/l)	Energy density (kWeh/kg)	Special properties
Primary battery			**Not rechargeable**
Li/MnO$_2$	0.535	0.230	
Li/SO$_2$	0.415	0.260	
Secondary battery			**Rechargeable**
Nickel-cadmium	0.100	0.035	
Nickel-metal hydride	0.240	0.075	
Lithium-ion	0.400	0.150	
Chemical fuels			**Thermal energy**
Methanol	4.384	5.600	
Butane	7.290	12.600	
Iso-octane	8.680	12.340	

The need to reduce the overall system' weight and volume, to increase the operational lifetime, and to reduce manufacturing costs has led to intensified research on hydrocarbon-fueled microreactors for power production. As shown in Table 11.1, the specific energy density of hydrocarbons is significantly higher than that of different battery types. Furthermore, liquid hydrocarbons are easy to transport and relatively safe, rendering them a suitable alternative for commercial products. Table 11.1 draws on ideas in Linden et al. [16].

Catalytic microreactors cover a broad range of applications: catalytic microburners are used for direct chemical to thermal energy conversion in downscaled thermal engines or in combination with thermoelectric (TE), piezoelectric and thermophotovoltaic (TVP) generators for electrical power generation [17–20]. They are further used for fuel reforming in micro-solid oxide fuel cells [21–23], while catalytic microthrusters are promising for space applications [24]. For the fabrication of these high-precision microdevices, advanced manufacturing techniques such as electro-discharge machining (EDM), laser beam machining (LBM) or focused ion beam machining (FIBM) are employed. Materials and techniques similar to those used in the semiconductor industry, for example micro electromechanical systems (MEMS), rapid prototyping, and batch-manufacturing techniques, allow the fabrication of devices in the millimeter range. In most cases noble metal catalysts are applied since they provide a good activity for the oxidation and the reforming of hydrocarbons at modest reactor temperatures [25].

11.2.1 MICROCOMBUSTORS

Microcombustors can be used in different ways for power generation. One approach is to combine them with piezoelectric, TE, or TVP energy conversion devices for electrical power generation. Alternatively, they can be used as fuel reformers for fuel cells or for follow-up catalytic reactors operating on reformates (see Section 11.2.2). The major advantage of these devices is the lack of moving parts, thus mitigating friction losses and reducing maintenance costs.

However, thermal management is a major challenge for micro-scale reactors since the temperature (which is affected by the heat losses) greatly impacts the combustion stability and efficiency of the complete system. TE devices, for example, require for high efficiency a significant temperature gradient, which is difficult to establish due to the small scale of the system and the generally enhanced thermal conductivity of the good electrical conductor materials.

In order to increase flame stability and extend quenching limits, especially for very lean mixtures, Weinberg and co-workers developed a mesoscale burner which uses heat recirculation from the combustion products to preheat the fresh reactants [26, 27] (Figure 11.3). These burners are often referred to as "excess enthalpy" burners, since the enthalpy of the incoming reactants is increased by preheating. To reduce the available surface and thus heat losses to the ambient, the burner can be wrapped around to form the "Swiss roll" configuration, where combustion takes place at the center of the burner (see Figure 11.4). A planar "Swiss roll" combustor still has significant heat losses in the lateral direction. It is conceivable to roll the quasi two-dimensional burner in the lateral direction to create a three-dimensional toroidal shape (Figure 11.5); however, when designing such microscale combustors appropriate fabrication techniques are required [28]. Another way to reduce heat losses is to apply catalytic combustion.

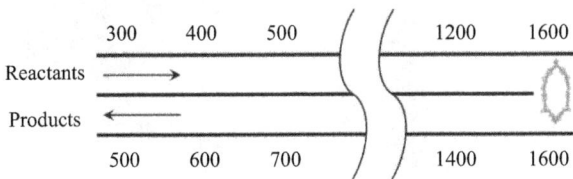

Figure 11.3. Schematic design of excess enthalpy burner with typical temperatures in Kelvin.

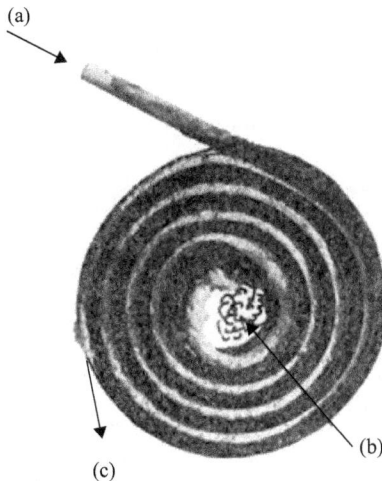

Figure 11.4. Model of an original "Swiss roll" burner with (a) reactant inlet, (b) combustion chamber, and (c) product outlet (reproduced by permission from Lloyd and Weinberg [27]).

Since homogeneous or catalytic combustion is a flameless surface reaction with a significant lower activation energy barrier compared to that of gas-phase chemical reactions, it is possible to sustain stable combustion at lower operational temperatures. Therefore, the minimum inlet temperature at which stable catalytic combustion can be sustained has been introduced in a manner analogous to the flammability limits for gas-phase combustion.

At the University of Southern California, the original Swiss roll concept has been further developed and macro- and meso-scale two-dimensional and toroidal three-dimensional "Swiss roll" burners have been constructed [17, 30]. The channel supplying the unburned mixture and the exhaust gas channel are coiled around each other in a spiral. Combustion takes place in the center of the burner. The heat of the combustion is transferred across the channel walls and used to preheat the incoming fresh mixture and thus increases the total reactant enthalpy. In order to support combustion at low temperatures ($T \approx 380$ K) and low Reynolds numbers ($R_e \approx 1$), the channel walls at the burner center are plated with a platinum catalyst. The effects of catalyst coating on the reactor extinction limits are shown in Figure 11.6. The results are replicated

Figure 11.5. Three-dimensional catalytic burner [29].

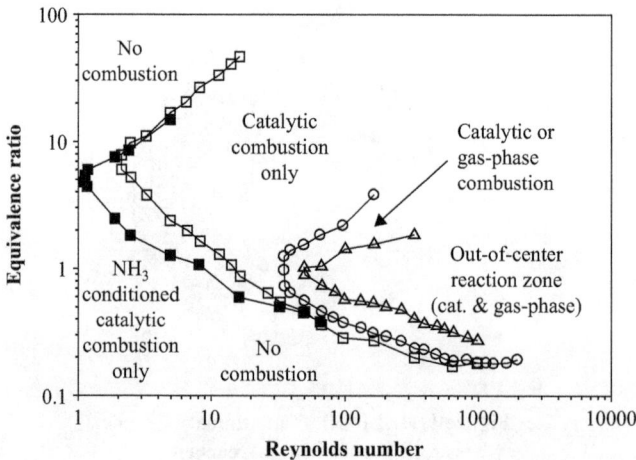

Figure 11.6. Extinction limit map for catalytic and gas-phase combustion in an inconel Swiss roll burner [17].

from Ahn et al. [17]. To improve the thermal efficiency, a thermoelectrically active layer has been embedded into the channel walls separating cold fresh mixture and hot exhaust gas [31]. However, integrating and efficiently operating TE elements in such burners is still difficult and necessitates advances on fabrication techniques and material properties.

At the Yale Center for Combustion Studies, a mesoscale catalytic combustor for power generation applications using liquid fuels (JP-8 jet fuel) has been developed [18, 32]. Since the vaporization of thin films of heavy hydrocarbons at hot surfaces is often difficult to achieve in microscale applications due to coking or plugging of the small channels, electrosprayed atomization of the fuel is employed. This technique is applicable if the fuel comes with a sufficiently high electric conductivity and a moderate surface tension. When feeding the fuel through nozzles kept at a high electric potential relative to a ground electrode, the fuel jet breaks into charged, monodisperse droplets. As a result of the net charge, coagulation of the droplets is prevented and mixing with the oxidant stream is enhanced. For this burner a catalyst substrate design (microlith) [33] has been used, consisting of several catalytically coated grids or screens (Figure 11.7). The results are replicated from Kyritsis et al. [18]. This design provides a high active surface-to-volume ratio that allows converting more fuel at a given pressure drop compared to the classical monolithic designs containing arrays of small channels. It is thus possible to build very compact burners. For best results with regard to combustion efficiency and CO emissions reduction, a platinum/palladium bimetallic catalyst doped with cerium and nickel for better sulfur resistance has been used. The thermal power of the system is in the order of 100 W

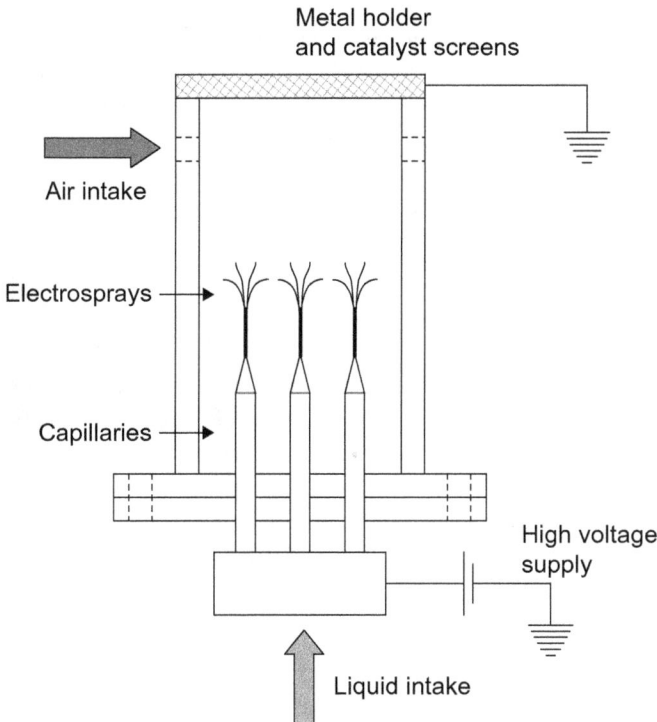

Figure 11.7. Principle of burner using electrosprayed fuel injection and catalytic combustion on a catalytic screen [18].

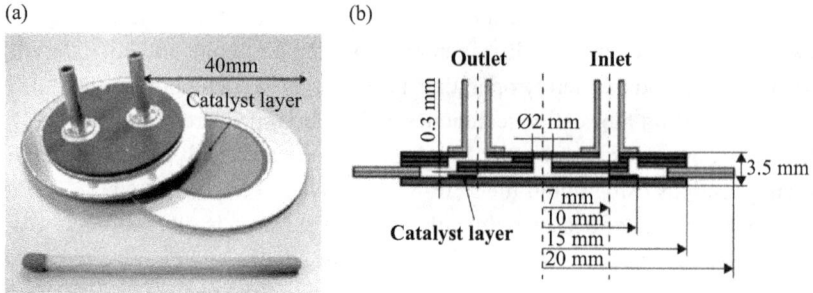

Figure 11.8. (a) Catalytic radial flow combustor. (b) Schematic of the design (reproduced by permission from Kamijo et al. [19]).

at a JP-8 fuel flow rate of 10 mL/h and a fuel-to-air equivalence ratio ranging between 0.35 and 0.70. Further miniaturization of the catalyst and the fuel atomizer by means of microfabrication techniques has led to an increased volumetric heat release rate of 270 MW/m^3, which is in the range of conventional gas turbines [34]. This is a very encouraging result since the developed combustor has a total volume of only 0.22 cm^3 and, despite its large surface-to-volume ratio that favors higher heat loss to the ambient, stable combustion could be sustained.

At the Turbulence and Heat Transfer Laboratory (Nobuhide Kasagi's Laboratory) of the University of Tokyo, a micro TPV system using a catalytic micro ceramic combustor has been demonstrated [19, 35, 36]. For combustor fabrication high-precision tape-casting [37] MEMS approaches were used, which were originally developed for semiconductor ceramic packages. Platinum and palladium on a nano-porous alumina support were evaluated as catalysts. A schematic of the burner design is shown in Figure 11.8. It has been found that a palladium catalyst is better suited for the developed combustor since it can be used up to 900°C and has a larger active surface than platinum at temperatures above 600°C [19]. Using n-butane as fuel, a maximum wall temperature of 850°C and a heat generation density of 2–5×10^8 W/m^3 were reported in the combustor prototype. The latter was designed for high-temperature applications such as TVP power generation.

Another catalytic butane combustor used in combination with TE power generator was presented by Yoshida et al. [20]. The TE module consists of a set of 34 BiTe TE elements, which are directly bonded to the combustor as shown in Figure 11.9. A platinum catalyst on a TiO$_2$ support was used to maintain combustion below the quenching distance (2 mm for butane) in a small combustion chamber ($8 \times 8 \times 0.4$ mm^3) and at low temperatures (200–400°C). The burner itself without connected TE module was able to ignite electrically and self-sustain combustion with almost 100 percent efficiency. However, with connected TE module, combustion was not possible due to increased heat losses. Moreover, when using hydrogen instead of butane, a maximum electric power output of 184 mW (theoretical maximum electrical power of 6.6 W) was obtained.

Mircoturbine-based microreactors have also been investigated. Concepts for 10–20 W electric power have been proposed [38]. Production of net power at such scales is very challenging; however, mesoscale approaches (electrical powers of 1–5 kW) using microturbines appear to be feasible. Figure 11.10 provides the components (meso-scale compressor and turbine, meso-scale electrical generator and combustor) of the Swiss Federal Institute of Technology, Zurich (ETHZ) and Paul Scherrer Institute (PSI) (ETHZ/PSI) effort of a recuperative mesoscale

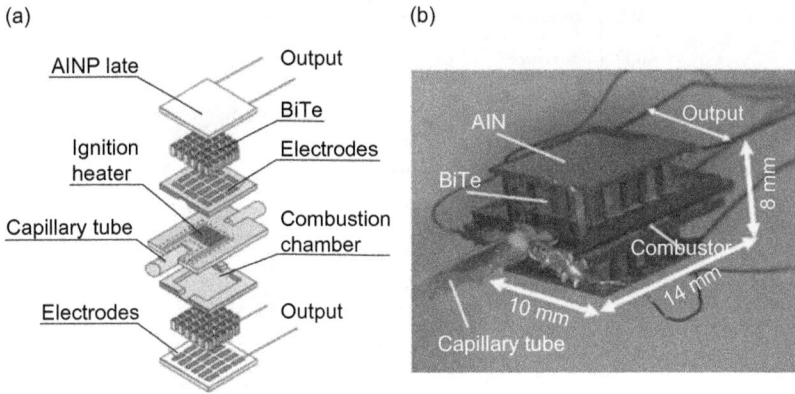

Figure 11.9. (a) Schematic of microcombustor-TE assembly, and (b) prototype of TE generator [20].

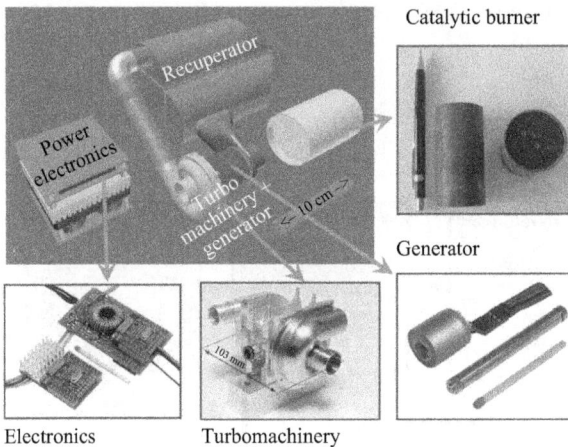

Figure 11.10. Component layout of mesoscale, turbine-based, portable power generation system [39].

turbine systems, aiming at 1 kW electric power [39]. A catalytic mesoscale reactor was constructed, consisting of an FeCr-alloy metallic honeycomb structure coated with platinum (see Figure 11.10). The combustion performance of this reactor is illustrated in Figure 11.11 [40], and the set goal of 3.1 kW thermal power at a flow rate of 6.7 g/s was exceeded.

11.2.2 CATALYTIC MICROREFORMERS

For power generation in portable devices such as notebooks or cell phones, hydrogen-based proton exchange membrane fuel cells (PEM) [21] or solid oxide fuel cells (SOFC) (the latter can also run with hydrocarbon or syngas fuels) are an alternative to batteries [41]. However, due to safety, cost, and weight requirements, the storage of hydrogen in portable devices is difficult and expensive. Instead, a catalytic reactor can be used to reform an easier-to-handle hydrocarbon fuel to hydrogen or syngas. Advantages of this concept compared to conventional state-of-the-art

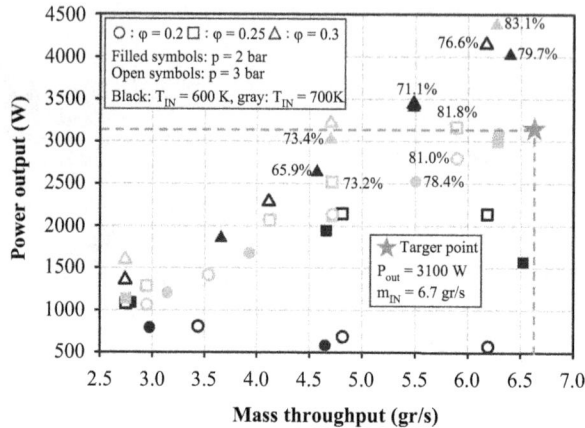

Figure 11.11. Performance of catalytic reactor in the concept of Figure 11.10 (reproduced by permission from Karagiannidis et al. [40]).

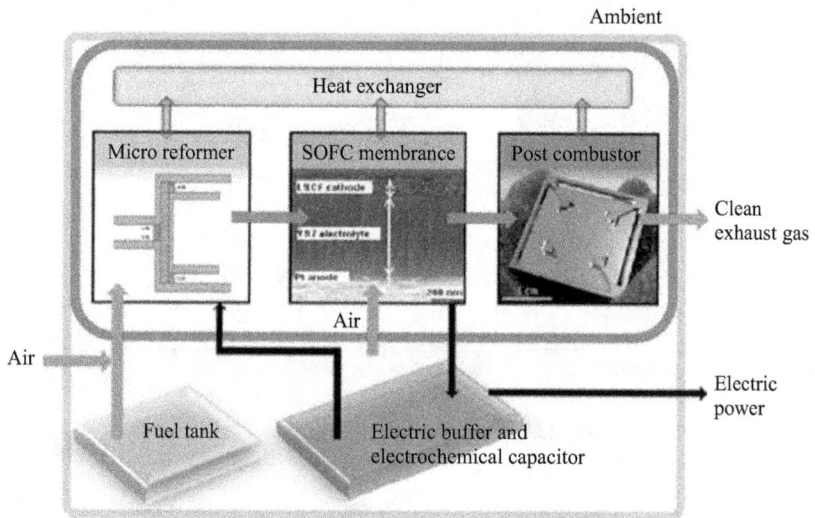

Figure 11.12. Layout of the ONEBAT micro SOFC system with catalytic microreformer (reproduced by permission from Santis-Alvarez et al. [23]© 2011 by Royal Society of Chemistry).

Li-ion batteries are the significantly higher energy densities and the fast "recharging" with liquid fuels for which a worldwide supply infrastructure exists.

Example of such a system is the ONEBAT [22, 23], a micro-SOFC system developed at the Swiss Federal Institute of Technology, Zurich (ETHZ). A schematic of the system layout is given in Figure 11.12. This system contains a fuel cell membrane, a gas processing unit (GPU) consisting of a catalytic microreformer and a post-combustor for exhaust gas cleaning, a heat exchanger, a fuel tank, and an electrochemical buffer for start-up. The system is fueled with *n*-butane, which is commercially available in small cartridges. Rhodium-doped ceria/zirconia ($Ce_{0.5}Zr_{0.5}O_2$) catalytic nanoparticles [42] are used to partially oxidize the *n*-butane to syngas

(H_2 and CO). The heat generated in this exothermic reaction is used to sustain the working temperature of the fuel cell between 500°C and 700°C.

A major issue for microreactors is the start-up, especially when used in mobile devices where a frequent switch-on and switch-off is required. When starting up, neither the catalytic reactor nor the fuel cell work at their design temperatures, resulting in reduced fuel conversion rates and increased pollutant generation, reduced efficiency and even potential detrimental conditions for the system. Catalytic ignition under reforming conditions will be elaborated in Section 11.4.3. In larger systems, catalytic ignition is usually achieved by preheating the inlet gas flow or the reactor with an external flame. However, it is not feasible for small-scale systems to implement such an additional heating system. Instead, internal electric heating has to be used. To reduce the start-up time of the ONEBAT system, a hybrid start-up method combining electric heating and total catalytic oxidation is used [23, 43]. This approach, which uses the micro-reformer for heating up the system, is significantly faster and energetically more efficient than electric heating alone. Future system developments are expected to further decrease start-up times by using a post combustor, which is designed for complete fuel conversion, by carrying out total catalytic oxidation of the remaining fuel.

11.2.3 CATALYTIC MICROTHRUSTERS

In recent years there has been increased interest in developing small spacecraft [44]; to this end, new classes considering mass, size, and power for such devices were established. Notwithstanding the lack of official convention, artificial satellites with a wet mass (mass including fuel) between 10 and 100 kg are often referred to as "microsatellites," satellites with a wet mass between 1 and 10 kg "nanosatellites" and satellites with a wet mass between 0.1 and 1 kg "pico-satellites." Sometimes several small satellites are designed to work in a network, potentially together with a larger "mother satellite," which is responsible for network coordination and communication with ground control [44]. Propulsion systems for positioning and maneuvering such small satellites must be capable of providing low levels of thrust with very high accuracy in terms of magnitude and firing duration. Furthermore, they have to be small, light-weight, and finally fuel- and cost-efficient. Among micro-solid rockets [45] and bipropellant rocket engines [46], monopropellant thrusters using catalyzed fuel decomposition are very promising because of their robust structure and their mechanical simplicity.

Microthrusters for "nanosatellites" using the catalyzed chemical decomposition of high-concentration hydrogen peroxide (H_2O_2) as a propulsion mechanism were developed at NASA Goddard Space Flight Center [24]. The prototype of a monopropellant MEMS thruster is shown in Figure 11.13. This thruster uses liquid high-purity hydrogen peroxide as a propellant, which is fed from a reservoir through an injector producing the necessary pressure drop into a catalytic chamber whereby the propellant undergoes a rapid chemical decomposition and phase change. The decomposition rate of hydrogen peroxide is highly dependent on temperature, purity, concentration, and surface activity (non-catalytic decomposition is too slow to be used in a thruster). An important design parameter is the length of the catalyst chamber and thus the critical residence time of the propellant to achieve complete decomposition. However, with an increasing in length, the pressure drops and, more importantly, the heat loss increases appreciably. Therefore, it is crucial to increase the active surface. Different surface geometries have been evaluated in the mentioned study. The photograph in Figure 11.13 shows

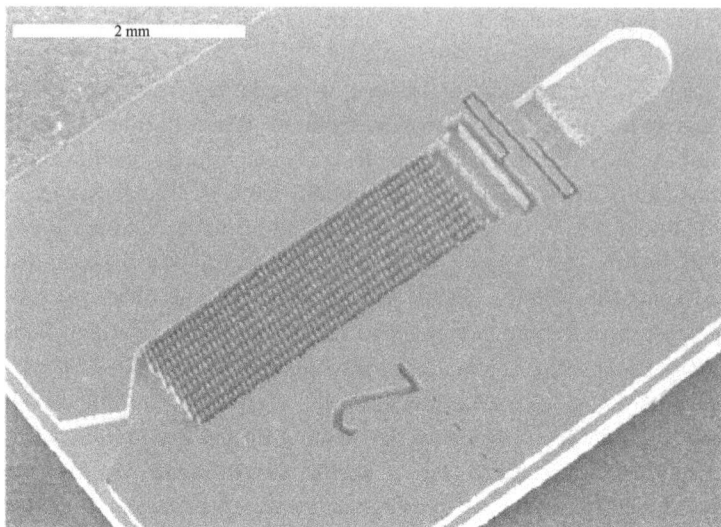

Figure 11.13. SEM photograph of a microthruster composed of an inlet plenum, an injector, a catalyst chamber, and a supersonic nozzle. The catalyst chamber is 2178 μm long, 1103 μm wide and 100 μm deep. The scale bar is ~2 mm (reproduced by permission from Hitt et al. [24]).

a catalyst chamber filled with diamond-shaped pillars. The surface is coated with high-purity silver, which acts as a catalyst for the propellant decomposition. This is a common catalyst in space flight applications for macroscale satellites and other monopropellants like hydrazine. The product gases (O_2 and water) are fed through a converging/diverging nozzle where the flow is accelerated to supersonic velocities and thus the desired thrust is attained.

For fundamental microthruster investigations, simple geometries are preferred as they can be manufactured easily and are amenable to numerical simulations. Thus, catalytic microtubes were often used in several studies aiming at the development of microthrusters for space-based applications [47–50]. Initial studies with platinum microtubes having diameters 0.4 mm and 0.8 mm and H_2/O_2 combustion [47] showed that these devices can achieve 1–10 mN thrust and a specific impulse of 300 s. The combustion in the catalytic tube can be self-sustained; however, a critical temperature must be exceeded for ignition. Therefore, an additional electric heating is required, but the estimated power consumption (~1 W) of such a microthruster is still significantly lower than that of a comparable electric propulsion device [47].

For longer missions in space the storage of methane as fuel is much easier than that of hydrogen. Further studies [48] with the above-mentioned platinum microtubes and also reduced ambient pressure (down to 0.0136 atm) showed that methane is a feasible fuel. The achieved thrust was again between 1 and 10 mN, while the specific impulse varied between 180 and 190 s. However, the critical temperature to ignite the CH_4/O_2 mixture is significantly higher. Ignition characteristics depend mainly on the mass flow rate, the ambient pressure, and the imposed heat flux, which can be reduced by adding a small amount H_2 [48]. However, hydrogen addition has a very strong impact on the thrust. If the volumetric flow rate of the thruster is kept constant, the addition of hydrogen will cause a decrease of thrust because of the decrease of exit pressure and the mass flow rate. The specific impulse depends inversely on the mass flow rate and the molecular weight of the products (which decreases with increasing hydrogen addition), so it will increase. For constant mass flow rates in the thruster, the picture is the opposite and thrust increases with the addition of hydrogen because of the higher fuel inlet velocities [50].

11.3 BASIC THEORY OF CATALYTIC COMBUSTION

Heterogeneous fuel conversion is usually accomplished in ceramic or metallic reactors, which are coated with a catalytically active material and provide an adequate surface-to-volume ratio. In gas turbines for large-scale power generation, honeycomb reactors are typically used, consisting of a multitude of catalytically active channels, each with a hydraulic diameter of 0.5–2 mm. This is also the size of typical microreactors, which may also assume the geometry of one or few catalytic channels. Nevertheless, in both large-scale and micro-scale power generation, the same physicochemical processes occur inside each catalytic channel. These processes are depicted in Figure 11.14.

Preheated fuel–air premixtures are admitted in each catalytic channel at inflow velocities that guarantee, for microreactor applications, laminar flows. Fuel and oxidizer diffuse transversely to the catalytically active channel surfaces. Catalytic ignition (light-off) is attained at a certain distance from the entry, which depends not only on the operating conditions (inflow velocity and temperature, pressure, fuel-to-air stoichiometry, fuel type), and the catalytic reactivity, but also on key in-channel heat transfer mechanisms such as heat conduction in the solid walls and surface radiation heat transfer. At the light-off location, the surface temperature is sufficiently high such that the heterogeneous conversion shifts from the kinetically controlled to the transport-controlled regime [51]. Subsequently, the fuel and oxidizer react vigorously at the catalyst surface forming a degenerate diffusion reaction sheet [52]. This terminology reflects the fact that both reactants diffuse from the same side of the sheet and the position of the reaction sheet is fixed in space. Heat and reaction products diffuse back to the main flow and homogeneous (gas-phase) combustion can be established inside the channel, if conditions are appropriate. Given the lower activation energy of the catalytic reaction pathway compared to

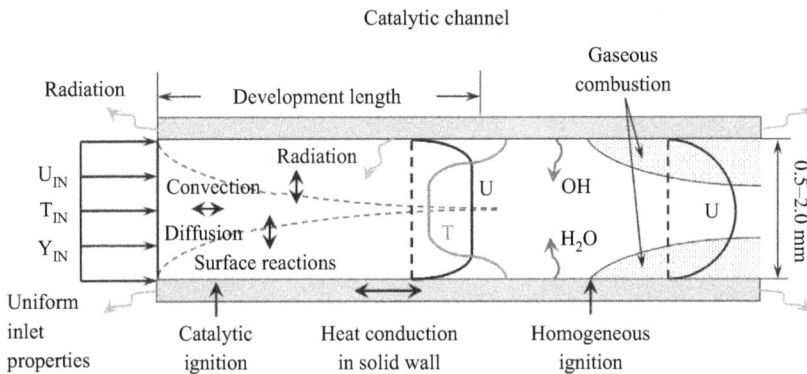

Figure 11.14. Physical and chemical processes in a catalytic channel.

that of the gaseous pathway, homogeneous ignition, when present, is typically initiated down-stream of the light-off position. Fundamentally, there are four main coupling routes between the two reaction pathways affecting gaseous combustion. The catalytically induced near-wall fuel depletion inhibits homogeneous ignition, whereas the heat transfer from the hot catalytic walls to the reacting gas promotes ignition. Additionally, gas-phase ignition is mildly inhibited by the recombination of radicals on the catalyst [53, 54]. Finally, homogeneous ignition can either be inhibited or promoted by heterogeneously produced major species (notably H_2O), depending on the fuel type.

For power-generation systems usually noble metals, which are supported by a porous metal oxide layer (e.g., Al_2O_3 or ZrO_2), are used as catalysts. The support layer provides a high surface area, a uniform dispersion of the catalytic active material, and features good adhesion to the ceramic or metallic structure of the reactor.

11.3.1 MATHEMATICAL FORMULATION

The governing equations for a hetero-homogeneous reaction system are given below. Although valid for any complex reactor geometry, the high computational cost makes them attractive only for single-channel simulations. These models resolve all relevant spatiotemporal scales and are also referred to as direct numerical simulation (DNS). The following complete set of equations, or a simplified version of it, will be used in the forthcoming sections.

Continuity equation:

$$\frac{\partial \rho}{\partial t} + \nabla \cdot (\rho \, \vec{u}) = 0. \tag{11.1}$$

Momentum equations:

$$\frac{\partial(\rho \vec{u})}{\partial t} + \nabla \cdot (\rho \, \vec{u}\vec{u}) + \nabla p - \nabla \cdot \mu \left[\nabla \vec{u} + (\nabla \vec{u})^T - \frac{2}{3}(\nabla \cdot \vec{u}) \underline{\underline{I}} \right] = 0. \tag{11.2}$$

Total enthalpy equation:

$$\frac{\partial(\rho h)}{\partial t} + \nabla \cdot (\rho \vec{u} h) + \nabla \cdot \left(\sum_{K=1}^{K_g} \rho Y_k h_k \vec{V}_k - \lambda \nabla T \right) = 0. \tag{11.3}$$

Gas-phase species equations:

$$\frac{\partial(\rho Y_k)}{\partial t} + \nabla \cdot \rho Y_k (\vec{u} + \vec{V}_k) - \dot{\omega}_k W_k = 0, k = 1,...., K_g. \tag{11.4}$$

Surface species equations:

$$\frac{\partial \theta_m}{\partial t} = \sigma_m \frac{B \dot{s}_m}{\Gamma}, m = 1,..., K_s. \tag{11.5}$$

Finally, the ideal and caloric gas laws close the system of equations:

$$p = \rho R^\circ T / \bar{W} \quad \text{and} \quad h_k = h_k^0(T_0) + \int_{T_0}^{T} c_{p,k} \, dT, k = 1, \ldots, K_g. \tag{11.6}$$

The diffusion velocities $\vec{V}_k$ in Equations (11.3) and (11.4) can be computed using either the full multicomponent transport equations:

$$\nabla X_k = \sum_{\ell=1}^{K_g} \frac{X_k X_\ell}{D_{k\ell}} (\vec{V}_\ell - \vec{V}_k) + (Y_k - X_k) \frac{\nabla p}{p} + \sum_{\ell=1}^{K_g} \frac{X_k X_\ell}{\rho D_{k\ell}} \left(\frac{D_\ell^T}{Y_\ell} - \frac{D_k^T}{Y_k} \right) \frac{\nabla T}{T}, \tag{11.7}$$

or the more commonly used mixture-average diffusion augmented with thermal diffusion:

$$\vec{V}_k = -D_{km} \nabla \left[\ln \left(Y_k \bar{W} / W_k \right) \right] + \left[D_k^T W_k / (\rho Y_k \bar{W}) \right] \nabla (\ln T). \tag{11.8}$$

The interfacial gas-phase species boundary conditions become

$$\left[\rho Y_k (\vec{V}_k + \vec{u}_{st}) \right]_+ \cdot \vec{n}_+ = B W_k \dot{s}_k, \tag{11.9}$$

with $\vec{n}_+$ the outward-pointing normal to the catalytic walls, and $\vec{u}_{st}$ the Stefan velocity, having a magnitude:

$$\left| u_{st} \right| \equiv (1 / \rho) B \sum_{k=1}^{K_g} W_k \dot{s}_k. \tag{11.10}$$

In Equation (11.10) "+" denotes gas properties just above the gas–wall interface. Multidimensional heat transfer in the solid wall is also considered, given the rise of significant temperature gradients in the normal to the surface direction during transient microreactor operation:

$$\frac{\partial (\rho_s c_s T_s)}{\partial t} - \nabla \cdot (k_s \nabla T_s) = 0, \tag{11.11}$$

with the interfacial energy boundary condition:

$$\left[-\lambda \nabla T + \vec{q}_{rad} \right]_+ \cdot \vec{n}_+ + (k_s \nabla T_s)_- \cdot \vec{n}_+ + \sum_{k=1}^{K_g + M} B h_k \dot{s}_k W_k = 0. \tag{11.12}$$

The Stefan velocity in Equation (11.10) is identically zero for catalytic systems at steady state, since there is no net mass deposition or etching of the surface.

Intraphase transport (inside the porous catalyst layer) may also be needed in modeling of microreactors having sufficiently thick washcoats and well-dispersed loading within the washcoat volume. High temperatures, in particular, enhance diffusion limitations through the porous catalyst structure. For certain combinations of fuels/catalysts, it has been shown [55, 56] that intraphase diffusion limitations may become important in washcoats as thin as 50 μm and temperatures as low as 700 to 800 K. Intraphase modeling approaches are reviewed in [57] and include reaction–diffusion models for multidimensional porous structures, 1D (normal to the wall) reaction–diffusion models, and the effectiveness factor methodology. By neglecting convection, a set of transient transport reaction–diffusion equations for the surface and gaseous species inside the washcoat can be developed [58]. Spatial discretization of the resulting equations leads to a significant increase in the computational cost, such that this approach is not typically used for transient multidimensional flow models; there, the simpler effectiveness factor model is preferred.

For transient applications, the quasi-steady assumption for the gas-phase and surface chemistry can be invoked. In this case the only transient term kept is that in the solid heat conduction Equation (11.11). Conditions for the validity of the quasi-steady assumption have been discussed in [59] for methane-fueled microreactors.

11.3.2 MICROREACTOR THERMAL MANAGEMENT

Thermal management is of great concern in the design of microreactors since small devices suffer from relatively large surface-to-volume ratios and thus significant heat losses to the ambient. However, such large ratios favor the application of a catalytic chemical processes. By converting the fuel on the catalytic wall, heat is produced on a larger area compared to pure gas-phase combustion where the heat release is confined into a typically thin flame zone. This allows for a more uniform heating of the reactor structure. On the other hand, the distributed heat generation via heterogeneous reactions on a relatively large surface area increases external heat losses. A detailed discussion about these mechanisms and the impact on ignition characteristics and microreactor combustion stability is given in Section 11.4.1.

To reduce external heat losses, many microreactor designs implement heat-recirculation. These excess enthalpy burners [27, 28] may not necessarily use a catalyst, but they usually benefit from it, depending on the operating conditions [17]. In such a heat recirculation reactor the hot reaction products are used to preheat the cold incoming mixture (see Figure 11.3). This heat exchange increases combustion stability. One important parameter is the heat conductivity of the wall separating the cold incoming mixture and the hot exhaust gases. Materials with low thermal conductivity have an insulating effect and hinder heat transfer. However, different numerical studies have shown that a very high thermal conductivity does not result in a most favorable heat transfer [60, 61]. In this case, the heat transfer efficiency of the counter-flow approaches is lower than the one of a parallel-flow heat exchanger [62]. Depending on the geometrical and operational parameters (e.g., steady versus transient conditions), the optimal thermal heat conductivity can vary over several orders of magnitudes. Nonetheless, numerical simulations [62] indicate that ceramics are suitable materials for a wide range of operating conditions.

Experiments with non-catalytic reactors show very different performance. For the reactor in Figure 11.15, combustion in the catalytic mode is feasible at significantly lower mass flow rates. The schematic is replicated from Ahn et al. [17]. Furthermore, it is possible to sustain combustion at much lower temperatures in the catalytic reactor (~350 versus 920 K in the non-catalytic reactor) [17]. With regard to long-term operation, lower temperatures are advantageous since thermal stress in the reactor materials is reduced. By using a catalyst it is possible to better define the position of the reaction zone and thus the location of the heat release. Typically this zone should be located at the center of the reactor, to make the inlet and the exhaust

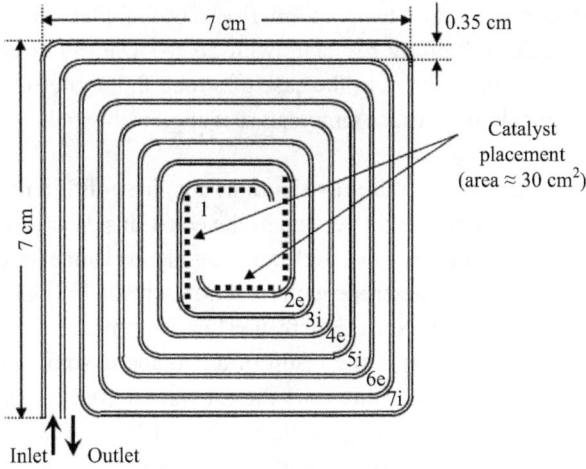

Figure 11.15. Schematic plan view of a heat-recirculation ("Swiss roll") burner. Platinum strips can be placed optionally in the center. Thermocouples 1–7 in the inlet (i) and exhaust (e) turns are indicated by numbers [17].

channels equally long and allow for sufficient heat transfer from the hot exhaust gases to the cold reactants [63]. In addition, the catalyst usually allows anchoring of the gaseous flame in the desired zone by lowering the stoichiometry of the fuel/air mixture and mitigating "flashback." This is usually more difficult for non-catalytic reactors.

The diffusional imbalance (non-unity Lewis number) of the limiting reactant has also a profound impact on the surface temperature of catalytic reactors and has to be carefully considered in thermal management concepts. The unequal heat and mass diffusivities can cause super- or under-adiabatic wall temperatures. The wall temperature in a case where fuel is the deficient reactant is given by [64]

$$T_W = T_G + Le_F^{\beta-1}(Y_{F,G} - Y_{F,W})q/c_p, \tag{11.13}$$

where T_G and $Y_{F,G}$ are the bulk gas temperature and the bulk fuel mass fraction, respectively, $Y_{F,W}$ is the fuel mass fraction at the wall, Le_F is the Lewis number of the limiting reactant (in this case fuel), q is the heat release per unit mass of fuel, c_p is the mixture heat capacity, and $\beta = 0$ for fully developed or $\beta = 1/3$ for developing channel flows.

Because of the moderate channel length-to-diameter ratio, the relatively high inlet velocities and the flow acceleration due to combustion, fully developed flows are generally not attained in microreactors. Thus the maximum wall temperature is achieved under mass-transport limited operation conditions (i.e., when $Y_{F,W} = 0$) and Equation (11.13) reduces to

$$T_W = T_G + Le_F^{-2/3}\Delta T \tag{11.14}$$

with

$$\Delta T = T_{ad} - T_G = Y_{F,G}q/c_p \tag{11.15}$$

denoting the adiabatic combustion temperature rise. For infinitely fast surface chemistry (transport-limited operation), limiting reactants with $Le < 1$ ($Le > 1$) lead to superadiabatic (underadiabatic) surface temperatures. In the case of strong finite-rate surface chemistry ($Y_w/Y_{F,G} >> 5$ percent), the maximum surface temperature is appreciably reduced because of incomplete combustion.

The calculated surface temperature distributions for these different cases are given in Figure 11.16. Simulations (using the 2D version of the equations described in Section 11.3.1) were carried out for an H_2/air mixture in a 300-mm long plane channel coated with platinum. For all simulations an elementary heterogeneous reaction scheme [65] has been used. Curves (a) and (b) refer to a fuel-lean mixture where the Lewis number of the limiting reactant (hydrogen) is $Le_{H2} \approx 0.3$. Curve (a) is the result of an adiabatic simulation with artificially imposed infinitely fast chemistry. In this case the maximum wall temperature approaches the theoretical value (≈ 2100 K for $Le_{H2} \approx 0.3$) obtained by Equation (11.14). The temperature profile (b) is the result of a non-adiabatic simulation whereby, apart from finite-rate chemistry, surface radiation heat transfer and heat conduction in the solid are considered. In both cases superadiabatic temperatures ($T_{ad} = 1148$ K) are achieved near the channel entry; however, the heat losses and finite-rate chemistry effects lead to significantly lower surface temperatures for (b). Superadiabaticity is more pronounced at the upstream region of the channel due to the higher amount of available fuel, while the temperatures approach the adiabatic equilibrium temperature at the channel rear upon complete hydrogen conversion. The predicted peak temperatures in either (a) or (b) are too high and would damage the catalyst and cause reactor meltdown.

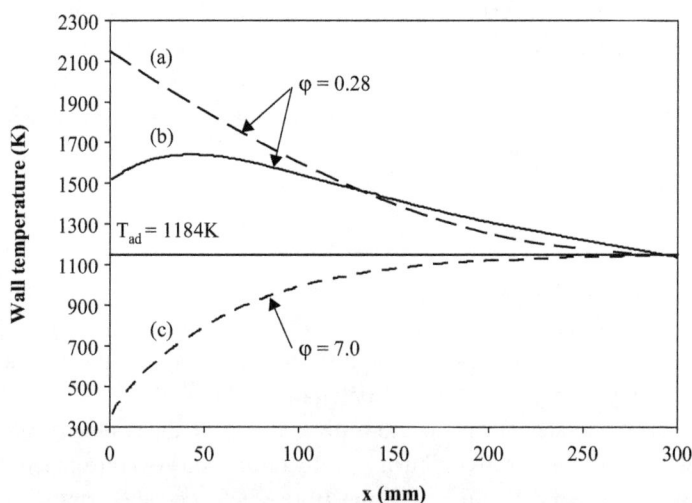

Figure 11.16. Predicted catalyst temperature distribution in H_2/air channel flow catalytic combustion: $U_{IN} = 2$ m/s, $T_{IN} = 310$ K, and $\varphi = 0.28$. The adiabatic flame temperature is 1148 K and the Lewis number of hydrogen is $Le_{H2} \approx 0.3$. (a) Adiabatic calculations with mass-transport limited catalytic fuel conversion. (b) Non-adiabatic simulations with an elementary heterogeneous reaction scheme, surface radiation, and heat conduction in the solid. (c) Fuel-rich ($\varphi = 7.0$) simulation with same adiabatic flame temperature; oxygen is the limiting reactant ($Le_{O2} \approx 2.3$).

For simulation (c) an H_2/air mixture with a fuel-rich stoichiometry ($\varphi = 7.0$) is considered, which results in the same adiabatic equilibrium temperature as in the lean cases (a) and (b). In this case oxygen is the limiting reactant, having a Lewis number $Le_{O_2} \approx 2.3$. In contrast to cases (a) and (b), the wall temperature now approaches the adiabatic equilibrium value from underadiabatic values. Of course, this behavior is similar for fuel-lean stoichiometries when the deficient fuel has $Le > 1$, such as the case of $C2^+$ hydrocarbons.

Since the residence time in typical microreformers is relatively short (a few milliseconds in short-contact-time CPO reactors), another type of superadiabaticity may occur due to the presence of multiple and/or competing reaction pathways. An example is illustrated in Figure 11.17 providing axial profiles of the predicted wall temperature (T_w), the mean gas temperature (T_{gas}), and methane and hydrogen mole fractions during CPO of methane in a 75-mm long channel with a diameter of 1.2 mm; the channel is coated with rhodium only over the axial length 10 mm $< x <$ 65 mm. Predictions were carried out using the 2D version of the model equations in Section 11.3.1 and the detailed surface reaction mechanism [66] for the CPO of methane over rhodium. In the same figure, the measured temperatures along the reactor, the calculated adiabatic equilibrium temperature, and the CH_4 and H_2 mole fractions are shown [8]. The predicted wall temperatures exceed the adiabatic equilibrium temperature (T_{ad}) by as much as 200 K, while the mean gas temperatures surpass T_{ad} for $x >$ 27 mm. This is a result of chemical non-equilibration due to the short convective time scales and the presence of multiple reaction pathways with different time scales (fast oxidation reactions $CH_4 + 2O_2 = CO_2 + 2H_2O$ and $2CH_4 + O_2 = 2CO + 4H_2O$ and slow steam reforming $CH_4 + H_2O = CO + 3H_2$).

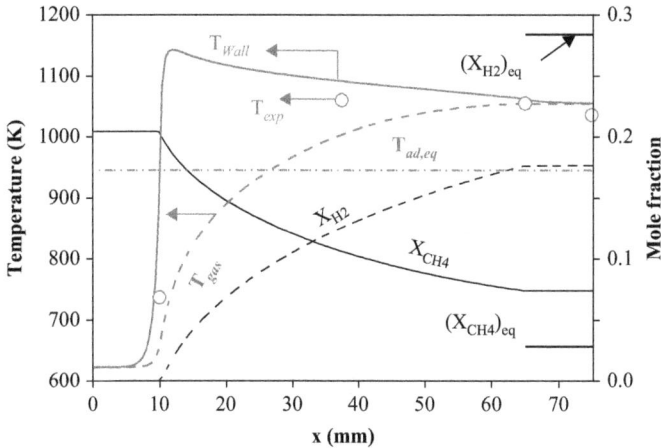

Figure 11.17. Profiles of temperature and species mole fractions in catalytic partial oxidation of methane. Computations: wall temperature T_{Wall}, mean gas temperature T_{gas}, mean CH_4 mole fraction (X_{CH4}), mean H_2 mole fraction (X_{H2}). Adiabatic equilibrium calculations: gas temperature ($T_{ad,eq}$) and CH_4 and H_2 mole fractions (($X_{CH4})_{eq}$ and ($X_{H2})_{eq}$). Measurements: temperature (open circles). The coated section of the reactor extends over 10 mm $< x <$ 65 mm [8].

11.4 OPERATION OF MICROREACTORS FOR POWER GENERATION

The numerical model of Section 11.3.1 will be herein applied to illustrate key issues of microreactor steady-state or transient operation.

11.4.1 STEADY OPERATION OF CATALYTIC MICROREACTORS

Key issue in the design and operation of catalytic and non-catalytic microreactors is the mapping of the regimes where stable combustion can be sustained and the parameters affecting combustion stability. Two-dimensional numerical simulations using a one-step reaction of pure gas-phase combustion of methane/air and propane/air mixtures in channels with sub-millimeter gap sizes and walls without radical quenching have delineated the stable operating regimes [67, 68]. Heat transfer mechanisms, especially heat conduction inside the reactor walls and external heat losses, have a strong influence on the stability envelope of the flame. The aforementioned studies have also shown that propane/air flames are more stable and tolerant to heat loss and heat conduction than methane/air flames, since the propane's lower ignition temperature causes the reaction front to stabilize farther upstream. In terms of catalytic microreactors, for pure heterogeneous methane combustion over platinum combustion stability was investigated numerically in a cylindrical tubular reactor with 1 mm diameter [69]. The walls of the reactor were either adiabatic or non-adiabatic and were considered thermally thin (heat conduction in the solid was neglected). In this study a detailed heterogeneous reaction scheme [65] was used. Extinction limits for catalytic and non-catalytic combustion of propane were also studied experimentally in a heat recirculation "Swiss roll" burner [17]. All aforementioned studies were conducted under atmospheric pressure.

Residence times and wall temperatures in catalytic microreactors can be sufficiently high so as to ignite homogeneous combustion. Thus, in many practical systems gas-phase reactions cannot be neglected, such that validated homogeneous and heterogeneous reaction schemes are necessary for designing and simulating catalytic microreactors. The gas-phase reaction scheme must be able to capture the homogeneous ignition characteristics in the presence of the catalytic reaction pathway. The anchoring position of the flame in turn affects the heat transfer mechanisms in the reactor wall and thus the extinction limits. An experimental methodology using *in situ* Raman measurements of major gas-phase species and laser-induced fluorescence (LIF) of gaseous radical species [70–72, 54] has been used to validate heterogeneous and homogeneous kinetics over a wide range of operating conditions. Using this methodology, an appropriate elementary gas-phase reaction scheme was identified and refined, which reproduced the measured gas-phase combustion characteristics of fuel-lean ($\varphi \leq 0.5$) methane/air mixtures over a platinum at pressures 1 bar $\leq p \leq$ 16 bar [54]. The lower pressure range (1–5 bar) is relevant for gas-turbine-based microreactors. This homogeneous reaction scheme has been used in combination with a detailed heterogeneous reaction scheme (validated over the same pressure range 1–16 bar in [71]) in a 2D full-elliptic steady model to investigate the stability of a methane-fueled platinum-coated microreactor. The model also included heat conduction in the reactor walls, surface radiation heat transfer, and imposed external heat losses $h(T_w - T_o)$ with h a heat transfer coefficient. This model delineated the combustion stability envelop of such reactors [73].

By varying independently the external heat transfer coefficient h, the inlet velocity U_{IN}, and the wall thermal conductivity k_s, combustion stability maps (see Figure 11.18) could be constructed. This figure presents stability diagrams for 1 bar and 5 bar at $T_{IN} = 700$ K and for $p = 5$ bar at $T_{IN} = 600$ K. The wall thermal conductivity is $k_s = 2$ W/mK and the catalyst surface emissivity $\varepsilon = 0.6$. Two different velocity scales are used in Figure 11.18 in order to facilitate the comparison of the stability limits at the two different pressures of 1 and 5 bar for a given mass throughput. The stability limits at low U_{IN} define extinction due to heat loss while at large U_{IN} denote blowout caused by insufficient residence time.

In addition to the inlet mass flow, the thermal conductivity of the solid k_s plays a key role on the microreactor combustion stability. Typical thermal properties for two practical materials used in the simulations are given in Table 11.2. For low thermal conductivity materials such as cordierite ($k_s \sim 2$ W/mK), heat is not transferred effectively upstream to promote catalytic ignition and thus a blowout may be caused. With increasing thermal conductivity, combustion becomes more stable; however, values above $k_s = 30$ W/mK have only a very low impact on

Table 11.2. Properties for cordierite and FeCr-alloy channel wall material

Material	k_s (W/mK)	ρ_s (kg/m^3)	c_s (J/kgK)	$\rho_s c_s$ (kJ/m^{3}K)
Cordierite	2	2600	1464	3806
FeCr alloy	16	7200	615	4428

Source: Karagiannidis et al. [59].

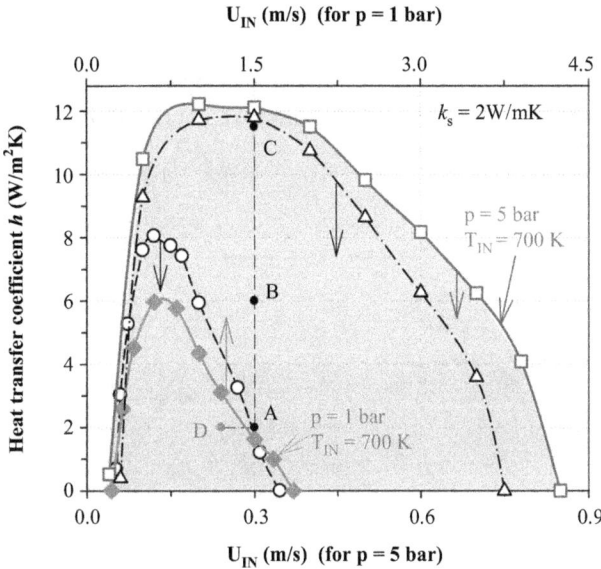

Figure 11.18. Stability diagrams of methane/air (equivalence ratio 0.4) catalytic planar microchannels (channel height and length of 1 mm and 10 mm, respectively) in terms of critical coefficient h and inlet velocity U_{IN}: $p = 5$ bar (squares: $T_{IN} = 700$ K, triangles: $T_{IJN} = 700$ K without gas-phase chemistry, circles: $T_{IN} = 600$ K), $p = 1$ bar (filled diamonds: $T_{IN} = 700$ K). The stable regimes for the 700 K cases are shown by the shaded areas [73].

combustion stability. Figure 11.19 illustrates this behavior by providing stability limits for two different inlet velocities at $p = 5$ bar and $T_{IN} = 700$ K as a function of the solid thermal conductivity. For low k_s the stability limits are narrower at higher inlet velocities since there the limits are defined by blowout characteristics. At higher k_s, however, stability limits are wider for higher velocities because of heat-loss-induced extinction [73].

Another way to increase the amount of upstream transferred heat and thus stabilize combustion is to increase the wall thickness. Besides an increased thermal stability, this measure also causes an improvement of mechanical stability which is of interest for microreactors used for mobile applications. On the other hand, it will of course increase the size and the weight of the system.

Compared to pure gas-phase combustion [67], catalytic systems show some remarkable differences. While the stability limit reaches its maximum at ~20 W/mK and does not change significantly by further increasing the solid thermal conductivity, in pure gas-phase combustion extinction limits reach a peak at approximately $k_s = 5$ W/mK [67]. A further increase of k_s reduces the stable regime considerably because in this case heat transferred from the localized flame may lead to extinction. At low k_s and for pure gas-phase combustion, the blowout limit is almost independent of h and only extends over the narrow range 0.4 W/m $< k_s <$ 0.8 W/mK. The reason is that only limited heat is transferred from the flame region and thus the wall temperature upstream of the flame remains relatively modest. Since the heat generation is located in a very narrow region, external heat losses become less effective in removing heat from the walls. On the other hand, in catalytic systems heat is generated via the heterogeneous pathway along an extended surface area, resulting in a better preheat of the incoming mixture but at the expense of increased external heat losses. This has to be considered when designing a catalytic microreactor with low thermal conductivity walls. Nonetheless, catalytic microreactors are able to sustain stable combustion at significantly lower k_s compared to non-catalytic microreactors: their stability limits extend to $k_s <$ 0.1 W/mK; however, in practice there is little practical interest for materials with such properties [73].

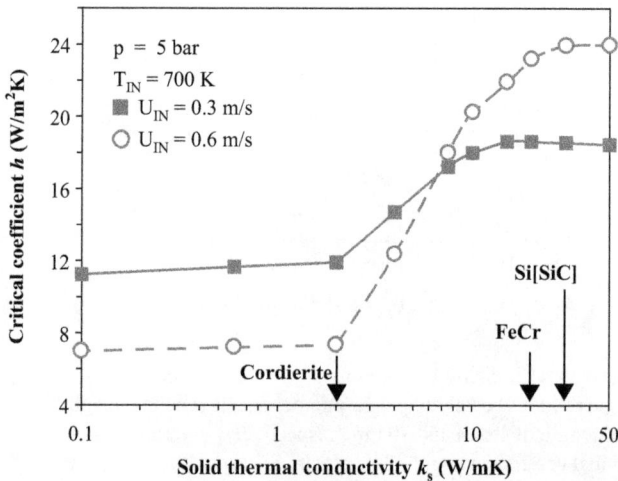

Figure 11.19. Stability limits of methane catalytic microchannels. Critical heat transfer coefficient versus solid thermal conductivity [73].

The stable combustion envelope of a methane-fueled microreactor can be substantially extended by increasing the pressure. This is depicted in Figure 11.18, when comparing the stability limits at 1 and 5 bar for T_{IN} = 700 K and the same mass throughput ($\rho_{IN}U_{IN}$). The increase of pressure almost doubles the allowable critical heat transfer coefficient, while the maximum mass flow before the blowout limit drastically increases. The reason for this behavior is the positive $p^{+0.47}$ dependence of the catalytic reactivity of methane over Pt [71], while the corresponding gas-phase reactivity scales as $p^{+1.0}$ [74].

Besides heat conductivity in the solid, thermal radiation is another energy transfer mechanism, which can have a strong influence on the stability limits and the combustion processes inside the stability envelope, especially for reactors made of low thermal conductivity materials. Because of the short distance of a microreactor, the resulting viewing factors between the hot catalytic walls and the significantly colder entry section become relatively large. A detailed energy balance of the solid walls reveals that thermal radiation is a net heat loss mechanism [73]. Computed surface temperatures at different surface radiation emissivities for a case located well-inside the stability envelope in Figure 11.18 are depicted in Figure 11.20. Compared to the non-radiating case ($\varepsilon = 0.0$) the predicted surface temperature for an $\varepsilon = 0.6$ case is significantly lower, that is by ~240 K at the inlet and 110 K at the channel rear. Moreover, with increasing emissivity the temperature peak is shifted downstream. The aforementioned characteristics of the temperature profiles agree qualitatively with simulations in longer catalytic channels [75].

When the stability limits in Figure 11.18 are approached, the role of radiation changes substantially. Even though the contribution of radiation to the overall solid energy balance is small [73], close to the stability limit its impact can be decisive. Since close to the blow-out limit the catalyst surface temperature is low at the channel entry, the resulting radiative heat losses to the inlet section are low. On the other hand heat is transferred from the rear section to the front, and this in-channel energy redistribution has a stabilizing effect on combustion. For

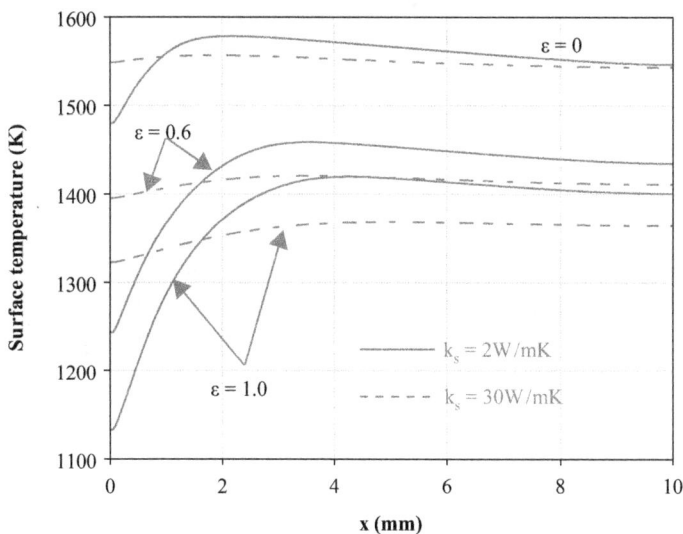

Figure 11.20. Surface temperatures in methane catalytic microchannels for a case located well-inside the stability envelope of Figure 11.18 (point A in Figure 11.18) for three catalyst surface emissivities. The inlet temperature is 700 K [73].

materials with higher thermal conductivity, the effect of radiation becomes less pronounced as can be seen in Figure 11.20.

Goal in microreactor development is miniaturization. A key question arises whether the homogeneous pathway still plays a role with further increased confinements. Numerical simulations indicate that it is possible to sustain gas-phase combustion in catalytic microreactors at sub-millimeter scales [73]. In Figure 11.21 the two-dimensional distribution of the OH mass fraction and the gaseous heat release rate are shown for two planar channels with half-heights of 0.3 mm and 0.15 mm, respectively. These cases lie well inside the stability envelope in Figure 11.18 and, apart from the half-height, all parameters are kept constant. With increasing confinement (smaller channel half-height) the gas-phase combustion intensity decreases as seen from the absolute OH levels and heat release magnitudes in Figure 11.21. Moreover, the flame moves upstream with increasing confinement. Because of the reduced viewing factors from the hot catalytic walls to the colder entry section, the radiative heat loss decreases and thus the surface temperature in the first 2 mm of the channel increases. This also enhances combustion stability on the extinction branch allowing for a higher value of the critical heat transfer coefficient. However, a blowout of the flame will occur at mass throughputs lower, when compared to a larger-height channel, because of an insufficient residence time [76].

While in many experimental and numerical studies of hydrocarbon-fueled microreactors methane is used, high hydrocarbons or other organic compounds, for example alcohols [77] are of prime commercial interest. Propane is a fuel of special interest for micro-energy conversion systems because it liquefies at room temperature under moderate pressures, simplifying storage and distribution. However, with increasing C atoms in the hydrocarbon fuel, the number of species and elementary reactions involved in the hetero-/homogenous schemes is growing enormously. The development of adequate reduced detailed hetero- and homogeneous reaction schemes valid over a wide range of operating conditions is still an important topic in research [78, 79]. The comparison of the stability maps for methane and propane has been elaborated in [76]. Results illustrate that propane is more robust on the extinction branch. In this regime of lower inlet velocities, which is equivalent to longer residence times, the hetero-/homogeneous reactivity [80, 81] of propane is higher than the one of methane and controls combustion stability [76]. On the blowout branch, however, methane is more robust to heat losses. This can

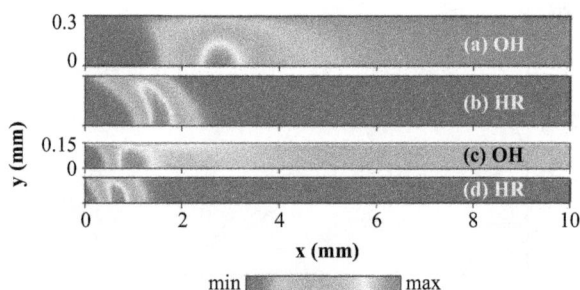

Figure 11.21. Two-dimensional distributions of OH mass fraction and gaseous heat release (W/cm^3) for two half-height planar channels: (a and b) 0.3 mm and (c and d) 0.15 mm. Methane/air combustion ($\varphi = 0.4$), all other parameters as in point A of Figure 11.18. The color bar indicates values from zero to (a) 5.2×10^{-5}, (b) 198, (c) 7.1×10^{-6}, and (d) 134 [73].

be explained by the diffusive transport of the fuel to the catalytic surface. In a lean methane/air mixture, the Lewis number of the fuel is $Le_{CH4} \approx 0.97$, which leads to a significantly higher transverse transport towards the catalytic wall compared to propane, which has a Lewis number $Le_{C3H8} \approx 1.82$. When the residence time in the catalytic channel becomes of the same order of magnitude as the transverse fuel diffusion time, transport becomes the controlling factor of fuel conversion and combustion stability instead of chemical reaction times [76].

Because of the increased surface-to-volume ratio, homogeneous reactions in a microreactor are often neglected in numerical simulations especially since the inclusion of a detailed reaction scheme is computationally expensive. However, neglecting gas-phase chemistry can cause appreciable underprediction of the combustion stability limits, especially at elevated pressures. In Figure 11.18 the stability maps of methane at 5 bar and $T_{IN} = 700$ K are computed with and without the inclusion of homogeneous chemical reactions. It is evident that gas-phase reactions increase stability especially at the blowout limits. This enhancement is not a sole outcome of the total oxidation of methane via the homogeneous pathway. It has been shown [73] that this result largely reflects a coupled hetero-/homogeneous reaction route: methane is oxidized incompletely to CO via gas-phase reactions, followed by the main exothermic oxidation of the formed CO to CO_2 via the catalytic pathway [73]. Of course, the effect of gas-phase chemistry decreases for a smaller catalytic channel because of the shorter transverse diffusion times, but even in sub-millimeter channels its impact is noticeable [76].

11.4.2 TRANSIENT OPERATION OF CATALYTIC MICROREACTORS

Light-off under various operating conditions is one of the most important parameters for the success of a given microreactor design, since ignition determines the required start-up time and the emission of pollutants during this period. As many mobile devices require a frequent power-on and power-off, it is desirable to minimize the elapsed start-up time. Another reason for short start-up times is the incomplete fuel conversion and thus a significantly increased emission of pollutants, mainly CO and unburned hydrocarbons, during this time.

Compared to the number of studies on steady operation, research on the transient microreactor behavior is limited. In [82] a start-up strategy for a propane-fueled platinum-coated microreactor has been developed and demonstrated in an experimental setup. To facilitate homogeneous ignition of propane/air mixtures, hydrogen has been added. For a similar reactor fueled with methanol/air mixtures, start-up characteristics have been investigated in [83]. This study showed the feasibility of a self-igniting microreactor, which was connected to a TE power generation system. As several factors influence performance, numerical studies have been used to expose the controlling parameters and hence to aid in the design of microreactors for specific applications.

Since materials with a higher heat capacity ($\rho_s c_s$) require a higher energy input to raise their temperature, it is evident that the reactor solid wall plays an important role during the heat-up phase. Furthermore, heat transport via radiation and conduction in the reactor walls affect the start-up time. To demonstrate the impact of these parameters, transient simulations (using the model of Section 11.3.1.) for the start-up of two microreactors made out of cordierite and FeCr alloy (for material properties see Table 11.2) have been performed [59]. The microreactor is a plane channel with a length of 10 mm and a height of 1 mm; the initial 1 mm length is chemically inert while the remaining 9 mm is coated with platinum. Detailed heterogeneous [84] and homogeneous reaction mechanisms are used in the simulations. The influence of the solid material and the inlet pressure is summarized in Figure 11.22 for two methane-fueled microreactors made of cordierite and FeCr alloy, respectively, at different inlet pressures. The mass inflow $(\rho_{IN}U_{IN})$ is constant for all cases in Figure 11.22. The cordierite reactor ignites and also reaches steady state in a shorter time period when compared to the FeCr-alloy reactor. Because of the lower thermal conductivity of this material, heat generated on the catalytic surface cannot diffuse away from the reaction front. This leads to a spatially confined reaction zone, which increases the fuel consumption rate and leads to a faster catalytic ignition. These results are in strong contrast with observations made for steady-state operations of the microreactor discussed in Section 11.4.1. While a ceramic material with a low thermal conductivity such as cordierite has an advantage during the start-up, metallic microreactors provide extended steady-state stability limits and thus more robust combustion (see Section 11.4.1.). Moreover, it can be shown [59] that the FeCr-alloy microreactor has advantages in terms of lower attained steady wall temperatures.

It was shown in Section 11.4.1. that elevated pressures increase the stability limits of steady methane combustion on platinum. The increase in catalytic reactivity with rising pressure also strongly affects the start-up process. An earlier homogeneous ignition of the microreactor is hence facilitated, particularly in low thermal conductivity materials for which the initial reaction zone is spatially confined. However, the elapsed time between light-off and steady state is only marginally affected by the increased pressure (see differences between ignition and steady times in Figure 11.22). This indicates that the approach to steady state is controlled by the chemical power input once the reactor is ignited [59]. For the design of practical microreactors, the previous observations, in conjunction with the extended stability limits at steady state, clearly show the advantage of moderate operating pressures.

An increase of the fuel-to-air equivalence ratio can significantly reduce the time required for catalytic ignition and also for steady state. Numerical studies of a methane-fueled microre-

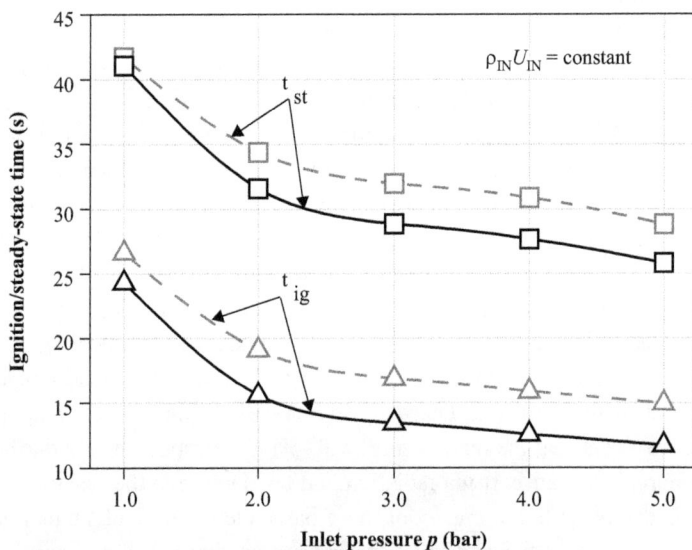

Figure 11.22. Ignition times (t_{ig}, triangles) and steady-state times (t_{st}, squares) versus inlet pressure for a cordierite (solid lines) and an FeCr-alloy (dashed lines) microchannel platinum-coated reactor. The fuel is methane with equivalence ratio $\varphi = 0.4$. The mass inflow $(\rho_{IN}U_{IN})$ is constant in all cases [59].

actor have shown that an increase of the equivalence ratio from $\varphi = 0.4$ to $\varphi = 0.6$ (at 5 bar) can decrease the catalytic ignition time by more than 50 percent [59]. The equivalence ratio and thus the catalytic reactivity reduce the impact of the solid material properties: with increased equivalence ratio the difference in ignition time and steady-state times becomes less pronounced for a cordierite and an FeCr-alloy microreactor.

Radiation heat transfer is an important energy transfer mechanism for catalytic reactors, which cannot be neglected [73, 75, 85]. Because of the short distance and the resulting apreciable viewing factors between the hot walls and the colder inlet section, microreactors suffer from significant heat losses via surface radiation. Depending on the reactor material and the equivalence ratio, around 30 percent of the chemical energy can be lost via surface radiation heat transfer [59]. Surface radiation heat transfer can have direct influence on ignition times, especially for reactors with low thermal conductivity material walls: because of radiative heat transport away from the initially spatially confined catalytic reaction zone, catalytic ignition is hindered and thus the ignition time becomes longer. However, after light-off, radiation is the most effective heat transfer mechanism in the channel and facilitates the upstream propagation of the reaction front. Due to this energy transfer from the rear to the front of the catalytic channel, the steady-state time is reduced compared to that of a similar case without inclusion of radiation [59].

Rapid start-up is a key requirement for practical microreactors. One strategy to reduce start-up times of hydrocarbon-fueled microreactors is to add a small amount of hydrogen to the fuel. Furthermore, with this concept it is possible to ignite hydrocarbon fuels without additional ignition devices thus aiding the further miniaturization of microreactor-based power generation devices. Pure hydrogen/air mixtures are known to self-ignite over platinum at very fuel lean stoichiometries [86–88] and at room temperatures in microchannel geometries [82]. Since hydrocarbons, especially methane, are difficult to ignite catalytically, the heat release from the combustion of hydrogen facilitates hydrocarbon ignition [84]. For hydrogen addition, different strategies are possible. A ready-to-use hydrocarbon/hydrogen blend can be combusted continuously. On the other hand, it is possible to add hydrogen only during the start-up process to facilitate hydrocarbon ignition. Since hydrogen is the main target of reformers for fuel cell applications, it is conceivable to store small amounts of hydrogen during normal operation to use it for the next start-up phase. Figure 11.23 shows the temperature history 1.6 cm downstream of the inlet of a propane-fueled catalytic microreactor, whereby different strategies of hydrogen-assisted start-up procedures have been tested [82]. The first method (solid line) entails feeding the reactor with a 90 percent propane and 10 percent hydrogen mixture, while in the second method (dashed line) the reactor is initially preheated with a pure hydrogen/air mixture before the propane supply is switched on. For both cases, the hydrogen supply is switched off when a specific temperature has been reached and propane combustion is self-sustaining. The insert shows the effect of hydrogen concentration: with only 4 percent hydrogen content, the start-up phase is significantly longer.

A numerical study of a methane-fueled catalytic microreactor has shown that the advantage of a hydrogen-assisted start-up is attenuated at elevated pressures because of the increased catalytic reactivity of methane with rising pressure [89]. The main advantage of hydrogen addition is its immediate heat release during the start-up period. Results are shown in Figure 11.24 for a catalytic plane channel with a length of 10 mm and a height of 1 mm; the channel is coated with platinum over the last 9 mm length. Simulations were firstly performed with the inclusion of catalytic chemistry only. The reaction rate progress of the methane and hydrogen fuels differs significantly throughout the heat-up process. For an initially high wall temperature of

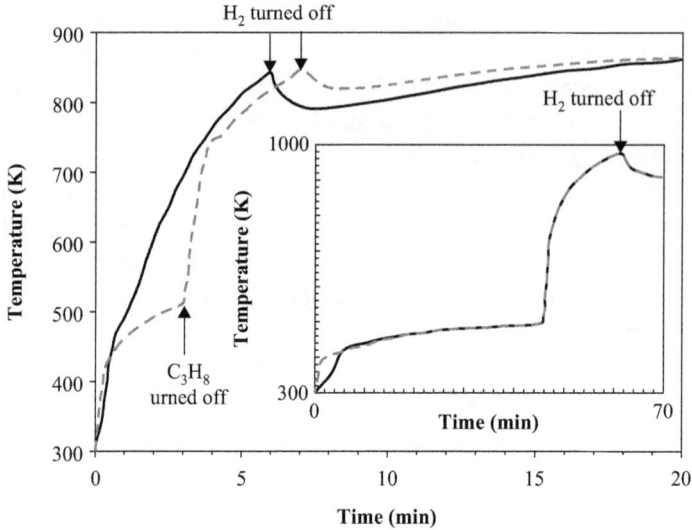

Figure 11.23. Measured temperature time histories 1.6 cm downstream of the reactor inlet for H_2/air/N_2/C_3H_8 mixtures using different ignition procedures. Solid lines correspond to simultaneous feeding of C_3H_8 and H_2; dashed lines correspond to mode with initial pure H_2 feeding and subsequent introduction of propane. Hydrogen concentration is 10 percent in the main figure, and 4 percent in the insert [82].

$T_w = 850$ K, the hydrogen conversion rate peaks already at $t \sim 0.0$ s at the beginning of the catalytic section ($x = 1.0$ mm).

In this phase, heat is mainly generated from hydrogen conversion at the catalytic walls. Methane in contrast maintains a low reaction rate in the early stage of the pre-ignition period and overtakes that of hydrogen just before ignition. After light-off and until steady state is reached, the contribution of hydrogen to the total heat generation is very low compared to that of methane. Since most of the unburned fuel or CO emissions are formed when the temperature is relatively low and thus methane is not converted completely, the temperature increase due to hydrogen conversion during the pre-ignition period reduces emissions drastically (Figure 11.25).

The impact of gaseous chemistry is subsequently studied. It was shown in Section 11.4.1. that homogeneous chemistry enhances combustion stability, especially at elevated pressures [73] and steady-state operation. By comparing a numerical simulation of the transient start-up process with full hetero-/homogeneous chemistry to the previously presented one with deactivated homogeneous reactions it can be shown that gas-phase reactions have only minor impact on the ignition time since surface chemistry is the dominant pathway before ignition. Homogeneous combustion influences the emissions of a microreactor. Since homogeneous reactions consume CH_4 either via complete oxidation or via the incomplete oxidation to CO, which is oxidized subsequently on the catalyst (as also discussed in the end of Section 11.4.1.), the unburned fuel emissions during the start-up phase are typically overpredicted when gas-phase chemistry is neglected in a numerical simulation. On the other hand, CO emissions are significantly higher when gas-phase reactions are included. When

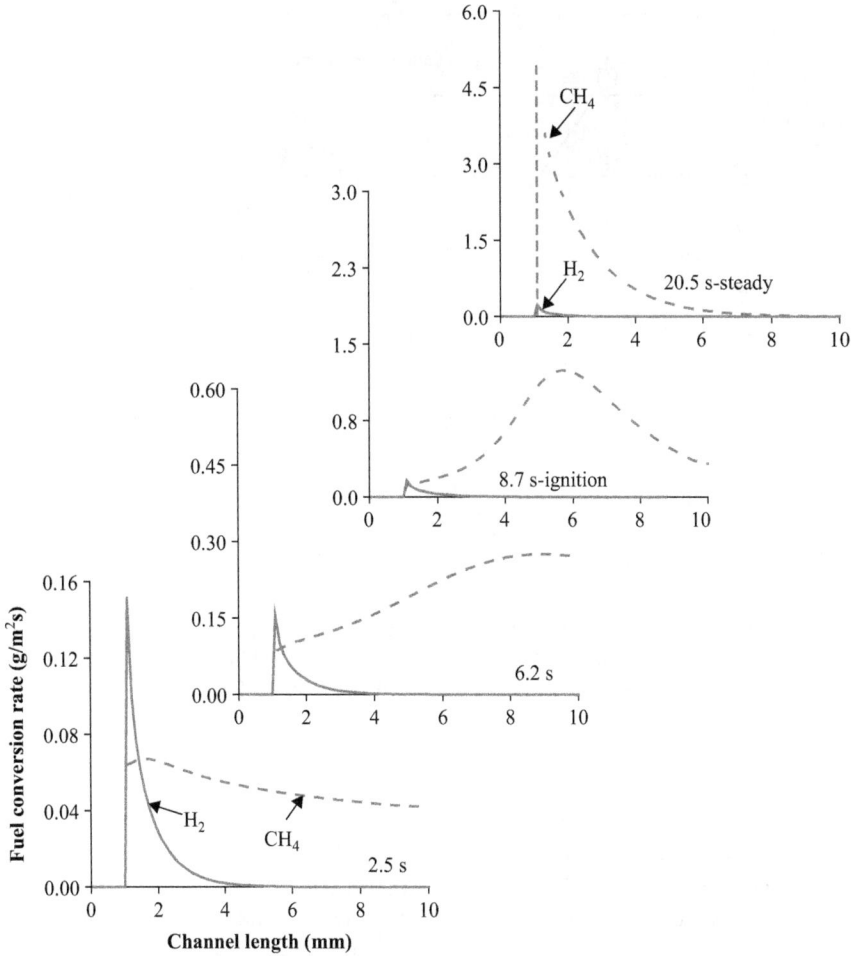

Figure 11.24. Hydrogen (solid lines) and methane (dashed lines) catalytic conversion rates along a cordierite microchannel reactor during the start-up phase at four time instances: ignition, steady-state, and two pre-ignition time instances. Inlet pressure $p = 5$ bar, $\varphi = 0.37$ with 90 percent CH_4 and 10 percent H_2 (by volume); the platinum catalyst coating starts at $x = 1.0$ mm [89].

designing a catalytic microreactor with a numerical model, which neglects homogeneous chemistry, only the ignition time can be predicted accurately while the steady-state time will be, especially for high-thermal-conductivity materials, underpredicted. These conclusions apply for pure hydrocarbon/air mixtures as well as for hydrogen-assisted start-up of methane-fueled (and higher hydrocarbons) microreactors [89].

11.4.3 CATALYTIC MICROREFORMERS—STEADY AND TRANSIENT OPERATION

Experimental and numerical studies of transient as well as steady catalytic methane reforming were performed in an FeCr-alloy honeycomb reactor coated with Rh/ZrO_2 [10]. The reactor

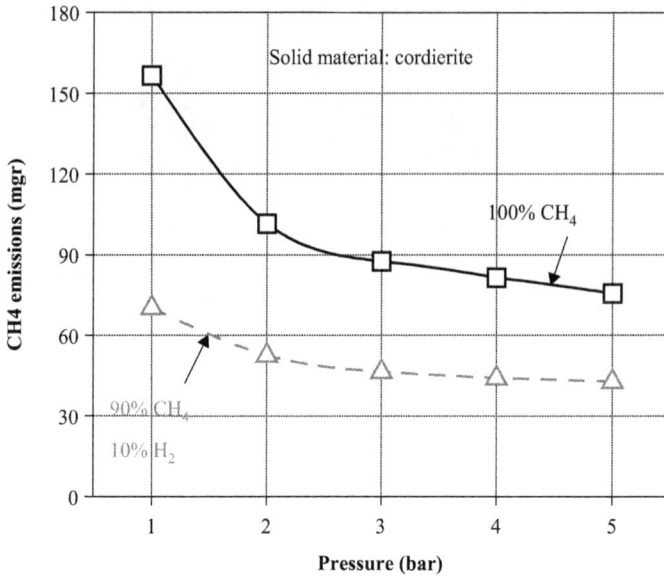

Figure 11.25. Cumulative microreactor emissions of unburned CH_4 versus pressure for a methane-hydrogen blend (dashed line) and the corresponding cases computed for pure methane in the fuel stream (solid line) [89].

has a length of 75 mm (with only the central 55 mm length is catalytically coated), diameter of 35 mm, and individual channel hydraulic diameter of 1.2 mm (see the inset in Figure 11.26). Operating conditions are $p = 5$ bar, $\varphi = 4.0$, exhaust gas dilution of 46.3 percent H_2O and 23.1 percent CO_2 per volume, inlet temperature $T_{IN} = 680$ K, and inlet velocity $U_{IN} = 5.6$ m/s. Of particular interest are the catalytic ignition transient characteristics during reforming of methane to syngas and also the steady-state combustion stability. The temperatures inside and at the exit of the reactor are monitored with thermocouples, while gas chromatography provides the exhaust gas composition. A 2D transient model (see Section 11.3.1) is used to simulate a single channel of the reactor. The model accounts for heat conduction in the solid, thermal radiation heat transfer between the inner channel surfaces, detailed transport, and elliptic flow description. At the moderate pressure of 5 bar, gas-phase chemistry is altogether negligible [8, 10].

Simulated and measured temporal histories of temperature are depicted in Figure 11.26 at four locations marked *B* to *E* positioned inside the honeycomb structure. Predicted axial profiles of the wall temperatures at selected times are further shown in Figure 11.27. In Figure 11.26 simulations are shown for both the surface and the radially averaged gas temperatures. The predictions in Figure 11.26 capture, at all positions, the measured elapsed times for the onset of abrupt temperature rise, the temporal extent of the main transient event, and the steady-state temperatures. The measured temperatures at late times ($t \geq 10$ s), whereby steady state has been practically achieved, are in good agreement with the computed surface temperatures. An exception is location *B*, with the measurements lying in-between the predicted surface and mean gas temperatures. However, this can be attributed to the very steep spatial temperature gradients at *B*, which is located at the beginning of the catalytically coated section (see also Figure 11.27).

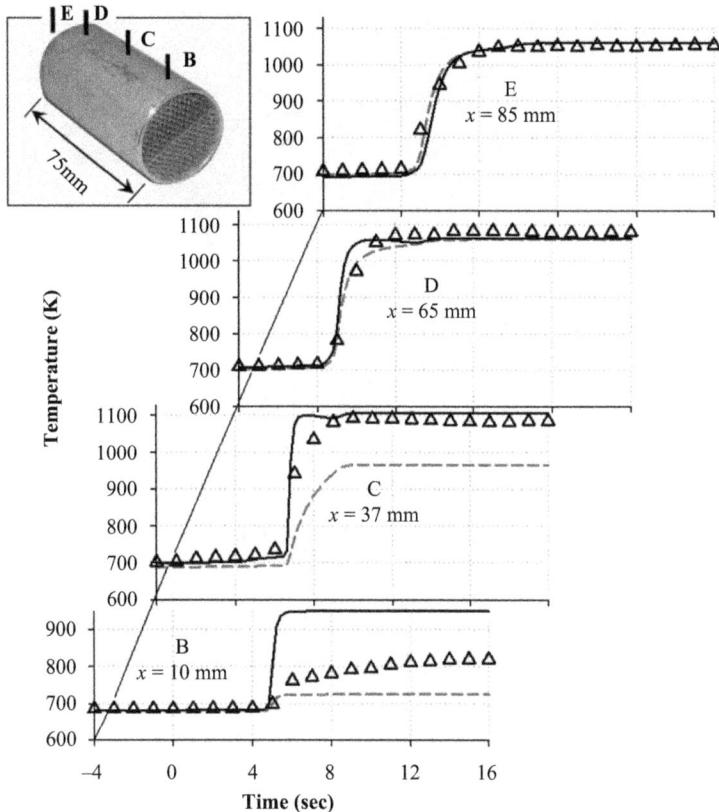

Figure 11.26. Catalytic reforming of methane over Rh/ZrO_2, $p = 5$ bar, $T_{IN} = 680$ K, $\varphi = 4.0$, with 46.3 percent H_2O and 23.1 percent CO_2 vol. dilution. Predicted (lines) and measured (symbols) temperatures at axial positions B through E in a honeycomb reactor (adapted from [10]). The reactor (shown in the insert top left) is 75 mm long, with catalyst applied at 10 mm $\leq x \leq$ 65 mm. The predictions refer to the wall temperature (solid lines) and the mean gas temperature (dashed lines). The origin of the time scale is arbitrary.

Additional computations with $T_{IN} \leq 670$ K do not yield light-off (in the sense of a vigorous burning solution). This is in good agreement with the experimentally assessed minimum inlet temperature for ignition ($T_{IN} = 680$ K).

Extinction was investigated by lowering slowly the inlet temperature and monitoring the exhaust gas composition and the reactor temperature. Measurements and predictions in Figure 11.28 are in good agreement with each other, clearly showing that combustion can be sustained at inlet temperatures at least as low as 473 K in CPO with exhaust gas dilution. The model captures the increase in total oxidation products and the decrease in syngas yields with decreasing inlet temperature. Additional simulations with CH_4/air mixtures and without H_2O and CO_2 addition have shown that combustion stability can be maintained (by reducing the inlet temperature after stable combustion is achieved) at inlet temperatures as low as 290 K [10]. This large hysteresis in the ignition/extinction characteristics shows that catalytic hydrocarbon microreformers have extended combustion stability limits when compared to total oxidation catalytic hydrocarbon microreactors.

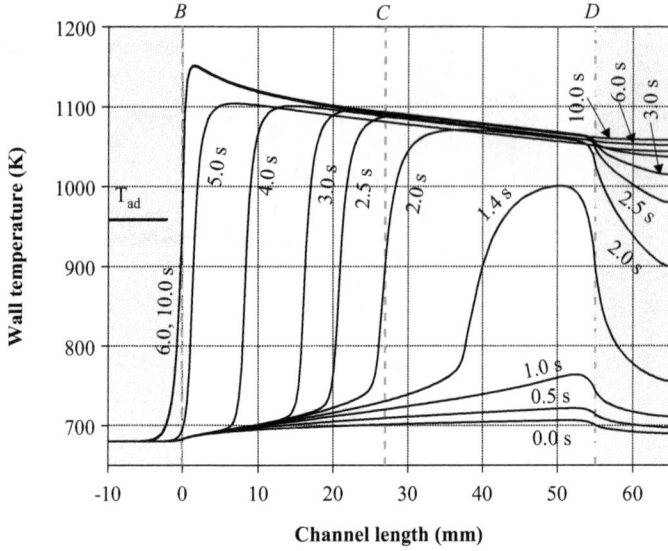

Figure 11.27. Computed axial profiles of wall temperature at various times for the catalytic reforming case in Figure 11.26. The shaded areas denote the non-catalytic part of the reactor. The vertical dashed lines at B, C, and D indicate the thermocouple locations inside the reactor in Figure 11.26 and the horizontal mark denotes the adiabatic equilibrium temperature T_{ad} [10].

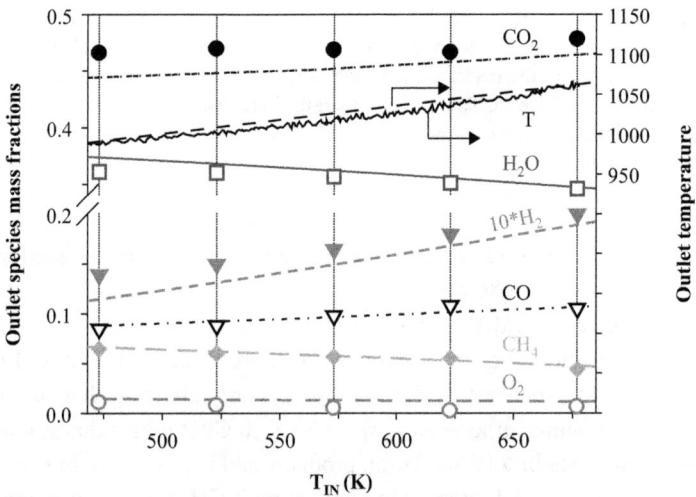

Figure 11.28. Predicted (lines) and measured (symbols) outlet species mass fractions in catalytic reforming of methane, as a function of T_{IN}; the other parameters are as in Figure 11.26. On the same graph, the predicted (dashed line) and measured (solid line) outlet gas temperatures are also given [10].

ACKNOWLEDGMENTS

Support has been provided by the European Union project HRC-Power.

REFERENCES

[1] W.C. Pfefferle, Catalytically supported thermal combustion, U.S. Patent No. 3,928,961, May 8, 1973.

[2] K.W. Beebe, K.D. Cairns, V.K. Pareek, S.G. Nickolas, J.C. Schlatter, T. Tsuchiya, Development of catalytic combustion technology for single-digit emissions from industrial gas turbines, *Catalysis Today*, 59 (2000) 95–115. doi: http://dx.doi.org/10.1016/S0920-5861(00)00276-5.

[3] H. Karim, K. Lyle, S. Etemad, L. Smith, W. Pfefferle, P. Dutta, K. Smith, Advanced catalytic pilot for low NO_x industrial gas turbines, ASME Conference Proceedings, Paper No. GT20-30083, pp. 483–490, Amsterdam, The Netherlands, June 3–6, 2002.

[4] C. Appel, J. Mantzaras, R. Schaeren, R. Bombach, A. Inauen, Turbulent catalytically stabilized combustion of hydrogen/air mixtures in entry channel flows, *Combustion and Flame*, 140 (2005) 70–92. doi: http://dx.doi.org/10.1016/j.combustflame.2004.10.006.

[5] S. Kajita, S. Tanaka, J. Kitajima, Evaluation of a catalytic combustor in a gas turbine generator unit, ASME, ASME paper 90-GT-89, June 11–14, Brussels, 1990.

[6] R. Carroni, T. Griffin, J. Mantzaras, M. Reinke, High-pressure experiments and modeling of methane/air catalytic combustion for power-generation applications, *Catalysis Today*, 83 (2003) 157–170. doi: http://dx.doi.org/10.1016/S0920-5861(03)00226-8.

[7] L.L. Smith, H. Karim, M.J. Castaldi, S. Etemad, W.C. Pfefferle, Rich-catalytic lean-burn combustion for fuel-flexible operation with ultra low emissions, *Catalysis Today*, 117 (2006) 438–446. doi: http://dx.doi.org/10.1016/j.cattod.2006.06.021.

[8] A. Schneider, J. Mantzaras, P. Jansohn, Experimental and numerical investigation of the catalytic partial oxidation of CH_4/O_2 mixtures diluted with H_2O and CO_2 in a short contact time reactor, *Chemical Engineering Science*, 61 (2006) 4634–4649. doi: http://dx.doi.org/10.1016/j.ces.2006.02.038.

[9] G. Veser, M. Ziauddin, L.D. Schmidt, Ignition in alkane oxidation on noble-metal catalysts, *Catalysis Today*, 47 (1999) 219–228. doi: http://dx.doi.org/10.1016/S0920-5861(98)00302-2.

[10] A. Schneider, J. Mantzaras, S. Eriksson, Ignition and extinction in catalytic partial oxidation of methane-oxygen mixtures with large H_2O and CO_2 dilution, *Combustion Science and Technology*, 180 (2008) 89–126. doi: http://dx.doi.org/10.1080/00102200701487087.

[11] T. Griffin, D. Winkler, M. Wolf, C. Appel, J. Mantzaras, Staged catalytic combustion method for the advanced zero emissions gas turbine power plant, ASME Conference Proceedings, Paper No. GT2004-54101, pp. 705–711, Vienna, Austria, June 14–17, 2004.

[12] L.L. Smith, H. Karim, M.J. Castaldi, S. Etemad, W.C. Pfefferle, V. Khanna, K.O. Smith, Rich-catalytic lean-burn combustion for low-single-digit NO_x gas turbines, *Journal of Engineering for Gas Turbines and Power*, 127 (2005) 27–35. doi: http://dx.doi.org/10.1115/1.1787510.

[13] S.K. Alavandi, S. Etemad, B.D. Baird, Low single digit NO_x emissions catalytic combustor for advanced hydrogen turbines for clean coal power systems, ASME Paper Proceedings, Paper No. GT2012-68128, pp. 53–62, Copenhagen, Denmark, June 11–15, 2012.

[14] F. Bolãnos, D. Winkler, F. Piringer, T. Griffin, R. Bombach, J. Mantzaras, Study of a rich/lean staged combustion concept for hydrogen under gas turbine relevant conditions, ASME Proceedings, Paper No. GT2013-94420, pp. V01AT04A031, San Antonio, Texas, June 3–7, 2013.

[15] Y.F. Tham, J.Y. Chen, Numerical study of rich-quick-lean (RQL) combustion of syngas, The 4th Joint Meeting of the U.S. Sections of the Combustion Institute, Pittsburgh, PA, March 20–23, 2005.

[16] D. Linden, T.B. Reddy. Handbook of Batteries (3rd Edition), McGraw-Hill, 2002.

[17] J. Ahn, C. Eastwood, L. Sitzki, P.D. Ronney, Gas-phase and catalytic combustion in heat-recirculating burners, *Proceedings of the Combustion Institute*, 30 (2005) 2463–2472. doi: http://dx.doi.org/10.1016/j.proci.2004.08.265.

[18] D.C. Kyritsis, B. Coriton, F. Faure, S. Roychoudhury, A. Gomez, Optimization of a catalytic combustor using electrosprayed liquid hydrocarbons for mesoscale power generation, *Combustion and Flame*, 139 (2004) 77–89. doi: http://dx.doi.org/10.1016/j.combustflame.2004.06.010.

[19] T. Kamijo, Y. Suzuki, N. Kasagi, T. Okamasa, High-temperature micro catalytic combustor with Pd/nano-porous alumina, *Proceedings of the Combustion Institute*, 32 (2009) 3019–3026. doi: http://dx.doi.org/10.1016/j.proci.2008.06.118.

[20] K. Yoshida, S. Tanaka, S. Tomonari, D. Satoh, M. Esashi, High-energy density miniature thermoelectric generator using catalytic combustion, *Journal of Microelectromechanical Systems*, 15 (2006) 195–203. doi: http://dx.doi.org/10.1109/JMEMS.2005.859202.

[21] K. Yoshida, S. Tanaka, H. Hiraki, M. Esashi, A micro fuel reformer integrated with a combustor and a microchannel evaporator, *Journal of Micromechanics and Microengineering*, 16 (2006) S191. doi: http://dx.doi.org/10.1088/0960-1317/16/9/S04.

[22] A. Bieberle-Hütter, D. Beckel, A. Infortuna, U.P. Muecke, J.L.M. Rupp, L.J. Gauckler, S. Rey-Mermet, P. Muralt, N.R. Bieri, N. Hotz, M.J. Stutz, D. Poulikakos, P. Heeb, P. Müller, A. Bernard, R. Gmür, T. Hocker, A micro-solid oxide fuel cell system as battery replacement, *Journal of Power Sources*, 177 (2008) 123–130. doi: http://dx.doi.org/10.1016/j.jpowsour.2007.10.092.

[23] A.J. Santis-Alvarez, M. Nabavi, N. Hild, D. Poulikakos, W.J. Stark, A fast hybrid start-up process for thermally self-sustained catalytic *n*-butane reforming in micro-SOFC power plants, *Energy & Environmental Science*, 4 (2011) 3041–3050. doi: http://dx.doi.org/10.1039/c1ee01330k.

[24] D.L. Hitt, C.M. Zakrzwski, M.A. Thomas, MEMS-based satellite micropropulsion via catalyzed hydrogen peroxide decomposition, *Smart Materials and Structures*, 10 (2001) 1163. doi: http://dx.doi.org/10.1088/0964-1726/10/6/305.

[25] M. Aryafar, F. Zaera, Kinetic study of the catalytic oxidation of alkanes over nickel, palladium, and platinum foils, *Catalysis Letters*, 48 (1997) 173–183. doi: http://dx.doi.org/10.1023/A:1019055810760.

[26] A.R. Jones, S.A. Lloyd, F.J. Weinberg, Combustion in heat exchangers, *Proceedings of the Royal Society of London. A. Mathematical and Physical Sciences*, 360 (1978) 97–115. doi: http://dx.doi.org/10.1098/rspa.1978.0059.

[27] S.A. Lloyd, F.J. Weinberg, A burner for mixtures of very low heat content, *Nature*, 251 (1974) 47–49. doi: http://dx.doi.org/10.1038/251047a0.

[28] S.A. Lloyd, F.J. Weinberg, Limits to energy release and utilization from chemical fuels, *Nature*, 257 (1975) 367–370. doi: http://dx.doi.org/10.1038/257367a0.

[29] L. Sitzki, E. Borer, E. Schuster, P.D. Ronney, Combustion in microscale heat-recirculating burners, The Third Asia-Pacific Conference on Combustion, Seoul, Korea, June 24–27 2001.

[30] L. Sitzki, K. Borer, S. Wussow, E. Schuster, K. Maruta, P. Ronney, A. Cohen, Combustion in microscale heat-recirculating burners, 38th AIAA Space Sciences and Exhibit, Reno, NV, 2001.

[31] A.L. Cohen, P. Ronney, U. Frodis, L. Sitzki, E. Meiburg, S. Wussow, Microcombustor and combustion-based thermoelectric microgenerator, U.S. Patent No. US6,613,972, January 5, 2001.

[32] D.C. Kyritsis, I. Guerrero-Arias, S. Roychoudhury, A. Gomez, Mesoscale power generation by a catalytic combustor using electrosprayed liquid hydrocarbons, *Proceedings of the Combustion Institute*, 29 (2002) 965–972. doi: http://dx.doi.org/10.1016/S1540-7489(02)80122-5.

[33] R.N. Carter, P. Menacherry, W.C. Pfefferle, G. Muench, S. Roychoudhury, Laboratory evaluation of ultra-short metal monolith catalyst, SAE Technical Paper 980672, February 1998.

[34] W. Deng, J.F. Klemic, X. Li, M.A. Reed, A. Gomez. Liquid fuel microcombustor using microfabricated multiplexed electrospray sources, *Proceedings of the Combustion Institute*, 31 (2007) 2239–2246. doi: http://dx.doi.org/10.1016/j.proci.2006.08.080.

[35] Y. Suzuki, J. Saito, N. Kasagi, Development of micro catalytic combustor with Pt/Al$_2$O$_3$ thin films, *JSME International Journal Series B, Fluids and Thermal Engineering*, 47 (2004) 522–527. doi: http://dx.doi.org/10.1299/jsmeb.47.522.

[36] T. Okamasa, G.-G. Lee, Y. Suzuki, N. Kasagi, S. Matsuda, Development of a micro catalytic combustor using high-precision ceramic tape casting, *Journal of Micromechanics and Microengineering*, 16 (2006) S198. doi: http://dx.doi.org/10.1088/0960-1317/16/9/S05.

[37] Y. Suzuki, Y. Horii, N. Kasagi, S. Matsuda, Micro catalytic combustor with tailored porous alumina, Proceedings 17th IEEE International Micro Electro Mechanical Systems Conference (MEMS), (2004) 312–315.

[38] A.H. Epstein, S.D. Senturia, G. Anathasuresh, A. Ayon, K. Breuer, K.-S. Chen, F. Ehrich, G. Gauba, R. Ghodssi, C. Groshenry, S. Jacobson, J. Lang, C.-C. Mehra, J.M. Mur, S. Nagle, D. Orr, E. Piekos, M. Schmidt, G. Shirley, S. Spearing, C. Tan, Y.-S. Tzeng, I. Waitz, Power MEMS and microengines, Proceedings International Solid State Sensors and Actuators Transducers '97 Conference, Vol. 2, pp. 753–756, Chicago, IL, June 1997.

[39] B. Schneider, M. Bruderer, D. Dyntar, C. Zwyssig, M. Diener, K. Boulouchos, R.S. Abhari, L. Guzzella, J.W. Kolar, Ultra-high-energy-density converter for portable power, Proceedings Power MEMS, pp. 81–84, Tokyo, Japan, November 28–30, 2005.

[40] S. Karagiannidis, K. Marketos, J. Mantzaras, R. Schaeren, K. Boulouchos, Experimental and numerical investigation of a propane-fueled, catalytic mesoscale combustor, *Catalysis Today*, 155 (2010) 108–115. doi: http://dx.doi.org/10.1016/j.cattod.2010.04.030.

[41] S.K. Kamarudin, W.R.W. Daud, A.M. Som, M.S. Takriff, A.W. Mohammad, Y.K. Loke, Design of a fuel processor unit for PEM fuel cell via shortcut design method, *Chemical Engineering Journal*, 104 (2004) 7–17. doi: http://dx.doi.org/10.1016/j.cej.2004.07.007.

[42] N. Hotz, M.J. Stutz, S. Loher, W.J. Stark, D. Poulikakos, Syngas production from butane using a flame-made Rh/Ce$_{0.5}$Zr$_{0.5}$O$_2$ catalyst, *Applied Catalysis B: Environmental*, 73 (2007) 336–344. doi: http://dx.doi.org/10.1016/j.apcatb.2007.01.001.

[43] M.J. Stutz, N. Hotz, D. Poulikakos, Optimization of methane reforming in a microreactor–effects of catalyst loading and geometry, *Chemical Engineering Science*, 61 (2006) 4027–4040. doi: http://dx.doi.org/10.1016/j.ces.2006.01.035.

[44] M.M. Micci, A.D. Ketsdever, Micropropulsion for Small Spacecraft, AIAA, Reston, Virginia, 2000.

[45] D.W. Youngner, S.T. Lu, E. Choueiri, J.B. Neidert, R.E. Black III, K.J. Graham, R. Lucas, X. Zhu. MEMS mega-pixel micro-thruster arrays for small satellite stationkeeping, Proceedings of the 14th Annual/USU Conference on Small Satellites, Paper No. SSC00-X-2, North Logan, UT, August 21–24, 2000.

[46] A.P. London, A.A. Ayon, A.H. Epstein, S.M. Spearing, T. Harrison, Y. Peles, J.L. Kerrebrock, Microfabrication of a high pressure bipropellant rocket engine, *Sensors and Actuators A: Physical*, 92 (2001) 351–357. doi: http://dx.doi.org/10.1016/S0924-4247(01)00571-4.

[47] G.A. Boyarko, C. Sung, S.J. Schneider, Catalyzed combustion of hydrogen-oxygen in platinum tubes for micro-propulsion applications, *Proceedings of the Combustion Institute*, 30 (2005) 2481–2488. doi: http://dx.doi.org/10.1016/j.proci.2004.08.203.

[48] S.J. Volchko, C.-J. Sung, Y. Huang, S.J. Schneider, Catalytic combustion of rich methane/oxygen mixtures for micropropulsion applications, *Journal of Propulsion and Power*, 22 (2006) 684–693. doi: http://dx.doi.org/10.2514/1.19809.

[49] C. Mento, C.-J. Sung, A. Ibarreta, S. Schneider, Catalytic ignition of methane/hydrogen/oxygen mixtures for microthruster applications, Joint Propulsion Conferences, July 2006.

[50] C.A. Mento, C.-J. Sung, A.F. Ibarreta, S.J. Schneider, Catalyzed ignition of using methane/hydrogen fuel in a microtube for microthruster applications, *Journal of Propulsion and Power*, 25 (2009) 1203–1210. doi: http://dx.doi.org/10.2514/1.42592.

[51] W.C. Pfefferle, L.D. Pfefferle, Catalytically stabilized combustion, *Progress in Energy and Combustion Science*, 12 (1986) 25–41. doi: http://dx.doi.org/10.1016/0360-1285(86)90012-2.

[52] F. A. Williams, Combustion Theory, Benjamin Cummings, Menlo Park, CA, 1985.

[53] C. Appel, J. Mantzaras, R. Schaeren, R. Bombach, A. Inauen, B. Kaeppeli, B. Hemmerling, A. Stampanoni, An experimental and numerical investigation of homogeneous ignition in catalytically stabilized combustion of hydrogen/air mixtures over platinum, *Combustion and Flame*, 128 (2002) 340–368. doi: http://dx.doi.org/10.1016/S0010-2180(01)00363-7.

[54] M. Reinke, J. Mantzaras, R. Bombach, S. Schenker, A. Inauen, Gas phase chemistry in catalytic combustion of methane/air mixtures over platinum at pressures of 1 to 16 bar, *Combustion and Flame*, 141 (2005) 448–468. doi: http://dx.doi.org/10.1016/j.combustflame.2005.01.016.

[55] R.E. Hayes, S.T. Kolaczkowski, Mass and heat transfer effects in catalytic monolith reactors, *Chemical Engineering Science*, 49 (1994) 3587–3599. doi: http://dx.doi.org/10.1016/0009-2509(94)00164-2.

[56] S.T. Kolaczkowski, S. Serbetcioglu, Development of combustion catalysts for monolith reactors: a consideration of transport limitations, *Applied Catalysis A: General*, 138 (1996) 199–214. doi: http://dx.doi.org/10.1016/0926-860X(95)00296-0.

[57] N. Mladenov, J. Koop, S. Tischer, O. Deutschmann, Modeling of transport and chemistry in channel flows of automotive catalytic converters, *Chemical Engineering Science*, 65 (2010) 812–826. doi: http://dx.doi.org/10.1016/j.ces.2009.09.034.

[58] H. Zhu, G.S. Jackson, Transient modeling for assessing catalytic combustor performance in small gas turbine applications, ASME Turbo Expo 2001, Paper No. 2001-GT-0520, New Orleans, LA, June, 2001.

[59] S. Karagiannidis, J. Mantzaras, Numerical investigation on the start-up of methane-fueled catalytic microreactors, *Combustion and Flame*, 157 (2010) 1400–1413. doi: http://dx.doi.org/10.1016/j.combustflame.2010.01.008.

[60] W. Bier, W. Keller, G. Linder, D. Seidel, K. Schubert, H. Martin, Gas to gas heat transfer in micro heat exchangers, *Chemical Engineering and Processing: Process Intensification*, 32 (1993) 33–43. doi: http://dx.doi.org/10.1016/0255-2701(93)87004-E.

[61] S. Hasebe, Design and operation of micro-chemical plants—bridging the gap between nano, micro and macro technologies, *Computers & Chemical Engineering*, 29 (2004) 57–64. doi: http://dx.doi.org/10.1016/j.compchemeng.2004.07.020.

[62] T. Stief, O.-U. Langer, K. Schubert, Numerical investigations of optimal heat conductivity in micro heat exchangers, *Chemical Engineering & Technology*, 22 (1999) 297–303. doi: http://dx.doi.org/10.1002/(SICI)1521-4125(199904)22:4<297::AID-CEAT297>3.3.CO;2-P.

[63] J.A. Federici, E.D. Wetzel, B.R. Geil, D.G. Vlachos, Single channel and heat recirculation catalytic microburners: an experimental and computational fluid dynamics study, *Proceedings of the Combustion Institute*, 32 (2009) 3011–3018. doi: http://dx.doi.org/10.1016/j.proci.2008.07.005.

[64] J. Mantzaras. Interplay of transport and hetero-/homogeneous chemistry, In S.Z. Jiang, editor, Focus on Combustion Research. Nova Publishers, New York, NY, 2006.

[65] O. Deutschmann, R. Schmidt, F. Behrendt, J. Warnatz, Numerical modeling of catalytic ignition, *International Symposium on Combustion*, 26 (1996) 1747–1754.

[66] R. Schwiedernoch, S. Tischer, C. Correa, O. Deutschmann, Experimental and numerical study on the transient behavior of partial oxidation of methane in a catalytic monolith, *Chemical Engineering Science*, 58 (2003) 633–642. doi: http://dx.doi.org/10.1016/S0009-2509(02)00589-4.

[67] D.G. Norton, D.G. Vlachos, Combustion characteristics and flame stability at the microscale: A CFD study of premixed methane/air mixtures, *Chemical Engineering Science*, 58 (2003) 4871–4882. doi: http://dx.doi.org/10.1016/j.ces.2002.12.005.

[68] D.G. Norton, D.G. Vlachos, A CFD study of propane/air microflame stability, *Combustion and Flame*, 138 (2004) 97–107. doi: http://dx.doi.org/10.1016/j.combustflame.2004.04.004.

[69] K. Maruta, J. Takeda, K. Ahn, K. Borer, L. Sitzki, P.D. Ronney, O. Deutschmann, Extinction limits of catalytic combustion in microchannels, *Proceedings of the Combustion Institute*, 29 (2002) 957–963. doi: http://dx.doi.org/10.1016/S1540-7489(02)80121-3.

[70] M. Reinke, J. Mantzaras, R. Schaeren, R. Bombach, W. Kreutner, A. Inauen, Homogeneous ignition in high-pressure combustion of methane/air over platinum: Comparison of measurements and detailed numerical predictions, *Proceedings of the Combustion Institute*, 29 (2002) 1021–1029. doi: http://dx.doi.org/10.1016/S1540-7489(02)80129-8.

[71] M. Reinke, J. Mantzaras, R. Schaeren, R. Bombach, A. Inauen, S. Schenker, High-pressure catalytic combustion of methane over platinum: In situ experiments and detailed numerical predictions, *Combustion and Flame*, 136 (2004) 217–240. doi: http://dx.doi.org/10.1016/j.combustflame.2003.10.003.

[72] C. Appel, J. Mantzaras, R. Schaeren, R. Bombach, A. Inauen, N. Tylli, M. Wolf, T. Griffin, D. Winkler, R. Carroni, Partial catalytic oxidation of methane to synthesis gas over rhodium: In situ Raman experiments and detailed simulations, *Proceedings of the Combustion Institute*, 30 (2005) 2509–2517. doi: http://dx.doi.org/10.1016/j.proci.2004.08.055.

[73] S. Karagiannidis, J. Mantzaras, G. Jackson, K. Boulouchos, Hetero-/homogeneous combustion and stability maps in methane-fueled catalytic microreactors, *Proceedings of the Combustion Institute*, 31 (2007) 3309–3317. doi: http://dx.doi.org/10.1016/j.proci.2006.07.121.

[74] C.K. Westbrook, F.L. Dryer, Simplified reaction mechanisms for the oxidation of hydrocarbon fuels in flames, *Combustion Science and Technology*, 27 (1981) 31–43. doi: http://dx.doi.org/10.1080/00102208108946970.

[75] A.L. Boehman, Radiation heat transfer in catalytic monoliths. *AIChE Journal*, 44 (1998) 2745–2755. doi: http://dx.doi.org/10.1002/aic.690441215.

[76] S. Karagiannidis, J. Mantzaras, K. Boulouchos, Stability of hetero-/homogeneous combustion in propane- and methane-fueled catalytic microreactors: channel confinement and molecular transport effects, *Proceedings of the Combustion Institute*, 33 (2011) 3241–3249. doi: http://dx.doi.org/10.1016/j.proci.2010.05.107.

[77] D.A. Behrens, I.C. Lee, C.M. Waits, Catalytic combustion of alcohols for microburner applications, *Journal of Power Sources*, 195 (2010) 2008–2013. doi: http://dx.doi.org/10.1016/j.jpowsour.2009.10.001.

[78] R. Quiceno Gonzalez, Evaluation of a Detailed Reaction Mechanism for Partial and Total Oxidation of C_1 - C_4 Alkanes. PhD thesis, Universität Heidelberg, May 2008.

[79] M. Hartmann, L. Maier, H.D. Minh, O. Deutschmann, Catalytic partial oxidation of iso-octane over rhodium catalysts: an experimental, modeling, and simulation study, *Combustion and Flame*, 157 (2010) 1771–1782. doi: http://dx.doi.org/10.1016/j.combustflame.2010.03.005.

[80] S. Karagiannidis, J. Mantzaras, R. Bombach, S. Schenker, K. Boulouchos, Experimental and numerical investigation of the hetero-/homogeneous combustion of lean propane/air mixtures over platinum, *Proceedings of the Combustion Institute*, 32 (2009) 1947–1955. doi: http://dx.doi.org/10.1016/j.proci.2008.06.063.

[81] T.F. Garetto, E. Rincon, C.R. Apesteguia, Deep oxidation of propane on Pt-supported catalysts: drastic turnover rate enhancement using zeolite supports, *Applied Catalysis B: Environmental*, 48 (2004) 167–174. doi: http://dx.doi.org/10.1016/j.apcatb.2003.10.004.

[82] D.G. Norton, D.G. Vlachos, Hydrogen assisted self-ignition of propane/air mixtures in catalytic microburners, *Proceedings of the Combustion Institute*, 30 (2005) 2473–2480. doi: http://dx.doi.org/10.1016/j.proci.2004.08.188.

[83] A.M. Karim, J.A. Federici, D.G. Vlachos, Portable power production from methanol in an integrated thermoelectric/microreactor system, *Journal of Power Sources*, 179 (2008) 113–120. doi: http://dx.doi.org/10.1016/j.jpowsour.2007.12.119.

[84] O. Deutschmann, L.I. Maier, U. Riedel, A.H. Stroemman, R.W. Dibble, Hydrogen assisted catalytic combustion of methane on platinum, *Catalysis Today*, 59 (2000) 141–150. doi: http://dx.doi.org/10.1016/S0920-5861(00)00279-0.

[85] S.-T. Lee, R. Aris, On the effects of radiative heat transfer in monoliths, *Chemical Engineering Science*, 32 (1977) 827–837. doi: http://dx.doi.org/10.1016/0009-2509(77)80068-7.

[86] M. Fassihi, V.P. Zhdanov, M. Rinnemo, K.E. Keck, B. Kasemo, A theoretical and experimental study of catalytic ignition in the hydrogen–oxygen reaction on platinum, *Journal of Catalysis*, 141 (1993) 438–452. doi: http://dx.doi.org/10.1006/jcat.1993.1153.

[87] M. Rinnemo, M. Fassihi, B. Kasemo, The critical condition for catalytic ignition. H_2/O_2 on Pt, *Chemical Physics Letters*, 211 (1993) 60–64. doi: http://dx.doi.org/10.1016/0009-2614(93)80052-Q.

[88] N.E. Fernandes, Y.K. Park, D.G. Vlachos, The autothermal behavior of platinum catalyzed hydrogen oxidation: Experiments and modeling, *Combustion and Flame*, 118 (1999) 164–178. doi: http://dx.doi.org/10.1016/S0010-2180(98)00162-X

[89] S. Karagiannidis, J. Mantzaras, Numerical investigation on the hydrogen-assisted start-up of methane-fueled, catalytic microreactors, *Flow, Turbulence and Combustion*, 89 (2012) 215-230. doi: http://dx.doi.org/10.1007/s10494-011-9343-2

NOMENCLATURE

b	Channel half-height
B	Ratio of catalytically-active to geometrical surface area
c_p, c_s	Specific heat of gas at constant pressure, specific heat of solid
D_{kl}	Binary diffusion coefficient between species k and ℓ
D_{km}	Mixture average diffusion coefficient of k-th species
h, h_k	Total enthalpy, enthalpy of gaseous species k
$\underset{=}{I}$	Unity matrix
k_s	Thermal conductivity of solid
K_g	Total number of gas-phase species
L	Channel length
Le	Lewis number
M	Total number of surface species
p	pressure
$\dot{q}_{rad}$	Radiative flux
R°	Universal gas constant
Re	Reynolds number
$\dot{s}_k$	Heterogeneous molar production rate of k-th species
T	Temperature
t	Time
$\vec{u}$	Velocity vector
$\vec{V}_k$	Diffusion velocity vector for k-th species
W_k	Molecular weight of k-th species
$\bar{W}$	Average mixture molecular weight
X_k	Mole fraction of k-th gaseous species
Y_k	Mass fraction of k-th gaseous species

GREEK SYMBOLS

Γ	Surface site density
ε	Surface emissivity
θ_m	Coverage of m-th surface species
λ	Thermal conductivity of the gas

μ	Gas viscosity
ρ	Gas density
φ	Equivalence ratio
$\dot{\omega}_k$	Gas-phase species molar production rate

SUBSCRIPTS

ad	Adiabatic
ig	Ignition
IN	Inlet
k, m	Index for gas-phase species, index for surface species
s	Surface, solid

CHAPTER 12

MICROREACTOR WITH A TEMPERATURE GRADIENT

Kaoru Maruta

12.1   WEAK FLAME IN A TEMPERATURE GRADIENT

As discussed in Chapter 1, flame–structure thermal coupling induces flame bifurcations, which exhibit weak flames. Such flames are also observed under the gas–solid thermal coupling in particle-laden flames [1] and strained flames under the combined effects of radiation and unequal diffusive-thermal transport (Lewis number effect) in counterflow flames in microgravity [2]. Temperature and reactivity of weak flames are thought to be low in general since they are stabilized at the near-limit condition or even beyond the limit due to those coupling effects.

Weak flames in microcombustion systems exhibit various novel aspects since the micro-combustion device is under the strong thermal coupling in nature. In this chapter, characteristics of weak flames in a meso-scale channel with a prescribed or controlled temperature profile by an external heat source [3] and their application to the study on chemical kinetics for combustion and ignition characteristics are discussed.

Thermal coupling of flames in a meso-scale channel having prescribed temperature profile is investigated using a fine, straight quartz tube [3, 4] and an external heat source. The stationary monotonic positive temperature gradient by the heater is formed along the inner surface of the tube wall in the axial direction. Measured flame responses for different mean flow velocities at the tube inlet are shown in Figure 12.1 [3]. Flames stabilized in the low-velocity region are normal weak flames (hereafter weak flames) while those in the high- and intermediate-velocity regions respectively correspond to the normal flames and *flames with repetitive extinction and ignition* (FREI) [3, 4], where the latter was discussed in detail in Chapter 3. The overall flame response was also examined analytically and shown to be S-shaped response (see Figure 3.6 in Chapter 3).

Since the tube wall temperature is prescribed by an external heater, weak-flame extinction induced by heat loss is not likely to occur. However, the lower limit of weak flame was identi-fied experimentally at very low mixture flow velocity [5] of methane/air mixture. Although a weak flame at a mean flow velocity of 0.2 cm/s was observed, no flame was observed below

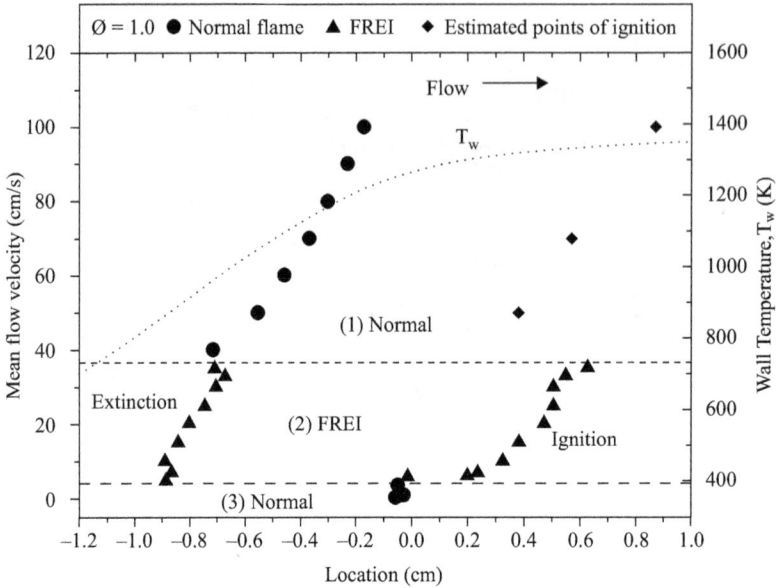

Figure 12.1. Measured flame response for different mixture velocities. Flame positions and extinction and ignition points of FREI are indicated for methane/air mixture at an equivalence ratio of 1.0. Estimated ignition locations in the upper normal flame regime are also indicated (reproduced by permission from Maruta et al. and Tsuboi et al. [3, 5]).

that flow velocity even with longer exposure time recording. This indicates the existence of a lower limit of weak flame, even with heat compensation by external heating. For further understanding, gas-phase and wall temperature measurements were carefully conducted. The temperature difference between the flame and the tube wall at the flame position, that is, temperature increase in the weak flame, is shown in Figure 12.2 [5]. The figure shows the gradual decrease of temperature difference with the decrease of the mean flow velocity. The temperature difference is almost zero at a mean flow velocity of 0.2 cm/s, where the wall temperature is around 1225 K regardless of mixture equivalence ratios. Based on these two characteristics of weak flame, that is, (1) small temperature increase at weak flame and (2) flame location being close to the ignition points of FREI, wall temperature at the lower limit of weak flame is considered to correspond to the ignition temperature of the given mixture at the given residence time. Although ignition has been considered to be strongly transient phenomenon in general, the present weak flame established as a stationary propagating flame near the ignition point can represent the initial stage of ignition of the given mixture.

Computations for one-dimensional plug-flow with detailed chemistry and transport properties were conducted for further examining the mechanism of the lower limit of weak flame [5]. A steady-state flame code [3] based on PREMIX with GRI-mech 3.0 was used. The energy equation, which has a convective heat transfer between the wall with prescribed temperature profile and gas, is as follows:

$$\dot{M}\frac{dT}{dx} - \frac{1}{c_p}\frac{d}{dx}\left(\lambda A \frac{dT}{dx}\right) + \frac{A}{c_p}\sum_{k=1}^{K}\rho Y_k V_k c_{pk}\frac{dT}{dx} + \frac{A}{c_p}\sum_{k=1}^{K}\dot{\omega}_k h_k W_k - \frac{A}{c_p}\frac{4\lambda Nu}{d^2}(T_W - T) = 0 \qquad (12.1)$$

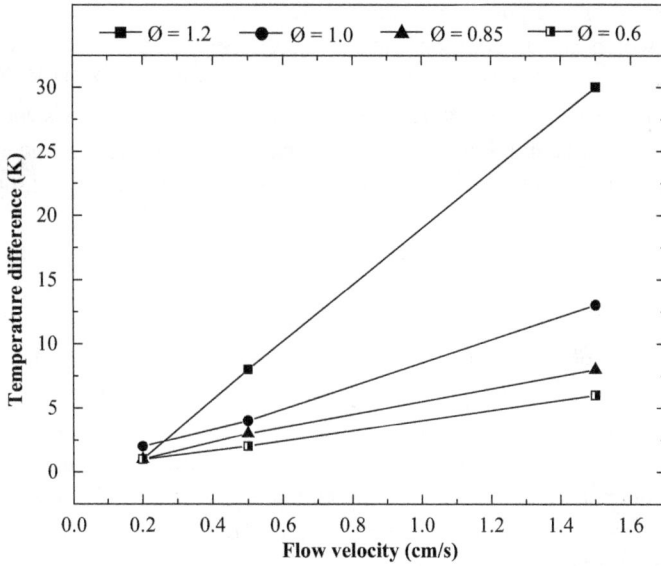

Figure 12.2. Measured temperature difference between the flame and the tube wall at the flame position as a function of the inlet mean flow velocity (reproduced by permission from Tsuboi et al. [5]).

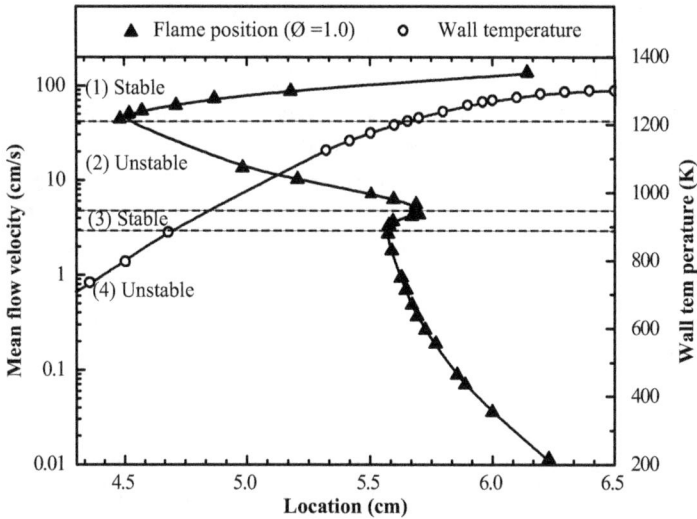

Figure 12.3. Computed flame position as a function of the inlet mean flow velocity in log scale ($\phi = 1.0$) (reproduced by permission from Tsuboi et al. [5]).

Because of the fully developed flow in a circular tube for the case between constant wall heat flux and constant wall temperature, the *Nusselt number* of the inner wall surface is assumed to be constant ($Nu = 4$). Figure 12.3 [5] shows the computed flame response to the mean flow velocity at equivalence ratio $\phi = 1$. The location of the CH peak is considered to be the flame position. The solid line with open circles in the figure is the prescribed wall temperature profile.

The computational flame response to the mean inlet flow velocity in the log scale showed an ε-shaped curve, which has an additional lower velocity branch with the previously reported S-shaped curve [3, 4]. Based on the stability analysis for the S-shaped curve, the lowest velocity branch here is considered to be unstable. If this assumption is true, the existence of two stable and two unstable solutions in the four regimes can be inferred. Therefore, the existence of the lower limit of the experimental weak flame can be interpreted as the lower limit of the region (3) in Figure 12.3 [5].

Computational temperature difference between the flame and the wall temperature at the weak flame was quite consistent with the experimental results. It becomes almost zero (Tg - Tw < 1 K) at a mean flow velocity of 0.1 cm/s when wall temperature is 1230 K, which agrees well with the experimental results (1225 K). The lower limit of stable weak flame is considered to be at the midpoint of the lowest velocity branch (regime (4)) in Figure 12.3. The boundary between the stable and unstable branches on the curve is unresolved. However, it should be noted that the extremely small temperature increase at the weak flame does not directly correspond to flame quenching in the low-velocity regime; nevertheless, the conventional reaction with intense heat release no longer occurs in regime (4), even though the heat loss from the flame zone is compensated by the external heating.

Since thermal quenching of weak flame does not occur due to the external heating in the present system, it was assumed to be the cause of lower limit of weak flame being related to the dominant diffusive mass dissipation over the convective mass transfer in the extremely low-velocity regime. Based on this assumption, the effect of diffusion on the lower limit of weak flame was examined by comparing the convective and diffusive mass fluxes of OH in the cases of the high- and low-velocity regimes. OH was chosen because it is one of the key radicals for the initiation of chain branching reactions as well as its high diffusivity. Mass transport by convection is dominant in the case of the high-velocity regime (not shown), which is similar to the conventional flame. Meanwhile, diffusive mass flux is dominant at a mean flow velocity of 1.82 cm/s in the low-velocity region as shown in Figure 12.4 [5]. This indicates that the effect of diffusion on total mass transport increases at the extremely low flow velocity. Furthermore, the contribution of negative total mass flux in the low-velocity regime is significantly larger than that in the high-velocity regime, and the absolute value of the negative total mass flux is larger than that of the positive total mass flux in the same figure. Note that the negative value means that OH moves to the upstream direction. Since the radical species represented by OH, which are vital for the chain branching reaction, diffuse out from the reaction zone in both downstream and upstream directions, the number of collisions and the rate of production of chain carriers are considered to be low. Thus, the flame temperature may become lower with the decrease of mean flow velocity. The existence of the lower limit of weak flame and its mechanism was also investigated by flames under the reduced pressure [5, 6].

13.2 MULTISTAGE OXIDATION STUDY USING WEAK FLAME IN A TEMPERATURE GRADIENT

Weak flames observed in a meso-scale channel with a prescribed temperature profile can be considered to represent the initial stage of ignition phenomena of the given mixture. In particular, wall temperature at the lower limit of weak flame can be regarded as ignition temperature of the mixture at the given condition. It is also noted that the wall temperatures at weak flames

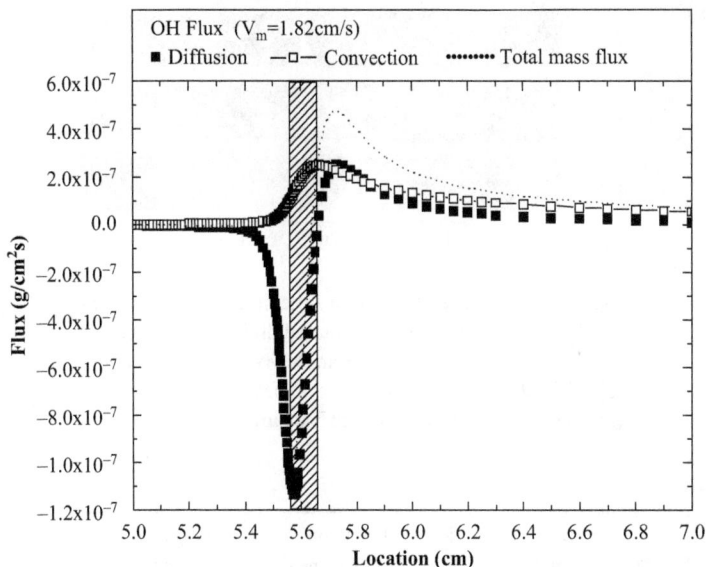

Figure 12.4. Distributions of convective, diffusive, and total mass fluxes of OH at a mean flow velocity of 1.87 cm/s (reproduced by permission from Tsuboi et al. [5]).

were quite insensitive to the change in the mean flow velocity as can be seen in Figure 12.1 (Refer also Figure 12.6). This is due to the significant effect of mass diffusions in such a low flow velocity regime as discussed in the previous section. Thus, more practically, ignition properties can be investigated based on the wall temperatures at the weak-flame locations.

In this way, the microflow reactor with controlled temperature profile was applied for chemical kinetics studies of several fuels with multistage oxidations to date. Oshibe et al. [7] examined ignition and combustion characteristics of stoichiometric DME/air mixture using the microflow reactor. Results demonstrated that the existence of the steady double luminous weak flames. Gas sampling analysis by GC and one-dimensional computation with detailed chemistry and transport indicated that these double weak flames are the separated high temperature oxidation (HTO) of the DME/air mixture [7]. Gas sampling showed that the existence of low-temperature oxidation (LTO) in the upstream region while this LTO cannot be detected as observable luminous weak flame.

Further study for a stoichiometric *n*-heptane/air mixture identified the separated, stationary triple luminous weak flames [8]. Figure 12.5 shows three kinds of flame responses observed in the microflow reactor [8]. Figures 12.5 (a) and (b) are normal flames at a mean inlet flow velocity of 40 cm/s and FREI at 20 cm/s, and the characteristic feature of the weak flame with triple luminous zones is shown in Figure 12.5 (c). Gas sampling analysis and computation with detailed chemistry clarified that the first weak flame corresponds to the cool flame, which represents LTO of *n*-heptane/air mixture. The second and third weak flames are the separated HTO of the mixture as similar to the results of DME/air mixture. It is supposed that LTO could be observed in *n*-heptane/air weak flame due to the higher carbon number in *n*-heptane rather than that of DME.

Figure 12.6 [8] shows observed flame responses for the stoichiometric *n*-heptane/air mixture. While the behaviors of normal flames and FREI are similar to those for the methane/air

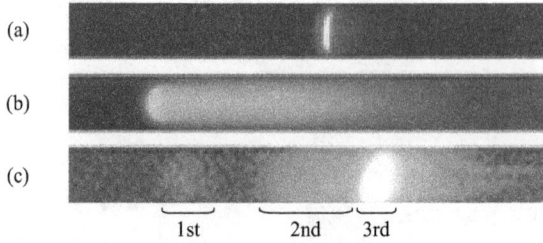

Figure 12.5. Flame images of a stoichiometric *n*-heptane/ air mixture recorded by a CH-filtered digital still camera. (a) Normal flame at an inlet mean flow velocity of 40 cm/s, (b) FREI at 20 cm/s and (c) weak flames at 3.0 cm/s (reproduced by permission from Yamamoto et al. [8]).

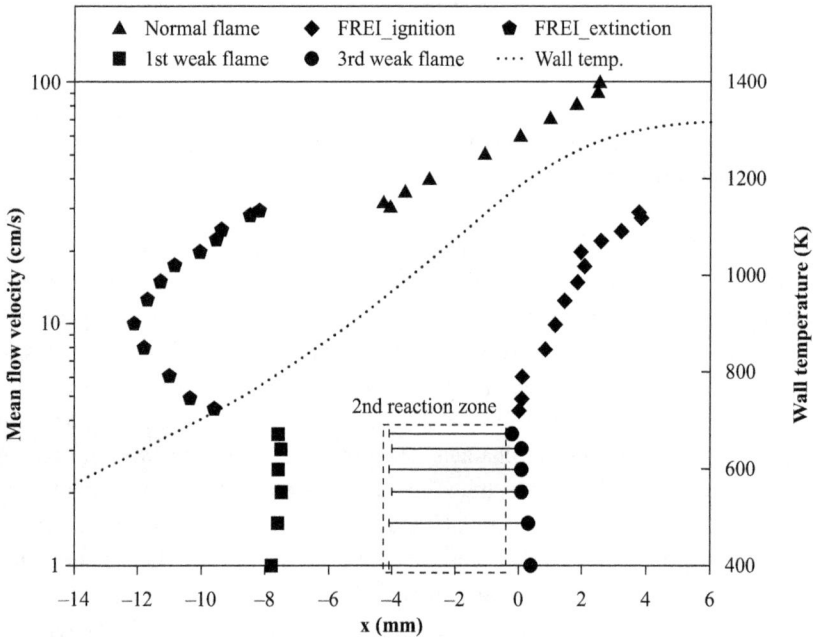

Figure 12.6. Measured flame response for stoichiometric *n*-heptane/air mixture. Flame positions for normal and multiple weak flames and extinction and ignition points of FREI as a function of the mean flow velocity are indicated. Dashed line depicts the wall temperature profile (reproduced by permission from Yamamoto et al. [8]).

flame case, the existence of three weak flames and their insensitivities to the mixture flow velocities are noted. The insensitiveness of the weak flames to the mixture flow velocity is due to the dominant mass diffusion effect discussed above, and it is advantageous for examining ignition properties from the appearances of those weak flames. Figure 12.7 [8] shows computational mole fraction profiles of major species, which agreed with measured species profiles based on gas sampling analysis. Vertical three dotted lines are the first, second, and third weak flames from left to right. The figure shows significant productions of CH_2O and H_2O_2 in the

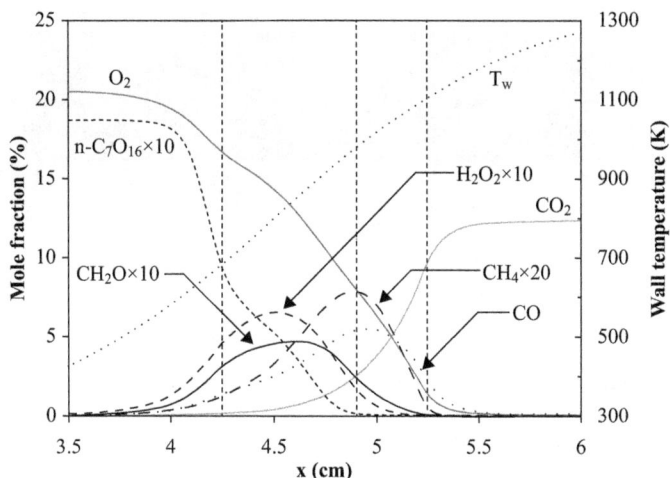

Figure 12.7. Computational mole fraction profiles of major species at a mean flow velocity of 2.0 cm/s. Vertical dotted lines depict the locations of three weak flames (reproduced by permission from Yamamoto et al. [8]).

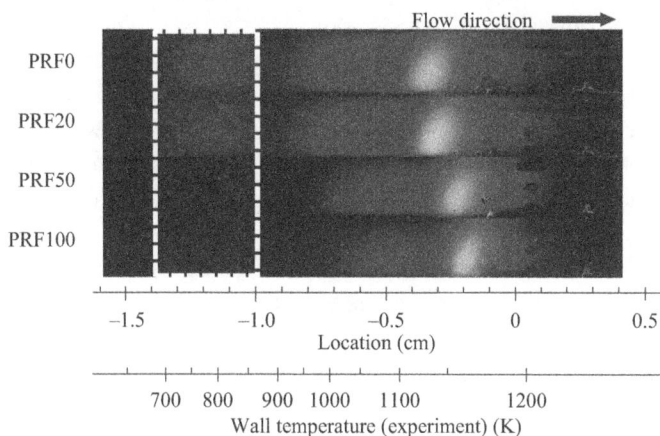

Figure 12.8. Weak flame images for RON 0, 20, 50, and 100. Chemiluminescence from the first weak flames (cool flames) can only be seen in RON 0 and 20 (reproduced by permission from Hori et al. [9]).

temperature between 600 K and 700 K, their consumption at the higher temperature around 900–1000 K, and the total oxidation of all the intermediates at the temperature higher than 1100 K. Hence, those three weak flames are LTO (first weak flame) and separated HTOs till CO (second weak flame) and total oxidation (third weak flame).

Further examinations were conducted for primary reference fuel for gasoline, that is, the mixture of n-heptane and iso-octane, which are used to define the research octane number (RON). Figure 12.8 [9] shows that the images of multiple weak flames for PRF xx, which defines the RON, where xx is a mole fraction of iso-octane in percent of the mixed fuel of

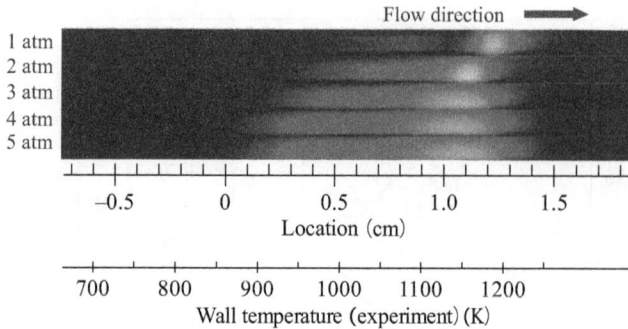

Figure 12.9. Weak flame images of RON 100 at elevated pressures. Chemiluminescence from the first weak flame is invisible at any pressures (reproduced by permission from Hori et al. [9]).

n-heptane and iso-octane. The figure shows that weak flames for LTO (cool flame) were only seen in PRF 0 and 20, and their luminosities are weaker in the larger RON. Separated dual weak flames for HTO shifted to the higher temperature side, and the second weak flames are weaker in the larger RON. Those weak-flame characteristics roughly correspond to the ignition properties of those fuels, implying that the microflow reactor would be contributing to, for example, chemical kinetics study, developments of reaction control agents, and flexible combustion devices for various alternative fuels.

Pressure dependences of multiple weak flames for PRF 0, 50, and 100 were also addressed by elevating operating pressures up to 5 atm. Results for RON 100 are shown in Figure 12.9 [9]. It is noted that weak luminosities were also observed for the first weak flames at elevated pressures for RON 0 and 50 but not for RON 100. Relevant computations showed that the contributions of each weak flame to the heat release significantly shifted to the flames at lower temperature side at higher pressures.

Furthermore, soot precursor formation and cetane number dependences are addressed. It is expected that stationary multiple weak flames in the microflow reactor with the prescribed temperature profile can be utilized for examining general combustion studies from the different viewpoints in addition to the matured methodologies such as shock tube, rapid compression machine, and ordinary flow reactor.

REFERENCES

[1] T. Mitani, and T. Niioka, *Extinction phenomenon of premixed flames with alkali metal compounds.* Combustion and Flame, 1984. **55**(1): pp. 13–21. doi: http://dx.doi.org/10.1016/0010-2180(84)90145-7.

[2] Y. Ju, H. Guo, K. Maruta, and F. Liu, *On the extinction limit and flammability limit of non-adiabatic stretched methane-air premixed flames.* Journal of Fluid Mechanics, 1997. **342**(1): pp. 315–334. doi: http://dx.doi.org/10.1017/S0022112097005636.

[3] K. Maruta, T. Kataokaa, N. Il Kima, S. Minaevb, and R. Fursenko, *Characteristics of combustion in a narrow channel with a temperature gradient.* Proceedings of the Combustion Institute, 2005. **30**(2): pp. 2429–2436. doi: http://dx.doi.org/10.1016/j.proci.2004.08.245.

[4] K. Maruta, J. K. Parc, K. C. Oh, T. Fujimori, S. S. Minaev, and R. V. Fursenko, *Characteristics of microscale combustion in a narrow heated channel.* Combustion, Explosion and Shock Waves, 2004. **40**(5): pp. 516–523. doi: http://dx.doi.org/10.1023/B:CESW.0000041403.16095.a8.

[5] Y. Tsuboi, T. Yokomori, and K. Maruta, *Lower limit of weak flame in a heated channel.* Proceedings of the Combustion Institute, 2009. **32**(2): pp. 3075–3081. doi: http://dx.doi.org/10.1016/j.proci.2008.06.151.

[6] Y. Tsuboi, T. Yokomori, and K. Maruta, *Extinction characteristics of premixed flame in heated microchannel at reduced pressures.* Combustion Science and Technology, 2008. **180**(10–11): pp. 2029–2045. doi: http://dx.doi.org/10.1080/00102200802269723.

[7] H. Oshibe, H. Nakamura, T. Tezuka, S. Hasegawa, and K. Maruta, *Stabilized three-stage oxidation of DME/air mixture in a micro flow reactor with a controlled temperature profile.* Combustion and Flame, 2010. **157**(8): pp. 1572–1580. http://dx.doi.org/10.1016/j.combustflame.2010.03.004.

[8] A. Yamamoto, H. Oshibe, H. Nakamura, T. Tezuka, S. Hasegawa, K. Maruta, *Stationary three-stage oxidation of a gaseous n-heptane/air mixture in a microflow reactor with a controlled temperature profile.* Proceedings of the Combustion Institute, 2011. **33**(2): pp. 3259–3266. http://dx.doi.org/10.1016/j.proci.2010.05.004.

[9] M. Hori, A. Yamamoto, H. Nakamura, T. Tezuka, S. Hasegawa, and K. Maruta, *Study on octane number dependence of PRF/air weak flames at 1–5 atm in a microflow reactor with a controlled temperature profile.* Combustion and Flame, 2012. **159**(3): pp. 959–967. http://dx.doi.org/10.1016/j.combustflame.2011.09.020.

CHAPTER 13

CHEMICAL MICROPROPULSION

Ming-Hsun Wu, Richard A. Yetter, and Vigor Yang

Micropropulsion has received significant attention as a promising application of microcombustion, and different types of microthrusters have been developed, aiming to meet the primary propulsion and attitude control needs of microspacecraft. The following list provides several examples of the use of microthrusters in small spacecraft.

- Orbit change and attitude control of microspacecraft
- Precise positioning control of spacecraft constellations (formation flying) for interferometry missions (gravity wave detection, planet searches around distant solar systems), communications, and earth observations
- Small spacecraft propulsion for inspection and repair of large spacecraft
- Small spacecraft propulsion for decoys and active protection of large spacecraft
- Sample return missions from asteroids
- Insertion into final orbits when microspacecraft are launched as a secondary payload

Micropropulsion technology has applications beyond microthrusters, however, and can be applied to any process requiring small quantities of directed gas flow. For example, the technology necessary for the successful development of microthrusters has been applied to microgas generators for usage in airbags, microactuators, and fluid pumps.

In this chapter, the field of chemical micropropulsionis reviewed. First, some of the design considerations regarding fundamental combustion, heat transfer, and fluid mechanics are discussed. This discussion is followed by a brief description of manufacturing processes for micropropulsion systems. Finally, the field is reviewed through presentation of examples of micropropulsion systems from the literature, using propellant type (solid, liquid, and gaseous) as the method of classification.

13.1 MICROPROPULSION AND SCALING

The effects of miniaturization on microthrusters is important to both combustion dynamics and nozzle flow, which are coupled via heat transfer through the body structure. Many of the

371

important issues in combustion, heat transfer, and fluid dynamics in microthrusters have been discussed earlier in this book and elsewhere [1–8]. Here, we briefly discuss some of the effects of decreasing the characteristic length scale of thrusters.

As compared to macroscale systems, micropropulsion devices operate at lower Re and Pe numbers. Consequently, viscous and diffusive effects dominate. The flows are less turbulent, and laminar conditions generally prevail. Boundary conditions, which are usually not influential in large-scale systems, play a more significant role at smaller scales. Diffusive processes, which in micron-sized channels can be fast, will largely dominate species mixing, but in millimeter sized combustion chambers can be too slow to be effective. Although the greater viscous forces will result in larger pumping requirements, certain advantages, such as the reduced leakage of gases through micron-sized joints, may be achieved. Higher viscous losses also imply that larger Reynolds (Re) numbers may be needed at smaller length scales to achieve a given desired thrust [4]. One approach to achieve high Re numbers in microthrusters is to operate at high pressure. As viscosity becomes more important with continued reduction in size, the simple isentropic expansion expressions used in conventional thruster design are less applicable. For example, measured I_{sp} of rocket micronozzles was found to decrease by a factor of 10 when the nozzle Re was reduced from 4000 to 400 by decreasing the size [4, 9]. Navier–Stokes and direct simulation Monte Carlo (DSMC) calculations on 2D planar and 2D axisymmetric nozzles have shown that flow in 2D flat nozzles is actually 3D and produced ~20 percent less thrust than a 2D axisymmetric nozzle, where the surface-to-cross-sectional area ratio is minimized [8].

Lower specific impulses are also expected, due to lower combustion chamber temperatures resulting from enhanced heat transfer losses as the ratio of heat loss to heat generation rates is inversely proportional to length scale. Depending upon the dominant mode of heat transfer at the external surface, the magnitude of the scaling may vary, but downsizing without effective means to recover lost energy results in lower combustion temperatures and hence lower I_{sp}. As the length scale is decreased, velocity, temperature, and species gradients at boundaries tend to increase, producing larger momentum, energy, and mass transport rates near surfaces, making it difficult to maintain large differences between the wall and bulk flow variables. Biot numbers associated with microstructures will generally be much less than unity, resulting in nearly uniform body temperatures, and Fourier numbers will result in small thermal response times of structures. Transient DSMC coupled thermal and fluid analyses of microthruster flows have shown that the silicon thruster body quickly attained a uniform temperature after start-up, and that heat was transferred from the combustion chamber to the nozzle diffuser flow through the thruster body [10]. Temperatures of the thruster with an adiabatic external surface and initial body temperature of 300 K were predicted to reach 1200 K after ~13 s of operation. Because view factors increase with decreasing characteristic length, radiative heat transfer also plays an important role in chemical thrusters.

For efficient combustion, residence times must be greater than chemical times. Residence times, however, decrease with decreasing length scale, and therefore, to sustain combustion, chemical times must also be reduced. As previously mentioned, increased surface-to-volume ratios make this difficult, because of the increase in heat loss relative to heat generation rates. While the high surface-to-volume ratio and small length scales favor catalytic combustion, it is generally slower than gas-phase combustion and deposits energy directly into the thruster body. For efficient gas-phase combustion, high inlet, wall, and combustion temperatures for increased kinetic rates, operation with stoichiometric mixtures, oxygen enrichment, excess enthalpy combustors, and use of highly energetic fuels and propellants are all approaches to enhance combustion at the microscale. The times of physical processes, such as mixing and liquid fuel evaporation, must also be reduced with length scale, if efficient combustion is to be achieved. For sprays, droplet sizes must be reduced to shorten evaporation times, which implies greater pressure and energy requirements for atomization. Microelectrospray atomizers as well as film-cooling techniques have been considered as means to introduce liquid fuels into the combustion chamber [11–13].

Clearly, there are many opportunities for combustion research at the microscale to facilitate the design of microthrusters. Although not discussed here, the research, design, and manufacturing of other components of microthruster systems, such as valves and pumps, remain equally important science and technology areas [2]. In fact, the availability of high pressure, leak-tight, micro valves that consume minimal power is currently one of the limiting factors in chemical microthruster development.

13.2 MATERIALS, FABRICATION, AND SYSTEM INTEGRATION

Chemical energy of the propellant is converted into thermal energy in the combustion chamber of a microchemical propulsion system. The high temperature and very often high pressure in the system pose challenges in selection of materials and fabrication techniques, requiring processes uncharacteristic of other microfluidic devices. Materials ranging from stainless steel to ceramic tapes have been used in the fabrication of the combustion chamber of chemical microthrusters. Silicon and Pyrex are traditionally the most common substrate materials for fabricating micro-electro mechanical systems (MEMS). Due to the readily available microfabrication processes, these materials have been applied to the fabrication of microchemical propulsion systems since the early stages of development [14–20]. Reaction chambers, feeding channels, and nozzles are directly patterned and etched on silicon wafer substrates using either wet etching or reactive ion etching. Conductive wirings and microheaters can be embedded using surface processing techniques. A typical example of a microthruster fabricated using the silicon MEMS process is the "digital" microthruster developed by TRW and CalTech [21]. For this microthruster, the expansion angle of the nozzle was restricted to 35.3° (from the centerline to one side wall) due to the nature of the KOH selective etching along the <100> planes of the silicon substrate, and consequently the angle could not be varied to determine the optimal expansion angle for a microscale nozzle. Micronozzles with smaller expansion angles have been fabricated by the deep reactive ion etch (DRIE) technique. The liquid bi-propellant microthruster developed at the Massachusetts Institute of Technology [17, 22] is probably the most sophisticated, from a fabrication perspective. This bi-propellant rocket engine consists of six individually etched single crystal silicon wafers. The layers were precisely aligned and then fusion bonded to form the 18 mm long, 13.5 mm wide, and 2.9 mm thick microrocket engine. A 2D converging diverging nozzle with throat cross-sectional dimensions of 0.5 mm × 1.4 mm was integrated in the rocket engine with the thrust chamber.

There is also a recent trend in the microfluidic community to fabricate fluidic devices on unconventional MEMS substrates, such as plastic, polymers, and even papers, but these materials are unlikely to survive at the temperatures characteristic of microcombustion chambers. Highly reactive fuels, oxygen enrichments, and excess enthalpy combustion are often applied to enhance the self-sustainability of the flame and to reduce the quenching distance in a microscale combustion chamber; adiabatic flame temperatures become higher when any of these approaches is applied. With the larger surface-to-volume ratio, a larger proportion of heat released by the flame is deposited to the combustor body in a microscale combustion chamber. The temperature distribution in the microscale combustor may be more homogeneous than in a macroscale combustor, due to the smaller Biot number in microsystems. Hence, bulk materials for fabricating a microcombustor must sustain temperatures over 1000 K in an oxidizing environment, and the material should be able to maintain good mechanical strength at high temperatures. Compatibility of the substrate materials with the propellants applied should also be taken into account. Moreover, thermal expansion matching of the materials in the operating temperature range has to be considered when more than one material is utilized and packaged together or the bonding of substrates may be compromised when the thermal expansion coefficients mismatch. Other than the properties of materials, the availability of corresponding microfabrication technology is another important factor to be considered in the selection of the substrate for developing a microchemical propulsion system.

Conventional conductive materials, such as stainless steel, aluminum, and other metal alloys, can also be used to fabricate mesoscale channels and chambers in chemical propulsion systems using electro-discharge machining (EDM). Wu et al. [23, 24] developed a mesoscale Inconel combustor using EDM. Unlike conventional machining methods, in which the material is removed physically by cutting and milling, EDM utilizes the erosive action of electrical discharges (sparks) to cut the material. With this high-energy electro-thermal erosion (instead of mechanical cutting forces), EDM is capable of machining mechanically difficult-to-cut materials, such as hardened steels, carbides, high strength alloys, and even ultra-hard conductive materials like polycrystalline diamond and ceramics. The surface roughness of an EDM-machined surface is typically on the order of microns, which is considerably higher than that obtained using the silicon microfabrication process. Furthermore, the materials compatible with EDM are limited to conductive and semiconductive materials, and the integration of components fabricated using EDM with other subcomponents into a complete miniaturized system may be costly and time consuming. For example, it is difficult to integrate the ignition electrodes, sensors, and valve units into a submillimeter scale stainless steel combustion chamber. Gas sealing and electrical insulation are also technically challenging. The fabrication technique is nonetheless very useful for fabricating mesoscale prototypes for concept validation during the early development stage.

Figure 13.1 shows an example of stainless steel channels fabricated using EDM, along with magnified views at the intersection of the crossed channels and at one end of the smaller channel. The channel width and length are 0.3 mm and 1 mm, respectively, while the thickness of the piece is 1 mm.

The minimum width of the channel is limited by the wire diameter used in the EDM machine. Microscopic pictures of the channel show irregularities at the edges. The amount of material removed is proportional to the energy applied across the discharge gap. Higher energy results in a higher removal rate but a rougher surface. As a result, the key in micro-EDM is to limit the energy in the discharge. In order to make microproducts with high accuracy and good surface finish, the energy per single discharge should be minimized (on the order of 10^{-6} J to 10^{-7} J) and discharge frequency kept high. Holes as small as 5 μm can be drilled using this technique, but 10–50 μm are more typical [25]. The largest achievable aspect ratio is around five for feature size on the order of 100 μm [26].

Ceramic has drawn increasing interest as the material for microcombustion and chemical propulsion systems. Alumina (Al_2O_3)-based ceramics such as mullite (60% Al_2O_3 and 40% SiO_2)

Figure 13.1. A microcounter flow burner fabricated using EDM.

have properties including good high-temperature strength, good thermal shock resistance, excellent thermal stability, and resistance to most chemical attack, all making it an excellent material for high-temperature microsystems. Ceramic stereolithography is among the handful of fabrication techniques capable of fabricating alumina parts at the submillimeter scale. In stereolithography, a laser beam scans over a liquid monomer resin to cure it into a solid polymer layer by layer. This class of materials, originally developed for the printing and packaging industries, quickly solidifies wherever the laser beam strikes the surface of the liquid. Once one layer is completely traced, the part is lowered a small distance into a vat and a second layer is traced right on top of the first. The self-adhesive property of the material causes the layers to bond to one another and a complete three-dimensional object will consist of layers of various contours. Ceramic parts are fabricated from both the aqueous and nonaqueous alumina suspensions in liquid photosensitive monomers. The cured solid polymer is burned out during the sintering process afterwards. Inhomogeneity in the alumina solution, and absorption and scattering in the dense ceramic colloid solution, are the major challenges for ceramic stereolithography.

Figure 13.2 shows a nozzle and combustion chamber fabricated using three-dimensional stereolithography [27]. The process starts with the design of the part to be made as a three-dimensional virtual solid in a computer-aided design program. The solid is then sliced finely by another computer program (Maestro) along the Z-axis, creating a build file that consists of a stack of X–Y layers. The build file is transferred to the build station used for the manufacture of the part (3-D Systems Inc. SLA-250, Valencia, CA). During the build, each layer is formed by raster scanning an ultra-violet laser across the surface of a photo-curable bath of liquid resin. The solidified layer is moved down by one layer thickness inside the liquid bath so that the resin flows across the previously cured layer and the patterned curing process is repeated to form the next layer on top of the preceding layer. The resin is a highly concentrated colloidal dispersion prepared by dispersing alumina powder in an aqueous solution of ultraviolet curable polymers. The ceramic green body is subsequently dried, the photo-curable binder is burned out, and the part is sintered to full density.

Ceramic microchemical propulsion devices have also been fabricated using low-temperature cofired ceramic (LTCC) tapes [28–30]; LTCC is a glass-ceramic composite material. Figure 13.3 shows a gaseous bi-propellant microthruster with integrated cooling microchan-

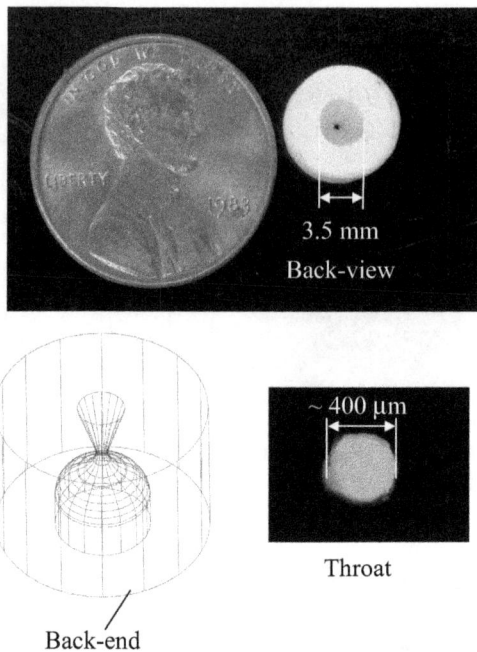

Figure 13.2. Alumina thruster body fabricated using ceramic stereolithography (top), wire-frame of thruster body (bottom left), and cross-section of the throat (bottom right).

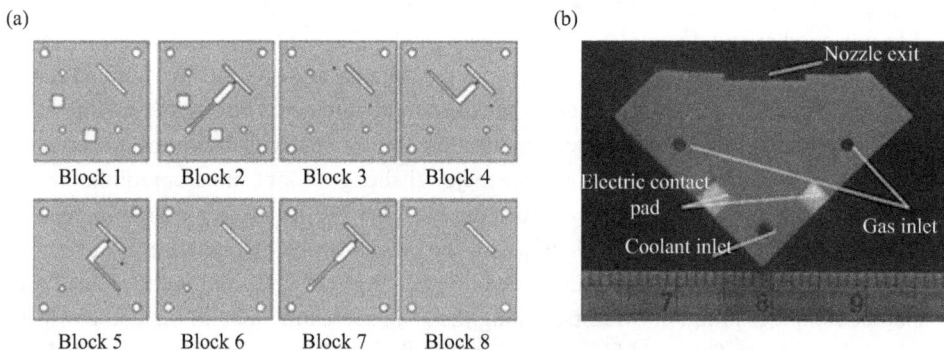

Figure 13.3. (a) The layer-by-layer layout of a low temperature cofired ceramic (LTCC) bi-propellant microthruster, and (b) photograph of the microthruster after cofiring.

nels. The ceramic filler is usually alumina, Al_2O_3, and the composition also includes a glass frit binder to lower processing temperature and render the material compatible with thick film technology. A third component of the composite is an organic compound for binding and viscosity control of the tape before sintering. Typical LTCC tape shrinks after cofiring, so the shrinkage ratio has to be compensated for during the design stage. One of the most intriguing features of LTCC tape technology is the capability of embedding electrical elements, such as resistors and conductors, on the tapes using a thick film screen printer [31]. Integration of optical windows and fiber optics, which provide optical access for in-situ diagnostics, has also been shown to be feasible. Recently, Sun et al. [32] have demonstrated the feasibility of fabricating a hybrid microcombustor for microthruster application with an LTCC chamber body and a nozzle plate made of high-temperature cofired ceramic (HTCC) tapes.

LTCC tapes are commercially produced in various thicknesses, typically in the range of 100–400 μm. Thicker tapes are obtained by laminating multiple tapes using an isostatic laminator or hot press. Each layer is processed green (before firing) separately with features such as vias, cavities, channels, and electrical elements. Vias, cavities, and channels can be cut out with high-speed mechanical punching machines, lasers, or computer numerical control (CNC) milling. It is worth noting that hot embossing has also been investigated to fabricate microchannels on both LTCC tapes and mullite. The smallest size that a punching machine can generate is ~100 μm. The smallest feature that CNC milling can create is slightly larger than the punching method, but CNC milling is more versatile and allows fabrication of shallow cavities and thin membranes by milling off a portion of the tape thickness. Laser machining has the ability to shape features as small as 50 μm on both fired and green tapes. The ability to sculpt the fired LTCC facilitates dimension control. After all layers are processed to the desired profiles, the tapes are collated, that is, they are stacked in sequence on a platform with alignment pins. The layer-to-layer alignment tolerance is around 25 μm. The next step is to laminate the stacked blanks by applying pressure (~3000 psi) and heat (~70°C) using a press. A holding time at peak temperature of typically 3–10 min is required to reach a homogeneous temperature profile in the stack. The applied pressure can also be achieved in an isostatic press, where the stacked tapes are vacuum-packaged and pressed in hot water. The thermo-compression squeezes the particles into the neighboring layers, so that the layered stack merges into a single block after sintering.

However, the compression process causes problems in the fabrication of internal cavities and channels. The cavities and channels collapse and deform during this thermo-compression lamination process because mass flow is created as the particles from the un-sintered tapes interpenetrate into each other. Two approaches are available to eliminate the problem: cold low pressure lamination [33] and lamination using organic fluids. In the cold low pressure lamination method, the green tapes are stuck together using double-sided polyacrylate-based adhesive tapes 12–25 μm thick at room temperature under low pressures. The lamination is achieved through a low viscosity melt derived from the polyester carrier film of the adhesive tape, when the temperature reaches around 350°C during the binder burnout process. This liquid phase, existing at the interface of the porous microstructure of the green tapes, causes capillary forces. These forces pull the neighboring ceramic layers toward each other and finally yield a rearrangement of the particles in the interface of the tapes. The binders of both the LTCC green tapes and the adhesive layers are already burned out, but the sintering phase has not started. The movement of the particles in the presence of the melt results in an interpenetration of the particles of adjacent tapes. The following sintering process transfers this structure to defect-free joining. The organic fluid approach uses a similar lamination mechanism, but it has the advantage of being able to deal with complex shapes and non uniform surfaces. One of the organic fluids successfully applied is a honey-water solution.

Sagging of suspended structures, such as the top wall of a wide channel, is another common problem when fabricating combustion chambers using LTCC. Several approaches are developed to reduce the sagging effect. For a long-span bridge structure, a shrinkage matched paste can be screen printed on top of the bridge layer to provide the tensile stresses upon sintering.

Graphite powders mixed with high viscosity organic fluid can also be filled into the internal channels to prevent the top and bottom layers of a buried void from sagging [34]. The graphite mixture is oxidized into carbon dioxide or carbon monoxide during the sintering cycle. Materials such as a lead bi-silicate frit that can be preferentially etched away after sintering have also been explored as a sacrificial layer to eliminate the sagging problem.

The last step of the LTCC fabrication process is sintering. The laminated LTCC stack is placed in an oven with a specific programmed multistep temperature profile. The typical profile starts with a slow temperature increase of about 2–5°C per minute up to around 450°C, and holds for ~2 hours for the organic binder to be burned out. The temperature is then increased up to between 850 and 875°C with a dwell time of about 10–15 min. The sintering takes place at this temperature.

For most microdevices, communication and interconnection with other macroscale components are unavoidable. For example, even if the pump and valves are integrated with the chemical microthruster on the same chip, propellant tanks at a relatively larger scale will still be required for liquid or gaseous microthrusters. The interconnection of components at different length scales is a challenge. Micro-to-macro interconnections, good sealing, and temperature-sensitive materials have been identified as the three major difficulties that have to be resolved in microfluidic systems [35]. The most common method for fluidic micro-to-macro interconnection is attaching the microfabricated microfluidic chips to aligned manifold blocks machined by traditional techniques using commercially available compression fittings (such as Swagelok). Reliable gas tight and temperature resisting interconnect adapters are essential for microcombustion systems. Among the microfabrication technologies discussed, ceramic tape technology has the potential to serve as a core technology for highly integrated high-temperature micro-thermal fluidic applications. Such microsystems are able to integrate, for example, electrical, mechanical, and microfluid mechanic functions. The relatively lower conductivity of ceramic tape is also advantageous for microcombustion applications. This technique may also serve as a bridge between fabricated parts from silicon devices and high-temperature ceramics.

13.3 SOLID PROPELLANT THRUSTERS

There have been numerous studies on microthrusters using solid propellants, either in the form of a single thruster or as arrays of thrusters [20,21,28,39,41,43,44]. In all designs, a grain of solid propellant is ignited, usually as an end burning grain, and the evolved gases are directed through a nozzle. The differences between the systems are the types of propellants, the igniter design, the fabrication materials and methods, the size of the thruster and consequently the thrust levels and impulse bit magnitudes, and in the case of arrays, the addressing technology. The designs generally consist of a layered approach where the flow of the gas is either planar to these layers or perpendicular to the layers. Different materials such as silicon (monocrystalline, porous), glass (Pyrex, quartz, Foturan), ceramics, and polymers have been used.

For solid propellants, as well as for liquids and gases, high combustion rates are essential to overcome heat losses and quenching effects of high surface-to-volume ratios. Ease of ignition is also a criterion, particularly for minimizing the ignition energies of small quantities of propellants. Various types of solid energetic materials have been used in microthrusters (or gas-generating devices), including lead styphnate, glycidylazide polymer (GAP) mixed with ammonium perchlorate (AP) and doped with zirconium, double-base propellants with black powder, black powder alone, high nitrogen compounds such as BTATz and DAATO 3.5, potassium nitrate with boron, azobisisobutyronitrile (AIBN), hetero-metallic compounds, and nanothermites with nanoaluminum and copper oxide. Porous silicon, with various perchlorates and nitrates as oxidizers filling the pores, are also currently under consideration. To assist ignition of these solids, a small quantity of a primer is often included in the propellant design. Thermite reactants have been considered by several research groups as energetic igniters, for example, in contrast to electrical resistors.

The first reporting of arrays of solid propellant microthrusters in 1999 by Rossi et al. [20], was quickly followed by a detailed publication by Lewis et al. [21]. The thruster system consisted of a three-layer sandwich containing microresistors, thrust chambers, and rupture diaphragms. The sandwich of silicon and glass layers was fabricated to contain an array of small plenums, each sealed with a rupturable diaphragm on one side. The plenums were loaded with combustible propellants, which were ignited and reacted to form a high-pressure, high-temperature working fluid. Once the pressure exceeded the burst pressure of the diaphragm, the diaphragm ruptured, and an impulse was imparted as the fluid was expelled from the plenum. Thus, each plenum delivered one impulse bit. The size of the impulse was determined by the size of the plenum and the mass of the propellant. Typical of solid rocket motors, each microthruster unit had one firing and was therefore consumable. However, as an array, the thrusters could be fired separately or together for thrust vectoring until all the thrust elements have been consumed. The three-layer structure is shown in Figure 13.4. The top layer contained an array of thin square diaphragms (which were 0.5 μm thick silicon nitride, 190, 290 or 390 μm on a side, that remained after an anisotropic KOH wet etch through a <100> silicon wafer). The middle layer (consisting of 1.5 mm thick photosensitive glass) contained an array of 300, 500, or 700 μm diameter through-holes that represented the combustion chambers and were loaded with propellant. The bottom layer contained a matching array of polysilicon microresistors. There sistors were fabricated on top of a 3 μm SiO_2 insulating layer. Power supplied to the resistors for ignition was about 50W (0.5 A and 100 V for a period of 50 μs, or ~2.5 mJ). The bottom two layers were bonded together using cyanoacrylate, then the chambers were filled with propellant, and lastly, the top layer was bonded also using cyanoacrylate to complete the assembly. When the resistor was energized, the propellant ignited, raising the pressure in the chamber, and the diaphragm was ruptured (burst pressure ~10 atm). An impulse was imparted as the high-pressure

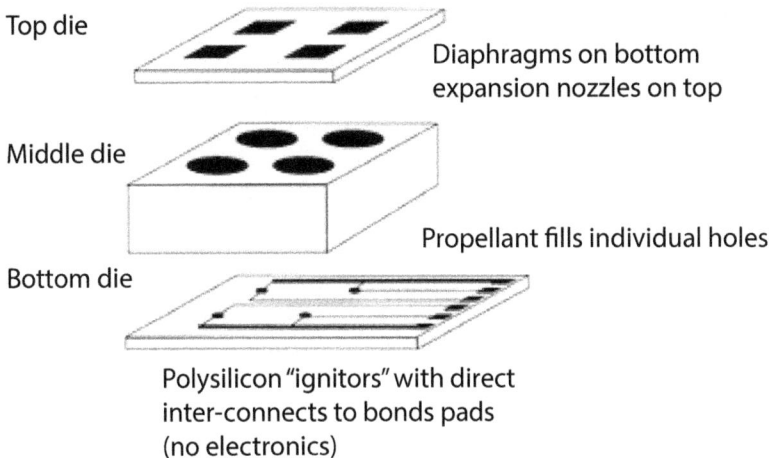

Figure 13.4. Schematic of digital microthruster array (from Lewis et al. [21]).

fluid was expelled from the chamber. Three by five thruster arrays were tested. The initial tests, using lead styphnate as the propellant, produced 10^{-4} N s of impulse and about 100 W of power. The duration of the thrust impulse from each chamber was about 1 ms.

Rossi et al. [20,36–38] have published a series of papers on the development of solid propellant thruster arrays. Their design has overcome several of the problems associated with the TRW, Aerospace, and Caltech design [21]. In that design (Figure 13.4), the igniter is at the opposite end of the motor from the nozzle, and, since the grain is solid and not perforated, gases buildup behind the grain and push a significant fraction of propellant out the nozzle un reacted. Also, the proximity of combustion chambers resulted in cross-talk between the chambers, as a result of heat transfer from one chamber to another, and when one thruster was fired, it often resulted in the firing of adjacent thrusters. Furthermore, misfires were common with the polysilicon resistor igniters. The design of the thruster array from Rossi et al. is shown in Figure 13.5.

The array consisted of 100 addressable microthrusters with three main micromachined layers. The first silicon layer contained the micronozzles. The second silicon layer held the addressing and heating elements. The third layer, either photoetchable glass or silicon (with insulating groves), held the main propellant grain. An intermediary silicon chamber was included between the igniter wafer and the propellant reservoir to allow sufficient heating time of the main grain prior to excessive pressure build-up in the chamber, to assure reliable ignition of the main grain. The propellant was a composite consisting of GAP mixed with AP and zirconium particles. The minimum diameter for deflagration of the grain was found to be ~1 mm, and consequently combustion chamber dimensions were chosen to be 1.5 mm × 1.5 mm in cross-section when etched in Foturan wafers 1 mm thick. Nozzle throats were 250 μm and 500 μm in diameter. To assist ignition, a primer consisting of zirconium perchlorate potassium (ZPP) was included in the igniter wafer. The layers were assembled using various gluing techniques. The main grain was loaded using screen printing under vacuum while the primer was manually filled. With polysilicon resistor igniters, ignition success was 100 percent with an input power of 250 mW. Thrusts generated varied from ~0.3–2.3 mN depending on the nozzle size. An example of the thrust profile is shown in Figure 13.6. As can be seen from the figure, the thrust achieved from the primer is not negligible, and further research would be needed to optimize the grain from being over driven.

More recently, Lee et al. [39] have also reported on solid propellant thruster arrays. One difference between this system and that of Rossi et al. [20] was that the microigniter was developed

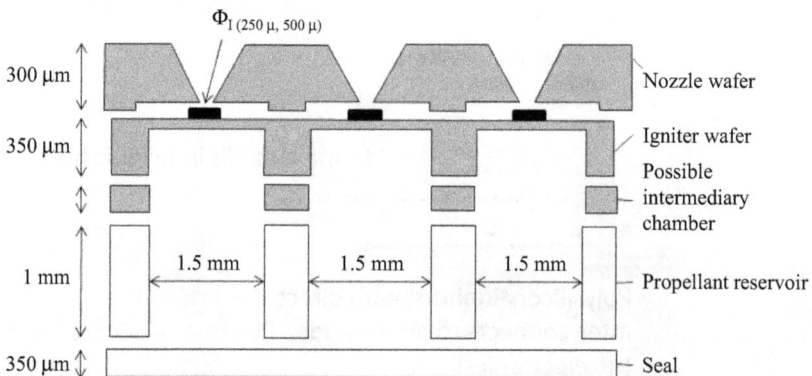

Figure 13.5. Schematic of the Rossi et al. microthruster array (from Rossi et al. [20]).

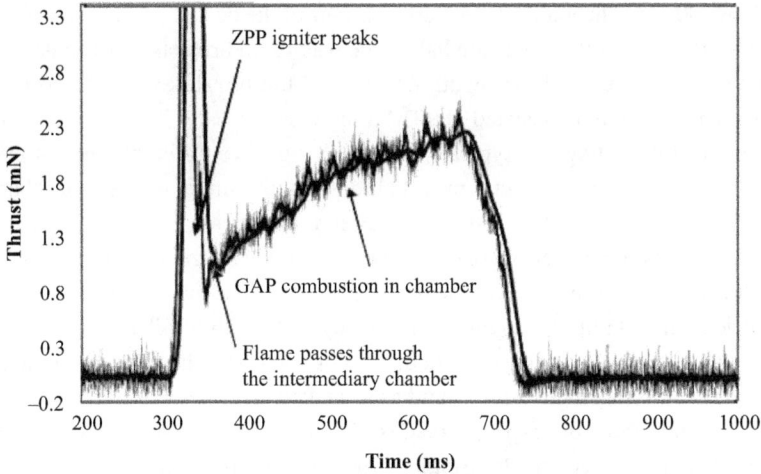

Figure 13.6. Thrust profile from a microthruster with a throat diameter of 500 μm (from Rossi et al. [20]).

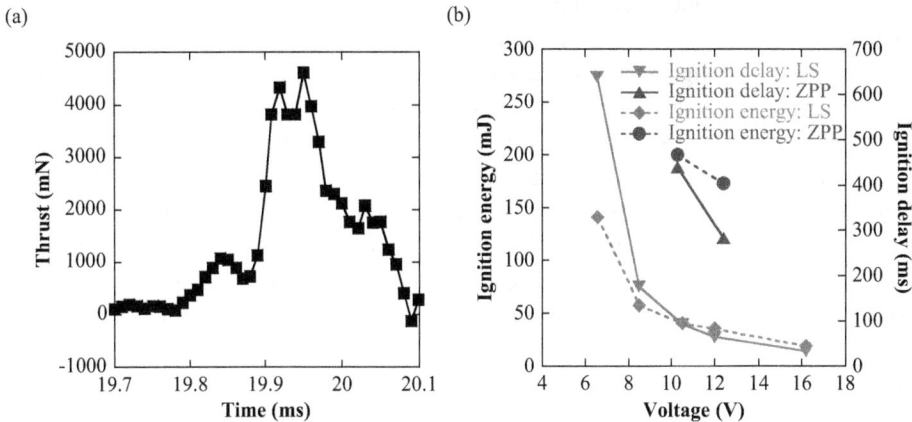

Figure 13.7. (a) Thrust profile for lead styphnate propellant and (b) ignition delay and energies for both lead styphnate and ZPP propellants as a function of voltage (from Lee et al. [39]).

on a photosensitive glass wafer instead of on a dielectric membrane. Lee et al. [39] report that the stability of the Pt igniter (with a 200 Angstrom thick Ti layer to improve the adhesion of Pt to glass) was improved by using the glass membrane with a thickness of 35 μm, which required a pressure of ~1531 kPa to burst the membrane. The propellant grains were either lead styphnate or ZPP, which was the primer used by Rossi et al. [20]. The electric power required to reach the ignition temperature of lead styphnate (260°C) was about 340 mW. The minimum ignition energy was 19.3 mJ with an ignition delay of 27.5 ms. The average maximum thrust and total impulse for lead styphnate were 3619 mN and 0.381 mN·s, respectively. The average specific impulse was 62.3 s. Examples of a thrust profile, ignition energies and delays are shown in Figure 13.7.

In addition to the vertical array microthruster arrangements described before, Zhang et al. [16, 40–42] have developed stacked vectors of planar thrusters to form the arrays. The thrusters were fabricated from silicon with a Pyrex cover. The combustion chambers were 1 μm in length with a depth of 400 µm. The widths were varied as part of the design process (the nominal value tested was 1 mm). The nozzle expansion half angle was 12°. Thrusters were constructed to vary the chamber width to nozzle throat width (A_c/A_t) and the nozzle exit width to nozzle throat width (A_e/A_t). The propellant consisted of 90% gunpowder, 6% AP, 3% aluminum, and 1% iron oxide. The propellant loading was typically about 0.8 mg. Two different types of igniters were tested. The first was a 25 µm diameter nickel chromium aluminum copper wire that protruded into the flow and was tested at the throat, midway into the combustion chamber and at the end of the combustion chamber. The second design was an Au/Ti igniter deposited on the Pyrex surface and located near the exit end of the propellant grain. The latter design was more suitable for batch fabrication and improved ignition efficiency and reliability. The location of the igniter also produced an end burning grain without disturbing the flow. The thruster configurations are shown in Figure 13.8 as a single thruster and as a stacked array.

The thrust trends show the expected results of a larger burning surface and a larger expansion ratio producing greater thrusts. Single thrusters have produced 2.11×10^{-5} to 1.15×10^{-4} Ns of total impulse and 2.68 to 14.68 s of specific impulse at sea level and 3.52×10^{-5} to 2.22×10^{-4} N s of total impulse and 4.48 to 28.29 s of specific impulse in vacuum. Figure 13.9 shows the repeatability of thrust profiles with the Au/Ti igniter, and comparisons of thrust profiles for different A_e/A_t and A_c/A_t ratios.

Zhang et al. [28] have also produced thrusters essentially identical to the aforementioned silicon devices, but using LTCC tape technology. As discussed previously, some advantages of LTCC include: (1) small coefficient of thermal expansion, (2) low tolerance in dielectric constant, (3) good electrical conductivity of metal paste embedded in LTCC, (4) fabrication of multilayer circuits, (5) ability to integrate RLC elements between layers, (6) airtight seals, and (7) low cost for mass production. The microthruster was fabricated by lamination of 10 individual layers of green tapes, each of which was individually processed. The microthruster had a cavity (combustion chamber), a convergent–divergent nozzle, and a resistor embedded inside the cavity. The resistor was connected to catch pads on the top of the thruster through electrical vias. An isometric view and cross-section of the LTCC microthruster are shown in Figure 13.10.

Figure 13.11 shows typical thrust results at sea level for the LTCC microthruster with a combustion chamber width (W_c) equal to 1500 µm, a nozzle throat width (W_t) of 370 µm, a nozzle diffuser length (L) of 600 µm, and a nozzle half divergence angle of 16°. The peak value of the thrust produced was about 0.38 N. Ignition occurred with an input voltage of 37 V.

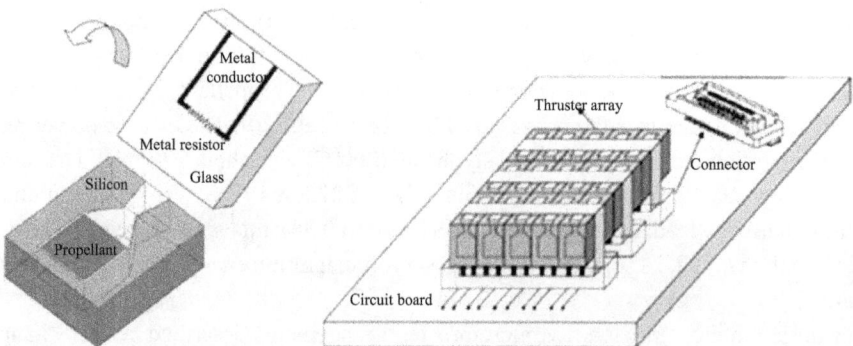

Figure 13.8. The planar microthruster arrays (from Zhang et al. [41]).

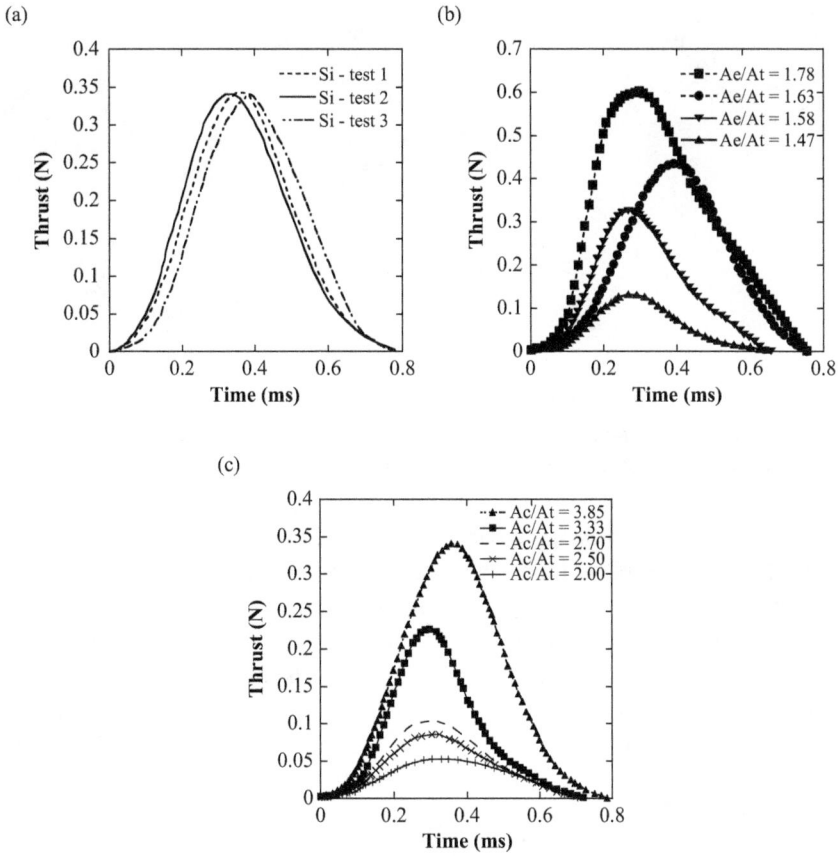

Figure 13.9. (a) Thrust profile repeatability, (b) thrust profiles for various A_e/A_t, and (c) thrust profiles for various A_c/A_t (from Zhang et al. [41]).

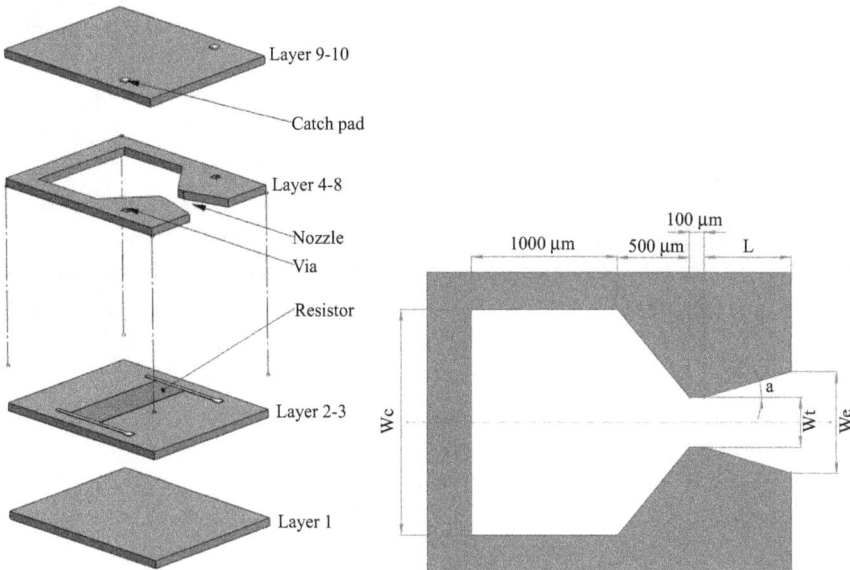

Figure 13.10. Isometric view and cross-section of the LTCC microthruster (from Zhang et al. [28]).

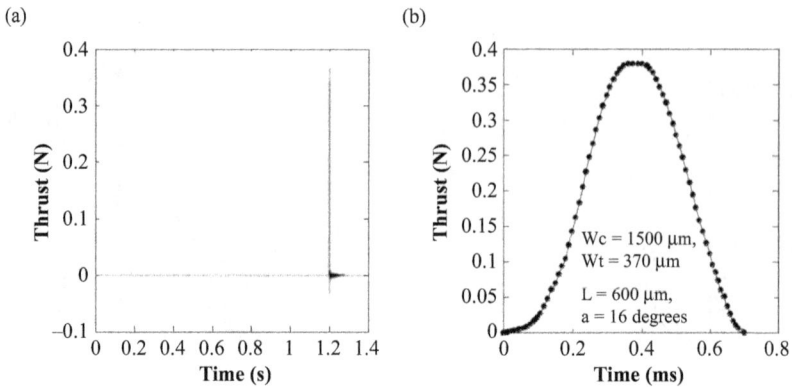

Figure 13.11. (a) Complete thrust profile of LTCC microthruster testing showing ignition delay and (b) expanded time about the peak thrust (from Zhang et al. [28]).

Although current was not measured directly, a direct current power supply output of 36 mA was observed at the time of ignition. The ignition power was estimated to be ~1.3 W. The ignition delay was 1.211 s with ignition energy estimated tobe 1.613 J. Expanding the time scale about the thrust peak shows that the duration of combustion was ~0.70 ms and the peak value of thrust was ~0.38 N. For this firing of the LTCC microthruster, the total impulse was 1.27×10^{-4} Ns. The thrusts produced by the LTCC microthrusters were found to be slightly higher than those produced by the similarly designed and sized silicon-based microthrusters. This trend was attributed to the better thermal insulation characteristics of the ceramic tape. The combustion thermal loss through the wall was lower, producing higher thrust. Specific impulses produced by the LTCC microthrusters were also higher than those generated by the silicon-based microthrusters (14.41 s versus 10.74 s vacuum; 10.28 s versus 6.56 s sea level), again probably the result of the better thermal insulation characteristics of LTCC.

More recently, Apperson et al. [43] have examined the thrust-generating characteristics of copper oxide/aluminum nanothermites in cylindrical motors with combustion chamber diameters of 1.59 mm. The propellant mixture was tested in various quantities (9–38 mg) by pressing the material over a range of densities. The testing was done in two different types of thrust motors: one with no nozzle and one with a convergent–divergent nozzle. As the packing density was varied, it was found that the material exhibited two distinct impulse characteristics. At low packing pressure, the combustion was in the fast regime, and the resulting thrust forces were ~75 N with durations of less than 50 μs full width at half-maximum. At high density, the combustion was relatively slow and the thrust forces were 3–5 N with a duration 1.5–3 ms. The change in density results in a change of burning rate as a result of a change in the propagation mechanism from a conductive mode at high densities to a convective mode at low densities. In both regimes, the specific impulse generated by the material was 20–25 s. Figure 13.12 shows the thrust as a function of propellant density and as a function of time for both the low and high burning rate modes of combustion. The motors were tested with and without a converging-diverging nozzle at the combustion chamber exit. For the slow propagation mode (conductive), the addition of the nozzle increased the thrust as a result of accelerating the gases to supersonic conditions; during the high propagation rate mode, however, the nozzle had little effect on the thrust, decreasing it slightly, most likely because the gases were already at a Mach number close to unity and due to the added drag and friction caused by the entrained condensed phase products.

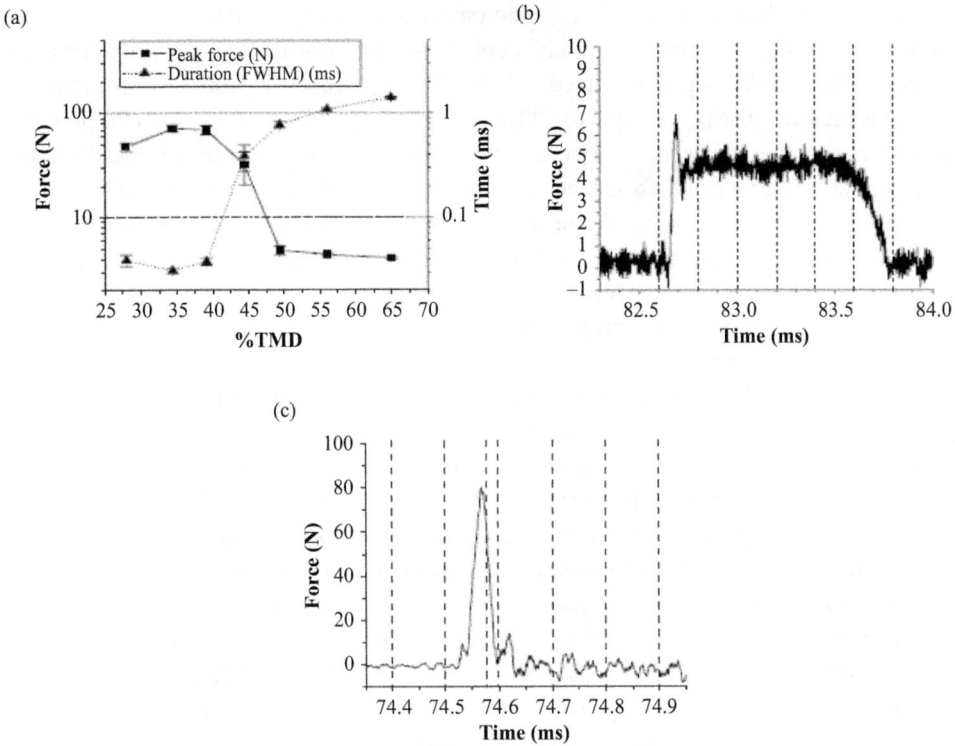

Figure 13.12. (a) peak thrust and duration, (b) thrust for the high density low burning rate mode, and (c) thrust for the low density and high burning rate mode (from Apperson et al. [43]).

Figure 13.13. Planar microthruster array using a Macor ceramic combustion chamber (from Chaalane et al. [44]).

Various other types of propellants and igniters have been investigated for solid propellant microthrusters. Chaalane et al. [44] have used propellant mixtures consisting of double-base propellants in a planar structure design fabricated from layers of silicon, glass, and Macor Corning ceramic. The Macor combustion chamber had the advantage of a coefficient of expansion close to that of silicon, a low thermal conductivity (1.46 W/m-K), and a high operating temperature (1273 K). The drawbacks were its higher weight and difficulty of machining. The width of the combustion chamber was ~1.5 mm, the length was ~5.3 mm, the throat width was ~0.15 mm, and the nozzle exit width was ~0.6 mm. The nozzle had a half angle of 10° and the combustion chamber volume with the Malcor covering was ~20 mm^3 and held ~7 mg of propellant (Figure 13.13).

With a double-base propellant only, ignition success was < 20 percent using polysilicon resistors. Ignition quality was improved by adding black powder to the double-base propellant. With > 10 percent by mass black powder mixed to the double-base propellant, 100 percent ignition success was achieved with a power input of 600 mW. To assure good contact between the propellant and igniter, high thermal conductivity epoxy glue was positioned between the propellant grain and the igniter to reduce contact resistance. Ignition delays were ~150 ms, while combustion rates varied from ~3.1 to 4.6 mm/s when the black powder fraction was increased from 10 to 30 percent. The average thrust, combustion times, and total impulse varied from 0.1 to 1 mN, 1.7 to 1.15 s, and 0.17 to 1.13 mN·s, respectively, for this same variation in propellant composition. An example of a thrust profile for a propellant composition of double base and 30 percent black powder is given in Figure 13.14.

Tanaka et al. [45] have considered boron potassium nitrate propellants with B/Ti multilayer igniters in solid microthruster arrays. The thrusters were constructed from three layers of silicon and two glass layers. Various loading designs of the solid propellant in the combustion chamber were studied to examine the effect on ignition quality. For ignition by resistive heating, good adhesion between the solid propellant and the igniter was found to be necessary. Despite much effort to ensure adhesion, the success rate of the ignition still remained ~90 percent at the maximum. To improve ignition quality, a reactive B/Ti multilayer igniter was developed, which generated sparks to a distance of several millimeters or more, for the noncontact ignition of the solid propellant. B/Ti multilayer igniters in three sizes were fabricated and tested in six configurations of the solid propellant. Although one rocket motor with the ignition charge was ignited successfully, the improvement of ignition by the noncontact B/Ti igniter was not confirmed. The B/T imultilayer igniter itself generated a small impulse of 10^{-6} N·s, however, which is suitable for attitude control of microsatellites, and incidentally illustrates the use of intermetallics as a propellant.

De Groot et al. [46] have considered the heterocyclic high nitrogen material 3,6-Diamino-1,2,4,5-trtrazine 1,4-dioxide ($C_2H_4N_6O_2$) with the trade name LAX–112 as a solid propellant for microthrusters. LAX–112 was chosen because of its excellent safety and environmental characteristics, low combustion temperature (less thermal stress imposed on the thruster chamber and better combustion stability) and competitive specific impulse. However, because of the incomplete combustion characteristics of LAX–112, upstream filters were required to prevent

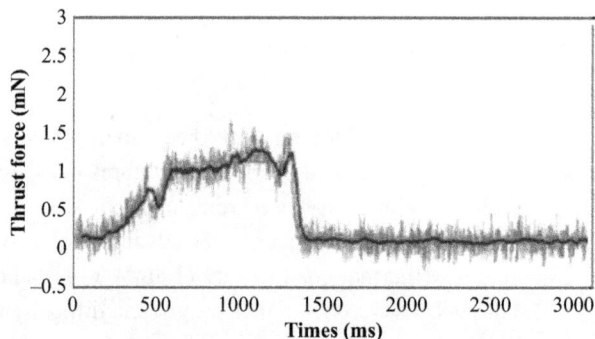

Figure 13.14. Thrust versus time for a microthruster using double base and 30 percent black powder propellant (from Chaalane et al. [44]).

solid particulates from clogging the nozzle. Compounding the problem was the slow burn rate of LAX–112, which necessitated a small nozzle throat to attain sufficiently high chamber pressures for stable combustion. LAX–112 also proved to have nonrepeatable ignition temperatures and was unable to be easily and repeatedly ignited by the chosen diode laser. More recently, Ali et al. [47] have considered more advanced high nitrogen, solid–gas-generating materials for microthrusters. These compounds included 3,6-bis(1H–1,2,3,4-tetrazol-5-ylamino)-s-tetrazine (BTATz) and mixed N-oxides of 3,3-azobis(6-amino-1,2,4,5-tetrazine) (DAATO3.5 where the 3.5 indicates the average oxide content) (Figure 13.15). DAATO3.5 has the highest burning rate of any known organic solid, although BTATz also has a fairly high burning rate. The high nitrogen compounds also yield significant hydrogen. Experiments were conducted in cylindrical quartz microthrusters. The DAATO3.5 and BTATz were mixed with 3 percent Hyemp 4051CG (Zeon Chemical, Inc.) binder and pressed into 3 mm pellets and then inserted through the back of the quartz tube thruster. Ignition was attained from an Nd: YAG laser.

The thrusters were 3 mm in diameter and ~25 mm in length. The mass of propellant varied from ~60 to 75 mg and had a density of 1.3 g/cm^3. Burning rates of the propellants generally varied from ~1 to 9 cm/s as the pressure was increased from 1 to 100 atm. The measured specific impulse ranged from 6 to 36 s, while the impulse ranged from 3 mN·s to 23 mN·s. The estimated chamber pressure ranged from ~2 to 19 atm. The I_{sp} efficiencies were about 14 percent. The low efficiencies were attributed to heat losses, flow losses, or combustion losses.

In addition to thrust, solid energetic materials at the microscale have also been used for actuation. Since these energetic materials differ from those typically used in microrockets, they are presented here for potential space application. Rodriguez et al. [48] have used the heterometallic compound $[Mn(NO_3)_4]_3[Co(NH_3)_6]_2$ as a gas-producing agent to deflect a PDMS membrane for microactuation. The material generates bio-compatible and nontoxic gases (N_2, O_2, and H_2O vapor) when the ignition temperature of 223°C is attained. The reaction proceeds according to the overall reaction

$$[Mn(NO_3)_4]_3[Co(NH_3)_6]_2 \rightarrow Co_2O_3 + 3MnO_2 + 18H_2O + 12N_2 + 9/2O_2 + 333J/g.$$

The actuator was small (<0.25 mm^2 × 100 μm), bio-compatible, and capable of generating sufficient overpressures (>10 kPa) under low electrical power (<100 mW), and could be used to eject a few nanoliters of fluid. The energetic material actuator device was fabricated using MEMS and microfluidic-compatible technology. Initial characterization indicated pressurization of 13 kPa and a membrane deformation of 46 μm for an electrical power of 90 mW (6.5 V, 13.9 mA).

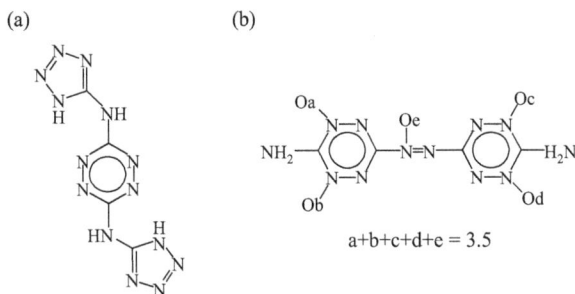

Figure 13.15. The chemical structure of (a) BTATz and (b) DAATO 3.5 high nitrogen compounds used as propellants.

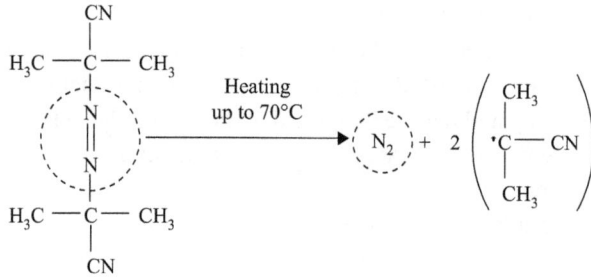

Figure 13.16. The decomposition of AIBN between 70 and 100°C (from Hong et al. [49]).

While these characteristics make such a microactuator well adapted for microfluidic applications and especially for the ejection of fluids contained in microchannels of a disposable lab-on-a-chip, they may also be useful for pumping liquid propellants.

Hong et al. [49] have used gas generation from solid chemical propellants to pump microfluids in applications of plastic-based disposable biochips or lab-on-a-chip systems. AIBN was used as the solid chemical propellant and was deposited on a microheater using a screen-printing technique, which heated the AIBN to 70°C, initiating decomposition and producing nitrogen gas. The overall decomposition of AIBN occurs as shown in Figure 13.16 for temperatures between ~70 and 100°C.

In the experiments, the output pressure of nitrogen gas generated from the solid chemical propellant could be adjusted to a desired pressure by controlling the input power of the heater. Using this chemical energy source, the generated pressure depended on the deposited amount of the solid chemical propellant and the temperature of the microheater. Experimental measurements showed that a pressure around 3 kPa was achieved when 189 mJ of energy was applied to heat 100 mg of AIBN. This pressure can drive 50 mL of water through a microfluidic channel 70 mm in length with a cross-sectional area of 100 μm × 50 μm.

Pi et al. [50] have developed a solid propellant microthruster based on the detonating agent diazodinitrophenol (DDNP), and the propellant black powder, for actuation of aremote-controlled capsule (RCC) for drug release. Ignition and combustion tests were performed to justify the feasibility and reliability of the system. The results demonstrated that 166 mW of power consumption led to successful combustion. Complete and rapid drug release was achieved when the propellant ranged from 16 mg to 20 mg. The RCC integrated with the microthruster could provide a promising alternative for site-specific drug delivery in human gastrointestinal tract.

13.4 LIQUID PROPELLANT THRUSTERS

Liquid propellants, both monopropellants and bi-propellants, have been much less studied at the microscale than solids. Molecular liquids like hydrazine and hydrogen peroxide have been studied. These monopropellants have been ignited with a catalyst (generally iridium for N_2H_4 and Ag or MnO_x for H_2O_2). Soluble catalysts like the $FeCl_2$ aqueous solution have also been proposed. Monopropellant liquids comprised of a concentrated aqueous solution of ionic oxidizer such as hydroxylammonium nitrate(HAN, $NH_3OH^+NO_3^-$), ammonium dinitramide

(ADN, $NH_4^+N(NO_2)_2^-$), and hydrazinium nitroformate (HNF, $N_2H_5^+C(NO_2)_3^-$) with a dissolved fuel have been studied with catalyst, as well as electrolysis, used for ignition.

Hitt et al. [51] have considered the catalyzed chemical decomposition of high-concentration hydrogen peroxide. The silicon planar thrusters consisted of an inlet of filters, a catalyst chamber, and a converging and diverging nozzle. The catalyst bed was formed from etched pillars that were silver coated. The combustion chambers were ~1 mm in width, 2 mm in length, and 50–300 μm in depth. Preliminary results indicated an incomplete reaction and a two-phase flow exiting from the nozzle. Plumlee et al. [52] also considered hydrogen peroxide microthrusters with a silver catalyst. Their design made use of low temperature co-fired ceramic tape structures to construct a nozzle and a silver-coated catalyst chamber in the substrate. Instead of a structural catalyst, Chen et al. [53] developed a homogeneous catalyst consisting of a water solution of $FeCl_2$, which results in a hypergolic ignition with H_2O_2. The solubility of $FeCl_2$ in water is 20.65 g/100 g H_2O at 20°C. The system operated like a bi-propellant with the two liquids injected into the combustion chamber as opposed flows to induce mixing. The thruster was fabricated from silicon and glass. The overall size was 5 mm × 5 mm × 3 mm with a weight of < 0.2 g. The nozzle profile was optimized with a throat size of 0.04 mm (width) × 0.2 mm (depth) and an exit of 0.75 mm (width) × 0.2 mm (depth). With 87 percent concentrated H_2O_2, and a flow rate ratio of 9:1 for hydrogen peroxide to the saturated solution of $FeCl_2$, the product temperature was predicted to be ~400°C, which yielded a theoretical vacuum I_{sp} of 163 s. Initial testing was performed with 30 percent concentrated H_2O_2, which yielded an impulse per controlling pulse of <80 μN s. The preliminary studies, however, showed excessive ignition delays.

Scharlemann and Tajmar [54] have reported on the development of a hydrogen peroxide microrocket engine that has the capability of delivering a specific impulse in vacuum over the range of 100–800 mN. The initial testing provided a $I_{sp,vac}$ of 153 s. The combustion chambers and nozzles were manufactured from nickel–chromium–cobalt alloy (Nimonic 115) using EDM techniques. This material was chosen for ease of manufacturing and welding, and because thermal stress calculations indicated that in some ceramic materials, even those characterized by high temperature resistance, low thermal conductivity generates excessive thermal stresses at the nozzle throat. Silicon carbide was also considered. The use of a monolithic catalyst [55] reduced the pressure loss across the catalyst bed significantly in comparison to the previously used pellet or gauze catalyst. Aqueous solutions of sodium permanganate ($NaMnO_4$) were used as the catalyst precursor. A decomposition efficiency of up to 99 percent was obtained. A maximum total load of nearly 1.2 kg was decomposed by one catalyst without showing performance deterioration tendencies.

Miniature liquid bipropellant rocket engines have also been developed using hydrogen peroxide (87.5 percent concentrated) to generate oxygen and ethanol as the liquid fuel [55, 56]. The original design made use of a glow plug to vaporize the ethanol, but because of power requirements, this was abandoned and the fuel was injected as a liquid. Additional modifications included replacement of the Nimonic 105 with stainless steel because of welding difficulties. The hydrogen peroxide was decomposed efficiently with a monolith catalyst, as previously described. The combustion chamber was designed to operate at 3 bar (0.3 MPa) with a diameter of 15 mm and a length of 10 mm. The residence times studied were found to be too short for successful mixing and ignition, and operating pressures were too low.

Wu et al. [24, 29] have reported on the fabrication and characterization of a microscale liquid monopropellant thruster using LTCC tape technology. Metal electrodes were deposited on the surfaces of the combustion chamber to investigate electrolytic ignition of HAN-based

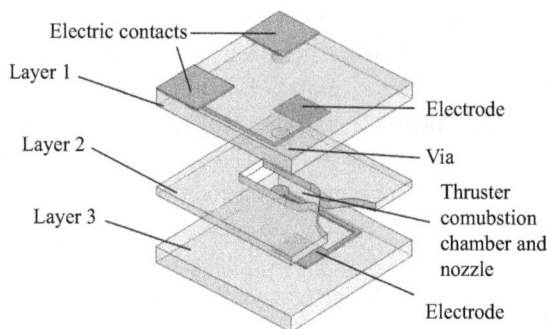

Figure 13.17. Exploded view of the LTCC electrolytic microthruster design.

propellants. In this concept, electrical energy was deposited directly into the liquid propellant (as opposed to the surrounding structure), which offers the potential to reduce energy losses and improve thermal management at the microscale. A digital (single-shot) microthruster was designed to demonstrate the feasibility of electrolytic ignition. The thruster design (Figure 13.17) was comprised of three major layers. The thrust chamber (Layer 2) was sandwiched between Layer 1 and Layer 3. Metal electrodes on the inner surfaces of Layer 1 and Layer 3 provided direct contacts for the electrolytic reaction of the HAN-based liquid monopropellant in the thruster combustion chamber. The electrodes were designed to cover the entire top and bottom surfaces of the chamber to maximize the contact area between the liquid propellant and the electrodes.

Metal wirings and vias were also embedded in the device to connect the electrodes to the electric pads on the outer surface of Layer 1. The thruster combustion chamber and nozzle were fabricated on Layer 2 using micropunches. Vias were punched on Layers 2 and 3 and filled with conductor paste. The vias and wiring patterns on Layer 1 and Layer 3 provided electric connections between the electrodes and the electric contacts on the outer surface of Layer 3. The dimensions of the whole thruster chip were ~25.4 × 12.7 × 1 mm. The volume of the internal combustion chamber was 0.82 mm^3. The half angle of the nozzle contraction was 54°, and 15° for the nozzle expansion.

Investigation of the evolutions of the current output (Figure 13.18) and the ignition process revealed a current spike of 14.5 A at the moment when ignition occurred. However, the current evolution showed that the current flow was not initiated immediately when the voltage was switched on, but after a lag of ~25 ms. The delay was due to the induction time for the initial ionic reaction. As the ionic flow was created in the liquid propellant, the overall current flow in the circuit increased initially to ~1 A, but rapidly decreased to a stable current level of ~0.2 A. These phenomena were explained by the less-efficient diffusive transport resulting from the reduction of effective liquid contact area on the electrodes due to gas bubbles forming on the electrode surfaces as the reaction developed. The current became zero after the spike at full ignition, even though a constant voltage output of 45 V was maintained at the power supply, indicating that all propellant was either consumed or propelled out of the combustion chamber. The total energy output of the power supply during the ignition process was estimated by integrating the power output from the power supply from zero time until the termination of the current spike at 0.25 s (see Figure 13.18), and was determined to be 1.9 J for the case shown.

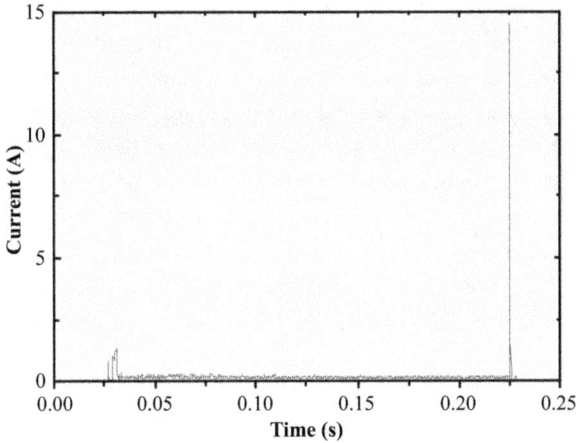

Figure 13.18. The evolution of the current flow during the ignition process.

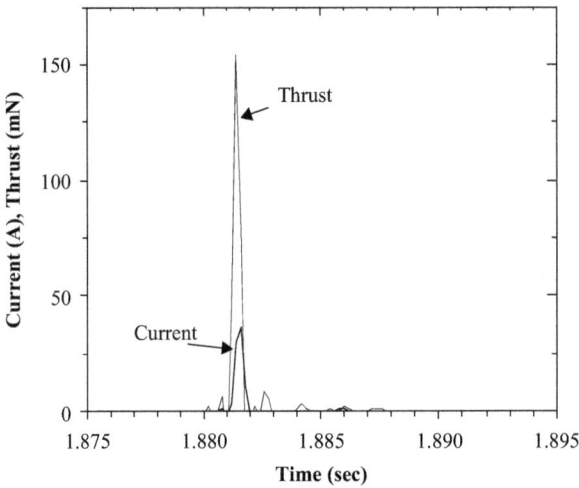

Figure 13.19. Current and thrust evolution of the HAN-based liquid monopropellant microthruster.

The ignition energy is ~40 percent of the energy released by the propellant through the combustion process.

Thrust measurements on the microthrusters (Figure 13.19) showed concurrent peak thrust output from the thruster and peak current output from the power supply. The firing shown in Figure 13.19 achieved a peak thrust of 150 mN. Peak thrust as high as 197 mN for a HAN-based liquid monopropellant microthruster has also been recorded. The impulse was 3.1×10^{-3} N·s. In some cases, the output current was found to be clipped at 50 V, which was the maximum current the power supply could provide. The current clipping indicates that the reactivity of the propellant becomes so high that the resistivity through ionic flow in the liquid propellant becomes negligible, such that the circuit draws all the current the power supply is able to provide (50 A).

Wu et al. [23, 24] have also reported the development of continuous vortex flow combustors for miniature rockets using HAN-based monopropellants and nitromethane. These thrusters were made from Inconel using EDM techniques. Combustion chambers were cylindrical (~5 mm diameter) and the volume was ~100 mm^3. Liquids were injected along the wall and because of the high surface-to-volume ratio, enough surface area was provided to vaporize the propellant. In addition, wall temperatures at small scale can be significant, and thus, the wall injection also provided some cooling advantages. For nitromethane injection, stable combustion was reached at a chamber pressure of 1.8 MPa; chamber pressures needed to be further increased to achieve over 99 percent chemical conversion (> 2 MPa). For HAN-based propellants, elevated pressures were still required, but the required pressures were somewhat lower.

13.5 GASEOUS PROPELLANT THRUSTERS

London et al. [17, 22] produced a high pressure microthruster manufactured by fusion bonding a stack of six individually etched single crystal silicon wafers. The device, originally designed for operation with liquid oxygen and ethanol, was tested as a gaseous bipropellant with methane and oxygen. The microthruster, 18 × 13.5 × 3 mm in size, was designed to operate at pressures as high as 125 atm. The design had internal cooling channels for liquid coolant and opposed injectors for fuel and oxidizer in the combustion chamber. The dies and assembled motor are shown in Figure 13.20. The injectors are located on the side walls of the combustor chamber (as a function of its length) and oppose each other across the chamber. The cooling jacket is also visible in the photographs. The design is an excellent example of the intricacy achievable at the microscale.

The microthruster operating at a pressure of 12.3 atm is shown in Figure 13.21. Because the maximum pressure was only 12.3 atm, separation in the nozzle was expected as shown in

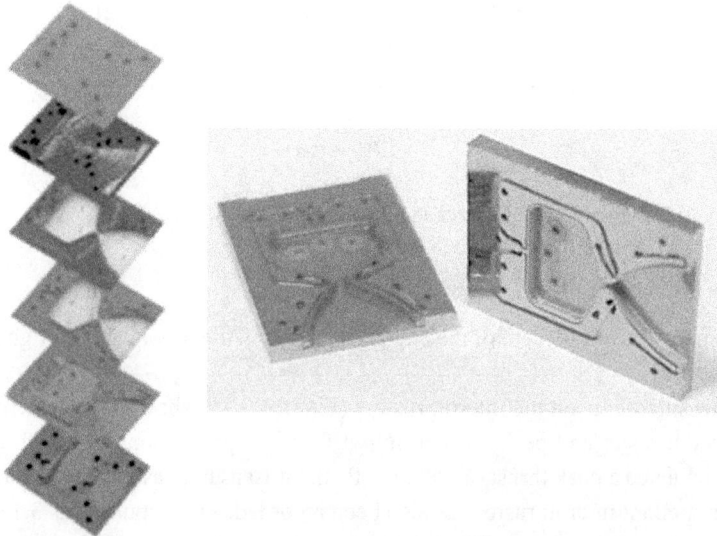

Figure 13.20. (left) Six dies of the MIT bi-propellant microrocket stack and (right) two halves of the assembled thrust chamber (from London et al. [17, 22]).

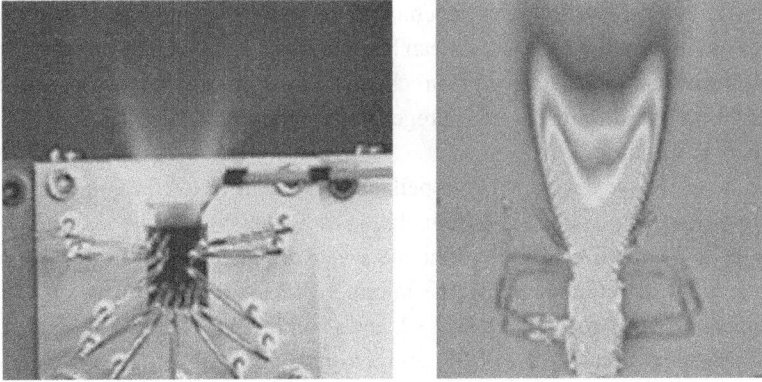

Figure 13.21. Visual and infrared images of the MIT bi-propellant microrocket motor operating at 12.3 atm (from London et al. [17, 22]).

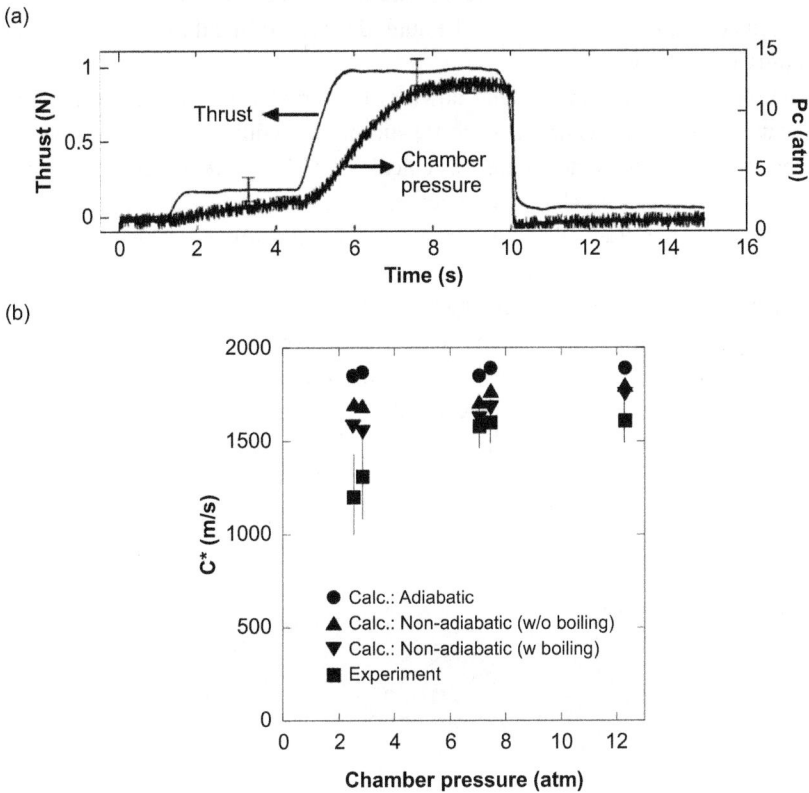

(a)

(b)

Figure 13.22. (a) Thrust profile and (b) comparison of experimental and calculated characteristic velocity (from London et al. [22]).

the figure. The bifurcation in the plume apparently results from the operation of the high aspect ratio rectangular nozzle under off-design conditions.

The thrust profile attained at the 12.3 atm chamber pressure is shown in Figure 13.22, which indicates a thrust of ~1 N for ~2 s of test time. As the chamber pressure was increased from 2 to 7 atm, measured characteristic velocities (C*) initially increased, but remained constant above ~7 atm. Chemical reaction rates generally increase as P^{1-2}, and therefore, the kinetics do not appear to be limiting the reaction. As can be seen from Figure 13.22, the measured C* are ~15 percent lower than those deduced from chemical equilibrium, suggesting that heat losses are responsible for the differences. With regenerative cooling, C* values should be close to those of larger scale motors.

Yetter et al. [27] have developed bipropellant thrusters in which the combustion process is stabilized in a cylindrical chamber with recirculation zones developed from vortex flows using the tangential injection of oxidizer. The microthrusters were designed and fabricated from aluminum oxide using ceramic stereolithography. The microthrusters had throat diameters ranging from ~700–300 µm, an exit or throat area ratio of ~2, and a combustion chamber volume of ~60 mm^3. Combustion tests were conducted with H_2–air mixtures without forced external cooling.

Ignition was obtained via a spark across two small electrodes. The device was run in continuous operation for ~1 hour, and many start-up and shut-down sequences were performed. Photographs of the thruster and the plume are shown in Figure 13.23. The thruster body in these photographs is reddish-orange. Surface temperatures of the ceramic thruster body from a thermocouple located on the top surface near the outer diameter of the thruster body (Figure 13.23) were generally near 600 K.

An example of the start-up and shut-down of the thruster is shown in Figure 13.24. Here the thruster was operated on a nonpremixed H_2–air mixture with an overall equivalence ratio of 1.6. The chamber adiabatic temperature was calculated to be ~2100 K. The Reynolds numbers of the inlet air, hydrogen, combustion chamber, and throat were 7500, 550, 330, and 2300, respectively. For this condition, the experimental C* was determined to be 970 m/s and the theoretical vacuum specific impulse was 214 s. The start-up procedure consisted of setting the air flow rate, then firing the spark, and then injecting the hydrogen. Ignition was rapidly achieved, as indicated by the jump in pressure and temperature around 35 s. The internal temperature measurement was located at the upstream end of the combustion chamber near the combustor wall, but not in contact with the wall. The slight increase in chamber temperature at 90 s was the result of a change in the hydrogen flow rate.

Figure 13.24 also presents C* measurements and C* efficiency as a function of equivalence ratio for operation under hot fire conditions on hydrogen–air mixtures. The characteristic velocity is observed to increase with increasing equivalence ratio and attain a maximum near

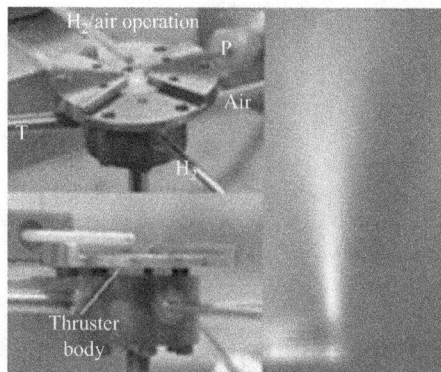

Figure 13.23. Thruster in hot fire operation with hydrogen air mixture. The thruster body in these pictures was reddish-orange in color (bottom-left), except for near the throat (top-left) where the surface was bright white. The image to the right is a schlieren image of the plume.

(a)

(b)

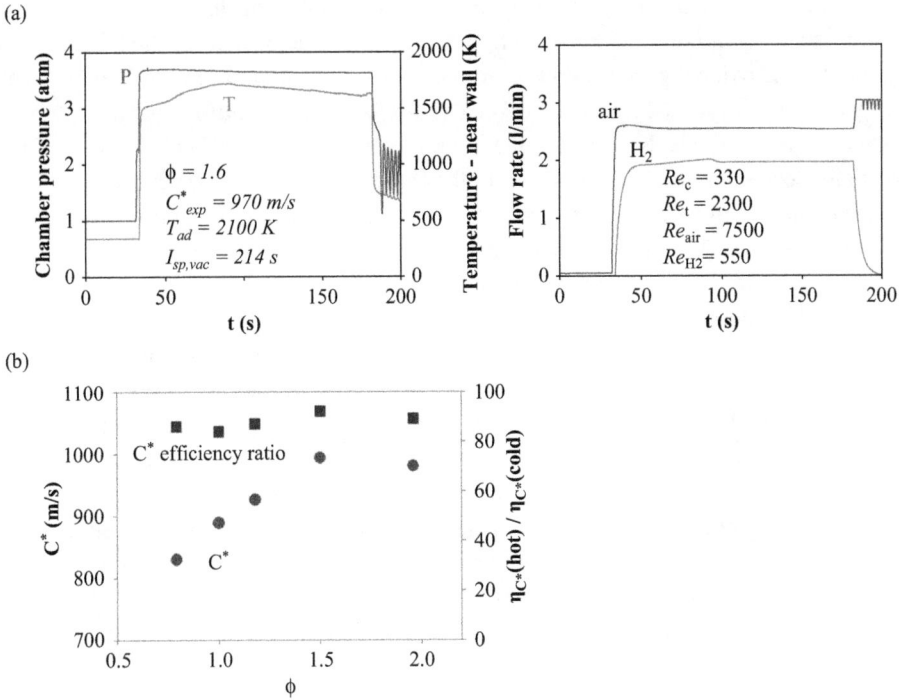

Figure 13.24. (a) Start-up and shutdown of the thruster under hot-fire conditions. (b) Characteristic velocity and ratio of C* efficiency under hot flow operation to C* efficiency under cold flow operation.

an equivalence ratio of 1.5. In the figure, the C* efficiency of the hot fire tests are plotted relative to the C* efficiency of the cold flow tests, to separate the inefficiencies of the cold flow from the combustion efficiencies. Since the low cold flow efficiencies are probably attributable to uncertainties in the throat area measurement, this presentation also removes this ambiguity. As can be seen from the figure, ratios of hot C* efficiencies relative to cold C* efficiencies are high, near 90 percent indicating good combustion efficiency and relatively small heat loss from the thruster. This research was continued by Sun et al. [32], using thrusters fabricated from both low and high temperature cofired ceramic tapes. The high temperature cofired ceramic tapes were required for the entire combustor or as a nozzle inserted into a low temperature cofired ceramic combustor body, because of the high temperatures in the throat region for extended periods of operation with hydrogen oxygen mixtures.

The multifunctional power production capabilities of propulsion systems may also be beneficial to space applications. Ceramic tapes enable the incorporation of solid oxide fuel cells (SOFC) directly into the combustion chambers of micro propulsion systems. Because thermal management is important in these systems, the combination of thruster and fuel cell may be an approach to recover thermal losses. Donadio et al. [57] have investigated the feasibility of partially decomposing a HAN-based liquid propellant to form gaseous oxidizers of nitrogen oxides and vaporized fuel (such as methanol) and water, since low temperature decomposition of such tri-component propellants is generally initiated by HAN decomposition with little consumption of the fuel. The passage of this low temperature gas to the cathode side of an solid oxide fuel cell (SOFC) with a catalyst preferential for the oxygen was considered; the remainder of the propellant was injected into the combustion chamber to complete the high temperature reaction and form a fuel-rich mixture with excess hydrogen as a source of fuel for the SOFC anode. A direct-flame solid oxide fuel cell (DFFC) composed of a 150 μm thick Hionic™ zirconia-based electrolyte, a nickel-yttria-stabilized zirconia (Ni-YSZ) anode, and a lanthanum strontium manganite (LSM) cathode, both 50 μm thick, was tested with various N_xO_y oxidizers (including partially decomposed 13 M HAN) fed to the cathode; combustion products from fuel-rich hydrogen combustion were fed to the anode. The electrodes had a 12.5 mm diameter and the entire button cell was 20 mm in diameter. Figure 13.25 shows the polarization curves for NO_2 and O_2. As can be seen from the figure, NO_2 performs as well as molecular oxygen in producing power. Very good performance was also obtained with NO and N_2O oxidizers.

Wu and Lin [30] have developed a gaseous bi-propellant chemical microthruster using LTCC tapes. A sapphire window and silver spark ignition electrodes were cofired in the LTCC microthruster. The dimensions of the thruster chamber were 5.22 mm × 1.6 mm × 1.19 mm, while the volume of the combustion chamber was 9.9 mm³. The thickness of the nozzle throat was the same as that of the chamber and equaled 268 μm, the cross-sectional area of the nozzle throat was 0.319 mm², and the contraction ratio of the nozzle was approximately six, with a contraction angle of 45° on the converging section and an expansion angle of 10° on the diverging section. Dimensions for the inlet ports were 0.7 mm × 0.595 mm. Figure 13.26 shows a schematic of the thruster designs with counter flow and swirling injector configurations, with the latter shown to enhance mixing from numerical simulations.

Thrust outputs between 0.2 mN and 1.97 mN were achieved using ethylene–argon/oxygen mixtures. Thrust was linearly proportional to the total mass flow rate of the inlet. The chemical energy input of the microthruster was ~37 W for a maximum measured thrust of 1.97 mN. The swirling inlets were demonstrated to improve mixing and therefore combustion.

There have also been a number of studies investigating reaction in microtubes with and without catalytic surfaces for application to microthrusters. Mento et al. [58] have studied catalyzed combustion of propellants in a microtube as a model of a microthruster. The effect of

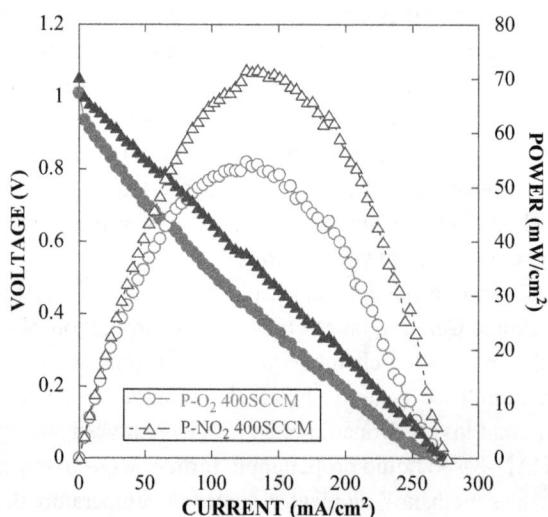

Figure 13.25. Polarization curves for NO_2 and O_2. The solid and open symbols represent voltage and power density values respectively (from Donadio et al. [57]).

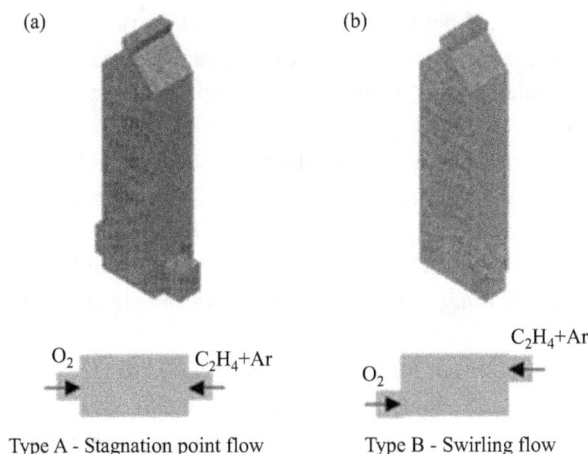

Type A - Stagnation point flow   Type B - Swirling flow

Figure 13.26. The inlet design and internal structure of the LTCC bipropellant microrockets (from Wu and Lin [30]).

hydrogen addition on fuel-rich methane/oxygen ignition within a 0.40 mm diameter platinum microtube was investigated. The tests were conducted in a vacuum chamber with an ambient pressure of 0.0136 atm to simulate high-altitude conditions. The results showed that the critical temperature needed to catalytically light off fuel-rich methane/oxygen mixtures was reduced by the addition of small amounts of hydrogen to the mixture. A two-stage ignition phenomenon was observed for low levels of hydrogen addition (2 to 7 percent by volume), with the first and second ignition conditions corresponding to the reactions of hydrogen and methane, respectively. The effects of changing flow rate (residence time), equivalence ratio, and amount of hydrogen addition on the critical ignition temperature were investigated. The ability of the catalyst to sustain chemical reactions once the input power was turned off was also explored and, for most cases, self-sustainability was realized. A thrust of 1–13 mN was obtained while specific impulses were between 125 and 300 s for the reacting flow. In a similar study, but for rich methane/oxygen mixtures, Volchko et al. [59] indicated that a microthruster could be developed with as little as 1 W of power for ignition and that the reactions could be self-sustained. Gamezo and Oran [60] have numerically studied flame acceleration in narrow channels and suggest that the interaction of the flame development with the boundary growth can be applied to micropropulsion applications as the combustion process allows the efficient use of fuel and gradual development of thrust. Wu and Lu [61] have recently developed a chemical microthruster based on gaseous pulsed detonation. Stable valveless operation, at frequencies as high as 200 Hz, was achieved in a 100 mm long micro channel with 1 mm × 0.6 mm cross-section. The propagation velocity of reaction waves near the exit of the channel was close to Chapman–Jouguet velocity, according to the signals from ion probes embedded in the LTCC pulsed detonation microthruster.

13.6   CONCLUSION AND FUTURE WORK

Significant progress has been made in the development of chemical microthrusters over the last 15 years. Many of the proposed concepts are based on scaling of larger systems. Because thrust levels (millinewtons and below) and impulse bit levels (tens of micronewtons–seconds and below) required by many microscale applications can be considerably smaller than those attainable from macroscale systems, however, novel approaches have also been studied at the microscale.

Solid propellant thrusters have received the most attention at the microscale because of their simplicity. The use of solid propellants with high burning rates has been successful in overcoming heat losses and radical quenching at walls. Liquid and gaseous propellant thrusters have also been studied, with somewhat less success due to greater difficulty in ignition, thermal management with longer burning times, and the need for integrated and connected peripheral flow control modules.

To further the development of microthrusters, continued fundamental research is needed on fluid flow in microchannels, combustion in small volumes, coupled heat and mass transfer between flows and structures, high temperature materials with low thermal conductivity, integrated microscale design and fabrication techniques, and microscale diagnostics. Equally important to these fundamental studies are systems level research and manufacturing development.

ACKNOWLEDGMENT

This work was supported by the U.S. Air Force Office of Scientific Research under Contract AFOSR F49620-01-1-0376. The authors gratefully acknowledge the support from Dr. Mitat Birkan, contract monitor of the program.

REFERENCES

[1] Fernandez-Pello, A.C. (2002) Micropower Generation using Combustion: Issues and Approaches, Proceedings of the Combustion Institute 29, 883–899. doi: http://dx.doi.org/10.1016/S1540-7489(02)80113-4.

[2] Micci, M.M., and Ketsdever, A.D. (Eds.) (2000) Micropropulsion for Small Spacecraft, Progress in Astronautics and Aeronautics, Vol. 187, AIAA, Reston, VA.

[3] Bruno, C. (2001) Chemical Microthrusters: Effects of Scaling on Combustion, AIAA 2001-3711, 37th AIAA/ASME/SAE/ASEE Joint Propulsion Conference, Salt Lake City, UT.

[4] Bruno, C., Giacomazzi, E., and Ingenito, A. (2003) Chemical Microrocket: Scaling and Performance Enhancement, Final Report, SPC 02-4034, Contract No. FA8655-02-M034, AFRL.

[5] Ketsdever, A.D., and Mueller, J. (1999a) Systems Considerations and Design Options for Microspacecraft Propulsion Systems, AIAA 99-2723, 35th AIAA / ASME / SAE / ASEE Joint Propulsion Conference and Exhibit, Los Angeles, CA.

[6] Ketsdever, A.D., Wadsworth, D.C., Wapner, P.G., Ivanov, M.S., and Markelov, G.N. (1999b) Fabrication and Predicted Performance of Conical DeLaval Micronozzles, 35th AIAA/ASME/SAE/ASEE Joint Propulsion Conference, LA.

[7] Ketsdever, A.D. (2003) Microfluidics Research in MEMS Propulsion Systems, AIAA 2003-0783, 41st Aerospace Sciences Meeting & Exhibit, Reno, NV.

[8] Alexeenko, A.A., Gimelshein, S.F., Levin, D.A., and Collins, R.J. (2000) Numerical Modeling of Axisymmetric and Three-Dimensional Flows in MEMS Nozzles, AIAA 2000-3668, 36th AIAA/ASME/SAE/ASEE Joint Propulsion Conference, Huntsville, AL.

[9] Janson, S.W., Helvajian, H., Hansen, W.W., and Lodmell, J. (1999) Batch-Fabricated CW Micro-thrusters for Kilogram-Class Spacecraft, AIAA 99-2722, 35th AIAA/ASME/SAE/ASEE Joint Propulsion Conference, LA.

[10] Alexeenko, D.A., Levin, D.A., Fedosov, D.A., Gimelshein, S.F., and Collins, R.J. (2003) Coupled Thermal and Fluid Analysis of Microthruster Flows, AIAA 2003-0673, 41st Aerospace Sciences Meeting and Exhibit, Reno, NV.

[11] Kyritsis, D.C., Guerrero-Arias, I., Roychoudhury, S., and Gomez, A. (2002) Mesoscale Power Generation by a Catalytic Combustor Using Electrosprayed Liquid Hydrocarbons, Proceedings of the Combustion Institute 29, 965–972. doi:http://dx.doi.org/10.1016/S1540-7489(02)80122-5.

[12] Gemci, T., and Chigier, N. (2003) Electrodynamic Atomization for MEMS Combustion Systems, AIAA 2003-675, 41st Aerospace Sciences Meeting and Exhibit, Reno, NV.

[13] Stanchi, S., Dunn-Rankin, D., and Sirignano, W.A. (2003) Combustor Miniaturization with Liquid Fuel Filming, AIAA 2003-1163, 41st Aerospace Sciences Meeting and Exhibit, Reno, NV.

[14] Arana, L.R., Schaevitz, S.B., Franz, A.J., Schmidt, M.A., and Jensen, K.F. (2003) A Micro-fabricated Suspended-Tube Chemical Reactor for Thermally Efficient Fuel Processing. Journal of Microelectromechanical Systems 12, 600–612. doi: http://dx.doi.org/10.1109/JMEMS.2003.817897.

[15] Tanaka, S., Yamada, T., Sugimoto, S., Li, J.F., and Esashi, M. (2003) Silicon Nitride Ceramic-Based Two-Dimensional Microcombustor. Journal of Micromechanics and Microengineering 13, 502–508. doi: http://dx.doi.org/10.1088/0960-1317/13/3/321.

[16] Zhang, K.L., Chou, S.K., and Ang, S.S. (2004b) Development of a Solid Propellant Microthruster with Chamber and Nozzle Etched on a Wafer Surface, Journal of Micromechanics and Microengineering 14, 785–792. doi: http://dx.doi.org/10.1088/0960-1317/14/6/004.

[17] London, A.P., Epstein, A.H., and Kerrebrock, J.L. (2001a) High-Pressure Bipropellant Microrocket Engine, Journal of Propulsion and Power 17, 780–787. doi: http://dx.doi.org/10.2514/2.5833.

[18] Waitz, I.A., Gauba, G., and Tzeng, Y.S. (1998) Combustors for Micro-Gas Turbine Engines, Journal of Fluids Engineering-Transactions of the ASME 120, 109–117. doi: http://dx.doi.org/10.1115/1.2819633.

[19] Yang, W.M., Chou, S.K., Shu, C., Li, Z.W., and Xue, H. (2004) A Prototype Microthermophotovoltaic Power Generator. Applied Physics Letters 84, 3864–3866. doi: http://dx.doi.org/10.1063/1.1751614.

[20] Rossi, C., Briand, D., Dumonteuil, M., Camps, T., Pham, P.Q., and de Rooij, N.F. (2006a) Matrix of 10 × 10 Addressed Solid Propellant Microthrusters: Review of the Technologies, Sensors and Actuators A 126, 241–252, 2006. doi: http://dx.doi.org/10.1016/j.sna.2005.08.024.

[21] Lewis, D.H., Janson, S.W., Cohen, R.B., and Antonsson, E.K., (2000) Digital Micropropulsion, Sensors and Actuators 80, 143–154. doi: http://dx.doi.org/10.1016/S0924-4247(99)00260-5.

[22] London, A.P., Ayon, A.A., Epstein, A.H., Spearing, S.M., Harrison, T., Peles, Y., and Kerrebrock, J.L. (2001b) Microfabrication of a High Pressure Bipropellant Rocket Engine, Sensors and Actuators A 92, 351–357, 2001. doi: http://dx.doi.org/10.1016/S0924-4247(01)00571-4.

[23] Wu, M.H., Wang, Y., Yang, V., and Yetter, R.A. (2007) Combustion in Mesoscale Vortex Chambers, Proceedings of the Combustion Institute 32, 3091–3098.

[24] Wu, M.H., Wang, Y., Yetter, R.A., and Yang, V. (2009b) Liquid Monopropellant Combustion in Mesoscale Vortex Chamber, Journal of Propulsion and Power 24, 829–832. doi: http://dx.doi.org/10.2514/1.40662.

[25] Schoth, A., Forster, R., and Menz, W. (2005) Micro Wire EDM for High Aspect Ratio 3D Microstructuring of Ceramics and Metals, Microsystem Technologies 11, 250–253. doi: http://dx.doi.org/10.1007/s00542-004-0399-y.

[26] Benavides, G.L., Bieg, L.F., Saavedra, M.P., and Bryce, E.A. (2002) High Aspect Ratio Meso-Scale Parts Enabled by Wire Micro-EDM, Microsystem Technologies 8, 395–401. doi: http://dx.doi.org/10.1007/s00542-002-0190-x.

[27] Yetter, R.A., Yang, V., Wang, Z., and Wang, Y. (2003) Development of Meso and Micro Scale Propellant Thrusters, the 41st AIAA Aerospace Sciences Meeting and Exhibit, Reno, NV.

[28] Zhang, K.L., Chou, S.K., and Ang, S.S. (2005b) Development of a Low-Temperature Co-Fired Ceramic Solid Propellant Microthruster, Journal of Micromechanics and Microengineering 15, 944–952, 2005. doi: http://dx.doi.org/10.1088/0960-1317/15/5/007.

[29] Wu, M.H., and Yetter, R.A. (2009a) A Novel Electrolytic Ignition Monopropellant Microthruster Based on Low Temperature Co-fired Ceramic Tape Technology, Lab on a Chip 9, 910–916. doi: http://dx.doi.org/10.1039/b812737a.

[30] Wu, M.H., and Lin, P.S. (2010) Design, Fabrication and Characterization of a Low-Temperature Co-fired Ceramic Gaseous Bi-Propellant Microthruster, Journal of Micromechanics and Microengineering 20, 1–11. doi: http://dx.doi.org/10.1088/0960-1317/20/8/085026.

[31] Gongora-Rubio, M.R., Espinoza-Vallejos, P., Sola-Laguna, L., and Santiago-Aviles, J.J. (2001) Overview of Low Temperature Co-Fired Ceramics Tape Technology for Meso-System Technology (MsST), Sensors and Actuators a-Physical 89, 222–241. doi: http://dx.doi.org/10.1016/S0924-4247(00)00554-9.

[32] Sun, W., Cai, G., Danadio, N., Kuo, B.H., Lee J., and Yetter, R.A. (2011) Development of Meso-Scale Co-Fired Ceramic Tape Axisymmetric Combustors, International Journal of Applied Ceramic Technology 9, 833–846. doi: http://dx.doi.org/10.1111/j.1744-7402.2011.00698.x.

[33] Roosen, A. (2001) New Lamination Technique to Join Ceramic Green Tapes for the Manufacturing of Multilayer Devices, Journal of the European Ceramic Society 21, 1993–1996. doi: http://dx.doi.org/10.1016/S0955-2219(01)00158-3.

[34] Birol, H., Maeder, T., and Ryser, P. (2006) Processing of Graphite-Based Sacrificial Layer for Microfabrication of Low Temperature Co-Fired Ceramics (LTCC), Sensors and Actuators a-Physical 130, 560–567. doi: http://dx.doi.org/10.1016/j.sna.2005.12.009.

[35] Pattekar, A.V., and Kothare, M.V. (2003) Novel Microfluidic Interconnectors for High Temperature and Pressure Applications, Journal of Micromechanics and Microengineering 13, 337–345. doi: http://dx.doi.org/10.1088/0960-1317/13/2/324.

[36] Rossi, C., Orieux, S., Larangot, B., Do Conto, T., and Estève, D. (2002) Design, Fabrication and Modeling of Solid Propellant Microrocket-Application to Micropropulsion, Sensors and Actuators A 99, 125–133. doi: http://dx.doi.org/10.1016/S0924-4247(01)00900-1.

[37] Rossi, C., Larangot, B., Lagrange, D., and Chaalane, A. (2005) Final Characterizations of MEMS-Based Pyrotechnical Microthrusters, Sensors and Actuators A 121, 508–514. doi: http://dx.doi.org/10.1016/j.sna.2005.03.017.

[38] Rossi, C., Larangot, B., Pham, P. Q., Briand, D., de Rooij, N.F., Puig-Vidal, M., and Samitier, J. (2006b) Solid Propellant Microthrusters on Silicon: Design, Modeling, Fabrication, and Testing, Journal of Micro Electro Mechanical Systems 15, 1805–1815. doi: http://dx.doi.org/10.1109/JMEMS.2006.880232.

[39] Lee, J., Kim, K., and Kwon, S. (2010) Design, Fabrication, and Testing of MEMS Solid Propellant Thruster Array Chip on Glass Wafer, Sensors and Actuators A 157, 126–134, 2010. doi: http://dx.doi.org/10.1016/j.sna.2009.11.010.

[40] Zhang, K., Chou, S.K., and Ang, S.S. (2004a) MEMS-Based Solid Propellant Microthruster Design, Simulation, Fabrication, and Testing, Journal of Micro Electro Mechanical Systems 13, 165–175. doi: http://dx.doi.org/10.1109/JMEMS.2004.825309.

[41] Zhang, K.L., Chou, S.K., Ang, S.S., and Tang, X.S. (2005a) A MEMS-Based Solid Propellant Microthruster with Au/Ti Igniter, Sensors and Actuators A 133, 113–123. doi: http://dx.doi.org/10.1016/j.sna.2005.04.021.

[42] Zhang, K., Chou, S.K., and Ang, S.S. (2007) Investigation on the Ignition of a MEMS Solid Propellant Microthruster Before Propellant Combustion, Journal of Micromechanics and Microengineering 17, 322–332. doi: http://dx.doi.org/10.1088/0960-1317/17/2/019.

[43] Apperson, S.J., Bezmelnitsyn, A.V., Thiruvengadathan, R., Gangopadhyay, K., and Gangopadhyay, S. (2009) Characterization of Nanothermite Material for Solid-Fuel Microthruster Applications, Journal of Propulsion and Power 25, 1086–1091. doi: http://dx.doi.org/10.2514/1.43206.

[44] Chaalane, A., Rossi, C., and Estève, D. (2007) The Formulation and Testing of New Solid Propellant Mixture (DB + x%BP) for a New MEMS-Based Microthruster, Sensors and Actuators A 138, 161–166. doi: http://dx.doi.org/10.1016/j.sna.2007.04.029.

[45] Tanaka, S., Kondo, K., Habu, H., Itoh, A., Watanabe, M., Hori, K., and Esashi, M. (2008) Test of B/Ti Multilayer Reactive Igniters for a Micro Solid Rocket Array Thruster, Sensors and Actuators A 144, 361–366. doi: http://dx.doi.org/10.1016/j.sna.2008.02.015.

[46] De Groot, W.A., Reed, B.D., and Brenizer, M. (1998) Preliminary Results of Solid Gas Generator Micropropulsion, AIAA 34th Joint propulsion Conference, Reston, VA.

[47] Ali, A.N., Son, S.F., Hiskey, M.A., and Naud, D.L. (2004) Novel High Nitrogen Propellant Use in Solid Fuel Micropropulsion, Journal of Propulsion and Power 20, 120–126. doi: http://dx.doi.org/10.2514/1.9238.

[48] Rodríguez, G.A., Suhard, S., Rossi, C., Estève, D., Fau, P., Sabo-Etienne, S., Mingotaud, A.F., Mauzac, M., and Chaudret, B. (2009) A Microactuator Based on the Decomposition of an Energetic Material for Disposable Lab-on-Chip Applications: Fabrication and Test, Journal of Micromechanics and Microengineering 19, 015006 (8pp).

[49] Hong, C.C., Murugesan, S., Kim, S., Beaucage, G, Choi, J.W., and Ahn, C.H. (2003) A Functional On-Chip Pressure Generator using Solid Chemical Propellant for Disposable Lab-on-a-Chip, Lab on a Chip 3, 281–286. doi: http://dx.doi.org/10.1039/b306116g.

[50] Pi, X., Lin, Y., Wei, K., Liu, H., Wang, G., Zheng, X., Wen, Z., and Li, D. (2010) A Novel Micro-Fabricated Thruster for Drug Release in Remote Controlled Capsule, Sensors and Actuators A 159, 227–232, 2010. doi: http://dx.doi.org/10.1016/j.sna.2010.03.035.

[51] Hitt, D.L., Zakrzwski, C.M., and Thomas, M.A. (2001) MEMS-Based Satellite Micropropulsion via Catalyzed Hydrogen Peroxide Decomposition, Smart Materials and Structures 10, 1163–1175. doi: http://dx.doi.org/10.1088/0964-1726/10/6/305.

[52] Plumlee, D., Steciak, J., and Moll, A. (2007) Development and Simulation of an Embedded Hydrogen Peroxide Catalyst Chamber in Low-Temperature Co-Fired Ceramics, International Journal of Applied Ceramic Technology 4, 406–414, 2007. doi: http://dx.doi.org/10.1111/j.1744-7402.2007.02161.x.

[53] Chen, X., Li, Y., Zhou, Z., and Fan, R. (2003) A Homogeneously Catalyzed Micro-Chemical Thruster, Sensors and Actuators A 108, 149–154. doi: http://dx.doi.org/10.1016/S0924-4247(03)00376-5.

[54] Scharlemann, C., and Tajmar, M. (2007) Development of Propulsion Means for Microsatellites, AIAA-2007-5184, 43rd Joint Propulsion Conference, July, Cincinnati, OH.

[55] Scharlemann, C., Schiebl, M., Marhold, K., Tajmar, M., Miotti, P., Kappenstein, C., Batonneau, Y., Brahmi, R., and Hunter, C. (2006a) Development and Test of a Miniature Hydrogen Peroxide Monopropellant Thruster, AIAA-2006-4550, 42nd Joint Propulsion Conference, July, Sacramento, CA.

[56] Miotti, P., Tajmar, M., Guraya, C., Perennes, F., Marmiroli, B., Soldati, A., Campolo, M., Kappenstein, C., Brahmi, R., and Lang, M. (2004) Bi-propellant Micro-Rocket Engine, 40th Joint Propulsion Conference, July, AIAA Paper 2004-3690, Ft. Lauderdale, FL.

[57] Donadio N.M., Lee, J.G., and Yetter, R.A. (2013) The Performance of a Direct-Flame Solid Oxide Fuel Cell Operating on Various Nitrogen-Based Oxidizers, in preparation.

[58] Mento, C.A., Sung, C.J., and Ibarreta, A.F. (2009) Catalyzed Ignition of Using Methane/Hydrogen Fuel in a Microtube for Microthruster Applications, Journal of Propulsion and Power 25, 1203–1210. doi: http://dx.doi.org/10.2514/1.42592.

[59] Volchko, S.J., Sung, C.J., and Huang, Y. (2006) Catalytic Combustion of Rich Methane/Oxygen Mixtures for Micropropulsion Applications, Journal of Propulsion and Power 22, 684–693. doi: http://dx.doi.org/10.2514/1.19809.

[60] Gamezo, V.N., and Oran, E.S. (2006) Flame Acceleration in Narrow Channels: Applications for Micropropulsion in Low-Gravity Environments, AIAA Journal 44, 329–336. doi: http://dx.doi.org/10.2514/1.16446.

[61] Wu, M.H., and Lu, T.H. (2012) Development of a Chemical Microthruster based on Pulsed Detonation, Journal of Micromechanics and Microengineering, 22, 105040. doi: http://dx.doi.org/10.1088/0960-1317/22/10/105040.

CHAPTER 14

MICRO-ROTARY ENGINE POWER SYSTEM

A. Carlos Fernandez-Pello

This chapter presents a project overview and a summary of research conducted for the Micro-Rotary Engine Power System project at the University of California at Berkeley. The research motivation for the project was to develop a microrotary engine that would benefit from the high specific energy density of hydrocarbon fuels and would be used for small-scale portable power generation. When compared with the energy density of batteries, hydrocarbon fuels may have as much as 50 to 100 times more energy; thus a combustion system such as an engine even with low efficiency is an atractive solution for portable power generation. However, the technical challenge is the conversion of hydrocarbon fuel to electricity in an efficient and clean micro-engine. This was the objective of the project described here, specifically the development of small-scale rotary engine (Wankel type) capable of providing portable power. The rotary engine was selected because of its simplicity and favorable geometry for micro-fabrication. The project had two tasks, one the development of an electrical discharge machining (EDM) fabricated mesoscale rotary engine (10 mm scale) capable of delivering power of the order of 100 W and the other the development of a micro electro-mechanical systems (MEMS) fabricated microscale rotary engine (1 mm scale) capable of delivering around 100 mW power. During the project development, several distinct designs were produced to characterize the effect of engine size, port timing, fuel type, and sealing on power generation. The results of the research conducted during the development of the two scales rotary engines is described and the implications toward the potential implementation of small-scale combustion engines discussed.

14.1 INTRODUCTION

Advances in device miniaturization have led to opportunities in reduced-scale, distributed power generation [1]. Particularly interesting is the production of power from internal combustion engines using liquid hydrocarbon fuels. Liquid hydrocarbons have a very large chemical energy density and consequently are particularly suitable for portable power generation. In comparison with rechargeable batteries, with an energy density of around 0.5 MJ/kg for lithium-ion batteries, hydrocarbon fuels have 45 MJ/kg of stored chemical energy. Therefore, if this chemical energy can be converted to electricity with an internal combustion engine with efficiency as low as even 5 percent, then portable devices could be developed that would replace batteries for a number of applications. This is the primary reason why energy conversion systems using combustion are very attractive for portable, small-scale, power generation. A review of the issues related to combustion device development at the small scale has been previously reported [2,3]. Several of the heat engines and thermochemical approaches in development are addressed in the above two reviews and the other chapters in this book.

Leveraging the inherent advantages in storage and energy density of liquid hydrocarbon fuels and the well-established operation of internal combustion engines to convert that energy into mechanical/electrical power is the prime motivation for the project described here. With this objective in mind, a research project was undertaken at the University of California Berkeley to develop an internal combustion rotary engine with sizes that are well below those commercially available. The rotary engine design of the Wankel type was originally chosen because of its simplicity (Figure 14.1) and specifically for an MEMS-fabricated engine that would benefit from the planar design of the rotary engine [4]. The planar design allows for high precision electrical discharge machining (EDM) and MEMS fabrication and the mechanical shaft output can be directly coupled to an electric motor to produce electrical power. Furthermore, the rotary engine is still well suited for small-scale power generation because it operates on a four-stroke cycle, which increases maximum efficiency and consequently has a high specific power. It also has a low cost due to a minimum number of moving parts because it is naturally aspirated without the use of intake and exhaust valves (it has no valves) and transmission is not required for operation. The power output is in the form of rotary motion of the shaft and has a high specific power relative to other internal combustion engines. The emissions of rotary engines are lower than two-stroke engines due to better intake and exhaust scavenging that prevents contamination of fresh charge with exhaust, although sealing problems may reduce this benefit. Also, rotary engines are inherently capable of fuel flexible operation with glow plug ignition, which is a preferable mode of operation in small-scale engines. This is due to the motion of the rotor whereby the compression stroke is separated from the glow plug location, which prevents premature initiation of the combustion reaction.

Although the ultimate goal of the project was to develop an MEMS-based micro-engine capable of delivering power on the order of milliwatts, realization of the major challenges that the project involved led to the parallel development of a micro-machining-based engine that would be easier to test and develop while still having significant portable power applications. The MEMS task of the project aimed at developing a "micro-rotary engine" with a rotor diameter of 2.4 mm, which was deemed as the largest size that at the time could be fabricated using the available MEMS fabrication techniques. Initial modeling and testing efforts indicated that the micro-engine would suffer from MEMS manufacturing capabilities

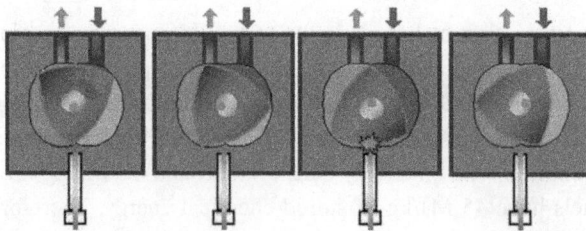

Figure 14.1. Rotary engine operation.

(the device was too big and complicated) together with excessive heat transfer from the reaction zone. Furthermore, in the absence of diagnostics, the larger 2.4 mm engine development proved to be a challenging research effort. The "mini-rotary engine" task of the project aimed at the development of a series of rotary engines with sizes larger than 10 mm rotor diameters, which at the time were the minimum size that could be fabricated using EDM. These engines were fabricated and tested to investigate combustion efficiency, engine operation behavior, and design issues. These series of larger engines were fabricated from hardened stainless steel by EDM. It was expected that these engines would develop up to 100 W power. A test bench for the mini-rotary engine was also developed and experiments were conducted with gaseous-fueled mini-rotary engines to examine the effects of sealing, ignition, design, and thermal management on efficiency. Since there were no available test benches in the small scale required, a custom-based bench had to be developed. Testing was performed initially using gaseous H_2/air mixtures to facilitate ignition and burning, and as the sizes of the engines were increased tests were conducted with a wide variety of liquid fuels (methanol, butane, and gasoline). In addition a range of spark and glow plug designs were tested as the fuel/air mixture ignition source. Iterative design and testing of the mini-engine led to improved sealing designs that helped improve volumetric intake and compression. Exhaust recirculation was also studied to reduce heat losses from the combustion chamber. Several distinct designs were produced to characterize the effect of engine size, port timing, fuel type, and sealing on power generation.

The MEMS engine task proved challenging. Even with the simple design of the Wankel engine, the engine parts were quite complicated for the MEMS fabrication process. Furthermore, the small size effects on combustion forced the design of the engine to be the largest possible, which at the time were maximum thicknesses of the Si wafer of the order of 0.9 mm. This presented significant challenges to produce precise straight surfaces. The results were significant advances in the MEMS fabrication of complex parts, although the project ended before a complete operating micro-engine was produced and tested.

In the sections below the different engine designs and results will be summarized and some of the encountered challenges discussed. For more detailed information, the reader is referred to the different documents published from this project listed in the reference section [4–24]. The project was a significant research effort, with several faculty members, research staff, technical staff, graduate students, and undergraduate students contributing to it. Thus the publications have a variety of authors depending on their particular expertise and contribution.

14.2   MESO-SCALE "MINI-ROTARY" ENGINE

A series of mesoscale "mini-rotary" engines of increasing scales were constructed from steel using electrodischarge machining (EDM) to investigate design and combustion issues related to the small scale of the engine. Four engines were built and tested over the course of this development project to examine the impact of engine design and construction on performance. Engine design characteristics and operating conditions are given in Table 14.1. Photographs of these engines are given below in their respective descriptions. Changes in manufacturing processes, sealing design, porting methods, and combustion pocket design, as well as increases in engine size, led to improvements in overall engine performance. A brief description of the design and testing results with these engines is given in this section.

Table 14.1. Engines design characteristics and operating conditions

Name	P78	P348	S367	S1500
Displacement (mm^3)	78	348	367	1500
Porting	Peripheral	Peripheral	Side	Side
Rotor length (mm)	9.5	12.5	17	25
Rotor depth (mm)	3.6	9	6.3	10
Pocket depth (mm)	0.46	0.67	1.3	1.5
Tolerance (mm)	±0.2	±0.02	±0.003	±0.003
Fuel type	Gaseous	Gas/liquid	Liquid	Liquid
Air induction	Supercharged	Supercharged	Naturally aspirated	Naturally aspirated
Speed range (RPM)	2500–5000	3000–9000	1000–6000	8000–15000
Housing temperature (°C)	25–45	45–75	45–100	90–115
Ignitor location	Minor axis	Minor axis	Off axis	Off axis

Figure 14.2. Test bench for mini-engine.

14.2.1 TEST APPARATUS

An engine test stand was designed and fabricated to test the mini-engine operation. A schematic of the test apparatus is shown in Figure 14.2. The test bench consists of an electric motor/dynamometer, an optical tachometer, and an ignition system. The characteristics for the test bench changed slightly from the first two engines tested to the third and fourth. It is worth noting that no commercial testing product existed at this small scale so one had to be custom designed and built. This also proved to be a significant challenge and development effort.

The fuel supply system varied depending on the type of engine tested. The two peripherally ported engines with the smallest combustion chamber depths were tested with hydrogen as fuel because the high reaction rates of hydrogen allows for combustion in smaller chambers.

Hydrogen and air at elevated pressure were mixed upstream of the engine in a T-junction and individually controlled by mass flow controllers. The two side-ported engines were tested with a liquid fuel blend, which consisted of methanol (70% by vol), nitromethane (10 percent), and castor oil as a lubricant (20 percent). A micro-dispensing valve was used as a fuel injector, and a function generator controlled the fuel flow rate over the range of 10–100 mg/s. The maximum pressure of the fuel injection system was limited to 2 atm. Air was inducted by natural aspiration and mixed with fuel in the intake manifold, which consisted of 3 mm diameter tubing with 6 and 8 mm lengths.

A spark ignition system that utilized a Hall effect sensor was attached to a rotary dial to control the ignition and spark timing. A customized igniter was made from a ceramic tube with a tungsten electrode. In addition to a spark plug, a modified glow plug was used on the P348 engine. The ignition system for these engines was improved by controlling the ignition delay with a function generator. The function generator was actuated with a signal from an optical sensor that detects the position of the crankshaft. To ensure that spark power was consistent at higher operation speeds (limited by capacitor charging time), an automotive ignition system from MSD was implemented for the later engine testing. The S367 engine used the same customized spark plug as the peripheral-ported engines; however, for the S1500 engine a commercial ¼"-32 spark plug was used.

Engine speed for all engines was measured using a Monarch Instruments ACT-3 tachometer with ROS-5W remote optical sensor. The engines were rigidly coupled to the dynamometer via a steel shaft. The engine power for all engines except the S1500 was determined by measuring the electrical power generated from a dynamometer. The electrical power was measured with a Maxon brushless electronically commutated motor with a rectifier. Power generated by the engine spun the motor, which acted as a generator and produced electrical power. The rectifier circuit converted the motor's three-phase output to a DC voltage potential. Rheostats were used to apply a resistance to the motor and could be adjusted to produce the appropriate load, based on the engine being tested. For the S1500 engine, the mechanical power was measured with a 235 mm long torque arm and a load cell that measured up to 2.5N. The torque arm was attached to the motor, and the motor shaft was mounted so that the motor could rotate freely on the shaft axis. Temperature measurements were made on the housing near the combustion chamber using type K thermocouples. These results were verified by infrared imaging. It is noted that at the scale of these engines, the Biot number is estimated at 0.05, and therefore the temperature in the housing is fairly uniform, although thermal stresses was an important issue in the operation of the engines.

Improvements in design and manufacturing techniques produced a steady increase in engine performance over the course of the project. Slight differences in the power measurement technique were used, though in general an electric motor was used to both motor the engine with no combustion and measure the power delivered by the engine while operating with combustion.

A summary of the engines' design characteristics and results relating to combustion and power production for each engine is presented here.

14.2.2 PERIPHERAL PORTED P78ENGINE

The P78 engine was the first engine developed. It had a displacement of 78 mm³ (rotor length 9.5 mm), which is a 1/64th scale version of the smallest commercially available rotary engine, the 5 cc Graupner/O.S. 49-PI. An initial version of the engine (Figure 14.3) did not have apex

Figure 14.3. P78 engine without apex seals.

Figure 14.4. P78 engine with apex seals.

seals, because it was expected that the viscous forces at this small scale would help the sealing. Good sealing at the rotor apex and sides is important to obtain a high compression ratio of the engine and reduce the escape of combustion gases from the combustion chamber. It is worth mentioning that initially it was thought that the problem with the viscous forces would be in the loss of power in the engine, but that this problem would be counteracted by the simplicity of not using apex seals. However initial testing showed that the compression of the fuel/air mixture was very low, that blow by of gases was significant, and that consequently the use of sealing would be necessary. This led to an improved version of the P78 that incorporated apex seals (Figure 14.4). The engine consisted of 15 parts: front plate, epitrochoid housing, back plate, rotor, internal gear, spur gear, shaft and three apex seals, and three leaf springs (Figure 14.5). The apex seals consisted of tabs of brass or steel between 0.5 and 1.0 mm thick. The seals were backed with metal leaf springs to maintain contact between the seals and engine housing. The performance of these seals was a continuous problem because being so small they broke continuously. Two bearings were also added in this design. They were mounted in the front and back plates and positioned the shaft. The engine was made from steel with a tolerance of 0.2 mm. The location of the peripheral ports is the same as the Graupner/O.S. 49-PI, with the intake mixture and exhaust gases entering and exiting the engine through radial passages in the

Figure 14.5. P78 engine components.

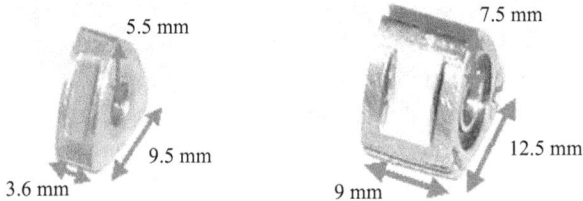

Figure 14.6. Rotors of engines P78 and P348.

engine housing. The combustion pocket is comprised of a simple rectangular recess made on the side of the rotor as seen in Figure 14.6 and the spark plug is located on the minor axis of the engine housing.

The P78 engine was tested with stoichiometric hydrogen–air and propane–air mixtures. The engine did not generate a net power output, but with combustion there was a reduction in the amount of power needed by the electric motor to turn the engine. One problem encountered during the tests was the lack of sensors that would indicate the operating conditions of the engine and would help adjusting engine parameters, such as fuel/air mixture, spark timing, and so forth. Attempts to use the smallest commercially available transducer to measure the pressure in the combustion chamber failed because its dead volume was bigger than the engine combustion chamber causing an unacceptable reduction in compression ratio. To get qualitative information about the combustion characteristics inside the engine, a few tests were run with a transparent polycarbonate front plate so that infrared images of combustion could be obtained.

14.2.3 PERIPHERAL PORTED P348 ENGINE

The low performance of the P78 engine appeared to be caused in part by excessive heat losses to the surrounding environment and in part by poor sealing at the rotor. One obvious solution to reduce the heat loss problem was to increase the width of the combustion chamber, that is, the rotor, to reduce the percentage of the heat losses through the front plate of the engine. Furthermore, it was clear that the engine size was at the scale limit of EDM fabrication and that increasing the size somewhat would help improve the fabrication tolerances and reduce the sealing problem. This led to the development of the P348 engine, with a rotor 2.5 times thicker than the P78 and a slightly longer rotor length (12.5 mm). A photograph of the two rotors is shown in Figure 14.6. The P348 had a displacement of 348 mm, that is, 4.5 times larger than the P78 engine. Heat-treated 17-4 PH900 stainless steel was used to reduce part wear and improve part tolerance from 0.2 mm to 0.02 mm. The same fifteen-part engine design with peripheral porting, rectangular combustion pocket, and minor axis igniter location was used. Two types of face seals were developed for the P348: a one-piece graphite seal backed with a circular wavy spring and three-piece steel springs backed with individual steel leaf springs as shown in Figure 14.7. Although sealing was improved with the face seals, it had no major effect on output power compared to the apex-seal only designs. Therefore, the simple and relatively effective apex-seal only design was eventually selected for the engine. Sealing of the rotor face was improved by having a close fit between the rotor and the end plates.

A typical testing procedure for the P348 consisted of bringing the engine up to a speed of ~3000 RPM through the use of the pressurized air to overcome friction. Then the spark system was energized and hydrogen was introduced until sustained combustion was obtained. With the spark energized, speed increases of 2000 to 4000 RPM were noted and the power increased from one to three watts. By subtracting the amount of power generated with the spark off from the power generated with the spark on, the net power generated by the engine was calculated. The results of power tests over a range of engine speeds are shown in Figure 14.8. Notice the linear rise in power with speed with no drop in power at high speeds, indicating that the residence time is still greater than the chemical reaction time for all engine speeds tested. The equivalence ratio for the tests in Figure 14.8 is in the range of 0.37–0.42. Higher equivalence

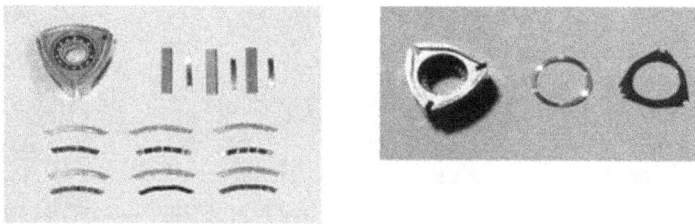

Figure 14.7. Rotors and seals of engine P348.

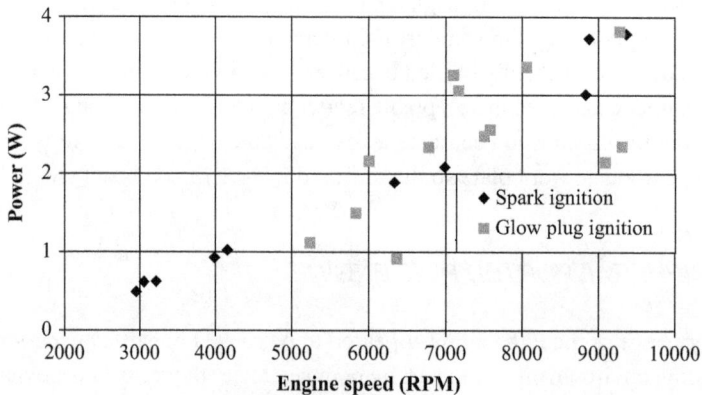

Figure 14.8. Net power output of the P348 engine.

ratios would result in more power, but flashback at the exhaust pipe occurred frequently, preventing the experiment from achieving safe operation at elevated equivalence ratios. The flashback was attributed to excess porting overlap, which allows the flame from the exhaust to enter the intake. The spark timing is zero degrees BTDC for all conditions, with similar results being obtained with a glow plug. Using the values of the chemical energy rate in and the electrical power out, the efficiency can be calculated as the ratio of the electrical power out divided by the chemical energy rate in. For the conditions tested, the efficiency at the maximum output power of 3.8 W was calculated as 0.27 percent.

14.2.4 SIDE PORTED S367 ENGINE

The limitation of not being able to vary the mixture composition in the P348 engine due to flashback problems motivated us to investigate an engine with side porting rather than peripheral porting. Peripheral ports are advantageous at high speeds for maximum power output because of an increase in volumetric efficiency. However, peripheral ported engines have increased rotor port overlap, which occurs when the intake and exhaust ports are open at the same time, contaminating the intake charge with exhaust gases and potentially causing flashback at the exhaust pipe. Side port designs alleviate contamination of the intake charge by exhaust gases, as the ports are designed to prevent any port overlap. This is the design approach followed for the S367 engine. The size of the S367 engine is approximately the same as the P348 with a displacement of 367 mm^3, but with side-ported design instead of peripheral ports. A photograph of the S367 engine is shown in Figure 14.9. The displacement changes are due to slight changes in housing and combustion pocket geometry. The rotor and rotor housing are fabricated with M2 tool steel to improve tolerances to 0.003 mm. The improved tolerances were viewed as necessary to reduce leaking at the rotor apex, which was still viewed as a major cause of low engine efficiency. The side plates were fabricated with aluminum. These end plates maintain assembly tolerance, but do not see direct contact with rotating components. Wear plates that contain the side ports are also made of M2 tool steel and are used between the aluminium end plate and rotor to minimize damage incurred by wear.

The rotors of the S367 engine are thinner compared to the P348 to induce flame front motion in the direction of the applied torque rather than toward the end plates, more consistent with operating rotary engines. It has been shown in the previous studies that the peak pressure

Figure 14.9. Side-ported engines S367 and S1500.

during combustion in closed and semi-closed chambers on the order of 1 mm is reduced as the chamber height and width are reduced. For this reason, a teardrop-shaped rotor recess was used to increase the depth of the combustion chamber to increase combustion efficiency by allowing heat losses to the ignition kernel to be reduced (Figure 14.10). The inverted teardrop shape was designed to also increase turbulence, and in turn the reaction rate and mixing through a "squish flow" effect. Tests on the S367 were conducted with a methanol/nitromethane mixture (50 percent methanol, 30 percent nitromethane, and 20 percent castor oil) as fuel. For the combustion tests, the engine was driven by the electric motor and excess fuel was supplied to the engine with the ignition system on. The fuel rate was reduced until combustion was found to occur and then reduced further until no combustion is detected. The tests did not produce positive power, although combustion and a significant reduction in the motoring power required were observed.

14.2.5 SIDE PORTED S1500 ENGINE

Figure 14.11 is a photograph of the S1500 engine assembled. Analysis of the performance of the engines tested to that point indicated that low efficiency of the engines was a combination of low compression ratios, high heat losses, and poor combustion efficiencies. The low compression ratio was caused primarily by leaking through the apex seals and side walls. The small

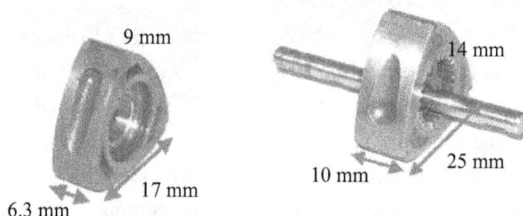

Figure 14.10. Rotors of the S367 and S1500 engines.

Figure 14.11. S1500 engine assembled.

size of the engines made difficult not only their fabrication with very small tolerances, but the seals were very small and fragile reducing their effectiveness and breaking continuously. Thus it was concluded that it was necessary to increase the size of the engines. Obviously a larger size would not only improve relative tolerances but also reduce the heat losses and increase the combustion efficiency. It was decided to fabricate the S1500 engine, which was four times larger than the S367 with a displacement of 1500 mm³. Although larger, the engine was manufactured using the same method and materials as the S367. The same peripheral porting, teardrop combustion pocket, and off minor axis igniter location were used.

The S1500 generated positive power with naturally aspirated, liquid-fuelled operation. A fuel blend with 50 percent methanol, 30 percent nitromethane, and 20 percent castor oil was used with a flow rate of 47 ± 3mg/s. To motor the engine at 10,000 RPM with no spark, 77 W of input power was needed. With the spark on, the engine speed increased to 11,000 RPM and the input power required dropped to 2.6 W. When power to the electric motor was turned off, the engine speed was reduced to 8600 RPM, and the engine continued to operate. To load the engine, a bank of rheostats was put in series with the rectified output of the drive motor. The spark timing was adjusted to find the optimum operating range, and it was observed that the engine operated best with retarded spark timing, ranging from 5°ATDC to 25° ATDC. The maximum sustained mechanical power during these tests was determined to be 33 W (Figure 14.12). The conditions were a speed of 10,500 RPM and 15°ATDC spark timing with 45 ± 3 mg/s of the fuel blend indicated above. Using the positive power generated and estimating a chemical energy of the fuel mixture of 19 MJ/kg, the efficiency was calculated to be 3.9 percent. With an engine mass of 210 g, this results in a peak-specific power of 158 W/kg. Further, assuming a 1 L fuel storage vessel, the resulting specific energy of 157 W-hr/kg is achieved when only the mass of the engine block and fuel are considered. Fuel storage, delivery and control as well as power conditioning and conversion will add significant mass and reduce the system-specific energy.

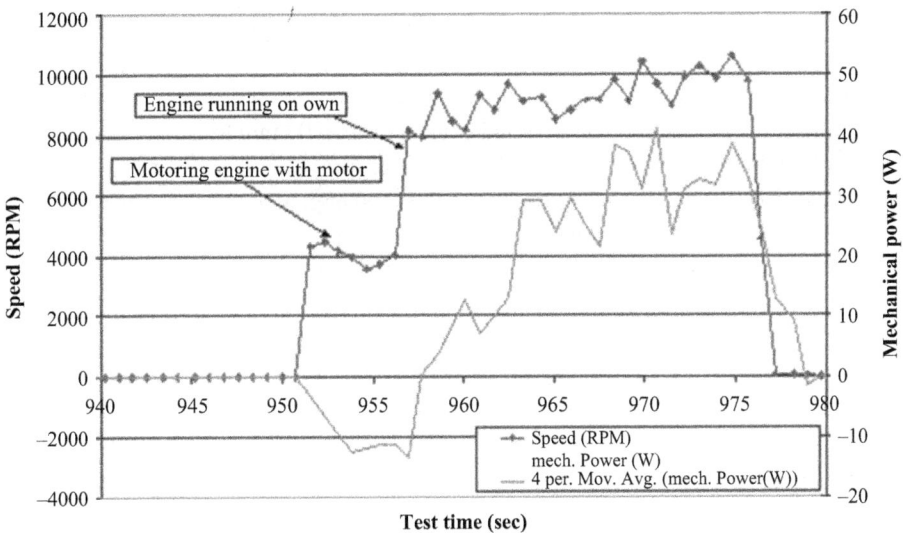

Figure 14.12. S1500 engine test, 15° ATDC spark timing, and 45 percent cycle fuel rate.

14.2.6 DISCUSSION OF RESULTS

The comparison of the performance of the different engines, and in particular the P348 and S1500 engines, shows that there is an increase in power and efficiency between these engines. It is notable that although the displacement is 4.3 times larger in the S1550 engine, the power output is greater by a factor of nine than that of the P348. The improved efficiency is attributed to several factors, including better sealing, no port overlap, deeper combustion pockets, and reduced blowby. As it was mentioned above, one of the problems with the diagnosis of the operation of these small engines is the lack of instruments that would provide the information needed to understand the engine operation, particularly the combustion performance. For these reasons, a number of peripheral tests and analyses were conducted in an attempt to quantify the apparent problems in the engines' operation. Some of these are summarized below, more detailed descriptions can be found in the referred publications for this project.

14.2.6.1 Effect of Engine Seals

Sealing of the rotor at its apex and sides is probably one of the important factors affecting the efficiency of a small rotary engine. In the first place, sealing is critical in maintaining a good compression ratio, which affects the thermodynamic efficiency of the engine. But maybe what is more important is that it affects the combustion efficiency through the increase in the reaction rate with pressure and the loss of unburned fuel through the exhaust. It was clear in the testing that reduction of leakage from the combustion chamber was a major factor in improving the efficiency of the engines.

Improvements in sealing from the P348 engine to the S1500 were accomplished through three mechanisms: reduced gap clearance due to better tolerance, higher spring force on the apex seals, and increased viscosity of fuel. Breakage due to cyclic loading was found to be a problem with early spring designs using brass and 316 stainless steel. The longest reliable performance was achieved with the use of 17-4 stainless steel springs with slight design changes to prevent cyclic failure. The 17-4 stainless steel springs used in the S1500 engine appeared to provide sufficient force to keep the apex seals in contact with the engine housing. The improvement in fabrication tolerance of the rotor and housing from 0.02 mm to 0.003 mm also allowed for a closer fit between the rotor face and the endplates, providing a smaller leakage pathway. The use of a continuous supply of lubricant in the S1500 engine, which filled the gap between the rotor face and endplates, also increased the sealing performance of this design. Still, the measured value for the maximum pressure of the S1500 while being motored at 10,000 RPM was only 6.0 atm, which indicated that still there was a high level of leakage. This and other tests and calculations led us to conclude that there was a necessity for further improvements in sealing in order to increase engine efficiency.

14.2.6.2 Effect of Port Timing

Port overlap occurs when both the intake and exhaust ports are open to the combustion chamber at the same time. This allows combustion products to be drawn into the intake from the exhaust port. These gases decrease the volumetric efficiency since the exhaust gases take the place of a fresh intake charge. As with blowby, the compression efficiency is also reduced. Port overlap particularly affects peripheral ported engines, the peripheral ports on P348 are 3.5 mm and 4.5 mm diameter holes located at ±18° off the minor axis. This results in a large amount of overlap of approximately one quarter of the rotor revolution, which contributes to the low efficiency of the P348 engine. Since the ports on the S1500 are located on the side of the end-plates, the port opening and closing can be adjusted independently. This allows for zero degrees of port overlap since the intake can be opened after the exhaust is closed. These improvements reduced contamination of the fresh charge with exhaust gases and contributed to the increase in fuel efficiency.

14.2.6.3 Effect of Combustion Chamber Design

A portion of the total engine efficiency improvement was the result of improvement of the combustion efficiency. Quenching effects were reduced with the design of the S1500 combustion pocket due to the smaller heat loss to the housing wall. Because the combustion pocket depth in the S1500 is 1.5 mm compared to 0.67 mm for the P348, the deeper pocket increased conduction resistance and consequently reduced heat losses. The increase of thermal resistance reduced heat loss to the engine, and the combustion efficiency was improved. However, the sampling of the exhaust gases indicated that the combustion was incomplete, most likely due to heat losses to the combustion chamber wall. In an attempt to reduce heat losses further, a housing design was fabricated where the exhaust gases were recirculated past the combustion chamber walls. This design was never fully tested because of poor tolerances of that specific engine, and concerns that the pressure losses through the exhaust may jeopardize further the engine efficiency.

14.2.6.4 Effect of Blowby

Blowby occurs when high-pressure exhaust gases travel past the apex seal from the combustion chamber to the following chamber undergoing compression. The high-pressure exhaust gases contaminate the fresh charge with combustion products, leading to a reduction in compression and combustion efficiency. The former is due to the preheating of the incoming gases by the hot exhaust gases that hampers the compression of the mixture in the compression stroke. The later effect is due to contamination of the fresh charge. Although the hot combustion products help ignition by expanding the fuel flammability limits, they reduce the amount of fuel in the combustion chamber reducing the heat release rate. In the case of small-scale engines, with already reduced gas temperatures due to heat losses, the impact is compounded and can lead to reduced power and instability of the combustion event. This blowby generally occurs at the spark plug recess since a direct path is opened from the combustion chamber to the compression chamber when the apex crosses the recess. This problem was minimized in the S1500 by decreasing the size of the spark plug recess using a surface gap spark plug to reduce the area available for blowby. In addition, the spark plug location was set at 15° off the minor axis so that the exhaust port was open at the time the apex seal crossed the spark plug. With the exhaust port open, the pressure in the chamber is low and less exhaust gas will pass into the chamber undergoing compression.

14.3 MEMS-SCALE "MICRO-ROTARY" ENGINE

The final objective the project was the development of a MEMS-based micro rotary engine in the mm scale. The overall concept of the engine system integration is shown graphically in Figure 14.13. Being aware that combustion at the mm scale was difficult to achieve because of quenching issues, and that consequently it was essential to develop as large an MEMS engine as possible, the MEMS component of the project was centered on the fabrication of the individual components of the engine, particularly the housing, rotor, and shaft. It should be noted that at the time this project was initiated, normal MEMS scales were of the order of 0.1 mm for component thickness, thus expanding this scale to the order of mm was essential for the success of the project. Achieving control over the overall dimensions of these components, their quality and their interaction with each other were an important first step toward an operational power-generation system. Furthermore, certain components of the parts, like the apex seals, required complex fabrication, which led us to seek designs for these components that were more amenable to MEMS fabrication. Because of these issues, the initial objective of the project was somewhat changed to that of developing fabrication techniques and approaches that would produce the required engine components. Since the major emphasis in this task was on MEMS fabrication developments, which is beyond the scope of this chapter, only a summary of this task is presented here. The interested reader is referred to the publications indicated in the reference section for further details.

The MEMS scale micro-rotary engine parts were fabricated primarily of silicon (Si), silicon carbide (SiC), and silicon dioxide (SiO_2). All parts subjected to high temperatures and stresses were built or planned to be built either using molded SiC or a Si substrate with a thin SiC coating. Two primary micro-rotary engines were fabricated at the UC Berkeley Microfabrication

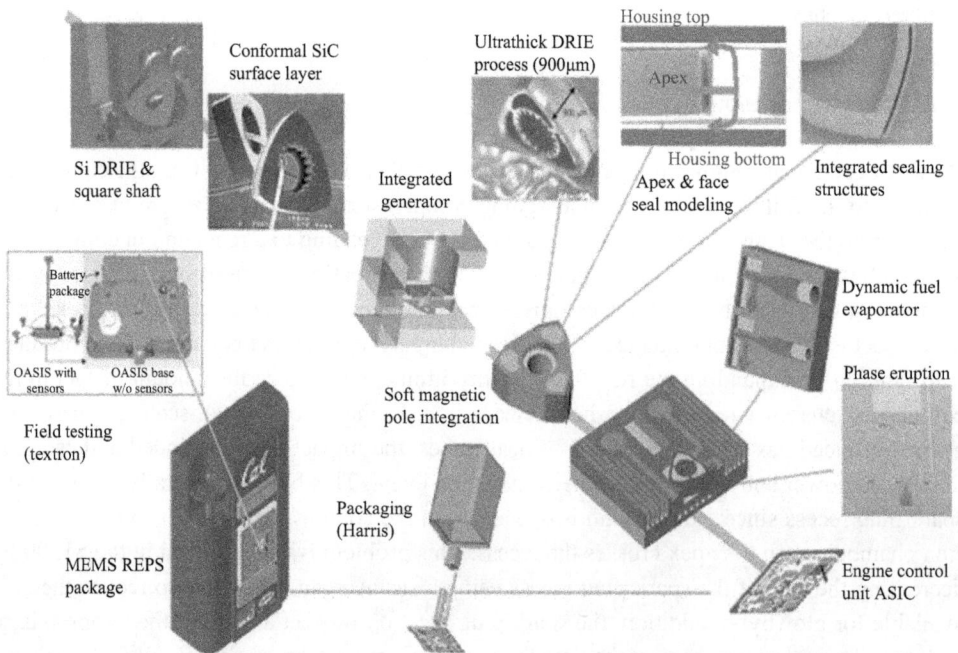

Figure 14.13. Integration concept of the MEMS micro-power generator.

Laboratory: a 0.08 mm^3 displacement engine with a 1-mm rotor diameter, and a 1.2 mm^3 displacement engine with a 2.4-mm rotor diameter.

The 1 mm rotor diameter engine was used to develop the basic MEMS fabrication processes, by investigating what fabrication tolerances were possible in larger-scale MEMS devices. After the fabrication process was proven, the larger 2.4 mm rotor diameter engine was designed and its parts fabricated. The engine fabrication processes consisted of multiple mask, deep reactive ion etching (DRIE) with wafer-to-wafer bonding steps.

The 1 mm engine components consisted of a housing wafer, rotor, and square shaft. A photograph of the engine housing and of the rotor is shown in Figure 14.14. The housing wafer featured inlet/exhaust channels, the epitrochoidal housing bore, and a spur gear. The rotors and housings were coated with a novel low temperature (650°C) SiC coating, which showed excellent structural detail transfer. Initial measurement of the fabricated 1 mm engine pieces indicated that the fabrication process had produced viable engine parts for an initial investigation into sealing and material characteristics. The housing spur gear and rotor through-holes were fabricated using a self-masking method to produce structures 150 μm in diameter and 75–100 μm in height. Issues with the fabrication process included significant lateral etching of the spur gear profile and non-uniform etch rates across the wafer that reduced the number of useable parts fabricated during the manufacturing process. Modifications to the fabrication process and STS etch recipe were implemented to reduce the Si etch uncertainties. The improved fabrication process produced acceptable engine parts including housings and rotors with a 100:1 sidewall straightness, but at the expense of low yield.

The knowhow acquired during the fabrication of the 1 mm rotor engine was used to improve the design and fabrication of the 2.4 mm rotor diameter (Figure 14.15). Some design improvements included an integrated apex sealing system, rotor pockets for soft magnetic material fill, and waste heat channels. Changes to the base fabrication process were introduced to eliminate DRIE over etch and increase overall component tolerance. Several design features that were considered necessary for the operation of the engine were also included, but were in the early stages of development at the end of the project. It was planned to test this first-generation micro-rotary engine by compressed air first to determine engine leakage and wear characteristics. The objective of these tests was to investigate the long-term wear, lubricity, and toughness of the materials. Thermal management approaches were also considered and some of them were implemented in the design and fabrication process of the parts. These included: recirculating the exhaust around the combustion section of the housing and packaging the engine in an aerogel/vacuum container to reduce heat losses, using catalytic surfaces to sustain combustion at low temperatures, and using the exhaust heat to evaporate the liquid fuel.

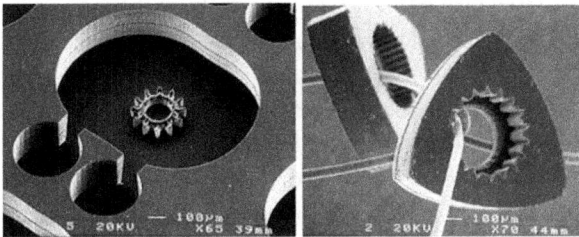

Figure 14.14. 1 mm engine components. Rotor 1 mm diameter and 300 μm thick.

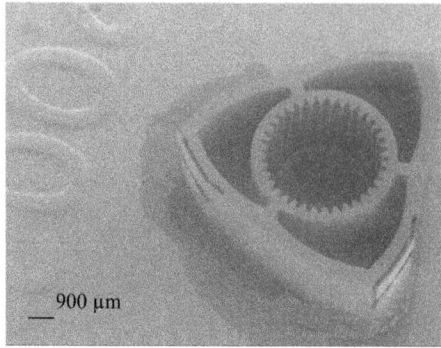

Figure 14.15. 2.4 mm engine rotor with open space for electroplated soft magnetic poles and cantilever apex seals.

14.4 CONCLUSION AND FUTURE WORK

When this project started, micro-power generation was an emerging field in mechanical engineering. As a result, its development encountered a number of unexpected problems that had to be resolved as the project progressed. Although the final goals of the project were not met, we feel that it was not because the goal was impossible to reach, but because the means to reach it were not available in terms of technical support, funding, and time. A research project in an academic institution has a number of other limitations that begin with the fact that it has a teaching component and that the staff working on the project (primarily students) are learning with the project. Obviously, the technical support and facilities are also not comparable to those available in industrial settings. Nevertheless, we feel that the project has helped us understanding a number of fundamental problems related to the development of a small-scale engine. Most of the obstacles encountered during the project were more technological rather than fundamental; we don't think that there are fundamental impediments for micro-combustion in the mm scale although thermal management is essential for its implementation. Although internal combustion engines using hydrocarbon fuels have the attractive of higher energy densities than other devices, the fact that an engine has moving parts presents a significant challenge for the development of a small-scale engine at least at the mm scale. Fabrication tolerances at the meso-scale (10 mm) and fabrication of complex parts at the MEMS-scale (1 mm) is a serious challenge. Also there is a big difference in the tractability of the thermal management problem at each scale. Thermal issues at the 1 mm scale are very difficult to overcome unless we are able to come up with a new material that is able to support very large thermal gradients (C. Cadou pers. comm.) [2]. The lack of diagnostic tools at mm scale is also a major drawback.

The approach that was followed in this project was to improve system performance based on successes and failures of earlier engine testing. For example to reduce the heat losses and to reduce leakage from the combustion chamber the approach was to increase the overall size of the mini-engine. Although the approach worked, it was a quick fix. With more resources, other more fundamental solutions could have been followed, such as exhaust recirculation, improved chamber design and timing, and so forth. The MEMS scale the fabrication challenges were very important at the scale we were working (1 mm). When the author first proposed the project, he thought that maintaining combustion at the micro-scale was the primary problem. However, although combustion at the micro-scale was undoubtedly an important problem, as the project progressed, it became clear that fabrication tolerances of the mini-engine and MEMS fabrication limitations of the micro-engines were also a serious challenge. Maybe time is needed so that better micro-fabrication techniques and new materials capable of maintaining large temperature gradients are developed before a small-scale engine as those attempted in this project can be commercially built.

Although the project did not reach the initially proposed goals, the author feels that the project contributed to the development of the field and to the understanding of the issues and limitations of micro-scale power generation. Furthermore, the project being developed in an academic research institution it had an educational component, which the author believes it was achieved. In this regard the author would also like to quote Professor Christopher Cadou's comment regarding the project (C. Cadou pers. comm.), which I agree with "It is inevitable that certain lessons must be learned before such an ambitious dream can be realized and it is academia's job to ferret them out. A good way to do this is to pursue very ambitious programs like this one."

A final thought worth mentioning is that the approach followed in this project of downsizing conventional engines may not be the best approach for successful micro-power generation. Upsizing biological systems or nano-engineering systems could be a better approach, but of course it too would need thinking "out of the box."

ACKNOWLEDGMENTS

This work was supported by DARPA through grant number DABT 63-98-1-0016 and by research grants from ChevronTexaco and the Industry University Cooperative Research Program (IUCRP) at UC Berkeley. Professor A. P. Pisano and Professor D. Liepmann were primary Co-PI of the project. Dr. D. C. Walther was in charge of the day-to-day operation of the project; his contribution was essential to the development of the project. Graduate students that participated in the project include B.A. Cooley, M. Swanger, A.N. Chen, M. Wasilik, A. Cardes, C. McCoy, F. Martinez, J. Dirner, K. Fu, A. Knobloc, Y. Tsuji, B. Sprague, S-W. Park. Visiting Scholars that also contributed critically to the project were Drs. K. Miyasaka, K. Maruta and L. Inaoka. The project would not have reached the state of development that it did without their contributions. The author would also like to thank Professor. C. Cadou for his helpful comments and suggestions to this chapter. The support to the author's work from the Almy C. Maynard and Agnes Offield Maynard Endowment fund is also acknowledged.

REFERENCES

[1] D. Dunn-Rankin, E. Leal, and D.C. Walther, "Personal Power Systems," Progress in Energy and Combustion Science 31, 422–465, 2005. doi: http://dx.doi.org/10.1016/j.pecs.2005.04.001.

[2] R.B. Peterson, "Size Limits for Regenerative Heat Engines," Microscale Thermophysical Engineering 2, 121–131, 1998. doi: http://dx.doi.org/10.1080/108939598200033.

[3] A.C. Fernandez-Pello, "Micro-scale Power Generation Using Combustion: Issues and Approaches," Proceedings of the Combustion Institute 29, 883–899, 2002. doi: http://dx.doi.org/10.1016/S1540-7489(02)80113-4.

[4] K. Fu, A.J. Knobloch, F.C. Martinez, D.C. Walther, A.C. Fernandez-Pello, A.P. Pisano, D. Liepmann, K. Maruta, and K. Miyasaka, "Design and Experimental Results of Small-Scale Rotary

Engines," Proceedings of 2001 ASME International Mechanical Engineering Congress and Exposition (IMECE), ASME publication IMECE/MEMS-23924, New York, NY, November 11–16, 2001.

[5] B.A. Cooley, D. Walther, and A.C. Fernandez-Pello, "Exploring the Limits of Microscale Combustion," 1999 Fall Technical Meeting, Western States Section/Combustion Institute, Irvine, CA, October 25, 26, 1999.

[6] K. Fu, D. Walther, D. Liepmann, A.C. Fernandez-Pello, and K. Miyasaka, "Preliminary Investigation of a Small-scale Rotary Internal Combustion Engine," 1999 Fall Technical Meeting, Western States Section/Combustion Institute, Irvine, CA, October 25, 26, 1999.

[7] K. Fu, A.J. Knobloch, B.A. Cooley, D.C. Walther, A.C. Fernandez-Pello, D. Liepmann, and K. Miyasaka, "Microscale Combustion Research for Applications to MEMS Rotary IC Engine," Proceedings of the 35th ASME 2001 National Heat Transfer Conference, ASME publication NHTC2001-20089, Anaheim, CA, 2001.

[8] K. Fu, A.J. Knobloch, F.C. Martinez, D.C. Walther, A.C. Fernandez-Pello, A.P. Pisano, and D. Liepmann, "Design and Fabrication of a Silicon-Based MEMS Rotary Engine," Proceedings of 2001 ASME International Mechanical Engineering Congress and Exposition, (IMECE), ASME publication IMECE/MEMS-23925, New York, NY, November 11–16, 2001.

[9] D.C. Walther and A.C. Fernandez-Pello, "Microscale Combustion: Issues and Opportunities," 2001 Technical Meeting, pp. 33–40, ESS/CI, Hilton Head, NC, December 1–5, 2001.

[10] P. Pisano, K. Fu, D. Walther, A. Knobloch, F. Martinez, and A.C. Fernandez-Pello, "MEMS Rotary Engine Power System," Special Issue on Power MEMS of The Sensors and Micromachines Associated Society of the Institute of Electrical Engineers of Japan (T. IEE-J), Vol. 122-B, No. 11, 2002.

[11] F.C. Martinez, N. Chen, M. Wasilik, and A.P. Pisano, "Optimized Ultra-DRIE for the MEMS Rotary Engine Power System," European Micro and Nano Systems Conference, TIMA, 2004.

[12] A.J. Knobloch, M. Waskilik, A.C. Fernandez-Pello, and A.P. Pisano, "Micro, Internal-Combustion Engine Fabrication with 900 micron Deep Features via DRIE," IMECE2003-42558, Proceedings of 2003 International Mechanical Engineering Congress and Exposition (IMECE), Berkeley, CA, November 2003.

[13] M.B.J. Wijesundara., D.C. Walther, C.R. Stoldt, K. Fu, D. Gao, C. Carraro, A.P. Pisano, and R. Maboudian, "Low Temperature CVD SiC Coated Si Microcomponents for Reduced Scale Engines," IMECE2003-42558, Proceedings of 2003 International Mechanical Engineering Congress and Exposition (IMECE), Washington, DC,November15–21, 2003.

[14] B.M. Swanger, D.C. Walther, A.P. Pisano, and A.C. Fernandez-Pello, "Small-Scale Rotary Engine Power System Development Status," 2004 Spring Technical Meeting, Western States Section/ Combustion Institute, Davis, CA, March 29, 30, 2004.

[15] B. Sprague, Y. Tsuji, D.C. Walther, A.P. Pisano, and A.C. Fernandez-Pello, "Observations of Flame Speed and Shape in Small Combustion Chambers," 2004 Spring Technical Meeting, Western States Section/Combustion Institute, Davis, CA, March 29, 30, 2004.

[16] B.E Haendler, J.M. Rheaume, D.C. Walther, and A.P. Pisano, "The Technical Arguments for Micro Engines," PowerMEMS, Berkeley, CA, 2004.

[17] Y. Tsuji, B. Sprague, D. Walther, A. Pisano, and A.C. Fernandez-Pello, "Flame Propagation in 2-D Channels," Asian Conference on Propulsion and Power 2005, Kita-Kyusyu, Japan, January 2005.

[18] Y. Tsuji, S.B. Sprague, D.C. Walther, A.P. Pisano, and A.C. Fernandez-Pello, "Effect of Chamber Width on Flame Characteristics in Small Combustion Chambers," 43rd AIAA Aerospace Science Meeting and Exhibit, publication AIAA-2005-0943, Reno, NV, January 2005.

[19] A. Cardes, C. McCoy, L. Inaoka, D.C. Walther, A.P. Pisano, and C. Fernandez-Pello, "Characterization of Fuel Flexibility in a 4.97 cm^3 Rotary Engine," Proceedings of the Fourth Mediterranean Combustion Symposium, paper X6, Lisbon, Portugal, October 2005.

[20] S.-W. Park, D.C. Walther, A.P. Pisano, and A.C. Fernandez-Pello, "Development of Liquid Fuel Injection System for Small Scale Rotary Engines," 44th AIAA Aerospace Science Meeting and Exhibit, publication AIAA-2006-1345,Reno, NV, January 2006.

[21] S.W. Park, S.B. Sprague, D.C. Walther, A.P. Pisano, and A.C. Fernandez-Pello, "Improved Power Generation of a Small-Scale, Naturally Aspirated and Liquid Fuel Injected Rotary Engine," Power MEMS 2007, Freiburg, Germany, November 28–29, 2007.

[22] S.B. Sprague, and C. Fernandez-Pello, "Development of Small-Scale Internal Combustion Rotary Engines," Seventh International Symposium on Advanced Fluid Information (AFI/TFI 2007), Sendai, Japan, December 14–15, 2007.

[23] S.B. Sprague, A. Cardes, D.C. Walther, S.-W. Park, A.P. Pisano, and A.C. Fernandez-Pello, "Effect of Leakage on Optimal Compression Ratio for Small-Scale Rotary Engines," 45th AIAA Aerospace Science Meeting and Exhibit, AIAA publication AIAA-2007-0578, Reno, NV, January 2007.

[24] S.B. Sprague, and A.C. Fernandez-Pello, "Characterization of Small-Scale Internal Combustion Rotary Engines," 4th International Symposium on Innovative Aerial/Space Flyer Systems, Tokyo, Japan, January 14–15, 2008.

CHAPTER 15

SMALL-SCALE RECIPROCATING ENGINES

Shyam Menon and Christopher Cadou

Reciprocating internal combustion (IC) engines are a very well-established technology for converting the chemical potential energy stored in liquid or gaseous fuels into useful mechanical work. While they come in a wide range of sizes, configurations, and complexity levels, and can implement several different thermodynamic cycles, all use pistons sliding in cylinders to extract p-dV work from an exothermically reacting fuel–air mixture. Most also use a connecting rod and crank mechanism to convert the piston's reciprocating motion into rotary motion of a shaft. Mechanical power can be extracted directly from the shaft or electrical power can be produced by attaching a generator.

Reciprocating IC engines offer a number of important advantages over other energy conversion technologies: they are able to consume a wide variety of energy dense liquid hydrocarbon fuels, achieve reasonable levels of overall thermodynamic efficiency (>20 percent except at the smallest scales), achieve relatively high power densities (>2 kW/kg), and have excellent transient response. They are a mature and thus rather reliable technology that can be built at relatively low cost. Compared to gas turbines, IC engines offer lower power densities but better transient response and significantly lower cost per kW. Fuel cells offer direct fuel to electric power conversion and significantly higher efficiencies at the stack level (35 to 55 percent), but parasitic losses associated with the balance of plant (pumps, thermal management, power management, fuel reforming, de-sulfurization, etc.) usually erode all or most of this advantage. The balance of plant components also drive up fuel cell systems' mass and volume generally leading to significantly lower power densities.[1] Finally, fuel cells' higher mechanical complexity also makes them more expensive per kW and results in poorer transient response.

Figure 1.5 that a commercially available glow fueled piston engine comes closer to meeting current Department of Defense (DOD) performance targets than any other sub-kilogram technology. The fact that it is an "off-the-shelf" engine that has not been optimized for efficiency suggests that even a modest investment in research and development could move performance

---

[1] This is not to say that balance of plant issues are unimportant in IC engine-based energy conversion systems: All engines require fuel tanks, pumps, controls, and packaging—all of which become increasingly significant fractions of overall system weight/volume as scale is reduced. However, the problem is much worse with fuel cell systems because many more components are usually required.

Figure 15.1. Photographs of miniature piston engines: (a) Evolution EVOE 26 GX engine (79), (b) AP Yellow jacket engine, and (c) PAW 049 engine.

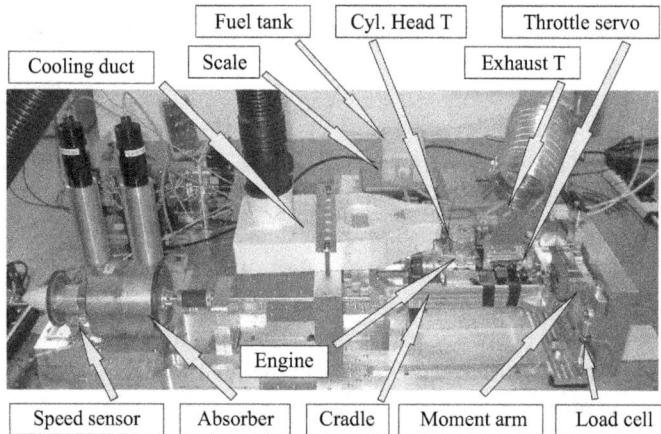

Figure 15.2. Photograph of the small engine dynamometer system at the University of Maryland with geared transmission.

into the DOD target range. Therefore, miniature IC engines represent a particularly promising solution to the microscale power-generation problem.

The objectives of this chapter are to develop a fundamental understanding of the current state of the art in miniature engine performance, how it scales with size and why, and to identify the research steps necessary to improve performance. This will be accomplished by briefly reviewing some of the research efforts focused on building more efficient miniature reciprocating engines before moving on to a somewhat more detailed discussion of the performance of the smallest reciprocating IC engines available commercially today.

15.1 SURVEY OF MINIATURE RESEARCH ENGINES

There have been a number of efforts to build kilogram and sub-kilogram scale energy conversion systems based on various types of reciprocating IC engines. Many of these are oscillating free piston designs with integral electrical generators that are intended as direct battery replacements. One of the earliest attempts was a collaborative effort between Honeywell and the University of Minnesota. The original idea was to microfabricate a planar "knock" engine with a 1 mm bore and 1–4 mm stroke based on a concept by Yang et al. [1] and integrate it with a linear alternator. However, difficulties with microfabrication ultimately led to the adoption of a somewhat larger 3 mm cylindrical design manufactured using conventional techniques. Piston-induced homogeneous charge compression ignition (HCCI) was achieved without external heating with strokes in the 3–4 mm range [2] but heat transfer imposed time and length-scale limitations on miniaturization [3, 4]. A liquid-piston engine was proposed as a way to reduce blow-by losses without incurring unacceptable friction penalties. A new approach using the linear engine to power a compressor device for orthotic applications (weighing 235 g and at a much larger scale of about 4 inches) has yielded a power output of about 10 W but with an overall efficiency of only 0.6 percent [5].

A similar effort at Georgia Tech focused on building a free piston engine/generator with a planar geometry that is suitable for microfabrication [6]. It is a two-stroke engine that has a spring-loaded piston, is spark ignited, and operates on a mixture of propane and air. A 58.8 mm wide × 6.35 mm piston with a 42.4 mm stroke oscillated at 31 Hz and produced 15 W of electrical power. Leakage around the piston was a major development challenge.

A smaller-scale oscillating free piston device with a rectangular piston (1 mm × 2 mm) was fabricated by Lee et al. [7] using photosensitive glass and pyrex instead of silicon. Glass was chosen for its dielectric properties that enabled the spark electrodes to be embedded directly through the structure. A combustion study was performed first to determine an appropriate combustor volume and then a prototype was designed and fabricated using standard reactive ion etching and wafer bonding techniques. Ignition of an H_2–air premixture in a combustion chamber volume of a few mm^3 was achieved and was able to drive a piston at engine-appropriate speeds. However, a fully working engine was never realized. Again, the leakage of hot gases around the piston was found to be a critical issue.

The most commercially advanced linear free piston engine with integrated alternator is Aerodyne Inc.'s linearly oscillating miniature internal combustion engine (MICE). It operates in a glow-plug-assisted HCCI mode on propane and Jet-A. A 10 W version operating on propane weighs 15 g and produces 10 W at an overall chemical to electrical conversion efficiency of ~7.5 percent [8]. A larger 3.1 kg/185 W device achieved 15 percent overall efficiency while operating on Jet-A [9]. The most recent version of the device produces 250–300 W of electrical power on Jet-A with an overall efficiency of 24 percent and weighs 2.6 kg. The principal challenges at the larger 300 W scale are low scavenging and fuel atomization. Additional challenges at the smaller scale include piston leakage and lower than expected combustion efficiency.

The rotational analog of the linear free piston engine is the micro-internal combustion swing engine (MICSE) developed at the University of Michigan. A two-bladed swing arm serves as a piston which is driven back and forth by alternate combustion events in four chambers that form between the piston blades and the base structure. The swing arm's shaft drives a generator to produce electric power. A 65 W bench-scale version operating on butane achieved an overall thermal efficiency of 21 percent [10]. A 20 W version weighing about 170 g is in development.

D-star engineering takes a somewhat more conventional approach that uses a small two-stroke diesel piston engine to drive a permanent magnet starter/generator. The device produces about 40 W of electrical power with an overall efficiency of ~10 percent [11]. While the prime mover looks like a "standard" model aircraft engine, it is not. It operates on JP-8 without fouling, uses ceramic materials, and operates at a lower compression ratio to reduce weight. The advantage of this engine is that it can be disconnected from the generator and used to drive mechanical loads (like propellers) directly. The smallest proven engine has a displacement of 0.82 cm^3 and operates on a mixture of JP-8 and two-stroke oil (3 percent). It produces 90 W with a SFC of 0.75 kg/kW-hr or efficiency of 11 percent [12].

Gomez et al. developed a meso-scale free piston Stirling engine equipped with a multiplexed electrospray capable of dispersing heavy fuels like JP-8 [13]. A recuperator included in the design improved system efficiency. The combustor coupled with a free-piston Stirling engine could produce up to 42.5 W of power with an overall efficiency of 22 percent. A fully packaged system delivered 34.5 W net electric power with 18 percent overall efficiency and weighed 2100 g. Challenges associated with making the technology sufficiently affordable have prevented its commercialization to date.

Finally, a number of studies have focused on miniaturizing Stirling engines for use in actuators and heat pumps. Nakajima et al. [14] at the University of Tokyo studied the scaling of a Stirling engine and developed a centimeter-scale engine that produced mechanical power in the range of 10 mW with the a hot side temperature of 373 K. White [15] at the University of Washington has developed a 0.71 cc engine producing 5 W of power for artificial heart applications.

15.2 MINIATURE COMMERCIAL ENGINES

In contrast to the handful of miniature research engines, there are hundreds of sub-kilogram scale engines that are produced to power model aircraft and cars. They come in displacements as low as 0.3 cm^3 and consume a variety of fuels ranging from "glow fuel," to "model diesel," to gasoline. Some examples are provided in Figure 15.1. While such engines are not optimized for power density and efficiency nor do they generally consume militarily appropriate fuels, one can learn a lot about the influence of scale on engine performance by studying them.

We will use detailed performance measurements of nine such small engines to investigate engine performance scaling. All are two-stroke, glow fueled, loop-scavenged designs. The measurements were made on the small engine dynamometer illustrated in Figure 15.2. Detailed descriptions of the dynamometer and how these data were acquired are available elsewhere [16, 17]. The basic parameters of these engines are presented in Table 15.1.

Table 15.1. Specifications of the miniature engines considered in this study.

	Make/model	Displacement		Mass	Speed range	Bore	Stroke
		in^3	cm^3	(g)	(kRPM)		(mm)
A	OS46FX	0.46	7.54	488	2.5–17	22.0	19.6
B	OS40FX	0.4	6.55	386	2–17	20.5	19.6
C	OS25FX	0.25	4.10	248	2.5–19	18.0	16.0
D	AP Yellow jacket	0.15	2.46	150	3.5–18.0	15.5	12.0
E	AP Hornet	0.09	1.47	128	4–18	12.5	12.0
F	AP Wasp	0.06	1.00	54	3–25	11.3	10.0
G	Cox 049	0.05	0.80	42	10–20+	10.2	9.6
H	Cox 020	0.02	0.33	25	10–20+	7.5	7.5
I	Cox 010	0.01	0.16	15	10–20+	6.0	5.7

15.3 QUANTIFYING ENGINE PERFORMANCE

The principal performance metrics of an IC engine are power output and overall efficiency. Power is the product of torque and rotational speed in radians/s. It is given by:

$$P = \frac{2\pi N}{60} \Gamma \tag{15.1}$$

where N is rotational speed in revolutions per minute. Overall efficiency is the ratio of the useful power output to the rate of chemical energy input [18, p. 84]:

$$\eta_o = \frac{P}{\dot{m}_f Q_r} \tag{15.2}$$

In this expression $\dot{m}_f$ is the fuel mass flow rate and Q_R is its lower heating value. Efficiency is also reported in terms of the brake-specific fuel consumption (BSFC), which is given by [18, p. 209]:

$$BSFC = \frac{\dot{m}_f}{P} \tag{15.3}$$

As a "normalized" performance parameter, BSFC (usually reported in kg/kW-hr or lb/Hp-hr) is nominally independent of size and therefore useful for comparing the performance of engines of different displacements. There are a variety of other normalized parameters that are also useful for comparing engines.

In naturally aspirated engines, brake mean effective pressure (BMEP) measures how well an engine is able to ingest air and react it with fuel to do useful work.[2] It is given by:

$$BMEP = \frac{P n_R}{V_d N} \tag{15.4}$$

where n_R=1 for two-stroke engines and 2 for four-stroke engines. "Power per piston area" (P_{parea}) measures how well the engine designer has been able to use the available piston area to extract work and "normalized power" (P_{norm}) is useful in understanding how engine performance scales with size [19]

$$P_{parea} = \frac{P}{A_p} \tag{15.5}$$

$$P_{norm} = \frac{P}{S_{mean}} \tag{15.6}$$

Mean piston speed (S_{mean}) is related to the design's ability to handle inertial and frictional loads. It is a function of stroke length (L) and speed (N):

$$S_{mean} = 2L(2\pi N / 60) \tag{15.7}$$

---

[2] In turbocharged and supercharged engines, BMEP reflects the engine's ability to withstand higher gas temperatures and pressures.

Power density measures how much power the engine produces per kilogram of engine mass:

$$\rho_{pwr} = \frac{P}{m_e} \tag{15.8}$$

The fuel–air ratio (F/A) plays an important role in determining engine performance and overall efficiency:

$$F/A = \frac{\dot{m}_f}{\dot{m}_a} \tag{15.9}$$

The equivalence ratio (ϕ) is the ratio of the actual fuel–air mixture ratio to the fuel–air ratio associated with stoichiometric combustion:

$$\phi = \frac{F/A}{(F/A)_{stoichiometric}} \tag{15.10}$$

15.3.1 QUANTIFYING LOSSES

Quantifying losses begin with a control volume analysis. Since good models for the transfer of charge from the crankcase to the cylinder volume (scavenging) and for charge leakage around the piston are not available, it is most convenient to draw the control volume around the entire engine as shown in Figure 15.3. The various quantities shown in the figure are explained next.

The "delivery ratio" (DR) describes the ability of a two-stroke scavenged engine to transfer fresh charge to the cylinder [18, p. 237]. It is defined as [18, p. 237]:

$$DR = \frac{\dot{m}_a}{\dot{m}_{a,i}} \tag{15.11}$$

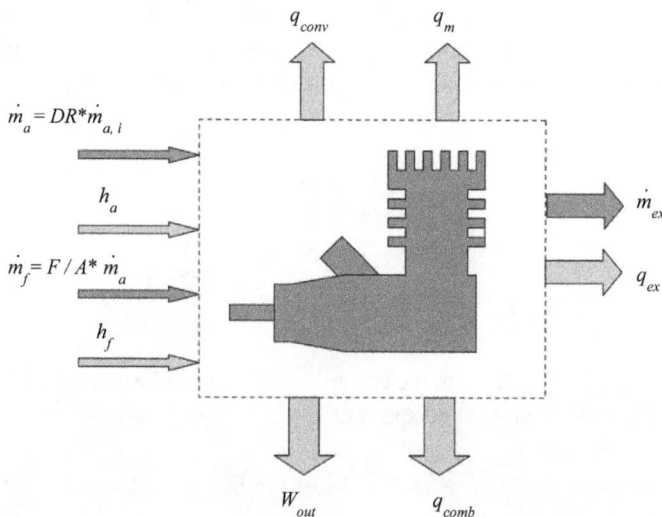

Figure 15.3. Energy balance in a two stroke engine with control volume drawn around the entire engine.

where the numerator is the actual air flow and the denominator is the ideal air flow associated with complete filling of the cylinder with fresh charge at ambient conditions. Similar to volumetric efficiency in a four-stroke engine, the major factors influencing DR are engine speed, charge heating, choking of the intake passages, and fluid friction through the engine flow path [20]—all of which become more important as scale is reduced. Therefore, on this basis alone one should expect miniature engines to burn less fuel and make less power per unit cylinder volume than conventional-scale engines. The mass of air entering the engine at any operating speed is given by

$$\dot{m}_a = DR\,\rho_a V_d \frac{N}{60} \tag{15.12}$$

Combining Equations 15.9 and 15.12 gives the fuel mass flow rate:

$$\dot{m}_f = DR\,\rho_a V_d \frac{N}{60}\left(\frac{F}{A}\right) \tag{15.13}$$

The total rate of enthalpy addition from the fuel–air mixture entering the engine is given by

$$q_{IN} = \dot{m}_a h_a + \dot{m}_f h_f \tag{15.14}$$

Since the enthalpy of the incoming air is negligible compared to the heat released by the fuel, the total rate of enthalpy addition from the fuel–air mixture is well approximated by

$$q_{A\,vailable} = DR\,\rho_a V_d \frac{N}{60}\left(\frac{F}{A}\right)Q_R \tag{15.15}$$

This has already been mentioned. Equation 15.15 gives the total amount of energy available for release inside the control volume. Part of this energy is converted into useful work at the engine shaft while the rest is lost to the environment through various processes including the discharge of unburned fuel.

The overall energy balance on the control volume is given by

$$q_{A\,vailable} = \dot{W}_{out} + q_{l,tot} \tag{15.16}$$

where W_{out} is the engine shaft power and $q_{l,tot}$ is the sum of thermal, mechanical, and combustion losses:

$$q_{l,tot} = q_{th} + q_m + q_{comb} \tag{15.17}$$

The thermal loss includes convective heat transfer to the walls and sensible enthalpy lost in the exhaust:

$$q_{th} = q_{comv} + q_{ex} \tag{15.18}$$

Frictional losses (associated with fluid flow and rubbing at the piston–cylinder interface and bearings) account for ~24 percent of energy released from the combustible mixture in conventional–scale IC engines [21]. The frictional power loss is the product of the torque required to overcome friction (Γ_F not the engine torque) and speed:

$$q_m = \frac{2\pi N}{60} \Gamma_F \qquad (15.19)$$

Since thermal and mechanical losses' relative importance increase with increasing surface-to-volume ratio, they are also expected to increase as engine size is reduced. The overall conversion efficiency of the engine is the shaft work output divided by the total energy input:

$$\eta_o = \frac{\dot{W}_{out}}{q_{Available}} = \frac{q_{Available} - q_{l,tot}}{q_{Available}} \qquad (15.20)$$

Combining Equations 15.15 and 15.20 give the following expression for the engine power output:

$$\dot{W}_{out} = \eta_o DR \rho_a V_d \frac{N}{60} \left(\frac{F}{A}\right) Q_R \qquad (15.21)$$

The overall conversion efficiency defined earlier can be expressed as the product of component efficiencies corresponding to the different loss mechanisms:

$$\eta_o = \eta_{th} \eta_m \eta_{comb} \qquad (15.22)$$

Inserting into 15.21 gives:

$$\dot{W}_{out} = \eta_{th} \eta_m \eta_{comb} DR \rho_a V_d \frac{N}{60} \left(\frac{F}{A}\right) Q_R \qquad (15.23)$$

where the component efficiencies are given by

$$\eta_{comb} = \frac{q_{Available} - q_{comb}}{q_{Available}} \qquad (15.24)$$

$$\eta_{th} = \frac{q_{Available} - q_{comb} - q_{th}}{q_{Available} - q_{comb}} \qquad (15.25)$$

$$\eta_{th} = \frac{q_{Available} - q_{comb} - q_{th} - q_m}{q_{Available} - q_{comb} - q_{th}} \qquad (15.26)$$

The first four terms in Equation 15.23 represent the various loss processes occurring in the engine. These times the product of the remaining terms (which is the rate of chemical energy addition through the fuel) give the total power output of the engine. The major losses are associated with incomplete filling of the cylinder with each stroke (DR), energy loss via heat transfer and sensible enthalpy in the exhaust (η_{th}), energy loss due to incomplete combustion (η_{comb}), and energy loss due to fluid and mechanical friction (η_m). Again, one should expect all of these efficiencies to decrease with decreasing scale.

Finally, engine performance also varies with the number of oxygen molecules per unit volume in the atmosphere and hence the local pressure, temperature, and humidity. As a result, engine performance is usually corrected to standard atmospheric conditions so that measurements made on different days may be compared. The methods for accomplishing this are well known [22] and will not be repeated here. All comparisons made in this chapter are based on corrected power.

15.4   PERFORMANCE MEASUREMENTS

Measuring performance in miniature engines is difficult because of the high operating speeds, low torques, small physical size, and large 1/rev disturbances associated with single cylinder engines. The pulsatile nature of the flow entering the engine makes it difficult to make good air flow measurements and it is even harder to measure the fuel flow rate as there are very few commercial instruments in the proper range. Therefore, it is worth saying a few words about how basic performance measurements are made.

Instruments for measuring engine performance are called "dynamometers" and consist of five basic components: a physical structure that supports the engine and other components, an "absorber" that is attached to the engine (sometimes via a transmission) that applies a load to the engine, a torque-measuring system, a control system that maintains engine speed either by controlling the load or throttle position, and an additional sensor system that measures whatever other parameters are of interest. While the first four components exist in some form in virtually all engine testing facilities, the additional sensor systems can vary widely.

15.4.1   ABSORBERS

There are many different types of absorbers. The simplest is a propeller. It has the advantage of being inexpensive, attaching directly to the engine with no need for a transmission, and supplying cooling air to the engine. The principal disadvantage is that the relationship between speed and power consumption is fixed. This means that a large number of experiments with different propellers are needed to measure the full operating map of an engine. Fixed pitch propellers have been used in many investigations of miniature engine performance [23–30].

Examples of controllable absorbers are water brakes, electric motors, eddy current brakes, and hysteresis brakes. Permanent magnet electric motors run as generators have been used successfully in several miniature engine testing programs [31–35]. Eddy current brakes are more expensive but offer the advantage of being able to produce the same braking torque at any speed. They are very common in conventional-scale dynamometers and have also been used successfully with small engines [16, 36, 37, p. 65].

15.4.2   TORQUE

Most conventional-scale dynamometers measure torque at the absorber and a number of small engine investigations take this approach [32, 35, 36, 38]. A disadvantage of this scheme is that losses in the drive/transmission influence the torque measurement. While this may be a relatively small problem at larger scales, it can be very important with small engines that run at high speed and produce very small torques. As a result, a number of innovative techniques have been developed for small engines.

Gierke [25] mounted the engine on a bearing with a moment arm held in place by a spring scale. The scale reading times the moment arm length gave the torque. Raine et al. [31] used a pendulum-based torque sensor where the engine mount was also free to rotate about the crankshaft but the center of mass lay below the axis of rotation so that gravity provided a constant restoring force. Torque was proportional to the deflection angle in this system. The commercial miniature dynamometer used by Mengitsu [39] supports the engine in a cradle that is free to rotate about the engine's crankshaft and that is restrained by a load cell. This is also the approach taken by Menon [16]. The challenge in these systems is controlling the dynamic behavior of the cradle, which acts with the load cell as a lightly damped spring-mass oscillator. Great care needs to be taken to ensure that the system stays out of resonance. Finally, a torque balance can be avoided entirely if the absorber is a permanent magnet DC motor where the torque is a known function of the armature current [40]. In this case, engine torque is determined by applying an electrical load to the generator, measuring the armature current, and multiplying by the effective gear ratio between the engine and the absorber. This is the method used by Papac et al. [33].

15.4.3   SPEED

Speed is a relatively easy measurement. Strobe-based tachometers [25] and Hall Effect sensors using magnetic "pulser" discs [16] or single magnetic elements attached to the shaft have all been used successfully. Optical encoders producing a pulsed output corresponding to the optical pattern on a disc coupled to the engine shaft have been used [35, 36, 41, 42]. The advantage of the optical sensor is that it also provides crank angle information, which can be used to estimate displacement volume corresponding to cylinder pressure measurements. Moulton [38] and Raine [43] used optical interrupter switches that detect the passage of notches on a disk mounted on the engine shaft.

15.4.4   FUEL FLOW RATE

The fuel flow rate is a challenging measurement because rates are approximately a few cm^3/min, which is a range served by few commercial instruments. In addition, even small pressure losses associated with in-line instruments can impede engine operation. Therefore, some innovation is also required here. Gierke [26] and Roberts et al. [32] used calibrated burettes and a stopwatch. Schauer et al. [36] used a Max Machinery piston flow meter with a range of 1–1800 cc/min in their studies. Mengitsu [39] used a turbine flow meter capable of measuring flow rates as small as 11 cc/min. We have used a variety of methods including a nutating flow meter with a range of 1–250 cc/min made by DEA Engineering and a gravimetric method where the weight of the fuel in the fuel tank is recorded as a function of time [37, p. 80]. The latter usually works best and is the only one suitable for use with the smallest engines. Similar techniques were used by Moulton [38], Pompa [35], and Papac [33].

15.4.5   AIR FLOW RATE

Unsteadiness caused by the opening and closing of valves in the engine can cause rotameters and other conventional instruments to over-predict air flow. The problem was noted by Roberts [32] and has reappeared in other studies. It can be corrected by placing a large plenum equipped with screens between the engine and the flow meter, although care must be taken to ensure that the pressure loss through the system is relatively small. This approach has been followed by Pompa [35] and Menon [37, pp. 188–194].

15.4.6 CYLINDER PRESSURE

The principal challenge is finding a sensor that is small enough to be installed in the cylinder without significantly altering the clearance volume. Schauer et al. [36] used a PCB Model 112B1 with a spark plug mount (Model 65A). This was possible because the engine they studied was not that small (35 cc). Menon [37, p. 226] used an Optrand model M3 × 0.5 fiber-optic pressure transducer installed in the cylinder head. This is the smallest commercially available pressure transducer that is easily installable. A similar transducer of slightly larger diameter was used by Papac [33] and Pompa [35]. Manente [44] has utilized a Kistler Type 6052 C piezoresistive pressure transducer with a charge amplifier to measure cylinder pressure in a 4 cc glow ignition engine. A similar sensor made by Endevco (Model 8541-100) has been used by Disseau [45] to measure combustion pressure in a free-piston engine.

15.4.7 TRANSMISSIONS

While not a measurement system, determining which combination of transmission, couplings, and absorber to use with which engine is a critical aspect of dynamometer testing. Unfortunately, it is largely a trial and error process (even at the conventional scale) and is one of the most challenging aspects of making reliable measurements in miniature engines. The task is much simpler if it can be done without regard for its influence on the torque measurement system so we favor measuring torque on the engine side of the transmission. We have found that smaller engines (< 1 cm^3 displacement) work best in direct drive arrangements with small hysteresis brakes. Geared arrangements have worked well for intermediate size engines (4 cm$^3 <$ displacement < 10 cm^3) and belt drive arrangements are better for certain larger-scale engines.

15.4.8 MEASURING COMPONENT EFFICIENCIES

Only brief overviews of the methods used to measure each component efficiency are provided here. More extensive discussions are available elsewhere [37, p. 157], [46].

15.4.8.1 Intake Losses (DR)

DR (Equation 15.11) is computed directly from measurements of air flow at each operating condition. The principal challenge, as described earlier, is reducing intake air velocity fluctuations.

15.4.8.2 Mechanical Efficiency

The torque required to overcome mechanical friction is determined by "motoring" the engine that is driving it with an electric motor w/o combustion and measuring the torque at the engine shaft. Speed is swept through a range appropriate for the engine and fuel is provided in order to maintain proper lubrication. Menon [46], Shin [47], and Roberts [32] have used motoring techniques to measure frictional power loss in miniature engines.

15.4.8.3 Thermal Efficiency

There are two main contributors to thermal loss: heat transfer from the hot combustion products to the cylinder walls (q_{conv}) and flow of hot combustion products out the exhaust port (q_{ex}). Determining the latter is relatively straightforward—it is computed directly from measurements of the fuel flow rate, air flow rate, and exhaust gas temperature:

$$q_{ex} = (\dot{m}_a + \dot{m}_f)C_{p,ex}T_{ex} \tag{15.27}$$

$C_{p,ex}$ depends on the composition and temperature of the exhaust and so is determined from equilibrium calculations performed on a series of mixtures whose compositions ($1 < \phi < 3$; burned and unburned) and temperatures (300 K $< T < 800$ K)[3] span the range of conditions expected in the engines. We have done this using CANTERA [48] for a 70 percent methanol, 10 percent nitromethane, and 20 percent oil mixture (by volume). The oil is treated as a non-reacting species.

Computing the heat loss to the cylinder (q_{conv}) is much more challenging. Many different approaches have been developed for modeling heat transfer from the reacting gases to the inner cylinder wall [49–52], but the problem with all is that they require detailed in-situ measurements that are already difficult to perform in conventional-scale engines and would be extremely difficult in miniature engines. Therefore, the approach we have followed is to use the concept of an "apparent" mean gas temperature and convection process where conductive, convective, and radiative processes are lumped together [18, p. 274]. The advantage of this approach is that it can be implemented directly on the dynamometer without the need for detailed in-situ measurements of the gas–wall interaction. The net result is that the convective heat transfer is given by

$$q_{conv} = 10.4\frac{k_g}{b}\left(\frac{Gb}{\mu}\right)_g^{0.75} A_p(\bar{T}_g - \bar{T}_c) \tag{15.28}$$

where $\bar{T}_g$ is the "mean" hot gas temperature representing the average temperature within the cylinder volume over the course of an engine cycle, $\bar{T}_c$ is the mean temperature of the cooling air, and G is the mass flow rate through the engine per piston area. The mean gas temperature concept, how it is measured, and details of the actual implementation of this method are described in detail elsewhere [46].

15.4.8.4 Combustion Efficiency

"Combustion efficiency" is defined as the fraction of chemical energy supplied by the fuel that is released during the combustion process [53] per Equation 15.24. Determining it directly from

---

[3] Note that temperatures need only span the range of possible mean gas temperatures, not combustion temperatures.

exhaust gas composition measurements is impractical in small two-stroke engines because the fraction of residual burned gases carried over from the previous cycle (i.e., the scavenging efficiency) is usually not known. An additional complicating factor—as will be shown in a later section—is that Damköhler numbers approach 1 in these small engines and chemical reaction may not be complete when fluid exits the exhaust port. In light of these challenges, we take an indirect approach where the engine power output and all other losses are measured directly and the value of the combustion efficiency is inferred to satisfy Equation 15.23.

15.5 FUELS

Miniature engines operate on a variety of fuels, including gasoline, kerosene (diesel), and methanol. Most require oil to be added to the fuel since they do not have separate oil systems and many also require the addition of nitromethane, which is a fuel-borne oxidizer. Increasing the nitromethane fraction increases the energy release per unit of combustion air and therefore can offset problems with low volumetric efficiency but decreases the overall energy density of the mixture.

15.6 SAMPLE PERFORMANCE DATA FROM THREE MINIATURE ENGINES

This section presents the results of detailed performance measurements in engines A, D, and E in Table 15.1 made on the small engine dynamometer illustrated in Figure 15.2. Detailed descriptions of the dynamometer and how these data were acquired are available elsewhere [16, 17]. The objective is to give the reader a general idea of what the operating maps of these engines can look like and to highlight the aspects of small engine performance that differ from what is typically observed in conventional-scale engines. The two parameters varied in the experiments are the fuel–air mixture ratio and engine speed. The throttle is always wide open. In the next section, these data along with others [17, 37, p. 100] are used to develop scaling relationships for peak power and efficiency.

15.6.1 AP HORNET 09

Figure 15.4 is an example of a complete set of basic engine performance measurements for engine E in Table 15.1. The full operating speed range of this engine is 4,000–18,000 revolutions per minute (RPM) but the peaks in torque, power, and overall efficiency occur between 10,000 and 14,500 RPM so the x-axis limits have been set to these values. While there appears to be an "optimum" mixture ratio at which torque and power are maximum and fuel consumption is minimum, it is not well-defined. The fuel–air mixture ratio decreases linearly as the fuel flow rate is reduced. Power output peaks at equivalence ratios ~1.3. This is somewhat richer than in conventional-scale gasoline engines where power peaks at equivalence ratios between 1 and 1.1 [18, p. 830] and is a preliminary indication of low volumetric efficiency.

Figure 15.5 is derived from the data in Figure 15.4. It shows peak torque and power as a function of engine speed with corresponding values of overall efficiency, DR, and fuel–air mixture ratio. Overall efficiency increases at higher engine speeds where F/A is lower and peaks at an equivalence ratio of less than 1. This is similar to what is seen in conventional-scale engines

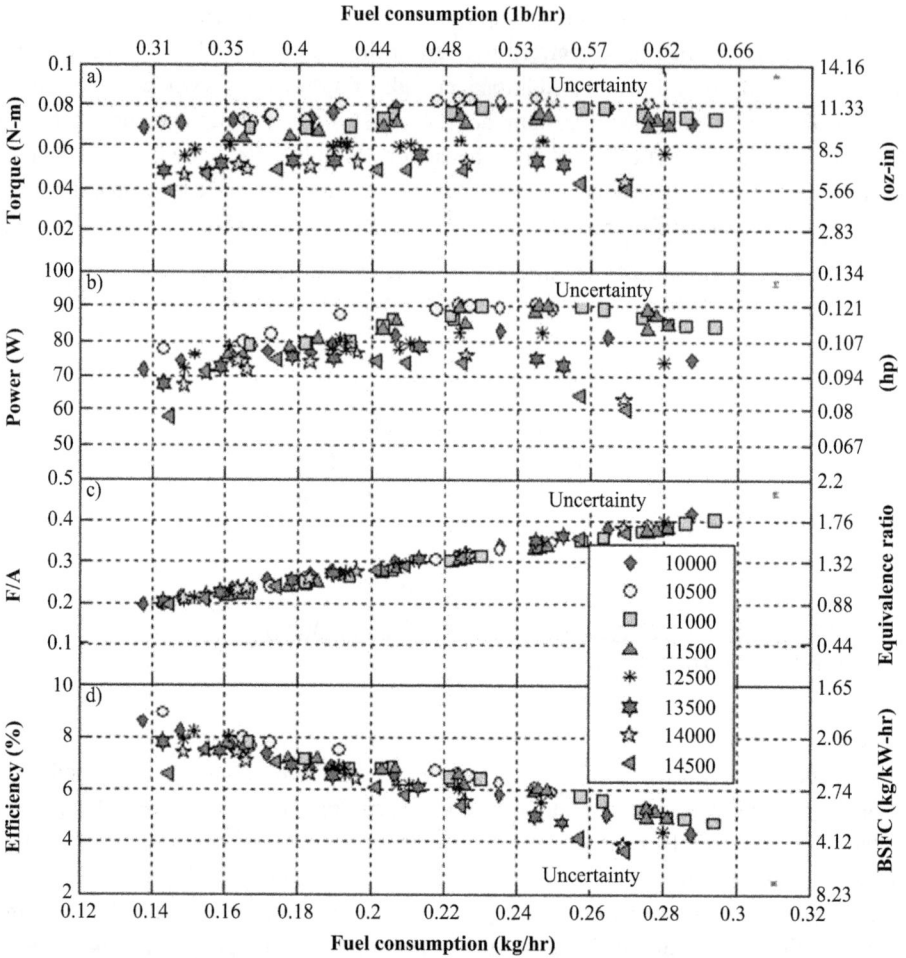

Figure 15.4. Performance map for engine E (the 1.47 cc AP Hornet). Markers represent different engine speeds.

where fuel conversion efficiency generally decreases as $1/\phi$ when the mixture is rich ($\phi > 1$) and increases linearly with a decrease in ϕ for lean mixtures ($\phi < 1$) [18, p. 831].

Peak torque is 0.083 N–m at 10,500 RPM and decreases with further increases in speed. Peak power is 91 W at 10,500 RPM but does not decrease significantly until 11,500 RPM. The efficiency at peak power is ~6.5 percent, whereas the overall peak efficiency is 7 percent and occurs at 14,500 RPM. This translates to a BSFC of 2.35 kg/kW–hr. Peak *DR* is 64 percent at 10,000 RPM and decreases monotonically (and substantially) with increasing speed.

Peak power and efficiency occur at different operating points because of the particular ways power and fuel–air mixture ratio vary with speed in this engine. While power output decreases with further increases in speed beyond its peak at 10,500 RPM, the fuel/air ratio remains approximately constant until ~12,500 RPM. This causes efficiency to drop between 10,500 and 12,500 RPM because more fuel is being consumed while less power is being made. However, beyond 12,500 RPM the fuel/air ratio decreases faster than the power output and efficiency increases again. The net result is that peak efficiency in this engine occurs at peak RPM.

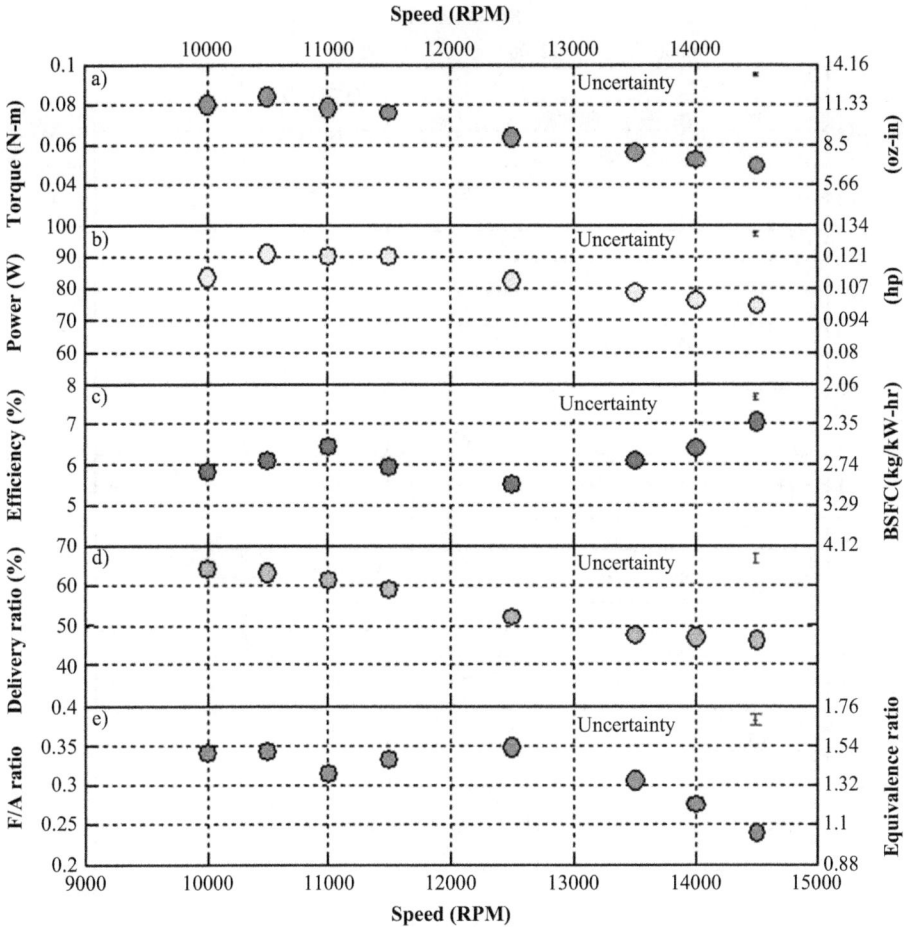

Figure 15.5. Peak performance of the 1.47 cc AP Hornet engine as a function of engine speed.

This is unlike most other engines where efficiency tends to peak at low RPM where frictional and combustion losses are also minimum.

15.6.2   OS 46FX

Figure 15.6 shows the peak operating points for engine A in Table 15.1 as a function of engine speed. Peak torque decreases with increasing engine speed until about 13,000 RPM. This is probably because the DR also decreases with increasing speed in this interval thereby reducing the amount of fuel that can be burned. Peak power remains approximately constant in this interval because the decrease in torque is offset by the increase in speed. With the amount of fresh charge and the F/A ratio decreasing in this interval, the efficiency also increases. Beyond 13,000 RPM, the DR increases again allowing more fuel to be burned. The net result is increased torque and power output but decreased efficiency. The minimum in the DR that occurs in the middle of the engine's operating range has also been observed in other investigations of crankcase scavenged two-stroke engines [17, 54, 55].

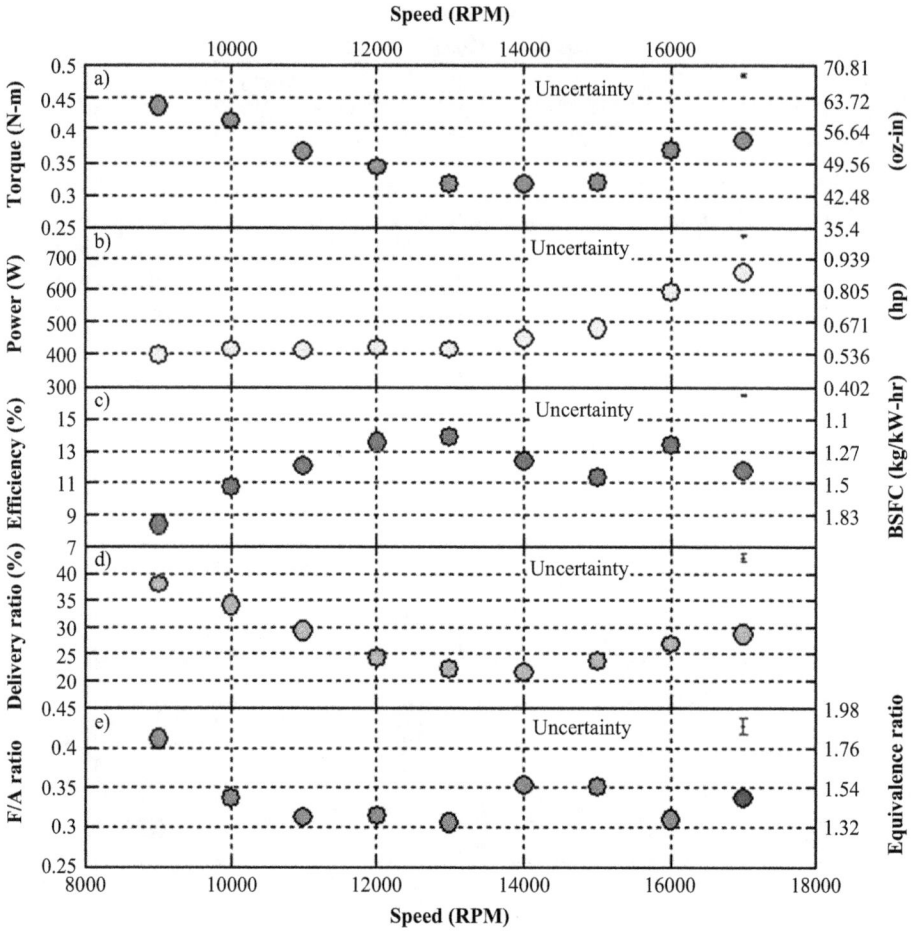

Figure 15.6. Peak performance of 7.54 cc OS 46 FX engine as a function of engine speed.

15.6.3 COX 049

Figure 15.7 shows the peak operating points for engine G in Table 15.1 at each engine speed. The *DR* also goes through a minimum but is generally much higher than the other engines. This is probably because it does not have a muffler. Peak torque does not change much between 9,000 and 17,000 RPM but drops rapidly thereafter. Peak power and peak efficiency are achieved at 17,000 RPM at much lower equivalence ratios than the other engines because of this engine's superior volumetric efficiency.

15.7 SCALING OF ENGINE PERFORMANCE

15.7.1 PERFORMANCE MEASUREMENTS

Data from the authors' own investigations of small engines [17, 37, p. 100] and data for larger (>26 cc) engines compiled from various textbooks, publications, and light aircraft-operating

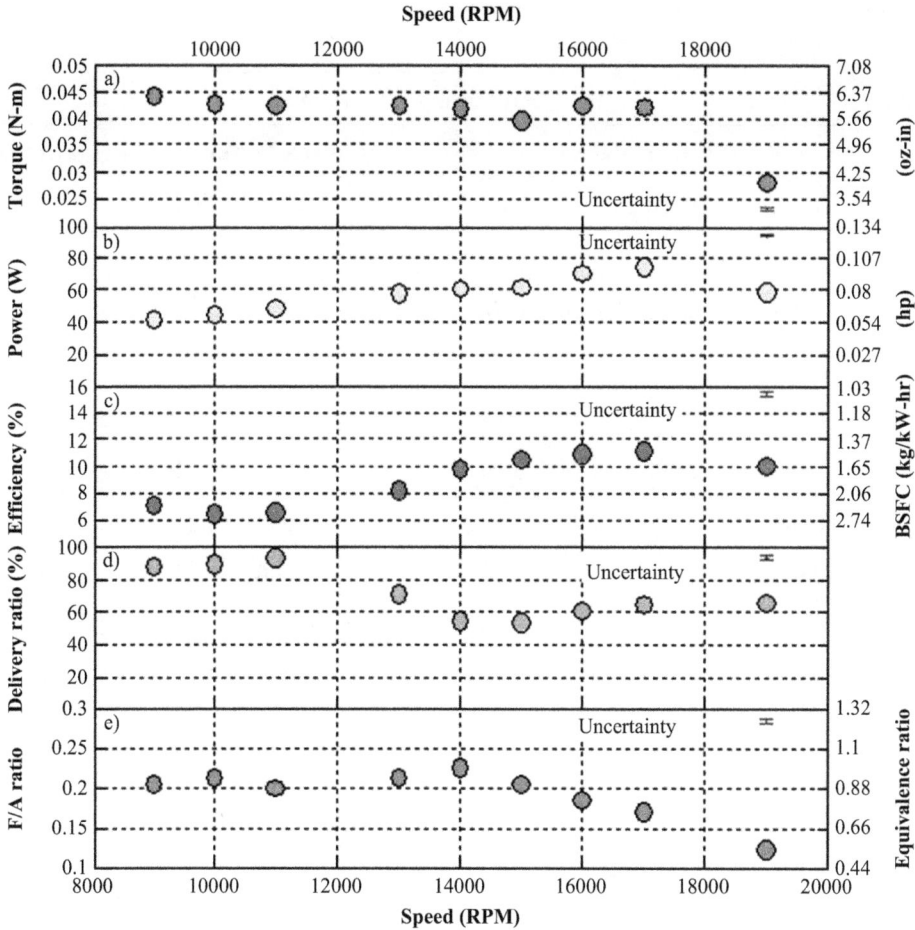

Figure 15.7. Peak performance of the 0.16 cc Cox 049 engine as a function of engine speed.

manuals are used to develop scaling relations for some of the performance metrics presented in Section 15.4 and to estimate size limits for miniaturization of heat engines based on current technology. The results of our measurements of small engine performance are summarized in Table 15.2. The other data sources used to develop the scaling relations are presented elsewhere [17].

The straight line in the log–log space of Figure 15.8 shows that engine mass follows fairly well-defined power laws with displacement. Four-stroke engines for ground applications follow a slightly different scaling than aircraft engines—presumably because of the increased importance of weight in aircraft. Figures 15.9 and 15.10 show that peak torque and power also follow power law scaling with engine displacement over a wide range of engine types and sizes although a different scaling of peak power emerges for two-stroke versus four-stroke versus diesel. Power per piston area (Equation 15.5) is often used as an indicator of the quality of an engine's design—that is how effectively it produces power from its "allocation" of piston area [18, p. 57]. Figure 15.11 shows that two distinct trends for two-stroke and four-stroke designs are apparent. The "crossover" occurs at $V_d \sim 1000$ cc with two-stroke designs having higher power/area

Table 15.2. Summary of performance characteristics for the engines tested on the University of Maryland small engine dynamometer

Engine	Operating Condition								
	Peak Torque				Peak Power				Peak *DR*
	Torque	Speed	F/A	η_o	Power	Speed	F/A	η_o	
A	0.437	9,000	0.41	8.39	656	17,000	0.35	11.80	38.4
B	0.445	7,000	0.46	6.88	677	18,000	0.35	11.88	57.3
C	0.299	8,500	0.33	6.49	311	11,000	0.36	9.12	80.6
D	0.133	11,000	0.40	7.48	158	12,500	0.38	8.45	68.7
E	0.083	10,500	0.34	6.10	91	10,500	0.34	5.14	63.8
F	0.029	12,500	0.30	4.50	40	13,000	0.26	5.14	52.6
G	0.044	9,000	0.21	7.06	75	17,000	0.17	11.12	93.6
H	0.016	13,000	0.48	3.08	24	20,000	0.37	4.93	81.6
I	0.005	14,500	0.29	2.95	8	16,000	0.32	2.76	80.4
	N-m	RPM		%	W	RPM		%	%

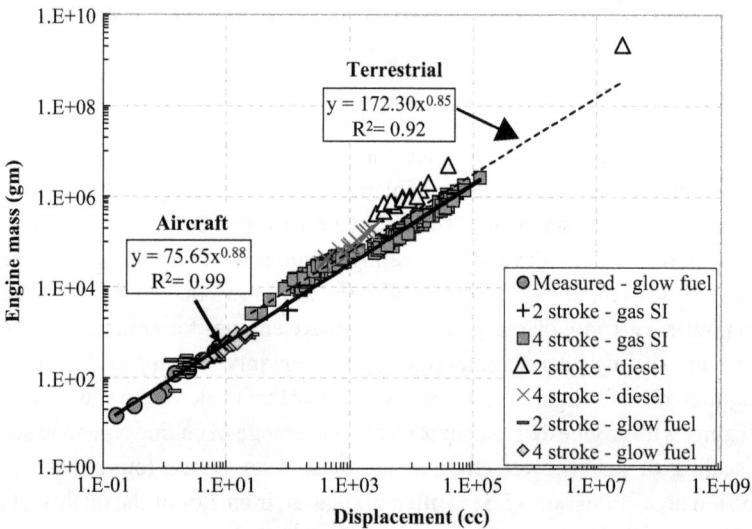

Figure 15.8. Variation of engine mass with engine displacement. Aircraft engines follow a different (higher) power law.

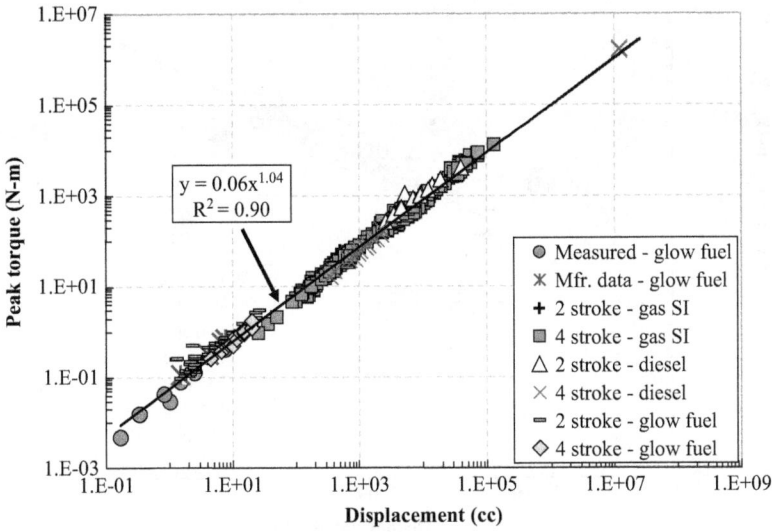

Figure 15.9. Peak torque output as a function of engine displacement.

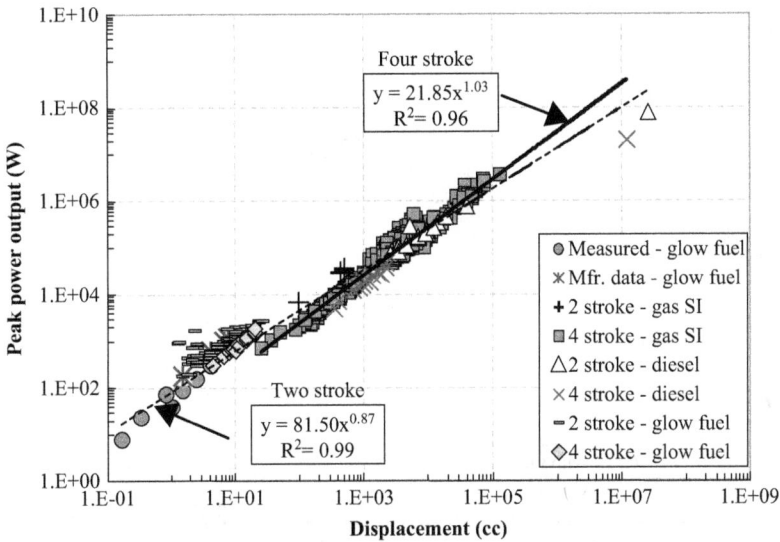

Figure 15.10. Peak power output as a function of engine displacement.

below this point and four-stroke designs having higher power/area above it. Considering these plots together, nothing seems to suggest that small engines are fundamentally different than larger ones.

However, the story is very different when it comes to efficiency and BMEP. Figure 15.12 shows not only that a single power law does not capture the scaling of efficiency very well for engines in general, but that the trend changes abruptly with scale in the vicinity of 10 cm^3 displacement. Above 10 cm^3, the data cluster about different simple power laws (straight lines

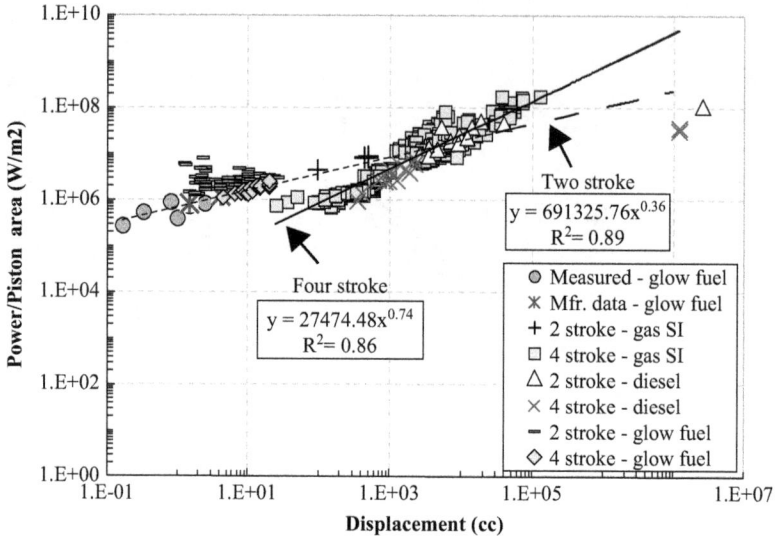

Figure 15.11. Variation of power per unit piston area with engine displacement.

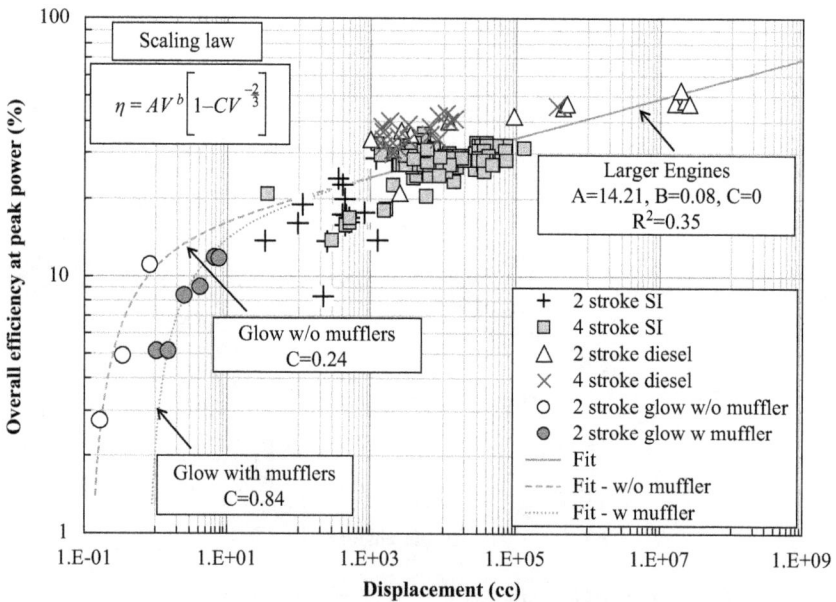

Figure 15.12. Variation of overall efficiency at peak power with engine displacement.

in log–log space) based on engine type. This reflects the strong influence of engine type on overall efficiency. Below 10 cm^3 displacement, however, something else is going on. Not only are these engines much more sensitive to scale, they are also sensitive to the presence/absence of a muffler as engines without mufflers (G–I in Table 15.2) seem to be able to maintain reasonable levels of efficiency at smaller scales better than those with mufflers (A–F in Table 15.2). Similar configuration sensitivity is observed in the scaling of BMEP presented in Figure 15.13.

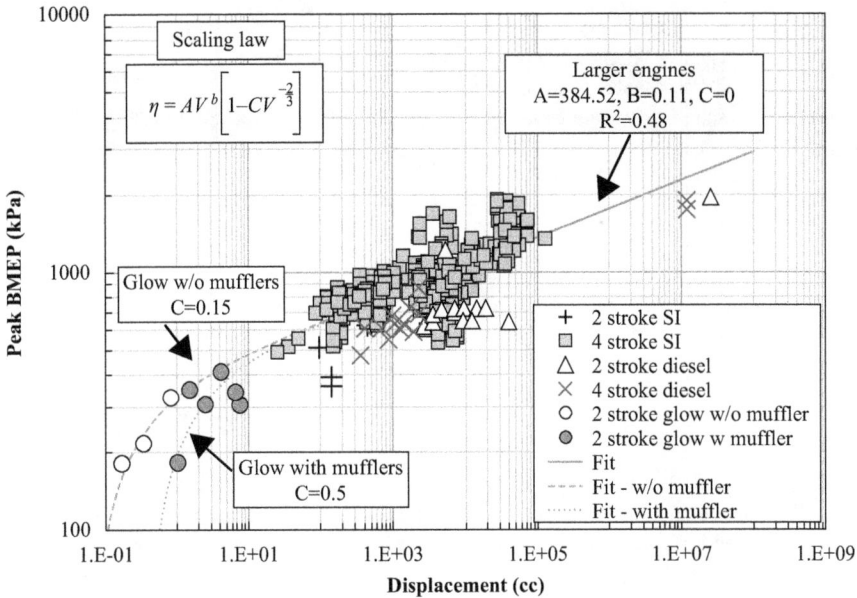

Figure 15.13. Variation of BMEP at peak power with engine displacement.

This suggests that flow losses through the engine are an important driver of performance at the smallest scales and that reducing them is one way to offset thermal losses as the scale is reduced. Eventually, however, decreasing engine size leads to drastic efficiency decreases in both groups of engines in a manner that is very similar to that predicted by Figure 1.6 in the Introduction to the volume.

While no universal laws for the scaling of efficiency or BMEP with displacement are evident, the following modified power law with three fitting parameters (A, B, and C) works better than a simple power law:

$$\eta = A V_d^B (1 - C V_d^{-2/3}) \tag{15.29}$$

A and B are determined by fitting a simple power law to the data for the larger engines (>100 cc where it appears to hold). C is determined by minimizing errors between the fit and the data at the smallest scales and takes on different values depending on whether or not a muffler is present. The values of these coefficients can be found on the respective efficiency and BMEP plots.

15.7.2 COMPONENT LOSSES AND EFFICIENCIES

Figure 15.14 summarizes how the various energy loss rates scale with engine size at constant speed (10,000 RPM) and equivalence ratio (~1). The smallest losses are associated with mechanical friction. These are followed by sensible enthalpy loss in the exhaust, thermal losses, and incomplete combustion. Incomplete combustion losses are three times larger than the next largest loss process and so are plotted on a separate y-axis. Figure 15.15 shows the ratio of each type of loss to the total engine power output under the same operating conditions as in Figure 15.14.

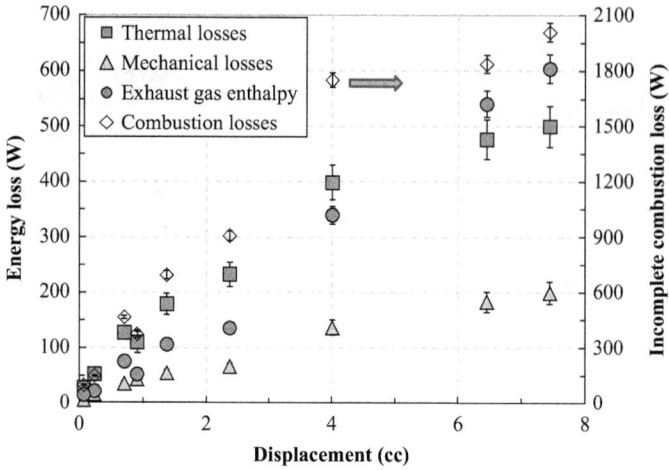

Figure 15.14. Component enthalpy losses for all miniature engines listed in Table 15.1 as a function of displacement at a constant speed (10,000 RPM).

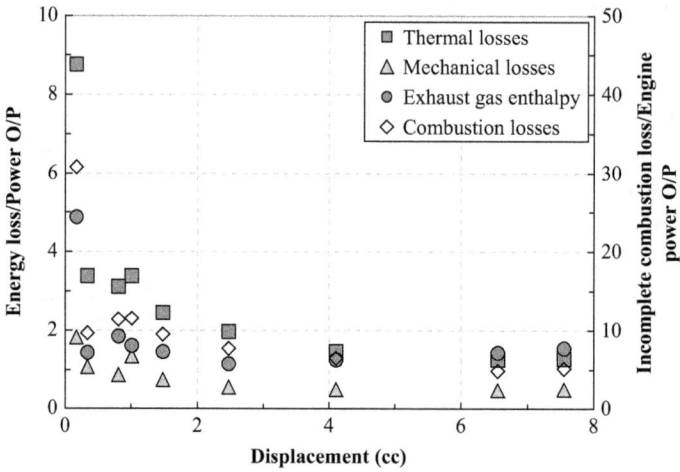

Figure 15.15. Magnitude of various energy loss mechanisms relative to the net engine power output as a function of engine displacement at 10,000 RPM and equivalence ratio ~1.

While all types of losses consume ever-increasing fractions of the total power as the size of the engine is reduced, incomplete combustion remains the largest consuming between 5 and 30 times the net power output of the engine.

Figure 15.16 shows how the individual component efficiencies in Equation 15.22 scale with engine size at an operating speed of 10,000 RPM and an equivalence ratio of about 1. The overall efficiency is not shown in this figure. We could change the plot and show the overall efficiency on the secondary y-axis if required. While all efficiencies decrease with scale, the figure shows that mechanical and thermal efficiencies are the most scale-dependent. The mechanical efficiency begins to drop at 4 cm^3. The thermal efficiency does not start to drop

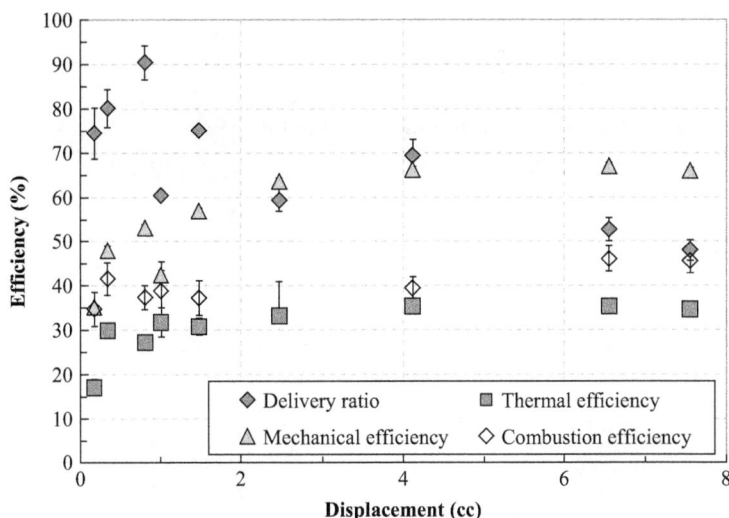

Figure 15.16. Scaling of the various contributors to the overall efficiency at 10,000 RPM and an equivalence ratio of ~1.

significantly until about 1 cm^3 but when it does, its fall is more precipitous. Meanwhile, the combustion efficiency does not experience a precipitous drop at all but decreases more or less smoothly from ~45 percent at 8 cm^3 to 35 percent at 0.16 cm^3. Efficiencies follow similar trends at 12,000 RPM although the values are slightly lower.

Taken together, the data indicate that the dramatic decrease in overall efficiency with reduced scale observed in Figure 15.12 is mainly driven by decreases in thermal and mechanical efficiencies. However, this does not mean that combustion efficiency is unimportant in the scaling of engine performance. In fact, it is really the most important factor as incomplete combustion is the primary energy loss in miniature engines.

To see this, consider Figure 15.17, which presents a series of pie charts comparing how the energy available in the fuel is distributed between useful work and losses in miniature versus conventional-scale engines. The top two pie charts show the distributions in engines I and A, which lie at opposite ends of the scale range investigated here. Below these are similar charts for conventional-scale gasoline and diesel engines. The figure shows that combustion efficiency–or the lack thereof–is the primary differentiator between conventional and miniature-scale engines: incomplete combustion accounts for <4 percent of the fuel power in conventional-scale engines but accounts for >50 percent of the fuel power in all of the miniature engines tested here. While frictional losses consume comparable proportions of fuel power, the waste of chemical potential energy associated with the unburned fuel is enormous. This distorts the flow of thermal energy making it difficult to make meaningful comparisons between the other efficiency metrics. For example, the low degree of fuel conversion probably reduces the mean and exhaust gas temperatures. These, in turn, reduce thermal losses and raise thermal efficiency. Therefore, the extremely low combustion efficiency has the effect of making miniature engines look more thermally efficient than they really would be if they were properly (i.e., nearly completely) utilizing their fuel like larger engines do.

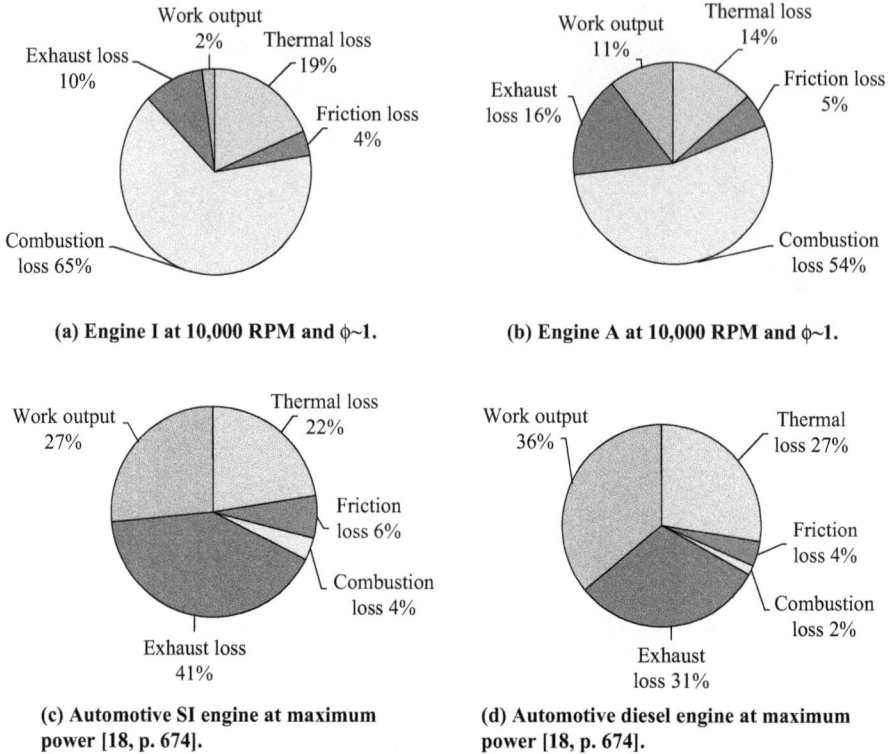

(a) Engine I at 10,000 RPM and φ~1.

(b) Engine A at 10,000 RPM and φ~1.

(c) Automotive SI engine at maximum power [18, p. 674].

(d) Automotive diesel engine at maximum power [18, p. 674].

Figure 15.17. Comparison of fuel energy consumption in miniature versus conventional-scale engines.

15.7.3 MINIATURIZATION LIMITS

The results of the previous two sub-sections can be used to infer the minimum size of a thermo-dynamically viable heat engine—that is one that operates but with net power output = 0.[4] The size limit is the value of V_d that makes η_b or BMEP equal to zero. The results are summarized in Table 15.3 and compared to minimum size estimates made by other researchers. The estimates span almost three orders of magnitude. The largest are those associated with actual measurements in engines while the smallest are associated with idealized "first principles" analyses of heat transfer [56] and flame propagation in micro-channels [57]. Between these extrema lie the various theoretical estimates of Sher et al. [58, 59] that model the two-stroke engine cycle and consider various fuels (gaseous methane or propane) and representations of losses (heat transfer, blowby, etc.).

The ordering of the estimates is hardly surprising: theoretical estimates should predict the smallest scales because they are based on "canonical" geometries and rudimentary representations of losses. Improving the fidelity of the models by accounting for more loss processes produces results that are closer to measured values in engines. Taken together, the results indicate that the minimum displacement of a piston engine based on present technology is ~0.1 cm³.

---

[4] Note that a thermodynamically viable engine is not necessarily a *useful* engine.

Table 15.3. Engine miniaturization limits determined from various sources

Scaling criterion		Length scale (mm)	Displacement (cc)
Overall efficiency at peak power	Without muffler	5.48	0.12
	With muffler	10.25	0.77
Peak BMEP	Without muffler	4.33	0.06
	With muffler	7.91	0.35
Sher (2009) [59]		7.49	0.3
Sher (2005)—Methane [58]		2.14	0.007
Sher (2005)—Propane [58]		3.59	0.033
Peterson [56]		1	7e-4
Leach [57]		0.75	3e-4

While they do not indicate in any way that building a smaller engine is impossible, they do suggest that such an engine would have to be considerably different from those we have today.

15.8 SMALL TWO-STROKE PISTON ENGINE COMBUSTION

The previous section has shown that incomplete combustion is the primary driver of inefficiency in miniature two-stroke engines. Exactly why this is so remains unclear but there are several good reasons to expect the combustion process to be different in miniature engines:

- Higher operating speeds could lead to proportionally lower Damköhler numbers.
- Cylinder bores <10 mm place the engines in the meso-scale combustion regime suggesting that wall interactions/quenching could be important/dominant.
- Crank case-scavenged designs lead to large variations in fuel–air and charge–exhaust mixedness.
- Existing engines operate with excess fuel (rich) to control operating temperatures.

As a result, work has begun to investigate the combustion process in miniature engines. In-cylinder pressure measurements by Papac concluded that combustion occurred in a combination of diesel and partially premixed modes with fuel vapors from the walls forming a fuel-rich zone next to it resulting in a diffusion driven process [33]. Similar measurements by Ball investigating cycle-to-cycle cylinder pressure variations in a miniature glow ignition IC engine attributed variations to the presence of trapped radicals from the previous engine cycle [60]. The most detailed photographic information regarding ignition and combustion phenomena in miniature IC engines is reported by Manente who described a "dual mode combustion" process where the charge sometimes ignites and burns homogeneously and sometimes burns in a premixed flame mode [34].

The objective of this section is to explore the influence of scale on the combustion process occurring in miniature engines. It will be accomplished by considering the heat release rate in three different size engines, the role of combustion intermittency, in-situ images of the combustion process, and the effect of scale on an engine's location on a combustion regime diagram.

15.8.1 HEAT RELEASE RATE MEASUREMENTS

The heat release rate in engines A, B, and C is computed from measurements of cylinder pressure as a function of crank angle using standard methods that are based on a single-zone model of the cylinder volume [18, p. 386]. Figure 15.18 is an example of a typical cylinder pressure trace (averaged over 50 cycles) in engine C of Table 15.1. The shape of the trace is similar to that observed in conventional-scale two-stroke engines: the energy release portion of the cycle appears to occur mostly at constant volume as in the idealized Otto cycle. Similar results are obtained in engines A and B.

Figure 15.19 shows the heat release profile inferred from the pressure trace in Figure 15.18. The details of how this is accomplished are presented elsewhere [18, 41]. The locations of exhaust port opening (EPO) and closing (EPC) are also marked in the plot. Scatter is significant

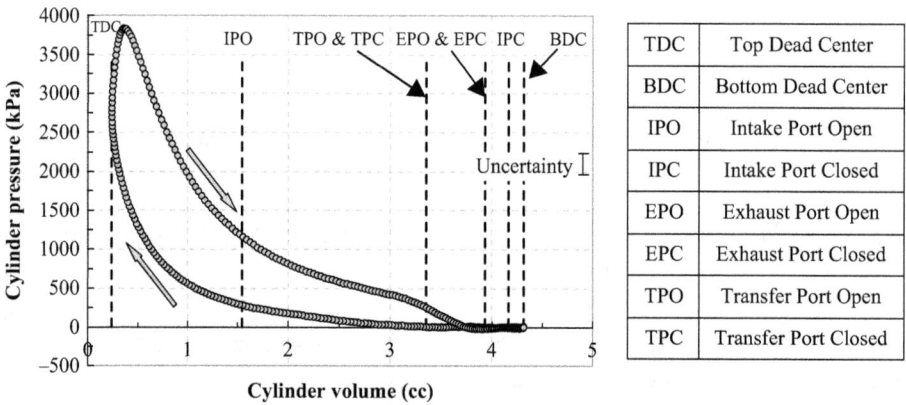

Figure 15.18. Cylinder pressure (absolute) versus volume in engine C operating at 10,000 rpm and an equivalence ratio of 1.05. Average of 50 cycles.

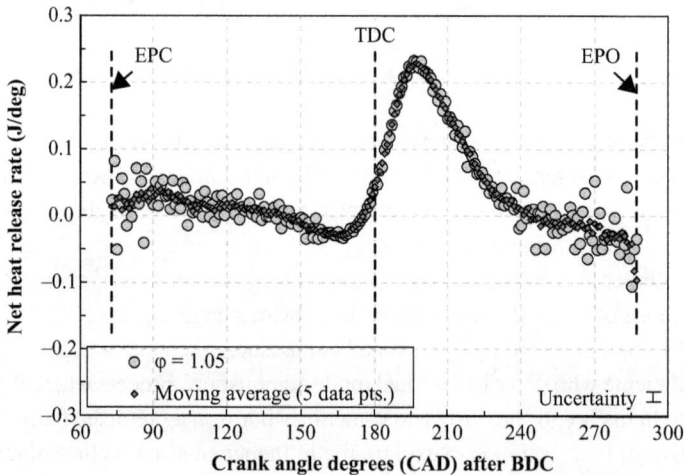

Figure 15.19. Net heat release rate as a function of crank angle in engine C operating at 10,000 RPM and an equivalence ratio of 1.05.

before ignition and after about 240 degrees after Bottom Dead Center (BDC). A 5 point moving average removes some of the scatter and gives a clearer picture of the overall engine heat transfer processes [61]. The net heat release rate begins a slow decrease at ~100 degrees after BDC and continues to drop until ~160 degrees after BDC. Presumably, this is due to the combined effects of heat transfer from the compressed charge to the cylinder walls and blowby from the cylinder to the crankcase. The point at which the net heat release rate goes positive indicates the start of combustion. Heat release rate peaks at about 200 degrees after BDC. Beyond this point, the heat release rate remains positive but begins to decrease and eventually becomes negative at ~270 degrees after BDC. The exhaust port opens at 287 degrees after BDC.

Figure 15.20 compares five point moving averaged net heat release rates at 10,000 RPM in engine C at 7 different equivalence ratios. The data show that changing the equivalence ratio only affects the combustion portion of the cycle and that the peak in the heat release rate occurs slightly rich of stoichiometric. This is in contrast with the cylinder pressure, which peaks at an equivalence ratio of 0.83. This mismatch is caused by the retarding of the peak in the heat release rate as the mixture moves from a lean to rich setting. In lean mixtures, early ignition leads to more rapid heat release prior to Top Dead Center (TDC) and a greater pressure rise because the charge is still being compressed by the piston's upward motion. In richer mixtures, the magnitude of the heat release is higher but ignition is delayed so that most of the heat release occurs past TDC resulting in a lower overall pressure rise.

Figure 15.21 shows that engine speed has an interesting effect on the average heat release rate: at high speeds (10,000 and 11,000 RPM), the average heat release rate has a single peak similar to those observed in Figure 15.20. However, at 9,000 RPM and below, it develops two peaks. The first occurs well before TDC (160 crank angle degrees [CAD] after BDC) and shifts right while decreasing in strength as engine speed increases. The second peak occurs after TDC in the 190–200 degree after BDC range and grows in amplitude but remains approximately in the same place as speed increases. The minimum between the peaks occurs between 175 and 185 degrees after BDC. Figure 15.22 suggests that this phenomenon may also depend on the engine scale as it appears only in the smallest of the three engines for which cylinder pressure data are available.

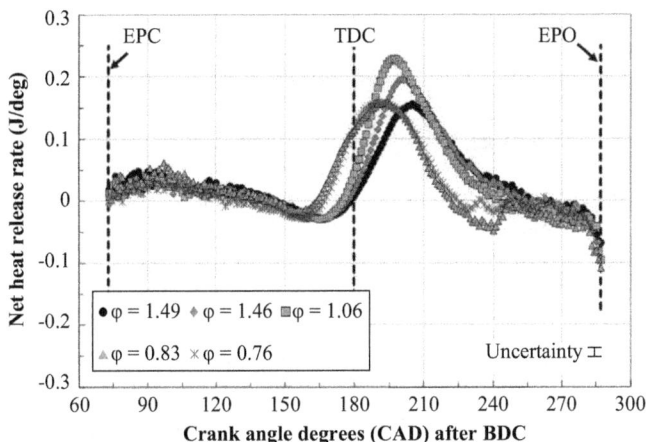

Figure 15.20. Effect of equivalence ratio on five point moving average heat release rate (engine C).

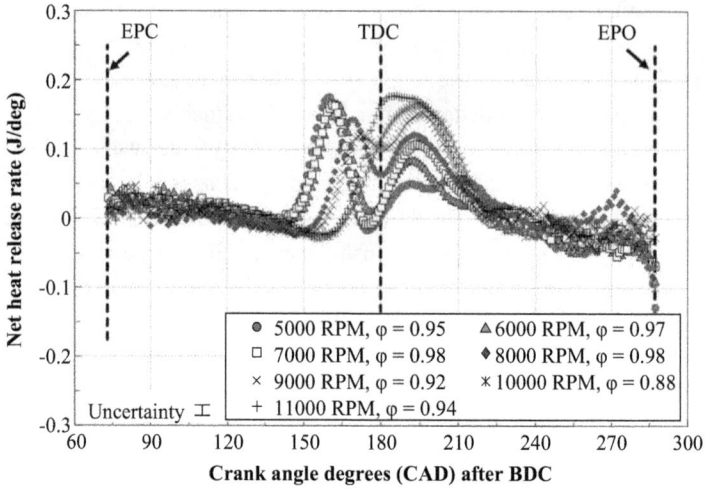

Figure 15.21. Effect of engine speed on five point moving average net heat release in engine C at an equivalence ratio of ~0.98.

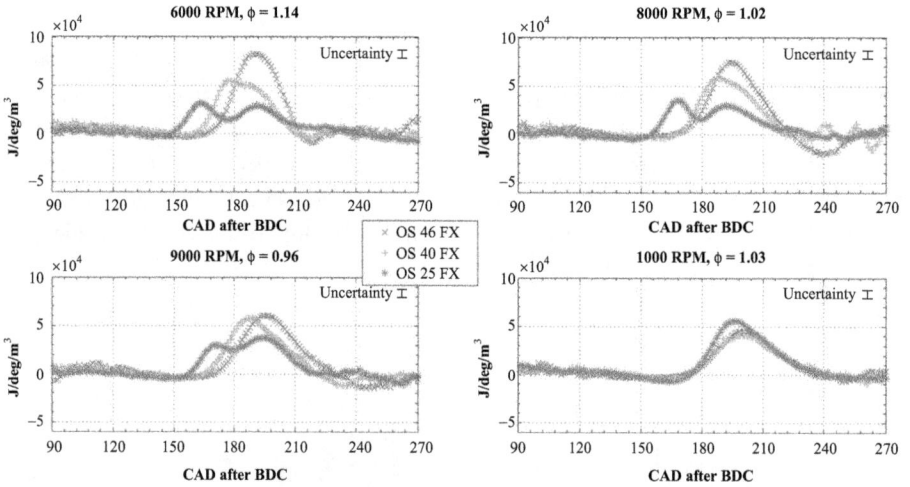

Figure 15.22. Net heat release rates in engines A, B, and C operating at different speeds and near stoichiometric conditions.

Two-stage combustion processes are not unprecedented in engines. For example, a double peak in the net heat release rate is a common characteristic of direct injected diesel engines where a portion of the fuel that has mixed with air and vaporized during the ignition delay phase burns in a premixed manner [62]. This is followed by a mixing-controlled phase, which depends on the duration of fuel injection. Annen et al. observed secondary peaks in pressure traces during startup of the MICE engine [9], which also operates in a glow-plug-assisted HCCI mode. They attribute the dual peaks to re-ignition during the expansion stroke and their disappearance after ignition to the buildup of the residual radical pool in the cylinder that increases the reaction rate. Finally, Manente's images of the cylinder volume in a small (4.1 cm³) engine also suggest a "dual mode combustion" process where the charge sometimes ignites and burns homogeneously and sometimes burns in a premixed flame mode [34].

The reasons for why this two-stage combustion process occurs remain unclear. One possible explanation is that glow engines operate in a partially premixed mode where the combustion process begins with a turbulent flame that is initiated at the glow plug and propagates through the lean premixed portion of the charge. This is followed some time later (possibly after a second ignition delay) by an HCCI event facilitated by combustion radicals left over from the previous cycle or a diffusion-controlled burning of fuel droplets and oil attached to the cylinder walls. Increasing the engine speed increases the Reynolds number (i.e., turbulence levels) and hence the reaction rate while decreasing the engine's size has the opposite effect on the reaction rate by decreasing the Reynolds number. Such a mechanism would be consistent with suggestions by Papac [63] that combustion occurs in a film attached to the cylinder walls but more work would be required to verify this. Another factor may be variations in the concentrations of trapped radicals.

15.8.2 INTERMITTENCY

Cycle-to-cycle variations in the combustion process are widely observed in conventional-scale IC engines and more recently in miniature-scale ones [60]. The process manifests itself as a random variation in the flame spread through the combustion chamber resulting in variations in cylinder pressure rise and associated heat release over many engine cycles [64]. This variability is typically quantified using the coefficient of variation (COV) of the indicated mean effective pressure (IMEP) [18, p. 417].

IMEP can be estimated by integrating the area within the pressure–volume diagram constructed using the pressure trace from each engine cycle [18, p. 50]:

$$IMEP = \frac{\int pdV}{V_d} \tag{15.30}$$

The COV is defined by

$$COV = \frac{\sigma}{\bar{X}} * 100 \tag{15.31}$$

where

$$\bar{X} = \frac{1}{n}\sum_{i=1}^{n} x_i \tag{15.32}$$

is the mean, and

$$\sigma = \left(\frac{1}{n}\sum_{i=1}^{n}(x_i - \bar{X})^2\right)^{1/2} \tag{15.33}$$

Heywood attributes cycle-to-cycle variations in single cylinder engines with premixed charges to two main factors: variation in gas motion during combustion from one cycle to another and variation in mixture composition within the cylinder and especially near the ignition source [18, p. 419]. Ball attributes cycle-to-cycle variations in miniature glow-fueled engines to the presence of trapped radicals from the previous engine cycle [60], which would be consistent with the second mechanism.

Figure 15.23 shows the COV of IMEP for engine A in Table 15.1 (i.e., the OS 46 FX) as a function of operating speed and equivalence ratio. It shows that cycle-to-cycle variability generally increases with engine speed and that variability is much greater when the mixture is rich. Similar trends are also observed in engine B. The increase in COV as one deviates from stoichiometric conditions is consistent with observations in larger engines where cyclic variation increases as the burn duration accounts for a bigger fraction of the total cycle duration [65]. Other sources of variation could be fluctuating turbulence levels in the vicinity of the glow plug and variations in the scavenging process that lead to fluctuations in temperature and species concentration near the glow plug or within the cylinder volume itself. Regardless of the mechanism, the most important finding is summarized in Figure 15.24, which shows that an increase in cycle-to-cycle variation is correlated with a decrease in overall efficiency. This suggests that a main source of inefficiency is loss of unreacted charge through the exhaust port.

15.8.3 IN-CYLINDER IMAGING

A variety of methods have been used to visualize the combustion process in miniature engines using flame luminescence. All are difficult because of the relatively small volume, geometric constraints, and relatively large temperatures and pressures. Pompa [35] used an optical fiber installed in the engine cylinder head with a photo multiplier tube (PMT) to capture natural combustion luminescence in a 4.9 cc engine. Therkelsen [42] used multiple fiber optic cables

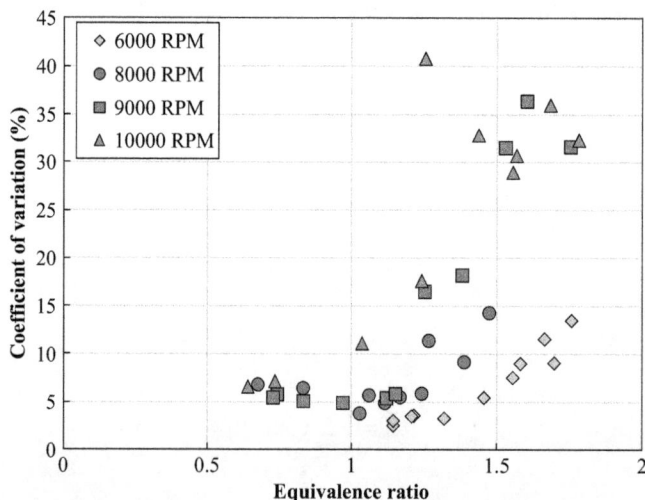

Figure 15.23. Coefficient of variation of IMEP as a function of equivalence ratio for the OS 46 FX engine operating at different speeds.

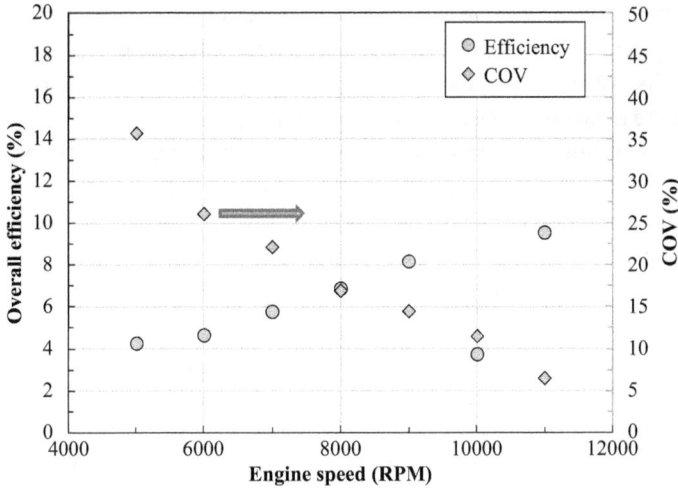

Figure 15.24. Overall efficiency and COV as a function of engine speed for the datasets presented earlier in Figures 15–21.

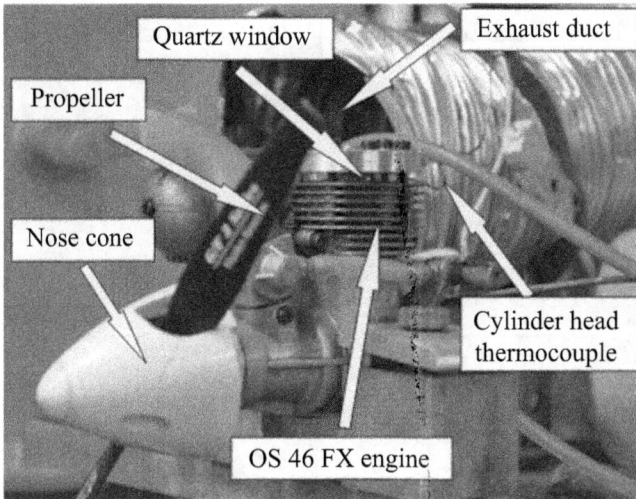

Figure 15.25. Photograph of engine A on test stand with quartz disc installed. A propeller provides the engine load.

with PMT's primarily to distinguish between propagating flame and auto-ignition phenomena in a 25 cc engine. Manente acquired traditional 2D images of flame propagation across the piston face in a 4.1 cm^3 engine but doing so required extensive modifications to the cylinder head that included relocating the glow plug to near the bottom of the squeeze gap so that it would not obstruct the images. It is not clear to what extent these modifications alter the combustion process occurring in the engine.

The images presented in this section were acquired by replacing the stock cylinder head with a custom-machined quartz disc with a hole in its center to admit a glow plug. This has the advantage of providing optical access to the squeeze volume (through the edge of the disc as illustrated in Figure 15.25 or through the top via a series of ports) without changing the shape or dimensions of the combustion chamber. A high-speed camera (Phantom v12.1 by Vision Research) acquired images of natural flame luminescence in the 350–950 nm range. More extensive discussions of the modifications to the cylinder head and the particulars of the imaging system are available elsewhere [37, p. 291], [41].

Figure 15.26 is a side-view (in false color) through the edge of the quartz disc of natural luminescence from the top of the combustion chamber (i.e., the squeeze volume) in engine A of Table 15.1 over two consecutive cycles. The speed is 8,000 RPM and the fuel–air mixture ratio is 0.21, which corresponds to an equivalence ratio of about 1. The white lines show the approximate edges of the combustion chamber "bowl" and the text in the upper left hand corner of each image shows the crank angle. The piston crown never breaks the bottom plane of the quartz disc so it is not visible in any of the images. This also means that only a portion of the total luminescence occurring in the cylinder is being captured by the camera. The exhaust port is located to the left of the bowl center. The bright spot in the center of the images that remains so throughout the engine cycle is the glow plug. The images show an increase in background luminescence that spreads throughout the chamber and then recedes. The luminescence begins at about 100 CAD after TDC and ends at about 180 CAD after TDC suggesting that the

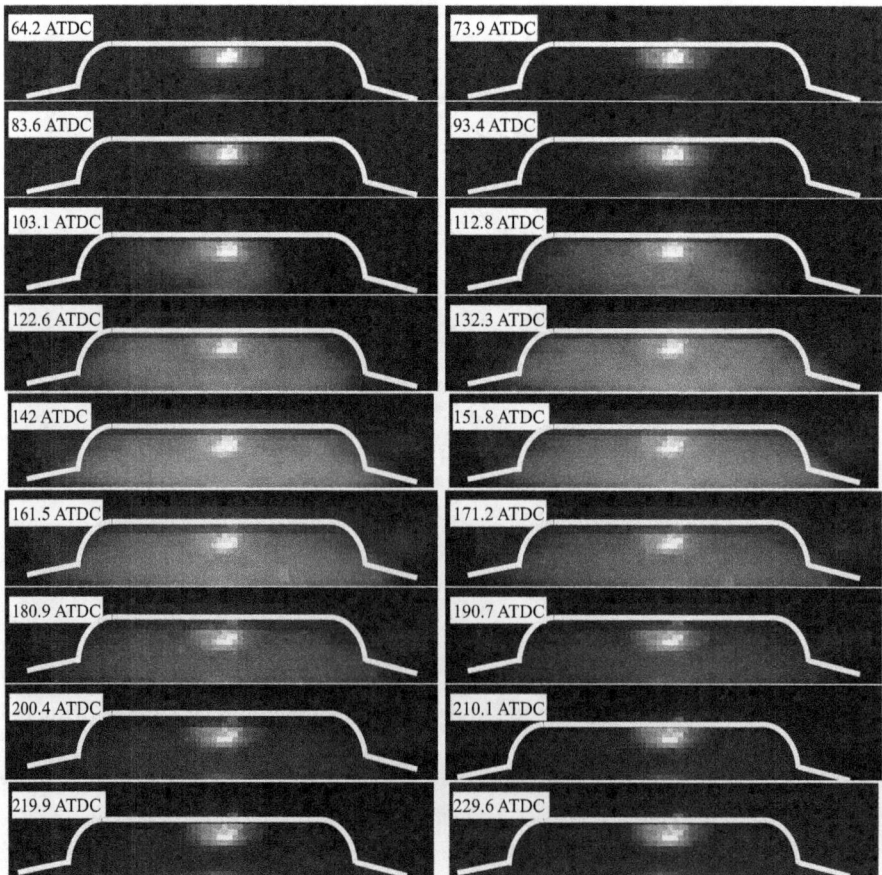

Figure 15.26. High speed false color images of the combustion process in engine A at 8,000 rpm and an equivalence ratio of ~1. Uncertainty in the crank angle location is about 40 CAD.

combustion event occurs within this interval. The images also suggest that the portion of the mixture to the left of the glow plug ignites the first and a flame front propagates outward. (See 93.4–112.8 CAD in Figure 15.26).

Since the images are spectrally broad and probably contain a lot of black body radiation from soot particles, there is no way to infer the location of the flame from them. This plus the relatively coarse phase information makes it difficult to determine whether the combustion process is characteristic of a premixed turbulent flame or of an HCCI-type event. However, the fact that the luminous region seems to spread from the exhaust port throughout the bowl suggests a propagation type of phenomenon (possibly initiated by hot products near the exhaust port that are left over from the previous cycle), which would be more typical of a premixed flame than of an HCCI combustion event.

Figure 15.27 shows another set of high-speed images at 8,000 RPM and an equivalence ratio of approximately 1. This set is interesting because it shows several types of irregularities that occur during some engine cycles. A soot particle or burning fuel droplet is visible to the left of the glow plug from 122.6 after top dead center (ATDC) to 151.8 ATDC and appears to travel some distance across the face of the bowl to the left. Other isolated areas of luminosity at different locations in the cylinder volume suggest that combustion does not always occur via a propagating flame front. For example, a re-ignition event seems to begin in the lower left of the image at 151.8 ATDC. This could be evidence of the two-stage combustion process observed in the cylinder pressure measurements. However, more measurements are needed to understand the combustion process.

15.8.4 COMBUSTION REGIME ANALYSIS

A regime diagram is a non-dimensional representation of how the ratios of various flow length and time scales (see the nomenclature section) depend on two fundamental dimensionless parameters: the Turbulence Reynolds number (horizontal axis) and the Damköhler number (vertical axis) [66, p. 458]. Contours of constant $\dot{v}_{rms}/S_L$, l_k/δ_L and l_0/δ_L divide the space into different regimes of turbulent combustion: flamesheet, flamelet-in-eddy, distributed reaction, and so forth. Abraham and Williams [67] used a regime diagram to show that combustion in conventional-scale four-stroke gasoline and diesel engines generally occurs in a wrinkled laminar flame regime. This result was important for understanding what types of flame structures to expect and how to model them. The objective of this section is to do the same for miniature engines and in so doing develop insight into the influence of scale on the way the overall combustion process takes place.

The procedure for placing miniature engines on this chart is somewhat complex and so will only be described briefly here. Details are available elsewhere [37, p. 315], [41]. In brief, it involves collecting information on various engines' physical dimensions, operating speeds, and fuels. Laminar flame speeds and transport properties are computed using Cantera [48] where combustion of methanol/nitromethane/air mixtures is represented using a 77 species, 484 reaction mechanism developed by Bendtsen, Glarborg, and Dam–Johansen [68] for reactions occurring in hydrocarbon–nitrogen mixtures.

The "space" occupied by miniature engines on the regime diagram is determined by computing Da and Re while varying five different engine parameters. Table 15.4 lists the parameters and the ranges over which each is varied. The "baseline value" refers to the value the parameter

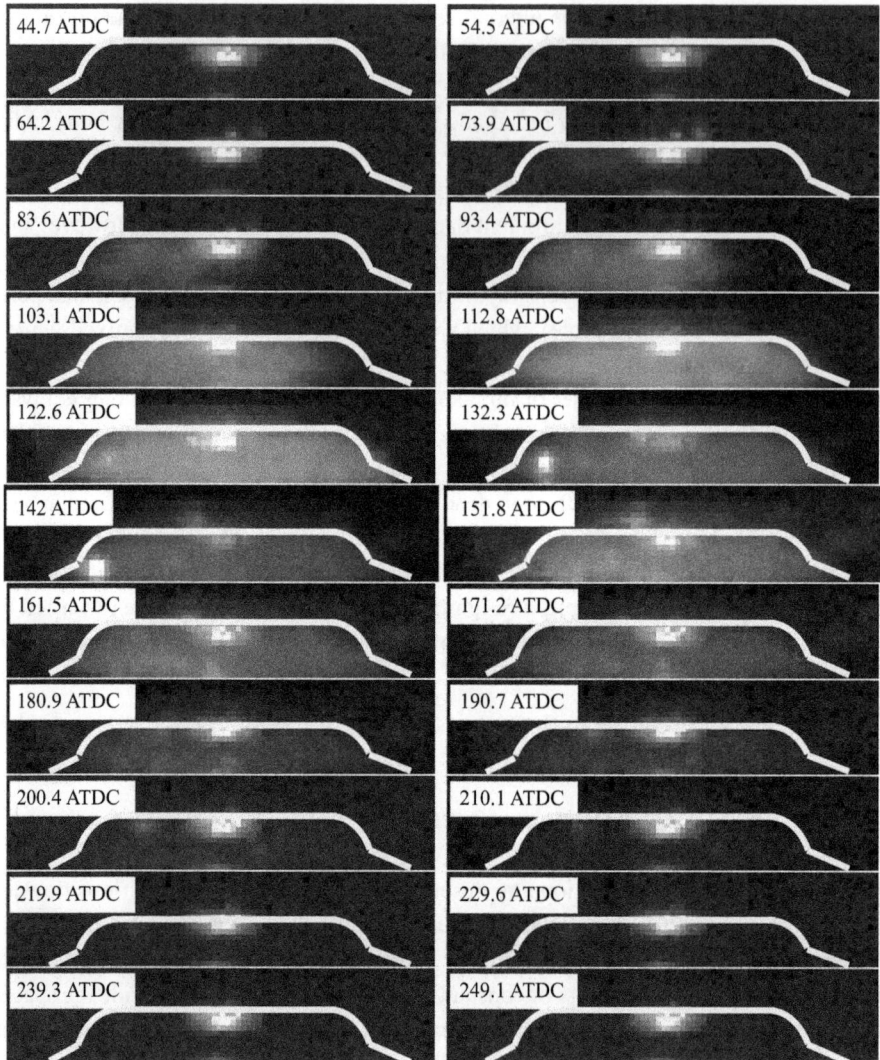

Figure 15.27. High-speed false color images of the combustion process in the OS 46 FX engine at 8,000 rpm and an equivalence ratio of ~1. The images appear to show the formation of soot and possibly a thin film of fuel burning on the walls of the quartz window. Uncertainty in the crank angle location is ~40 CAD.

takes when another parameter is being varied. The range explored reflects the range of conditions observed in all of the small engines tested on our small engine dynamometer. The charge dilution range is more speculative as it has not been measured. The fuel is "glow fuel" in all cases.

The results of the combustion regime analysis are summarized in Figure 15.28. The open symbols correspond to the engines initially investigated by Abraham and Williams [67] while the solid symbols correspond to the nine miniature engines at the baseline conditions listed in Table 15.4. In all cases, decreasing engine size shifts operating points down and to the left in the figure. This is because decreasing the length scale decreases both the Damköhler and turbulent Reynolds numbers. Different boxes bound the regions occupied by this set of miniature engines

Table 15.4. Parameters associated with miniature two-stroke IC engines used in the combustion regime analysis

Parameter	Range explored	Baseline value	Units
Displacement	0.16–7.54	N/A	cc
Equivalence ratio	0.6–2.2	1.4	
Speed	5,000–15,000	10000	rpm
Ignition timing	10 BTDC–5 ATDC	5 BTDC	CAD
Charge dilution	0–45	30	%

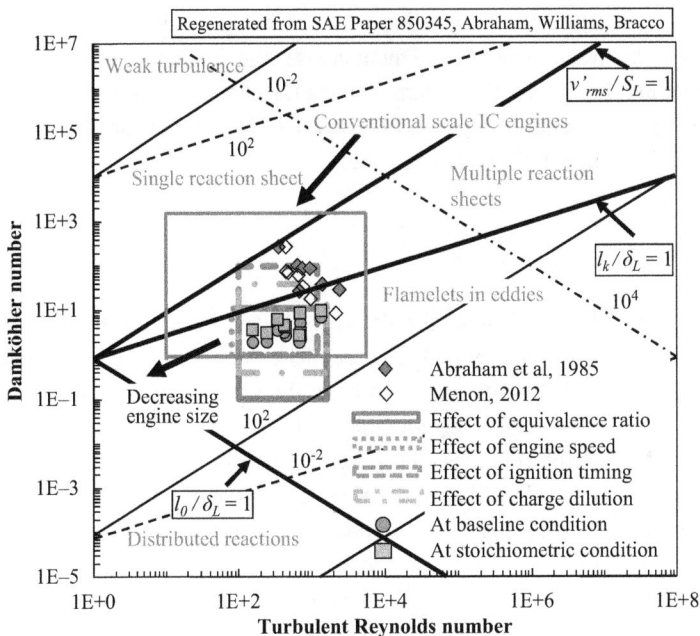

Figure 15.28. Operating points for methanol fueled two-stroke miniature internal combustion engines showing the effect of variation in engine size and equivalence ratio.

as each of the parameters in Table 15.4 is varied. The effects of each variation are discussed below in more detail.

15.8.4.1 Effect of Equivalence Ratio

Operating off stoichiometric on the rich or lean side shifts the operating point down in the regime diagram. The effect can be large for very rich or lean mixtures and can drive the engine well into the distributed reaction zone regime. The shift is due primarily to the large decrease in flame speed as the equivalence ratio moves away from stoichiometric conditions.

15.8.4.2   Effect of Engine Speed

The principal effect of increasing engine speed is to increase the mean piston speed. This increases the turbulence intensity which, in turn, increases the turbulent Reynolds number, and decreases the Damköhler number. The net effect is to move the operating point lower and to the right in the figure, but it is not as strong as that of equivalence ratio. While increasing engine speed drives the combustion regime outside the range associated with conventional-scale engines, realistic changes in speed will not drive it completely out of the flamelet in eddies regime.

15.8.4.3   Effect of Ignition Timing

Retarding ignition timing from the baseline value of 5 degrees BTDC increases the pressure and temperature at the point of ignition. This increases the flame speed resulting in an increase in the Damköhler number without any change in the turbulent Reynolds number. This shifts the operating point up in the regime diagram, but the effect is relatively small and the combustion mode remains in the flamelets-in-eddies regime.

15.8.4.4   Effect of Charge Dilution

Changing the charge dilution changes the flame speed and hence the Damköhler number but without changing the turbulent Reynolds number. The net effect is to shift the engines up or down in the space. Charge dilution seems to have the strongest effect on the location of the engines in the regime diagram after the equivalence ratio. Finally, it should be noted that while the loop-scavenged two-stroke engines considered here must operate with relatively high levels of charge dilution, reducing the dilution to near zero percent as in conventional four-stroke engines (possibly by changing the valving scheme) would return small engines to the same general combustion regime as conventional-scale engines.

Taken together, the results of the combustion regime analysis indicate that miniature engines usually operate within the lower boundaries of the space associated with conventional-scale engines by Bracco and Williams. However, operating miniature engines at the sub-stoichiometric equivalence ratios that are required to achieve acceptable levels of efficiency will push the engines out of the conventional regime to the point where flamelet in eddy combustion models may be more appropriate than the reaction sheets. Reducing the dilution to near zero percent as in conventional four-stroke engines (possibly by changing the valving scheme or switching to a four-stroke spark ignited cycle) could return small engines to the lower part of the Da-Re space associated with conventional-scale engines but is probably not a practical strategy for two-stroke engines.

15.9   CONCLUSION AND FUTURE WORK

Reciprocating IC engines that consume energy dense liquid hydrocarbon fuels are an extremely promising technology for meeting the power and energy density requirements for battery replacements and miniature vehicles. While noise and system-level thermal management pose other challenges, technological solutions are available today. By far the most important problem is the loss of overall thermodynamic efficiency as the scale is reduced. This chapter has presented a systematic study of the smallest commercially available engines in order to develop insight into the fundamental physical processes that determine how engine performance scales with size. Performance measurements on a series of engines ranging in displacement from 0.16 to 7.54 cm^3 show that while most parameters follow the same power law scalings with displacement that equivalent larger engines do, overall thermodynamic efficiency and BMEP do not and instead became progressively more sensitive to displacement as the scale is reduced. Extrapolating these curves to zero efficiency (or BMEP) suggests that the minimum thermodynamically viable size of a reciprocating heat engine *based on current technology* is ~ 0.12 cm^3. This corresponds to a length scale ~5 mm. More detailed measurements show that the principal factors responsible for low thermodynamic efficiency are thermal losses and incomplete combustion. The former is a function of surface-to-volume ratio and necessarily gets worse as the scale is reduced. Preliminary evidence suggests that cycle-to-cycle variations in the rate of reaction (i.e., combustion intermittency) is an important factor driving the low combustion efficiency but it is not the only one as the effects of residence time, charge quality/mixing, and ignition timing may also be important. All of these processes require more study using non-intrusive optical diagnostics. A first-principle scaling analysis shows that miniature piston engines operate at lower overall Damköhler numbers than conventional-scale engines. This puts them entirely in the "flamelet in eddy" regime whereas conventional-scale piston engines lie mostly in the flame sheet regime. Reducing engine scale drives the Damköhler number close to 1 where a distributed reaction model may be required.

REFERENCES

[1] W. Yang, U. Bonne and B. R. Johnson, "Microcombustion Engine/Generator," US Patent 6276313, 2001.

[2] H. T. Aichlmayr, D. B. Kittelson and M. R. Zachariah, "Micro-HCCI Combustion: Experimental Characterization and Development of a Detailed Chemical Kinetic Model with Coupled Piston Motion," Combustion and Flame, vol. 135, no. 3, pp. 227–248, 2003. doi: http://dx.doi.org/10.1016/S0010-2180(03)00161-5.

[3] H. T. Aichlmayr, D. B. Kittelson and M. R. Zachariah, "Miniature Free-Piston Homogeneous Charge Compression Ignition Engine–Compressor Concept - Part I: Performance Estimation and Design Considerations Unique to Small Dimensions," Chemical Engineering Science, vol. 57, no. 19, pp. 4161–4171, 2002. doi: http://dx.doi.org/10.1016/S0009-2509(02)00256-7.

[4] H. T. Aichlmayr, D. B. Kittelson and M. R. Zachariah, "Miniature Free Piston Homogeneous Charge Compression Ignition Engine-Compressor Concept - Part II: Modeling HCCI Combustion in Small Scales With Detailed Homogeneous Gas Phase Chemical Kinetics," Chemical Engineering Science, vol. 57, no. 19, pp. 4173–4186, 2002. doi: http://dx.doi.org/10.1016/S0009-2509(02)00257-9.

[5] L. Tian, D. Kittelson and W. Durfee, "Miniature HCCI Free Piston Engine Compressor for Orthosis Application," in Small Engine Technology Conference, Penang, Malaysia, 2009.

[6] J. Jagoda, "The Development and Investigation of a Small High Aspect Ratio Two-Stroke Engine," Ph.D, Georgia Institute of Technology, 2003.

[7] D. H. Lee, D. E. Park, B. J. Yoon, S. Kwon and E. Yoon, "Fabrication and Test of a MEMS Combustor and Reciprocating Device," Journal of Micromechanics and Microengineering, vol. 12, no. 1, pp. 26–34, 2002. doi: http://dx.doi.org/10.1088/0960-1317/12/1/305.

[8] K. Annen, D. B. Stickler and J. Woodroffe, "Linearly oscillating miniature internal combustion engine (MICE) for portable electric power," in AIAA 2003-1113, 2003.

[9] K. Annen, D. Stickler and J. Woodroffe, "Miniature Internal Combustion Engine-Generator for High Energy Density Portable Power," in Proceedings of the Army Science Conference (26th), Orlando, FL, December 1–4 2008.

[10] W. J. A. Dahm, J. Ni, K. Mijit, R. Mayor, G. Qiao, A. Benjamin, Y. Gu, Y. Lei and M. Papke, "Micro Internal Combustion Swing Engine (MICSE) for Portable Power Generation Systems," in 40th AIAA Aerospace Sciences Meeting, Reno, NV, 2002.

[11] N. S. Lewis, "Portable Energy for the Dismounted Soldier," JASON, McLean, Virginia, June, 2003.

[12] D.-S. Engineering, "0.1 hp/75 Watt Heavy Fuel Engine," D-STAR Engineering, March 2007. [Online]. Available: http://www.dstarengineering.com/piston-pdfs/Engines_75W.pdf [Accessed January 26,2013].

[13] A. Gomez, J. J. Berry, S. Roychoudhury, B. Coriton and J. Huth, "From Jet Fuel to Electric Power Using a Mesoscale, Efficient Stirling Cycle," Proceedings of the Combustion Institute, vol. 31, pp. 3251–3259, 2007. doi: http://dx.doi.org/10.1016/j.proci.2006.07.203.

[14] N. Nakajima, K. Ogawa and I. Fujimasa, "Study on Micro Engines–Miniaturizing Stirling Engines for Actuators and Heat Pumps," in Micro Electro Mechanical Systems, Proceedings, An Investigation of Micro Structures, Sensors, Actuators, Machines, and Robots, 1989.

[15] M. White, "Miniature Stirling Engines for Artificial Heart Power," in Intersociety Energy Conversion Engineering Conference, Orlando, FL, 1983.

[16] S. Menon, N. Moulton and C. Cadou, "Development of a Dynamometer for Measuring Small Internal Combustion Engine Performance," AIAA Journal of Propulsion and Power, vol. 23, no. 1, pp. 194–202, 2007.

[17] S. Menon and C. Cadou, "Scaling of Miniature Piston Engine Performance - Part I : Overall Engine Performance," AIAA Journal of Propulsion and Power, vol. 29, no. 4, pp. 774–787, 2013.

[18] J. B. Heywood, "Internal Combustion Engine Fundamentals," Singapore: McGraw-Hill Book Company, 1988.

[19] D. M. Chon and J. B. Heywood, "Performance Scaling of Spark-Ignition Engines: Correlation and Historical Analysis of Production Engine Data," in 2000 SAE World Congress and Exposition, Detroit, MI, March 6–9, 2000.

[20] J. B. Heywood, "Internal Combustion Engine Fundamentals," Singapore: McGraw-Hill Book Company, 1988, p. 217.

[21] M. Skjoedt, R. Butts, D. N. Assanis and S. V. Hohac, "Effects of Oil Properties on Spark-Ignition Gasoline Engine Friction," Tribology International, vol. 41, no. 6, pp. 556–563, June 2008. doi: http://dx.doi.org/10.1016/j.triboint.2007.12.001.

[22] SAE, "Engine Power Test Code—Spark Ignition and Compression Ignition–Gross Power Rating," Society of Automotive Engineers, June, 1995.

[23] P. G. F. Chinn, "Import Review," Model Airplane News, p. 28, 1957.

[24] D. Gierke, "We Test 10.60 Engines: Which Is Right for You?," Model Airplane News, vol. 131, no. 5, p. 28, 2003.

[25] D. Gierke, "Part I Dynamometer and Engine Performance Analysis," Flying Models, pp. 21–25, June 1973.

[26] D. Gierke, "Part II Dynamometer and Engine Performance Analysis," Flying Models, pp. 43–51, July 1973.

[27] D. Gierke, "Part III Dynamometer and Engine Performance Analysis," Flying Models, pp. 38–47, July 1973.

[28] D. Gierke, "Two-Stroke Glow Engines for R/A Aircraft," Air Age Inc., 1994, p. 48.

[29] V. Manente, P. Tunestal and B. Johansson, "Influence of Inlet Temperature and Hot Residual Gases on the Performances of a Mini High Speed Glow Plug Engine," in Small Engine Technology Conference, San Antonio, TX, 2006.

[30] V. Manente, "A Study of a Glow Plug Ignition Engine by Chemiluminescence Images," in Fuel and Lubricant Meeting, Kyoto, Japan, 2007.

[31] R. Raine, K. Moyle and G. Otte, "A Cost Effective Teaching and Research Dynamometer for Small Engines," International Journal of Engineering Education, vol. 18, no. 1, pp. 50–57, 2002.

[32] C. P. Roberts, R. G. Salter, R. G. Smith and K. Y. Tang, "Study of Miniature Engine-Generator Sets," Ohio State University Research Foundation, Columbus, 1953.

[33] J. Papac and D. Dunn-Rankin, "Characteristics of Combustion in a Miniature Four-Stroke Engine," Journal of Aeronautics, Astronautics and Aviation, Series A, vol. 38, no. 2, pp. 77–88, 2006.

[34] V. Manente, P. Tunestal and B. Johansson, "Influence of the Compression Ratio on the Performance and Emissions of a Mini HCCI Engine Fueled with Diethyl Ether," in Fuel and Lubricant Meeting, Chicago, IL, 2007.

[35] J. Pompa, S. Karnani and D. Dunn-Rankin, "Performance Characterization and Combustion Analysis of a Centimeter-Scale Internal Combustion Engine," Journal of Aeronautics, Astronautics and Aviation, Series A, vol. 40, no. 4, pp. 205–216, 2008.

[36] F. Schauer, "The Effects of Varied Octane Rating on a Small Spark Ignition Internal Combustion Engine," in 48th AIAA Aerospace Sciences Meeting, Orlando, Florida, 2010.

[37] S. Menon, The Scaling of Performance and Losses in Miniature Internal Combustion Engines, Ph.D Thesis, University of Maryland, College Park, 2010.

[38] N. Moulton, "Performance Measurement and Simulation of a Small Internal Combustion Engine," University of Maryland, College Park, 2007.

[39] I. Mengitsu, "Small Internal Combustion Engine Testing for a Hybrid-Electric Remotely-Piloted Aircraft," Biblioscholar, 2011.

[40] M. B. Histand and D. G. Alciatore, Introduction to Mechatronics and Measurement Systems, New York, NY: WCB/McGraw-Hill, 1999.

[41] S. Menon and C. Cadou, "Investigation of Combustion Processes in Miniature Internal Combustion Engines," Combustion Science and Technology, vol. 185, no. 11, pp. 1667–1695, 2013.

[42] P. Therkelsen and D. Dunn-Rankin, "Small-scale HCCI Engine Operation," Combustion Science and Technology, vol. 183, no. 9, pp. 928–946, 2011. doi: http://dx.doi.org/10.1080/00102202.2011.561819.

[43] H. Ma, K. Kar, R. Stone, R. Raine and H. Thorwarth, "Analysis of Combustion in a Small Homogeneous Charge Compression Assisted Ignition Engine," International Journal of Engine Research, vol. 7, no. 3, pp. 237–253, 2006. doi: http://dx.doi.org/10.1243/146808705X60834.

[44] V. Manente, "Characterization of Glow Plug and HCCI Combustion Processes in a Small Volume at High Engine Speed," Division of Combustion Engines, Department of Energy Sciences, Lund University, Faculty of Engineering, LTH, 2007.

[45] M. Disseau, D. Scarborough and J. Jagoda, "The Development and Investigation of a Small High Aspect Ratio, Two-Stroke Engine," in 41st Aerospace Sciences Meeting and Exhibit, Reno, NV, 2003.

[46] S. Menon and C. Cadou, "Scaling of Miniature Piston Engine Performance - Part II: Engine loss Mechanisms," AIAA Journal of Propulsion and Power, vol. 29, no. 4, pp. 788–799, 2013.

[47] Y. Shin, S.-H. Chang and S.-O. Koo, "Performance Test and Simulation of a Reciprocating Engine for Long Endurance Miniature Unmanned Aerial Vehicles," Proceedings of the Institute of Mechanical Engineers, Part D: Journal of automobile engineering, vol. 219, no. 4, pp. 573–581, 2005.

[48] D. Goodwin, N. Malaya, H. Moffat and R. Speth. "Cantera. An Object-Oriented Software Toolkit for Chemical Kinetics, Thermodynamics, and Transport Processes," Version 2.1a1, available at https://code.google.com/p/cantera/

[49] C. A. Finol and K. Robinson, "Thermal Modeling of Modern Engines: A Review of Empirical Correlations to Estimate the In-cylinder Heat Transfer Coefficient," Proceedings of the Institution of Mechanical Engineers, Part D: Journal of Automotive Engineering, vol. 220, no. 12, pp. 1765–1781, 2006. doi: http://dx.doi.org/10.1243/09544070JAUTO202.

[50] C. F. Taylor and T. Y. Toong, "Heat Transfer in Internal Combustion Engines," in ASME 57-HT-17, 1957.

[51] W. J. D. Annand, "Heat Transfer in the Cylinders of Reciprocating Internal Combustion Engines," Proceedings of the Institution of Mechanical Engineers, vol. 177, no. 36, pp. 973–990, 1963. doi: http://dx.doi.org/10.1243/PIME_PROC_1963_177_069_02.

[52] T. LeFeuvre, P. S. Myers and O. A. Uyehara, "Experimental Instantaneous Heat Fluxes in a Diesel Engine and their Correlation," SAE Transactions, vol. 78, 1969.

[53] R. Bishop, "Combustion Efficiency in Internal Combustion Engines," Thesis report, Massachusetts Institute of Technology, Cambridge, MA, 1985.

[54] K. Komotori and E. Watanabe, "A Study of the Delivery Ratio Characteristics of Crankcase-Scavenged Two-Stroke Cycle Engines," SAE transactions, vol. 78, pp. 608–636, 1969.

[55] Y. Motoyama and T. Gotoh, "The Effect of Higher Compression Ratio in Two-Stroke Engines," SAE PAPER 931512, 1993.

[56] R.B. Peterson, "Size Limits for Regenerative Heat Engines," Microscale Thermophysical Engineering, vol. 2, pp. 121–131, 1998. doi:http://dx.doi.org/10.1080/108939598200033.

[57] T. Leach and C. Cadou, "The Role of Structural Heat Exchange and Heat Loss in the Design of Efficient Silicon Micro-Combustors," Proceedings of the Combustion Institute, vol. 30, no. 2, pp. 2437–2444, January 2005. doi: http://dx.doi.org/10.1016/j.proci.2004.08.229.

[58] E. Sher and D. Levinzon, "Scaling Down of Miniature Internal Combustion Engines: Limitations and Challenges," Heat Transfer Engineering, vol. 26, no. 8, pp. 1–4, October 2005. doi: http://dx.doi.org/10.1080/01457630591004780.

[59] I. Sher, D. Levinzon-Sher and B. Sher, "Miniaturization Limits of HCCI Internal Combustion Engines," Applied Thermal Engineering, vol. 29, pp. 400–411, 2009. doi: http://dx.doi.org/10.1016/j.applthermaleng.2008.03.020.

[60] J. Ball, R. Raine and R. Stone, "Combustion Analysis and Cycle-by-Cycle Variations in Spark Ignition Engine Combustion. Part 1," Proceedings of the Institution of mechanical engineers, Part D: Journal of automobile engineering, vol. 212, no. D5, pp. 381–399, 1998. doi: http://dx.doi.org/10.1243/0954407981526046.

[61] P. Lakshminarayanan, Y. Aghav, A. Dani and P. Mehta, "Accurate Prediction of the Rate of Heat Release in a Modern Direct Injection Diesel Engine," Proceedings of the institution of mechanical engineers, Part D: Journal of automobile engineering, vol. 216, no. 8, pp. 663–675, 2002. doi: http://dx.doi.org/10.1177/095440700221600805.

[62] S. Jindal, B. Nandwana and N. Rathore, "Comparative Evaluation of Combustion Performance and Emissions of Jatropha Methyl Ester and Karanj Methyl Ester in a Direct Injection Engine," Energy Fuels, vol. 24, no. 3, pp. 1565–1572, 2010. doi: http://dx.doi.org/10.1021/ef901194z.

[63] J. Papac and D. Dunn-Rankin, "Combustion in a Centimeter Scale Four Stroke Engine," in Proceedings of the Western States Section of the Combustion Institute, Spring 2004 meeting, University of California, Davis, CA, 2004.

[64] M. Rashidi, "The Nature of Cycle-by-Cycle Variation in the S.I. Engine from High Speed Photographs," Combustion and Flame, vol. 42, pp. 111–122, 1981. doi: http://dx.doi.org/10.1016/0010-2180(81)90150-4.

[65] J. Daily, "Cycle to Cycle Variations: A Chaotic Process?," Combustion Science and Technology, vol. 57, no. 4–6, pp. 149–162, 1988. doi: http://dx.doi.org/10.1080/00102208808923950.

[66] S. Turns, An Introduction to Combustion, Second ed., New York, NY: McGraw-Hill Book company, 2000.

[67] J. Abraham, F. Williams and F. Bracco, "A Discussion of Turbulent Flame Structures in Premixed Charges," SAE Paper no.850345, 1985.

[68] A. Bendtsen, P. Glarborg and K. Dam-Johansen, "Low Temperature Oxidation of Methane: The Influence of Nitrogen Oxides," Combustion Science and Technology, vol. 151, no. 1, pp. 31–71, 2000. doi: http://dx.doi.org/10.1080/00102200008924214.

NOMENCLATURE

$\dot{W}_{out/p}$	Power output
Γ	Torque
N	Shaft speed in revolutions/minute
η	Efficiency
$\dot{m}$	Mass flow rate
$BSFC$	Brake-specific fuel consumption
V	Volume
n_R	Number of strokes/power stroke
BMEP	Brake mean effective pressure
A	Area
S_{mean}	Mean piston speed
L	Stroke length
ρ	Density (either mass/volume or power/mass)
m	Mass
F/A	Fuel-to-air ratio
φ	Equivalence ratio
DR	Delivery ratio
h	Enthalpy per unit mass
q	Heat flux
p	Pressure
C_p	Specific heat at constant pressure
k	Thermal conductivity
b	piston diameter
G	Mass flow per piston area
μ	dynamic viscosity
Re	Reynolds number
Da	Damköhler number
l_k	Taylor length scale
l_0	Integral length scale
δ_L	Flame speed
v'_{rms}	Turbulence intensity

SUBSCRIPTS

o	Overall
f	Fuel
a	Air
$available$	Available
d	Displacement
p	Piston
$parea$	Piston area
$norm$	Normalized
pwr	Power

e	Engine
i	Ideal
IN	In
l	Loss
tot	Total
th	Thermal
m	Mechanical
$conv$	Convective
ex	Exhaust
$comb$	Combustion
g	Gas
F	Fuel

CHAPTER 16

COMBUSTORS FOR MICROGAS TURBINE ENGINES

C.M. Spadaccini and I.A. Waitz

16.1   INTRODUCTION

The study of microcombustion science and engineering as a practical endeavor began to develop in the mid 1990s with the advent of the microgas turbine engine concept. Prior to this, combustion at small scales was largely an academic topic revolving around the question of quenching diameter. With the rapid development of microelectromechanical systems (MEMS) and the advancement of microfabrication techniques, realizing a gas turbine-based micropower system became possible. This eventually spawned a host of sub-disciplines including power-MEMS, micropropulsion, and microreactor technologies. Devices in each of these areas require a heat source, typically a combustor due to the high energy density of liquid hydrocarbon fuels. Thus, combustion in small volumes has become an important area of research with impact across a range of applications.

As a result of the variety of microcombustion applications, several types of combustors have been developed and studied. They can be divided into two categories: homogeneous gas-phase microcombustors and heterogeneous catalytic microcombustors. The first type is a flow-through system similar to the combustors that are found in conventional gas turbines, and the second is more characteristic of a chemical reactor with a solid surface catalyst to increase reaction rates. The type of combustor chosen for a given application will depend on the required fuel, materials limitations, the thermodynamic cycle (if being used for a power application), and available fabrication techniques. The governing physics and limitations of these combustors are largely the same.

This chapter will provide a general understanding of microcombustion fundamentals and challenges most directly associated with a microgas turbine engine. Additionally, we will show microcombustor operational data and provide some design guidelines. Initially, we briefly introduce the concept of the microgas turbine engine and provide a cursory introduction to basic combustion principles and definitions. Next, the challenges specific to combustion in microengines

465

will be presented. This will be followed by a discussion of the two primary strategies for operating a microcombustor: homogenous gas phase systems and heterogeneous catalytic systems. These sections include both simplified analytical modeling and experimental results for fabricated devices designed for integration with a rotating structure. A section dedicated to the fabrication of these devices is then presented and followed by a conversation regarding future trends.

16.2 MICROGAS TURBINE ENGINES

Engines with rotating machinery typically operate on the Brayton cycle, the cycle that governs the performance of gas turbine engines. An ideal Brayton cycle consists of an isentropic compression step (1–2), followed by constant pressure heat addition (2–3), isentropic expansion (3–4), and heat rejection (4–1).These individual processes correspond to the compressor (1–2), combustor (2–3), turbine (3–4), and exhaust (4–1) as shown in Figure 16.1. The Brayton cycle was selected for the micro-engine in an attempt to realize the high power densities achieved by conventional-scale gas turbine engines at the micro-scale.

In this chapter, we will focus on the miniaturization and performance of the constant pressure heat addition component of the Brayton cycle. A combustor for a gas turbine is a flow-through device (hence the constant pressure assumption) where high pressure air from a compressor enters a chamber, mixes with fuel that is injected from a separate source, ignites, and burns. This raises the enthalpy and temperature of the working fluid. The high temperature combustion products exit the chamber and enter a turbine where they expand and work is extracted. Figure 16.2 shows a conventional combustor found in a gas turbine engine.

At the micro-scale, combustion for the purpose of driving rotating machinery must satisfy similar requirements. A microcombustor must accept high pressure air from a microcompressor, inject and adequately mix fuel, burn the mixture efficiently, retain enthalpy in the fluid, and provide this fluid to the turbine for work extraction; however, there are additional constraints due the small scale. These include very short time scales, increased heat loss via fluid–structure coupling, and fabrication and material challenges.

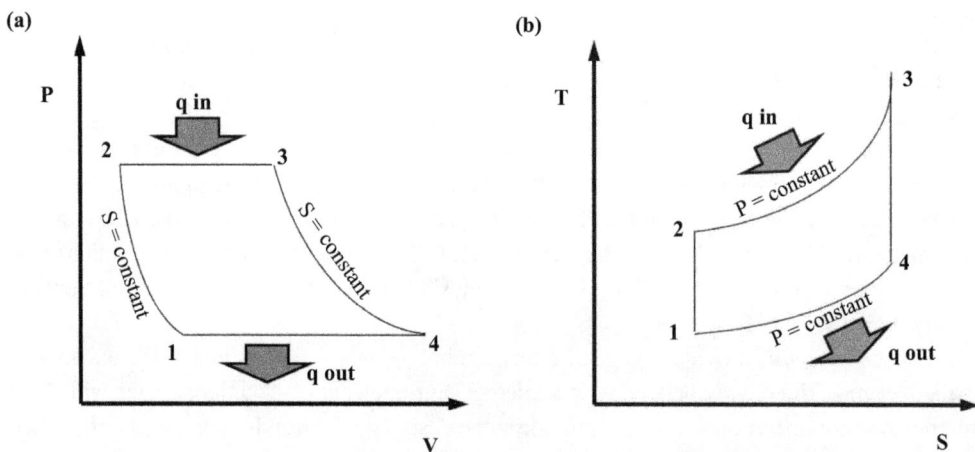

Figure 16.1. Ideal Brayton cycle: (a) pressure–volume coordinates and (b) temperature–enthalpy coordinates.

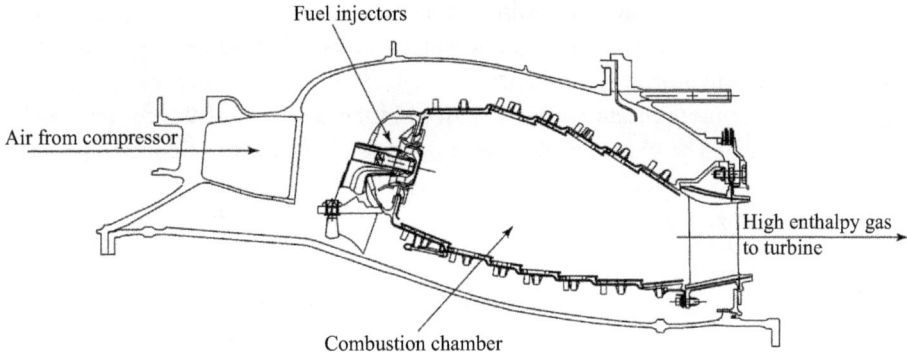

Figure 16.2. Conventional gas turbine combustor.

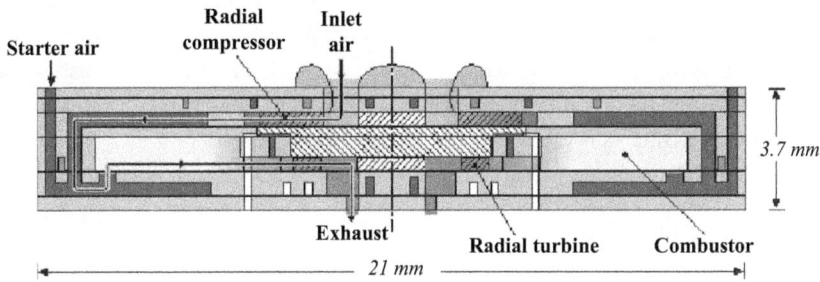

Figure 16.3. The MIT microgas turbine engine.

Despite these challenges, there are tremendous advantages to a micro-scale gas turbine engine. A small-scale power system with high power density can have broad application as both a power source when coupled to a generator or as a propulsion system for microair vehicles. An example of an initial design of a microgas turbine first developed by researchers at MIT is shown in Figure 16.3.

16.3 BASIC COMBUSTION CONCEPTS

Before discussing microcombustor concepts and issues, it is useful to familiarize the reader with several basic combustion concepts. These include the fundamental chemical equations of combustion reactions, the concept of equivalence ratio (normalized fuel–air ratio), reaction rate and Arrhenius expressions, and various definitions of efficiency which will be used in the context of combustors for microgas turbine engines.

16.3.1 COMBUSTION CHEMISTRY

Combustion consists of a chemical reaction between a fuel (typically hydrogen or a hydrocarbon species) and an oxidizer leading to heat release in the fluid. In general, the overall chemical equation for complete combustion of a hydrocarbon with air can be written as

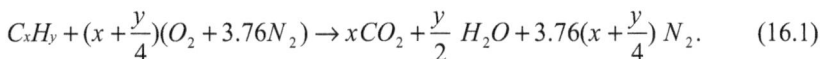

$$C_xH_y + (x + \frac{y}{4})(O_2 + 3.76N_2) \rightarrow xCO_2 + \frac{y}{2} H_2O + 3.76(x + \frac{y}{4}) N_2. \quad (16.1)$$

Typical fuels for gas turbines are long chain hydrocarbon liquids with carbon numbers between 12 and 15 primarily due to their high energy densities. For the microcombustors discussed here, gaseous fuels such as hydrogen (H_2), ethylene (C_2H_4), and propane (C_3H_8) are most commonly used [1–6]; however more recent work has moved in the direction of the conventional long chain liquid fuels [7]. The choice of fuel often depends on fuel storage constraints, reaction rates, and emissions requirements.

The balanced chemical reactions of hydrogen and propane with air respectively are:

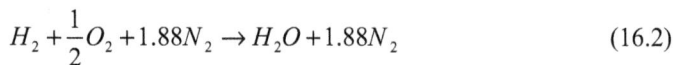

$$H_2 + \frac{1}{2}O_2 + 1.88N_2 \rightarrow H_2O + 1.88N_2 \tag{16.2}$$

$$C_3H_8 + 5O_2 + 18.8N_2 \rightarrow 3CO_2 + 4H_2O + 18.8N_2 \tag{16.3}$$

If the reaction proceeds to completion, the only products are water and carbon dioxide (in the ideal case nitrogen is inert); however in reality this is not the case. Trace quantities of pollutants such as oxides of nitrogen (NO_x) and carbon monoxide (CO) can be present as well as some unburned fuel. Incomplete combustion occurs if the reactants are not well mixed or if there is insufficient residence time as is often the case with microcombustors.

16.3.2 EQUIVALENCE RATIO

If fuel and oxidizer, propane and oxygen in air in this case, are present in a combustion system in a 23.8:1 molar ratio, the mixture is said to be stoichiometric. In this example, all fuel and oxygen would be consumed in the reaction if it proceeds to completion. If there is excess fuel, the reaction is considered to be *rich*, whereas with excess air or oxidizer, it is *lean*. The fuel/air ratio is usually given on a mass basis which can be obtained by scaling the molar ratio by the molecular weights of the reactants. In a combustor, the mass-based fuel/air ratio is the ratio of the fuel and air mass flow rates:

$$fa = \frac{\dot{m}_{fuel}}{\dot{m}_{air}} \tag{16.4}$$

The fuel/air ratio can be normalized by the stoichiometric ratio to yield a quantity known as the equivalence ratio:

$$\phi = \frac{fa_{actual}}{fa_{stoichiometric}} \tag{16.5}$$

This is the most commonly used measure of mixture ratio. ϕ equal to one indicates that the fuel/air mixture is stoichiometric; ϕ is less than unity for a lean mixture; and ϕ is greater than unity if it is rich. Combustion reactions typically peak in temperature around an equivalence ratio of one and are reduced the more lean or rich the mixture. This is due primarily to the unreacted components absorbing some of the heat released from the reaction. The theoretical temperature achieved with an adiabatic boundary is known as the adiabatic flame temperature which is an important reference point for measuring the completeness of a combustion process.

16.3.3 KINETICS

Chemical kinetics is the study of the rate at which a reaction occurs and is generally a function of reactant concentration, temperature, and the presence of a catalyst. This rate can be expressed as the rate of change in concentration of the reactants per unit time. This can be equated to a rate constant multiplied by the concentrations of the reactant species raised to the power of their respective stoichiometric coefficients. For example, a simple reaction can be written as

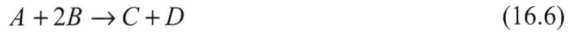

$$A + 2B \rightarrow C + D \tag{16.6}$$

where the reaction rate is

$$-R = \frac{d[A]}{dt} = -k[A][B]^2 \tag{16.7}$$

with brackets denoting concentration and k being the aforementioned rate constant.

The temperature dependence of reaction rate is contained within the rate constant k and is usually expressed using the Arrhenius equation

$$k = A \exp\left(\frac{-E_a}{RT}\right) \tag{16.8}$$

where E_a is the activation energy for the reaction, R is the gas constant, T is the temperature, and A is the kinetic pre-exponential factor.

The activation energy represents an energy barrier which must be overcome to begin the reaction which subsequently progresses from reactants to products. This energy is usually thermal energy, especially in a combustion process, and once it is provided, the reaction rapidly proceeds. Figure 16.4 notionally shows this as a function of reaction coordinate for a typical

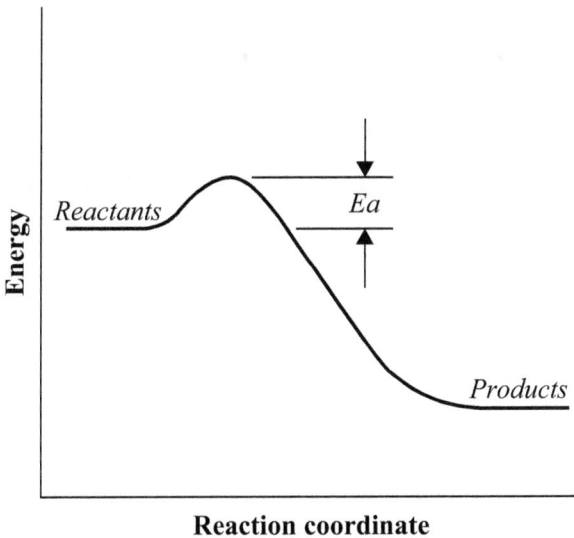

Reaction coordinate

Figure 16.4. Energy as a function of reaction coordinate for a generic exothermic reaction.

exothermic combustion reaction. With a catalyst present, the activation energy is reduced and the reaction proceeds more easily.

16.3.4 MICROCOMBUSTOR EFFICIENCY DEFINITIONS

In the context of combustors for microgas turbines, overall combustor efficiency is based upon an enthalpy balance across the combustor and is defined as

$$\eta_c = \frac{(\dot{m}_a + \dot{m}_f)\, h_2 - \dot{m}_a h_1)}{(\dot{m}_f h_f)} \tag{16.9}$$

where station (1) refers to the combustor inlet and station (2) is the exit. The combustor efficiency can be thought of as the product of two component efficiencies which are a chemical efficiency and a thermal efficiency. They can be written as

$$\eta_{chem} = \frac{(\dot{m}_a + \dot{m}_f)\, h_2 - \dot{m}_a h_1) + Q_{loss}}{(\dot{m}_f h_f)} = \frac{\text{(total enthalpy released)}}{\text{(maximum enthlapy release possible)}} \tag{16.10}$$

$$\eta_{therm} = \frac{(\dot{m}_a + \dot{m}_f)\, h_2 - \dot{m}_a h_1)}{(\dot{m}_a + \dot{m}_f)\, h_2 - \dot{m}_a h_1) + Q_{loss}} = \frac{\text{(enthalpy rise of fluid)}}{\text{(total enthlapy released)}} \tag{16.11}$$

where Q_{loss} is the heat lost to the structure.

16.3.5 COMBUSTION METHODOLOGIES

There are two combustion modes which can be used in microcombustors for microgas turbines: homogeneous gas-phase combustion and heterogeneous catalytic combustion. Homogenous gas-phase combustion is the mixing and reacting of the fuel and oxidizer species while both are gaseous and is the most common combustion process in gas turbines. Heterogeneous catalytic combustion involves the use of a solid-phase catalyst which is usually a noble metal such as platinum and/or palladium. In this type of combustor, the solid catalyst surface provides local sites at which the fuel and oxidizer species will bind and react. Figure 16.5 is a schematic illustration of this combustion process. The reaction occurring at the catalytic sites propagates at a significantly faster rate than in the gas phase; hence the reaction is "catalyzed." This is the result of the surface site lowering the activation energy of the reaction. The addition of a catalyst adds

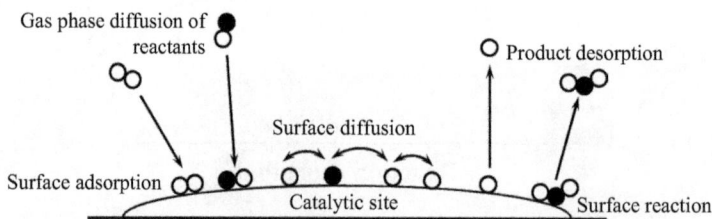

Figure 16.5. Schematic illustration of a heterogeneous combustion process.

complexity to the system as well as the fabrication process. The primary physical limitations include:

1. transport of the fuel and oxidizer species to the active surface sites,
2. minimizing gas to surface transport time while also minimizing pressure loss,
3. start-up and stability of the reaction process, and
4. heat loss through the solid catalyst surface.

16.4 CHALLENGES OF MICROGAS TURBINE COMBUSTORS

There are fundamental challenges unique to a microcombustor designed for integration with other microgas turbine engine components. These include residence-time constraints, material constraints, fluid–structure thermal coupling, the overall engine thermodynamic cycle efficiency requirements, and fabrication and packaging.

16.4.1 COMPETING TIME SCALES

Among the most fundamental physical restrictions for a microcombustor are the significantly reduced time-scales available for reactions to take place. For energy conversion applications, power density is a critical performance metric. For gas turbine combustors, the differences in power density across length scales are shown in Table 16.1. The high power density of a microcombustor directly results from high mass flow per unit volume. It is these small volumes that

Table 16.1. Comparison of operating parameters and requirements for a microgas turbine engine combustor with those estimated for a conventional GE90 gas turbine combustor [2]

	Conventional gas turbine combustor	Microengine combustor
Length	0.2 m	0.001 m
Volume	0.073 m^3	6.6×10^{-8} m^3
Cross-sectional area	0.36 m^2	6×10^{-5} m^2
Inlet total pressure	37.5 atm	4 atm
Inlet total temperature	870 K	500 K
Mass flow rate	140 kg/s	1.8×10^{-4} kg/s
Residence time	~7 ms	~0.5 ms
Efficiency	>99%	>90%
Pressure ratio	>0.95	>0.95
Exit temperature	1800 K	1600 K
Power density	1960 MW/m^3	3000 MW/m^3

Note: Residence times are calculated using inlet pressure and an average flow temperature of 1000 K.

increase the power density. However, time scales associated with chemical reaction kinetics do not vary with mass flow rate or combustor volume. Only temperature, pressure, and species concentrations can affect the kinetics. As a result, the realization of high power density is contingent upon completing the combustion process within a significantly shorter combustor through-flow time which results from the small volume.

One unusual aspect of the microgas turbine engine is that the combustor is also the structure of the engine so its basic design and dimensions must be finalized quickly in order to permit development of the rest of the engine. As a result, time was not available to perform the usual detailed investigations and the combustor had to be designed largely on the basis of the extremely simple time-scale analyses described below.

The homogeneous Damköhler number is the non-dimensional parameter which best quantifies the effects of reduced residence time. It is defined as the ratio of the residence time to the characteristic chemical reaction time:

$$Da_h = \frac{\tau_{residence}}{\tau_{reaction}} \qquad (16.12)$$

A Da_h of unity or greater is required in order to achieve "complete" combustion. This constraint is met either by increasing the flow residence time or decreasing the chemical reaction time; however this can be difficult in cases where the system geometry, thermo-fluid requirements, and fuel type are fixed.

The characteristic combustor residence time can be estimated as

$$\tau_{residence} \approx \frac{\text{volume}}{\text{volumetric flow rate}} = \frac{VP}{\dot{m}_{RT}}, \qquad (16.13)$$

and the corresponding power density is given by

$$\text{Power density} \propto \frac{\dot{m}_f LHV}{V} \propto \frac{\rho_f}{\tau_{residence}} \propto \frac{\rho_f}{Da_h \tau_{reaction}}. \qquad (16.14)$$

Chemical reaction times can be estimated using an Arrhenius expression for the reaction rate.

$$\tau_{reaction} \approx \frac{[\text{fuel}]_0}{A[\text{fuel}]^a[O_2]^{b_e}{}^{-Ea/RT_0}} \qquad (16.15)$$

Achieving high power density requires high mass flow rates through small volumes but increasing the mass flow rate drives the Damköhler number down. This leads to a tradeoff between power density and residence time (or Damköhler number) as shown in Equation 16.14. One way to increase power density without lowering the overall Damköhler number is to use a catalyst to reduce the chemical reaction time.

The performance of catalytic microcombustors depends upon several additional time-scales: the time required for reactants to be transported to the surface (often by diffusion) and adsorb to it, the time for them to diffuse across/through the surface to come into contact with an active site, and the time for products to diffuse away from the active site and desorb from the surface. In many catalytic systems, the time for transport from the gas phase to the surface is the rate limiting process. Thus, a second Damköhler number based on the ratio of diffusion to reaction time also becomes important.

$$Da_2 = \frac{\tau_{\text{diffusion}}}{\tau_{\text{reaction}}} \tag{16.16}$$

The final ratio of time-scales to consider is the ratio of diffusion time to residence time which is referred to as a Peclet number. For large Peclet numbers (>>1) all reactants do not have time to diffuse to the active catalytic surface before exiting the system:

$$Pe = \frac{\tau_{\text{diffusion}}}{\tau_{\text{reaction}}} \tag{16.17}$$

In this case, the chemical reaction is said to be "diffusion limited" because the diffusion time is much greater than the chemical time. The opposite is true at low Peclet numbers (<<1) where the chemical reaction time at the active site may control the combustion process. If both Damköhler numbers are of the same order, then the microcombustor operates in a "diffusion controlled" mode, where the overall reaction rate is controlled by the Peclet number and power density can be rewritten as

$$\text{Power density} \propto \frac{\rho_f P_e}{\tau_{\text{diffusiom}}}. \tag{16.18}$$

Hence, there is a basic tradeoff between power density and diffusion time. The Peclet number can also be written as

$$Pe = \frac{Lv}{D} \tag{16.19}$$

where L is the reactor length and v is the velocity. Consequently, power density becomes

$$\text{Power density} \propto \frac{\rho L v}{D \tau_{\text{diffusiom}}}. \tag{16.20}$$

Equation 16.20 shows that for a given operating pressure, temperature, and reactant species, reducing diffusion time increases power density. Reducing the diffusion time scale can be achieved by increasing catalyst surface area-to-volume ratio; however, this comes at the expense of pressure loss.

The sensitivity of Brayton cycle's efficiency to combustor pressure loss makes it a critical design parameter. Minimizing pressure loss is challenging in microcombustors due to the inherently small size scales of the structure. In the case of a catalytic device, this is even more difficult to mitigate. The need for a high surface area to promote the surface reaction scales inversely with pressure drop resulting in a fundamental tradeoff. In the case of a catalytic microcombustor where catalyst material is distributed throughout the combustion chamber, pressure loss can be estimated using the Darcy friction factor,

$$f = \frac{64}{Re_d} \tag{16.21}$$

and

$$-\frac{dP}{dz} = f\left(\frac{L}{d_h}\right)\left(\frac{1}{2}\rho v^2\right)$$

(16.22)

for single channels within a monolithic multi-channel geometry. The Ergun equation can be used for a porous type distribution of material:

$$-\frac{dP}{dz} = \frac{v}{L}\frac{(1-\alpha)}{\alpha^3}\left[\frac{150\mu(1-\alpha)}{L}+1.75\rho v\right].$$

(16.23)

16.4.2 HEAT TRANSFER AND FLUID–STRUCTURE COUPLING

Energy loss due to heat transfer is usually neglected in conventional-scale gas turbine combustors but cannot be in microcombustors. Waitz et al. [5] have shown that the ratio of heat lost to that generated scales with the hydraulic diameter as

$$\frac{Q_{loss}}{Q} \propto \frac{1}{d_h^{12}}.$$

(16.24)

Therefore, while heat transfer losses are usually neglected in conventional-scale gas turbine combustors whose diameters are approximately tens of centimeters, they cannot be neglected in microcombustors whose diameters are ~1 mm. This large thermal loss has a direct impact on the combustor's thermal efficiency (Equation 16.11) but also reduces the chemical efficiency (Equation 16.10) by lowering the reaction temperature which increases kinetic times according to Equations 16.7 and 16.8. This narrows flammability limits and exacerbates residence time constraints.

Several factors make the heat loss problem especially problematic in catalytic devices: one is the need for higher surface area to volume ratios to facilitate transport to the active surface. Another is the fact that the heat release occurs on a solid, highly conductive surface with a direct conduction path through the structure to the exterior of the device. Unlike gas-phase microcombustors, no thermal boundary layer between the hot reacting gases and the wall is available to provide insulation and limit heat loss. In fact, the thermal boundary layer has the opposite effect in a catalytic device: it isolates the hot wall from the relatively cooler free-stream flow.

16.4.3 MATERIAL CONSTRAINTS

Choice of materials for microcombustors is often limited by the availability of fabrication techniques which can produce the appropriate feature sizes. For the microgas turbine upon which this discussion is primarily based, semiconductor and MEMS microfabrication methods were used. Deep Reactive Ion Etching (DRIE) in silicon wafers and aligned fusion bonding are one means of achieving two-dimensional extruded geometries which serve as combustion chambers, fuel/air mixers, fuel injectors, and other components. These techniques are not well developed for other materials; hence all of the microcombustion systems discussed in the remainder of this chapter are silicon-based.

For silicon microcombustion systems, the most critical requirement is to limit wall temperature to less than ~950 K which is the point beyond which single crystal silicon will plastically deform under some stress [8]. While this temperature is relatively low, silicon also has high thermal conductivity which helps to conduct heat away and reduce wall temperature. Of course the resulting conductive heat loss also degrades combustor performance as noted previously. Silicon-based microcombustors are usually nearly isothermal. This is a result of the combination of the high thermal conductivity of silicon and the short conduction paths in a microsystem. This is quantified by the heat transfer Biot number

$$B_i = \frac{hL}{k} = \frac{\text{resistance to conductive heat transfer}}{\text{resistance to convective heat transfer}} \tag{16.25}$$

which is typically $\ll 1$ for these systems.

The use of catalytic materials introduces additional constraints. For example, noble metals like platinum may agglomerate in thin layers at temperatures greater than 1000 K [10,11] thus reducing the available surface area. This can also put limits on fabrication methods like wafer bonding and various packaging techniques which often occur at high temperatures.

16.4.4 DESIGN SPACE

While time-scale, heat transfer, and material constraints are important, frequently in microgas turbines the most limiting constraints arise from the thermodynamic cycle used to produce the useful work. Figure 16.6 shows a typical design space for a hydrogen-fueled micro-scale gas turbine combustor as a function of equivalence ratio and heat rejected [2]. A turbine inlet temperature between 1600 and 1800 K is required for the cycle proposed for this particular engine. The flame stability limit represents stable operation of a non-adiabatic, perfectly stirred reactor while the thermal stress constraint indicates the material limits of an all silicon structure. The design space is further bounded by the desire for lean burning and the flammability limits of the fuel–air mixture.

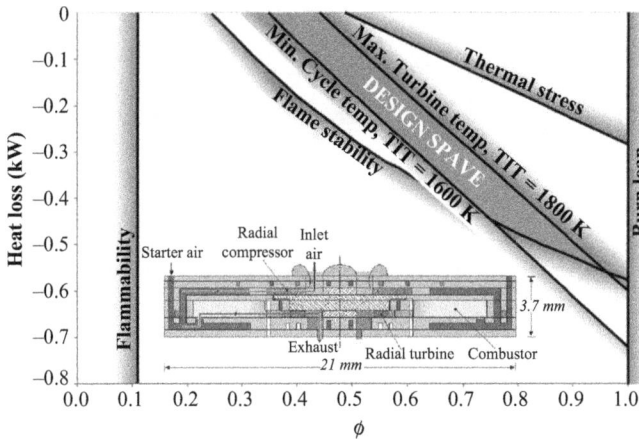

Figure 16.6. Microcombustor design space [2].

16.5   HOMOGENEOUS GAS-PHASE MICROCOMBUSTORS

The development of homogeneous gas-phase microcombustors began in the mid 1990s with simple metal flame tube experiments. This later evolved into a six silicon wafer, axisymmetric, dual-zone combustor designed for integration with a complete microgas turbine engine. Here we will discuss the highlights of this work.

16.5.1   REVIEW OF EARLY MICROGAS TURBINE COMBUSTORS

Early experiments by Waitz et al. and Tzeng [5, 12] used a flame tube to map the flammability boundaries of hydrogen-air mixtures in a small diameter tube. The low equivalence ratio hydrogen-air mixtures were found to be limited by heat loss in millimeter-scale tubes. Using the knowledge gained from the flame tube experiments, the first microcombustor designed for a microgas turbine engine was developed. This combustor was 0.13 cm^3 and was machined out of steel. Lean premixed hydrogen–air combustion was stabilized in this system. Measurements made in this device laid the foundation for the development of the first silicon microfabricated combustors [12]. Mehra and Waitz went on to develop the first silicon microcombustor [13]. This three-wafer combustor was 0.066 cm^3 and was designed for integration with the first microgas turbine engine. A schematic and a scanning electron micrograph (SEM) image of the device's cross section are shown in Figures 16.7 and 16.8 respectively [13].

This three-wafer system was tested over a range of equivalence ratios from 0.4 to 1.6 for a fixed mass flow rate of 0.045 g/s and atmospheric pressure resulting in a residence time of approximately 0.5 ms. Exit gas temperatures in excess of 1800 K were achieved with overall combustor efficiencies of up to 70 percent for premixed hydrogen–air. Chemical efficiency was estimated to be nearly 100 percent so it was concluded that heat transfer from the combustion chamber to the surroundings was mostly responsible for the low overall efficiency. The power density of the device was approximately 1200 MW/m^3 and operated successfully for tens of hours [13]. Mehra went on to design and fabricate a six silicon wafer microcombustor based on the engine layout shown in Figures 16.3 and 16.9.

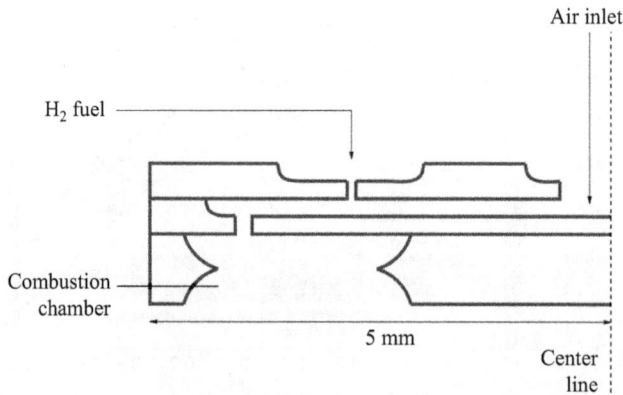

Figure 16.7.   Schematic of three-wafer silicon microcumbustor [13].

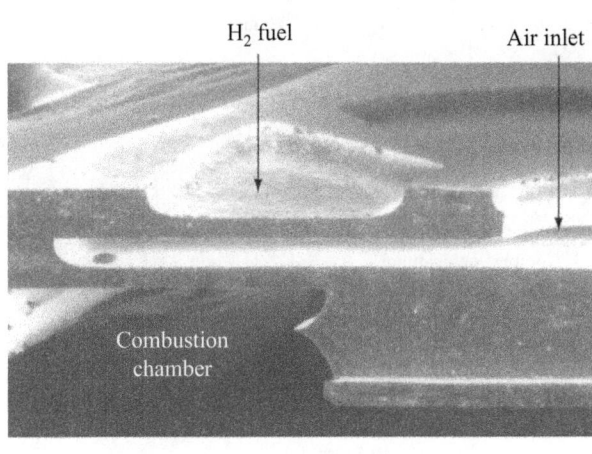

Figure 16.8. Scanning electron micrograph (SEM) of the three-wafer device [13].

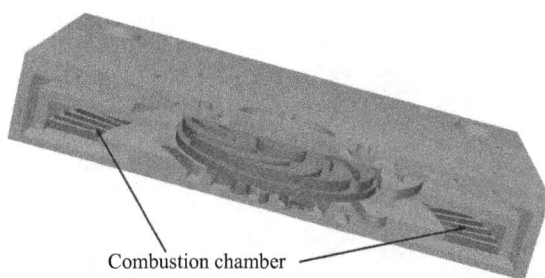

Figure 16.9. 3D schematic of microgas turbine engine.

16.5.2 BASELINE SIX SILICON WAFER GAS-PHASE MICROCOMBUSTOR

A 191 mm^3 diameter radial inflow axisymmetric microcombustor fabricated from six silicon wafers is shown in Figures 16.10 and 16.11. It includes a recirculation jacket which serves as both a thermal management system and fuel/air mixer, multiple fuel injection ports, and can be fabricated with various flame holding inlet geometries.

Experiments were performed with premixed hydrogen–air in this device. Figures 16.12 and 16.13 show combustor exit temperature and efficiency as functions of mass flow rate. The mass flow rate was controlled by varying the inlet pressure for a fixed exhaust area. Exit gas temperatures in excess of 1600 K were achieved for a mass flow rate of 0.11 g/s and an operating pressure of 1.13 atm. The overall efficiency exceeded 90 percent and the power density was approximately 1100 MW/m^3. The break in the $\phi = 0.5$ and $\phi = 0.6$ exit temperature and efficiency curves is due to the approximately 1600 K temperature limit of the type K thermocouples used for these measurements. Above this temperature, type K thermocouples become unreliable and rapidly cease to give any reading at all. The $\phi = 0.7$ curve terminates at a mass flow rate of 0.015 g/s due to upstream burning in the recirculation jacket resulting in destruction of the silicon structure [2, 4, 13].

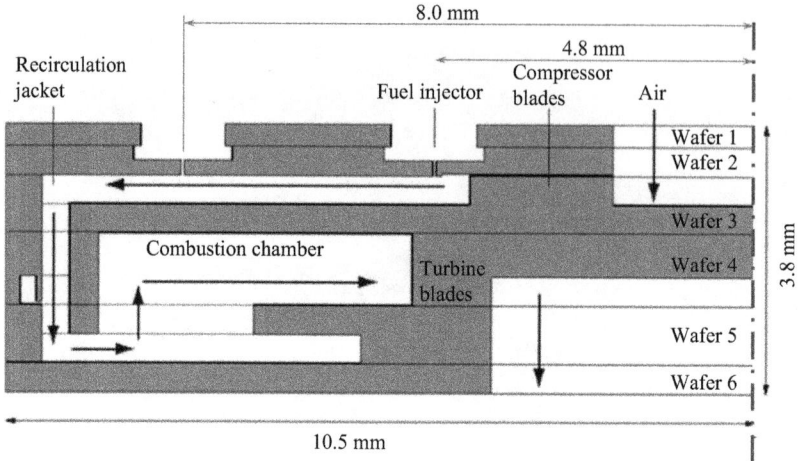

Figure 16.10. Schematic of baseline six wafer microcombustor [13].

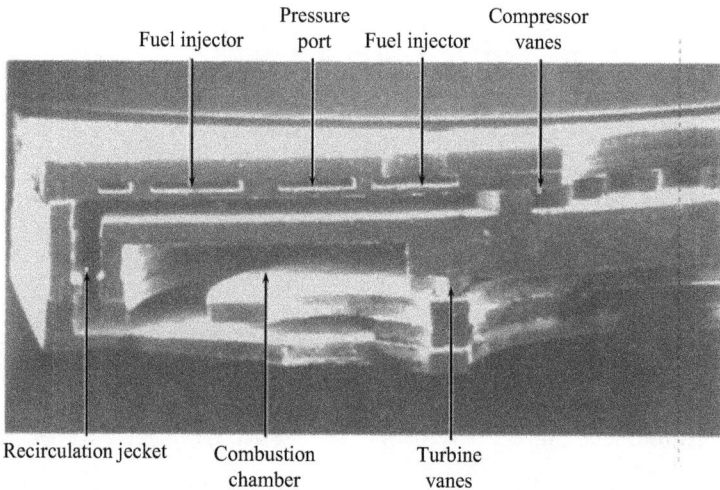

Figure 16.11. Scanning electron micrograph (SEM) of baseline six wafer microcombustor [13].

The lean blowout characteristics of this microcombustor can be explained by examining the change in Da_h over a constant equivalence ratio operating line. Using Equation (16.13) for residence time, and a simple one step hydrogen–air reaction mechanism,

$$\tau_{\text{reaction}} \approx \frac{[H_2]_0}{A[H_2]^a[O_2]e^{-Ea/RT}}, \tag{16.26}$$

With $E_a = 10950$ cal/mole and $A = 1.62 \times 10^{18}$ (cm³/mole)² sec, to calculate Da_h [14, 15]. Figure 16.14 shows a plot of reaction time, residence time, and Da_h versus mass flow rate for a constant temperature and equivalence ratio as an example in order to illustrate these tradeoffs.

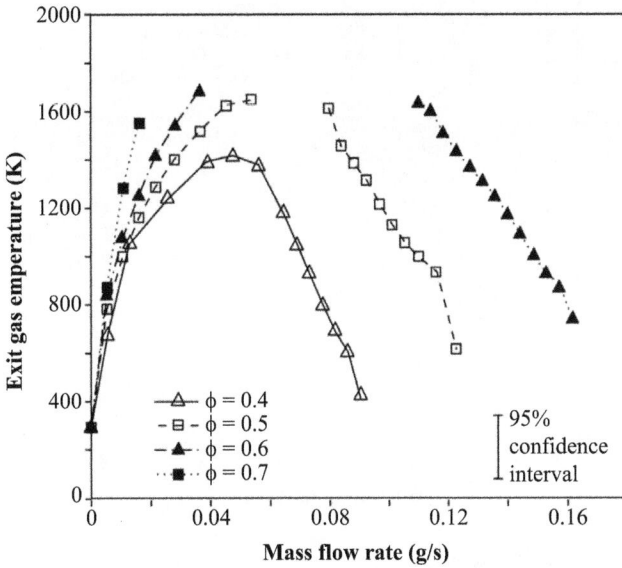

Figure 16.12. Exit gas temperature versus mass flow rate for baseline six wafer microcombustor [13].

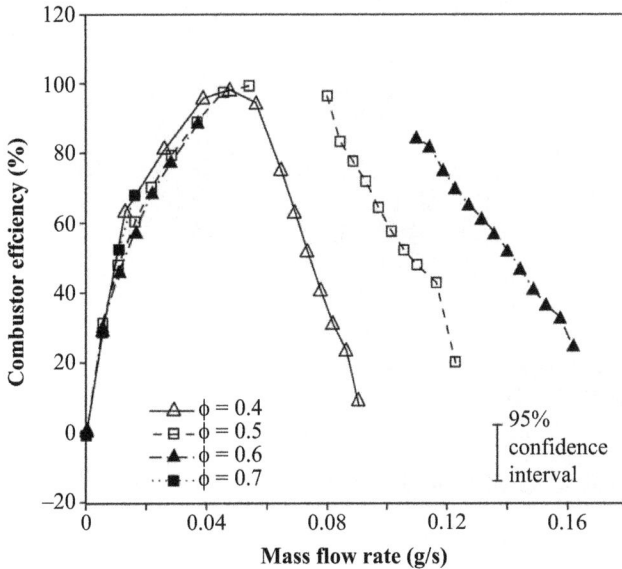

Figure 16.13. Overall combustor efficiency versus mass flow rate for baseline six wafer microcombustor [13].

Initially, the residence time decreases more rapidly than the chemical reaction time as mass flow increases. Consequently, Da_h drops below unity indicating that residence time is less than reaction time, likely causing chemical inefficiency and flame blowout. Although Da_h will begin to rise again as pressure rises with flow rate and the device enters the choked flow regime and chemical time-scales fall, in a practical system, the flame may blowout prior to achieving this condition.

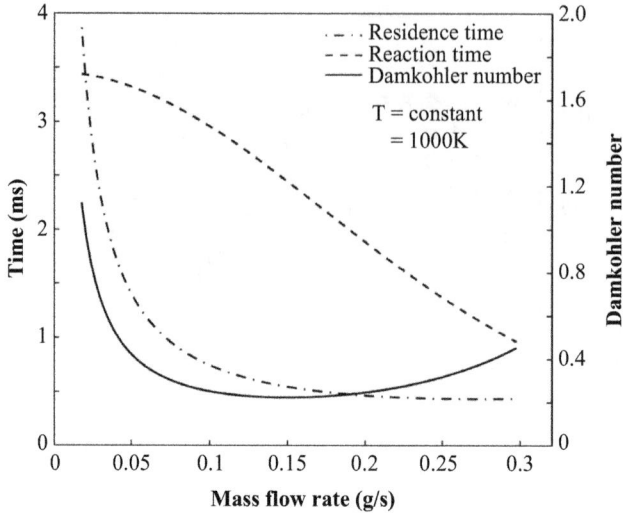

Figure 16.14. Reaction time, residence time, and Da_h versus mass flow rate for baseline six wafer microcombustor [13].

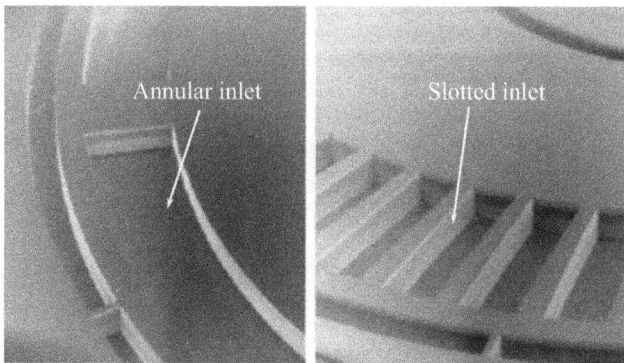

Figure 16.15. SEMs of annular (left) and slotted (right) inlet geometries [13].

16.5.3 RECIRCULATION ZONES AND FLAME STABILITY

Fluid recirculation zones and local eddies are useful in stabilizing flames. Two types of combustion chamber inlets were used to investigate the effect of different flameholding geometries: a 1.2 mm wide annular inlet and an array of 60 slots 2.2 mm long. The multi-slot geometry was intended to create multiple small recirculation zones for more rapid and uniform ignition of the incoming flow. SEMs of both types of inlets are shown in Figure 16.15 [13].

Combustor efficiency for the slotted-inlet geometry is plotted in Figure 16.16. Performance is similar to that of the annular-inlet combustor (Figures 16.10 and 16.11); however, the slotted-inlet produced higher exit temperatures and efficiencies in the high mass flow part of the operating space. This can be attributed to the presence of multiple recirculation zones which more rapidly and uniformly ignite the incoming mixture. The sharp drop in performance for the

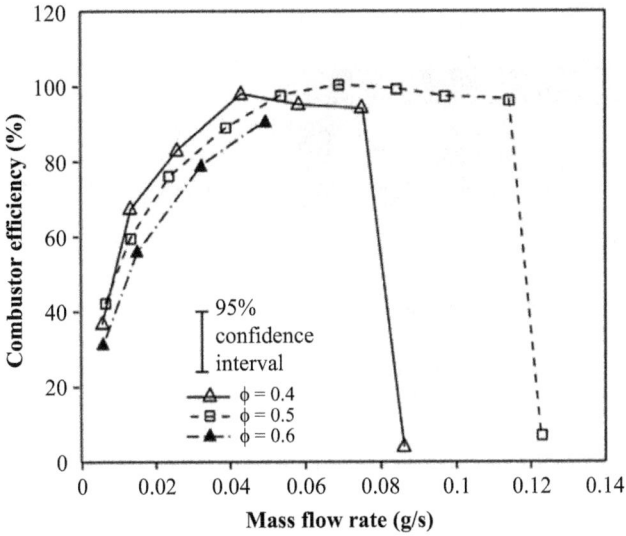

Figure 16.16. Overall combustor efficiency versus mass flow rate for a six wafer combustor with slotted inlets [13].

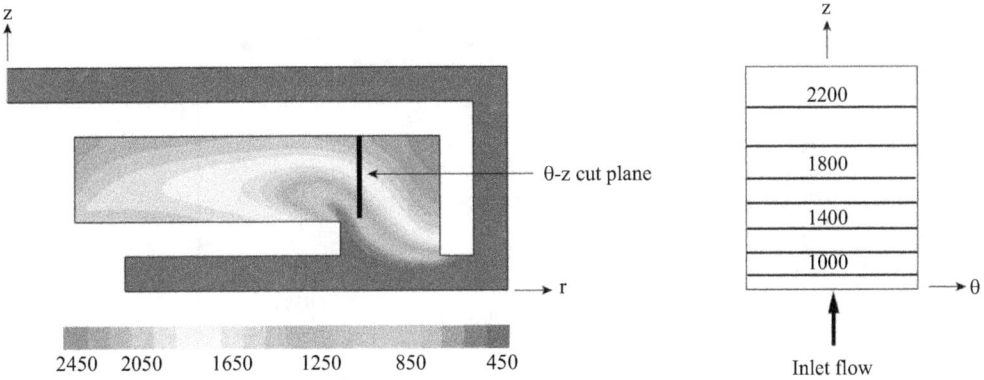

Figure 16.17. Temperature contours in the r–z and z–θ planes for the annular inlet six-wafer microcombustor [13].

slotted combustor is due to the rapid extinction of these small ignition zones. While the annular inlet geometry with a single large ignition zone stays lit over a broader range of mass flow rates, it does not ignite the incoming reactants as uniformly and completely [13].

Without internal measurements or flow visualization, numerical simulation is the only tool available to study the combustion process. Figure 16.17 shows temperature contours generated from computational fluid dynamics (CFD) calculations for the annular inlet geometry. The temperature gradient from the bottom of the combustor to the top (in the r–z plane) indicates that the primary ignition zone is in the upper right corner of the combustor. Figure 16.16 shows the same contour plots for a combustor with a slotted inlet. Note that temperatures are significantly higher overall and in the lower regions of the combustion chamber. The shape of the contours in the z–θ plane also indicate that there are smaller, hot ignition zones near the slotted inlet as expected (Figure 16.18) [1, 2].

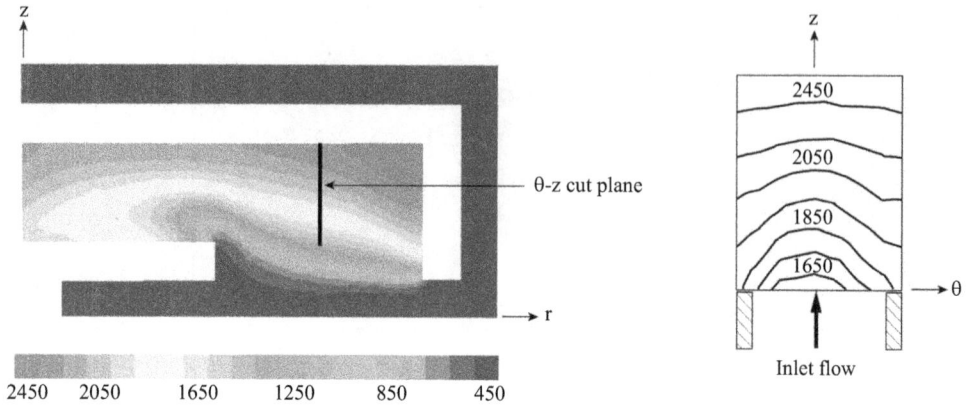

Figure 16.18. Temperature contours in the r–z and z–θ planes for the slotted inlet six-wafer micro-combustor [13].

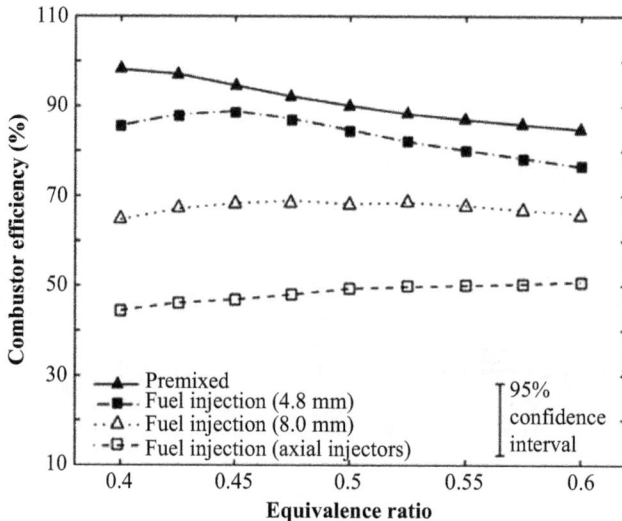

Figure 16.19. Overall combustor efficiency for each set of fuel injector versus equivalence ratio [13].

16.5.4 FUEL INJECTION AND MIXING

The six wafer microcombustor was also used to evaluate fuel injection schemes and mixing distances. Three sets of fuel injectors were fabricated into the device at different locations within the recirculation jacket to examine the effect of mixing length scales. By comparing overall combustor efficiencies for each fuel injector location, the required mixing length can be observed. This can also be compared to the "ideal" performance associated with pre-mixed operation. Figure 16.19 plots overall efficiency for each set of fuel injectors versus equivalence ratio at a constant mass flow rate of 0.045 g/s. The injectors were located at radii of 4.8 mm and 8.0 mm into the recirculation jacket (see Figures 16.10 and 16.11) as well as in the sidewall at the base of the recirculation jacket (axial injectors) just upstream from the combustion chamber. The figure indicates that performance decreases as the injectors get closer to the combustion chamber due to the decreased mixing time and length [13].

16.5.5 DUAL-ZONE MICROCOMBUSTORS

Large-scale combustors normally utilize a multi-zone, or staged combustion strategy where an initial primary zone consisting of high-temperature, near-stoichiometric combustion is followed by dilution jets of air. These jets form a secondary zone with a leaner mixture and lower temperature. A dual-zone microcombustor which operates in a similar fashion was fabricated by adding a circumferential array of holes through the upper combustor wall. These connect the upper recirculation jacket to the combustion chamber allowing a portion of the inlet air to be diverted downstream into the combustion chamber and splitting it into two zones. This air also serves to dilute the hot combustion products from the initial primary zone reducing their temperature to the required turbine inlet temperature. As a result of the desire to divert only air through these dilution holes, fuel must be injected downstream in the recirculation jacket. This results in reduced mixing length-scales and inherently non-premixed operation. In the primary zone, the fuel–air mixture is burned near stoichiometric conditions with higher temperatures increasing kinetic rates. In turn, this decreases reaction time, effectively increasing Da_h. A three-dimensional, adiabatic, reacting flow CFD solution (Figure 16.20) shows the effect of the dual-zone configuration on the temperature distribution [1, 2].

Operational data shown in Figure 16.19 indicate that the dual-zone combustor configuration exhibits slightly lower exit gas temperatures and overall efficiencies than the baseline six wafer device; however, its operating range was broader than either single zone combustor. Combustion

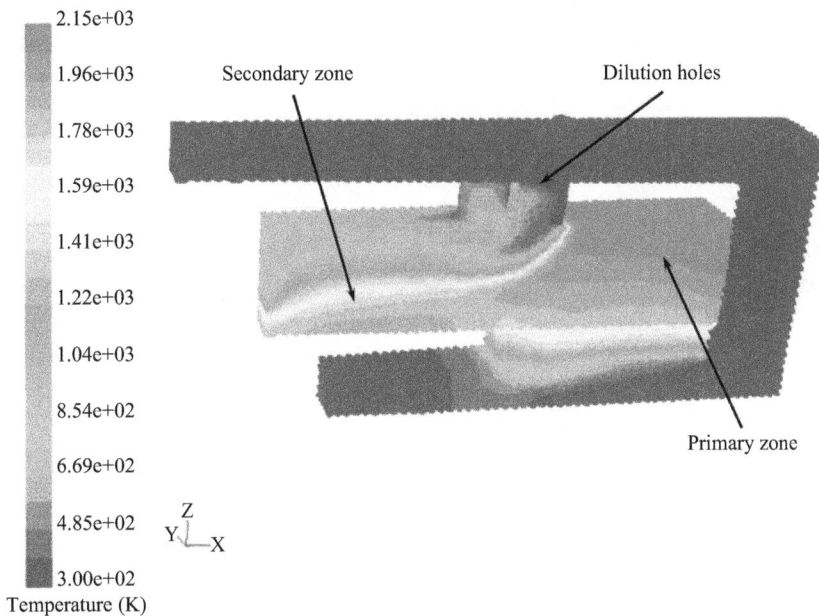

Figure 16.20. Temperature contours in a "dual-zone" microcombustor [1, 2].

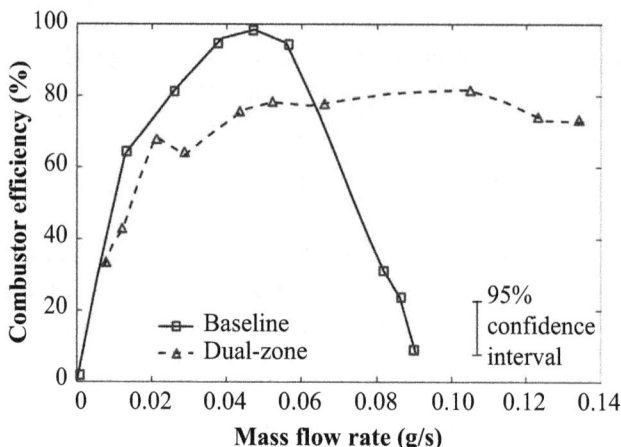

Figure 16.21. Comparison of dual-zone to baseline (unslotted) microcombustor for hydrogen–air operation at $\phi = 0.4$ [1, 2].

was stabilized at equivalence ratios as low as 0.2 and mass flow rates as high as 0.2 g/s [1, 2]. The extension in the operating range is largely due to the hot, stable primary zone. For the same overall efficiency, the dual-zone combustor achieved a 100 percent increase in the mass flow rate before blowout when compared to the standard six wafer device. The ability to operate at the low overall equivalence ratio is a result of a higher local equivalence ratio in the primary zone.

16.5.6 HYDROCARBON–AIR OPERATION

Hydrocarbon fuels were tested in both the baseline and dual-zone versions of the microgas turbine combustor. Fuels which can be easily stored as a liquid such as ethylene (C_2H_4) and propane (C_3H_8) are desirable for a microgas turbine engine due to their high energy densities. Results from combustion tests of the baseline, unslotted configuration are shown in Figures 16.21 and 16.22. Ethylene–air mixtures achieved maximum power density at an equivalence ratio of 0.9 and ~1 atm pressure, with exit temperature exceeding 1400 K. This results in an overall efficiency of 60 percent and a power density of ~500 MW/m^3. Propane reacts significantly slower than ethylene and as a result, combustion could only be stabilized at an equivalence ratio of 0.8 as shown in Figure 16.23. Exit temperatures of 1200 K were achieved at ~1 atm with an efficiency of 55 percent and a power density of ~140 MW/m^3. Residence times for these fuels were 1.6 ms and 2.8 ms for ethylene–air and propane–air respectively and differ largely due to the relatively higher efficiency and temperature associated with the ethylene–air mixture. Reaction time-scales are 0.5–1 ms for ethylene–air and 1–2 ms for propane–air, much longer than hydrogen–air reaction times which are approximately 0.2 ms.

The dual-zone configuration did not result in an extended range of mass flow rates for hydrocarbon-fueled operation largely because premixed fuel–air operation was not possible [1, 13]. Due to the dilution hole location, fuel was injected downstream of these holes in the recirculation jacket. While for H$_2$-air operation this results in adequate mixing time, for hydrocarbons this is not the case and the fuel–air mixture was not well mixed. Table 16.2 summarizes the maximum power densities achieved by both devices with various fuels [1].

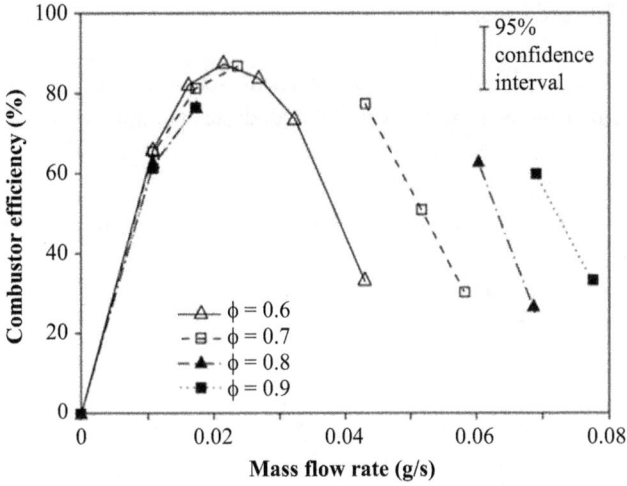

Figure 16.22. Overall combustor efficiency for baseline (unslotted) six wafer microcombustor with various ethylene/air mixtures [1, 13].

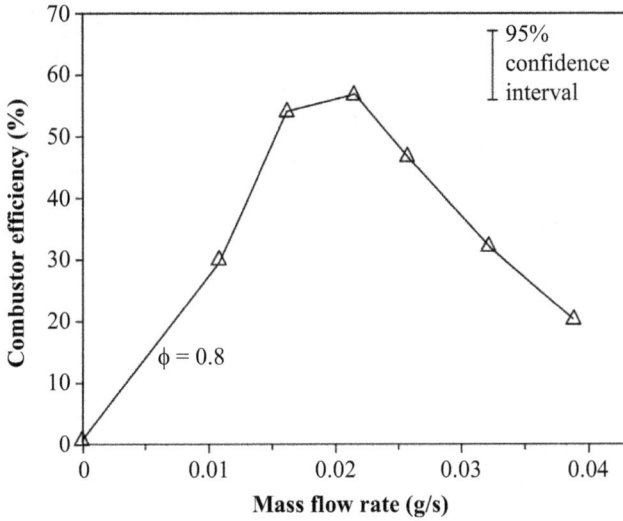

Figure 16.23. Overall combustor efficiency for baseline (unslotted) six wafer microcombustor with various propane/air mixtures [1, 13].

Table 16.2. Maximum power densities and efficiencies for microcombustors [1].

Fuel	Baseline Microcombustor		Dual-zone Microcombustor	
H_2	1150 MW/m^3	>99%	1400 MW/m^3	92%
C_2H_4	500 MW/m^3	85%	100 MW/m^3	38%
C_3H_8	140 MW/m^3	50%	75 MW/m^3	29%

16.5.7 OPERATING SPACE

Figure 16.24 plots the empirically identified operating space for the gas-phase microcombustors operating with hydrogen–air mixtures. The boundaries include:

1. The lean blowout limit.
2. The upstream burning limit where combustion in the cooling jacket occurs and the device is destroyed.
3. The structural limitations.
4. The 1600 K required turbine inlet temperature associated with the microengine thermodynamic cycle.

The light gray area represents the operating range for the baseline, unslotted device while the dark gray area represents the extension due to the dual-zone configuration. The upstream burning limit is a result of flashback into the recirculation zone. At very low mass flow rates, total heat release, and temperature rise are low enough that flashback does not occur but as temperature rises at relatively low flow rates and hence, longer residence times, flashback becomes more likely. There are two nominal design mass flow rates of 0.2 and 0.35 g/s. Figure 16.24 illustrates the difficulty in achieving the desired operating conditions in the baseline design while remaining within these limits. The dual-zone combustor has a broader operating space for hydrogen–air operation due to its stable, hot primary zone with a lower lean blowout limit and the absence of an upstream burning limit due to its inherent non-premixed operation.

Microcombustor performance is a function of both chemical and thermal losses which can be quantified in terms of the non-dimensional power density. It is defined here as power density normalized by the maximum possible power density (that which is achieved with 100 percent chemical and thermal efficiency) at given conditions. If it is assumed that there is adequate fuel-air mixing and rapid, uniform ignition, the non-dimensional power density

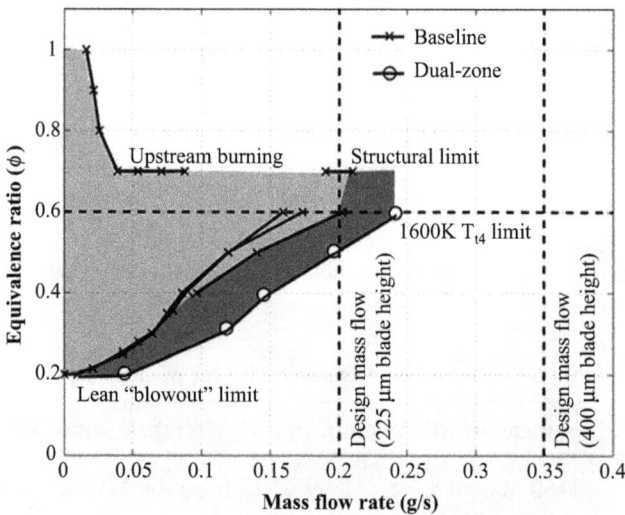

Figure 16.24. Operating space for six wafer and dual-zone microcombustors [1].

is a function of only two parameters: the Damköhler number and the non-dimensional heat loss. These can be considered proxies for chemical and thermal efficiencies respectively that are more easily estimated a priori. The Damköhler number can be estimated from residence time as previously described and a suitable reaction kinetics software package with detailed reaction mechanisms for chemical time estimates. Heat loss can be cast in non-dimensional terms by normalizing the heat lost from the device by the maximum possible heat that can be generated. The heat lost can be estimated using a lumped capacitance model due to the inherent isothermal nature of silicon microcombustors. Higher-fidelity thermal models could also be used to estimate this quantity.

Figure 16.25 plots this non-dimensional power density for several microcombustors in three-parameter space. This includes multiple fuels: hydrogen, ethylene, and propane. The heat loss parameter and homogenous Damköhler number are on the x- and y-axes, respectively, and a second order, least squares surface fit was used to generate contours of non-dimensional power density (the fit explains 65 percent of the variance in the data) [1, 2].

Thermal efficiency in silicon microgas turbine combustors is inversely proportional to heat loss whereas chemical efficiency is directly proportional to the Damköhler number. As a result, optimum performance is achieved at low levels of heat loss and high Damköhler number. At low mass flow, performance is limited by heat loss. Conversely, at high mass flow, the Damköhler number is lower and performance is limited by chemical losses. The highest power densities are achieved at moderate mass flow rates where these two competing effects are optimized.

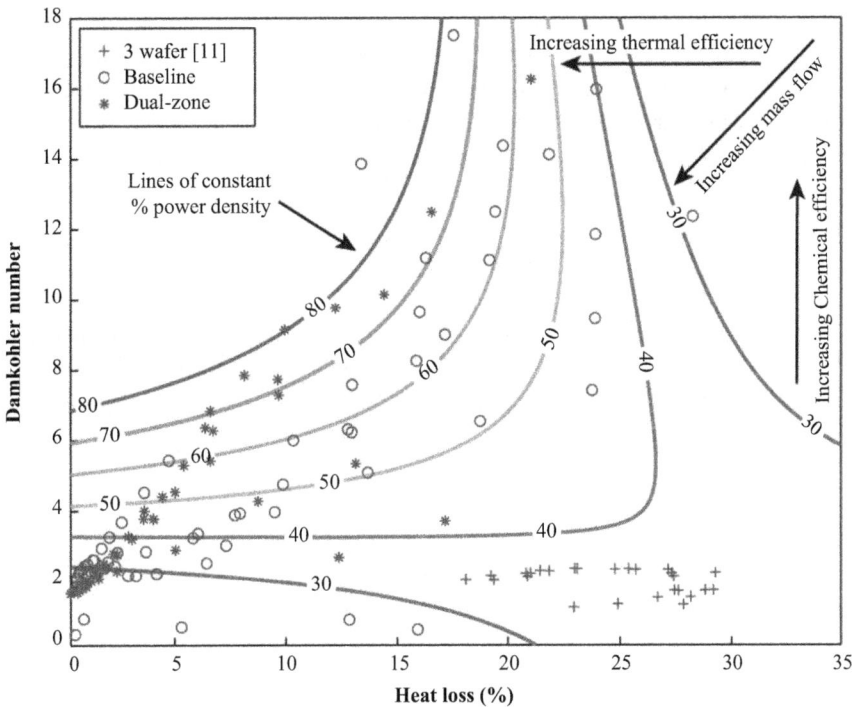

Figure 16.25. Non-dimensional operating space for six wafer and dual-zone microcombustors [1, 2].

16.6 HETEROGENEOUS (CATALYTIC) MICROCOMBUSTORS

We now have an understanding of the how the volume required for a practical gas-phase device depends on requirements for mass flow, exhaust gas temperature, and fuel type. Additionally, we have reviewed the details of specific microcombustor elements and physics. This has provided an understanding of the operating behavior and performance of a microcombustor with homogeneous gas-phase hydrocarbon-air mixtures. Results and analyses previously presented has led to the conclusion that operating with a more practical hydrocarbon fuel like propane will require at least a five times larger volume than a device operating with hydrogen for the same nominal performance. The difference is primarily a result of the difference in chemical kinetic rates of the two fuels.

For most practical applications of microcombustors, operation with storable liquid hydrocarbon fuels such as propane or ethylene is required. Chemical conversion rates must be increased in order to utilize these fuels without significantly increasing combustor volume or temperature, or decreasing power density. Surface catalysis is a means of increasing reaction rates by lowering activation energy. Heterogeneous reactions of hydrocarbon–air mixtures over noble metal catalysts are faster than those in the gas-phase which leads us to the development of catalytic microcombustors as an attractive alternative.

A typical large-scale catalytic combustor consists of a monolithic structure of parallel channels made from a refractory material like alumina. The surface of these channels is coated with a ceramic "washcoat" to which nodules of noble metals like platinum or palladium are adsorbed. There are also more exotic and proprietary metal oxide catalysts which have been developed. The reactants flow through the channels and diffuse to the catalytic surface where the fuel and air molecules are then adsorbed. This is followed by transport to active "sites" on the surface and rapid reaction and heat release. Subsequently, the reaction products are desorbed and flow out of the channel. The heat liberated by the reaction is transferred both conductively through the solid and convectively to the fluid through a thermal boundary layer. Eventually, any remaining gas phase reactant which did not reach the catalyst surface may be ignited by heat provided by the surface reaction and burn homogeneously in the gas phase. Figure 16.26 shows a schematic of a typical catalytic combustor.

A micro-catalytic combustor was constructed by filling the combustor volume shown in Figure 16.8 with a porous substrate made of nickel foam. A photograph of the substrate along with an SEM image of the pores are shown in Figure 16.27 (a) and 16.27 (b). The porosity of the substrate was 95 percent where porosity is defined as open volume normalized by total geometric volume:

$$\alpha = \frac{V_{open}}{V_{total}}. \qquad (16.27)$$

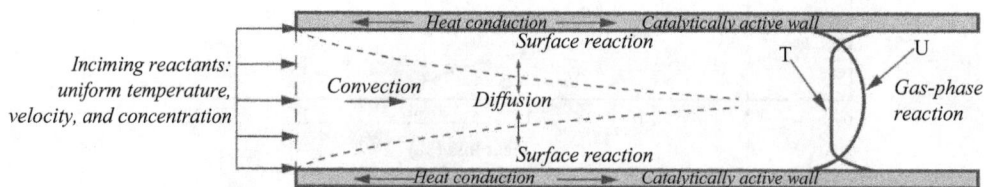

Figure 16.26. Schematic illustration of processes occurring in a "typical" catalytic combustor.

(a)   (b)

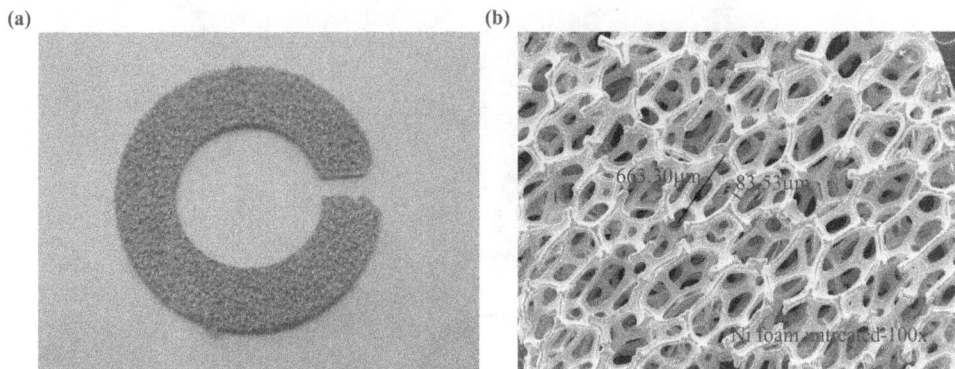

Figure 16.27. Nickel foam substrate material (~95% porosity): (a) photograph; (b) SEM (×50) [1].

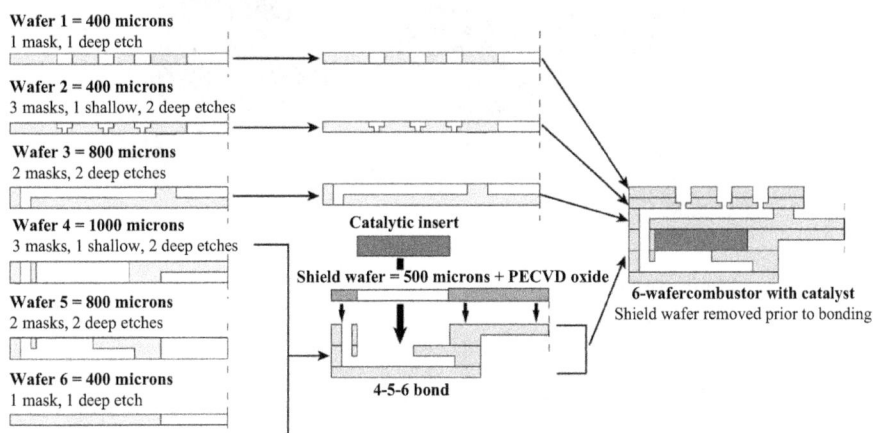

Figure 16.28. Fabrication, assembly, and bonding process for six-wafer catalytic microcombustor [1].

The porous substrates were conformally coated with platinum which acted as the active catalytic surface. Coating was accomplished by soaking the substrates in a metal salt solution (H_2PtCl_6 and deionized water here) and then baking in a reducing furnace. Water evaporates in the furnace and H_2 strips the remaining Cl into the gas phase to form HCl which is carried away leaving only platinum on the surface. The nickel foam substrates coated using this technique typically increased in weight by approximately 3 to 5 percent.

Trapping the catalytic inserts within the combustor volume was a challenging manufacturing process that required the development of a new process. First, a "shield wafer" was fabricated to protect the bonding surfaces, which must remain uncontaminated in order to achieve a fusion bond. This wafer was etched through with the shape of the inserts and silicon dioxide was deposited on its surface via plasma-enhanced chemical vapor deposition (PECVD) to prevent the shield from sticking to other wafer layers. Second, the fourth, fifth, and sixth wafer levels were fusion bonded to form an open combustion chamber. Then, the shield wafer was aligned and contacted with these three levels. While in contact, the catalytic pieces were pressed manually through the shield waver into the bonded 4–5–6 wafer stack. Finally, the protective shield wafer was carefully removed. A schematic illustration of the assembly process is shown in Figure 16.28 and a photograph of the installation of the catalytic inserts is shown in Figure 16.29.

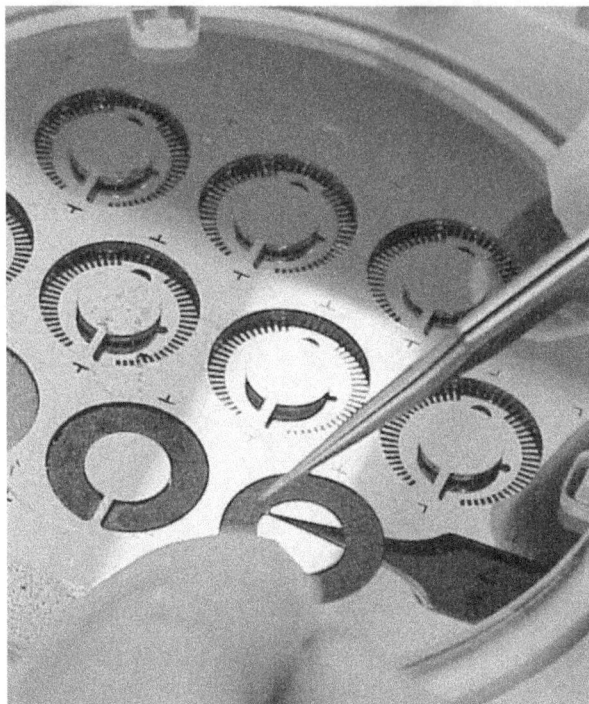

Figure 16.29. Catalytic pieces being inserted into the six-wafer microcombustor during fabrication [1].

16.6.1 IGNITION TECHNIQUES AND CHARACTERISTICS

In order to achieve auto-thermal operation, the catalyst has to be heated to a suitable ignition temperature by either:

1. Beginning with hydrogen–air operation (which will ignites at lower temperatures) to heat the catalyst and then transitioning to a hydrocarbon fuel once the catalyst has exceeded ~500 K.
2. Externally heat the entire microcombustor and begin flow of the hydrocarbon–air mixture when the required pre-heating temperature is achieved.
3. Heat the catalytic element resistively until the required pre-heat temperature is achieved.

A hybrid approach for ignition was used here. First, the entire device was brought to start-up temperature via an external electrical heater while a hydrogen–air mixture was flowing. Catalytic ignition occurred at approximately 20 to 30 percent of maximum heater power, or 80°C to 100°C wall temperature. After removing the external heater the combustor continued to operate auto-thermally. At this point, the hydrogen fuel was slowing phased out while hydrocarbon fuel, such as propane, was directed into the system.

16.6.2 PERFORMANCE OF A CATALYTIC MICROCOMBUSTOR

Plots of exit gas temperature and overall combustor efficiency for a range of mixture ratios are shown in Figures 16.30 and 16.31, respectively. The maximum exit gas temperature for this

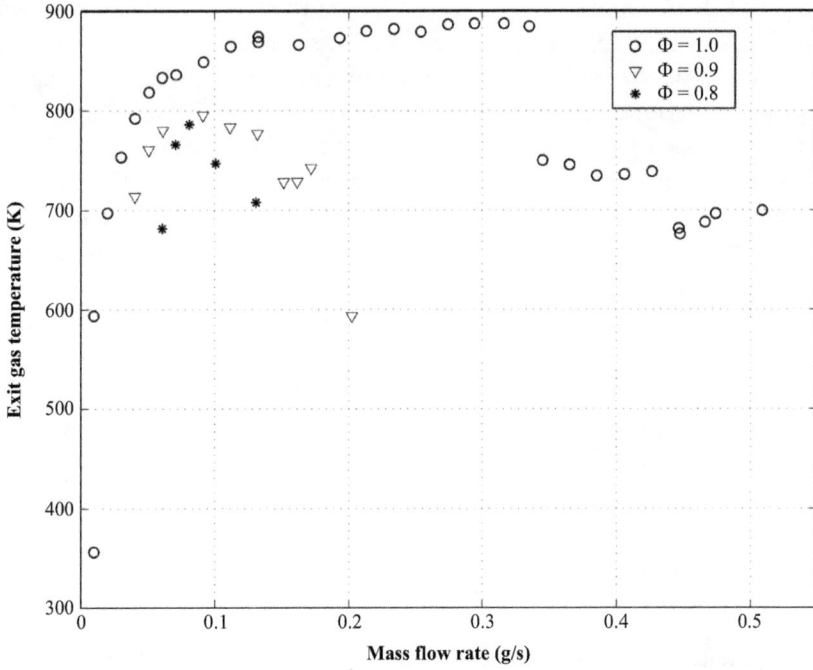

Figure 16.30. Exit gas temperatures versus mass flow rate for a catalytic microcombustor with a 95 percent porous platinum catalyst and propane fuel [1, 3].

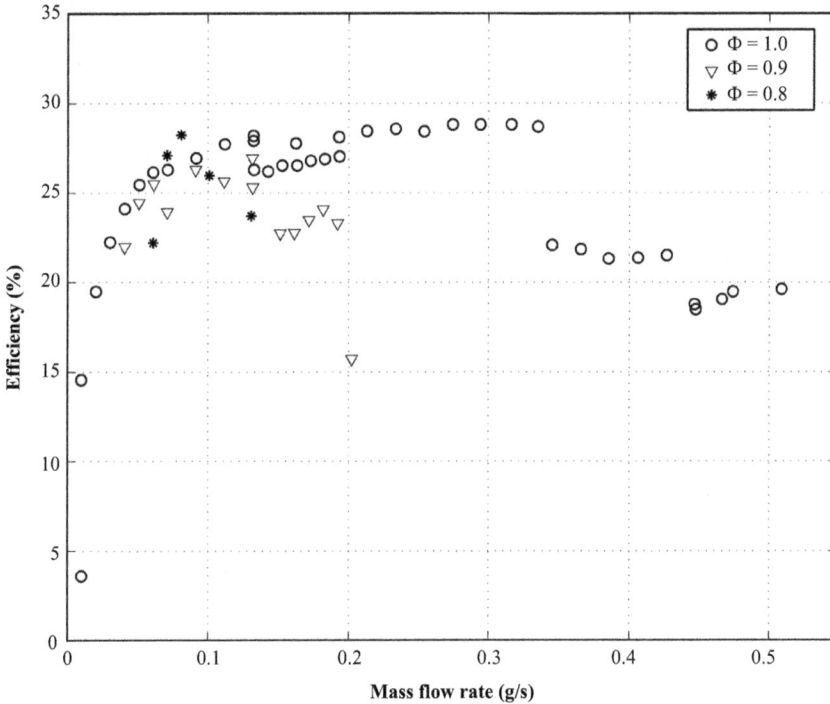

Figure 16.31. Overall combustor efficiency for a catalytic microcombustor with a 95 percent porous platinum catalyst and propane fuel [1, 3].

device was 850 K to 900 K over a wide range of flow rates. Overall combustor efficiencies, as defined by Equation 16.9, approached 30 percent. The sharp drop in performance at 0.35 g/s was due to errors associated with the mass flow controllers in the experimental setup. Although these temperatures and efficiencies are quite low, the mass flow rates achieved were well beyond those measured for the baseline gas-phase device and combustion could be stabilized at flow rates approaching 0.4 g/s resulting in a maximum power density of 1050 MW/m^3. This represents a 7.5 fold increase compared to the gas-phase device with propane–air mixtures [1, 3].

Losses in the catalytic combustor can be identified by separating the overall combustor efficiency into its thermal and chemical components. Wall temperature measurements combined with a simple 1-D heat transfer model [13, p.55] for estimating heat loss, reveals that thermal losses are greater than in the gas-phase case. However, this is not the dominant loss mode. Inefficiency in this device is dominated by unreacted chemical components indicating that most of the reactants pass through the device without ever coming into contact with the active surface. Figure 16.32 shows an estimate of these sub-efficiencies in the device for a stoichiometric mixture ratio.

16.6.3 DIFFUSION AND TIME SCALES

The relevant physical time-scales can be evaluated to determine which phenomena govern the operation of a catalytic microcombustor. These time-scales are reaction time, residence time, diffusion time of the fuel species, and diffusion time of the oxidizer species. Residence time can be estimated using Equation 16.13 and the reaction rate can be obtained from an Arrhenius type

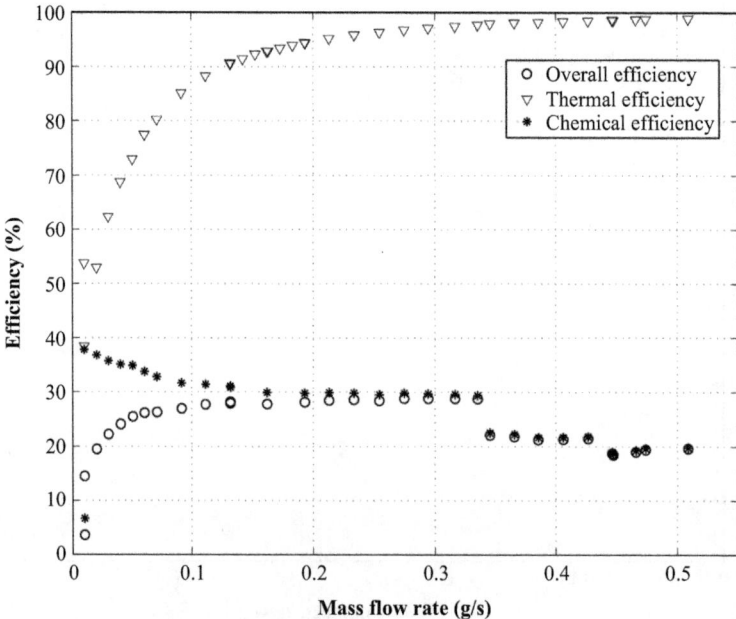

Figure 16.32. Efficiency analysis for a catalytic microcombustor with 95 percent porous platinum and propane fuel at $\phi = 1.0$ [1, 3].

rate expression. For a propane–air reaction on a platinum catalyst, the following single-step mechanism can be used [16]:

$$-R_{C3H8} = -k_s[C_3H_8]$$

(16.28)

where the rate constant is

$$k_s = 2.4 \times 10^5 \exp\left(\frac{-1.08 \times 10^4}{T}\right).$$

(16.29)

Molecular diffusion coefficients for propane and oxygen in air can be estimated using [16]

$$D_{AB} = \frac{(1.013 \times 10^{-2} T^{1.75}) \left(\frac{1}{M_A} + \frac{1}{M_B}\right)^{1/2}}{P\left[(\Sigma v_i)A^{1/3} + (\Sigma v_i)B^{1/3}\right]^2}.$$

(16.30)

In this expression, M_A and M_B are the molecular weights of the reactant species and v_i is the diffusion volume of each atom in the molecule. Diffusion volumes for carbon, hydrogen, and oxygen are 16.5, 1.98, and 5.48 cm^3 respectively [16].

The key non-dimensional parameters relating these time scales are summarized in Table 16.3. The results indicate that the diffusion time is the longest time-scale in the system and thus the diffusion of reactants from the gas phase to the surface is the rate-limiting process. The large values of Da_2 and Pe also suggest that one should expect significant fractions of unreacted fuel in the exhaust.

The Peclet number associated with flow through a porous element is estimated by modeling each pore as a cylindrical plug flow reactor. Although this does not adequately represent the porous catalyst of the experiments discussed previously, it does provide interesting general trends. For the purposes of this simple analysis, Figure 16.33 shows the Peclet number based on fuel or oxidizer diffusion as a function of pore diameter at $P = 2$ atm, $T = 1000$ K, and a mass flow rate = 0.3 g/s. The Peclet number is greater than one for diameters greater than ~100 μm suggesting that incomplete combustion will be a problem for the Nickel foams used in this work whose pore diameters measure in the hundreds of μm. The figure also shows that propane diffuses more slowly than oxygen. This is expected due to its larger molecular size and implies that propane diffusion to the active surface is the rate limiting process.

Table 16.3. Summary of non-dimensional parameters [1]

	Non-dimensional parameter	Range
Damköhler number for residence time	$Da_h = \dfrac{\tau_{residence}}{\tau_{reaction}}$	~1–5
Damköhler number for diffusion time	$Da_2 = \dfrac{\tau_{diffusion}}{\tau_{reaction}}$	~5–20
Peclet number	$Pe = \dfrac{\tau_{diffusion}}{\tau_{reaction}}$	~1–20

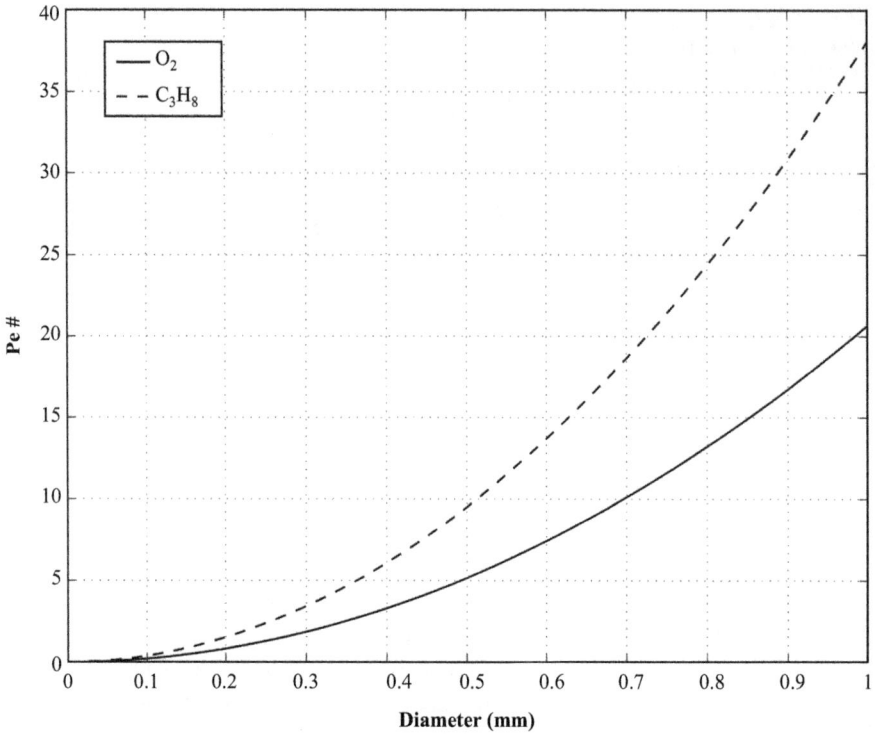

Figure 16.33. Peclet number versus catalytic insert pore diameter for oxygen and propane [1].

16.6.4 ISOTHERMAL PLUG FLOW REACTOR MODEL

In many cases, catalytic microcombustors can be modeled as isothermal plug flow reactors [16]. A simple analytical model can be derived from a control volume analysis as illustrated in Figure 16.34. The objective is to create a design tool that predicts trends in fuel conversion and bulk gas temperature rise through a catalytic combustor as functions of flow conditions and geometry. The initial Peclet number analysis has shown that the diffusion of fuel species is likely the rate limiting process. For this reason, only the mass transport and consumption of the fuel are taken into account here. Homogeneous gas-phase reactions are neglected for simplicity and the surface reaction is modeled using the single-step mechanism shown in Equations 16.28 and 16.29.

A steady-state gas-phase mole balance across the control volume yields [3]

$$C_b v \frac{dY_{f,b}}{dz} + a_v k_m C_b (Y_{f,b} - Y_{f,s}) = 0 \, . \tag{16.31}$$

The surface area-to-volume ratio is

$$a_v = \frac{4(1-\alpha)}{w} \tag{16.32}$$

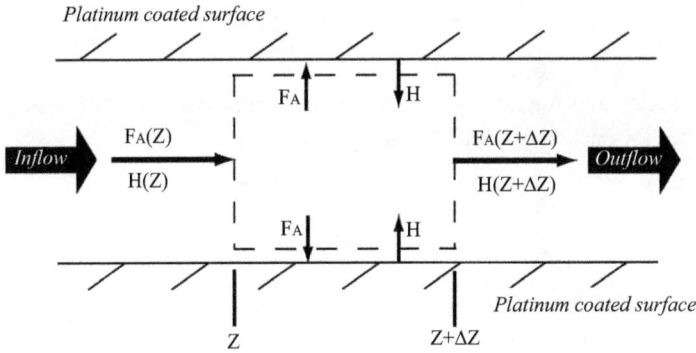

Figure 16.34. Control volume for a fluid element in a catalytic micro-combustor from Spadaccini [1] and Spadaccini, Peck, and Waitz [3]).

where α is the porosity and w is the thickness of a fiber in a porous foam catalyst support such as those used in the devices presented here. A similar balance between the moles of fuel transported to the surface and the moles reacted on the surface:

$$k_m C_b Y_{f,b} - Y_{f,S} = (-R_f)S\ (1-\alpha)\tag{16.33}$$

The bulk gas temperature is determined by performing an energy balance on the fluid element.

$$-\rho C_p v \frac{dT_b}{dz} + a_v h(T_S - T_b) = 0.\tag{16.34}$$

Since the wall temperature is a design constraint, the analysis treats it as a parameter. This is a significant simplification (and computational time saver) as a fourth energy balance across a differential element of the combustor wall would be needed to determine the wall temperature which would no longer be spatially uniform.

Correlations for heat and mass transport in packed bed reactors provide a straightforward approximation for heat and mass transfer in porous substrates. Thus, the heat and mass transfer coefficients obtained from these correlations are used in the above control volume analysis. Non-dimensional j factors, j_D for mass transfer and j_H for heat transfer, can be related to Schmidt and Prandtl numbers respectively [16]:

$$j_D = \frac{k_m}{v} Sc^{2/3}\tag{16.35}$$

$$j_H = \frac{h}{C_p \rho v} Pr^{2/3}\tag{16.36}$$

The following can be used with the appropriate Reynolds number to find j_D and j_H[16]:

$$j_D = j_H = 0.91 Re_j^{-0.51} S_f,\ Re_j < 50\tag{16.37}$$

$$j_D = j_H = 0.91 Re_j^{-0.41} S_f \quad Re_j > 50 \tag{16.38}$$

S_f is the shape factor and for this analysis a value of approximately 0.8 was used [16]. The Reynolds number for packed beds can be calculated using

$$Re = \frac{\rho v w}{6(1-\alpha)\mu S_f}. \tag{16.39}$$

Equations 16.35, 16.37, and 16.39 result in k_m which is then used in Equation 16.31. Equations 16.36, 16.38, and 16.39 give h which is subsequently used in Equation 16.34. The boundary conditions for this model include the inlet bulk gas temperature, the inlet fuel mole fraction, and the inlet surface mole fraction. The catalyst surface temperature must be specified and is constant due to the isothermal assumption resulting from the low Biot number. A typical set of boundary and flow conditions is shown in Table 16.4.

Figures 16.35 and 16.36 show predicted temperature and fuel concentration profiles respectively using the above boundary conditions assuming a substrate porosity of 95 percent, an average foam fiber thickness of 90 μm, and an equivalence ratio of unity.

Figure 16.37 shows trends in fuel conversion for various surface area-to-volume ratios, also at an equivalence ratio of unity, indicating that higher surface area-to-volume ratios (usually a result of lower porosity for a constant fiber thickness) will significantly improve fuel conversion. However, this comes at the expense of pressure loss.

16.6.5 OPERATING SPACE

This 1-D isothermal plug flow analytical model is used to visualize the catalytic microcombustor's operating space. Note that, this assessment is from the perspective of the combustor only and not the overall engine thermodynamic cycle. Figure 16.38 shows contours of maximum power density as a function of total pressure loss in the system and catalyst temperature at an equivalence ratio of one. The maximum power density is computed from the model's predictions of exhaust gas temperature. The figure shows that for constant surface temperature, maximum power density and thus combustor performance increase initially and then decrease

Table 16.4. Typical boundary and flow conditions for the reactor inlet in 1-D isothermal plug flow model

Parameter	Value
T_b	500 K
T_S	1000 K
$Y_{fuel,b}$ (from ϕ)	0.04 ($\phi = 1.0$)
$Y_{fuel,S}$	0.00
Pressure	2 atm
Mass flow	0.3 g/s

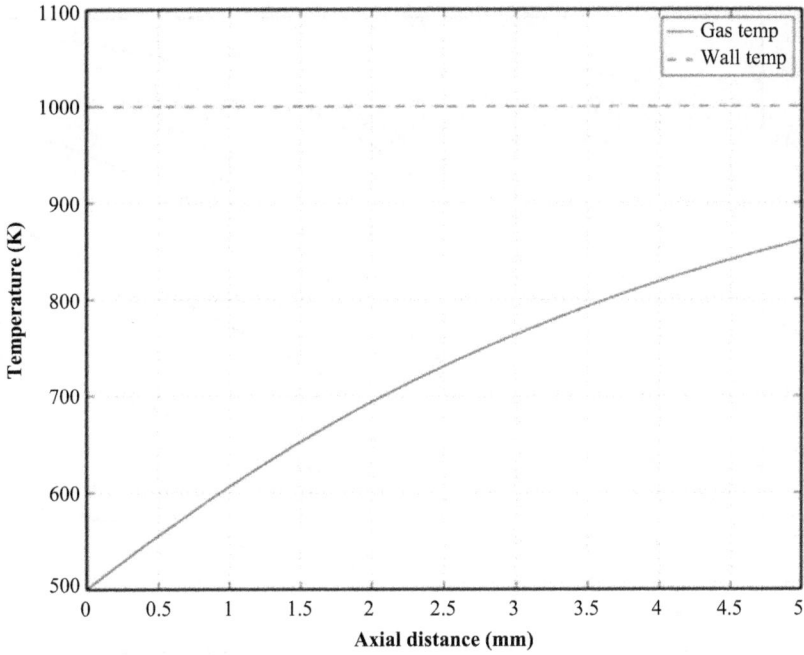

Figure 16.35. Axial temperature profile along porous media plug flow reactor $\phi = 1.0$ (from Spadaccini [1] and Spadaccini, Peck, and Waitz [3]).

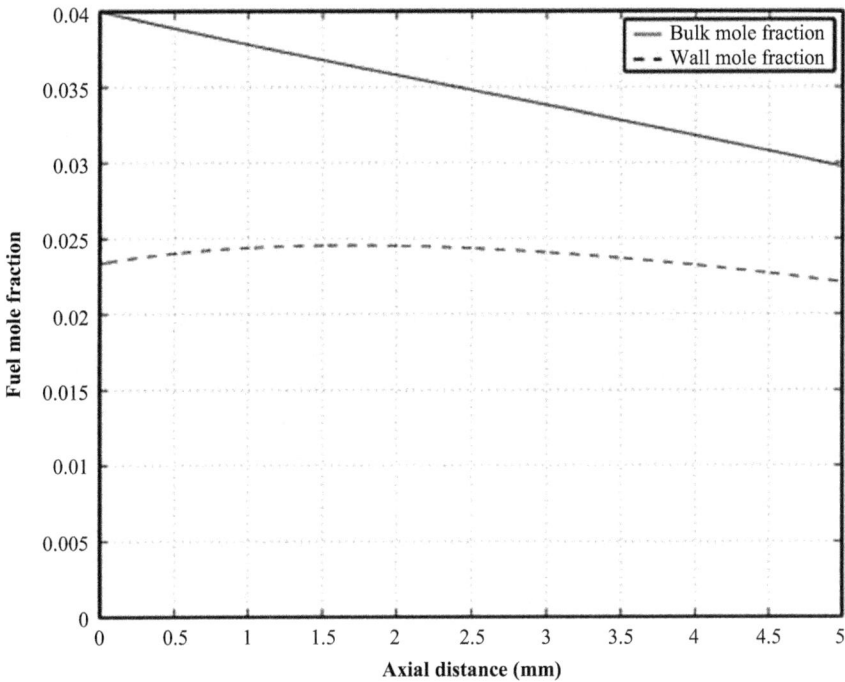

Figure 16.36. Axial fuel concentration profile along a porous media plug flow reactor $\phi = 1.0$ (from Spadaccini [1] and Spadaccini, Peck, and Waitz [3]).

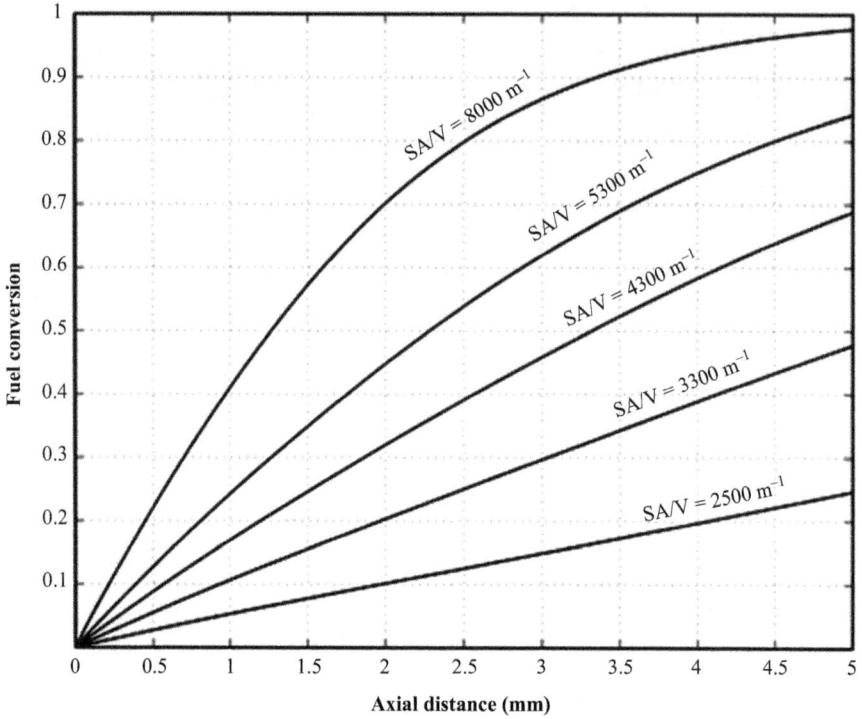

Figure 16.37. Fuel conversion profiles for various surface area-to-volume ratios in a porous media plug flow reactor $\phi = 1.0$ (from Spadaccini [1] and Spadaccini, Peck, and Waitz [3]).

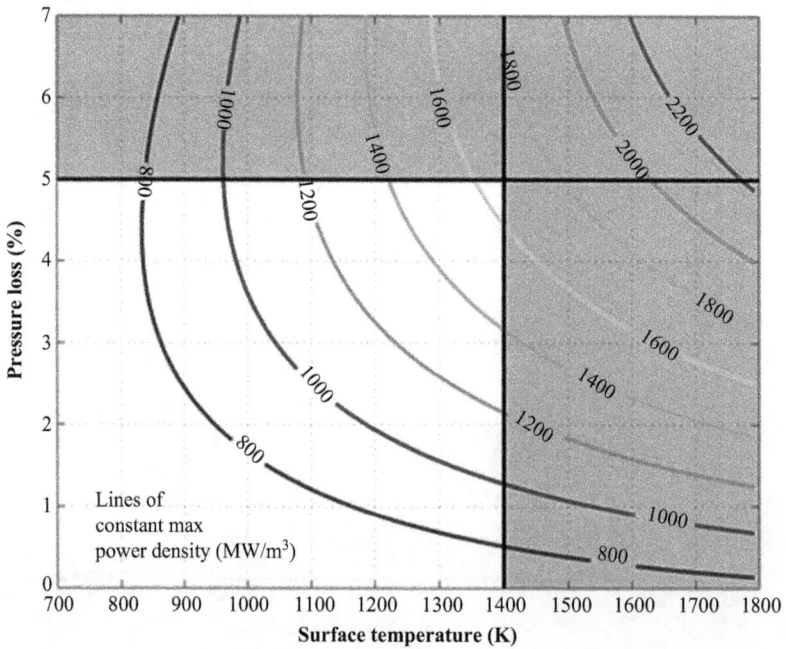

Figure 16.38. Operating space of a catalytic microcombustor at an overall equivalence ratio of 1 (from Spadaccini [1] and Spadaccini, Peck, and Waitz [3])

as pressure loss increases. The counter intuitive increase in power density with pressure loss is a result of the fact that the increased pressure loss is associated with increased surface area and decreased porosity (more catalytic surface). With the more surface area, there are more sites for fuel and oxidizer to react. At some point, the effect diminishes as the surface area is no longer the limiting factor and the parasitic effects of pressure loss start to dominate. The figure shows that higher catalyst temperatures and higher pressure losses (resulting from higher surface area-to-volume ratios) improve catalytic microcombustor performance (power density) at nearly all points within the pressure (<5 percent) and temperature (<1400 K) limits. The 1400 K limit represents an approximate catalyst agglomeration temperature which could be extended with advanced catalyst materials [9].

16.6.6   CONCLUSIONS

1. Catalytic microcombustors must operate in a diffusion-limited regime in order to achieve high power density. Thus, searching for more active catalyst materials is not required.
2. Power density is maximized by maximizing operating temperature. This can be achieved by:
   (a) Using the most thermally durable catalytic materials.
   (b) Using the highest surface area-to-volume ratio substrate available which does not violate the system pressure loss constraint and relaxing the total pressure loss constraint as much as possible.

16.7   CONCLUSION AND FUTURE WORK

Although significant progress has been made in understanding combustion at the micro-scale, there is still work to be done. Microcombustors for small-scale gas turbines must become smaller and more efficient in order for them to be practical. A hybrid homogeneous/heterogeneous microcombustor may be the optimal solution for achieving the high power densities required for microgas turbine engines. The combustor would consist of an initial low temperature catalytic section which ignites the mixture and raises the temperature and enthalpy of the fluid. The remaining reactants would ignite in the gas-phase to complete combustion and achieve the appropriate combustor exit conditions in a minimum volume.

Other future areas of work will include developing microcombustors that consume logistic fuels like gasoline, JP-8, and kerosene. This will require new fuel injection concepts that either atomize and mix the fuel or pre-vaporize the liquid. It also will likely require the use of a catalyst to increase reaction rates.

Finally, another advancement which would result in more practical combustion-based micropower systems is fuel flexibility. A small-scale gas turbine capable of operating without significant performance change on a variety of fuels would be extremely valuable for a variety of civilian and military applications.

REFERENCES

[1] Spadaccini C M 2004 Combustion systems for power-MEMS applications. Ph.D. Thesis, Massachusetts Institute of Technology.

[2] Spadaccini C M, Mehra A, Lee J, Zhang X, Lukachko S, and Waitz I A 2003 High power density silicon combustion systems for microgas turbine engines. *ASME Journal of Engineering for Gas Turbine and Power*, Vol. 125, pp. 709–719.

[3] Spadaccini C M, Peck J, Waitz I A 2007 Catalytic combustion systems for micro-scale gas turbine engines. *ASME Journal of Engineering for Gas Turbines and Power*, Vol. 129, pp. 49–60.

[4] Mehra A, Zhang X, Ayon A A, Waitz I A, Schmidt M A, and Spadaccini C M, 2000 A six-wafer combustion system for a silicon microgas turbine engine. *IEEE/ASME Journal of Microelectromechanical Systems*, Vol. 9, pp. 517–527.

[5] Waitz I A, Gauba G, and Tzeng Y-S 1998 Combustors for microgas turbine engines. *ASME Journal of Fluids Engineering*, Vol. 20, pp. 109–117

[6] Arana L R 2003 High-temperature microfluidic systems for thermally-efficient fuel processing. Ph.D. Thesis, Massachusetts Institute of Technology.

[7] Peck J J 2008 Development of a liquid fueled micro-combustor. Ph.D. Thesis, Massachusetts Institute of Technology.

[8] Rabier J and Demenet J L 2000 Low temperature, high stress plastic deformation of semiconductors: The silicon case. *Physica Status Solidi*, Vol. 222, pp. 63–74.

[9] Wang J, Zhao D G, Sun Y P, Duan L H, Wang Y T, Zhang S M, Yank H, Zhou S, and Wu M 2003 Thermal annealing behaviour of Pt on n-GaN Schottky contacts. *Journal of Physics D: Applied Physics*, Vol. 36, pp. 1018–1022

[10] Goralski C T Jr, and Schmidt L D 1996 Lean catalytic combustion of alkanes at short contact times. *Catalysis Letters*, Vol. 42, pp. 15–20.

[11] Veser G and Schmidt L D 1996 Ignition and extinction in catalytic oxidation of hydrocarbons. *AIChE Journal*, Vol. 42, pp. 1077–1087.

[12] Tzeng Y-S 1997 An investigation of microcombustion thermal phenomena. Master's Thesis, Massachusetts Institute of Technology.

[13] Mehra A 2000 Development of a high power density combustion system for a silicon microgas turbine engine. Ph.D. thesis, Massachusetts Institute of Technology.

[14] Coffee T P, Kotler A J, and Miller M S 1983 The overall reaction concept in premixed, laminar, steady-state flames: I. Stoichiometries. *Combustion and Flame*, Vol. 54, pp. 155–169.

[15] Coffee T P, Kotler A J, and Miller M S 1984 The overall reaction concept in premixed, laminar, steady-state flames: II. Initial temperatures and pressures. *Combustion and Flame*, Vol. 58, pp. 59–67.

[16] Hayes R E and Kolaczkowski S T 1997 *Introduction to Catalytic Combustion*, pp. 289–313, 389–394, Gordon and Breach Science Publishers, Canada.

NOMENCLATURE

Roman

A	Arrhenius pre-exponential factor
A	Arrhenius exponent
a_v	surface area-to-volume ratio
B	Arrhenius exponent
Bi	Biot number for heat transfer
C_b	molar density (mol/m^3)
C_p	constant pressure specific heat (J/kg K)

D	diffusion coefficient (cm^2/s)
Da_h	homogeneous Damköhler number
Da_2	diffusion-based Damköhler number
d_h	hydraulic diameter (m)
E_a	activation energy
F	Darcy friction factor
Fa	fuel/air ratio
h	convective heat transfer coefficient (W/m^2 K) or enthalpy
j_D	j-factor for mass transport
j_H	j-factor for heat transfer
k	thermal conductivity (W/m K), or rate constant
k_m	mass transport coefficient (m/s)
L	length (m)
LHV	lower heating value
M	molecular weight
$\dot{m}$	mass flow rate (kg/s)
P	static pressure (N/m^2)
Pe	Peclet number
Pr	Prandtl number
Q	heat, (W)
R	gas constant (J/kgK) or reaction rate
Re	Reynolds number
Sc	Schmidt number
S_f	shape factor
T	temperature (K)
V	volume
v	Velocity (m/s) or diffusion volume (cm^3)
W	thickness (m)
Y	mole fraction
Z	axial location (m)

Greek

α	porosity
ϕ	equivalence ratio
η	efficiency
μ	viscosity (Ns/m^2)
ρ	density (kg/m^3)
τ	characteristic time (s)

Subscripts

a	air
b	bulk flow
chem.	chemical
d	diameter (m)
f	fuel
S	surface
therm	thermal
0	initial
1	inlet
2	exit

Index

THIS TITLE IS FROM OUR MECHANICAL ENGINEERING GROUP COLLECTION.
OTHER TITLES OF INTEREST MIGHT BE...

Automotive Sensors
By John Turner, Joe Watson

Centrifugal and Axial Compressor Control
By Gregory K. McMillan

Virtual Engineering
By Joe Cecil

Reduce Your Engineering Drawing Errors:
Preventing the Most Common Mistakes
By Ronald Hanifan

Chemical Sensors: Fundamentals of Sensing
Materials Volume 2: Nanostructured Materials
By Ghenadii Korotcenkov

Biomedical Sensors
By Deric P. Jones, Joe Watson

Chemical Sensors: Comprehensive Sensor
Technologies Volume 4: Solid State Devices
By Ghenadii Korotcenkov

Acoustic High-Frequency Diffraction Theory
By Frederic Molinet

Chemical Sensors: Comprehensive Sensor
Technologies Volume 5: Electrochemical
and Optical Sensors
By Ghenadii Korotcenkov

Chemical Sensors: Comprehensive Sensor
Technologies Volume 6: Chemical Sensors
Applications
By Ghenadii Korotcenkov

Bio-Inspired Engineering
By Chris Jenkins

Chemical Sensors: Simulation and Modeling
Volume 1: Microstructural Characterization
and Modeling of Metal Oxides
By Ghenadii Korotcenkov

The Essentials of Finite Element Modeling and
Adaptive Refinement: For Beginning Analysts to
Advanced Researchers in Solid Mechanics
By John O. Dow

Chemical Sensors: Simulation and Modeling
Volume 2: Conductometric-Type Sensors
By Ghendaii Korotcenkov

Aerospace Sensors
By Alexander Nebylov

Chemical Sensors: Simulation and Modeling
Volume 3: Solid-State Devices
By Ghenadii Korotcenkov

Chemical Sensors: Simulation and Modeling
Volume 4: Optical Sensors
By Ghenadii Korotcenkov

Classical and Modern Engineering Methods in Fluid
Flow and Heat Transfer: An Introduction
for Engineers and Students
By Abram Dorfman

PEM Fuel Cells: Thermal and Water
Management Fundamentals
By Yun Wang, Ken S. Chen, Sun Chan Cho

Chemical Sensors: Simulation and Modeling
Volume 5: Electrochemical Sensors
By Ghenadii Korotcenkov

Announcing Digital Content Crafted by Librarians

Momentum Press offers digital content as authoritative treatments of advanced engineering topics by leaders in their fields. Hosted on ebrary, MP provides practitioners, researchers, faculty, and students in engineering, science, and industry with innovative electronic content in sensors and controls engineering, advanced energy engineering, manufacturing, and materials science.

Momentum Press offers library-friendly terms:

- perpetual access for a one-time fee
- no subscriptions or access fees required
- unlimited concurrent usage permitted
- downloadable PDFs provided
- free MARC records included
- free trials

The **Momentum Press** digital library is very affordable, with no obligation to buy in future years.

For more information, please visit **www.momentumpress.net/library** or to set up a trial in the US, please contact **mpsales@globalepress.com**.

www.ingramcontent.com/pod-product-compliance
Lightning Source LLC
Chambersburg PA
CBHW062010190326
41458CB00009B/3027